Group Theory in a Nutshell for Physicists

Group Theory in a Nutshell for Physicists

A. Zee

PRINCETON UNIVERSITY PRESS · PRINCETON AND OXFORD

Copyright © 2016 by Princeton University Press

Published by Princeton University Press, 41 William Street,
Princeton, New Jersey 08540
In the United Kingdom: Princeton University Press,
6 Oxford Street, Woodstock, Oxfordshire OX20 1TR

press.princeton.edu

Cover image: Modification of *Der Tiger* by Franz Marc (1912);
Städtische Galerie im Lenbachhaus und Kunstbau, München.

Library of Congress Cataloging-in-Publication Data
Names: Zee, A., author
Title: Group theory in a nutshell for physicists / A. Zee.
Other titles: In a nutshell (Princeton, N.J.)
Description: Princeton, New Jersey : Princeton University Press, [2016] |
    © 2016 | Series: In a nutshell | Includes bibliographical references and
    index.
Identifiers: LCCN 2015037408 | ISBN 9780691162690 (hardcover : alk. paper) |
    ISBN 0691162697 (hardcover : alk. paper)
Subjects: LCSH: Group theory.
Classification: LCC QC20.7.G76 Z44 2016 | DDC 512/.2—dc23 LC record available at
http://lccn.loc.gov/2015037408

British Library Cataloging-in-Publication Data is available

This book has been composed in Scala LF with ZzTEX
by Cohographics and Windfall Software.

Printed on acid-free paper ∞

Printed in the United States of America

10 9 8 7

To ZYH, the once and future

Contents

VIII   Part VIII: The Expanding Universe

IX   Part IX: The Gauged Universe

Preface

Trust me, it's not that hard

> In 1951, I had the good fortune of listening to Professor
> Racah's lecture on Lie groups at Princeton. After attending
> these lectures, I thought, "This is really too hard. I cannot learn
> all this . . . too damned hard and unphysical."
>
> —A. Salam, 1979 Nobel laureate in physics[1]

Trust me, it's not that hard. And as Salam's own Nobel-winning work helped show, group theory is certainly relevant to physics. We now know that the interweaving gauge bosons underlying our world dance to the music[2] of Lie groups and Lie algebras.

This book is about the use of group theory in theoretical physics. If you are looking for a mathematics book on group theory complete with rigorous proofs, the abstract[3] modern definition[4] of tensors and the like, please go elsewhere. I will certainly prove every important statement, but only at a level of rigor acceptable to most physicists. The emphasis will be on the intuitive, the concrete, and the practical.

I would like to convince a present day version of Salam that group theory is in fact very physical. With due respect to Racah, I will try to do better than him pedagogically. My goal is to show that group theoretic concepts are natural and easy to understand.

Elegant mathematics and profound physics: Honor your inheritance

> In great mathematics, there is a very high degree of unexpectedness, combined with inevitability and economy.
>
> —G. H. Hardy[5]

Group theory is a particularly striking example of what Hardy had in mind. For me, one of the attractions of group theory is the sequence of uniqueness theorems, culminating in Cartan's classification of all Lie algebras (discussed in part VI). Starting from a few innocuous sounding axioms defining what a group is, an elegant mathematical structure emerges, with many unexpected theorems.

My colleague Greg Huber pointed out to me that group theory is an anagram for rough poetry. Rough? I've always thought that it's close to pure poetry.

Although group theory is certainly relevant for nineteenth-century physics, it really started to play an important role with the work of Lorentz and Poincaré, and became essential with quantum mechanics. Heisenberg opened up an entire new world with his vision of an internal symmetry, the exploration of which continues to this very day in one form or another. Beginning in the 1950s, group theory has come to play a central role in several areas of physics, perhaps none more so than in what I call fundamental physics, as we will see in parts V, VII, VIII, and IX of this book. There are of course some areas[6] of physics that, at least thus far, seem not[7] to require much of group theory.

I understand that group theory has also played a crucial role in many areas of mathematics, for example, algebraic topology, but that is way outside the scope of this book. As a one-time math major who saw the light, while I do not know what mathematicians know about groups, I know enough to know that what I cover here is a tiny fraction of what they know.

This is a book about a branch of mathematics written by a physicist for physicists. One immediate difficulty is the title: the disclaimer "for physicists" has to be there; also the phrase "in a nutshell" because of my contractual obligations to Princeton University Press. The title "Group Theory for Physicists in a Nutshell" would amount to a rather lame joke, so the actual title is almost uniquely determined.

This is my third Nutshell book. As for my motivation for writing yet another textbook, Einstein said it better than I could: "Bear in mind that the wonderful things that you learn in your schools are the work of many generations. All this is put into your hands as your inheritance in order that you may receive it, honor it, add to it, and one day faithfully hand it on to your children."[8]

Advice to the reader

Some advice to the reader, particularly if you are self-studying group theory in physics. The number one advice is, of course, "Exercise!" I strongly recommend doing exercises as you read along, rather than waiting until the end of the chapter, especially in the early chapters. To the best of my knowledge, nobody made it into the NBA by watching the sports channels. Instead of passively letting group theory seep into your head, you should frequently do the mental equivalent of shooting a few baskets. When given a theorem, you should, in the spirit of doubting Thomas, try to come up with counterexamples.

I am particularly worried about the readers who are shaky about linear algebra. Since this is not a textbook on linear algebra, I did not provide lots of exercises in my coverage (see below) of linear algebra. So, those readers should make up their own (even straightforward) exercises, multiplying and inverting a few numerical matrices, if only to get a sense of how matrices work.

For whom is this book written

This brings me to prerequisites. If you know linear algebra, you can read this book. For the reader's convenience, I had planned to provide a brief appendix reviewing linear algebra. It grew and grew, until it became essentially self-contained and clamored to move up front. I ended up covering, quite completely, at least those aspects of linear algebra needed for group theory. So, yes, even if you don't know linear algebra, you could still tackle this book.

Several blurbers and reviewers of my Quantum Field Theory in a Nutshell[9] have said things along the line of "This is the book I wish I had when I was a student."[10] So that's roughly the standard I set for myself here: I have written the book on group theory I wished I had when I was a student.[11]

My pedagogical strategy is to get you to see some actual groups, both finite and continuous, in action as quickly as possible. You will, for example, be introduced to Lie algebra by the third chapter. In this strategy, one tactic is to beat the rotation group to death early on. It got to the point that I started hearing the phrase "beat rotation to death" as a rallying cry.

Group theory and quantum mechanics

I was not entirely truthful when I said "If you know linear algebra, you can read this book." You have to know some quantum mechanics as well. For reasons to be explained in chapter III.1, group theory has played much more of a role in quantum mechanics than in classical mechanics. So for many of the applications, I necessarily have to invoke quantum mechanics. But fear not! What is needed to read this book is not so much quantum mechanics itself as the language of quantum mechanics. I expect the reader to have heard of states, probability amplitudes, and the like. You are not expected to solve the Schrödinger equation blindfolded, and certainly those murky philosophical issues regarding quantum measurements will not come in at all.

For some chapters in parts V, VII, and IX, some rudimentary knowledge of quantum field theory will also be needed. Some readers may wish to simply skip these chapters. For braver souls, I try to provide a gentle guide, easing into the subject with a review of the Lagrangian and Hamiltonian in chapters III.3 and IV.9. The emphasis will be on the group theoretic aspects of quantum field theory, rather than on the dynamical aspects, of course. I believe that readers willing to work through these chapters will be rewarded with a deeper understanding of the universe. In case you run into difficulties, my advice is to muddle through and provisionally accept some statements as given. Of course, I also hope that some readers will go on and master quantum field theory (smile).

Applications of group theory

My philosophy here is not to provide a compendium of applications, but to endow the reader with enough of an understanding of group theory to be able to approach individual problems. The list of applications clearly reflects my own interests, for instance, the Lorentz group and its many implications, such as the Weyl, Dirac, and Majorana equations. I think that this is good. What is the sense of my transporting calculations from some crystallography and materials science textbooks (as some colleagues have urged me to do "to broaden the market"), when I do not feel the subject in my bones, so to speak? In the same way, I do not expect existing books on group theory in solid state physics to cover the Majorana fermion. To be sure, I cover some standard material, such as the nonexistence of crystals with 5-fold symmetry. But, judging from recent advances at the frontier of condensed matter theory, some researchers may need to get better acquainted with Weyl and Majorana rather than to master band structure calculations.

I try to give the reader some flavor of a smattering of various subjects, such as Euler's function and Wilson's theorem from number theory. In my choice of topics, I tend to favor those not covered in most standard books, such as the group theory behind the expanding universe. My choices reflect my own likes or dislikes.[12] Since field theory, particle physics, and relativity are all arenas in which group theory shines, it is natural and inevitable that this book overlaps my two previous textbooks.

The genesis of this book

This book has sat quietly in the back of my mind for many years. I had always wanted to write textbooks, but I am grateful to Steve Weinberg for suggesting that I should write popular physics books first. He did both, and I think that one is good training for the other. My first popular physics book is Fearful Symmetry,[13] and I am pleased to say that, as it reaches its thirtieth anniversary, it is still doing well, with new editions and translations coming out. As the prospective reader of this book would know, I could hardly talk about symmetry in physics without getting into group theory, but my editor at the time[14] insisted that I cut out my attempt to explain group theory to the intelligent public. What I wrote was watered down again and again, and what remained was relegated to an appendix. So, in some sense, this book is a follow-up on Fearful, for those readers who are qualified to leap beyond popular books.

Physics students here at the University of California, Santa Barbara, have long asked for more group theory. In an interesting pedagogical year,[15] I taught a "physics for poets" course for nonphysics majors, discussing Fearful over an entire quarter, and during the following quarter, a special topics course on group theory addressed to advanced undergraduates and graduate students who claimed to know linear algebra.

Teaching from this book and self-studying

As just mentioned, I taught this material (more than once) at the University of California, Santa Barbara, in a single quarter course lasting ten weeks with two and a half hours' worth of lectures per week. This is too short to cover all the material in the book, but with some skipping around, I managed to get through a major fraction. Here is the actual syllabus.

Week 1: Definition and examples of groups, Lagrange's theorem, constructing multiplication tables, direct product, homomorphism, isomorphism

Week 2: Finite group, permutation group, equivalence classes, cycle structure of permutations, dihedral group, quarternionic group, invariant subgroup, simple group

Week 3: Cosets, quotient group, derived subgroup, rotation and Lie's idea, Lie algebra

Week 4: Representation theory, unitary representation theorem, orthogonality theorem, character orthogonality

Week 5: Regular representation; character table is square; constructing character table

Week 6: Tray method; real, pseudoreal, and complex; crystals; Fermat's little theorem (statement only);[16] group theory and quantum mechanics

Week 7: $SO(N)$: why $SO(3)$ is special, Lie algebra of $SO(3)$, ladder operators, Casimir invariants, spherical harmonics, $SU(N)$

Week 8: $SU(2)$ double covers $SO(3)$, $SO(4)$, integration over group manifolds

Week 9: $SU(3)$, roots and weights, spinor representations of $SO(N)$

Week 10: Cartan classification, Dynkin diagrams

Thus, the single quarter course ends with part VI.

Students were expected to do some reading and to fill in some gaps on their own. Of course, instructors may want to deviate considerably from this course plan, emphasizing one topic at the expense of another. It would be ideal if they could complement this book with material from their own areas of expertise, such as materials science. They might also wish to challenge the better students by assigning the appendices and some later chapters. A semester would be ideal.

Some notational alerts

Some books denote Lie groups by capital letters, for example, $SU(2)$, and the corresponding algebras by lower case letters, for example, $su(2)$. While I certainly understand the need to distinguish group and algebra, I find the constant shifting between upper and lower case letters rather fussy looking. Most physicists trust the context to make clear whether the group or the algebra is being discussed. Thus, I will follow the standard physics usage and write $SU(2)$ for both group and algebra. An informal survey of physics students indicates that most agree with me. Of course, I am careful to say that, for example, $SU(2)$ and $SO(3)$ are only locally isomorphic and that one covers the other (as explained in detail in part IV).

In general, I vote for clarity over fussiness; I try not to burden the reader with excessive notation.

Parting comments: Regarding divines and dispensable erudition

A foolish consistency is the hobgoblin of little minds, adored by
little statesmen and philosophers and divines.
                                            —Ralph Waldo Emerson

I made a tremendous effort to be consistent in my convention, but still I have to invoke Emerson and hope that the reader is neither a little statesman nor a divine. At a trivial level, I capriciously use "1, 2, 3" and "x, y, z" to denote the same three Cartesian axes. Indeed, I often intentionally switch between writing superscript and subscript (sometimes driven by notational convenience) to emphasize that it doesn't matter. But eventually we come to a point when it does matter. I will then explain at length why it matters.

In Zvi Bern's Physics Today review of QFT Nut, he wrote this lovely sentence: "The purpose of Zee's book is not to turn students into experts—it is to make them fall in love with the subject."[17] I follow the same pedagogical philosophy in all three of my textbooks. This echoes a sage[18] who opined "One who knows does not compare with one who likes, one who likes does not compare with one who enjoys."

As I have already said, this is not a math book, but a book about math addressed to physicists. To me, math is about beauty, not rigor, unexpected curves rather than rock hard muscles.

Already in the nineteenth century, some mathematicians were concerned about the rising tide of rigor. Charles Hermite, who figures prominently in this book, tried to show his students the simple beauty of mathematics, while avoiding what Einstein would later refer to as "more or less dispensable erudition."[19] In this sense, I am Hermitean, and also Einsteinian.

Indeed, Einstein's aphorism, that "physics should be made as simple as possible, but not any simpler," echoes throughout my textbooks. I have tried to make group theory as simple as possible.[20]

Acknowledgments

I thank Dillon Allen, Yoni BenTov, Shreyashi Chakdar, Joshua Feinberg, David Gross, Christine Hartmann, Greg Huber, Boris Kayser, Choonkyu Lee, Henri Orland, Joe Polchinski, Rafael Porto, Srinivas Raghu, Skyler Saleebyan, Andrew Smith, Dam Thanh Son, Arkady Vainshtein, Miles Vinikow, Brayden Ware, Jake Weeks, Alexander Wickes, Yong-Shi Wu, and Tzu-Chiang Yuan for reading one or more chapters and for their comments. I particularly thank Alex Wickes for help in the early stages of this book and for getting the entire project rolling. As with my two previous books, I thank Craig Kunimoto for his invaluable help in taming the computer.

I wrote the bulk of this book in Santa Barbara, California, but some parts were written while I visited the Academia Sinica in Taipei, the Republic of China, Tsinghua University in Beijing, and Jiao Tong University in Shanghai, People's Republic of China. I very much appreciate their warm hospitality. My editor at Princeton University Press, Ingrid Gnerlich, has always been a pleasure to talk to and work with. I am delighted to have Cyd Westmoreland, for the fourth time, as my copy editor. I appreciate the work of the entire production team, including Laurel Muller, Karen Fortgang, Eric Henney, Paul Anagnostopoulos, Joe Snowden, and MaryEllen Oliver. I also thank Bjorn Burkel and Seth Koren for assistance with the final proofreading. Last but not least, I am deeply grateful to my wife Janice for her loving support and encouragement throughout the writing of this book.

Notes

1. Quoted in J-Q. Chen, Group Representation Theory for Physicists, p. 1.
2. I will get up to a "faint echo" in the closing parts of this book.
3. George Zweig, who independently discovered the notion of quarks, had this to say about the abstract approach: "Mathematics in the US was taught in a very formal manner. I learned algebra from a wonderful algebraist, Jack McLaughlin, but the textbook we used was Jacobson's two-volume set, 'Lectures in Abstract Algebra,' and abstract it was! It seemed like there were as many definitions as results, and it was impossible to see how Mr. Jacobson actually thought. The process was hidden, only polished proofs were presented." Hear, hear!
4. In contrast to the concrete "archaic" definition that physicists use.
5. G. H. Hardy, A Mathematician's Apology, Cambridge University Press, 1941.
6. I am often surprised by applications in areas where I might not expect group theory to be of much use. See, for example, "An Induced Representation Method for Studying the Stability of Saturn's Ring," by S. B. Chakraborty and S. Sen, arXiv:1410.5865. Readers who saw the film Interstellar might be particularly interested in this paper.
7. To paraphrase Yogi Berra, if some theoretical physicists do not want to learn group theory, nobody is going to make them.
8. Albert Einstein, speaking to a group of school children, 1934.
9. Henceforth, QFT Nut.
10. See, for example, F. Wilczek on the back cover of the first edition of QFT Nut or the lead page of the second edition.
11. Indeed, it would have been marvelous if I had had something like this book after I had learned linear algebra in high school.
12. Or even other people's dislikes. For instance, my thesis advisor told me to stay the heck away from Young tableaux, and so I have ever since.
13. Henceforth, Fearful.
14. He had heard Hawking's dictum that every equation in a popular physics book halves its sale.
15. At the urging of the distinguished high energy experimentalist Harry Nelson.
16. That is, the full proof is given here, but was not entered into when I taught the course.

17. Niels Bohr: "A expert is someone . . . who goes on to know more and more about less and less, and ends up knowing everything about nothing."

18. This represents one of the few instances in which I agree with Confucius. Alas, I am often surrounded by people who know but do not enjoy.

19. In fact, Einstein was apparently among those who decried "die Gruppenpest." See chapter I.1. I don't know of any actual documentary evidence, though.

20. While completing this book, I came across an attractive quote by the mathematician H. Khudaverdyan: "I remember simple things, I remember how I could not understand simple things, this makes me a teacher." See A. Borovik, Mathematics under the Microscope, p. 61.

A Brief Review of Linear Algebra

Algebra is generous; she often gives more than is asked of her.
—Jean-Baptiste le Rond d'Alembert (1717–1783)

Linear algebra is a beautiful subject, almost as beautiful as group theory. But who is comparing?

I wrote this originally as a review for those readers who desire to learn group theory but who need to be reminded of some key results in linear algebra. But then the material grew, partly because I want to have a leisurely explanation of how the basic concepts of matrix and determinant arise naturally. I particularly want to give a step-by-step derivation of Cramer's formula for the matrix inverse rather than to plop it down from the sky. So then in the end, I decided to put this review at the beginning.

This is of course not meant to be a complete treatment of linear algebra.* Rather, we will focus on those aspects needed for group theory.

Coupled linear equations

As a kid, I had a run-in with the "chicken and rabbit problem." Perhaps some readers had also? A farmer has x chickens and y rabbits in a cage. He counted 7 heads and 22 legs. How many chickens and rabbits are in the cage? I was puzzled. Why doesn't the farmer simply count the chickens and rabbits separately? Is this why crazy farmers have to learn linear algebra?

In fact, linear algebra is by all accounts one of the most beautiful subjects in mathematics, full of elegant theorems, contrary to what I thought in my tender years. Here I will take an exceedingly low-brow approach, eschewing generalities for specific examples and building up the requisite structure step by step.

---

* Clearly, critics and other such individuals looking for mathematical rigor should also look elsewhere. They should regard this as a "quick and dirty" introduction for those who are unfamiliar (or a bit hazy) with linear algebra.

Instead of solving $x + y = 7$, $2x + 4y = 22$, let us go up one level of abstraction and consider

$$ax + by = u \tag{1}$$

$$cx + dy = v \tag{2}$$

Subtracting b times (2) from d times (1), we obtain

$$(da - bc)x = du - bv \tag{3}$$

and thus

$$x = \frac{du - bv}{ad - bc} = \frac{1}{ad - bc} (d, -b) \begin{pmatrix} u \\ v \end{pmatrix} \tag{4}$$

Note that the scalar product* of a row vector with a column vector naturally appears. Given a row vector $\vec{P}^T = (p, q)$ and a column vector $\vec{Q} = \begin{pmatrix} r \\ s \end{pmatrix}$, their scalar product is defined to be $\vec{P}^T \cdot \vec{Q} = (p, q) \begin{pmatrix} r \\ s \end{pmatrix} = pr + qs$. (The superscript T on $\vec{P}$ will be explained in due time.)

Similarly, subtracting a times (2) from c times (1), we obtain

$$(cb - ad)y = cu - av \tag{5}$$

and thus

$$y = -\frac{cu - av}{ad - bc} = \frac{1}{ad - bc} (-c, a) \begin{pmatrix} u \\ v \end{pmatrix} \tag{6}$$

(With $a = 1$, $b = 1$, $c = 2$, $d = 4$, $u = 7$, $v = 22$, we have $x = 3$, $y = 4$, but this is all child's play for the reader, of course.)

Matrix appears

Packaging (4) and (6) together naturally leads us to the notion of a matrix:[†]

$$\begin{pmatrix} x \\ y \end{pmatrix} = \frac{1}{ad - bc} \begin{pmatrix} du - bv \\ -cu + av \end{pmatrix} = \frac{1}{ad - bc} \begin{pmatrix} d & -b \\ -c & a \end{pmatrix} \begin{pmatrix} u \\ v \end{pmatrix} \tag{7}$$

The second equality indicates how the action of a 2-by-2 matrix on a 2-entry column vector is defined. A 2-by-2 matrix acting on a 2-entry column vector produces a 2-entry column vector as follows. Its first entry is given by the scalar product of the first row of the matrix, regarded as a 2-entry row vector, with the column vector, while its second entry is given by the scalar product of the second row of the matrix, regarded as a 2-entry row vector, with the column vector. I presume that most readers of this review are already familiar with

---

* Also called a dot product.
† "Matrix" comes from the Latin word for womb, which in turn is derived from the word "mater." The term was introduced by J. J. Sylvester.

how a matrix acts on a column vector. As another example,

$$\begin{pmatrix} a & b \\ c & d \end{pmatrix} \begin{pmatrix} x \\ y \end{pmatrix} = \begin{pmatrix} ax + by \\ cx + dy \end{pmatrix} \tag{8}$$

At this point, we realize that we could write (1) and (2) as

$$\begin{pmatrix} a & b \\ c & d \end{pmatrix} \begin{pmatrix} x \\ y \end{pmatrix} = \begin{pmatrix} u \\ v \end{pmatrix} \tag{9}$$

Define the matrix $M = \begin{pmatrix} a & b \\ c & d \end{pmatrix}$ and write $\vec{x} = \begin{pmatrix} x \\ y \end{pmatrix}$ and $\vec{u} = \begin{pmatrix} u \\ v \end{pmatrix}$. Then we can express (9) as

$$M\vec{x} = \vec{u} \tag{10}$$

Thus, given the matrix M and the vector $\vec{u}$, our problem is to find a vector $\vec{x}$ such that M acting on it would produce $\vec{u}$.

Turning a problem around

As is often the case in mathematics and physics, turning a problem around[1] and looking at it in a novel way could open up a rich vista. Here, as some readers may know, it is fruitful to turn (10) around into $\vec{u} = M\vec{x}$ and to look at it as a linear transformation of the vector $\vec{x}$ by the matrix M into the vector $\vec{u}$, conceptualized as $M : \vec{x} \to \vec{u}$, rather than as an equation to solve for $\vec{x}$ in terms of a given $\vec{u}$.

Once we have the notion of a matrix transforming a vector into another vector, we could ask what happens if another matrix N comes along and transforms $\vec{u}$ into another vector, call it $\vec{p}$:

$$\vec{p} = N\vec{u} = NM\vec{x} = P\vec{x} \tag{11}$$

The last equality defines the matrix P. At this point, we may become more interested in how two matrices N and M could be multiplied together to form another matrix $P = NM$, and "dump"[2] the vectors $\vec{p}$, $\vec{u}$, and $\vec{x}$ altogether, at least for a while.

The multiplication of matrices provides one of the central themes of group theory. We will see presently that (11) tells us how the product NM is to be determined, but first we need to introduce indices.

Appearance of indices and rectangular matrices

If we want to generalize this discussion on 2-by-2 matrices to n-by-n matrices, we risk running out of letters. So, we are compelled to use that marvelous invention known as the index.

Write $M = \begin{pmatrix} M_{11} & M_{12} \\ M_{21} & M_{22} \end{pmatrix}$. Here we have adopted the totally standard convention of denoting by M_{ij} the entry in the ith row and jth column of M. The reader seeing this for the first time should make sure that he or she understands the convention about rows and columns by writing down M_{31} and M_{23} in the following 3-by-3 matrix:

$$M = \begin{pmatrix} a & b & c \\ d & e & f \\ g & h & i \end{pmatrix} \tag{12}$$

The answer is given in an endnote.[3]

Starting with the chicken and rabbit problem, we were led to square matrices. But we could just as well define m-by-n rectangular matrices with m rows and n columns. Indeed, a column vector could be regarded as a rectangular matrix with m rows and only one column. A row vector is a rectangular matrix, with one row and n columns.

Rectangular* matrices could be multiplied together only if they "match"; thus an m-by-n rectangular matrix can be multiplied from the right by an n-by-p rectangular matrix to form an m-by-p rectangular matrix.

Writing $\vec{x} = \begin{pmatrix} x_1 \\ x_2 \end{pmatrix}$ and $\vec{u} = \begin{pmatrix} u_1 \\ u_2 \end{pmatrix}$ (which amounts to regarding a vector as a rectangular matrix with two rows but only one column), we could restate (10) (or in other words, (1) and (2), the equations we started out with) as $u_i = M_{i1}x_1 + M_{i2}x_2 = \sum_{j=1}^2 M_{ij}x_j$, for $i = 1, 2$.

Multiplying matrices together and the Kronecker delta

Now the generalization to n-dimensional vectors and n-by-n matrices is immediate. We simply allow the indices i and j to run over $1, \cdots, n$ and extend the upper range in the summation symbol to n:

$$u_i = \sum_{j=1}^n M_{ij}x_j \tag{13}$$

The rule for multiplying matrices then follows from (11):

$$p_i = \sum_{j=1}^n N_{ij}u_j = \sum_{j=1}^n \sum_{k=1}^n N_{ij}M_{jk}x_k = \sum_{k=1}^n P_{ik}x_k \tag{14}$$

Hence $P = NM$ means

$$P_{ik} = \sum_{j=1}^n N_{ij}M_{jk} \quad \text{(multiplication rule)} \tag{15}$$

We now define the identity matrix I by $I_{ij} = \delta_{ij}$, with the Kronecker delta symbol δ_{ij} defined by

$$\delta_{ij} = \begin{cases} 1 & \text{if } i = j \\ 0 & \text{if } i \neq j \end{cases} \tag{16}$$

---

* We will seldom encounter rectangular matrices other than vectors; one occasion would occur in the proof of Schur's lemma in chapter II.2.

In other words, I is a matrix whose off-diagonal elements all vanish and whose diagonal elements are all equal to 1. In particular, for $n = 2$, $I = \begin{pmatrix} 1 & 0 \\ 0 & 1 \end{pmatrix}$.

It follows from (15) that $(IM)_{ik} = \sum_{j=1}^{n} \delta_{ij} M_{jk} = M_{ik}$. Similarly, $(MI)_{ik} = M_{ik}$. In other words, $IM = MI = M$.

If the reader feels a bit shaky about matrix multiplication, now is the time to practice with a few numerical examples.[4] Please also do exercises 1–5, in which the notion of elementary matrices is introduced; we will need the results later.

I also mention here another way of looking at matrix multiplication that will be useful later. Regard the n columns in M as n different column vectors $\vec{\psi}_{(k)}$, $k = 1, \cdots, n$, where by definition the jth component of $\vec{\psi}_{(k)}$ is equal to M_{jk}. (The parenthesis is a bit pedantic: it emphasizes that k labels the different column vectors.) Thus, schematically,

$$ M = \left(\vec{\psi}_{(1)}, \cdots, \vec{\psi}_{(k)}, \cdots, \vec{\psi}_{(n)} \right) \tag{17} $$

For example, for the 3-by-3 matrix in (12), $\vec{\psi}_{(1)} = \begin{pmatrix} a \\ d \\ g \end{pmatrix}$, $\vec{\psi}_{(2)} = \begin{pmatrix} b \\ e \\ h \end{pmatrix}$, and $\vec{\psi}_{(3)} = \begin{pmatrix} c \\ f \\ i \end{pmatrix}$.

Similarly, regard the n columns in P as n different column vectors $\vec{\phi}_{(k)}$, $k = 1, \cdots, n$, where by definition the jth component of $\vec{\phi}_{(k)}$ is equal to P_{jk}. Looked at this way, (15) is telling us that $\vec{\phi}_{(k)}$ is obtained by acting with the matrix N on $\vec{\psi}_{(k)}$:

$$ P = \left(\vec{\phi}_{(1)}, \cdots, \vec{\phi}_{(n)} \right) = NM = \left(N\vec{\psi}_{(1)}, \cdots, N\vec{\psi}_{(n)} \right) \tag{18} $$

Einstein's repeated index summation

Let us now observe that whenever there is a summation symbol, the index being summed over is repeated. For example, in (15) the summation symbol instructs us to sum over the index j, and indeed the index j appears twice in $N_{ij} M_{jk}$, in contrast to the indices i and k, which appear only once each and are sometimes called free indices.

Notice that the free indices also appear in the left hand side of (15), namely in P_{ik}. This is of course as it should be: the indices on the two sides of the equation must match. In contrast, the index j, which has been summed over, must not appear in the left hand side.

As you can see (and as could Einstein), the summation symbol is redundant in many expressions and may be omitted if we agree that any index that is repeated, such as j in this example, is understood to be summed over. In the physics literature, Einstein was among the first to popularize this repeated index summation, which many physicists regard as one of Einstein's important contributions.[5] We will adopt this convention and hence write (15) simply as $P_{ik} = N_{ij} M_{jk}$.

The index being summed over (namely, j in this expression) is sometimes called the dummy index, in contrast to the free indices i and k, which take on fixed values we are free to assign. Here is a self-evident truth seemingly hardly worth mentioning, but yet it sometimes confuses some abecedarians: it does not matter what we call the dummy index.

For example, $N_{1j}M_{j3} = N_{1m}M_{m3} = N_{1h}M_{h3} = N_{11}M_{13} + N_{12}M_{23} + \cdots + N_{1n}M_{n3}$ if we are dealing with n-by-n matrices.

Not commutative, but associative

Matrix multiplication is not commutative; that is, in general $NM \neq MN$: no reason for the two sums $\Sigma_j N_{ij}M_{jk}$ and $\Sigma_j M_{ij}N_{jk}$ to be equal. You could verify this with a few examples. For instance, define the matrices $\sigma_3 = \left(\begin{smallmatrix} 1 & 0 \\ 0 & -1 \end{smallmatrix} \right)$ and $\sigma_1 = \left(\begin{smallmatrix} 0 & 1 \\ 1 & 0 \end{smallmatrix} \right)$; then $\sigma_3\sigma_1 = -\sigma_1\sigma_3 \neq \sigma_1\sigma_3$.

Matrix multiplication is, however, associative: $(AB)C = A(BC)$. We show this by brute force: $((AB)C)_{ik} = (AB)_{ij}C_{jk} = A_{il}B_{lj}C_{jk}$, while $(A(BC))_{ik} = A_{ij}(BC)_{jk} = A_{ij}B_{jl}C_{lk}$, but these two expressions are the same, because in both the index j and index l are summed over according to Einstein's convention. They are dummies and can be renamed.

Keep in mind that

$$NM \neq MN \quad \text{but} \quad (AB)C = A(BC) \tag{19}$$

Transpose

We now go back to the superscript T slipped in earlier when we wrote $P^T \cdot Q$. The transpose of a matrix M, denoted by M^T, is defined by interchanging the rows and columns in M. In other words, $(M^T)_{ij} = M_{ji}$. The transpose of a column vector is a row vector, and vice versa.

A small* but important theorem: The transpose of a product is the product of the transposes, but in reversed order:

$$(NM)^T = M^T N^T \tag{20}$$

Proof: $((NM)^T)_{ij} = (NM)_{ji} = N_{jk}M_{ki} = N^T_{kj}M^T_{ik} = M^T_{ik}N^T_{kj} = (M^T N^T)_{ij}$. Note that in three of these expressions the index k is summed over.

Trace

An important concept is the trace of a matrix, defined as the sum of the diagonal elements:

$$\text{tr } M = \sum_i M_{ii} = M_{ii} \tag{21}$$

In the second equality we invoke Einstein's summation convention. For example, $\text{tr} \left(\begin{smallmatrix} a & b \\ c & d \end{smallmatrix} \right) = a + d$.

Another important theorem states that although in general AB and BA are not equal, their traces are always equal:

$$AB \neq BA \quad \text{but} \quad \text{tr } AB = \text{tr } BA \tag{22}$$

---

* By the way, this shows that exercise 4 follows from exercises 1–3 by transposition.

Proof: $\operatorname{tr} AB = (AB)_{ii} = A_{ij}B_{ji}$, while $\operatorname{tr} BA = (BA)_{ii} = B_{ij}A_{ji}$. But since the index i and index j are both summed over, these two expressions are in fact the same. In other words, we could simply rename the indices: $B_{ij}A_{ji} = B_{ji}A_{ij} = A_{ij}B_{ji}$. The last equality just states that the multiplication of numbers is commutative.

Using this and associativity, we see the cyclicity of the trace:

$$\operatorname{tr} ABC = \operatorname{tr}(AB)C = \operatorname{tr} C(AB) = \operatorname{tr} CAB \tag{23}$$

Under a trace, a product of matrices could be cyclically permuted. In particular, the trace in (23) is also equal to $\operatorname{tr} BCA$.

A quick summary

Here is a quick summary of what we have done thus far. From a system of linear equations, we abstracted the notion of a matrix and then were led naturally to multiplying matrices together. We discovered that matrix multiplication was associative but not commutative. Operations such as transposition and trace suggest themselves.

But we have not yet solved the system of linear equations we started out with. We will do so shortly.

The inverse

Another simple observation is that an n-by-n matrix M reduces to just an ordinary number m for $n = 1$. If $m \neq 0$, then it has an inverse m^{-1} such that $m^{-1}m = 1$. Indeed, $m^{-1} = 1/m$.

We might hope that if a matrix M satisfies some condition analogous to $m \neq 0$, then it has an inverse M^{-1} such that $M^{-1}M = I$. (Call this an example of physicist intuition if you like.) We will presently discover what that condition is.

If M does have an inverse, then (10), the equation $M\vec{x} = \vec{u}$ that started this entire discussion (for example, the chicken and rabbit problem), could be solved immediately by multiplying the equation by M^{-1} from the left:

$$\vec{x} = I\vec{x} = M^{-1}M\vec{x} = M^{-1}\vec{u} \tag{24}$$

If we know M^{-1}, then we simply act with it on $\vec{u}$ to obtain $\vec{x}$.

We are talking as if the existence of M^{-1} is hypothetical, but in fact, comparing (24) with (7), we can immediately read off the inverse M^{-1} of any 2-by-2 matrix $M = \begin{pmatrix} a & b \\ c & d \end{pmatrix}$, namely

$$M^{-1} = \frac{1}{\mathcal{D}} \begin{pmatrix} d & -b \\ -c & a \end{pmatrix} \tag{25}$$

where we have defined

$$\mathcal{D} \equiv ad - bc \tag{26}$$

Check it! The all-important quantity $\mathcal{D}$ is known as the determinant of the matrix M, written as $\det M$.

Note that M^{-1} exists if and only if the determinant, $\mathcal{D} = \det M$, does not vanish. This condition, that $\mathcal{D} \neq 0$, generalizes, to 2-by-2 matrices, the condition $m \neq 0$ that you learned in school necessary for a number m to have an inverse.

Inverting a matrix

Surely you know that famous joke about an engineer, a physicist, and a mathematician visiting New Zealand for the first time. No?

Well then, an engineer, a physicist, and a mathematician were traveling by train in New Zealand when they saw a black sheep. The engineer exclaimed, "Look, the sheep in New Zealand are black." The physicist objected, saying "You can't claim that. But wait, if we see another black sheep, or maybe yet another, then we can say almost for sure that all the sheep in New Zealand are black." The mathematician smirked, "All you can say is that, of the sheep in New Zealand we could see from this train, their sides facing the train are black."

Our strategy here, now that we have shown explicitly that the generic 2-by-2 matrix has an inverse, is that we will try to find the inverse of the generic 3-by-3 matrix. If we could do that, then by the standards of theoretical physics, we would have pretty much shown (hear the derisive howls of the mathematicians?) that the generic n-by-n matrix has an inverse.

Actually, our approach of inverting 2-by-2 and 3-by-3 matrices and hoping to see a pattern is in the spirit of how a lot of research in theoretical physics proceeds (rather than by the textbook "stroke of genius" method).

To find the inverse of the generic 3-by-3 matrix, we adopt the poor man's approach of solving the 3-by-3 analog of (1) and (2):

$$ax + by + cz = u \tag{27}$$

$$dx + ey + fz = v \tag{28}$$

$$gx + hy + iz = w \tag{29}$$

Perhaps the most elementary method is to eliminate one of the unknowns, say z, by forming the combination $i(27) - c(29)$:

$$(ai - cg)x + (bi - ch)y = iu - cw \tag{30}$$

Also, the combination $i(28) - f(29)$ gives

$$(di - fg)x + (ei - fh)y = iv - fw \tag{31}$$

We have thus reduced the problem to a previously solved problem,[6] namely, the system of equations

$$a'x + b'y = u' \tag{32}$$

$$c'x + d'y = v' \tag{33}$$

with $a' = ai - cg$ and so on, with the solution given in (4), (6), and (7), namely, $x = \frac{d'u' - b'v'}{a'd' - b'c'}$ and so on. We note that as $i \to 0$ both the numerator and denominator vanish. For example, $a'd' - b'c' \to cgfh - chfg = 0$. Thus, the numerator and the denominator must have a factor of i in common. Canceling this factor, we obtain

$$x = \frac{(ei - fh)u - (bi - ch)v + (bf - ce)w}{aei - bdi - afh - ceg + cdh + bfg} \tag{34}$$

Compare with (4). I cordially request the reader seeing this for the first time to solve for y and z. I will wait for you.

With $M = \begin{pmatrix} a & b & c \\ d & e & f \\ g & h & i \end{pmatrix}$, $\vec{x} = \begin{pmatrix} x \\ y \\ z \end{pmatrix}$, and $\vec{u} = \begin{pmatrix} u \\ v \\ w \end{pmatrix}$, write the three linear equations (27), (28), and (29) as $M\vec{x} = \vec{u}$.

Next, write the solution (a piece of which is in (34)) we (that includes you!) have just found as $\vec{x} = M^{-1}\vec{u}$, and read off the desired inverse M^{-1}.

For example, in the matrix M^{-1}, the entry in the first row and third column is equal to $(M^{-1})_{13} = (bf - ce)/\mathcal{D}$. (Do you see why?) Here we have once again defined the determinant $\mathcal{D}$ of M as the denominator in (34):

$$\mathcal{D} = \det M = aei - bdi - afh - ceg + cdh + bfg \tag{35}$$

Since you have solved for y and z, you can now write down M^{-1} in its entirety.

The determinant and the Laplace expansion

By the way, we could of course also have found M^{-1} by brute force: in the 2-by-2 case write it as $\begin{pmatrix} p & r \\ q & s \end{pmatrix}$ and solve the four equations contained in $\begin{pmatrix} p & r \\ q & s \end{pmatrix}\begin{pmatrix} a & b \\ c & d \end{pmatrix} = \begin{pmatrix} 1 & 0 \\ 0 & 1 \end{pmatrix}$ for the four unknowns p, q, r, s. (For example, one of the equations is $pa + rc = 1$.) Similarly for the 3-by-3 case, in which we have to solve nine equations for nine unknowns.

Clearly, neither this brute force method nor the method we followed by repeatedly reducing the problem to a problem we have solved before generalizes easily to the n-by-n case.

Instead, stare at the determinant $\mathcal{D} = ad - bc$ of the matrix $M = \begin{pmatrix} a & b \\ c & d \end{pmatrix}$. Note that, remarkably, if we define $A = d$, $B = -c$, $C = -b$, $D = a$, we can write $\mathcal{D}$ in four (count them, four) different ways, namely

$$\mathcal{D} = aA + bB = cC + dD = aA + cC = bB + dD \tag{36}$$

You say, this is silly. We have simply rewritten a, b, c, d using capital letters and thrown in some signs.

Indeed, apparently so, but the important point to focus on is that $\mathcal{D}$ can be written in four different ways. This is known as the Laplace expansion, named after a very clever guy indeed.

Notice that

$$\begin{pmatrix} A & C \\ B & D \end{pmatrix} \begin{pmatrix} a & b \\ c & d \end{pmatrix} = \mathcal{D} \begin{pmatrix} 1 & Y \\ X & 1 \end{pmatrix} \tag{37}$$

where the symbols X, Y denote elements as yet unknown. That the diagonal elements in the matrix on the right hand side are equal to 1 is guaranteed by the third and fourth form of $\mathcal{D}$ in (36).

We now show that both X and Y vanish.

By matrix multiplication, we see that in (37) $X = aB + cD$. By comparing this with the fourth form of $\mathcal{D}$ in (36), $\mathcal{D} = bB + dD$, we see that this is the determinant of a matrix obtained from $\begin{pmatrix} a & b \\ c & d \end{pmatrix}$ by replacing $b \to a$ and $d \to c$, namely, $\begin{pmatrix} a & a \\ c & c \end{pmatrix}$. Invoking (26), we see that the determinant of this matrix equals $ac - ac$, which all educated[7] people agree vanishes. I leave it to you to argue that Y also vanishes. Thus, we have found the inverse $M^{-1} = \frac{1}{\mathcal{D}} \begin{pmatrix} A & C \\ B & D \end{pmatrix}$ in agreement with the result in (7) and (25).

By now, the reader may be chuckling that this has got to be the most longwinded method for finding the inverse of a 2-by-2 matrix. But in mathematics and theoretical physics, it often pays to wander around the bush for a while rather than zero in by brute force.

Let's see if the pattern in the preceding discussion continues to hold in the 3-by-3 case. For the matrix $M = \begin{pmatrix} a & b & c \\ d & e & f \\ g & h & i \end{pmatrix}$, we have $\mathcal{D} = aei - bdi - afh - ceg + cdh + bfg$ as in (35).

Notice that in the 2-by-2 case, in the Laplace expansion, we can write $\mathcal{D}$ in (36) in $2 + 2 = 4$ ways; namely, (i) of a linear combination of a, b, the two elements in the first row; (ii) of a linear combination of c, d, the two elements in the second row; (iii) of a linear combination of a, c, the two elements in the first column; and (iv) of a linear combination of b, d, the two elements in the second column. So, the poor man reasons, there ought to be $3 + 3 = 6$ ways of writing $\mathcal{D}$ in the 3-by-3 case.

Sure enough, in the 3-by-3 case, we can collect terms and write

$$\mathcal{D} = aA + bB + cC \tag{38}$$

as a linear combination of a, b, c, the three elements in the first row. This is the analog of way (i). Inspecting (35), we have

$$A \equiv (ei - fh), \quad B \equiv -(di - fg), \quad C \equiv (dh - eg) \tag{39}$$

Furthermore, we recognize $A \equiv (ei - fh)$ as the determinant of the 2-by-2 matrix $\begin{pmatrix} e & f \\ h & i \end{pmatrix}$ obtained by crossing out in M the row and column a belongs to:

$$\begin{pmatrix} \cancel{a} & \cancel{b} & \cancel{c} \\ \cancel{d} & e & f \\ \cancel{g} & h & i \end{pmatrix} \tag{40}$$

Similarly, $B \equiv -(di - fg)$ is the determinant of the 2-by-2 matrix obtained by crossing out the row and column b belongs to, but now multiplied by (-1). I leave it to you to work out what C is.

Thus, in this case, the determinant is obtained as follows.

1. Take an entry in a particular row (namely, the first row).

2. Cross out the row and column the entry belongs to.

3. Calculate the determinant of the resulting 2-by-2 matrix.

4. Multiply the entry by this determinant.

5. Repeat for each entry in the row.

6. Sum with alternating signs.

In other words, the problem of evaluating a 3-by-3 determinant reduces to the previously solved problem of evaluating a 2-by-2 determinant.

Iterative evaluation of determinants

This yields an efficient way to evaluate determinants by hand, for 3-by-3 or even larger matrices, depending on your physical stamina. One standard notation for the determinant involves replacing the parentheses around the matrix by two vertical bars. Thus, we have found

$$\begin{vmatrix} a & b & c \\ d & e & f \\ g & h & i \end{vmatrix} = a \begin{vmatrix} e & f \\ h & i \end{vmatrix} - b \begin{vmatrix} d & f \\ g & i \end{vmatrix} + c \begin{vmatrix} d & e \\ g & h \end{vmatrix} \tag{41}$$

For those readers who are seeing this for the first time, a numerical example is

$$\begin{vmatrix} 2 & 0 & 1 \\ 3 & 1 & 1 \\ 1 & 2 & 1 \end{vmatrix} = 2 \begin{vmatrix} 1 & 1 \\ 2 & 1 \end{vmatrix} + \begin{vmatrix} 3 & 1 \\ 1 & 2 \end{vmatrix} = 2(1-2) + (6-1) = 3 \tag{42}$$

Note that this way of evaluating determinants applies to the 2-by-2 case also.

Nothing special about the first row. No surprise, we can also write $\mathcal{D}$ in (35) as a linear combination of the elements in the second row:

$$\mathcal{D} = dD + eE + fF \tag{43}$$

with

$$D \equiv -(bi - ch), \quad E \equiv (ai - cg), \quad F \equiv -(ah - bg) \tag{44}$$

In other words just the analog of (41) above. For example, F is just minus the determinant of the 2-by-2 matrix left over after crossing out the second row and the third column:

$$\begin{pmatrix} a & b & \cancel{c} \\ \cancel{d} & \cancel{e} & \cancel{f} \\ g & h & \cancel{i} \end{pmatrix} \tag{45}$$

You should do the same for the third row, working it out before reading further to make sure that you have it. You should find

$$\mathcal{D} = gG + hH + iI \tag{46}$$

with

$$G \equiv (bf - ce), \quad H \equiv -(af - cd), \quad I \equiv (ae - bd) \tag{47}$$

Indeed, what is the difference between rows and columns? None, in this context. We could also Laplace expand $\mathcal{D}$ as a linear combination of the elements in the first column:

$$\mathcal{D} = aA + dD + gG \tag{48}$$

Again, by simply staring at (35), we write down without any further ado that

$$A \equiv (ei - fh), \quad D \equiv -(bi - ch), \quad G \equiv (bf - ce) \tag{49}$$

The expressions for A, D, and G are manifestly consistent with what we have already gotten. I will leave you the pleasure of working out the second and third columns.

Clearly then, for the method of evaluating determinants given in (41), we could Laplace expand in any row or column we choose. The wise person would choose a row or a column with the largest number of zeroes in it. Let's illustrate with the numerical example in (42), now expanding in the second column instead of the first row:

$$\begin{vmatrix} 2 & 0 & 1 \\ 3 & 1 & 1 \\ 1 & 2 & 1 \end{vmatrix} = \begin{vmatrix} 2 & 1 \\ 1 & 1 \end{vmatrix} - 2 \begin{vmatrix} 2 & 1 \\ 3 & 1 \end{vmatrix} = (2 - 1) - 2(2 - 3) = 3 \tag{50}$$

in agreement with (42).

We can now find M^{-1} by the same laborious method we followed in the 2-by-2 case:

$$\begin{pmatrix} A & D & G \\ B & E & H \\ C & F & I \end{pmatrix} \begin{pmatrix} a & b & c \\ d & e & f \\ g & h & i \end{pmatrix} = \mathcal{D} \begin{pmatrix} 1 & X_1 & X_2 \\ X_3 & 1 & X_4 \\ X_5 & X_6 & 1 \end{pmatrix} \tag{51}$$

Again, the 1s in the matrix on the right hand side are guaranteed by the various expansions of $\mathcal{D}$ we have. Next, we show that the elements denoted generically by X vanish. For example, by matrix multiplication, $X_1 = bA + eD + hG$. Looking at (48), we see that this is the determinant of the matrix that results from the replacement $(a, d, g) \rightarrow (b, e, h)$,

$$\begin{pmatrix} b & b & c \\ e & e & f \\ h & h & i \end{pmatrix} \tag{52}$$

which manifestly vanishes. (Just evaluate the determinant by Laplace expanding in the third column.)

Notice that the first and second columns are identical. In the spirit of the physicist observing the sheep in New Zealand, we might propose a lemma that when two columns of a matrix are identical, the determinant vanishes; similarly when two rows are identical.

(Indeed, the astute reader might recall that the explanation I gave for X and Y vanishing in (37) invokes this lemma in the 2-by-2 case.)

Thus, we find

$$M^{-1} = \frac{1}{\mathcal{D}} \begin{pmatrix} A & D & G \\ B & E & H \\ C & F & I \end{pmatrix} \tag{53}$$

The right inverse is equal to the left inverse

Thus far, we have found the left inverse. To find the right inverse, we use a simple trick. Take the transpose of $M^{-1}M = I$ to obtain $M^T(M^{-1})^T = I$. In other words, the right inverse of M^T is just $(M^{-1})^T$. Since the left inverse of $M = \begin{pmatrix} a & b \\ c & d \end{pmatrix}$ is $\frac{1}{\mathcal{D}}\begin{pmatrix} A & C \\ B & D \end{pmatrix}$ (and we might as well illustrate with the 2-by-2 case), it follows that the right inverse of $M^T = \begin{pmatrix} a & c \\ b & d \end{pmatrix}$ is $\frac{1}{\mathcal{D}}\begin{pmatrix} A & B \\ C & D \end{pmatrix}$. But these squiggles called letters are just signs left to us by the Phoenicians; thus, we simply rename the letters b and c, B and C, and conclude that the right inverse of $M = \begin{pmatrix} a & b \\ c & d \end{pmatrix}$ is $\frac{1}{\mathcal{D}}\begin{pmatrix} A & C \\ B & D \end{pmatrix}$, which is exactly and totally the left inverse of M. Similarly for the 3-by-3 case. Of course, you could have also checked that the left inverse also works as the right inverse by direct multiplication.

The inverse is the inverse, no need to specify left or right. Groups also enjoy this important property, as we shall see.

Determinant and permutation

After this "baby stuff" we are now ready for the n-by-n case.

That $\mathcal{D}$ has $2 = 2!$ terms for a 2-by-2 matrix and $6 = 3!$ terms for a 3-by-3 matrix suggests that $\mathcal{D}$ for an n-by-n matrix has $n!$ terms. Since the number of permutations of n objects is given by $n!$, this suggests that $\mathcal{D}$ consists of a sum over permutations.

Evidently, continuing to write letters $a, b, c, \ldots$ is a losing proposition at this point. We invoke the magic trick of using indices.

To warm up, let us (once again, it would be instructive for you to do it) write the known expressions for $\mathcal{D}$ in terms of the matrix elements M_{ij}. For $n = 2$,

$$\mathcal{D} = ac - bd = M_{11}M_{22} - M_{12}M_{21} \tag{54}$$

and for $n = 3$,

$$\begin{aligned} \mathcal{D} &= aei - bdi - afh - ceg + cdh + bfg \\ &= M_{11}M_{22}M_{33} + M_{13}M_{21}M_{32} + M_{12}M_{23}M_{31} \\ &\quad - M_{12}M_{21}M_{33} - M_{11}M_{23}M_{32} - M_{13}M_{22}M_{31} \end{aligned} \tag{55}$$

Stare at these expressions for $\mathcal{D}$ for a while, and see if you can write down the expression for general n. The pattern in (55) is clear: write down $M_{1()}M_{2()}M_{3()}$, and insert all 3!

permutations of 1, 2, 3 into the empty slots, and attach a + or − sign according to whether the permutation is even or odd.

I trust that you would see that the general expression is given by

$$\mathcal{D} = \sum_{P} \text{sign}(P) M_{1P(1)} M_{2P(2)} M_{3P(3)} \cdots M_{nP(n)} \tag{56}$$

This may look complicated, but is actually very simple to describe in words. The sum runs over the $n!$ permutations of n objects. The permutation P takes the ordered set $(123 \cdots n)$ to $(P(1)P(2)P(3) \cdots P(n))$. Each term is given by the product of n matrix elements, $M_{iP(i)}$ for $i = 1, \cdots, n$, multiplied by $\text{sign}(P) = \pm 1$ depending on whether the permutation P is even or odd, respectively. Check this for $n = 2, 3$.

General properties of the determinant

We can now deduce various properties of the determinant.

1. If we multiply all the elements in any one row of the matrix M by λ, $\mathcal{D} \to \lambda \mathcal{D}$.

   This is clear. Suppose we pick the fifth row. Then, in (56), $M_{5P(5)} \to \lambda M_{5P(5)}$, and hence the stated property of $\mathcal{D}$ follows. In particular, if we flip the sign of all the elements in any one row of M, then $\mathcal{D}$ flips sign. Note the emphasis on the word "all."

2. If we interchange any two rows, $\mathcal{D}$ flips sign.

   To be definite and to ease writing, suppose we interchange the first two rows. The elements of the new matrix M' are related to the old matrix M by (for $k = 1, \cdots, n$) $M'_{ik} = M_{ik}$ for $i \neq 1, 2$ (in other words, anything outside the first two rows is left untouched) and $M'_{1k} = M_{2k}$, $M'_{2k} = M_{1k}$. According to (56),

   $$\mathcal{D}(M') = \sum_{P} \text{sign}(P) M'_{1P(1)} M'_{2P(2)} M'_{3P(3)} \cdots M'_{nP(n)}$$

   $$= \sum_{P} \text{sign}(P) M_{1P(2)} M_{2P(1)} M_{3P(3)} \cdots M_{nP(n)}$$

   $$= -\sum_{Q} \text{sign}(Q) M_{1Q(1)} M_{2Q(1)} M_{3Q(3)} \cdots M_{nQ(n)}$$

   $$= -\mathcal{D}(M) \tag{57}$$

   For the next to last equality, we define the permutation Q by $Q = PX$, where X is the permutation that exchanges 1 and 2, leaving the other $(n - 2)$ integers unchanged. (The multiplication of permutations should be intuitively clear: we first perform one permutation X, and then the permutation P. In fact, this foreshadows group multiplication, as will be discussed in chapter I.1.)

3. If any two rows of a matrix are identical, the determinant of the matrix vanishes.

   This follows immediately from property 2. If we interchange the two identical rows, the matrix remains the same, but the determinant flips sign.

4. If we multiply any row (say, the jth row) by a number λ and then subtract the result from any other row (say, the ith row), $\mathcal{D}$ remains unchanged.

For the sake of definiteness, let $i = 3$, $j = 5$. We change the matrix by letting $M_{3k} \rightarrow M_{3k} - \lambda M_{5k}$, for $k = 1, \cdots, n$. Then according to (56),

$$\mathcal{D} \rightarrow \sum_P \text{sign}(P) M_{1P(1)} M_{2P(2)} (M_{3P(3)} - \lambda M_{5P(3)}) M_{4P(4)} M_{5P(5)} \cdots M_{nP(n)}$$

$$= \mathcal{D} - \lambda \sum_P \text{sign}(P) M_{1P(1)} M_{2P(2)} M_{5P(3)} M_{4P(4)} M_{5P(5)} \cdots M_{nP(n)}$$

$$= \mathcal{D} \tag{58}$$

The last equality follows from property 3, because the displayed sum is the determinant of a matrix whose third and fifth rows are identical.

5. All the preceding statements hold if we replace the word "row" by the word "column."

   A really simple proof: Just rotate your head by 90°.

The reader who did the exercises as we went along will recognize that the operations described here are precisely the elementary row and column operations described in exercises 1–5.

Since we can interchange rows and columns, it follows that

$$\det M = \det M^T \tag{59}$$

Evaluating determinants

In simple cases, these properties enable us to evaluate determinants rather quickly. Again, let us illustrate with the example in (42):

$$\begin{vmatrix} 2 & 0 & 1 \\ 3 & 1 & 1 \\ 1 & 2 & 1 \end{vmatrix} = \begin{vmatrix} 0 & 0 & 1 \\ 1 & 1 & 1 \\ -1 & 2 & 1 \end{vmatrix} = \begin{vmatrix} 1 & 1 \\ -1 & 2 \end{vmatrix} = 2 - (-1) = 3 \tag{60}$$

The first equality results when we multiply the third column by 2 and subtract it from the first column, and the second equality follows when we Laplace expand in the first row. The object is clearly to maximize the number of zeroes in a single row (or column). It goes without saying that the reader seeing this for the first time should practice this procedure. Make up your own numerical exercises. (Check your answers by evaluating the determinant in some other way or against a computer if you have one.)

This procedure could be systematized as follows. Given an arbitrary matrix M, we apply the various rules listed above to generate as many zeroes as possible, taking care not to change the determinant at each step. As we shall see presently, we could turn M into an upper triangular matrix, with all elements below the diagonal equal to 0.

So let's do it. Subtract the first row times (M_{21}/M_{11}) from the second row. (I will let you figure out what to do if $M_{11} = 0$.) The new matrix M' has $M'_{21} = 0$; that is, the element in the second row and first column of M' vanishes. Erase the prime on M'. Next, subtract the first row times (M_{31}/M_{11}) from the third row. The new matrix M' has $M'_{31} = 0$. Erase the prime on M'. Repeat this process until $M_{k1} = 0$ for $k > 1$, that is, until the entire first column has only one nonvanishing element.

Next, subtract the second row times (M_{32}/M_{22}) from the third row. The new matrix M' has $M'_{32} = 0$. Erase the prime on M'. Repeat this process until $M_{k2} = 0$ for $k > 2$, that is, until the entire second column has only one nonvanishing element.

Eventually, we end up with an upper triangular matrix of the form

$$
M = \begin{pmatrix}
* & * & * & \cdots & * \\
0 & * & * & \cdots & * \\
0 & 0 & * & \cdots & * \\
\vdots & \vdots & \vdots & \ddots & \vdots \\
0 & 0 & 0 & \cdots & *
\end{pmatrix}
\tag{61}
$$

where $*$ denotes elements that are in general not zero. We have succeeded in our stated goal of making all elements below the diagonal vanish.

Throughout this process, the determinant of M remains unchanged. Now we can evaluate the determinant by Laplace expanding in the first column, then in the second column, and so on. It follows that $\mathcal{D}(M)$ is given by the product of the diagonal elements in (61).

We illustrate the procedure described in this section with the same numerical example in (42):

$$
\begin{vmatrix}
2 & 0 & 1 \\
3 & 1 & 1 \\
1 & 2 & 1
\end{vmatrix}
=
\begin{vmatrix}
2 & 0 & 1 \\
0 & 1 & -\frac{1}{2} \\
0 & 2 & \frac{1}{2}
\end{vmatrix}
=
\begin{vmatrix}
2 & 0 & 1 \\
0 & 1 & -\frac{1}{2} \\
0 & 0 & \frac{3}{2}
\end{vmatrix}
= 2 \cdot 1 \cdot \frac{3}{2} = 3
\tag{62}
$$

The first equality follows from subtracting appropriate multiples of the first row from the second and from the third row. The second equality follows from subtracting an appropriate multiple of the second row from the third row. The result agrees (of course) with the result we already have obtained three times.

Evaluating the determinant by reducing the matrix to diagonal form

The readers who dutifully did exercises 1–3 will recognize that the manipulations we used to reach (61) are the three elementary row operations. For the purpose of evaluating the determinant, we might as well stop once we get to (61), but we could also continue using the three elementary column operations to knock off all the off-diagonal elements.

For the first step, multiply the first column by (M_{12}/M_{11}) and subtract it from the second column. The new matrix M' has $M'_{12} = 0$; that is, the element in the first row and second column of M', an element located above the diagonal, also vanishes. Erase the prime on M'. Subtracting the appropriate multiple of the first column and of the second column from the third column, we could set M_{13} and M_{23} to 0. I will let you carry on. Eventually, you end up with a diagonal matrix.

This is best illustrated by the example in (62). Here the first step is not needed since $M_{12} = 0$ already; we knock off M_{13} and M_{23} in two steps:

$$\begin{vmatrix} 2 & 0 & 1 \\ 0 & 1 & -\frac{1}{2} \\ 0 & 0 & \frac{3}{2} \end{vmatrix} = \begin{vmatrix} 2 & 0 & 0 \\ 0 & 1 & -\frac{1}{2} \\ 0 & 0 & \frac{3}{2} \end{vmatrix} = \begin{vmatrix} 2 & 0 & 0 \\ 0 & 1 & 0 \\ 0 & 0 & \frac{3}{2} \end{vmatrix} \tag{63}$$

We end up with a diagonal matrix, as promised. Note that the diagonal elements are unaffected in this procedure, and so the determinant is still equal to 3, of course.

In practice, few physicists these days would evaluate anything larger than a 3-by-3 matrix by hand, unless the matrix in question is full of zeroes.

Proof of the Laplace expansion

The general expression (56) for the determinant now allows us to prove the Laplace expansion. Write it (using what we all learned in elementary school) as

$$\mathcal{D} = \sum_P M_{1P(1)}[\text{sign}(P)M_{2P(2)}M_{3P(3)}\cdots M_{nP(n)}] \tag{64}$$

namely, a linear combination of the elements M_{1j}, $j = 1, \cdots, n$ in the first row with coefficients given by the square bracket. Probably many readers will have already seen that the expression in the square bracket is equal to, up to a sign, the determinant of an $(n - 1)$-by-$(n - 1)$ matrix, and so the case is more or less closed. I will try to describe the square bracket in words, but this may be one of those instances when a mathematical expression says it much more clearly than the words that describe the expression. But still, let me try.

For definiteness, let $n = 4$ and $P(1) = 3$. The coefficient of M_{13} in $\mathcal{D}$ is then $(-1)^{3-1}\{\sum_Q \text{sign}(Q)M_{2Q(2)}M_{3Q(3)}M_{4Q(4)}\}$, where Q is a permutation defined as follows. Since $P(1)$ is already nailed down to be 3, write $P(2) = \alpha$, $P(3) = \beta$, and $P(4) = \gamma$, with $\alpha, \beta, \gamma = 1, 2, 4$ in some order. Then the sum over Q runs over Q of the form $Q(2) = \alpha$, $Q(3) = \beta$, $Q(4) = \gamma$. The curly bracket is precisely the determinant of the 3-by-3 matrix obtained by crossing out in M the first row and the third column.

Indeed, define $\tilde{M}(j, i)$ as the $(n - 1)$-by-$(n - 1)$ matrix obtained by crossing out in M the jth row and the ith column. Then we can write (64) as

$$\mathcal{D} = \sum_P M_{1, P(1)}(-1)^{P(1)-1} \det \tilde{M}(1, P(1)) \tag{65}$$

This gives the Laplace expansion in the first row. I invite you to write the corresponding expression for Laplace expanding in an arbitrary row and in an arbitrary column.

We have thus derived in (56) the iterative structure of the determinant (as we have encountered in our examples above), allowing us to write the determinant of an n-by-n matrix in terms of the determinant of an $(n - 1)$-by-$(n - 1)$ matrix.

The determinant and the antisymmetric symbol

The determinant of an n-by-n matrix can be written compactly in terms of the anti-symmetric symbol $\varepsilon^{ijk\cdots m}$ carrying n indices, defined by the two properties:

$$\varepsilon^{\cdots l \cdots m \cdots} = -\varepsilon^{\cdots m \cdots l \cdots} \quad \text{and} \quad \varepsilon^{12 \cdots n} = 1 \tag{66}$$

Each index can take on only values $1, 2, \cdots n$. The first property listed says that ε flips sign upon the interchange of any pair of indices. It follows that ε vanishes when two indices are equal. (Note that the second property listed is then just normalization.) For example, for $n = 2$, $\varepsilon^{12} = -\varepsilon^{21} = 1$, with all other components vanishing. For $n = 3$, $\varepsilon^{123} = \varepsilon^{231} = \varepsilon^{312} = -\varepsilon^{213} = -\varepsilon^{132} = -\varepsilon^{321} = 1$, with all other components vanishing.

The determinant of any matrix M is then given by

$$\varepsilon^{pqr \cdots s} \det M = \varepsilon^{ijk \cdots m} M^{ip} M^{jq} M^{kr} \cdots M^{ms} \tag{67}$$

where the repeated indices $ijk \cdots m$ are summed over. You can readily see how this works. For example, set $pqr \cdots s = 123 \cdots n$. Then the sum over the repeated indices on the right hand side just reproduces (56) with rows and columns interchanged. The relation (67) will be of great importance when we discuss the special orthogonal groups in chapter IV.1.

Cramer's formula for the inverse

In 1750, the Swiss mathematician Gabriel Cramer published the formula for the inverse of an n-by-n matrix:

$$(M^{-1})_{ij} = \frac{1}{\mathcal{D}} (-)^{i+j} \det \tilde{M}(\hat{j}, \hat{i}) \tag{68}$$

The matrix $\tilde{M}(\hat{j}, \hat{i})$ was defined earlier as the $(n-1)$-by-$(n-1)$ matrix obtained by crossing out in M the jth row and the ith column. As usual, $\mathcal{D} = \det M$. Indeed, (25) furnishes the simplest example of Cramer's formula (for $n = 2$).

Note the transposition of i and j in the definition of $\tilde{M}(\hat{j}, \hat{i})$. In the 3-by-3 example in (53), the $(1, 3)$ entry of M^{-1} for instance (namely, the quantity G), is equal to $(bf - ce)$, the determinant of the 2-by-2 matrix obtained by crossing out in M the row and column the $(3, 1)$ (note $(3, 1)$, not $(1, 3)$!) entry, namely g, belongs to.

All this is a mouthful to say, but easy to show on a blackboard. Anyway, the diligent reader is cordially invited to verify Cramer's formula by calculating $M_{ki}(M^{-1})_{ij}$ (repeated index summation convention in force). You will need (56), of course.

Another interim summary

Again, an interim summary is called for, as we have covered quite a bit of ground. In our quest for the matrix inverse M^{-1}, we encounter the all-important concept of the determinant. We learned how to evaluate the determinant $\mathcal{D}(M)$ by Laplace expanding and discovered some properties of $\mathcal{D}(M)$. Any matrix can be transformed to an upper triangular form and to a diagonal form without changing the determinant. The entries of the inverse, M^{-1}, which sparked our quest in the first place, are given as ratios of determinants.

Hermitean conjugation

> There exists, if I am not mistaken, an entire world which is the
> totality of mathematical truths, to which we have access only
> with our mind, just as a world of physical reality exists, the one
> like the other independent of ourselves, both of divine creation.
>
> —Charles Hermite[8]

I totally agree with Hermite. No doubt, intelligent beings everywhere in the universe would have realized the importance of hermitean conjugation.

In our discussion, what is required of the entries of the matrices, which we denote by $a, b, c, d, \cdots$, is that we are able to perform the four arithmetical operations on them, namely, addition, subtraction, multiplication, and division. (These operations are needed to evaluate Cramer's formula, for example.) Entities on which these four operations are possible are called fields or division algebras. Real and complex numbers are of course the standard ones that physicists deal with.[9]

Therefore, in general, we should think of the entries in our matrices as complex numbers, even though our illustrative examples thus far favor simple real numbers.

Given a matrix M, its complex conjugate M^* is defined to be the matrix whose entries are given by $(M^*)_{ij} = (M_{ij})^*$. To obtain M^*, simply complex conjugate every entry in M.

The transpose of M^*, namely $(M^*)^T = (M^T)^*$ (verify the equality), is called the hermitean conjugate of M, in honor of the French mathematician Charles Hermite, and is denoted by $M^\dagger$. Hermitean conjugation turns out to be of fundamental importance in quantum physics.

A matrix whose entries are all real is called a real matrix. More formally, if $M^* = M$, then M is real.

A matrix such that $M^T = M$ is called symmetric.

A matrix such that

$$M^\dagger = M \tag{69}$$

is called hermitean. As we shall see, hermitean matrices have many attractive properties. A real symmetric matrix is a special case of a hermitean matrix.

Matrices that are not invertible

The appearance of $\mathcal{D} = \det M$ in the denominator in (68) is potentially worrisome; we've been told since childhood to stay away from vanishing denominators.

If $\mathcal{D} = 0$, then (68) informs us that $M^{-1} \propto 1/\mathcal{D}$ does not exist, and thus the solution $\vec{x} = M^{-1}\vec{u}$ of the problem $M\vec{x} = \vec{u}$ we started out with does not exist. An example suffices to show what is going on: suppose we are given $\begin{pmatrix} 2 & 1 \\ 6 & 3 \end{pmatrix} \begin{pmatrix} x \\ y \end{pmatrix} = \begin{pmatrix} u \\ v \end{pmatrix}$, that is, $2x + y = u$,

$6x + 3y = v$, to solve. The determinant of the relevant matrix M is equal to $2 \cdot 3 - 1 \cdot 6 = 0$, and thus its inverse M^{-1} does not exist. There is no solution for arbitrary u and v. (To see this, multiply the first equation by 3 and compare with the second equation. A solution exists only if v happens to be equal to $3u$.)

But what if $\vec{u}$ also vanishes? In other words, suppose we were asked to solve

$$M\vec{x} = \vec{0} \tag{70}$$

Then in the purported solution $\vec{x} = M^{-1}\vec{u}$ we have $0/0$ on the right hand side, and there is a fighting chance that a solution might exist. In the example just given, set $u = 0$, $v = 0$. Then $y = -2x$ solves it. In fact, an infinite number of solutions exist: if $\vec{x}$ solves $M\vec{x} = \vec{0}$, then $s\vec{x}$ is also a solution for any number s. (In the example, only the ratio of x and y is fixed.)

Finally, if $\mathcal{D} \neq 0$, then $M\vec{x} = \vec{0}$ does not have a nontrivial (that is, $\vec{x} \neq \vec{0}$) solution. This follows since in this case the inverse exists, which implies that $\vec{x} = M^{-1}\vec{0} = \vec{0}$.

The bottom line: $M\vec{x} = \vec{0}$ has a solution (other than $\vec{x} = \vec{0}$) if and only if $\det M = 0$.

Eigenvectors and eigenvalues

A matrix M acting on some arbitrary vector $\vec{x}$ will in general take it into some other vector $\vec{y} = M\vec{x}$, pointing in some direction quite different from the direction $\vec{x}$ points in.

An interesting question: Does there exist a special vector, call it $\vec{\psi}$, such that $M\vec{\psi} = \lambda\vec{\psi}$ for some complex number λ?

A vector with this special property is known as an eigenvector of the matrix M. The number λ is known as the eigenvalue associated with the eigenvector $\vec{\psi}$. In other words, M acting on $\vec{\psi}$ merely stretches $\vec{\psi}$ by the factor λ, without rotating it to point in some other direction. Evidently, $\vec{\psi}$ is a very special vector among all possible vectors.

Note that if $\vec{\psi}$ is an eigenvector, then $s\vec{\psi}$ is also an eigenvector for any number s: we are free to fix the normalization of the eigenvector.

Thus far, we speak of one eigenvector, but as we shall see presently, an n-by-n matrix in general has n eigenvectors, each with a corresponding eigenvalue.

So, let us now solve

$$M\vec{\psi} = \lambda\vec{\psi} \tag{71}$$

For a given M, both $\vec{\psi}$ and λ are unknown and are to be determined. The equation (71) is called an eigenvalue problem.

The first step is to rewrite (71) as $(M - \lambda I)\vec{\psi} = 0$, which we recognize as an example of (70). For $\vec{\psi} \neq 0$, the preceding discussion implies that

$$\det(M - \lambda I) = 0 \tag{72}$$

The matrix $(M - \lambda I)$ differs from M only along the diagonal, with the entry M_{ii} replaced by $M_{ii} - \lambda$ (here the repeated index summation convention is temporarily suspended). Laplace expanding, we see that, for M an n-by-n matrix, $\det(M - \lambda I)$ is an nth-degree polynomial in λ. The equation (72), with λ regarded as the unknown, will have n solutions according to the fundamental theorem of algebra.

For pedagogical clarity, let us first develop the theory for $n = 2$. For $M = \begin{pmatrix} a & b \\ c & d \end{pmatrix}$, the eigenvalue equation is a simple quadratic equation

$$\begin{vmatrix} a - \lambda & b \\ c & d - \lambda \end{vmatrix} = (a - \lambda)(d - \lambda) - bc = \lambda^2 - (a + d)\lambda + ad - bc = 0$$

with the two solutions: $\lambda_\pm = \frac{1}{2}\left[(a + d) \pm \sqrt{(a - d)^2 + 4bc}\right]$.

What about the eigenvectors? For each of the two λs, the corresponding eigenvector $\begin{pmatrix} x \\ y \end{pmatrix}$ is determined only up to a multiplicative constant (as remarked above) by $\begin{pmatrix} a - \lambda & b \\ c & d - \lambda \end{pmatrix}\begin{pmatrix} x \\ y \end{pmatrix} = 0$. By inspection, we see that the solution can be written as either $\begin{pmatrix} b \\ \lambda - a \end{pmatrix}$ or $\begin{pmatrix} \lambda - d \\ c \end{pmatrix}$. Indeed, the condition that these two vectors are proportional to each other, $b/(\lambda - a) = (\lambda - d)/c$, gives precisely the quadratic equation for λ above. Thus, we can take the two eigenvectors to be

$$\vec{\psi}_+ = \begin{pmatrix} b \\ \lambda_+ - a \end{pmatrix} = \begin{pmatrix} \lambda_+ - d \\ c \end{pmatrix} \quad \text{and} \quad \vec{\psi}_- = \begin{pmatrix} \lambda_- - d \\ c \end{pmatrix} = \begin{pmatrix} b \\ \lambda_- - a \end{pmatrix} \tag{73}$$

Understand clearly that there are two eigenvectors, $\vec{\psi}_+$ associated with λ_+ and $\vec{\psi}_-$ associated with λ_-. (Henceforth we will omit the arrow on top of the eigenvector.) Note that the overall normalization of the two eigenvectors is left for us to choose.

Hermitean and real symmetric matrices

Here a, b, c, and d are four complex numbers, and so the eigenvalues $\lambda_\pm$ are complex in general. Now notice that if M happens to be hermitean, that is, if $M = M^\dagger = M^{*\mathrm{T}}$, that is, if $\begin{pmatrix} a & b \\ c & d \end{pmatrix} = \begin{pmatrix} a^* & c^* \\ b^* & d^* \end{pmatrix}$, so that a and d are real and $c = b^*$, then $\lambda_\pm = \frac{1}{2}\left[(a + d) \pm \sqrt{(a - d)^2 + 4|b|^2}\right]$. The eigenvalues are guaranteed to be real! But note that the eigenvectors (73) are still complex.

Now that we understand the 2-by-2 case, we can readily prove the same theorem for the general n-by-n case. Since an nth-degree polynomial equation in general has n solutions, we label the eigenvalues λ_a and the corresponding eigenvectors ψ_a by an index $a = 1, \cdots, n$, satisfying the eigenvalue equation

$$M\psi_a = \lambda_a \psi_a \tag{74}$$

Taking the hermitean conjugate of the eigenvalue equation gives $\psi_a^\dagger M^\dagger = \psi_a^\dagger \lambda_a^*$, and thus $\psi_a^\dagger M^\dagger \psi_b = \lambda_a^* \psi_a^\dagger \psi_b$ upon multiplying by ψ_b from the right. In contrast, multiplying the eigenvalue equation from the left by $\psi_b^\dagger$ gives $\psi_b^\dagger M \psi_a = \lambda_a \psi_b^\dagger \psi_a$.

That's that in general, two equations, one involving M, the other $M^\dagger$. But now suppose that M is hermitean: $M = M^\dagger$. Relabel the indices $a \leftrightarrow b$ in the first of these two equations and subtract to obtain $0 = (\lambda_a - \lambda_b^*)\psi_b^\dagger \psi_a$.

Let $a = b$. Then $\lambda_a = \lambda_a^*$, which implies that λ_a is real (in agreement with the 2-by-2 case). All eigenvalues of a hermitean matrix are real.

Set $a \neq b$. In general, $\lambda_a \neq \lambda_b$, and we conclude that $\psi_b^\dagger \psi_a = 0$. For a hermitean matrix, different eigenvectors are complex orthogonal. (If for some reason, $\lambda_a = \lambda_b$, we have what is known as a degeneracy in quantum mechanics. See chapter III.1.)

We already mentioned that a real symmetric matrix is just a special case of a hermitean matrix. Note that for real symmetric matrices, not only are the eigenvalues real, but the eigenvectors also are real. You can see this explicitly in the 2-by-2 case. Prove it for the general case.

In summary, a hermitean matrix has real eigenvalues, and the eigenvectors are orthogonal: $\psi_b^\dagger \psi_a = 0$ for $a \neq b$. This is an important theorem for subsequent development. Since $\psi_a^\dagger \psi_a = \sum_{i=1}^n (\psi_a)_i^* (\psi_a)_i$ is a fortiori nonzero (for $\psi_a \neq 0$), we can always normalize the eigenvectors by $\psi_a \rightarrow \psi_a / (\psi_a^\dagger \psi_a)^{\frac{1}{2}}$, so that $\psi_a^\dagger \psi_a = 1$. Thus,

$$H = H^\dagger \Longrightarrow \lambda_a = \lambda_a^* \quad \text{and} \quad \psi_a^\dagger \psi_b = \delta_{ab} \tag{75}$$

Some readers may know that the concepts of eigenvalue and eigenvector are central to quantum mechanics. In particular, physical observables are represented by hermitean operators, or loosely speaking, hermitean matrices. Their eigenvalues are postulated to be measurable quantities. For example, the Hamiltonian is hermitean, and its eigenvalues are the energy levels of the quantum system. Thus, the theorem that the eigenvalues of hermitean matrices are real is crucial for quantum mechanics. Again, see chapter III.1.

Scalar products in complex vector space

When we first learned about vectors in everyday Euclidean spaces, we understood that the scalar product $\vec{v} \cdot \vec{w} = v^T w = \sum_{i=1}^n v_i w_i$ tells us about the lengths of vectors and the angle between them. For complex vectors, the theorem in (75) indicates that the natural generalization of the scalar product involves complex conjugation as well as transposition, that is, hermitean conjugation. The scalar product of two complex vectors ϕ and ψ is defined as $\phi^\dagger \psi = \phi^{*T} \psi = \sum_{i=1}^n \phi_i^* \psi_i$. The length squared $\psi^\dagger \psi$ of a complex vector is then manifestly real and positive.

A useful notation

At this point, let us introduce a notation commonly used in several areas of physics. Up to now, the indices carried by vectors and matrices have been written as subscripts. My only justification is that it would be sort of confusing for beginners to see superscripts right from the start. Anyway, I now decree that complex vectors carry an upper index: ψ^i. Next, for each vector carrying an upper index ϕ^i, introduce a vector carrying a lower index by defining

$$\phi_i \equiv \phi^{i*} \tag{76}$$

To be precise, the right hand side here is $(\phi^i)^*$, the complex conjugate of ϕ^i.

At the most pedestrian level, you could regard this as a cheap move merely to avoid writing the complex conjugation symbol $*$. But then it follows that the scalar product of two complex vectors is given by

$$\phi^\dagger \psi = \phi_i \psi^i \tag{77}$$

The right hand side is shorthand for $\sum_i \phi_i \psi^i = \sum_i \phi^{i*} \psi^i$.

To Einstein's repeated index summation convention we can now add another rule: an upper index must be summed with a lower index. We will never sum an upper index with an upper index, or sum a lower index with a lower index. With this rule, the matrix M would then have to be written as $M^i{}_j$ so that the vector $M\psi$ would be given by $(M\psi)^i = M^i{}_j \psi^j$. In some sense, then, the upper index is a row index, and the lower index a column index. This is consistent with the fact that in the scalar product (77) $\phi^\dagger \psi = \phi_i \psi^i$, the vector $\phi^\dagger$ is a row vector, that is, a 1-by-n rectangular matrix, and the lower index labels the n different columns.

Diagonalization of matrices

We can now show how to diagonalize a general n-by-n matrix M (not necessarily hermitean). Denote the n eigenvalues and eigenvectors of M by λ_a and ψ_a, respectively, for $a = 1, \cdots, n$. Define the n-by-n matrix

$$S = (\psi_1, \psi_2, \cdots, \psi_n) \tag{78}$$

In other words, the ath column of S is equal to ψ_a. (This notation was introduced earlier in (17), but now that we are more sophisticated, we drop the arrows and the small parentheses.) For example, for the 2-by-2 matrix we had before, $S = (\psi_+, \psi_-)$, with $\psi_\pm$ given by (73).

Write the inverse of the n-by-n matrix S in the form

$$S^{-1} = \begin{pmatrix} \phi_1 \\ \phi_2 \\ \vdots \\ \phi_n \end{pmatrix} \tag{79}$$

where each ϕ_a is to be regarded as a row vector. That $S^{-1}S = I$ implies that $\phi_a \psi_b = \delta_{ab}$: the row vectors ϕ_a and the column vectors ψ_b are orthonormal to each other.

With this notation, we see that the product of M and S is given by

$$MS = (M\psi_1, M\psi_2, \cdots, M\psi_n) = (\lambda_1 \psi_1, \lambda_2 \psi_2, \cdots, \lambda_n \psi_n) \tag{80}$$

The ath column of S is multiplied by the number λ_a. Multiplying from the left by S^{-1} gives

$$S^{-1}MS = \begin{pmatrix} \lambda_1 & 0 & 0 & \cdots & 0 \\ 0 & \lambda_2 & 0 & \cdots & 0 \\ 0 & 0 & \lambda_3 & \cdots & 0 \\ 0 & 0 & 0 & \ddots & \vdots \\ 0 & 0 & 0 & \cdots & \lambda_n \end{pmatrix} \equiv \mathrm{diag}\{\lambda_1, \lambda_2, \cdots, \lambda_n\} = \Lambda \tag{81}$$

(The self-evident notation diag$\{\cdots\}$ introduced here will clearly be useful.)

Given M, $S^{-1}MS$ defines the similarity transform of M by S. Here, a similarity transformation by the specific S in (78) has turned M into a diagonal matrix Λ with the eigenvalues of M along the diagonal. This makes total intuitive sense. The similarity transformation S takes the standard basis vectors $\begin{pmatrix} 1 \\ 0 \\ \vdots \\ 0 \end{pmatrix}, \begin{pmatrix} 0 \\ 1 \\ \vdots \\ 0 \end{pmatrix}$, and so on into the eigenvectors ψ_as. In the basis furnished by the ψ_as, the matrix M stretches each one of them by some factor λ_a. The matrix S^{-1} then takes the eigenvectors ψ_a back to the standard basis vectors. Thus, the net effect of $S^{-1}MS$ is described simply by the matrix Λ.

Note also that the eigenvectors ψ_a need not be normalized.

Occasionally, it does pay to be slightly more rigorous. Let us ask: Are there matrices that cannot be diagonalized? See if you can come up with an example before reading on. (Hint: Try some simple 2-by-2 matrices.) Scrutinize the construction given here for hidden assumptions. Indeed, we implicitly assumed that S has an inverse. This holds generically, but for some exceptional Ms, the determinant of S could vanish. This is explored further in exercise 9.

Diagonalizing hermitean matrices

The preceding discussion is for a general M. If M is hermitean, then we can say more. According to an earlier discussion, the eigenvalues λ_a are real and the eigenvectors are orthogonal: $\psi_a^\dagger \psi_b = 0$ for $a \neq b$.

Let us give S in the preceding section a new name: $U = (\psi_1, \psi_2, \cdots, \psi_n)$. By normalizing the eigenvectors, we can write $\psi_a^\dagger \psi_b = \delta_{ab}$; that is, $\phi_a = \psi_a^\dagger$. In other words,

$$S^{-1} = \begin{pmatrix} \psi_1^\dagger \\ \psi_2^\dagger \\ \vdots \\ \psi_n^\dagger \end{pmatrix}, \text{ namely, } U^\dagger. \text{ The identity } S^{-1}S = I \text{ translates to}$$

$$U^\dagger U = I \tag{82}$$

A matrix U that satisfies (82) is said to be unitary.

For later use, we mention parenthetically that a real matrix O that satisfies the condition

$$O^\mathsf{T} O = I \tag{83}$$

is said to be orthogonal. If U happens to be real, the condition (82) reduces to (83). In other words, a real unitary matrix is orthogonal.

Now repeat the argument in the preceding section: $MU = (\lambda_1 \psi_1, \lambda_2 \psi_2, \cdots, \lambda_n \psi_n)$, and $U^\dagger M U = \Lambda \equiv \mathrm{diag}\{\lambda_1, \lambda_2, \cdots, \lambda_n\}$. The input that M is hermitean has given us the additional information that Λ is real and that U is unitary.

We obtain another important theorem that we will use often. If M is hermitean, then it can be diagonalized by a unitary transformation; that is, for M hermitean, there exists a unitary matrix U such that

$$U^\dagger M U = \Lambda \tag{84}$$

is diagonal and real.

Again, a special case occurs when M is real symmetric. We learned earlier that not only are its eigenvalues real but also its eigenvectors are real. This implies that U is not only unitary, it is also orthogonal. To emphasize this fact, we give U a new name, O. For M real symmetric, there exists an orthogonal matrix O such that

$$O^T M O = \Lambda \tag{85}$$

is diagonal and real.

Applying the theorem in (22) about the trace we had before, we obtain

$$\mathrm{tr}\, S^{-1} M S = \mathrm{tr}\, S S^{-1} M = \mathrm{tr}\, M = \mathrm{tr}\, \Lambda = \sum_a \lambda_a \tag{86}$$

The trace of a matrix is the sum of its eigenvalues. As special cases, we have for M hermitean, $\mathrm{tr}\, U^\dagger M U = \mathrm{tr}\, M = \mathrm{tr}\, \Lambda$, and for M real symmetric, $\mathrm{tr}\, O^T M O = \mathrm{tr}\, M = \mathrm{tr}\, \Lambda$.

With the notation discussed earlier (see (77)), the matrix $U^\dagger M U$ is written as

$$(U^\dagger M U)^i{}_j = (U^\dagger)^i{}_k M^k{}_l U^l{}_j \tag{87}$$

When we discuss complex, hermitean, and unitary matrices, this notation, with upper and lower indices, will prove to be extremely useful, as we shall see in chapter II.2, for example. In contrast, when we discuss real and orthogonal matrices, we could stay with a notation in which all indices are subscripts.

Simultaneously diagonalizable

Given two matrices A and B, suppose we manage to diagonalize A. In other words, we find a matrix S such that $A^d = S^{-1} A S$ is diagonal. There is no reason that in the new basis B would also be diagonal. But this would be true if A and B commute.

First, let us define the commutator $[M, N] \equiv MN - NM$ for any two n-by-n matrices M and N. Thus, $[A, B] = 0$ if A and B commute.

To prove the assertion above, apply the similarity transformation to the vanishing commutator:

$$0 = S^{-1}[A, B]S = S^{-1}ABS - S^{-1}BAS = S^{-1}ASS^{-1}BS - S^{-1}BSS^{-1}AS$$
$$= A^d B^d - B^d A^d = [A^d, B^d] \tag{88}$$

In the fourth equality, we have merely defined $B^d \equiv S^{-1}BS$. But in fact the preceding equation tells us that B^d is diagonal. Simply write out the ij entry in this matrix equation:

$$0 = (A^d B^d)_{ij} - (B^d A^d)_{ij} = \sum_k A^d_{ik} B^d_{kj} - \sum_k B^d_{ik} A^d_{kj} = (A^d_{ii} - A^d_{jj})B^d_{ij} \tag{89}$$

Thus, for $i \neq j$, if $A^d_{ii} \neq A^d_{jj}$, then the off-diagonal elements $B^d_{ij} = 0$. In other words, B^d is diagonal.

What if $A^d_{ii} = A^d_{jj}$ for some specific i and j? For ease of exposition, relabel rows and columns so that $i = 1$, $j = 2$, and write $a = A^d_{11} = A^d_{22}$. Then the northwest corner of A^d is proportional to the 2-by-2 unit matrix I_2:

$$A^d = \begin{pmatrix} aI_2 & 0 & 0 & 0 & 0 \\ \hline 0 & A^d_{33} & 0 & 0 & 0 \\ 0 & 0 & A^d_{44} & 0 & 0 \\ 0 & 0 & 0 & \ddots & 0 \\ 0 & 0 & 0 & 0 & \ddots \end{pmatrix} \tag{90}$$

Assuming that A^d_{11} and A^d_{22} are the only two diagonal elements of A^d that are equal, we could conclude that B^d is diagonal except for the 2-by-2 block in its northwest corner. But now we could focus on this 2-by-2 matrix and diagonalize it by a similarity transformation. Since A^d is proportional to the identity matrix in that corner, it remains unchanged by this similarity transformation.

At this point, you could generalize to the case when three diagonal elements of A^d are equal, and so on.

If two matrices commute, then they are simultaneously diagonalizable. This important theorem, which we will use repeatedly in this book, is worth repeating. If $[A, B] = 0$, and if $A^d = S^{-1}AS$ is diagonal for some S, then $B^d = S^{-1}BS$ is also diagonal for the same S.

Count components

Let's count the independent components of different kinds of matrices. An n-by-n complex matrix M has n^2 complex components, that is, $2n^2$ real components. The hermiticity condition (69) actually amounts to $n + 2 \cdot \frac{1}{2}n(n-1) = n^2$ real conditions; the diagonal elements are required to be real, while the entries below the diagonal are determined by the entries above the diagonal. It follows that a hermitean matrix H has $2n^2 - n^2 = n^2$ real components. In particular, a general hermitean matrix has four real components for $n = 2$ and nine real components for $n = 3$.

Similarly, the unitary condition (82) amounts to n^2 real conditions. Thus, a unitary n-by-n matrix U also has n^2 real components.

In chapter IV.4, we will see that a general unitary matrix U can always be specified by a hermitean matrix H. This statement is indeed consistent with the counting done here.

Functions of matrices

Given a function $f(x) = \sum_{k=0}^{\infty} a_k x^k$ defined by its power series, we can define $f(M)$, for M some matrix, by the power series $f(M) = \sum_{k=0}^{\infty} a_k M^k$. For example, define $e^M \equiv \sum_{k=0}^{\infty} M^k/k!$; since we know how to multiply and add matrices, this series makes perfect sense. (Whether any given series converges is of course another issue.) We must be careful, however, in using various identities, which may or may not generalize. For example, the identity $e^a e^a = e^{2a}$ for a a real number, which we could prove by applying the binomial theorem to the product of two series (square of a series in this case) generalizes immediately. Thus, $e^M e^M = e^{2M}$. But for two matrices M_1 and M_2 that do not commute with each other, $e^{M_1} e^{M_2} \neq e^{M_1 + M_2}$.

Let M be diagonalizable so that $M = S^{-1} \Lambda S$ with $\Lambda = \text{diag}\{\lambda_1, \lambda_2, \cdots, \lambda_N\}$ a diagonal matrix. Since $M^k = (S^{-1} \Lambda S) \cdots (S^{-1} \Lambda S) = S^{-1} \Lambda^k S$, we have

$$f(M) = \sum_{k=0}^{\infty} a_k S^{-1} \Lambda^k S = S^{-1} f(\Lambda) S \tag{91}$$

But since Λ is diagonal, Λ^k is easily evaluated: $\Lambda^k = \text{diag}\{\lambda_1^k, \lambda_2^k, \cdots, \lambda_N^k\}$, so that $f(\Lambda)$ is just the diagonal matrix with the diagonal elements $f(\lambda_j)$. Functions of a matrix are readily evaluated after it is diagonalized.

Note also that

$$\text{tr } f(M) = \text{tr } S^{-1} f(\Lambda) S = \text{tr } f(\Lambda) \tag{92}$$

The function $f(x)$ does not have to be Taylor expanded around $x = 0$. For instance, to define $\log M$, the logarithm of a matrix, we could use the series $\log x = \log(1 - (1 - x)) = -\sum_{k=1}^{\infty} \frac{1}{n}(1 - x)^n$.

In particular, we learned that

$$\text{tr} \log M = \text{tr} \log \Lambda = \sum_a \log \lambda_a = \log \prod_a \lambda_a \tag{93}$$

More on determinants

To proceed further, we need to prove a theorem about determinants:

$$\det AB = \det A \det B \tag{94}$$

The determinant of a product is the product of the determinants.

To prove this, we have to go back to the elementary row and column operations discussed earlier in connection with the evaluation of determinants and in the exercises. Refer back to the discussion around (61). Denote the matrices effecting the elementary row and column operations by E generically. (The Es are sometimes called elementary matrices.) We will need a series of "small" lemmas, left for you to verify.

Lemma 1. The inverse of E is another E. For example, the inverse of $E = \begin{pmatrix} 1 & 0 \\ s & 1 \end{pmatrix}$ is $E = \begin{pmatrix} 1 & 0 \\ -s & 1 \end{pmatrix}$.

Lemma 2. We have $\det EM = \det E \det M$. For example, the E corresponding to multiplying a row by s has determinant equal to s, and indeed, the determinant of EM is equal to s times the determinant of M. Similarly, $\det ME = \det E \det M$ for the analogous column operations.

Lemma 3. As we have shown in the discussion around (61), any matrix M can be turned into a diagonal matrix D by the various elementary operations, thus $D = EE \cdots EME \cdots E$. The schematic notation should be clear: the Es denote a generic elementary matrix. But by lemma 1, this means $M = EE \cdots EDE \cdots E$.

Lemma 4. In lemma 3, D itself has the form of an E (that is, when it multiplies a matrix, from the right say, it multiplies each of the columns of the matrix by one of its diagonal elements). In other words, any matrix can be written as

$$M = EEE \cdots E \tag{95}$$

But then by repeated use of lemma 2, $\det M = \det E \det(EE \cdots E)$ is a product of $\det E$. By construction, the allowed Es have $\det E \neq 0$ (that is, we are not allowed to multiply a row by 0 on our way to (95)). The exceptional but trivial case is when D has some vanishing diagonal elements, in which case D does not have the form of an E, and $\det M = \det D = 0$.

We are now ready to prove the theorem. Let $C = AB$. Write $C = EE \cdots EB$. Then using the lemmas repeatedly, we obtain $\det C = \det E \det E \cdots \det(EB) = \det E \det E \cdots \det E \det B = \det(E \cdots) \det B = \det A \det B$. The theorem is proved.

Setting $B = A^{-1}$ in (94), we obtain

$$\det A^{-1} = (\det A)^{-1} \tag{96}$$

More generally, setting $B = A$ repeatedly, we obtain

$$\det A^n = (\det A)^n \tag{97}$$

The integer n can take on either sign.

Here is a slick plausibility argument. By the definition of the determinant, $\det AB$ is an elementary algebraic expression obtained by performing various arithmetical operations on the matrix elements of A and B. Think of it as a polynomial of the matrix elements of B. Suppose $\det B = 0$. Then there exists a vector $\vec{v}$ such that $B\vec{v} = 0$. It follows that $AB\vec{v} = 0$, and hence $\det AB = 0$ also. In other words, $\det AB$ vanishes when $\det B$ vanishes, whatever A might be. Hence $\det B$ must be a factor of $\det AB$; that is, $\det AB = \rho \det B$ for some ρ.

But from (59) and (20), we have det $AB = \det(AB)^T = \det B^T A^T$, and hence by the same argument, det $AB = \phi \det A^T = \phi \det A$ for some ϕ. It follows that det $AB = \det A \det B$. The matrices A and B have equal rights.

A useful identity for the determinant of a matrix

The theorem in the preceding section implies that det $S^{-1}MS = \det S^{-1}(MS) = \det(MS)S^{-1} = \det MSS^{-1} = \det M$. Thus, if M is diagonalizable, then det $M = \det \Lambda = \prod_{a=1}^{n} \lambda_a$. The determinant of a matrix is equal to the product of its eigenvalues.

A useful identity then follows from (93):

$$\det M = e^{\text{tr} \log M} \tag{98}$$

which is equivalently written as log det $M = \text{tr} \log M$.

Note that varying (98) in the manner of Newton and Leibniz, we obtain

$$\delta \det M = e^{\text{tr} \log M} \text{ tr } \delta(\log M) = (\det M) \text{ tr } M^{-1}\delta M = (\det M) \sum_{ij}(M^{-1})_{ij}\delta M_{ji} \tag{99}$$

This amounts to a formula for the ij element of the inverse M^{-1}, consistent of course with Cramer's formula.

The direct product of matrices

Given an n-by-n matrix C_{ab}, $a, b = 1, \cdots, n$, and a v-by-v matrix $\Gamma_{\alpha\beta}$, $\alpha, \beta = 1, \cdots, v$, define the direct product $M = C \otimes \Gamma$ as the nv-by-nv matrix given by

$$M_{a\alpha,b\beta} = C_{ab}\Gamma_{\alpha\beta} \tag{100}$$

(Think of the two indices of M as $a\alpha$ and $b\beta$, with the symbols $a\alpha$ and $b\beta$ running over nv values.) For example, given the 2-by-2 matrix $\tau = \begin{pmatrix} 0 & -i \\ i & 0 \end{pmatrix}$ and C some n-by-n matrix, then $M = C \otimes \tau$ is the $2n$-by-$2n$ matrix $\begin{pmatrix} 0 & -iC \\ iC & 0 \end{pmatrix}$.

Since $M_{a\alpha,b\beta}M'_{b\beta,c\gamma} = C_{ab}\Gamma_{\alpha\beta}C'_{bc}\Gamma'_{\beta\gamma} = (C_{ab}C'_{bc})(\Gamma_{\alpha\beta}\Gamma'_{\beta\gamma})$, it is particularly easy to multiply two direct product matrices together:

$$MM' = (C \otimes \Gamma)(C' \otimes \Gamma') = (CC' \otimes \Gamma\Gamma') \tag{101}$$

The direct product is a convenient way of producing larger matrices from smaller ones. We will use the direct product notation in chapters II.1, IV.8, and VII.1.

Dirac's bra and ket notation

I now introduce Dirac's bra and ket notation. Unaccountably, a few students would occasionally have difficulty understanding this elegant, and exceedingly useful, notation. Perhaps the association with a revered name in physics suggests something profound and mysterious, such as the Dirac equation* or the Dirac monopole. But heavens to Betsy, it is just an extremely appealing notation. Let me explain, adopting the lowest brow approach imaginable.

A glib way of explaining bras and kets is to appeal to the saying about "the tail that wags the dog." We will keep the tail and throw away the dog. I have already written the eigenvectors as ψ_a, $a = 1, \cdots, n$. Here a labels the eigenvector: it tells us which eigenvector we are talking about. Note that we are suppressing the index specifying the components of ψ_a. If we want to be specific, we write the jth component of ψ_a as ψ_a^j. One reason j is often suppressed is that the specific value of ψ_a^j is basis dependent. Our friend could be using a different basis related to our basis by a similarity transformation S, so that the jth component of her eigenvector $\psi_a'^j$, which for maximal clarity we might want to denote by a different Greek letter, ϕ_a^j, is given by $\psi_a'^j = \phi_a^j = S_{jl}\psi_a^l$.

After all this verbiage, you realize that the most important element in the symbol ψ_a^j is a! In particular, the letter ψ is really redundant, considerably more so than j; it is what I would call a coatrack to hang a on. In a flash of genius, Dirac said that we might as well write ψ_a^j as $|a\rangle$, known as a ket. The conjugate, $(\psi_a^j)^*$, regarded as a row vector, is written as $\langle a|$, known as a bra. Thus, under hermitean conjugation, $|a\rangle \leftrightarrow \langle a|$. By the way, Dirac invented the peculiar terms, bra and ket, by splitting the word bracket, as in the bracket $\langle \cdot \rangle$.

The scalar product $\sum_j (\psi_a^j)^* \psi_a^j$ is then written as $\langle a|a \rangle$, which is equal to 1 if ψ_a is properly normalized. Similarly, $\sum_j (\psi_b^j)^* \psi_a^j$ is written as $\langle b|a \rangle$ and orthonormality is expressed by

$$\langle b|a \rangle = \delta_{ba} \tag{102}$$

One tremendous advantage of Dirac's notation is that we could put whatever label into $|\rangle$ or $\langle|$ that we think best characterizes that ket or bra. For example, given a matrix M, we could denote the eigenvector with the eigenvalue λ by $|\lambda\rangle$: that is, define $|\lambda\rangle$ by $M|\lambda\rangle = \lambda|\lambda\rangle$.

Let us prove the important theorem we proved earlier about the eigenvalues of a hermitean matrix using Dirac's notation. Hermitean conjugating $M|\lambda\rangle = \lambda|\lambda\rangle$ gives $\langle\lambda| M^\dagger = \langle\lambda| \lambda^*$. Contracting with $|\rho\rangle$ then leads to $\langle\lambda| M^\dagger |\rho\rangle = \lambda^*\langle\lambda|\rho\rangle$. But contracting $M |\rho\rangle = \rho |\rho\rangle$ with $\langle\lambda|$ gives $\langle\lambda| M |\rho\rangle = \rho\langle\lambda |\rho\rangle$. If $M = M^\dagger$, then we can subtract one equation from another and obtain $0 = (\lambda^* - \rho)\langle\lambda|\rho\rangle$, which implies that λ is real if $\rho = \lambda$ (so that $\langle\lambda|\lambda\rangle \neq 0$) and that $\langle\lambda|\rho\rangle = 0$ if $\lambda^* \neq \rho$.

Of course this proof is exactly the same as the proof we gave earlier, but we have avoided writing a lot of unnecessary symbols, such as ψ, a, and b. Things look considerably cleaner.

Another advantage of the Dirac notation is that we can study a matrix M without committing to a specific basis. Let $|i\rangle$, $i = 1, 2, \cdots, n$ furnish an orthonormal complete set of kets, that is, $\sum_{i=1}^n |i\rangle \langle i| = I$. We often omit the summation symbol and write $I = |i\rangle \langle i|$. The specific components of M in this basis are then given by $\langle i| M |j\rangle = M^i_{\ j}$. Change of basis is then easily done. Let $|p\rangle$, $p = 1, 2, \cdots, n$ furnish another orthonormal complete

---

* See chapter VII.3, for example.

set of kets. Then $\langle p| M |q\rangle = \langle p |i\rangle \langle i| M |j\rangle \langle j| q\rangle$. (We merely inserted I twice, omitting the summation symbol; in other words, the left hand side here is simply $\langle p| I M I |q\rangle$.) This is just the similarity transformation $M'^p_{\ q} = (S^{-1})^P_{\ i} M^i_{\ j} S^j_{\ q}$, with the transformation matrix $S^j_{\ q} = \langle j| q\rangle$. (Note that $(S^{-1})^P_{\ i} S^i_{\ q} = \langle p |i\rangle \langle i| q\rangle = \langle p|q\rangle = \delta^p_{\ q}$.) We see the advantage of the Dirac notation: we don't have to bother putting a prime on M and to "waste" the letter S. These may seem like small things, but the notation, once you get used to it, renders a lot of equations more transparent just because of less clutter. We will be using this notation in chapter IV.2.

I might conclude this quick review of linear algebra by noting that what d'Alembert said was very French indeed. Still, I hope that you agree with what he said.

The reader being exposed to linear algebra for the first time could safely skip the following appendices, at least for now.

Appendix 1: Polar decomposition

Any matrix M can be written as $M = HU$, with H hermitean and U unitary.

This generalizes the usual polar decomposition of a complex number $z = re^{i\theta}$.

Proof: The matrix $MM^\dagger$ is hermitean and has real eigenvalues. These eigenvalues are not negative, since $MM^\dagger |\lambda\rangle = \lambda |\lambda\rangle$ implies $\lambda = \langle\lambda| MM^\dagger |\lambda\rangle = \langle\lambda| M |\rho\rangle \langle\rho| M^\dagger |\lambda\rangle = |\langle\lambda| M |\rho\rangle|^2 \geq 0$. In other words, there exists a unitary matrix V and a real diagonal matrix D^2 with non-negative elements such that

$$MM^\dagger = V D^2 V^\dagger = V D V^\dagger V D V^\dagger = (V D V^\dagger)(V D V^\dagger) = H^2 \tag{103}$$

where $H \equiv V D V^\dagger$ is hermitean. In a sense, we have taken the square root of $MM^\dagger$. (Note that we could also choose D to be non-negative real diagonal.) Define the matrix $U = H^{-1}M$, where we are assuming that H does not have zero eigenvalues and letting you deal with what happens otherwise. We now show that U is unitary by direct computation: $UU^\dagger = H^{-1}MM^\dagger H^{-1} = H^{-1}H^2 H^{-1} = I$. The theorem is proved.

We counted the real components of complex, hermitean, and unitary matrices earlier, and indeed $2n^2 = n^2 + n^2$. Note the composition is unique but not particularly attractive.

Appendix 2: Complex and complex symmetric matrices

The polar decomposition theorem (appendix 1) has a number of useful applications, in particle physics, for example.

Given a complex n-by-n matrix M, write it[10] in polar decomposition: $M = HU = V D V^\dagger U$. Thus we have

$$V^\dagger M W = D \tag{104}$$

where $W = U^\dagger V$ is a unitary matrix, and D is a positive real diagonal matrix. Let Ψ denote a phase matrix $\text{diag}\{e^{i\psi_1}, \cdots, e^{i\psi_n}\}$, then $\Psi^\dagger D \Psi = D$. Thus (104) determines the two unitary matrices V and W only up to $V \to V\Psi$ and $W \to W\Psi$.

Furthermore, let $M = M^T$ be symmetric.[11] Transpose $V^\dagger M W = D$ to get $W^T M^T V^* = W^T M V^* = D$. Now compare $V^\dagger M W = D$ and $W^T M V^* = D$.

You may be tempted to conclude that $W = V^*$, but first we have to deal with some phase degrees of freedom. Again, let Φ^2 be an arbitrary phase matrix (the square is for later convenience).

We replace D in the equation $W^T M V^* = D$ by $D = \Phi^{*2} D \Phi^2$ and conclude that $W = V^* \Phi^2$. Then the equation $V^\dagger M W = D$ becomes $\Phi^{*2} W^T M W = D$. Multiplying this equation by Φ from the left and Φ^* from the right, we obtain $W'^T M W' = D$ where $W' = W\Phi^*$. Dropping the prime, we have proved that for a symmetric complex matrix M, there always exists a unitary matrix W such that

$$W^T M W = D \tag{105}$$

with D positive real diagonal. Note that W^T, not $W^\dagger$, appears here.

It is worthwhile to remark that, in general, the diagonal elements of D are not the eigenvalues of M, since $\det(D - \lambda I) = \det(W^T M W - \lambda I) \neq \det(M - \lambda I)$. The inequality follows because W is unitary but not orthogonal. For the same reason, $\operatorname{tr} M \neq \operatorname{tr} D$.

Exercises

1 Let $E = \left(\begin{smallmatrix} 0 & 1 \\ 1 & 0 \end{smallmatrix} \right)$ and M be a 2-by-2 matrix. Show that the matrix EM is obtained from M by interchanging the first and second rows.

2 Let $E = \left(\begin{smallmatrix} s_1 & 0 \\ 0 & s_2 \end{smallmatrix} \right)$ and M be a 2-by-2 matrix. Show that the matrix EM is obtained from M by multiplying the elements in the first row by s_1 and the elements in the second row by s_2.

3 Let $E = \left(\begin{smallmatrix} 1 & 0 \\ s & 1 \end{smallmatrix} \right)$ and M be a 2-by-2 matrix. Show that the matrix EM is obtained from M by multiplying the first row by s and adding it to the second row. Similarly for the matrix $E = \left(\begin{smallmatrix} 1 & s \\ 0 & 1 \end{smallmatrix} \right)$: EM is obtained from M by multiplying the second row by s and adding it to the first row.

4 The three operations effected in exercises 1–3 are known as elementary row operations. The Es defined here are called elementary matrices. How do you effect the corresponding three elementary column operations on an arbitrary 2-by-2 matrix M? Hint: Multiply M by the appropriate E from the right.

5 Let $e = \left(\begin{smallmatrix} 1 & 0 & 0 \\ 0 & 0 & 1 \\ 0 & 1 & 0 \end{smallmatrix} \right)$. Show that multiplying a 3-by-3 matrix M from the right with e interchanges two columns in M, and that multiplying M from the left with e interchanges two rows in M.

6 Generalize the elementary row and column operations to n-by-n matrices.

7 Let $c = \left(\begin{smallmatrix} 0 & 0 & 1 \\ 1 & 0 & 0 \\ 0 & 1 & 0 \end{smallmatrix} \right)$ and d be a 3-by-3 diagonal matrix. Show that the similarity transformation $c^{-1}dc$ permutes the diagonal elements of d cyclically.

8 Show that the determinant of antisymmetric n-by-n matrices vanishes if n is odd.

9 Diagonalize the matrix $M = \left(\begin{smallmatrix} a & 1 \\ 0 & b \end{smallmatrix} \right)$. Show that M is not diagonalizable if $a = b$.

10 Write the polynomial equation satisfied by the eigenvalues of a matrix M that is diagonalizable as $\lambda^n + \sum_{k=1}^{n} c_k \lambda^{n-k} = 0$. Show that the matrix M satisfies the equation $M^n + \sum_{k=1}^{n} c_k M^{n-k} = 0$.

11 Show by explicit computation that the eigenvalues of the traceless matrix $M = \left(\begin{smallmatrix} c & a-b \\ a+b & -c \end{smallmatrix} \right)$ with a, b, c real have the form $\lambda = \pm w$, with w either real or imaginary. Show that for $b = 0$, the eigenvalues are real. State the theorem that this result verifies.

12 Let the matrix in exercise 11 be complex. Find its eigenvalues. When do they become real?

13 There were of course quite a few nineteenth-century theorems about matrices. Given A and B an m-by-n matrix and an n-by-m matrix, respectively, Sylvester's theorem states that $\det(I_m + AB) = \det(I_n + BA)$. Prove this for A and B square and invertible.

14 Show that $(M^{-1})^T = (M^T)^{-1}$ and $(M^{-1})^* = (M^*)^{-1}$. Thus, $(M^{-1})^\dagger = (M^\dagger)^{-1}$.

15 The n-by-n matrix M defined by $M_{ij} = x_j^{i-1}$ plays an important role in random matrix theory, for example. Show that $\det M$, known as the van der Monde determinant, is equal to (up to an overall sign) $\Pi_{i<j}(x_i - x_j)$. Hint: Write out M for $n = 2, 3$ to see what is going on.

16 Let the Cartesian coordinates of the three vertices of a triangle be given by (x_i, y_i) for $i = 1, 2, 3$. Show that the area of the triangle is given by

$$\text{Area of triangle} = \frac{1}{2} \det \begin{pmatrix} x_1 & x_2 & x_3 \\ y_1 & y_2 & y_3 \\ 1 & 1 & 1 \end{pmatrix}$$

Interestingly, this expression is the beginning of the notion of projective geometry and plays a crucial role in recent development in theoretical physics (see, for example, Henriette Elvang and Yu-tin Huang, Scattering Amplitudes in Gauge Theory and Gravity, p. 203).

Notes

1. Philip Roth told of a writer who wrote a sentence every day before lunch, and who would then turn the sentence around after lunch. See Fearful, p. 97.
2. This illustrates a basic principle of French haute cuisine made popular in the particle physics community by Murray Gell-Mann. See Fearful, p. 178. The matrices correspond to the pheasant, the vectors to the veal.
3. $M_{31} = g$, $M_{23} = f$.
4. Assuming that you can check your answers against those given by a friend or by a computer program capable of multiplying matrices together.
5. See Einstein Gravity in a Nutshell, p. 46, particularly the footnote.
6. About the reduction of a problem to a previously solved problem, there is also a famous joke about an engineer, a physicist, and a mathematician, but this time the joke is at the expense of the ethereal mathematician in favor of the practical engineer.
7. We are implicitly assuming that the entries in the matrices discussed here are either real or complex numbers, which commute on multiplication. Thus, some of the standard theorems of linear algebra do not hold for matrices whose elements are quarternions. Technically, real or complex numbers form a field, but quarternions do not.
8. Gaston Darboux. Eloges académiques et discours. Hermann, Paris, 1912, p. 142.
9. Some readers might know about quaternions, invented by W. R. Hamilton, as the next division of algebra after complex numbers. Complex numbers $a + bi$ are generalized to quarternionic numbers $a + bi + cj + dk$, with a, b, c, d real numbers and the "quarternionic units" postulated to satisfy $i^2 = j^2 = k^2 = -1$, and $ij = -ji = k$, $jk = -kj = i$, and $ki = -ik = j$. Note that $(a - bi - cj - dk)(a + bi + cj + dk) = a^2 + b^2 + c^2 + d^2$, which allows us to divide quarternionic numbers. The difficulty is that quarternionic numbers do not commute (this is conceptually to be distinguished from the noncommutativity of matrix multiplication) and hence many of the theorems we have proved do not hold for matrices with entries given by quarternionic numbers. For example, the determinant of $\begin{pmatrix} q_1 & q_1 \\ q_2 & q_2 \end{pmatrix}$ either vanishes or does not vanish, depending on whether we Laplace expand in the first row or the first column, respectively.
10. In particle physics, the M here stands for a quark or lepton mass matrix, and this discussion shows how to diagonalize such mass matrices. See chapter VII.4.
11. The neutrino Majorana mass matrix has this form. See chapter VII.4.

Part I Groups: Discrete or Continuous, Finite or Infinite

The notion of a group is defined and then illustrated with many examples. Finite groups are studied. The rotation group, as the canonical example of a continuous group, is approached in two different ways. The all-important concepts of Lie group and Lie algebra are introduced.

I.1 | Symmetry and Groups

Symmetry and Transformation

Symmetry plays a central role in modern theoretical physics.[1]

As the etymologist tells us, symmetry ("equal measure") originates in geometry ("earth measure"). We have a sense that an isosceles triangle is more symmetrical than an arbitrary triangle and that an equilateral triangle is more symmetrical than an isosceles triangle. Going further, we feel that a pentagon is more symmetrical than a square, a hexagon more symmetrical than a pentagon, and an $(n + 1)$-sided regular polygon is more symmetrical than an n-sided regular polygon. And finally, a circle is more symmetrical than any regular polygon.

The n-sided regular polygon is left unchanged by rotations through any angle that is an integer multiple of $2\pi/n$, and there are n of these rotations. The larger n is, the more such rotations there are. This is why mathematicians and physicists feel that the hexagon is more symmetrical than a pentagon: $6 > 5$, QED.

To quantify this intuitive feeling, we should thus look at the set of transformations that leave the geometrical figure unchanged (that is, invariant). For example, we can reflect the isosceles triangle across the median that divides it into equal parts (see figure 1a).

Call the reflection r; then the set of transformations that leave the isosceles triangle invariant is given by $\{I, r\}$, where I denotes the identity transformation, that is, the transformation that does nothing. A reflection followed by a reflection has the same effect as the identity transformation. We write this statement as $r \cdot r = I$.

In contrast, the equilateral triangle is left invariant not only by reflection across any of its three medians (figure 1b) but also by rotation R_1 through $2\pi/3 = 120°$ around its center, as well as rotation R_2 through $4\pi/3 = 240°$. The set of transformations that leave the equilateral triangle invariant is thus given by $\{I, r_1, r_2, r_3, R_1, R_2\}$. That this set is larger than the set in the preceding paragraph quantifies the feeling that the equilateral triangle is more symmetrical than the isosceles triangle. Note that $R_1 \cdot R_1 = R_2$ and that $R_1 \cdot R_2 = I$.

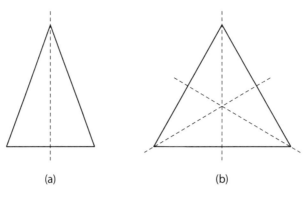

(a)   (b)

Figure 1

A circle is left invariant by an infinite number of transformations, namely, rotation $R(\theta)$ through any angle θ, and reflection across any straight line through its center. The circle is more symmetrical than any regular polygon, such as the equilateral triangle.

Symmetry in physics

In physics we are interested in the symmetries enjoyed by a given physical system. On a more abstract level, we are interested in the symmetries of the fundamental laws of physics. One of the most revolutionary and astonishing discoveries in the history of physics is that objects do not fall down, but toward the center of the earth. Newton's law of gravitation does not pick out a special direction: it is left invariant by rotations.

The history of theoretical physics has witnessed the discoveries of one unexpected symmetry after another. Physics in the late twentieth century consists of the astonishing discovery that as we study Nature at ever deeper levels, Nature displays more and more symmetries.[2]

Consider a set of transformations $T_1, T_2, \cdots$ that leave the laws of physics invariant. Let us first perform the transformation T_j, and then perform the transformation T_i. The transformation that results from this sequence of two transformations is denoted by the "product" $T_i \cdot T_j$. Evidently, if T_i and T_j leave the laws of physics invariant, then the transformation $T_i \cdot T_j$ also leaves the laws of physics invariant.[3]

Here we label the transformations by a discrete index i. In general, the index could also be continuous. Indeed, the transformation could depend on a number of continuous parameters. The classic example is a rotation $R(\theta, \varphi, \zeta)$, which can be completely characterized by three angles, as indicated. For example, in one standard parametrization,[4] the two angles θ and φ specify the unit vector describing the rotation axis, while the angle ζ specifies the angle through which we are supposed to rotate around that axis.

Groups

This discussion rather naturally invites us to abstract the concept of a group.

A group G consists of a set of entities $\{g_\alpha\}$ called group elements (or elements for short), which we could compose together (or more colloquially, multiply together). Given any two elements g_α and g_β, the product $g_\alpha \cdot g_\beta$ is equal to another element,[*] say, g_γ, in G. In other words, $g_\alpha \cdot g_\beta = g_\gamma$. Composition or multiplication[5] is indicated by a dot (which I usually omit if there is no danger of confusion). The set of all relations of the form $g_\alpha \cdot g_\beta = g_\gamma$ is called the multiplication table of the group.

Composition or multiplication (we will use the two words interchangeably) satisfies the following axioms:[†]

1. Associativity: Composition is associative: $(g_\alpha \cdot g_\beta) \cdot g_\gamma = g_\alpha \cdot (g_\beta \cdot g_\gamma)$.

2. Existence of the identity: There exists a group element, known as the identity and denoted by I, such that $I \cdot g_\alpha = g_\alpha$ and $g_\alpha \cdot I = g_\alpha$.

3. Existence of the inverse: For every group element g_α, there exists a unique group element, known as the inverse of g_α and denoted by g_α^{-1}, such that $g_\alpha^{-1} \cdot g_\alpha = I$ and $g_\alpha \cdot g_\alpha^{-1} = I$.

A number of comments follow.

1. Composition is not required to commute.[6] In general, $g_\alpha \cdot g_\beta$ is not equal to $g_\beta \cdot g_\alpha$. In this respect, the multiplication of group elements is, in general, like the multiplication of matrices but unlike that of ordinary numbers.

    A group for which the composition rule is commutative is said to be abelian,[‡] and a group for which this is not true is said to be nonabelian.[§]

2. The right inverse and the left inverse are by definition the same. We can imagine mathematical structures for which this is not true, but then these structures are not groups. Recall (or read in the review of linear algebra) that this property holds for square matrices: provided that the inverse M^{-1} of a matrix M exists, we have $M^{-1}M = MM^{-1} = I$ with I the identity matrix.

3. It is often convenient to denote I by g_0.

4. The label α that distinguishes the group element g_α may be discrete or continuous.

5. The set of elements may be finite (that is, $\{g_0, g_1, g_2, \cdots, g_{n-1}\}$), in which case G is known as a finite group with n elements. (Our friend the jargon guy[7] informs us that n is known as the order of the group.)

Mathematicians[8] of course can study groups on the abstract level without tying g_i to any physical transformation, but in some sense the axioms become clearer if we think of transformations in the back of our mind. For example, $gI = Ig = g$ says that the net effect of first doing nothing and then doing something is the same as first doing something and then doing nothing, and the same as doing something. Existence of the inverse says that the transformations of interest to physics can always[9] be undone.[10]

---

[*] This property, known as closure, is sometimes stated as an axiom in addition to the three axioms given below.

[†] See also appendices 1 and 2.

[‡] Named after the mathematician Niels Henrik Abel, one of the founders of group theory.

[§] As the reader might have heard, in contemporary physics, the theory of the strong, weak, and electromagnetic interactions are based on nonabelian gauge symmetries. See chapter IX.1.

Examples of groups

To gain a better understanding of what a group is, it is best to go through a bunch of examples. For each of the following examples, you should verify that the group axioms are satisfied.

1. Rotations in 3-dimensional Euclidean space, as already mentioned, form the poster child of group theory and are almost indispensable in physics. Think of rotating a rigid object, such as a bust of Newton. After two rotations in succession, the bust, being rigid, has not been deformed in any way: it merely has a different orientation. Thus, the composition of two rotations is another rotation.

    Rotations famously do not commute. See figure 2.

    Descartes taught us that 3-dimensional Euclidean space could be thought of as a linear vector space, coordinatized with the help of three unit basis vectors $\vec{e}_x = \begin{pmatrix} 1 \\ 0 \\ 0 \end{pmatrix}$, $\vec{e}_y = \begin{pmatrix} 0 \\ 1 \\ 0 \end{pmatrix}$, and $\vec{e}_z = \begin{pmatrix} 0 \\ 0 \\ 1 \end{pmatrix}$, aligned along three orthogonal directions traditionally named x, y, and z. A rotation takes each basis vector into a linear combination of these three basis vectors, and is thus described by a 3-by-3 matrix. This group of rotations is called $SO(3)$. We shall discuss rotations in great detail in chapter I.3; suffice it to mention here that the determinant of a rotation matrix is equal to 1.

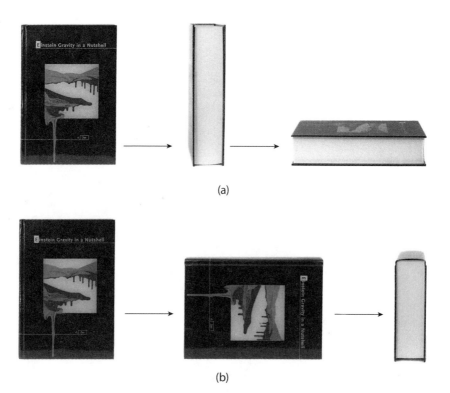

(a)

(b)

Figure 2

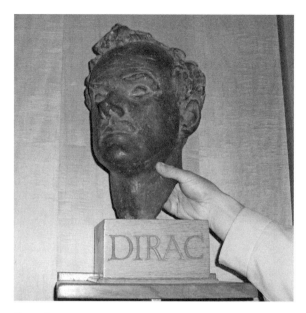

Figure 3

2. Rotations in 2-dimensional Euclidean space, namely a plane, form a group called $SO(2)$, consisting of the set of rotations around an axis perpendicular to the plane. Denote a rotation through angle ϕ by $R(\phi)$. Then $R(\phi_1)R(\phi_2) = R(\phi_1 + \phi_2) = R(\phi_2)R(\phi_1)$. These rotations commute. (See figure 3. I was surprised to discover that this bust of Dirac at St. John's College, Cambridge University, was not nailed down but could be rotated around the vertical axis. The photo depicts my attempt to give the bust a half-integral amount of angular momentum.)

3. The permutation group S_4 rearranges an ordered set of four objects, which we can name arbitrarily, for example, (A, B, C, D) or $(1, 2, 3, 4)$. An example would be a permutation that takes $1 \rightarrow 3, 2 \rightarrow 4, 3 \rightarrow 2$, and $4 \rightarrow 1$. As is well known, there are $4! = 24$ such permutations (since we have four choices for which number to take 1 into, three choices for which number to take 2 into, and two choices for which number to take 3 into). The permutation group S_n evidently has $n!$ elements. We discuss S_n in detail in chapter I.2.

4. Even permutations of four objects form the group A_4. As is also well known, a given permutation can be characterized as either even or odd (we discuss this in more detail in chapter I.2). Half of the 24 permutations in S_4 are even, and half are odd. Thus, A_4 has 12 elements. The jargon guy tells us that A stands for "alternating."

5. The two square roots of 1, $\{1, -1\}$, form the group Z_2 under ordinary multiplication.

6. Similarly, the three cube roots of 1 form the group $Z_3 = \{1, \omega, \omega^2\}$ with $\omega \equiv e^{2\pi i/3}$.

   Chugging right along, we note that the four fourth roots of 1 form the group $Z_4 = \{1, i, -1, -i\}$, where famously (or infamously) $i = e^{i\pi/2}$.

   More generally, the N Nth roots of 1 form the group $Z_N = \{e^{i2\pi j/N} : j = 0, \cdots, N-1\}$. The composition of group elements is defined by $e^{i2\pi j/N} e^{i2\pi k/N} = e^{i2\pi(j+k)/N}$.

   Quick question: Does the set $\{1, i, -1\}$ form a group?

7. Complex numbers of magnitude 1, namely $e^{i\phi}$, form a group called $U(1)$, with $e^{i\phi_1}e^{i\phi_2} = e^{i(\phi_1+\phi_2)}$. Since $e^{i(\phi+2\pi)} = e^{i\phi}$, we can restrict ϕ to range from 0 to 2π. At the level of physicist rigor, we can think of $U(1)$ as the "continuum limit" of Z_N with $e^{i2\pi j/N} \to e^{i\phi}$ in the limit $N \to \infty$ and $j \to \infty$ with the ratio held fixed $2\pi j/N = \phi$.

8. The addition of integers mod N generates a group. For example, under addition mod 5 the set $\{0, 1, 2, 3, 4\}$ forms a group: $2 + 1 = 3$, $3 + 2 = 0$, $4 + 3 = 2$, and so on. The composition of the group elements is defined by $j \cdot k = j + k$ mod 5. The identity element I is denoted by 0. The inverse of 2, for example, is 3, of 4 is 1, and so on. The group is clearly abelian. Question: Have you seen this group before?

9. The addition of real numbers form a group, perhaps surprisingly. The group elements are denoted by a real number u and $u \cdot v \equiv u + v$, where the symbol $+$ is what an elementary school student would call "add." You can easily check that the axioms are satisfied. The identity element is denoted by 0, and the inverse of the element u is the element $-u$.

10. The additive group of integers is obtained from the additive group of real numbers by restricting u and v in the preceding example to be integers of either sign, including 0.

11. As many readers know, in Einstein's theory of special relativity,[11] the spacetime coordinates used by two observers in relative motion with velocity v along the x-direction (say) are related by the Lorentz transformation (with c the speed of light):

$$ct' = \cosh\varphi\, ct + \sinh\varphi\, x$$
$$x' = \sinh\varphi\, ct + \cosh\varphi\, x$$
$$y' = y$$
$$z' = z \tag{1}$$

where the "boost angle" φ is determined by $\tanh\varphi = v$. (In other words, $\cosh\varphi = 1/\sqrt{1 - \frac{v^2}{c^2}}$, and $\sinh\varphi = \frac{v}{c}/\sqrt{1 - \frac{v^2}{c^2}}$.) Suppressing the y- and z-coordinates, we can describe the Lorentz transformation by

$$\begin{pmatrix} ct' \\ x' \end{pmatrix} = \begin{pmatrix} \cosh\varphi & \sinh\varphi \\ \sinh\varphi & \cosh\varphi \end{pmatrix} \begin{pmatrix} ct \\ x \end{pmatrix} \tag{2}$$

Physically, suppose a third observer is moving at a velocity defined by the boost angle φ_2 relative to the observer moving at a velocity defined by the boost angle φ_1 relative to the first observer. Then we expect the third observer to be moving at some velocity determined by φ_1 and φ_2 relative to the first observer. (All motion is restricted to be along the x-direction for simplicity.) This physical statement is expressed by the mathematical statement that the Lorentz transformations form a group:

$$\begin{pmatrix} \cosh\varphi_2 & \sinh\varphi_2 \\ \sinh\varphi_2 & \cosh\varphi_2 \end{pmatrix} \begin{pmatrix} \cosh\varphi_1 & \sinh\varphi_1 \\ \sinh\varphi_1 & \cosh\varphi_1 \end{pmatrix} = \begin{pmatrix} \cosh(\varphi_1+\varphi_2) & \sinh(\varphi_1+\varphi_2) \\ \sinh(\varphi_1+\varphi_2) & \cosh(\varphi_1+\varphi_2) \end{pmatrix} \tag{3}$$

The boost angles add.*

12. Consider the set of n-by-n matrices M with determinants equal to 1. They form a group under ordinary matrix multiplication, since as was shown in the review of linear algebra,

---

* To show this, use the identities for the hyperbolic functions.

the determinant of the product two matrices is equal to the product of the determinants of the two matrices: $\det(M_1 M_2) = \det(M_1) \det(M_2)$. Thus, $\det(M_1 M_2) = 1$ if $\det(M_1) = 1$ and $\det(M_2) = 1$: closure is satisfied. Since $\det M = 1 \neq 0$, the inverse M^{-1} exists. The group is known as $SL(n, R)$, the special linear group with real entries. If the entries are allowed to be complex, the group is called $SL(n, C)$. (Matrices with unit determinant are called special.)

From these examples, we see that groups can be classified according to whether they are finite or infinite, discrete or continuous. Note that a discrete group can well be infinite, such as the additive group of integers.

Concept of subgroup

In group theory, many concepts are so natural that they practically suggest themselves,[12] for example, the notion of a subgroup. Given a set of entities $\{g_\alpha\}$ that form a group G, if a subset $\{h_\beta\}$ also form a group, call it H, then H is known as a subgroup of G and we write $H \subset G$.

Here are some examples.

1. $SO(2) \subset SO(3)$. This shows that, in the notation $\{g_\alpha\}$ and $\{h_\beta\}$ we just used, the index sets denoted by α and β can in general be quite different; here α consists of three angles and β of one angle.

2. $S_m \subset S_n$ for $m < n$. Permuting three objects is just like permuting five objects but keeping two of the five objects untouched. Thus, $S_3 \subset S_5$.

3. $A_n \subset S_n$.

4. $Z_2 \subset Z_4$, but $Z_2 \not\subset Z_5$.

5. $SO(3) \subset SL(3, R)$.

Verify these statements.

Cyclic subgroups

For a finite group G, pick some element g and keep multiplying it by itself. In other words, consider the sequence $\{g, g^2 = gg, g^3 = g^2 g, \cdots\}$. As long as the resulting product is not equal to the identity, we can keep going. Since G is finite, the sequence must end at some point with $g^k = I$. The set of elements $\{I, g, g^2, \cdots, g^{k-1}\}$ forms a subgroup Z_k. Thus, any finite group has a bunch of cyclic subgroups. If k is equal to the number of elements in G, then the group G is in fact Z_k.

Lagrange's theorem

Lagrange[13] proved the following theorem. Let a group G with n elements have a subgroup H with m elements. Then m is a factor of n. In other words, n/m is an integer.

The proof is as follows. List the elements of H: $\{h_1, h_2, \cdots, h_m\}$. (Note: Since H forms a group, this list must contain I. We do not list any element more than once; thus, $h_a \neq h_b$ for $a \neq b$.) Let $g_1 \in G$ but $\notin H$ (in other words, g_1 is an element of G outside H). Consider the list $\{h_1 g_1, h_2 g_1, \cdots, h_m g_1\}$, which we denote by $\{h_1, \cdots, h_m\} g_1$ to save writing. Note that this set of elements does not form a group. (Can you explain why not?)

I claim that the elements on the list $\{h_1 g_1, h_2 g_1, \cdots, h_m g_1\}$ are all different from one another. Proof by contradiction: For $a \neq b$, $h_a g_1 = h_b g_1 \Longrightarrow h_a = h_b$ upon multiplication from the right by $(g_1)^{-1}$ (which exists, since G is a group).

I also claim that none of the elements on this list are on the list $\{h_1, \cdots, h_m\}$. Proof: For some a and b, $h_a g_1 = h_b \Longrightarrow g_1 = h_a^{-1} h_b$, which contradicts the assumption that g_1 is not in H. Note that H being a group is crucial here.

Next, pick an element g_2 of G not in the two previous lists, and form $\{h_1 g_2, h_2 g_2, \cdots, h_m g_2\} = \{h_1, h_2, \cdots, h_m\} g_2$.

I claim that these m elements are all distinct. Again, this proof follows by contradiction, which you can supply. Answer: For $a \neq b$, $h_a g_2 = h_b g_2 \Longrightarrow h_a = h_b$. I also claim that none of these elements are on the two previous lists. Yes, the proof proceeds again easily by contradiction. For example, $h_a g_2 = h_b g_1 \Longrightarrow g_2 = h_a^{-1} h_b g_1 = h_c g_1$, since H is a group, but this would mean that g_2 is on the list $\{h_1, h_2, \cdots, h_m\} g_1$, which is a contradiction.

We repeat this process. After each step, we ask whether there is any element of G left that is not on the lists already constructed. If yes, then we repeat the process and construct yet another list containing m distinct elements. Eventually, there is no group element left (since G is a finite group). We have constructed k lists, including the original list $\{h_1, h_2, \cdots, h_m\}$, namely, $\{h_1, h_2, \cdots, h_m\} g_j$ for $j = 0, 1, 2, \cdots, k-1$ (writing I as g_0).

Therefore $n = mk$, that is, m is a factor of n. QED.

As a simple example of Lagrange's theorem, we can immediately state that Z_3 is a subgroup of Z_{12} but not of Z_{14}. It also follows trivially that if p is prime, then Z_p does not have a nontrivial subgroup. From this you can already sense the intimate relation between group theory and number theory.

Direct product of groups

Given two groups F and G (which can be continuous or discrete), whose elements we denote by f and g, respectively, we can define another group $H \equiv F \otimes G$, known as the direct product of F and G, consisting of the elements (f, g). If you like, you can think of the symbol (f, g) as some letter in a strange alphabet. The product of two elements (f, g) and (f', g') of H is given by $(f, g)(f', g') = (ff', gg')$. The identity element of H is evidently given by (I, I), since $(I, I)(f, g) = (If, Ig) = (f, g)$ and $(f, g)(I, I) = (fI, gI) = (f, g)$. (If we were insufferable pedants, we would write $I_H = (I_F, I_G)$, since the identity elements I_H, I_F, I_G of the three groups H, F, G are conceptually quite distinct.)

What is the inverse of (f, g)? If F and G have m and n elements, respectively, how many elements does $F \otimes G$ have?

Evidently, the inverse of (f, g) is (f^{-1}, g^{-1}), and $F \otimes G$ has mn elements.

Klein's Vierergruppe V

A simple example is given by $Z_2 \otimes Z_2$, consisting of the four elements: $I = (1, 1)$, $A = (-1, 1)$, $B = (1, -1)$, and $C = (-1, -1)$. For example, we have $AB = (-1, -1) = C$. Note that this group is to be distinguished from the group Z_4 consisting of the four elements $1, i, -1, -i$. The square of any element in $Z_2 \otimes Z_2$ is equal to the identity, but this is not true of Z_4. In particular, $i^2 = -1 \neq 1$.

Incidentally, $Z_2 \otimes Z_2$, also known as Klein's Vierergruppe ("4-group" in German) and denoted by V, played an important historical role in Klein's program.

Note that the elements of F, regarded as a subgroup of $F \otimes G$, are written as (f, I). Similarly, the elements of G are written as (I, g). Clearly, (f, I) and (I, g) commute.

The direct product would seem to be a rather "cheap" way of constructing larger groups out of smaller ones, but Nature appears to make use of this possibility. The theory of the strong, weak, and electromagnetic interaction is based on the group* $SU(3) \otimes SU(2) \otimes U(1)$.

A teeny[14] bit of history: "A pleasant human flavor"

Historians of mathematics have debated about who deserves the coveted title of "the founder of group theory." Worthy contenders include Cauchy, Lagrange, Abel, Ruffini, and Galois. Lagrange was certainly responsible for some of the early concepts, but the sentimental favorite has got to be Évariste Galois, what with the ultra romantic story of him feverishly writing down his mathematical ideas the night before a fatal duel at the tender age of 20. Whether the duel was provoked because of the honor of a young woman named du Motel or because of Galois's political beliefs[†] (for which he had been jailed) is apparently still not settled. In any case, he was the first to use the word "group." Nice choice.

To quote the mathematician G. A. Miller, it is silly to argue about who founded group theory anyway:

> We are inclined to attribute the honor of starting a given big theory to an individual just as we are prone to ascribe fundamental theorems to particular men, who frequently have added only a small element to the development of the theorem. Hence the statement that a given individual founded a big theory should not generally be taken very seriously. It adds, however, a pleasant human flavor and awakens in us a noble sense of admiration and appreciation. It is also of value in giving a historical setting and brings into play a sense of the dynamic forces which have contributed to its development instead of presenting to us a cold static scene. Observations become more inspiring when they are permeated with a sense of development.[15]

---

* The notation $SU(n)$ and $U(n)$ will be explained in detail later in chapter IV.4.

† Galois was a fervent Republican (in the sense of being against the monarchy, not the Democrats).

While symmetry considerations have always been relevant for physics, group theory did not become indispensable for physics until the advent of quantum mechanics, for reasons to be explained in chapter III.1. Eugene Wigner,[16] who received the Nobel Prize in 1963 largely for his use of group theory in physics, recalled the tremendous opposition to group theory among the older generation (including Einstein, who was 50 at the time) when he first started using it around 1929 or so.[17] Schrödinger told him that while group theory provided a nice derivation of some results in atomic spectroscopy, "surely no one will still be doing it this way in five years." Well, a far better theoretical physicist than a prophet!

But Wigner's childhood friend John von Neumann,[18] who helped him with group theory, reassured him, saying "Oh, these are old fogeys. In five years, every student will learn group theory as a matter of course."[19]

Pauli[20] coined the term "die Gruppenpest" ("that pesty group business"), which probably captured the mood at the time. Remember that quantum mechanics was still freshly weird, and all this math might be too much for older people to absorb.

Multiplication table: The "once and only once rule"

A finite group with n elements can be characterized by its multiplication table,[21] as shown here. We construct a square n-by-n table, writing the product $g_i g_j$ in the square in the ith row and the jth column:

A simple observation is that, because of the group properties, in each row any group element can appear once and only once. To see this, suppose that in the ith row, the same group element appears twice, that is, $g_i g_j = g_i g_k$ for $j \neq k$. Then multiplying by g_i^{-1} from the left, we obtain $g_j = g_k$, contrary to what was assumed. It follows that each of the n elements must appear once to fill up the n slots in that row. We might refer to this as the "once and only once rule."

The same argument could be repeated with the word "row" replaced by "column," of course.

For n small, all possible multiplication tables and hence all possible finite groups with n elements can readily be constructed. Let us illustrate this for $n = 4$. For pedagogical reasons, we will do this in two different ways, one laborious,* the other "slick."

---

* An undergraduate in my class advised me to include also this laborious way as being the more instructive of the two ways. I agree with him that textbooks tend to contain too many slick proofs.

Finite groups with four elements: The slow way

First, we proceed very slowly, by brute force. Call the four elements I, A, B, and C.

1. By definition of the identity, the first row and first column can be filled in automatically:

	I	A	B	C
I	I	A	B	C
A	A			
B	B			
C	C			

2. We are to fill in the second row with I, B, and C. The first entry in that row is A^2. There are two possible choices: choice (a): $A^2 = B$, or choice (b): $A^2 = I$. (You might think that there is a third choice, $A^2 = C$, but that is the same as choice (a) upon renaming the elements. What you call C I will call B.)

   Let us now follow choice (a) and come back to choice (b) later.

3. The multiplication table now reads

	I	A	B	C
I	I	A	B	C
A	A	B	2	3
B	B	4	5	6
C	C			

   where for your and my convenience I have numbered some of the boxes yet to be filled in.

4. We have to put C and I into boxes 2 and 3. But we cannot put C into box 3, since otherwise the fourth column will break the "once and only once rule": C would appear twice:

	I	A	B	C
I	I	A	B	C
A	A	B	C	I
B	B	4	5	6
C	C			

5. Again by the "once and only once rule," box 4 can only be C or I. The latter choice would mean $BA = I$ and hence $B = A^{-1}$, but we already know from the second row of the multiplication table that $AB = C \neq I$. Thus, box 4 can only be C. Hence box 5 is I, and 6 is A.

6. Finally, the last three blank entries in the fourth column are fixed uniquely by the "once and only once rule." We obtain[22]

	I	A	B	C
I	I	A	B	C
A	A	B	C	I
B	B	C	I	A
C	C	I	A	B

Now that we have the multiplication table, we know everything about the group, and we can ask: What group is this? From the second row, we read off $A^2 = B$, $A^3 = AA^2 = AB = C$, $A^4 = AA^3 = AC = I$. The group is Z_4. Interestingly, we don't even have to finish constructing the entire multiplication table. In this simple case, by the time we had filled in the second row, we could have quit.

The rest of the table, however, provides us with a lot of consistency checks to ensure that we have not messed up. For example, from the last row, we have $CB = A$. But we know from the second row that $B = A^2$ and $C = A^3$, and hence the statement $CB = A$ says that $A^3A^2 = A^5 = A$, showing that indeed $A^4 = I$.

We now go back to choice (b): $A^2 = I$, so that

	I	A	B	C
I	I	A	B	C
A	A	I	2	3
B	B	4	5	6
C	C	7	8	9

1. We are to fill boxes 2 and 3 with C and B. By the "once and only once rule" in the third and fourth columns, these boxes can only be C and B in that order.

2. By the same reasoning, we can only fill boxes 4 and 7 with C and B, respectively. We thus obtain

	I	A	B	C
I	I	A	B	C
A	A	I	C	B
B	B	C		
C	C	B		

3. Now it looks like we could fill in the four remaining empty boxes with either

I	A
A	I

or

A	I
I	A

But the two choices amount to the same thing. We simply rename B and C. Thus, we obtain

	I	A	B	C
I	I	A	B	C
A	A	I	C	B
B	B	C	I	A
C	C	B	A	I

Again, what group is this? It is just $Z_2 \otimes Z_2$: $A^2 = I$, $B^2 = I$, $C = AB = BA$ (and hence also $C^2 = I$).

A quick way: Construct the cyclic subgroups

Here is an alternative to this laborious procedure of constructing the multiplication table step by step. We use the earlier observation that in a finite group, if we keep multiplying an element by itself, we will reach the identity I.

Given a group G of four elements $\{I, A, B, C\}$, we keep multiplying A by itself. If $A^4 = I$, then $G = Z_4$. By Lagrange's theorem, the possibility $A^3 = I$ is not allowed. If $A^2 = I$, then we multiply B by itself. Either B^2 or B^4 equals I. The latter is ruled out, so the only possibility is that $B^2 = I$, and $AB = BA = C$. Then $G = Z_2 \otimes Z_2$, with the four elements represented by $(1, 1)$, $(1, -1)$, $(-1, 1)$, and $(-1, -1)$.

If you are energetic and driven, you could try to construct all possible finite groups with n elements, and see how large an n you could get to.[23] A quick hint: It's easy if n is prime.

Presentations

For large groups, writing down the multiplication table is clearly a losing proposition. Instead, finite groups are defined by their properties, as in the examples listed above, or by their presentations,[24] which list the elements (sometimes called generators) from which all other elements can be obtained by group multiplication, and the essential relations the generators satisfy. Thus, in a self-evident notation, the groups Z_4 and $Z_2 \otimes Z_2$ are defined by their presentations as follows:

$$Z_4 : \langle A | A^4 = I \rangle \tag{4}$$

$$Z_2 \otimes Z_2 : \langle A, B \, | A^2 = B^2 = I, \, AB = BA \rangle \tag{5}$$

The two groups are clearly distinct. In particular, Z_4 contains only one element that squares to I, namely A^2.

Homomorphism and isomorphism

A map $f : G \to G'$ of a group G into the group G' is called a homomorphism if it preserves the multiplicative structure of G, that is, if $f(g_1) f(g_2) = f(g_1 g_2)$. Clearly, this requirement implies that $f(I) = I$ (more strictly speaking, the identity of G is mapped to the identity of G'). A homomorphism becomes an isomorphism if the map is one-to-one and onto.

Now we can answer the question posed earlier: the additive group of integers mod N is in fact isomorphic* to Z_N.

For a more interesting example, consider $Z_2 \otimes Z_4$. We use the additive notation here and thus write the elements as (n, m) and compose them according to $(n, m) \cdot (n', m') = (n + n' \bmod 2, m + m' \bmod 4)$. We start with $(0, 0)$ and add $(1, 1)$ repeatedly: $(0, 0) \overset{+(1,1)}{\longrightarrow} (1, 1) \to (0, 2) \to (1, 3) \to (0, 4) = (0, 0)$; we get back to where we started. Next, we start with $(0, 1)$ and again add $(1, 1)$ repeatedly: $(0, 1) \overset{+(1,1)}{\longrightarrow} (1, 2) \to (0, 3) \to (1, 0) \to (0, 1)$, getting back to where we started. Thus we can depict $Z_2 \otimes Z_4$ by a rectangular 2-by-4 discrete lattice on a torus (see figure 4).

Now we come in for a bit of a surprise. Consider $Z_2 \otimes Z_3$ consisting of (n, m), which we compose by $(n + n' \bmod 2, m + m' \bmod 3)$. Again, we start with $(0, 0)$ and add $(1, 1)$ repeatedly: $(0, 0) \to (1, 1) \overset{+(1,1)}{\longrightarrow} (2, 2) = (0, 2) \to (1, 3) = (1, 0) \to (2, 1) = (0, 1) \to (1, 2) \to (2, 3) = (0, 0)$. We are back where we started! In the process, we cycled through all six elements of $Z_2 \otimes Z_3$. We conclude that the six elements $(0, 0)$, $(1, 1)$, $(0, 2)$, $(1, 0)$, $(0, 1)$, and $(1, 2)$ describe Z_6.

Thus, $Z_2 \otimes Z_3$ and Z_6 are isomorphic; they are literally the same group. Note that this phenomenon, of a possible isomorphism between $Z_p \otimes Z_q$ and Z_{pq}, does not require p and q to be prime, only relatively prime. (Consider the example of $Z_4 \otimes Z_9$.)

As another example of isomorphism, the groups $SO(2)$ and $U(1)$ introduced earlier in the chapter are isomorphic. The map $f : SO(2) \to U(1)$ is defined simply by $f(R(\phi)) = e^{i\phi}$.

---

* That the additive group of integers mod N is also isomorphic to the multiplicative group Z_n foreshadows the confusion some students have between the addition and multiplication of angular momenta in quantum mechanics. We discuss this later in chapter IV.3.

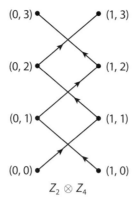

Figure 4

Appendix 1: Weakening the axioms

Two of the three axioms that define a group can in fact be weakened to the following:

2'. Existence of the left identity: A left identity I exists, such that for any element g, $Ig = g$.

3'. Existence of a left inverse: For any element g, there exists an element f, such that $fg = I$.

We now show that these imply axioms 2 and 3 given in the text. In other words, given the left identity and the left inverse, we are guaranteed that the right identity and the right inverse also exist.

Take the left inverse f of g. By 3', there exists an element k, such that $kf = I$. Multiplying this by g from the right, we obtain $(kf)g = Ig = g = k(fg) = kI$, where the second equality is due to 2', the third equality to associativity, and the fourth equality to 3'. Therefore $g = kI$. We want to show that $k = g$.

To show this, let us multiply $g = kI$ by I from the right. We obtain $gI = (kI)I = k(II) = kI = g$, where the second equality is due to associativity, and the third equality to 2', since I also qualifies as "any element." Thus, $gI = g$, so that I is also the right identity. But if I is also the right identity, then the result $g = kI$ becomes $g = k$. Multiplying by f from the right, we obtain $gf = kf = I$. Therefore, the left inverse of g, namely f, is also the right inverse of g.

Appendix 2: Associativity

Mathematically, the concept of a group is abstracted from groups of transformations. To physicists, groups are tantamount to transformation groups. In fact, if we are allowed to think of group elements as acting on a set of things $S = \{p_1, p_2, \cdots\}$, we can prove associativity. The "things" could be interpreted rather generally. For the geometrical examples given in this chapter, p_i could be the points in, for example, a triangle. Or for applications to fundamental physics, p_i could be some physical law as known to a particular observer, for example, an inertial observer in discussions of special relativity.

Suppose the group element g takes $p_1 \to p'_1$, $p_2 \to p'_2$, $\cdots$, so that the things in S are rearranged (as, for example, when a triangle is rotated). Suppose the group element g' takes $p'_1 \to p''_1$, $p'_2 \to p''_2$, $\cdots$, and the group element g'' takes $p''_1 \to p'''_1$, $p''_2 \to p'''_2$, $\cdots$, and so on.

Now consider the action of $g''(g'g)$ on S. The element $g'g$ takes p_j to p''_j, and then the element g'' takes p''_j to p'''_j. Compare this with the action of $(g''g')g$ on S. The element g takes p_j to p'_j, and then the element $g''g'$ takes p'_j to p'''_j. The final result is identical, and associativity is proved.

Most physicists I know would probably regard this kind of fundamental proof as painfully self-evident.

Appendix 3: Modular group

The modular group has become important in several areas of physics, for example, string theory and condensed matter physics. Consider the set of transformations of one complex number into another given by

$$z \to \frac{az+b}{cz+d} \tag{6}$$

with a, b, c, and d integers satisfying $ad - bc = 1$. The transformation (6) can be specified by the matrix

$$M = \begin{pmatrix} a & b \\ c & d \end{pmatrix} \quad \text{with } \det M = 1 \tag{7}$$

Clearly, M and $-M$ correspond to the same transformation in (6).

In the text, I introduced you to $SL(n, R)$, the special linear group of n-by-n matrices with real entries, and $SL(n, C)$, the special linear group of n-by-n matrices with complex entries. The matrices in (7) define the group $SL(2, Z)$, the special linear group of 2-by-2 matrices with integer entries.* The group that results upon identifying M and $-M$ in $SL(2, Z)$ is known as $PSL(2, Z)$ (the letter P stands for "projective"), otherwise known as the modular group.

The transformation in (6) can be generated by repeatedly composing (that is, multiplying together) the two generating transformations

$$S : z \to -\frac{1}{z} \tag{8}$$

and

$$T : z \to z + 1 \tag{9}$$

They correspond to the matrices $S = \begin{pmatrix} 0 & 1 \\ -1 & 0 \end{pmatrix}$ and $T = \begin{pmatrix} 1 & 1 \\ 0 & 1 \end{pmatrix}$, respectively.

Using the language of presentation introduced in the text, we can write

$$PSL(2, Z) : \langle S, T \mid S^2 = I, \ (ST)^3 = I \rangle \tag{10}$$

Incidentally, the modular group can be generalized to the triangular group $\mathcal{T}$, denoted by $(2, 3, n)$ and presented by

$$\mathcal{T} : \langle S, T \mid S^2 = I, \ (ST)^3 = I, \ T^n = I \rangle \tag{11}$$

The modular group is thus sometimes written as $(2, 3, \infty)$.

Exercises

1   The center of a group G (denoted by Z) is defined to be the set of elements $\{z_1, z_2, \cdots\}$ that commute with all elements of G, that is, $z_i g = g z_i$ for all g. Show that Z is an abelian subgroup of G.

2   Let $f(g)$ be a function of the elements in a finite group G, and consider the sum $\sum_{g \in G} f(g)$. Prove the identity $\sum_{g \in G} f(g) = \sum_{g \in G} f(gg') = \sum_{g \in G} f(g'g)$ for g' an arbitrary element of G. We will need this identity again and again in chapters II.1 and II.2.

---

* In mathematics, Z denotes the set of all integers, of either sign, including 0.

3   Show that $Z_2 \otimes Z_4 \neq Z_8$.

4   Find all groups of order 6.

Notes

1. See Fearful.
2. See parts VII and VIII.
3. We go into this in detail in chapter III.3.
4. See chapter IV.7.
5. Of course, we could also be more abstract and say that a group G is a structure endowed with the map $(G, G) \to G$ and so on and so forth.
6. In Strange Beauty, the biography of Murray Gell-Mann by G. Johnson, the following explanation about commutation is mentioned. When Gell-Mann was admitted only to MIT rather than the graduate school of his choice, he resolved to kill himself. But then he realized that killing himself and attending MIT do not commute, and so decided that he should go to MIT first and kill himself later, rather than the other way around.
7. He is one of several characters that populate my previous books Quantum Field Theory in a Nutshell and Einstein Gravity in a Nutshell. Hereafter QFT Nut and G Nut, respectively.
8. In the late nineteenth century, mathematicians felt that, with group theory, they had finally invented something of no use to the physicists. See p. v in R. Gilmore, Lie Groups, Lie Algebras, and Some of Their Applications.
9. Note the conceptual distinction between transformation and invariance. For example, the laws governing the weak interaction are famously not invariant under the interchange of left and right (known as a parity transformation P). But, regardless of whether a given law is invariant under parity, we still have $P \cdot P = I$.
10. This unfortunately is not true of many transformations in everyday life, such as cooking and aging.
11. See, for example, G Nut.
12. I once had a math professor who spoke of self-proving theorems. In the same sense, there are self-suggesting concepts.
13. Lagrange fell into a deep depression in his old age. Fortunately for him, the daughter of Lemonnier, an astronomer friend of Lagrange's, managed to cheer him up. Almost forty years younger than Lagrange, the young woman offered to marry him. Soon Lagrange was productive again. "Mathematicians Are People, Too," by L. Reimer and W. Reimer, p. 88.
14. "Teeny bit of history," because you can easily read your fill on the web.
15. "The Founder of Group Theory" by G. A. Miller, American Mathematical Monthly 17 (Aug–Sep 1910), pp. 162–165. http://www.jstor.org/stable/2973854.
16. My senior colleague Robert Sugar, who took a course on group theory at Princeton from Wigner, told me the following story. On the first day, Wigner asked the students whether they knew how to multiply matrices. Given Wigner's reputation of delivering long dull discourses, the students all said yes of course, and in fact, as graduate students at Princeton, they all knew how to do it. But Wigner was skeptical and asked a student to go up to the blackboard and multiply two 2-by-2 matrices together. The guy did it perfectly, but unfortunately, Wigner used a convention opposite to what was (and still is) taught in the United States. Wigner was convinced that the students did not know how to multiply matrices, and proceeded to spend a week tediously explaining matrix multiplication. If you look at the English edition of Wigner's group theory book, you would read that the translator had, with Wigner's permission, reversed all of his conventions.
17. The stories Wigner told about the early days of group theory used here and elsewhere in this book are taken from The Recollections of Eugene P. Wigner as told to Andrew Szanton, Plenum Press, 1992.
18. As you might have heard, the four Hungarians, Leo Szilard, Eugene Wigner, John von Neumann, and Edward Teller, all Jewish, formed a legendary group that had major impact on physics. Listed here in order of age, they were born within 10 years of one another. Wigner considered himself to be the slowest of the four, and anecdotal evidence suggests that this assessment is not due to exaggerated modesty; yet he is the only one of the four to have received a Nobel Prize.
19. Well, not quite—not even close.

20. This surprises me, since one of Pauli's famous contributions involves group theory. See the interlude to part VII. From what I have read, Pauli was brilliant but mercurial and moody, and always ready for a good joke.

21. As a child you memorized the standard 9-by-9 multiplication table; now you get the chance to construct your own.

22. Conspiracy nuts might notice that the acronym CIA appears not once, but four times, in this table.

23. Mathematicians have listed all possible finite groups up to impressively large values of n.

24. As in the rather old-fashioned and formal "May I present [Title] So-and-so to you?"

1.2 | Finite Groups

Let me first give you an overview or road map to this introduction to the theory of finite groups. We discuss various important notions, including equivalence classes, invariant subgroups, simple groups, cosets, and quotient groups. These notions are illustrated mostly with the permutation groups, which are the easiest to grasp and yet have enough structure for them to be highly nontrivial. We also introduce the dihedral groups and the quarternionic group.

Permutation groups and Cayley's theorem

The permutation group S_n and its natural subgroup A_n are sort of like the poster children of group theory, easy to define and to understand. Everybody knows how permutations work.

Furthermore, a theorem due to Cayley states that any finite group G with n elements is isomorphic (that is, identical) to a subgroup of S_n. (Try to figure this one out before reading on. Hint: Think about the multiplication table of G.)

List the n elements of G as $\{g_1, g_2, \cdots, g_n\}$ in the order pertaining to the row in the multiplication table corresponding to the identity element. Then in the row in the multiplication table corresponding to the element g_i we have, in order, $\{g_i g_1, g_i g_2, \cdots, g_i g_n\}$. By an argument* familiar from chapter I.1, this amounts to a permutation of $\{g_1, g_2, \cdots, g_n\}$. Thus, we can associate an element of S_n with g_i. This maps G into a subgroup of S_n. Note that Lagrange's theorem is satisfied.

For n large, we see that G is a tiny subgroup of S_n, which has $n!$ elements, as compared to n elements. In contrast, for n small, the situation is quite different, as shown by the following examples. The group Z_2 is in fact the same as S_2. (Check this; it's trivial.) But the group Z_3, with three elements, clearly cannot be the same as S_3, with $3! = 6$ elements, but it is the same as A_3. (Why?)

---

* Basically, the "once and only once rule."

Cycles and transpositions

As is often the case in mathematics and physics, a good notation is half the battle. To be specific, consider S_5. A "typical" element might be $g = \begin{pmatrix} 1 & 2 & 3 & 4 & 5 \\ 4 & 1 & 5 & 2 & 3 \end{pmatrix}$. This denotes a permutation that takes $1 \to 4$, $2 \to 1$, $3 \to 5$, $4 \to 2$, and $5 \to 3$, that is, a permutation that cyclically permutes $1 \to 4 \to 2 \to 1$ and interchanges $3 \to 5 \to 3$. A more compact notation suggests itself: write $g = (142)(35)$. In our convention, (142) means $1 \to 4 \to 2 \to 1$, and (35) means $3 \to 5 \to 3$.

The permutation $(a_1 a_2 \cdots a_k)$ is known as a cycle of length k and cyclically permutes $a_1 \to a_2 \to a_3 \to \cdots \to a_k \to a_1$. A cycle of length 2 is called a transposition, or more informally, an exchange. In the example above, (35) exchanges 3 and 5.

For brevity, we will call a cycle of length k a k-cycle. Clearly, the k numbers $(a_1 a_2 \cdots a_k)$ defining the k-cycle can be cyclically moved around without changing anything: for example, (35) and (53) are the same; (142), (421), and (214) are the same.

Any permutation P can be written as the product of cycles of various lengths, including cycles of length 1 (that is, consisting of an element untouched by P), with none of the cycles containing any number in common.[*] (An example is $g = (142)(35)$.) To see this, start by picking some integer between 1 and n, call it a_1, which is taken by P into some other number, call it a_2, which is in turn taken to a_3 by P, and so on, until we come back to a_1. This forms a cycle $(a_1 a_2 \cdots a_j)$ of some length, say j. If there are any numbers left over, pick one, call it b_1, which is taken by P into some other number, call it b_2, and so on, until we come back to b_1. We keep repeating this process until there aren't any numbers left. Then $P = (a_1 \cdots)(b_1 \cdots) \cdots (\cdots)$, consisting of n_j cycles of length j (with $\sum_j j n_j = n$, of course). By construction, the cycles do not have any number in common. For example, in the preceding discussion $g = (142)(35)$ with $n_2 = 1$, $n_3 = 1$.

Incidentally, the 1-cycle is trivial and does nothing. Hence it is usually omitted. For example, the permutation $g = \begin{pmatrix} 1 & 2 & 3 & 4 & 5 \\ 1 & 5 & 3 & 4 & 2 \end{pmatrix}$ could be written as $g = (25)(1)(3)(4)$ but is normally written as $g = (25)$.

Rules for multiplying permutations

Theorem: Any permutation can be written as a product of 2-cycles, that is, exchanges or transpositions.

This merely expresses the everyday intuition that a permutation can be performed in steps, exchanging two objects at a time. In some sense, exchanges are the "atoms" out of which permutations are built.

In our example, $g = (142)(35)$ is the product of a 3-cycle with a 2-cycle. Does this contradict the theorem?

---

[*] This is sometimes called resolving P into cycles.

No, as we will now show, (142) can itself be written as a product of exchanges. We write

$$(14)(42) = \begin{pmatrix} 1 & 2 & 4 \\ 4 & 2 & 1 \end{pmatrix} \begin{pmatrix} 1 & 2 & 4 \\ 1 & 4 & 2 \end{pmatrix} = \begin{pmatrix} 1 & 2 & 4 \\ 1 & 4 & 2 \\ 4 & 1 & 2 \end{pmatrix} = (142). \tag{1}$$

In the first equality, we merely go back to the more explicit notation, for example, $(14) = \begin{pmatrix} 1 & 2 & 4 \\ 4 & 2 & 1 \end{pmatrix}$. (Of course, 3 and 5 are not even in the game, so the upper row is written in the "canonical order" 124.) In the second equality, we invent on the spot a 3-tiered notation. The final equality is merely a simple way of representing the net effect of the two operations specified by the 3-tiered notation.

Thus, $g = (14)(42)(35)$ in accord with the theorem. Note that, when we resolve a permutation into cycles and write $g = (142)(35)$, the 3-cycle (142) and the 2-cycle (35) do not have any integer in common by construction. But there is no such restriction in the statement of the theorem. In our example, 4 appears in two separate 2-cycles.

We can readily develop some rules for multiplying 2-cycles:

1. If the two 2-cycles do not have a "number" in common, for example, (12) and (34), then they commute, and we have nothing more to say.

2. $(12)(23) = (123)$. (This was already shown earlier, if we simply rename the numbers; we had $(14)(42) = (142)$.) Note that since $(32) = (23)$, we can adopt the convention, when multiplying two 2-cycles, to match the head of one 2-cycle to the tail of the other 2-cycle.

3. We need hardly mention that $(12)(21) = I$.

4. $(12)(23)(34) = (12)(234) = \begin{pmatrix} 1 & 2 & 3 & 4 \\ 1 & 3 & 4 & 2 \\ 2 & 3 & 4 & 1 \end{pmatrix} = (1234)$.

5. $(123)(345) = (12)(23)(34)(45) = (12)(234)(45) = (12345)$.

And so it goes.

Indeed, we now see in hindsight that the preceding theorem is trivial.

Since any permutation can be decomposed into 2-cycles, these rules allow us to multiply permutations together.

As remarked earlier (without going into details), a permutation is either even or odd. The 2-cycle is clearly odd. (At the risk of being pedantic, let us observe that the 2-cycle (12) can be represented by the matrix $\begin{pmatrix} 0 & 1 \\ 1 & 0 \end{pmatrix}$, which has determinant $= -1$. We are anticipating a bit here by using the word* "represent.") The 3-cycle is even, since it is equal to the product of two 2-cycles. (We also note that (123) can be represented by $\begin{pmatrix} 0 & 0 & 1 \\ 1 & 0 & 0 \\ 0 & 1 & 0 \end{pmatrix}$, with determinant $= +1$.) A permutation is even or odd if it decomposes into the product of an even or odd number of exchanges (aka 2-cycles or transpositions), respectively.

---

* To be discussed in detail in chapter II.1.

Square root of the identity

The theory of finite groups is a rich subject with many neat theorems. You have already seen Lagrange's theorem. Here is another theorem for you to cut your teeth on.

Many of us were astonished to learn in school that there is another number besides 1 that would square to 1, namely, -1. Is there an analogous phenomenon for groups?

Theorem: Let G be a group of even order, that is, G has an even number of elements. There exists at least one element g, which is not the identity I, that also squares to the identity[1] $g^2 = I$.

You will prove this as an exercise. The alternating groups A_n for $n \geq 4$ and the permutation groups S_n for $n \geq 2$ are of even order, and the theorem holds for them. For example, in A_4, $(12)(34)$ squares to the identity. In contrast, A_3 has three elements and does not have any element other than the identity that squares to the identity.

Equivalence classes

Given a group G, distinct group elements are of course not the same, but there is a sense that some group elements might be essentially the same. The notion of equivalence class makes this hunch precise.

Before giving a formal definition, let me provide some intuitive feel for what "essentially the same" might mean. Consider $SO(3)$. We feel that a rotation through $17°$ and a rotation through $71°$ are in no way essentially the same, but that, in contrast, a rotation through $17°$ around the z-axis and a rotation through $17°$ around the x-axis are essentially the same. We could simply call the x-axis the z-axis.

As another example, consider S_3. We feel that the elements (123) and (132) are equivalent, since they offer essentially the same deal; again, we simply interchange the names of object 2 and object 3. We could translate the words into equations as follows: $(23)^{-1}(123)(23) = (32)(12)(23)(32) = (32)(21) = (321) = (132)$, where we use the rules just learned; for instance, in the first equality, we wrote $(123) = (12)(23)$. Note that at every step, we manipulate the 2-cycles so as to match head to tail. (Or, simply write $123 \rightarrow 132 \rightarrow 213 \rightarrow 312$.) A transformation using (23) has turned (123) and (132) into each other, as expected. Similarly, you would think that (12), (23), and (31) are essentially the same, but that they are in no way essentially the same as (123).

In a group G, two elements g and g' are said to be equivalent ($g \sim g'$) if there exists another element f such that

$$g' = f^{-1}gf \qquad (2)$$

The transformation $g \rightarrow g'$ is like a similarity transformation in linear algebra, and I will refer to it as such.

Since equivalence is transitive (friend of a friend is a friend)—that is, $g \sim g'$ and $g' \sim g''$ imply that $g \sim g''$—we can gather all the elements that are equivalent into equivalence classes.* The number of elements in a given equivalence class c, denoted by n_c, plays a crucial role in the representation theory to be discussed in part II.

Consider S_4 with $4! = 24$ elements. The even permutations form the subgroup A_4, with $4!/2 = 12$ elements. Given the preceding remarks, the even permutations fall into four equivalence classes:

$$\{I\}, \quad \{(12)(34), (13)(24), (14)(23)\}, \quad \{(123), (142), (134), (243)\}, \quad \text{and}$$
$$\{(132), (124), (143), (234)\} \tag{3}$$

For example, $((12)(34))^{-1}(123)(12)(34) = (43)(21)(12)(23)(12)(34) = (43)(234)(12) = (43)(34)(42)(21) = (421) = (142)$, where we used the various rules for multiplying 2-cycles repeatedly (for example, in the third equality, we write $(234) = (342) = (34)(42)$). As was explained earlier, we can also obtain the result more quickly by just performing the two exchanges $1 \leftrightarrow 2$ and $3 \leftrightarrow 4$: $(123) \rightarrow (213) \rightarrow (214) = (142)$.

The group S_4 is obtained by adjoining to A_4 the 12 odd permutations (12), (13), (14), (23), (24), (34), (1234), (1342), (1423), (1324), (1243), and (1432). Note that within S_4, (124) and (134) are equivalent, but within A_4, they are not: the element (23) is in S_4 but is not in A_4.

This example shows that, while permutations in the same equivalence class necessarily have the same cycle structure (more on this concept below), elements with the same cycle structure are not necessarily equivalent.

Three facts about classes

1. In an abelian world, everybody is in a class by himself or herself. Show this.

2. In any group, the identity is always proudly in its own private class of one. Show this.

3. Consider a class c consisting of $\{g_1, \cdots, g_{n_c}\}$. Then the inverse of these n_c elements, namely, $\{g_1^{-1}, \cdots, g_{n_c}^{-1}\}$, also form a class, which we denote by $\bar{c}$. Show this.

Cycle structure and partition of integers

We explained above that any permutation in S_n can be written as a product of n_j j-cycles with $\sum_j j n_j = n$. For example, a permutation with the cycle structure written schematically as $(xxxxx)(xxxxx)(xxxx)(xx)(xx)(xx)(x)(x)(x)(x)$ has $n_5 = 2, n_4 = 1, n_3 = 0, n_2 = 3$, and $n_1 = 4$ (and so $n = 24$) and is an element of S_{24}. As was remarked earlier, normally, the 1-cycles are not shown explicitly, a convention we have elsewhere followed.

Question: Given a cycle structure characterized by the integers n_j, determine the number of permutations in S_n with this cycle structure.

---

* The terms conjugate and conjugacy classes are also used, but physicists probably prefer to avoid these terms, since they often talk about complex conjugate and hermitean conjugate.

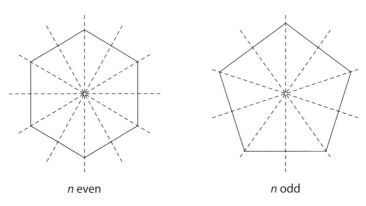

n even   n odd

Figure 1

Exercise! The desired number is

$$\mathcal{N}(n_1, \cdots, n_j, \cdots) = \frac{n!}{\Pi_j j^{n_j} n_j!} \tag{4}$$

For example, for A_4 (or S_4), the number of elements with the cycle structure $(xx)(xx)$ is (since $n_2 = 2$, with all other $n_j = 0$) $4!/(2^2 2!) = 3$, in agreement with what we wrote. Similarly, the number of elements with the cycle structure (xxx) is (since $n_3 = 1$, with all other $n_j = 0$) $4!/(3^1 1!) = 4 \cdot 2 = 8$, also in agreement with what we wrote.

From these examples, we also see that the cycle structures in the permutation group correspond to partitions of the integer n. For instance, for S_4, we have $1 + 1 + 1 + 1$, $2 + 2$, $3 + 1$, $2 + 1 + 1$, and 4. Note that the first three partitions appear in A_4.

The dihedral group D_n

There are of course many other finite groups besides the permutation groups. In chapter I.1, we already mentioned the set of transformations that leave the n-sided regular polygon invariant, forming a group known[*] as the dihedral group[2] D_n.

The group is generated by the rotation R through $2\pi/n$ and the reflection r through a median. For n odd (think equilateral triangle and pentagon), a median is a straight line going through the center of the polygon from one vertex to the midpoint of the opposite side. For n even (think square and hexagon), there are two types of median: a median is a straight line through the center of the polygon going from one vertex to another, or going from the midpoint of a side to another midpoint. There are always n medians (figure 1).

Clearly, $R^n = I$ and $r^2 = I$. Furthermore,[†] $rRr = R^{-1}$. Verify this! Thus, D_n has $2n$ elements, namely, $\{I, R, R^2, \cdots, R^{n-1}, r, Rr, R^2 r, \cdots, R^{n-1}r\}$. (Compare with the invariance group D_3 of the equilateral triangle in chapter I.1.)

---

[*] The terminology for finite groups can get quite confusing; this group is also known as C_{nv} in some circles.
[†] This just states the everyday fact that a rotation reflected in a mirror is a rotation in the opposite sense.

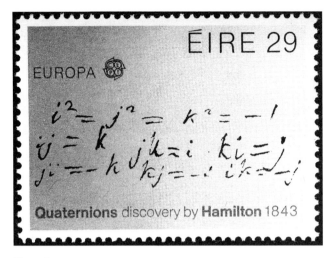

Figure 2

Using a language introduced in chapter I.1, we say that D_n is presented by

$$D_n = \langle R, r \mid R^n = I, r^2 = I, Rr = rR^{-1} \rangle \tag{5}$$

In exercise 11, you will work with D_n.

The quarternionic group $\mathcal{Q}$

Some readers may have heard that Hamilton generalized the imaginary unit i by adding two other units j and k satisfying the multiplication rules (see figure 2)

$$i^2 = j^2 = k^2 = -1 \quad \text{and} \quad ij = -ji = k, \ jk = -kj = i, \ ki = -ik = j \tag{6}$$

The quarternionic group $\mathcal{Q}$ consists of eight elements: $1, -1, i, -i, j, -j, k$, and $-k$, with group multiplication given by Hamilton's rules. As an exercise, show that $\mathcal{Q}$ forms a group.

Our long-time friend Confusio[3] looks a bit confused. He mutters, "In the review of linear algebra, I read in an endnote that the quarternionic numbers $a + bi + cj + dk$ form a division algebra."

No, we tell Confusio, quarternionic numbers are not to be confused with $\mathcal{Q}$. The eight-element quarternionic group $\mathcal{Q}$ is to quarternionic numbers as the four-element cyclic group Z_4 is to complex numbers.

Coxeter groups

One more example. A Coxeter group is presented by

$$\langle a_1, a_2, \cdots, a_k \mid (a_i)^2 = I, (a_i a_j)^{n_{ij}} = I, n_{ij} \geq 2, \ \text{with } i, j = 1, 2, \cdots, k \rangle \tag{7}$$

In other words, each generator squares to the identity (note that this does not mean that every group element squares to the identity), and for every pair of generators, there exists an integer $n_{ij} \geq 2$ such that $(a_i a_j)^{n_{ij}} = I$. The a_is correspond to reflections. As you can see, the Coxeter groups are inspired by the kaleidoscope.

Coxeter groups have many interesting properties. For example, we can readily show that $n_{ij} = n_{ji}$. Proof: Given $(ab)^n = I$, multiply from left and right by b. Using associativity, we have $I = b^2 = b(ab)^n b = b(abab \cdots ab)b = baba \cdots ba = (ba)^n$. QED.

Invariant subgroup

We know what a subgroup is, but now let us talk about a very special kind of subgroup, known as an invariant subgroup. Let H, consisting of elements $\{h_1, h_2, \cdots\}$, be a subgroup of G. Take any element g not in H. Then the set of elements $\{g^{-1}h_1 g, g^{-1}h_2 g, \cdots\}$ forms a subgroup (exercise!), which we naturally denote by $g^{-1}Hg$. In general, the subgroups H and $g^{-1}Hg$ are distinct.

But if H and $g^{-1}Hg$ are the same for all $g \in G$ (note the emphasis on "all") then H is called an invariant subgroup. In other words, H is invariant if the two lists $\{h_1, h_2, \cdots\}$ and $\{g^{-1}h_1 g, g^{-1}h_2 g, \cdots\}$ are the same* for any g. In other words, similarity transformations generated by the group elements of G leave H unchanged. (The jargon guy tells us that an invariant subgroup is also known as a normal subgroup. I prefer the term "invariant subgroup" as being more informative.)

An example is offered by A_4. The subgroup $\{I, (12)(34), (13)(24), (14)(23)\} = Z_2 \otimes Z_2 = V$ is invariant.[†] Verify this!

Another example: in a direct product group $G = E \otimes F$, E and F are invariant subgroups.[‡] Verify this as well.

Yet another example is $Z_4 = \{1, -1, i, -i\}$ with the invariant subgroup $Z_2 = \{1, -1\}$. One easy check: $(-i)\{1, -1\}i = \{1, -1\}$.

In fact, Z_4 is itself an invariant subgroup of the quarternionic group Q. (Please do check that it is invariant. For example, $j^{-1}\{1, -1, i, -i\}j = \{1, -1, i, -i\}$.)

We can have invariant subgroups inside invariant subgroups, like nested Russian dolls.

That G contains an invariant subgroup H is denoted by a funny-looking triangular symbol $G \rhd H$. Similarly, that H is an invariant subgroup of G is denoted by $H \lhd G$. These are evidently more restrictive forms of $G \supset H$ and $H \subset G$.

Derived subgroup

Given a group G, grab two elements a, b, and calculate

$$\langle a, b \rangle \equiv a^{-1}b^{-1}ab = (ba)^{-1}(ab) \tag{8}$$

---

* But of course the elements in these lists need not appear in the same order. The ordering $\{h_1, h_2, \cdots\}$ is arbitrary.

† Recall that V denotes Klein's Vierergruppe $Z_2 \otimes Z_2$ (see chapter I.1).

‡ More precisely, by E we mean $E \otimes I_F$.

First, note that $\langle a, a \rangle = I$ and $\langle a, b \rangle^{-1} = \langle b, a \rangle$. Denote by $\{x_1, x_2, \cdots\}$ the objects $\langle a, b \rangle$ as a and b range over all the elements in the group. These objects, together with the products of these objects with each other, that is, group elements of the form $x_i x_j \cdots x_k$, constitute a subgroup of G, known as the derived subgroup $\mathcal{D}$.

Note that the product $\langle a, b \rangle \langle c, d \rangle$ need not have the form $\langle e, f \rangle$ for some e and f. The derived subgroup $\mathcal{D}$ is not necessarily equal to the set of all objects of the form $\langle a, b \rangle$ as a and b range over G.

As an example, the derived subgroup of S_n is A_n. The objects $\langle a, b \rangle = a^{-1} b^{-1} ab$, being the product of four permutations, are necessarily even permutations.

A more involved example: the derived subgroup of A_4 is $V = Z_2 \otimes Z_2$. Calculate, for instance, that* $\langle (12)(34), (123) \rangle = (14)(23)$.

Our friend Dr. Feeling† strolls by. "Note that for an abelian group G, the derived subgroup is just the trivial group consisting of only the identity, since $\langle a, b \rangle = I$ for any a, b," he mumbles. "The object $\langle a, b \rangle = (ba)^{-1}(ab)$ measures how much ab differs from ba. Therefore, the derived subgroup tells us how nonabelian the group G is. The larger $\mathcal{D}$ is, the farther away G is from being abelian, roughly speaking."

Let's try out what he said using the quarternionic group Q. We have $\langle i, j \rangle = (-i)(-j)ij = (-i)kj = jj = -1$. Thus, $\mathcal{D} = Z_2$. In contrast, the derived subgroup of $Z_4 = \{1, -1, i, -i\}$, an abelian subgroup of Q, is manifestly just I.

Let us now show that $\mathcal{D}$ is an invariant subgroup of G. Use the convenient notation $\tilde{a} = g^{-1}ag$ (keeping in mind $\tilde{a}$ depends implicitly on g). Note that $(g^{-1}ag)(g^{-1}a^{-1}g) = g^{-1}aa^{-1}g = I$, which shows that $g^{-1}a^{-1}g = \tilde{a}^{-1}$. Now we simply calculate: $g^{-1}\langle a, b \rangle g = g^{-1}(a^{-1}b^{-1}ab)g = \tilde{a}^{-1}\tilde{b}^{-1}\tilde{a}\tilde{b} = \langle \tilde{a}, \tilde{b} \rangle$, which shows that the derived subgroup is an invariant subgroup. As an exercise, find the derived subgroup of the dihedral group and show that it is invariant.

In the example of Q, its derived subgroup $\mathcal{D} = Z_2$ is certainly an invariant subgroup. But we also know that Q contains the larger group Z_4 as an invariant subgroup.

I have to say a few words about terminology, but, to avoid interrupting the narrative flow, I have moved them to appendix 2.

A simple group does not contain a (nontrivial) invariant subgroup

In what follows, it is convenient to restrict the term "invariant subgroup" to mean proper invariant subgroup; we exclude G itself and the trivial subgroup consisting of only the identity.

A group is called simple[4] if it does not have any invariant subgroup.

---

* We have $(12)(34)(213)(12)(34)(123) = (12)(34)(21)(13)(12)(34)(12)(23) = (34)(13)(34)(23) = (34)(134)$ $(23) = (34)(341)(23) = (34)(34)(41)(23) = (41)(23)$. This calculation would go much faster using the matrices to be introduced in part II.

† Like Confusio and the jargon guy, Dr. Feeling has appeared previously in G Nut. To paraphrase a review of QFT Nut published by the American Mathematical Society, it is often more important to feel why something must be true rather than to prove that it is true. For all we know, Dr. Feeling might be a real person, rather than an imaginary friend from the author's childhood.

Dr. Feeling wanders by and explains: "We want to express the notion of a group being simple, of not containing smaller pieces. The naive first thought is that the group should not contain any subgroup, but subgroups are a dime a dozen. As we saw in chapter I.1, we could take any element g: it and its integer powers would form a cyclic subgroup. So, a garden variety cyclic subgroup does not count; any decent-sized group would be full of them. But an invariant subgroup is sort of special. Finding V inside A_4 is sort of like physicists finding quarks inside a hadron!"

As a physicist, I thought that was a bit of an exaggeration. But in any case, it provides a good mnemonic: not having an invariant subgroup makes a group simple.[5] Thus, Z_4, A_4,* and Q are all not simple.

By the way, given a group G, computing its derived subgroup is algorithmic, a task we can relegate to a computer or a student. If the derived subgroup is nontrivial, then we immediately realize that G is not simple.

Let f be a homomorphic[6] map of a group G into itself; in other words, the map is such that $f(g_1)f(g_2) = f(g_1 g_2)$. Show that the kernel of f, that is, the set of elements that are mapped to the identity, is an invariant subgroup of G. Exercise!

Invariant subgroup, cosets, and the quotient group

Let $G \triangleright H$. To repeat, this means that all the elements equivalent to the elements in the subgroup H are also in H, which makes H very special indeed, as we shall now see.

Having an invariant subgroup empowers us to form objects called cosets and construct another group called the quotient group.

For an element g, consider the set of elements $\{gh_1, gh_2, \cdots\}$, which we will denote by[†] gH. We have a whole bunch of such sets, $g_a H$, $g_b H$, $\cdots$.

We can naturally multiply two of these sets together: simply multiply each group element in the set $g_a H$ by every group element in the set $g_b H$ and look at the resulting set:

$$(g_a h_i)(g_b h_j) = g_a(g_b g_b^{-1})h_i g_b h_j = g_a g_b(g_b^{-1}h_i g_b)h_j = (g_a g_b)(h_l h_j) \tag{9}$$

In the third equality, we make crucial use of the fact that H is an invariant subgroup, so that $g_b^{-1}h_i g_b$ is some element h_l of H. (This step would not work if H is some garden variety subgroup of G.) Since H is a group, the product $h_l h_j$ is an element of H. Thus, $(g_a h_i)(g_b h_j) = (g_c h_k)$, where $g_c = g_a g_b$, and h_k depends on g_a, g_b, h_i, and h_j.

The objects gH, which our friend the jargon guy tells us are called left cosets, close under multiplication:

$$(g_a H)(g_b H) = (g_a g_b H) \tag{10}$$

The natural question is whether they form a group.

---

* A famous theorem states that A_n is simple for $n \geq 5$. We will prove in chapter II.3 that A_5 is simple.

† The set gH is definitely not to be confused with gHg^{-1}, which would be H itself, since H is an invariant subgroup. (Here g denotes a generic element not in H.) Indeed, unless $g = I$, the set gH is not a group; for one thing, the identity is not contained in gH.

Sure! Indeed, (10) maps the pair g_a and g_b to the product $g_a g_b$.

Thus, the identity of this group is $IH = H$, namely, H itself, since $(IH)(gH) = gH$. The inverse of gH is $g^{-1}H$, since $(g^{-1}H)(gH) = (gH)(g^{-1}H) = IH = H$. I will let you show associativity.

The left cosets form a group.

Thus, if a group G has an invariant subgroup H, then we can construct another group consisting of left cosets gH, a group known as the quotient group and written as $Q = G/H$. Why quotient? Well, if $N(G)$ denotes the number of elements in G and $N(H)$ the number of elements in H, each coset $\{g_a H\}$ contains $N(H)$ elements of G. Hence there can only be $N(Q) = N(G)/N(H)$ cosets. It is entirely reminiscent of how we first learned to divide, by putting, say, oranges into separate baskets.

The number of elements in Q is* $N(Q) = N(G)/N(H)$ (strong shades of Lagrange's theorem). In general, Q is not a subgroup of G.

There is nothing special about the left, of course. We could equally well have played with the right cosets, namely, the sets $\{h_1 g, h_2 g, \cdots\} = Hg$. Indeed, if $H \triangleleft G$, then the left cosets gH and right cosets Hg are manifestly the same.

As an example, consider the quarternionic group $\mathcal{Q}$, which has the invariant subgroup $Z_4 = \{1, -1, i, -i\}$, as we showed earlier. Construct the quotient group[†] $Q = \mathcal{Q}/Z_4$, which consists of only $2 = 8/4$ elements: $\{1, -1, i, -i\}$ and $j\{1, -1, i, -i\} = \{j, -j, k, -k\}$. (Note that $k\{1, -1, i, -i\} = \{j, -j, k, -k\}$, for example, does not give a different left coset.)

Indeed, Q is just the group Z_2. An easy check:

$$(jZ_4)(jZ_4) = \{j, -j, k, -k\}\{j, -j, k, -k\} = \{1, -1, i, -i\} = Z_4$$

namely, the identity of Z_2.

Given a group G, since its derived subgroup $\mathcal{D}$ is an invariant subgroup, we can always construct the quotient group $Q = G/\mathcal{D}$. This process can then be repeated with Q playing the role of G. Since taking the quotient G/H amounts to setting H to the identity, and since $\mathcal{D}$ measures how nonabelian G is, this process is known as abelianization.

A preview

In chapter I.3, we start discussing continuous groups, such as the rotation group. As you will see immediately, continuous groups are easier to deal with than finite groups in many respects: Newton and Leibniz come flying in with their wonderful concept of differentiation! We then have the concept of one group element being near another group element, in particular, the identity.

---

* The jargon guy tells us that $N(G)/N(H)$ is known as the index of H in G.

† The Q for the quotient group is not to be confused with the $\mathcal{Q}$ for the quarternionic group of course. Somewhat unfortunate; normally, not that many words start with the letter q.

Appendix 1 to this chapter gives you a tiny taste of finite group theory.[7] Indeed, the classification of all simple finite groups* is one of the crowning achievements of mathematics in modern times.[8]

Appendix 1: The composition series and the maximal invariant subgroup

Readers being exposed to group theory for the first time may safely skip this appendix.

Suppose we found an invariant subgroup H_1 of G. Then nothing prevents us from looking for an invariant subgroup H_2 of H_1. And so on. The sequence

$$G \triangleright H_1 \triangleright H_2 \triangleright \cdots \triangleright H_k \triangleright I \tag{11}$$

is called a composition series. By assumption, the Hs are invariant, and so we have a sequence of quotient groups $G/H_1 \supset H_1/H_2 \supset H_2/H_3 \supset \cdots \supset H_k$. The physics analogy might be that molecules contain atoms, atoms contain nuclei, nuclei contain nucleons, and nucleons contain quarks.

Again, use $\mathcal{Q}$, the eight-element quarternionic group, as an example. Then

$$\mathcal{Q} \triangleright Z_4 \triangleright Z_2 \triangleright I \tag{12}$$

where $Z_4 = \{1, i, -1, -i\}$, and $Z_2 = \{1, -1\}$. The quotient groups are $\mathcal{Q}/Z_4 = Z_2$ and $Z_4/Z_2 = Z_2$.

How do we know whether H_1 is the largest invariant subgroup of G?

Dr. Feeling strolls by. "In elementary school, we learned that we get a small number if we divide by a large number. So we might think that if the quotient group G/H is really small, then the invariant subgroup H should be the largest possible." This kind of intuition suggests the following theorem.

Given a group G and one of its invariant subgroups H, form the quotient group $Q = G/H$. Suppose that Q has no invariant subgroup. Then H is the maximal invariant subgroup.

The claim is that H is not contained in some larger invariant subgroup of G. Our intuitive feel is that if $Q = G/H$ is the smallest possible, then H is the largest possible, and if $Q = G/H$ does not contain an invariant subgroup, then it's kind of small.

Proof: We want to show that H is maximal. Assume to the contrary that H is not the maximal invariant subgroup. Then there exists an invariant subgroup F of G that contains H, that is, $G \triangleright F \supset H$. Let us list the elements of these various groups as follows: $H = \{h_1, h_2, \cdots\}$, $F = \{f_1, f_2, \cdots, h_1, h_2, \cdots\}$, $G = \{g_1, g_2, \cdots, f_1, f_2, \cdots, h_1, h_2, \cdots\}$. In other words, the fs are those elements in F but not in H, and the gs those elements in G but not in F. Note that since H is a group, one of the hs, say h_1, is the identity.

First, H is an invariant subgroup of F a fortiori, since it is an invariant subgroup of G. This implies that the quotient group $K = F/H$, consisting of $\{H, fH\}$ (the notation is compact but self-evident[†]) is a group. The quotient group $Q = G/H$ consists of $\{H, fH, gH\}$, and so, evidently, $K \subset Q$. But we will make the stronger claim that $K \lhd Q$. Simply check: for example, $(g^{-1}H)(fH)(gH) = (g^{-1}H)(fgH) = (g^{-1}fg)H = f'H$, where in the last step we use the fact that F is an invariant subgroup of G. This contradicts the assumption that Q has no invariant subgroup. But it does: namely, K. QED.[9]

To illustrate the theorem, use $\mathcal{Q}$, which has $Z_2 = \{1, -1\}$ as an invariant but not maximal subgroup. Then $\mathcal{Q}/Z_2$ is an $(8/2 = 4)$-element group with the elements $\{Z_2, iZ_2, jZ_2, kZ_2\}$. What is this group?

There are only two possibilities: Z_4 or the Viergruppe $V = Z_2 \otimes Z_2$. It is in fact the latter. We will work out presently the correspondence between the elements of $\mathcal{Q}/Z_2$ and $Z_2 \otimes Z_2$. To do this, first note that $(iZ_2)(iZ_2) = -Z_2 = Z_2$, and $(iZ_2)(jZ_2) = kZ_2$, so that we identify $Z_2 \leftrightarrow (+, +)$, $(iZ_2) \leftrightarrow (-, +)$, $(jZ_2) \leftrightarrow (+, -)$, and $kZ_2 \leftrightarrow (-, -)$. But V has an invariant subgroup consisting of the elements $(+, +)$ and $(-, -)$ (check that it is in fact invariant), which forms Z_2. Hence $\mathcal{Q}/Z_2$, which we have just shown is equal to V, also has an invariant subgroup Z_2. (This is admittedly a bit confusing. A cast of characters might help: $G = \mathcal{Q}$, $H = Z_2$, $Q = \mathcal{Q}/Z_2 = V$.) Since $V = \mathcal{Q}/Z_2$ does have an invariant subgroup, the invariant subgroup Z_2 is in fact not maximal.

---

* Including the discovery of the Monster group with $\sim 8 \otimes 10^{53}$ elements. See M. Ronan, *Symmetry and the Monster*.

[†] By fH we mean, of course, $f_1 H$, $f_2 H$, $\cdots$.

Appendix 2: Commutators and commutator subgroups

This appendix is devoted to the vexing issue of divergent terminology. I am aware that mathematicians call $\langle a, b \rangle$ the commutator of the two group elements a and b. I would like to avoid this terminology, because the term "commutator" is deeply ingrained in quantum physics and has a different meaning there. In physics, the commutator of A and B is defined* as $[A, B] = AB - BA$, which makes sense only if the subtraction symbol is defined (for matrices or operators in quantum mechanics, for example). When we discuss continuous groups in chapter I.3, then the concept of group elements near the identity makes sense. For $a \simeq I + A$, $b \simeq I + B$ near the identity (in the sense that A and B are small compared to I), then $\langle a, b \rangle = a^{-1}b^{-1}ab \simeq I + [A, B]$. Thus, for continuous groups, $\langle a, b \rangle$ and $[A, B]$ are intimately related but are still conceptually totally distinct.

In the text, I carefully refrained from giving $\langle a, b \rangle$ any name at all. This book is intended for physicists, and from my experience, calling $\langle a, b \rangle = a^{-1}b^{-1}ab$ a commutator invariably confuses some students. Also, a typical student statement is that it is really the same, since $\langle a, b \rangle = a^{-1}b^{-1}ab = I$ means that $ab = ba$ and thus $ab - ba = 0$, which ends up causing even more confusion, since the symbols $-$ and 0 do not exist in the definition of groups (think of the permutation group as an example). The student would then say something about rings, but we are not talking about rings here.

During one of my discussions with students about this issue, the name "grommutator" was suggested; I rather like it.[10]

Not surprisingly then, mathematicians usually call the derived subgroup the "commutator subgroup of G" and write $\mathcal{D} = [G, G]$, which looks odd at first sight to many physicists.

Denoting G by $G^{(0)}$, we can define the series $G^{(i+1)} = [G^{(i)}, G^{(i)}]$ with $i = 0, 1, \cdots$. This composition series of invariant subgroups may or may not end with the trivial group consisting of only the identity. If it does, G is known to mathematicians as solvable. If $G^{(1)} = G$, then G is known as perfect.[11]

Exercises

1   Show that for 2-cycles $(1a)(1b)(1a) = (ab)$.

2   Show that A_n for $n \geq 3$ is generated by 3-cycles, that is, any element can be written as a product of 3-cycles.

3   Show that S_n is isomorphic to a subgroup of A_{n+2}. Write down explicitly how S_3 is a subgroup of A_5.

4   List the partitions of 5. (We will need this later.)

5   Count the number of elements with a given cycle structure.

6   List the possible cycle structures in S_5 and count the number of elements with each structure.

7   Show that $\mathcal{Q}$ forms a group.

8   Show that A_4 is not simple.[†]

9   Show that A_4 is an invariant subgroup (in fact, maximal) of S_4.

10   Show that the kernel of a homomorphic map of a group G into itself is an invariant subgroup of G.

---

* As was already mentioned in the review of linear algebra.

[†] I can't resist mentioning here the possibly physically relevant fact that alone among all the alternating groups A_n, the group A_4 is not simple.

11 Calculate the derived subgroup of the dihedral group.

12 Given two group elements f and g, show that, while in general $fg \neq gf$, fg is equivalent to gf (that is, they are in the same equivalence class).

13 Prove that groups of even order contain at least one element (which is not the identity) that squares to the identity.

14 Using Cayley's theorem, map V to a subgroup of S_4. List the permutation corresponding to each element of V. Do the same for Z_4.

15 Map a finite group G with n elements into S_n a là Cayley. The map selects n permutations, known as "regular permutations," with various special properties, out of the $n!$ possible permutations of n objects.
(a) Show that no regular permutation besides the identity leaves an object untouched.
(b) Show that each of the regular permutations takes object 1 (say) to a different object.
(c) Show that when a regular permutation is resolved into cycles, the cycles all have the same length. Verify that these properties hold for what you got in exercise 14.

16 In a Coxeter group, show that if $n_{ij} = 2$, then a_i and a_j commute.

17 Show that for an invariant subgroup H, the left coset gH is equal to the right coset Hg.

18 In general, a group H can be embedded as a subgroup into a larger group G in more than one way. For example, A_4 can be naturally embedded into S_6 by following the route $A_4 \subset S_4 \subset S_5 \subset S_6$. Find another way of embedding A_4 into S_6. Hint: Think geometry!

19 Show that the derived subgroup of S_n is A_n. (In the text, with the remark about even permutations we merely showed that it is a subgroup of S_n.)

20 A set of real-valued functions f_i of a real variable x can also define a group if we define multiplication as follows: given f_i and f_j, the product $f_i \cdot f_j$ is defined as the function $f_i(f_j(x))$. Show that the functions $I(x) = x$ and $A(x) = (1-x)^{-1}$ generate a three-element group.[12] Furthermore, including the function $C(x) = x^{-1}$ generates a six-element group.

Notes

1. The jargon guy tells us that this is called an involution.
2. Dihedral means having two faces; in the context here, it means reflections are allowed. The root "hedra" means seat, bottom, base. Compare to polyhedron. Dihedrals occur quite often in everyday life, for example, in national emblems, such as the Ashok Chakra.
3. Some readers may know him as a much-loved personage who has appeared in QFT Nut and G Nut.
4. My preference for the term "invariant subgroup" over "normal subgroup" unfortunately deprives me of the pleasure of telling an ultra-nerd mathematical joke taken from Foolproof: A Sampling of Mathematical Folk Humor by Paul Renteln and Alan Dundes. But anyway, here it is. Question: What is purple and all of its offspring have been committed to institutions? Answer: A simple grape: it has no normal subgrapes.
5. Our friend the jargon guy won't give up; he tells us that furthermore a group is semi-simple if it does not have an abelian invariant subgroup. For example, S_3 is not semi-simple since it contains A_3, which is abelian.
6. Not to be confused with the homomorphic map in topology: homo from a Greek root meaning same, homeo meaning similar.
7. It used to be that finite groups appeared to be largely irrelevant for particle physics, but the situation may have changed. It is amusing to note that Sheldon Glashow, in his foreword to the 1982 first edition of Georgi's

Lie Algebras in Particle Physics, praised the book by saying that it "summarily dealt with finite group theory in the first seven pages."

8. For an introductory account, see, for example, Wikipedia. The Monster was found in 1981, and victory declared in 1983, but the full classification was only completed in 2004.

9. A naive physicist would just divide the number of elements to see whether $N(Q) = N(G)/N(H)$ is the smallest number "possible."

10. "Commuter" was also mentioned, but it sounds too similar to commutator. It also reminds me of the old nerd joke: what is purple and commutes? An abelian grape.

11. The smallest perfect group is A_5.

12. This exercise is from G. Hall.

1.3 | Rotations and the Notion of Lie Algebra

Two different approaches to rotations in the plane

If the permutation groups are the poster children for raising awareness of finite groups, then the rotation groups are surely the poster children for continuous groups. Given that we were all born into 3-dimensional Euclidean space, it is hardly surprising that the rotation group plays a crucial role in physics. Indeed, the very concept of a group was abstracted from the behavior of rotations.

My pedagogical strategy in this chapter[1] is to first take something you know extremely well, namely, rotations in the plane, present it in a way possibly unfamiliar to you, and go through it slowly in great detail—"beating it to death," so to speak. And then we will move up to rotations in 3-dimensional space and beyond.

I start with two different approaches to rotations in the plane, the first based on trigonometry, the second based on invariance considerations.

Cartesian coordinates and trigonometry

As you know, from a course on mechanics, we envisage either rotating the body we are studying or rotating the observer. Here we consistently rotate the observer.*

Put down Cartesian coordinate axes (see figure 1) so that a point P is labeled by two real numbers (x, y). Suppose another observer (call him Mr. Prime) puts down coordinate axes rotated by angle θ with respect to the axes put down by the first observer (call her Ms. Unprime) but sharing the same origin O. Elementary trigonometry tells us that the coordinates (x, y) and (x', y') assigned by the two observers to the same point P are related by[†]

$$x' = \cos\theta x + \sin\theta y, \qquad y' = -\sin\theta x + \cos\theta y \tag{1}$$

---

* This point of view is closer in spirit to the convention used in several advanced areas of theoretical physics, such as relativity and gravity.

[†] For example, by comparing similar triangles in the figure, we obtain $x' = (x/\cos\theta) + (y - x\tan\theta)\sin\theta$.

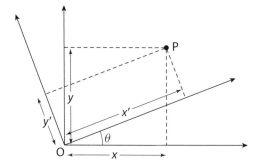

Figure 1

The distance from P to the origin O has to be the same for the two observers, of course. According to Pythagoras, this requires $\sqrt{x'^2 + y'^2} = \sqrt{x^2 + y^2}$, which you could check using (1).

Introduce the column vectors $\vec{r} = \begin{pmatrix} x \\ y \end{pmatrix}$ and $\vec{r}' = \begin{pmatrix} x' \\ y' \end{pmatrix}$ and the rotation matrix

$$R(\theta) = \begin{pmatrix} \cos\theta & \sin\theta \\ -\sin\theta & \cos\theta \end{pmatrix} \tag{2}$$

Then we can write (1) more compactly as $\vec{r}' = R(\theta)\vec{r}$.

We have already used the word "vector" in the linear algebra review. For the purpose of this chapter, a vector is a quantity (for example, the velocity of a particle in the plane) consisting of two real numbers, such that if Ms. Unprime represents it by $\vec{p} = \begin{pmatrix} p^1 \\ p^2 \end{pmatrix}$, then Mr. Prime will represent it by $\vec{p}' = R(\theta)\vec{p}$. In short, a vector is something that transforms like the coordinates $\begin{pmatrix} x \\ y \end{pmatrix}$ under rotation. The emphasis here is on the word "like."

Invariance under linear transformations

Given two vectors $\vec{p} = \begin{pmatrix} p^1 \\ p^2 \end{pmatrix}$ and $\vec{q} = \begin{pmatrix} q^1 \\ q^2 \end{pmatrix}$, the scalar or dot product is defined by $\vec{p}^T \cdot \vec{q} = p^1 q^1 + p^2 q^2$. Here the transposed vector $\vec{p}^T$ is the row vector (p^1, p^2).

According to Pythagoras, the square of the length of $\vec{p}$ is given by $\vec{p}^2 \equiv \vec{p}^T \cdot \vec{p} = (p^1)^2 + (p^2)^2$. By definition, rotations leave the length of $\vec{p}$, and hence $\vec{p}^2$, invariant. In other words, if $\vec{p}' = R(\theta)\vec{p}$, then $\vec{p}'^2 = \vec{p}^2$. Since this works for any vector $\vec{p}$, including the case in which $\vec{p}$ happens to be composed of two arbitrary vectors $\vec{u}$ and $\vec{v}$ (namely, $\vec{p} = \vec{u} + \lambda\vec{v}$ with λ an arbitrary real number), and since $\vec{p}^2 = (\vec{u} + \lambda\vec{v})^2 = \vec{u}^2 + \lambda^2\vec{v}^2 + 2\lambda\vec{u}^T \cdot \vec{v}$, rotation also leaves the dot product between two arbitrary vectors invariant: $\vec{p}'^2 = \vec{p}^2$ (where $\vec{p}' = \vec{u}' + \lambda\vec{v}' = R\vec{p} = R\vec{u} + \lambda R\vec{v}$) implies that $\vec{u}'^T \cdot \vec{v}' = \vec{u}^T \cdot \vec{v}$, since λ can be varied arbitrarily.

Since $\vec{u}' = R\vec{u}$ (to unclutter things, we often suppress the θ dependence in $R(\theta)$) and so $\vec{u}'^T = \vec{u}^T R^T$, we now have $\vec{u}^T \cdot \vec{v} = \vec{u}'^T \cdot \vec{v}' = (\vec{u}^T R^T) \cdot (R\vec{v}) = \vec{u}^T (R^T R)\vec{v}$. As this holds for any two vectors $\vec{u}$ and $\vec{v}$, we must have the matrix equation

$$R^T R = I \tag{3}$$

This imposes a condition on R.

Matrices that satisfy (3) are called orthogonal.[2] As was already mentioned in chapter I.1, these orthogonal matrices generate the group $O(2)$, with the multiplication law $R(\theta_1) R(\theta_2) = R(\theta_1 + \theta_2)$.

As promised, we followed two approaches to studying rotations. In the first approach, we used trigonometry, in the second, we insisted on linear transformations that leave the lengths of vectors unchanged.

At this point, we might pause to check that the two approaches are consistent by plugging in the explicit form of the rotation, given in (2), into the condition (3):

$$R(\theta)^T R(\theta) = \begin{pmatrix} \cos\theta & -\sin\theta \\ \sin\theta & \cos\theta \end{pmatrix} \begin{pmatrix} \cos\theta & \sin\theta \\ -\sin\theta & \cos\theta \end{pmatrix} = \begin{pmatrix} 1 & 0 \\ 0 & 1 \end{pmatrix} \tag{4}$$

Reflections

Recall from the linear algebra review that the determinant of a product of matrices is equal to the product of the determinants: $\det M_1 M_2 = (\det M_1)(\det M_2)$, and that the determinant of the transpose of a matrix is the same as the determinant of that matrix: $\det M^T = \det M$. Taking the determinant of (3), we obtain $(\det R)^2 = 1$, that is, $\det R = \pm 1$. The determinant of an orthogonal matrix may be -1 as well as $+1$. In other words, orthogonal matrices also include reflection matrices, such as $\mathcal{P} = \begin{pmatrix} 1 & 0 \\ 0 & -1 \end{pmatrix}$, a reflection flipping the y-axis.

To focus on rotations, let us exclude reflections by imposing the condition (since $\det \mathcal{P} = -1$)

$$\det R = 1 \tag{5}$$

Matrices with unit determinant are called special (as already mentioned in chapter I.1).

Note that matrices of the form $\mathcal{P} R$ for any rotation R are also excluded by (5), since $\det(\mathcal{P} R) = \det \mathcal{P} \det R = (-1)(+1) = -1$. In particular, a reflection flipping the x-axis $\begin{pmatrix} -1 & 0 \\ 0 & 1 \end{pmatrix}$, which is the product of $\mathcal{P}$ and a rotation through $90°$, is also excluded.

We define a rotation as a matrix that is both orthogonal and special, that is, a matrix that satisfies both (3) and (5). Thus, the rotation group of the plane consists of the set of all special orthogonal 2-by-2 matrices and is known as $SO(2)$.

In the linear algebra review, we mentioned in an endnote that it is good practice in physics and mathematics to turn equations and logical sequences around. Starting from (2), you can readily check (3) and (5). Verify that, given (3) and (5), you can also get to (2). Indeed, in the next section, we will learn how to do this in arbitrary dimensions, not just for the almost trivial 2-dimensional case.

Act a little bit at a time

The Norwegian physicist Marius Sophus Lie (1842–1899) had the almost childishly obvious but brilliant idea that to rotate through, say, 29°, you could just as well rotate through a zillionth of a degree and repeat the process 29 zillion times. To study rotations, it suffices to study rotation through infinitesimal angles. Shades of Newton and Leibniz! A rotation through a finite angle can always be obtained by performing infinitesimal rotations repeatedly. As is typical with many profound statements in physics and mathematics, Lie's idea is astonishingly simple. Replace the proverb "Never put off until tomorrow what you have to do today" by "Do what you have to do a little bit at a time."

A rotation through an infinitesimal angle θ is almost the identity I, that is, no rotation at all, and so can be written as

$$R(\theta) \simeq I + A \tag{6}$$

Here A denotes some infinitesimal matrix of order θ. The neglected terms in (6) are of order θ^2 and smaller.

Let us imagine Lie saying to himself, "Pretend that I slept through trigonometry class and I don't know anything about (2). Instead, I will define rotations as the set of linear transformations on 2-component objects $\vec{u}' = R\vec{u}$ and $\vec{v}' = R\vec{v}$ that leave $\vec{u}^T \cdot \vec{v}$ invariant. I will impose (3) $R^T R = I$ and derive (2). But according to my brilliant idea, it suffices to solve this condition for rotations infinitesimally close to the identity."

Following Lie, we plug $R \simeq I + A$ into $R^T R = I$. Since by assumption A^2, being of order θ^2, can be neglected relative to A, we have

$$R^T R \simeq (I + A^T)(I + A) \simeq (I + A^T + A) = I \tag{7}$$

Thus, this requires* $A^T = -A$, namely, that A must be antisymmetric.

But there is basically only one 2-by-2 antisymmetric matrix:

$$\mathcal{J} \equiv \begin{pmatrix} 0 & 1 \\ -1 & 0 \end{pmatrix} \tag{8}$$

In other words, the solution of $A^T = -A$ is $A = \theta \mathcal{J}$ for some real number θ, which, as we will discover shortly, is in fact the same as the angle θ we have been using. Thus, rotations close to the identity have the form[†]

$$R = I + \theta \mathcal{J} + O(\theta^2) = \begin{pmatrix} 1 & \theta \\ -\theta & 1 \end{pmatrix} + O(\theta^2) \tag{9}$$

---

* Note that this result, obtained by equating terms of order θ, is exact.

[†] An equivalent way of saying this is that for infinitesimal θ, the transformation $x' \simeq x + \theta y$ and $y' \simeq y - \theta x$ satisfies the Pythagorean condition $x'^2 + y'^2 = x^2 + y^2$ to first order in θ. (You could verify that (1) indeed reduces to this transformation to leading order in θ.) Or, write $x' = x + \delta x$, $y' = y + \delta y$, and solve the condition $x \delta x + y \delta y = 0$ that $\delta \vec{r}$ is perpendicular to $\vec{r}$.

The antisymmetric matrix $\mathcal{J}$ is known as the generator of the rotation group. We obtain, without knowing any trigonometry, that under an infinitesimal rotation, $x \to x' \simeq x + \theta y$, and $y \to y' = -\theta x + y$, which is of course consistent with (1). We could also obtain this result by drawing an elementary geometrical figure involving infinitesimal angles.

Now recall the identity $e^x = \lim_{N \to \infty} (1 + \frac{x}{N})^N$ (which you can easily prove by differentiating both sides). Then, for a finite (that is, not infinitesimal) angle θ, we have

$$R(\theta) = \lim_{N \to \infty} \left(R\left(\frac{\theta}{N}\right) \right)^N = \lim_{N \to \infty} \left(1 + \frac{\theta \mathcal{J}}{N} \right)^N = e^{\theta \mathcal{J}} \tag{10}$$

The first equality represents Lie's profound idea: we cut up the given noninfinitesimal angle θ into N pieces so that θ/N is infinitesimal for N large enough and perform the infinitesimal rotation N times. The second equality is just (9). For the last equality, we use the identity just mentioned, which amounts to the definition of the exponential.

As an alternative but of course equivalent path to our result, simply assert that we have every right, to leading order, to write $R(\frac{\theta}{N}) = 1 + \frac{\theta \mathcal{J}}{N} \simeq e^{\frac{\theta \mathcal{J}}{N}}$. Thus

$$R(\theta) = \lim_{N \to \infty} \left(R\left(\frac{\theta}{N}\right) \right)^N = \lim_{N \to \infty} \left(e^{\frac{\theta \mathcal{J}}{N}} \right)^N = e^{\theta \mathcal{J}} \tag{11}$$

In calculus, we learned about the Taylor or power series. Taylor said that if we gave him all the derivatives of a function $f(x)$ at $x = 0$ (say), he could construct the function. In contrast, Lie said that, thanks to the multiplicative group structure, he only needs the first derivative of the group element $R(\theta)$ near the identity. Indeed, we recognize that $\mathcal{J}$ is just $\frac{dR(\theta)}{d\theta}|_{\theta=0}$. The reason that Lie needs so much less is of course that the group structure is highly restrictive.

Finally, we can check that the formula $R(\theta) = e^{\theta \mathcal{J}}$ reproduces (2) for any value of θ. We simply note that $\mathcal{J}^2 = -I$ and separate the exponential series, using Taylor's idea, into even and odd powers of $\mathcal{J}$:

$$e^{\theta \mathcal{J}} = \sum_{n=0}^{\infty} \theta^n \mathcal{J}^n / n! = \left(\sum_{k=0}^{\infty} (-1)^k \theta^{2k} / (2k)! \right) I + \left(\sum_{k=0}^{\infty} (-1)^k \theta^{2k+1} / (2k+1)! \right) \mathcal{J}$$

$$= \cos\theta \, I + \sin\theta \, \mathcal{J} = \cos\theta \begin{pmatrix} 1 & 0 \\ 0 & 1 \end{pmatrix} + \sin\theta \begin{pmatrix} 0 & 1 \\ -1 & 0 \end{pmatrix} = \begin{pmatrix} \cos\theta & \sin\theta \\ -\sin\theta & \cos\theta \end{pmatrix} \tag{12}$$

which is precisely $R(\theta)$ as given in (2). Note that this works, because $\mathcal{J}$ plays the same role as i in Euler's identity $e^{i\theta} = \cos\theta + i \sin\theta$.

To summarize, the condition $R^T R = I$ determines the rotation matrix obtained previously by using trigonometry.

An old friend of the author's, Confusio, wanders by. He looks mildly puzzled. "How come we don't have to impose det $R = 1$?"

Ah, that's because we are looking at Rs that are continuously related to the identity I. The reflection $\mathcal{P}$, with det $\mathcal{P} = -1$, is manifestly not continuously related to the identity, with det $I = +1$.

Two approaches to rotation

To summarize, there are two different approaches to rotation.

In the first approach, applying trigonometry to figure 1, we write down (1) and hence (2).

In the second approach, we specify what is to be left invariant by rotations and hence define rotations by the condition (3) that rotations must satisfy. Lie then tells us that it suffices to solve (3) for infinitesimal rotations. We could then build up rotations through finite angles by multiplying infinitesimal rotations together, thus arriving also at (2).

It might seem that the first approach is much more direct. One writes down (2) and that's that. The second approach appears more roundabout. The point is that the second approach generalizes to higher dimensional spaces (and to other situations, for example in chapter IV.4 on the unitary groups) much more readily than the first approach does, as we shall see presently.

Distance squared between neighboring points

Before we go on, let us take care of one technical detail. We assumed that Mr. Prime and Ms. Unprime set up their coordinate systems to share the same origin O. We now show that this condition is unnecessary if we consider two points P and Q (rather than one point, as in our discussion above) and study how the vector connecting P to Q transforms. Let Ms. Unprime assign the coordinates $\vec{r}_P = (x, y)$ and $\vec{r}_Q = (\tilde{x}, \tilde{y})$ to P and Q, respectively. Then Mr. Prime's coordinates $\vec{r}'_P = (x', y')$ for P and $\vec{r}'_Q = (\tilde{x}', \tilde{y}')$ for Q are then given by $\vec{r}'_P = R(\theta)\vec{r}_P$ and $\vec{r}'_Q = R(\theta)\vec{r}_Q$. Subtracting the first equation from the second, we have $(\vec{r}'_P - \vec{r}'_Q) = R(\theta)(\vec{r}_P - \vec{r}_Q)$. Defining $\Delta x = \tilde{x} - x$, $\Delta y = \tilde{y} - y$, and the corresponding primed quantities, we obtain $\begin{pmatrix} \Delta x' \\ \Delta y' \end{pmatrix} = \begin{pmatrix} \cos\theta & \sin\theta \\ -\sin\theta & \cos\theta \end{pmatrix} \begin{pmatrix} \Delta x \\ \Delta y \end{pmatrix}$. Rotations leave the distance between the points P and Q unchanged: $(\Delta x')^2 + (\Delta y')^2 = (\Delta x)^2 + (\Delta y)^2$. You recognize of course that this is a lot of tedious verbiage stating the perfectly obvious, but I want to be precise here. Of course, the distance between any two points is left unchanged by rotations. (This also means that the distance between P and the origin is left unchanged by rotations; ditto for the distance between Q and the origin.)

Let us take the two points P and Q to be infinitesimally close to each other and replace the differences $\Delta x'$, Δx, and so forth by differentials dx', dx, and so forth. Indeed, 2-dimensional Euclidean space is defined by the distance squared between two nearby points: $ds^2 = dx^2 + dy^2$. Rotations are defined as linear transformations $(x, y) \to (x', y')$ such that

$$dx^2 + dy^2 = dx'^2 + dy'^2 \tag{13}$$

The whole point is that this now makes no reference to the origin O (or to whether Mr. Prime and Ms. Unprime even share the same origin).

The column $d\vec{x} = \begin{pmatrix} dx^1 \\ dx^2 \end{pmatrix} \equiv \begin{pmatrix} dx \\ dy \end{pmatrix}$ is defined as the basic or ur-vector, the template for all other vectors. To repeat, a vector is defined as something that transforms like $d\vec{x}$ under rotations.

From the plane to higher-dimensional space

The reader who has wrestled with Euler angles in a mechanics course knows that the analog of (2) for 3-dimensional space is already quite a mess. In contrast, Lie's approach allows us, as mentioned above, to immediately jump to N-dimensional Euclidean space, defined by specifying the distance squared between two nearby points as given by the obvious generalization of Pythagoras's theorem: $ds^2 = \sum_{i=1}^{N}(dx^i)^2 = (dx^1)^2 + (dx^2)^2 + \cdots + (dx^N)^2$. Rotations are defined as linear transformations $d\vec{x}' = Rd\vec{x}$ (with R an N-by-N matrix) that leave ds^2 unchanged.

The preceding discussion allows us to write this condition as $R^T R = I$. As before, we want to eliminate reflection and to focus on rotations by imposing the additional condition $\det R = 1$. The set of N-by-N matrices R that satisfy these two conditions forms the simple orthogonal group $SO(N)$, which is just a fancy way of saying the rotation group in N-dimensional space.

In chapter I.1, we already talked about the rotation group as if it is self-evident that rotations form a group. Let us make it official by checking that $SO(N)$ satisfies all the group axioms. The product of two rotations is a rotation: $(R_1 R_2)^T (R_1 R_2) = (R_2^T R_1^T)(R_1 R_2) = R_2^T (R_1^T R_1) R_2 = R_2^T R_2 = I$ (is that slow enough?), and $\det(R_1 R_2) = \det R_1 \det R_2 = 1$. Matrix multiplication is associative. The condition $\det R = 1$ guarantees the existence of the inverse.

Lie in higher dimensions

The power of Lie now shines through when we want to work out rotations in higher-dimensional spaces. All we have to do is satisfy the two conditions $R^T R = I$ and $\det R = 1$.

Lie shows us that the first condition, $R^T R = I$, is solved immediately by writing $R \simeq I + A$ and requiring $A = -A^T$, namely, that A be antisymmetric.

That's it. We could be in a zillion-dimensional space, but still, the rotation group is fixed by requiring A to be antisymmetric.

But it is very easy to write down all possible antisymmetric N-by-N matrices! For $N = 2$, there is only one, namely, the $\mathcal{J}$ introduced earlier. For $N = 3$, there are basically three of them:

$$\mathcal{J}_x = \begin{pmatrix} 0 & 0 & 0 \\ 0 & 0 & 1 \\ 0 & -1 & 0 \end{pmatrix}, \qquad \mathcal{J}_y = \begin{pmatrix} 0 & 0 & -1 \\ 0 & 0 & 0 \\ 1 & 0 & 0 \end{pmatrix}, \qquad \mathcal{J}_z = \begin{pmatrix} 0 & 1 & 0 \\ -1 & 0 & 0 \\ 0 & 0 & 0 \end{pmatrix} \qquad (14)$$

Any 3-by-3 antisymmetric matrix can be written as $A = \theta_x \mathcal{J}_x + \theta_y \mathcal{J}_y + \theta_z \mathcal{J}_z$, with three real numbers θ_x, θ_y, and θ_z. The three 3-by-3 antisymmetric matrices $\mathcal{J}_x$, $\mathcal{J}_y$, $\mathcal{J}_z$ are known as generators. They generate rotations, but are of course not to be confused with rotations, which are by definition 3-by-3 orthogonal matrices with determinant equal to 1.

One upshot of this whole discussion is that any 3-dimensional rotation (not necessarily infinitesimal) can be written as

$$R(\theta) = e^{\theta_x \mathcal{J}_x + \theta_y \mathcal{J}_y + \theta_z \mathcal{J}_z} = e^{\Sigma_i \theta_i \mathcal{J}_i} \tag{15}$$

(with $i = x, y, z$) and is thus characterized by three real numbers θ_x, θ_y, and θ_z. As I said, those readers who have suffered through the rotation of a rigid body in a course on mechanics surely would appreciate the simplicity of studying the generators of infinitesimal rotations and then simply exponentiating them.

To mathematicians, physicists often appear to use weird notations. There is not an i in sight, yet physicists are going to stick one in now. If you have studied quantum mechanics, you know that the generators $\mathcal{J}$ of rotation studied here are related to angular momentum operators. You would also know that in quantum mechanics observables are represented by hermitean operators or matrices. In contrast, in our discussion, the $\mathcal{J}$s come out naturally as real antisymmetric matrices and are thus antihermitean. To make them hermitean, we multiply them by some multiples of the imaginary[3] unit i. Thus, define[4] $J_x \equiv -i \mathcal{J}_x$, $J_y \equiv -i \mathcal{J}_y$, $J_z \equiv -i \mathcal{J}_z$, and write a general rotation as

$$R(\theta) = e^{i \Sigma_j \theta_j J_j} = e^{i \vec{\theta} \cdot \vec{J}} \tag{16}$$

treating the three real numbers θ_j and the three matrices J_j as two 3-dimensional vectors.

Now you should write down the generators of rotations in 4-dimensional space. At least count how many there are. See how easy it is to study rotations in arbitrarily high dimensional space? We will come back to this later in chapter IV.1.

Any student of physics knows that many physical situations exhibit spherical[5] symmetry, in which case the rotation group $SO(3)$ plays a central role.

Lie algebra

In chapter I.1, we mentioned that, in general, rotations do not commute. Following Lie, we could try to capture this essence of group multiplication by focusing on infinitesimal rotations.

Let $R \simeq I + A$ be an infinitesimal rotation. For an arbitrary rotation R', consider $RR'R^{-1} \simeq (I + A)R'(I - A) \simeq R' + AR' - R'A$ (where we have consistently ignored terms of order A^2). If rotations commute, then $RR'R^{-1}$ would be equal to R'. Thus, the extent to which this is not equal to R' measures the lack of commutativity. Now, suppose R' is also an infinitesimal rotation $R' \simeq I + B$. Then $RR'R^{-1} \simeq I + B + AB - BA$, which differs from $R' \simeq I + B$ by the matrix

$$[A, B] \equiv AB - BA, \tag{17}$$

known as the commutator* between A and B.

---

* And already mentioned in the review of linear algebra.

For $SO(3)$, for example, A is a linear combination of the J_is, which we shall call the generators of the Lie algebra of $SO(3)$. Thus, we can write $A = i \sum_i \theta_i J_i$ and similarly $B = i \sum_j \theta'_j J_j$. Hence $[A, B] = i^2 \sum_{ij} \theta_i \theta'_j [J_i, J_j]$, and so it suffices to calculate the commutators $[J_i, J_j]$ once and for all.

Recall that for two matrices M_1 and M_2, $(M_1 M_2)^T = M_2^T M_1^T$. Transposition reverses the order.* Thus, $([J_i, J_j])^T = -[J_i, J_j]$. In other words, the commutator $[J_i, J_j]$ is itself an antisymmetric 3-by-3 matrix and thus can be written as a linear combination of the J_ks:

$$[J_i, J_j] = i c_{ijk} J_k \tag{18}$$

The summation over k is implied by the repeated index summation convention. The coefficients c_{ijk} in the linear combination, with a factor of i taken out explicitly, are real (convince yourself of this) numbers. Evidently, $c_{ijk} = -c_{jik}$.

By explicit computation using (14), we find

$$[J_x, J_y] = i J_z \tag{19}$$

You should work out the other commutators or argue by cyclic substitution $x \to y \to z \to x$ that

$$[J_y, J_z] = i J_x \tag{20}$$

$$[J_z, J_x] = i J_y \tag{21}$$

The three commutation relations, (19), (20), and (21), may be summarized by

$$[J_i, J_j] = i \varepsilon_{ijk} J_k \quad \text{(sum over k implied)} \tag{22}$$

We define the totally antisymmetric symbol† ε_{ijk} by saying that it changes sign on the interchange of any pair of indices (and hence it vanishes when any two indices are equal) and by specifying that $\varepsilon_{123} = 1$. In other words, we found that $c_{ijk} = \varepsilon_{ijk}$. (Without further apology, I will often commit minor notational abuses, such as jumping back and forth between x, y, z and 1, 2, 3.)

The statement is that the commutation relations (22) tell us about the multiplicative structure of infinitesimal rotations. By exponentiating the generators J_i, as in (15), we then manage to capture the multiplicative structure of rotations not necessarily infinitesimal.

Note that the commutation relations fix the sign of J_i. In other words, the transformation $J_i \to -J_i, i = 1, 2, 3$ does not leave (22) invariant.

Structure constants

Lie's great insight is that the preceding discussion holds for any group whose elements $g(\theta_1, \theta_2, \cdots)$ are labeled by a set of continuous parameters such that $g(0, 0, \cdots)$ is the

---

* Note that this fact already played a crucial role in the section before last.
† Already introduced in the review of linear algebra.

identity I. (For example, the continuous parameters would be the angles θ_i, $i = 1, 2, 3$ in the case of $SO(3)$.)

For these groups, now known as Lie groups, this is what you do in four easy steps:

1. Expand the group elements around the identity by letting the continuous parameters go to zero: $g \simeq I + A$.

2. Write $A = i \sum_a \theta_a T_a$ as a linear combination of the generators T_a as determined by the nature of the group.

3. Pick two group elements near the identity: $g_1 \simeq I + A$ and $g_2 \simeq I + B$. Then $g_1 g_2 g_1^{-1} \simeq I + B + [A, I + B] \simeq I + B + [A, B]$. The commutator $[A, B]$ captures the essence of the group near the identity.

4. As in step 2, we can write $B = i \sum_b \theta'_b T_b$ as a linear combination of the generators T_b. Similarly, we can write $[A, B]$ as a linear combination of the generators T_c. (We know this because, for g_1 and g_2 near the identity, $g_1 g_2 g_1^{-1}$ is also near the identity.) Plugging in, we then arrive at the analog of (18) for any continuous group, namely, the commutation relations

$$[T_a, T_b] = i f_{abc} T_c \tag{23}$$

The commutator between any two generators can be written as a linear combination of the generators.

The commutation relations between the generators define a Lie algebra, with f_{abc} referred to as the structure constants of the algebra. The structure constants determine the Lie algebra, which essentially determines the Lie group.

This brief introduction to Lie algebra at this stage is necessarily somewhat vague, but we will go into more details soon enough. The key idea is that we can go a long way toward understanding a continuous group by studying its Lie algebra.

Note that, while a Lie group is characterized by multiplication, its Lie algebra is characterized by commutation.

Confusio said, "When I first studied group theory, I did not clearly distinguish between Lie group and Lie algebra. That they allow totally different operations did not sink in. I was multiplying the J_is together and couldn't make sense of what I got."

Absolutely, it is crucial to keep in mind that Lie group and Lie algebra are mathematically rather different structures. In a group, you multiply (or if one wants to pick nits, compose) two elements together to get another element. In the corresponding algebra (assuming of course that the group is continuous), you take two elements, and you commute them to get another element of the algebra. (Again, to keep the nitpickers at bay, it may be perhaps better to say two members of the algebra, since we have spoken often of group elements.)

The rotation group offers a good example. The members $\mathcal{J}$ of its algebra are real antisymmetric matrices,* but if you multiply two real antisymmetric matrices together, you certainly do not get a real antisymmetric matrix. The algebra does not close under

---

* Or hermitean matrices, if you prefer to talk about $J = -i\mathcal{J}$.

multiplication, only under commutation. Perhaps one confusing point for the beginner is that to calculate the commutator, one has to, in an intermediate step, multiply two real antisymmetric matrices together. Speaking somewhat colloquially, one has to first get out of the algebra before one can get back in.

We give presently a more mathematical formulation.

A Lie algebra is a linear space spanned by linear combinations $\sum_i \theta_i \mathcal{J}_i$ of the generators. In contrast, it makes no sense, in the rotation group, to form linear combinations of rotations. Given two rotations R_1 and R_2, the linear sum $R_s = c_1 R_1 + c_2 R_2$ certainly is not in general a rotation.

A mathematician might define a Lie algebra abstractly as a linear vector space V equipped with a map $f : V \otimes V \to V$ satisfying various properties, for instance, $f(A, B) = -f(B, A)$.

Historically, the relation between Lie group and Lie algebra was also hinted at by the Baker-Campbell-Hausdorff formula,[6] stating that $e^A e^B = e^C$, with $C = A + B + \frac{1}{2}[A, B] + \frac{1}{12}([A, [A, B]] + [B, [B, A]]) + \cdots$.

Rotations in higher-dimensional space

With your experience with (8) and (14), it is now a cinch for you to generalize and write down a complete set of antisymmetric N-by-N matrices.

Start with an N-by-N matrix with 0 everywhere. Stick a 1 into the mth row and nth column; due to antisymmetry, you are obliged to put a (-1) into the nth row and mth column. Call this antisymmetric matrix $\mathcal{J}_{(mn)}$. We put the subscripts (mn) in parentheses to emphasize that (mn) labels the matrix. They are not indices to tell us which element of the matrix we are talking about. As explained before, physicists like Hermite a lot and throw in a $-i$ to define the hermitean matrices $J_{(mn)} = -i \mathcal{J}_{(mn)}$. Explicitly,

$$(J_{(mn)})^{ij} = -i(\delta^{mi}\delta^{nj} - \delta^{mj}\delta^{ni}) \tag{24}$$

To repeat, in the symbol $(J_{(mn)})^{ij}$, which we will often write as $J_{(mn)}^{ij}$ for short, the indices i and j indicate, respectively, the row and column of the entry $(J_{(mn)})^{ij}$ of the matrix $J_{(mn)}$, while the indices m and n, which I put in parentheses for pedagogical clarity, indicate which matrix we are talking about. The first index m on $J_{(mn)}$ can take on N values, and then the second index n can take on only $(N-1)$ values, since, evidently, $J_{(mm)} = 0$. Also, since $J_{(nm)} = -J_{(mn)}$, we require $m > n$ to avoid double counting. Thus, there are only $\frac{1}{2}N(N-1)$ real antisymmetric N-by-N matrices $J_{(mn)}$. The Kronecker deltas in (24) merely say what we said in words in the preceding paragraph.

As before, an infinitesimal rotation is given by $R \simeq I + A$ with the most general A a linear combination of the $J_{(mn)}$s: $A = i \sum_{m,n} \theta_{(mn)} J_{(mn)}$, where the antisymmetric coefficients $\theta_{(mn)} = -\theta_{(nm)}$ denote $\frac{1}{2}N(N-1)$ generalized angles. (As a check, for $N = 2$ and 3, $\frac{1}{2}N(N-1)$ equals 1 and 3, respectively.) The matrices $J_{(mn)}$ are the generators of the group $SO(N)$.

Rotation around an axis or rotation in a plane

Confusio speaks up: "All this makes sense, but I am nevertheless puzzled. We physics students are used to rotating around an axis,* in particular, the x, y, and z axes. For $N = 3$, the case that I am most familiar with, the generators are labeled by only one index, to wit, J_x, J_y, and J_z. Strange that $J_{(mn)}$ is in general labeled by two indices."

We should thank Confusio. This notational peculiarity has indeed confused many students. The explanation is simple. Dear reader, please think: What does it mean to rotate around the fifth axis in, say, 8-dimensional Euclidean space? It makes no sense! However, it makes perfect sense to rotate in the (3-7) plane, for example: simply rotate the third axis and the seventh axis into each other. For $N = 3$, the indices m, n take on three values, and so we can write† $J_x = J_{23}$, $J_y = J_{31}$, and $J_z = J_{12}$. In 3-dimensional space, and only in 3-dimensional space, a plane is uniquely specified by the vector perpendicular to it. Thus, a rotation commonly spoken of as a rotation around the z-axis is better thought of as a rotation in the (12)-plane, that is, the (xy)-plane.‡

The Lie algebra for $SO(N)$

Our next task is to work out the Lie algebra for $SO(N)$, namely, the commutators between the $J_{(mn)}$s. You could simply plug in (24) and chug away. Exercise!

But a more elegant approach is to work out $SO(4)$ as an inspiration for the general case. First, $[J_{(12)}, J_{(34)}] = 0$, as you might expect, since rotations in the (1-2) plane and in the (3-4) plane are like gangsters operating on different turfs. Next, we tackle $[J_{(23)}, J_{(31)}]$. Notice that the action takes place entirely in the $SO(3)$ subgroup of $SO(4)$, and so we already know the answer: $[J_{(23)}, J_{(31)}] = [J_x, J_y] = i J_z = i J_{(12)}$. These two examples, together with antisymmetry $J_{(mn)} = -J_{(nm)}$, in fact take care of all possible cases. In the commutator $[J_{(mn)}, J_{(pq)}]$, there are three possibilities for the index sets (mn) and (pq): (i) they have no integer in common, (ii) they have one integer in common, or (iii) they have two integers in common. The commutator vanishes in cases (i) and (iii), for trivial (but different) reasons. In case (ii), suppose $m = p$ with no loss of generality, then the commutator is equal to $i J_{(nq)}$.

We obtain, for any N,

$$[J_{(mn)}, J_{(pq)}] = i(\delta_{mp} J_{(nq)} + \delta_{nq} J_{(mp)} - \delta_{np} J_{(mq)} - \delta_{mq} J_{(np)}) \tag{25}$$

This may look rather involved to the uninitiated, but in fact it simply states in mathematical symbols the last three sentences of the preceding paragraph. First, on the right hand side, a linear combination of the Js (as required by the general argument above)

---

* For example, dance around the maypole.
† We will, as here, often pass freely between the index sets (123) and (xyz).
‡ In this connection, note that the $\mathcal{J}$ in (8) appears as the upper left 2-by-2 block in $\mathcal{J}_z$ in (14).

is completely fixed by the first term by noting that the left hand side is antisymmetric under three separate interchanges: $m \leftrightarrow n$, $p \leftrightarrow q$, and $(mn) \leftrightarrow (pq)$. Next, all those Kronecker deltas just say that if the two sets (mn) and (pq) have no integer in common, then the commutator vanishes. If they do have an integer in common, simply "cross off" that integer. For example, $[J_{(12)}, J_{(14)}] = i J_{(24)}$ and $[J_{(23)}, J_{(31)}] = -i J_{(21)} = i J_{(12)}$. (Does this agree with what you got by the brute force approach? Surely you did the exercise!)

Confusio looks a bit puzzled. "But (25) does not look much like the canonical definition (23) of a Lie algebra."

We tell him that his confusion is merely notational. The generators are labeled by the peculiar symbol (mn) made out of two letters from a certain alphabet and parentheses. We might as well call it a, with $a = 1, 2, \cdots, \frac{1}{2}N(N-1)$, and write J_a for the generator $J_{(mn)}$. For an explicit example, consider $SO(4)$ with its six generators $J_{(mn)}$. Make a 1-to-1 map of the indices $(mn) \leftrightarrow a$ as follows: $(12) \leftrightarrow 1$, $(13) \leftrightarrow 2$, $(14) \leftrightarrow 3$, $(23) \leftrightarrow 4$, $(24) \leftrightarrow 5$, $(34) \leftrightarrow 6$. How the $\frac{1}{2}N(N-1)$ J_as are ordered does not matter as long as the correspondence $(mn) \leftrightarrow a$ is 1-to-1.

The right hand of (25) is just a linear combination of J_cs, and hence (25) can indeed be written as $[T_a, T_b] = i f_{abc} T_c$, as in (23).

We will come back to this point when we discuss the adjoint representation in chapter IV.1.

Appendix 1: Differential operators rather than matrices

Instead of using matrices, we can also represent the operators J_i by differential operators, acting on the coordinates (x, y, z). In particular, let $J_z = -i(x \frac{\partial}{\partial y} - y \frac{\partial}{\partial x})$, with similar expressions for J_x and J_y by cyclic permutation. In other words,

$$\vec{J} = -i \vec{x} \otimes \vec{\nabla} \tag{26}$$

Thus, $J_z x = -i(x \frac{\partial}{\partial y} - y \frac{\partial}{\partial x})x = iy$, $J_z y = -i(x \frac{\partial}{\partial y} - y \frac{\partial}{\partial x})y = -ix$, and $J_z z = -i(x \frac{\partial}{\partial y} - y \frac{\partial}{\partial x})z = 0$.

The important point is to verify the commutators defining the Lie algebra $SO(3)$. Simply note that $[\frac{\partial}{\partial z}, z] = 1$, and so forth, and that, for example, $[y \frac{\partial}{\partial z}, z \frac{\partial}{\partial x}] = y(\frac{\partial}{\partial z}z)\frac{\partial}{\partial x} = y\frac{\partial}{\partial x}$. Thus,

$$[J_x, J_y] = \left[-i\left(y \frac{\partial}{\partial z} - z \frac{\partial}{\partial y} \right), -i\left(z \frac{\partial}{\partial x} - x \frac{\partial}{\partial z} \right) \right] = i J_z \tag{27}$$

and its cyclic permutations, in agreement with (22).

I have already mentioned, for those who have studied some quantum mechanics, that the generators of rotation J_i are closely related to the angular momentum operators in quantum mechanics. The generators of rotation are of course dimensionless, while angular momentum has dimension of length times momentum, which is of course precisely the dimension of the new fundamental constant $\hbar$ ushered into physics by quantum mechanics. Indeed, the angular momentum operators (26) are obtained by applying the Heisenberg prescription $\vec{p} \to -i \vec{\nabla}$ to the classical expression for angular momentum $\vec{J} = \vec{x} \otimes \vec{p}$. See chapter IV.3 for more.

By the way, our ability to pass back and forth between matrices and differential operators is of course what underlies the equivalence between the Heisenberg and Schrödinger pictures in quantum mechanics.

Appendix 2: How the $SO(4)$ algebra falls apart

The group $SO(4)$ has some rather attractive features. Use (25) to write down its Lie algebra, or better yet, work it out ab initio. Write down the six 4-by-4 antisymmetric matrices $J_{12}, J_{23}, J_{31}, J_{14}, J_{24}, J_{34}$, and commute them explicitly.

The six matrices naturally divide into two sets: J_{12}, J_{23}, J_{31}, which generate the rotations in the 3-dimensional subspace spanned by the 1-, 2-, and 3-axes, and J_{14}, J_{24}, J_{34}, which rotate the 1-, 2-, and 3-axes, respectively, into the 4-axis. A convenient notation is to write $J_3 = J_{12}, J_1 = J_{23}, J_2 = J_{31}$ and $K_1 = J_{14}, K_2 = J_{24}, K_3 = J_{34}$. Verify that the Lie algebra takes on the attractive form

$$[J_i, J_j] = i\varepsilon_{ijk}J_k \tag{28}$$

$$[J_i, K_j] = i\varepsilon_{ijk}K_k \tag{29}$$

$$[K_i, K_j] = i\varepsilon_{ijk}J_k \tag{30}$$

The structure constants of $SO(4)$ can be expressed entirely in terms of the antisymmetric symbol ε.

The first set of commutation relations (28) restates that the J_is generate, as in (22), an $SO(3)$ algebra, a natural subalgebra of $SO(4)$.

The second set (29) tells us that the K_is transform like a vector under the $SO(3)$ subalgebra. As an exercise, show that, for example,

$$e^{-i\varphi J_3}K_1 e^{i\varphi J_3} = \cos\varphi\, K_1 + \sin\varphi\, K_2 \tag{31}$$

Under a rotation around the third axis, (K_1, K_2) transform like a vector in the (1-2) plane.

The third set (30) confirms that the algebra closes: the commutator of two Ks does not give us something new, which we know, since $SO(4)$ is a group.

Note also that the algebra is invariant under the discrete transformation $J_i \to J_i$ and $K_i \to -K_i$, which follows upon reflection of the fourth axis, $x^4 \to -x^4$ while leaving x^1, x^2, and x^3 unchanged. This implies that the K_ks cannot appear on the right hand side of (30). Alternatively, we could flip $x^i \to -x^i$, $i = 1, 2, 3$, leaving x^4 unchanged.

After staring at (28), (29), and (30) for a while, you might realize that the Lie algebra of $SO(4)$ actually falls apart into two pieces. Define $J_{\pm,i} = \frac{1}{2}(J_i \pm K_i)$. Verify that

$$[J_{+,i}, J_{-,j}] = 0 \tag{32}$$

The six generators $J_{\pm,i}$ divide into two sets of three generators each, the J_+s and the J_-s, that commute right past the other set. Furthermore,

$$[J_{+,i}, J_{+,j}] = i\varepsilon_{ijk}J_{+,k} \quad \text{and} \quad [J_{-,i}, J_{-,j}] = i\varepsilon_{ijk}J_{-,k} \tag{33}$$

In other words, the three J_+s, and the three J_-s, separately generate an $SO(3)$ algebra. The algebra of $SO(4)$ is the direct sum of two $SO(3)$ algebras.

The group $SO(4)$ is said to be locally* isomorphic with the direct product group $SO(3) \otimes SO(3)$, that is, the two groups, $SO(4)$ and $SO(3) \otimes SO(3)$ are identical near the identity. Note that under the discrete transformations mentioned just now, $J_+ \leftrightarrow J_-$, thus interchanging the two $SO(3)$ algebras.

One important consequence is that many results to be obtained later for $SO(3)$ could be immediately applied to $SO(4)$.

Here we discuss rotations in a 4-dimensional Euclidean space. As you have no doubt heard, Einstein combined space and time into a 4-dimensional spacetime. Thus, what you learn here about $SO(4)$ can be put to good use.[†]

---

* In chapters IV.5 and IV.7, we will discuss how two groups could be locally isomorphic without being globally isomorphic.

† Higher dimensional rotation groups often pop up in the most unlikely places in theoretical physics. For example, $SO(4)$ is relevant for a deeper understanding of the spectrum of the hydrogen atom. See part VII, interlude 1.

Appendix 3: Does exponentiating the Lie algebra cover the whole group?

For rotations, Lie's idea that we can build up finite rotations by repeating infinitesimal rotations clearly holds physically. For $SO(2)$ this was verified explicitly in (12). For $SO(3)$ we can simply call the rotation axis the z-axis, and then an arbitrary finite rotation reduces basically to an element of $SO(2)$. Similarly for $SO(N)$.

Is it always true that by exponentiating a general element of a Lie algebra, we can recover the Lie group that begot the Lie algebra? Think about this for a moment before reading on.

Cartan showed by a simple counterexample that this is not true. Consider the group $SL(2, R)$ defined in chapter I.1. Follow Lie, and write an element near the identity as $M \simeq I + A$. Using the identity $\det M = e^{\operatorname{tr} \log M}$ given in the review of linear algebra and expanding $\log(I + A) \simeq A$, we obtain $\det M \simeq 1 + \operatorname{tr} A$; thus, the S in $SL(2, R)$ implies that A is traceless. The Lie algebra consists of traceless matrices of the form $A = \begin{pmatrix} c & a-b \\ a+b & -c \end{pmatrix}$ with $a, b,$ and c real. From exercise 11 in the review of linear algebra, its eigenvalues have the form $\lambda = \pm w$, with w either real or imaginary. (In fact, this follows immediately from the tracelessness of A.)

Does $M = e^A$ range over the entire group? To disprove this, evaluate the trace

$$T \equiv \operatorname{tr} M = \operatorname{tr} e^A = \operatorname{tr} e^{S^{-1}AS} = e^w + e^{-w} \tag{34}$$

where S denotes the similarity transformation that diagonalizes A. This is equal to either $2 \cos \theta$ or $2 \cosh \varphi$, and in any case $T \geq -2$. On the other hand, the matrix $U = \begin{pmatrix} -u & 0 \\ 0 & -u^{-1} \end{pmatrix}$ for u real is manifestly an element of the group $SL(2, R)$. But $\operatorname{tr} U = -(u + u^{-1})$ is in general ≤ -2 for $u > 0$. Thus, not every element of the group $SL(2, R)$ can be written in the form e^A.

You might say that $O(3)$ provides another example: the reflection $(-I)$ is manifestly not connected to the identity I. Its trace is -3, but the trace of a rotation matrix is $2 \cos \theta + 1 \geq -2 + 1 = -1$. But this example is not quite satisfying, since we could think of $O(3)$ as formed by adjoining reflection to $SO(3)$. When we say "manifestly" above, we are appealing to physical experience: no matter how you rotate a rigid body, you are not going to turn it inside out. In essence, Cartan's argument is precisely a proof that the matrix U is manifestly not continuously connected to the identity.

Exercises

1 Suppose you are given two vectors $\vec{p}$ and $\vec{q}$ in ordinary 3-dimensional space. Consider this array of three numbers:

$$\begin{pmatrix} p^2 q^3 \\ p^3 q^1 \\ p^1 q^2 \end{pmatrix}$$

Prove that it is not a vector, even though it looks like a vector. (Check how it transforms under rotation!) In contrast,

$$\begin{pmatrix} p^2 q^3 - p^3 q^2 \\ p^3 q^1 - p^1 q^3 \\ p^1 q^2 - p^2 q^1 \end{pmatrix}$$

does transform like a vector. It is in fact the vector cross product $\vec{p} \otimes \vec{q}$.

2 Verify that $R \simeq I + A$, with A given by $A = \theta_x \mathcal{J}_x + \theta_y \mathcal{J}_y + \theta_z \mathcal{J}_z$, satisfies the condition $\det R = 1$.

3 Using (14), show that a rotation around the x-axis through angle θ_x is given by

$$R_x(\theta_x) = \begin{pmatrix} 1 & 0 & 0 \\ 0 & \cos \theta_x & \sin \theta_x \\ 0 & -\sin \theta_x & \cos \theta_x \end{pmatrix}$$

Write down $R_y(\theta_y)$. Show explicitly that $R_x(\theta_x) R_y(\theta_y) \neq R_y(\theta_y) R_x(\theta_x)$.

4 Use the hermiticity of J to show that the c_{ijk} in (18) are real numbers.

5 Calculate $[J_{(mn)}, J_{(pq)}]$ by brute force using (24).

6 Of the six 4-by-4 matrices $J_{12}, J_{23}, J_{31}, J_{14}, J_{24}, J_{34}$ that generate $SO(4)$, what is the maximum number that can be simultaneously diagonalized?

7 Verify (31).

Notes

1. This chapter is adapted from chapter I.3 of G Nut.
2. See the linear algebra review.
3. Nahin, An Imaginary Tale: The Story of $\sqrt{-1}$, Princeton University Press, 2010.
4. The signs agree with standard usage, for example QFT Nut 2, p. 114.
5. Fritz Zwicky, known for proposing dark matter, introduced the concept of the spherical bastard, defined as a person who is a bastard no matter which way you look at him (or her).
6. The matrix C is known only as a series, which can be obtained by brute force up to however many terms you want:

$$C = \log(e^A e^B) \simeq \log(1 + A + \tfrac{1}{2}A^2)(1 + B + \tfrac{1}{2}B^2) \simeq \log\{1 + A + B + \tfrac{1}{2}(A^2 + B^2) + AB\}$$

$$\simeq A + B + \tfrac{1}{2}(A^2 + B^2) + AB - \tfrac{1}{2}\{A + B\}^2 = A + B + \tfrac{1}{2}[A, B] + \cdots \tag{35}$$

(Try to get the next term!) The claim is that C is given entirely in terms of nested commutators of A and B. To physicists, this follows immediately from the fact that the product of rotations is a rotation (which in turn follows by rotating a rigid body in our mind's eye). The mathematical proof is quite involved. For application to lattice gauge theory, see chapter VII.1 in QFT Nut.

Part II | Representing Group Elements by Matrices

Group elements may be represented by matrices. Representation theory is full of delightful theorems and surprising results, such as the Great Orthogonality theorem. A favorite saying: character is a function of class. Indeed, some of us feel that constructing character tables is loads of fun.

The important concept of real, pseudoreal, and complex representations is studied, culminating in the construction of a reality checker. Using the character table, you can count the number of square roots of any element in a finite group.

The elegant theorem stating that crystals with five-fold symmetry are impossible is proven. After that, we relax and have fun with number theory, discussing various results associated with Euler, Fermat, Wilson, and Frobenius.

Jancsi considered my group theory problem for about half an hour's time. Then he said, "Jenö, this involves representation theory." Jancsi gave me a reprint of a decisive 1905 article by Frobenius and Schur. . . . He said, ". . . it's one of the things on which old Frobenius made his reputation. So it can't be easy."

—Wigner's autobiography[1]

What is a representation?

More than a century later, while representation theory is not exactly easy, it does not seem all that difficult, either.

The notion of representing group elements by matrices is both natural and intuitive. Given a group, the idea is to associate each element g with a $d \otimes d$ matrix $D(g)$ such that

$$D(g_1)D(g_2) = D(g_1g_2) \tag{1}$$

for any two group elements g_1 and g_2. The matrix[2] $D(g)$ is said to represent the element g, and the set of matrices $D(g)$ for all $g \in G$ is said to furnish or provide a representation of G. The size of the matrices, d, is known as the dimension of the representation.

The requirement (1) says that the set of matrices $D(g)$ "reflects" or "mirrors" the multiplicative table of the group. In words, the product g_1g_2 of two group elements g_1 and g_2 is represented by the product of the matrices representing g_1 and g_2 respectively. To emphasize this all-important concept of representation, let us write (1) "graphically" as

$$
\begin{array}{ccccc}
g_1 & \cdot & g_2 & = & g_1 \cdot g_2 \\
\downarrow & \downarrow & \downarrow & \downarrow & \downarrow \\
D(g_1) & \cdot & D(g_2) & = & D(g_1 \cdot g_2)
\end{array}
\tag{2}
$$

Note that the symbol $\cdot$ denotes two distinct concepts: in the top row, the composition, or more colloquially, the multiplication, of two group elements; in the bottom row, the multiplication of two matrices. (As already mentioned in chapter I.1, we often omit the dot, as in (1), for example.) All this will become clearer with the examples to be given here.

Consider S_4, the permutation group of four objects. Think of the four objects as the four vectors* $v_1 = \begin{pmatrix} 1 \\ 0 \\ 0 \\ 0 \end{pmatrix}$, $v_2 = \begin{pmatrix} 0 \\ 1 \\ 0 \\ 0 \end{pmatrix}$, $v_3 = \begin{pmatrix} 0 \\ 0 \\ 1 \\ 0 \end{pmatrix}$, and $v_4 = \begin{pmatrix} 0 \\ 0 \\ 0 \\ 1 \end{pmatrix}$. Then we can represent the element (2413), which takes $2 \to 4$, $4 \to 1$, $1 \to 3$, and $3 \to 2$, by the 4-by-4 matrix $D(2413) = \begin{pmatrix} 0 & 0 & 0 & 1 \\ 0 & 0 & 1 & 0 \\ 1 & 0 & 0 & 0 \\ 0 & 1 & 0 & 0 \end{pmatrix}$. By construction, $D(2413)v_2 = v_4$, $D(2413)v_4 = v_1$, and so on. The action of the matrix $D(2413)$ on the four vectors mirrors precisely the action of the permutation (2413) on the four objects labeled 1, 2, 3, and 4. Similarly, we have, for example, $D(34) = \begin{pmatrix} 1 & 0 & 0 & 0 \\ 0 & 1 & 0 & 0 \\ 0 & 0 & 0 & 1 \\ 0 & 0 & 1 & 0 \end{pmatrix}$.

According to what we have learned in chapter I.2, we have $(34)(2413) = (23)(14)$. Here, let us multiply the two matrices $D(34)$ and $D(2413)$ together. (Go ahead, do it!) We find $D(34)D(2413) = \begin{pmatrix} 0 & 0 & 0 & 1 \\ 0 & 0 & 1 & 0 \\ 0 & 1 & 0 & 0 \\ 1 & 0 & 0 & 0 \end{pmatrix}$, which is precisely $D((23)(14))$, as expected. This verifies (1), at least in this particular instance. Exercise! Write down a few more matrices in this 4-dimensional representation of S_4 and multiply them.

I presume that you are not surprised that we have found a 4-dimensional representation of S_4. In summary, the group S_4 can be represented by 24 distinct $4 \otimes 4$ matrices. Note that these are very special matrices, with 0 almost everywhere except for four 1s, with one single 1 in each column (and in each row). All this should be fairly self-evident: what is the difference between four vectors labeled v_1, v_2, v_3, v_4 and four balls labeled 1, 2, 3, 4?

Group elements and the matrices that represent them

In our example, the matrix $D(34)$ represents the permutation (34), but physicists might say that $D(34)$ is essentially what they mean by (34). In fact, physicists often confound group elements with the matrices that represent them. For example, when a physicist thinks of a rotation, he or she typically has in mind a 3-by-3 matrix. A mathematician, in contrast, might think of rotations as abstract entities living in some abstract space, defined entirely by how they multiply together. In practice, many of the groups used in theoretical physics are defined by the matrices representing them, for example, the Lorentz group described in chapter I.1.

Very roughly speaking, the representation of a group is like a photograph or a map of the group, to the extent that it preserves the multiplicative structure of the group. A photo or a map of a village is of course not the village itself, but it shows accurately how various buildings and geographical features are situated relative to one another.

---

* To lessen clutter, we omit the arrow on top of the vectors here.

On a practical level, it is much easier to tell a computer to multiply matrices together than to feed it the multiplication table of a group.

Note that (1) implies that $D(I) = I_d$. Here I have carefully distinguished the I on the left (denoting that special group element called the identity) from the I_d on the right (denoting the d-by-d identity matrix). The meaning of I_d depends on the particular representation being discussed. In contrast, the other I is an abstract entity fixed once the group is specified. More sloppily, I might have abused notation slightly and written the more mystifying but fairly self-evident $D(I) = I$.

To show that $D(I) = I_d$, observe that $D(I)D(g) = D(Ig) = D(g)$ for any g. Multiply by $D(g^{-1})$ from the right, and we obtain $(D(I)D(g))D(g^{-1}) = D(I)(D(g)D(g^{-1}))$, which, due to (1), reduces to $D(I)D(I) = D(I)$. Multiplying by $(D(I))^{-1}$, we obtain $D(I) = I_d$. This also tells us that $D(g^{-1}) = D(g)^{-1}$, as we would expect, since the representation is supposed to mirror the multiplicative structure of the group.

Introduction to representation theory

Now that we know what a representation is, we can naturally think of many questions. Does every group G have a representation? How many representations does it have? An infinite number, perhaps? What are some general properties of representations? How do we characterize these representations and distinguish among them?

How is your mathematical intuition? Do you feel that, the more sophisticated a group, the more representations it ought to have? But then, how would you measure the "sophistication" of a group? Is it merely the number of elements? Or, more intelligently, do you feel that sophistication would be measured more by the number of different types of elements? Recall the notion of equivalence classes from chapter I.2.

We have a partial answer to the first question. We learned in chapter I.2 that every finite group is isomorphic to a subgroup of S_n, and since S_n has a matrix representation, every finite group can be represented by matrices. As for continuous groups, in the list of examples given in chapter I.1, almost all groups—the rotation groups and the Lorentz group, for example—are defined in terms of matrices, so a fortiori they can be represented by matrices. An exception appears to be the additive group of real numbers. How in the world could addition be represented by multiplication?

You smile, since you already know the answer. Let $D(u)$ be the 1-dimensional matrix e^u: then $D(u)D(v) = D(u+v)$. Actually, we do not even have to invoke the exponential function. Consider the 2-dimensional matrix

$$D(u) = \begin{pmatrix} 1 & 0 \\ u & 1 \end{pmatrix} \tag{3}$$

Verify that $D(u)D(v) = D(u+v)$ and $D(0) = I$. (Note that 0 denotes the identity of the additive group.)

I bet you didn't know that addition could be represented by multiplying 2-by-2 matrices together. Let me also ask you, has the group described by (3) ever appeared in physics? For those of you who do not know the answer, just wait, and it will be revealed in chapter VII.2.

Indeed, the Lorentz group mentioned in chapter I.1 also defines a 2-dimensional representation of the additive group, since $D(\phi_1)D(\phi_2) = D(\phi_1 + \phi_2)$. (Recall that ϕ represents the boost angle.) You may have realized that this representation of addition secretly also involves the exponential function.

So by the physicist's laughable standard of rigor (recall the black sheep), it certainly seems that all the groups you are likely to encounter in physics can be represented by matrices.

To answer the second question that introduced this section, namely, how many representations a group might have, we are first obliged to mention that, in representation theory, the trivial representation $D(g) = 1$, for every $g \in G$, also counts as a perfectly valid representation. The basic requirement (1) of being a representation is certainly satisfied, since $D(g_1)D(g_2) = 1 \cdot 1 = 1 = D(g_1 g_2)$.

Some readers might chuckle: in our photo analogy, the entire village appears as a single dot. Yes indeed, this representation is trivial, hence the name. But as you will see, in the representation theory to be developed in this chapter and the next, it is important to include it. This is perhaps reminiscent of the introduction of the number 0 in the history of mathematics.

Here the notion of faithful versus unfaithful representations naturally suggests itself. To use a more mathematical language, we say that a d-dimensional representation is a map of the group G into some subgroup of* $GL(d, C)$. The requirement (1) merely says that the map is homomorphic. But if in addition the map is isomorphic, that is, one-to-one, then the representation is faithful. Otherwise, it is unfaithful.

As already mentioned, many of the groups given in chapter I.1 are defined in terms of matrices. For example, the rotation group $SO(3)$ is defined by 3-by-3 orthogonal matrices, as discussed in detail in chapter I.3. Naturally, these representations are known as defining or fundamental representations. The defining representation of $SO(3)$ is 3-dimensional.

As another example, the defining representation of Z_N is 1-dimensional, namely, the element $e^{i2\pi j/N}$ is represented by itself for $j = 0, \cdots, N - 1$. But interestingly, after some thought, we realize that we can also represent $e^{i2\pi j/N}$ by $e^{i2\pi kj/N}$ for some fixed integer k, which can take on any of the values $0, \cdots, N - 1$. Check that this indeed furnishes a representation: $D(e^{i2\pi j_1/N})D(e^{i2\pi j_2/N}) = e^{i2\pi kj_1/N}e^{i2\pi kj_2/N} = e^{i2\pi k(j_1+j_2)/N} = D(e^{i2\pi(j_1+j_2)/N})$. (I am being extra pedantic here.)

Thus, Z_N actually has N different 1-dimensional representations, labeled by the integer k. What is the $k = 0$ representation? (Yes, it is the trivial representation.)

More specifically, the group Z_3 has not only the 1-dimensional representation $\{1, \omega, \omega^2\}$ (where $\omega \equiv e^{2\pi i/3}$ is the cube root[3] of 1) but also the nontrivial 1-dimensional representation $\{1, \omega^2, \omega\}$. (Before reading on, you should verify that this indeed furnishes a representation. For example, $\omega^2\omega^2 = \omega^4 = \omega$.) Wait, you just read that Z_3 has three different irreducible representations. What is the third?[4]

---

* Here $GL(d, C)$ denotes the group consisting of invertible d-by-d matrices with complex entries. It contains the subgroup $SL(d, C)$ introduced in chapter I.1.

Character is a function of class

Now that we know that a given group can have many different representations, let us label the different representations by a superscript $r, s, \cdots$, and write $D^{(r)}(g)$ for the matrix representing the element g in the representation r.

Given a representation $D^{(r)}(g)$, define the important concept of the character $\chi^{(r)}$ of the representation by $\chi^{(r)}(g) \equiv \operatorname{tr} D^{(r)}(g)$. The character, as the name suggests, helps characterize the representation.

Nominally, the character depends on r and g. Recall from chapter I.2, however, that the elements of a group can be divided up into equivalence classes. Two elements g_1 and g_2 are equivalent ($g_1 \sim g_2$) if there exists another element f such that

$$g_1 = f^{-1} g_2 f \tag{4}$$

We then find

$$\chi^{(r)}(g_1) = \operatorname{tr} D^{(r)}(g_1) = \operatorname{tr} D^{(r)}(f^{-1} g_2 f) = \operatorname{tr} D^{(r)}(f^{-1}) D^{(r)}(g_2) D^{(r)}(f)$$
$$= \operatorname{tr} D^{(r)}(g_2) D^{(r)}(f) D^{(r)}(f^{-1}) = \operatorname{tr} D^{(r)}(g_2) D^{(r)}(I)$$
$$= \operatorname{tr} D^{(r)}(g_2) = \chi^{(r)}(g_2) \tag{5}$$

where in the third, fifth, and sixth equalities we used (1) and in the fourth equality we used the cyclicity of the trace. In other words, if $g_1 \sim g_2$, then $\chi^{(r)}(g_1) = \chi^{(r)}(g_2)$. Thus,

$$\chi^{(r)}(c) = \operatorname{tr} D^{(r)}(g) \quad \text{(for } g \in c) \tag{6}$$

Here c denotes the equivalence class of which the element g is a member. The trace on the right hand side does not depend on g as such, but only on the class that g belongs to. All the elements of a given equivalence class have the same character.

As an example, we learned in chapter I.2 that A_4 has four equivalence classes. For a given representation r of A_4, $\chi^{(r)}(c)$ is a function of c, a variable that takes on four values.

We can now proudly utter perhaps the most memorable statement in group theory: "Character is a function of class."

Equivalent representations

As for how many representations a group might have, we all agree, first of all, that two representations, $D(g)$ and $D'(g)$, are really the same representation (more formally, the two representations are equivalent) if they are related by a similarity* transformation[†]

$$D'(g) = S^{-1} D(g) S. \tag{7}$$

---

* Similarity transformations are discussed in the review of linear algebra.
[†] Not to be confused with (4); here we are talking about two different matrices representing the same group element.

As explained in the review of linear algebra, $D(g)$ and $D'(g)$ are essentially the same matrix, merely written in two different bases, with the matrix S relating one set of basis vectors to the other set. Put another way, given a representation $D(g)$, define $D'(g)$ by (7) with some S whose inverse exists. Then $D'(g)$ is also a representation, since $D'(g_1)D'(g_2) = (S^{-1}D(g_1)S)(S^{-1}D(g_2)S) = S^{-1}D(g_1)D(g_2)S = S^{-1}D(g_1g_2)S = D'(g_1g_2)$.

Note that it is the same S for all g. Think of it as follows. Suppose we have found a representation of a group G, for example, the 4-dimensional representation of S_4 described above. In other words, we list $24 = 4!$ 4-by-4 matrices $D(g)$ satisfying (1). Some ugly dude could come along, choose some ugly 4-by-4 matrix S, and use (7) to produce a list of 24 ugly 4-by-4 matrices $D'(g)$ satisfying (1). If he chooses a particularly messy S, the two sets of matrices $D'(g)$ and $D(g)$ could look very different.

If we are given two representations, how do we decide whether they are equivalent or not? Taking the trace of (7) and once again using the cyclicity of the trace, we obtain

$$\chi'(c) = \text{tr } D'(g) = \text{tr } SD(g)S^{-1} = \text{tr } D(g)S^{-1}S = \text{tr } D(g) = \chi(c) \tag{8}$$

where g is a member of the class c. Thus, if there exists some class c for which $\chi'(c) \neq \chi(c)$, we can conclude immediately that the two representations are in fact different. What if $\chi'(c) = \chi(c)$ for all c? If this holds for only one or two c, physicists of the "black sheep school of thought" might still admit that it could be a coincidence, but for all c? Most "reasonable" theoretical physicists would say that it is strong circumstantial evidence indicating that the two representations are in fact the same.

Indeed, physicist intuition is right. We will see in the next chapter that various theorems state that for two different representations r and s, the characters $\chi^{(r)}(c)$ and $\chi^{(s)}(c)$ are "more different than different": they are orthogonal in some well-defined sense.

Reducible or irreducible representation

Now we come to the all-important notion of whether a given representation is reducible or irreducible. For the sake of definiteness, focus on $SO(3)$. We have the trivial 1-dimensional representation $D^{(1)}(g) = 1$ and the 3-dimensional defining representation $D^{(3)}(g)$. Are there other representations? Think about it before reading on.

Can you give me an 8-dimensional representation?

"Sure," you say, "you want an 8-dimensional representation for $SO(3)$. I give you an 8-dimensional representation. Here it is:"

$$D(g) = \left(\begin{array}{c|c|c|c} D^{(1)}(g) & 0 & 0 & 0 \\ \hline 0 & D^{(1)}(g) & 0 & 0 \\ \hline 0 & 0 & D^{(3)}(g) & 0 \\ \hline 0 & 0 & 0 & D^{(3)}(g) \end{array} \right) \tag{9}$$

(The vertical and horizontal lines are merely to guide the eye. By the way, a matrix with this form is said to be block diagonal: it contains smaller matrices along its diagonal, with all other entries equal to zero. Note that the symbol 0 in (9) carries several different meanings: it could denote a 1-by-1 matrix with its single entry equal to zero, or a 1-by-3 rectangular matrix with all its entries equal to zero, or a 3-by-1 rectangular matrix with all its entries equal to zero, or a 3-by-3 square matrix with all its entries equal to zero.)

Each element g is represented by an 8-by-8 matrix $D(g)$ (8 since $1 + 1 + 3 + 3 = 8$). Show that this is indeed a representation; that is, it satisfies (1).

Ah, you have stacked two copies of $D^{(1)}(g)$ and two of $D^{(3)}(g)$ together. Indeed, by this cheap trick, you could construct representations with any dimension. You and I, being reasonable people, should agree that $D(g)$ does not count as a "new" representation.

The representation $D(g)$ is known as reducible, and usually written as a direct sum of the representations it reduces into: in our example, $D(g) = D^{(1)}(g) \oplus D^{(1)}(g) \oplus D^{(3)}(g) \oplus D^{(3)}(g)$.

Representations that are not reducible are called irreducible. Clearly, we should focus on irreducible, rather than reducible, representations.

It is clear as day that $D(g)$ in the form given above is reducible. But that ugly dude mentioned earlier could come along again and present you with a set of matrices $D'(g) = S^{-1}D(g)S$. If he chooses a particularly messy S, all those zeroes in (9) would get filled in, and we would have a hard time recognizing that $D'(g)$ is reducible.

Going back to the definition of the trivial representation $D^{(1)}(g) = 1$, you might have wondered why we used 1 and not the $k \otimes k$ identity matrix I_k for some arbitrary positive integer k. The answer is that the representation would then be reducible unless $k = 1$. The representation $D^{(1)}$ may be trivial, but it is not reducible.

One goal of representation theory is to develop criteria to determine whether a given representation is irreducible or not and to enumerate all possible irreducible representations. Since every group has an infinity of reducible representations, the real question is to figure out how many irreducible representations it has.

Restriction to a subgroup

A representation of a group G clearly also furnishes a representation of any of its subgroups. Denote the elements of the subgroup H by h. If $D(g_1)D(g_2) = D(g_1g_2)$ for any two group elements g_1 and g_2 of G, then a fortiori $D(h_1)D(h_2) = D(h_1h_2)$ for any two group elements h_1 and h_2 of H. We refer to this representation of H as the representation of G restricted to the subgroup H.

When restricted to a subgroup H, an irreducible representation of G will in general not be an irreducible representation of H. It will, in all likelihood, decompose, or fall apart, into a bunch of irreducible representations of H. The reason is clear: there may well exist a basis in which $D(h)$ is block diagonal (that is, has the form such as that shown in (9)) for all h in H, but there is no reason in general to expect that $D(g)$, for all g in G

but not in H, would also be block diagonal. Simply stated, there are, by definition, fewer elements in H than in G.

How an irreducible representation of a group G decomposes upon restriction of G to a subgroup H will be a leitmotif in this book.

Unitary representations

The all-important unitarity theorem states that finite groups have unitary representations, that is to say, $D^{\dagger}(g)D(g) = I$ for all g and for all representations.

In practice, this theorem is a big help in finding representations of finite groups. As a start, we can eliminate some proposed representations by merely checking if the listed matrices are unitary or not.

At this point, our friend Dr. Feeling strolls by. "Let's get an intuitive feel for this theorem," he says helpfully. Suppose the representation $D(g)$ is 1-by-1, that is, merely a complex number $re^{i\theta}$. Back in chapter I.1, we showed that in a finite group, if we keep multiplying g by itself, eventually it has to come back to the identity: $g^k = I$ for some integer k. But g^k is represented by $D(g^k) = D(g)^k = r^k e^{ik\theta}$. No way this could get back to 1 if $r \neq 1$. But if $r = 1$, then $D^{\dagger}(g)D(g) = e^{-i\theta}e^{i\theta} = 1$; that is, $D(g)$ is unitary. This, in essence, is why the theorem must be true.

Proof of the unitarity theorem

You worked out the rearrangement lemma in chapter I.1 as an exercise. Let me merely remind you what it says: Given a function on the group elements, we have, for any $g' \in G$,

$$\sum_{g \in G} f(g) = \sum_{g \in G} f(g'g) = \sum_{g \in G} f(gg') \tag{10}$$

The three sums are actually the same sum; they differ only by having the terms rearranged.

We are now ready to prove the unitarity theorem.

Suppose that a given representation $\tilde{D}(g)$ is nonunitary. Define

$$H = \sum_g \tilde{D}(g)^{\dagger}\tilde{D}(g) \tag{11}$$

where the sum runs over all elements $g \in G$. We note that, for any g',

$$\tilde{D}(g')^{\dagger}H\tilde{D}(g') = \sum_g \tilde{D}(g')^{\dagger}\tilde{D}(g)^{\dagger}\tilde{D}(g)\tilde{D}(g') = \sum_g (\tilde{D}(g)\tilde{D}(g'))^{\dagger}\tilde{D}(g)\tilde{D}(g')$$

$$= \sum_g (\tilde{D}(gg'))^{\dagger}\tilde{D}(gg') = H \tag{12}$$

The last equality holds because of the rearrangement lemma. The matrix H is remarkably "invariant."

Since H is hermitean, there exists a unitary matrix W such that $\rho^2 = W^{\dagger}HW$ is diagonal and real. We now show that in addition, the diagonal elements are not only real but also positive. (Hence the notation ρ^2: we can take the square root of the diagonal matrix ρ^2 to obtain the diagonal and real matrix ρ.) To show this, we invoke a theorem, cited in the review of linear algebra, that for any matrix M, the matrix $M^\dagger M$ has non-negative eigenvalues. Let ψ be the column vector with 1 in the jth entry and 0 everywhere else. Then

$$(\rho^2)_{jj} = \psi^\dagger \rho^2 \psi = \psi^\dagger W^\dagger H W \psi = \sum_g (\psi^\dagger W^\dagger) \widetilde{D}(g)^\dagger \widetilde{D}(g)(W\psi) = \sum_g \phi(g)^\dagger \phi(g) > 0 \qquad (13)$$

(Here we define $\phi(g) = \widetilde{D}(g)W\psi$.) Thus, the matrix ρ exists with the stated properties.

Define $D(g) \equiv \rho W^\dagger \widetilde{D}(g) W \rho^{-1}$. (I will let you worry about niceties, such as whether ρ^{-1} exists.) We now show that $D(g)$ is unitary. Simply calculate: $D^\dagger(g) = \rho^{-1} W^\dagger \widetilde{D}(g)^\dagger W \rho$, so that

$$\begin{aligned} D^\dagger(g)D(g) &= \rho^{-1} W^\dagger \widetilde{D}(g)^\dagger W \rho^2 W^\dagger \widetilde{D}(g) W \rho^{-1} \\ &= \rho^{-1} W^\dagger \widetilde{D}(g)^\dagger H \widetilde{D}(g) W \rho^{-1} \\ &= \rho^{-1} W^\dagger H W \rho^{-1} = \rho^{-1} \rho^2 \rho^{-1} = I \end{aligned} \qquad (14)$$

where the third equality holds because of (12).

The unitarity theorem is proved. It is instructive to see how the theorem works if $\widetilde{D}(g)$ is already unitary. Then $H = N(G)I$, with $N(G)$ the number of elements in the group, so that $W = I$, $\rho = \sqrt{N(G)}I$ is proportional to the identity matrix, and hence $D(g) = \rho \widetilde{D}(g)\rho^{-1} = \widetilde{D}(g)$.

Note that in almost all the examples given thus far, the representation matrices are real rather than complex. A real unitary matrix is orthogonal, of course;* in other words, for $D(g)$ real, $D^T(g)D(g) = I$ for all g. We will derive criteria for deciding whether a given irreducible representation is real or complex in chapter II.4.

Compact versus non-compact

To physicists, the natural thing to do is to check whether the unitarity theorem holds for groups other than finite groups. Well, it certainly works for $SO(2)$: $R(\theta)^T R(\theta) = I$, as was verified explicitly in chapter I.3 essentially by definition. Indeed, by the discussion there, it works for the rotation group in any dimension. To a sloppy physicist, finite, infinite, what's the difference? Assume everything converges unless proven otherwise. Shoot first, and ask questions later. So surely group representations have to be unitary in general.

The mathematician is aghast: you can't ignore rigor so blithely! What about the Lorentz group with $L(\varphi) = \begin{pmatrix} \cosh\varphi & \sinh\varphi \\ \sinh\varphi & \cosh\varphi \end{pmatrix}$? It is manifestly not true that $L(\varphi)^T L(\varphi)$ is equal to the identity. In fact, nothing as fancy as the Lorentz group is needed for a counterexample; how about the representation for the additive group mentioned earlier: $D(x) = \begin{pmatrix} 1 & 0 \\ x & 1 \end{pmatrix}$? Certainly, $D(x)^T D(x) = \begin{pmatrix} 1+x^2 & x \\ x & 1 \end{pmatrix} \neq I$.

---

* See the review of linear algebra if you need to be reminded of this point.

Indeed, the claimed "physicist theorem" also fails for discrete groups if we restrict x to the integers.[5] So the issue is not a question of whether the group is discrete or continuous. Rather the issue is whether the group is compact or not.

The notion of compactness arises naturally in the present context. The sloppy physicist would say that the theorem that representations are unitary should also hold for continuous groups; the jump from finite group to continuous group just amounts to replacing the sum $\sum_g$ over the elements of a finite group, following Newton and Leibniz, by some kind of integral $\int d\mu(g)$ over the continuous group (with $d\mu(g)$ some kind of integration measure).

Thus, if the relevant integral $\int d\mu(g)(\cdots)$ converges, the group is known to be compact, and the proof given above for finite groups formally goes through. The representations of compact groups are unitary. But what if the integral diverges?

Precisely, growls our mathematician friend, if an integral diverges, even physicists have to sit up and be on alert for possible danger.*

We will come back later to a precise definition of compactness, but for the moment let us simply note that this discussion makes a lot of sense. The rotation groups are parametrized by angles that typically are bounded between 0 and 2π, but for the Lorentz group the boost angle φ is not really an angle and ranges from $-\infty$ to $+\infty$. (Note that the issue is more intricate than this naive view. As mentioned in chapter I.1, we could also parametrize the Lorentz group using the more physical relative velocity v between inertial frames defined by $\tanh \varphi \equiv v$, with v ranging between -1 and $+1$. It really does matter what $d\mu(g)$ is.)

Product representation

Stacking representations on top of one another as in (9) gives an easy but mostly uninteresting way of producing a larger representation, which we might call a direct sum representation, out of smaller representations. The direct product of matrices discussed in the review of linear algebra provides a more interesting way of constructing a larger representation out of smaller ones.

Given two representations, of dimension d_r and d_s, with representation matrices $D^{(r)}(g)$ and $D^{(s)}(g)$, respectively, we can define the direct product representation defined by the direct product matrices $D(g) = D^{(r)}(g) \otimes D^{(s)}(g)$, namely, the $d_r d_s$-by-$d_r d_s$ matrix given by

$$D(g)_{a\alpha, b\beta} = D^{(r)}(g)_{ab} D^{(s)}(g)_{\alpha\beta} \tag{15}$$

We have intentionally used different letters to denote the indices on $D^{(r)}(g)$ and $D^{(s)}(g)$ to emphasize that they are entirely different beasts running over different ranges; in particular, $a, b = 1, \cdots d_r$, and $\alpha, \beta = 1, \cdots, d_s$.

---

* Incredulous at our insouciance, the mathematician demands, "Weren't divergent integrals the cause of the Stürm und Drang of quantum field theory in the early days, all that hand wringing over infinities and renormalization?"

The rule for multiplying direct product matrices together (derived in the review of linear algebra)

$$D(g)D(g') = (D^{(r)}(g) \otimes D^{(s)}(g))(D^{(r)}(g') \otimes D^{(s)}(g'))$$
$$= (D^{(r)}(g)D^{(r)}(g')) \otimes (D^{(s)}(g)D^{(s)}(g')) = D(gg') \tag{16}$$

shows explicitly that the product representation is indeed a representation. In general, however, there is no reason for this product representation to be irreducible. We will learn to determine how the product representation reduces to a direct sum of irreducible representations.

The character of the direct product representation is easily calculated by setting the index $a\alpha$ equal to the index $b\beta$ in (15) and summing:

$$\chi(c) = \sum_{a\alpha} D(g)_{a\alpha,a\alpha} = \left(\sum_a D^{(r)}(g)_{aa} \right)\left(\sum_\alpha D^{(s)}(g)_{\alpha\alpha} \right) = \chi^{(r)}(c)\chi^{(s)}(c) \tag{17}$$

As usual, c denotes the class that the group element g belongs to. The character of a direct product representation is the product of the characters of the representations that produce it. This result nicely parallels the statement mentioned earlier that the character of a direct sum representation is the sum of the characters of the representations that produce it.

Notice that nothing in this discussion requires that the representations r and s be irreducible.

In physics, it is often useful to think in terms of the objects furnishing the representations. Let ϕ_a, $a = 1, \cdots, d_r$, denote the d_r objects that transform into linear combinations of one another and thus furnish the representation r. Similarly, let ξ_α, $\alpha = 1, \cdots, d_s$, denote the d_s objects that furnish the representation s. Then, the $d_r d_s$ objects $\phi_a \xi_\alpha$ furnish the direct product representation $r \otimes s$. As we shall see in chapter III.1, these abstract mathematical objects are actually realized in quantum mechanics as wave functions.

Finally, let me quote what Wigner said about the article that von Neumann gave him: "Soon I was lost in the enchanting world of vectors and matrices, wave functions and operators. This reprint was my primary introduction to representation theory, and I was charmed by its beauty and clarity. I saved the article for many years out of a certain piety that these things create."

I hope that you find it equally enchanting.

Exercises

1 Show that the identity is in a class by itself.

2 Show that in an abelian group, every element is in a class by itself.

3 These days, it is easy to generate finite groups at will. Start with a list consisting of a few invertible d-by-d matrices and their inverses. Generate a new list by adding to the old list all possible pairwise products of these matrices. Repeat. Stop when no new matrices appear. Write such a program. (In fact, a student did write such a program for me once.) The problem is of course that you can't predict when the process will

(or if it will ever) end. But if it does end, you've got yourself a finite group together with a d-dimensional representation.

4   In chapter I.2, we worked out the equivalence classes of S_4. Calculate the characters of the 4-dimensional representation of S_4 as a function of its classes.

Notes

1. From "The Recollections of Eugene P. Wigner as told to Andrew Szanton." Jancsi (Johnny) and Jenö (Gene) are what John von Neumann and Eugene Wigner called each other in Hungarian.
2. D for the German word Darstellung, meaning "representation."
3. Of course, 1 is the most famous of them all, then comes -1, the square root of 1. By rights, ω, the cube root of 1, should be more famous than i, the fourth root of 1. Oh well, there is no justice in the world.
4. This recalls Eddington's famous response "Who is the third?" to a question. See G Nut, p. 369.
5. Note that the matrix called $H = \sum_{x=\text{integer}} D(x)^T D(x)$ does not exist.

II.2 Schur's Lemma and the Great Orthogonality Theorem

In this chapter[1] we prove a number of elegant theorems, considered by many to be among the most beautiful in mathematics. In chapter II.3, we will use these theorems to determine the irreducible representations of various finite groups. Those readers exposed to group theory for the first time might prefer to skip the detailed proofs and to merely absorb what these theorems say.

Schur's lemma

A crucial theorem in representation theory, known as Schur's lemma,* states the following: If $D(g)$ is an irreducible representation of a finite group G and if there is some matrix A such that $AD(g) = D(g)A$ for all g, then $A = \lambda I$ for some constant λ.

What does this mean?

If I give you a bunch of matrices $D_1, D_2, \cdots, D_n$, the identity matrix I commutes with all these matrices, of course. But it is also quite possible for you to find a matrix A, not the identity, that commutes with all n matrices. The theorem says that you can't do this if the given matrices $D_1, D_2, \cdots, D_n$ are not any old bunch of matrices you found hanging around the street corner, but the much-honored representation matrices furnishing an irreducible representation of a group.

To prove Schur's lemma, let's start with a small lemma to the lemma: A can be taken to be hermitean with no loss of generality.

To see this, recall that $D(g)$ is unitary according to the "unitary theorem" (see chapter II.1). Take the hermitean conjugate of $AD(g) = D(g)A$ to obtain $D(g)^\dagger A^\dagger = A^\dagger D(g)^\dagger$. Since $D(g)$ is unitary, we can write this as $D(g)^{-1}A^\dagger = A^\dagger D(g)^{-1}$, and hence $A^\dagger D(g) = D(g)A^\dagger$. Adding and subtracting, we obtain $(A + A^\dagger)D(g) = D(g)(A + A^\dagger)$ and $i(A - A^\dagger)D(g) = D(g)i(A - A^\dagger)$. The statement of Schur's lemma holds for the two hermitean

---

* As we shall see in chapter III.1, Schur's lemma amounts to an important statement in quantum mechanics.

matrices $(A + A^\dagger)$ and $i(A - A^\dagger)$. Thus, we might as well focus on each of them, and rename the original matrix H to emphasize its hermiticity.

Proof of Schur's lemma

We want to prove that if $HD(g) = D(g)H$ for all g, then $H = \lambda I$ for some constant λ.

Since H is hermitean, it can be diagonalized: $H = W^\dagger H' W$ with H' diagonal and W some unitary matrix. Transform to that basis: $D(g) = W^\dagger D'(g)W$. The statement of the theorem $HD(g) = D(g)H$ becomes $(W^\dagger H'W)(W^\dagger D'(g)W) = (W^\dagger D'(g)W)(W^\dagger H'W)$, which becomes, upon multiplication by W from the left and $W^\dagger$ from the right, $H'D'(g) = D'(g)H'$. Now drop the primes. In the statement of the theorem, H can be taken, not only to be hermitean, but also to be diagonal.*

Now take the ij-component of the statement $HD(g) = D(g)H$ (using the upper and lower indices explained in the review of linear algebra but suspending the repeated index summation convention for the moment): $(HD(g))^i{}_j = H^i{}_i D^i{}_j(g) = (D(g)H)^i{}_j = D^i{}_j(g)H^j{}_j$, which implies that $(H^i{}_i - H^j{}_j)D^i{}_j(g) = 0$. Note that there are many equations here, as i, j, and g run over their ranges.

We are almost there. For a given pair i, j, unless $D^i{}_j(g) = 0$ for all g (note the emphasis on "all" here), we can conclude $H^i{}_i = H^j{}_j$. We already know that H is diagonal; now we have shown that different diagonal elements are equal. Taking all possible i, j, we conclude that H is proportional to the identity matrix. This proves Schur's lemma.

The irreducibility of the representation is precisely to protect us against that "unless" clause in the preceding paragraph. Suppose that the representation reduces to a direct sum of a 3-dimensional and a 7-dimensional representation:

$$D(g) = \left(\begin{array}{c|c} D^{(3)}(g) & 0 \\ \hline 0 & D^{(7)}(g) \end{array} \right).$$

Then the element of $D(g)$ in the second row and fifth column, for example, vanishes for all g. In the proof, we could not show that $H^2_2 = H^5_5$. Hence we cannot conclude that H is proportional to the identity matrix, only[†] that

$$H = \left(\begin{array}{c|c} \mu I_3 & 0 \\ \hline 0 & \nu I_7 \end{array} \right)$$

is equal to two identity matrices I_3 and I_7 stacked together, with μ, ν two arbitrary real numbers.

---

* In the future, we will not go through the analog of these steps in detail again, but simply say, in this context, that we can go to a basis in which H is diagonal.

[†] This already provides us with very valuable information in physical applications. See chapters III.1 and III.2.

To repeat, the stipulation that the representation is irreducible is crucial. Otherwise, H could have the form just shown, for example, and is assuredly not proportional to the identity matrix.

The Great Orthogonality theorem

Now we are ready for the central theorem of representation theory: Given a d-dimensional irreducible representation $D(g)$ of a finite group G, we have

$$\sum_g D^\dagger(g)^i{}_j D(g)^k{}_l = \frac{N(G)}{d} \delta^i{}_l \delta^k{}_j \tag{1}$$

with $N(G)$ the number of group elements.

The heart of the theorem is the assertion that the sum on the left hand side is proportional to $\delta^i{}_l \delta^k{}_j$, which is either 0 or 1.

Dr. Feeling saunters by, muttering, "What this tells us is that when we sum over an entire group, any 'orientational' information is washed out. This is analogous to a common situation in physics: a physical result, after angles are integrated or averaged over,* cannot favor any particular direction."

Note also that we are again using the upper and lower index convention appropriate for unitary matrices, as explained in detail in the linear algebra review.

You need not remember the proportionality constant on the right hand side; that is easily determined by setting $j = k$ and summing over all d values that the index ranges over. The left hand side becomes $\sum_g \delta^i{}_l = N(G) \delta^i{}_l$, which fixes the constant to be $N(G)/d$.

Here is the proof of the theorem.

Form the matrix $A = \sum_g D^\dagger(g) X D(g)$ for some arbitrary matrix X. Observe that, for any g, $D^\dagger(g) A D(g) = D^\dagger(g)(\sum_{g'} D^\dagger(g') X D(g')) D(g) = (\sum_{g'} D^\dagger(g'g) X D(g'g)) = A$ because of the group axioms. (We invoked the rearrangement lemma again.) By Schur's lemma, $A = \lambda I_d$. Trace this to obtain tr $A = \lambda d = \sum_g$ tr $D^\dagger(g) X D(g) = \sum_g$ tr $X = N(G)$ tr X, which determines $\lambda = \frac{N(G)}{d}$ tr X.

Thus far, the discussion is for any X. We now choose it to be 0 everywhere except for the entry $X^j{}_k = 1$ in the jth row and kth column (for some specific j and k, for example, 3 and 11), which is set equal to 1. Thus[†] tr $X = \delta^k{}_j$. Now let us evaluate a specific entry $A^i{}_l$ in the matrix A, using what we have just learned: $A^i{}_l = \sum_g (D^\dagger(g) X D(g))^i{}_l = \sum_g D^\dagger(g)^i{}_j D(g)^k{}_l = \lambda \delta^i{}_l = \frac{N(G)}{d} \delta^i{}_l$ tr $X = \frac{N(G)}{d} \delta^i{}_l \delta^k{}_j$, which is precisely (1).

Clearly, (1) imposes a powerful constraint on the representation matrices, a constraint that we will exploit mercilessly to determine $D(g)$.

An important corollary is the following: If r and s are two inequivalent representations, then $\sum_g D^{(r)\dagger}(g)^i{}_j D^{(s)}(g)^k{}_l = 0$.

---

* Think of integrating over the group elements of the rotation group $SO(3)$.

[†] It may look strange that we have something without indices on the left hand side and $\delta^j{}_k$ on the right hand side, but in the present context, j, k are just two numbers we have arbitrarily chosen. In other words, X depends on j, k by construction.

One quick nonrigorous "physicist type" argument is that if the two irreducible representations are inequivalent, then the indices i, j and the indices k, l live in entirely different spaces, and there is no way to write Kronecker deltas on the right hand side. Strictly speaking, we should have written $\sum_g D^{(r)\dagger}(g)^\mu_\nu D^{(s)}(g)^k_l = 0$ to emphasize this point. Specifically, if r and s have different dimensions, then Kronecker deltas do not even make sense. The proof of the corollary involves several steps, and to avoid interrupting the narrative flow, I am relegating it to appendix 1.

With this corollary, we can then write (1) in the more complete form

$$\sum_g D^{(r)\dagger}(g)^i{}_j D^{(s)}(g)^k{}_l = \frac{N(G)}{d_r} \delta^{rs}\delta^i{}_l\delta^k{}_j \tag{2}$$

The Kronecker δ^{rs} is equal to 1 if the two irreducible representations are the same (that is, equivalent) and 0 if not.

Behold, the Great[2] Orthogonality theorem!

Character orthogonality

The representation matrices $D^{(r)}(g)$ are of course basis dependent: we can always make a similarity transformation $D^{(r)}(g) \to S^{-1}D^{(r)}(g)S$. Take the trace to get rid of the basis dependence. The character $\chi^{(r)}(c) \equiv \operatorname{tr} D^{(r)}(g)$ depends only on the class c that g belongs to, as discussed in chapter II.1.

Indeed, (2) contains so much information that we can afford to trace out some of it. Set $i = j$ and $k = l$ and sum. Invoke the celebrated saying "Character is a function of class!" and obtain $\sum_g (\chi^{(r)}(g))^*\chi^{(s)}(g) = \sum_c n_c(\chi^{(r)}(c))^*\chi^{(s)}(c) = N(G)\delta^{rs}$, with n_c denoting the number of elements belonging to class c. We have thus derived a statement about the characters:

$$\sum_c n_c(\chi^{(r)}(c))^*\chi^{(s)}(c) = N(G)\delta^{rs} \tag{3}$$

This result will turn out to be enormously useful.

We will study plenty of examples of (3) in chapter II.3, but for the time being note that this works nicely for Z_N. Indeed, the orthogonality in (3) for Z_N amounts essentially to the idea behind Fourier[3] analysis. For Z_N, the representations are labeled by an integer k that can take on the values $0, \cdots, N - 1$. The equivalence classes are labeled by an integer j that also can take on the values $0, \cdots, N - 1$. Since the irreducible representations are 1-dimensional, the characters are trivially determined to be $\chi^{(k)}(j) = e^{i2\pi kj/N}$. Character orthogonality (3) then says $\sum_{j=0}^{N-1} e^{-i2\pi lj/N}e^{i2\pi kj/N} = \sum_{j=0}^{N-1} e^{i2\pi(k-l)j/N} = N\delta^{lk}$, which is pretty much the basis of Fourier's brilliant idea.

Character table

For a given finite group, we can construct its "character table" displaying $\chi^{(r)}(c)$. Along the vertical axis, we list the different equivalence classes c, and along the horizontal axis,[*] the different irreducible representations r. To have something definite to wrap our minds around, I will give here, without derivation,[†] the character table for A_4 (with $\omega = e^{i2\pi/3}$, as was defined earlier):

A_4 n_c	c	1	1'	1''	3
1	I	1	1	1	3
Z_2 3	(12)(34)	1	1	1	-1
Z_3 4	(123)	1	ω	ω^*	0
Z_3 4	(132)	1	ω^*	ω	0

$$(4)$$

To the left of the vertical line, the third column lists the four different equivalence classes of A_4, as described in chapter I.2. Each equivalence class is identified by a "typical" member: I, (12)(34), (123), (132). The second column lists n_c, the number of elements belonging to each class: 1, 3, 4, 4. As always, the identity belongs to a "class of one." The first column indicates the subgroup generated by each of these classes. For example, $(123)^3 = I$, and so (123) generates Z_3.

The top row lists (to the right of the vertical line) the four different irreducible representations of A_4 (as will be derived in chapter II.3). They are named by their dimensions: three of them are 1-dimensional and known as 1, 1', 1'', and one is 3-dimensional, known as 3.

The irreducible representation 1 is just the trivial representation, representing every element of the group by the number 1. Thus, the first column in the table proper (just to the right of the vertical line) is fixed trivially.

The first row (again in the table proper; in the future this will be what we mean) gives $\chi^{(r)}(I) = d_r$, the dimension of the representation, since in the representation r the identity I, as explained earlier, is always represented by the $d_r \otimes d_r$ identity matrix.

One consequence follows immediately. Let us first define two numbers characteristic of the group G: $N(C) =$ the number of equivalence classes and $N(R) =$ the number of irreducible representations. (For A_4, they are both 4.) Consider, for each s, the array of the (in general) complex numbers $(n_c)^{\frac{1}{2}}\chi^{(s)}(c)$ as c ranges over its $N(C)$ possible values. Regard this array as a vector in an $N(C)$-dimensional complex vector space. We are told by (3) that these vectors, altogether $N(R)$ of them, are orthogonal to one another. In other words, the four columns in the table are orthogonal to one another. This is known as column orthogonality.

But in an $N(C)$-dimensional complex vector space, there are at most (do exercise 1) $N(C)$ such vectors. We have just proved that

$$N(R) \leq N(C) \tag{5}$$

---

[*] The opposite convention, listing the equivalence classes along the horizontal axis and the irreducible representations along the vertical axis, is also commonly used. I am used to the convention I was taught in school.

[†] In chapter II.3, we will use what we learn in this chapter to determine the character table for various finite groups, in particular the one given here.

The number of irreducible representations of a finite group is bounded above by the number of classes. This answers one of the questions raised in chapter II.2. Intuitively, you might not be surprised. We sense that the number of classes sort of measures how complicated the group is.

We will show later in this chapter that the inequality (5) is in fact an equality. Indeed, the equality holds for Z_N, which as we noted earlier, has $N(C) = N$ ("everybody in his or her own class") and $N(R) = N$. (OK, even physicists wouldn't call this a proof.)

We can go back to the orthogonality theorem (2) before we took the trace and play the same game. Clearly, we should expect to get more information. (See if you can do it before reading on.)

So, consider, for each triplet (s, k, l), the array of complex numbers $D^{(s)}(g)^k_l$ as g ranges over its $N(G)$ possible values. Regard this array as a vector in an $N(G)$-dimensional complex vector space. Since, for each s, the indices k and l each range over d_s values, there are altogether $\sum_s d_s^2$ of these vectors. We are told by (2) that they are orthogonal to one another. Thus, reasoning as before, we obtain

$$\sum_s d_s^2 \leq N(G) \tag{6}$$

Intuitively, just as above, you might not be surprised that there is some kind of upper bound. For a group of a certain "size" $N(G)$, the irreducible representations can't be "too big." (Reducible representations, in contrast, can be as big as you care to stack them.)

Later, we will show that the inequality (6) is actually also an equality.

A test for reducibility

Suppose a dubious looking character wants to sell you a used representation, swearing that it won't fall apart on you. Is there any way to tell if that's true?

Given a representation, how can we test whether it is reducible or not? Suppose it is reducible. Picture the representation matrices as consisting of a stack of irreducible representations:

$$D(g) = \begin{pmatrix} \ddots & 0 & 0 & 0 & 0 \\ 0 & D^{(r)}(g) & 0 & 0 & 0 \\ 0 & 0 & \ddots & 0 & 0 \\ 0 & 0 & 0 & D^{(s)}(g) & 0 \\ 0 & 0 & 0 & 0 & \ddots \end{pmatrix} \quad \text{for all } g \in G \tag{7}$$

In particular, the irreducible representation r with dimension d_r can appear n_r times possibly. (For example,* in (II.1.9), the irreducible representations 1 and 3 each appear

---

* Yes, (II.1.9) refers to the rotation group, while the discussion in this chapter is focused on finite groups. Note that the concept of the number of times an irreducible representation appears in a reducible representation applies to both continuous and finite groups.

twice.) Then this representation has characters given by $\chi(c) = \sum_r n_r \chi^{(r)}(c)$. According to (3),

$$\sum_c n_c \chi^*(c) \chi(c) = \sum_c n_c \sum_{r,s} n_r n_s \chi^{*(r)}(c) \chi^{(s)}(c) = N(G) \sum_{r,s} n_r n_s \delta^{rs} = N(G) \sum_r (n_r)^2 \qquad (8)$$

This is a powerful result that we will see in action repeatedly in chapter II.3; here let's see what it tells us.

Given a representation, all we have to do is to take some traces to compute $\chi(c)$, then a quick sum over c yields the quantity $\sum_r (n_r)^2$. If this quantity is equal to 1, we would know that one of the n_rs is equal to 1, with all the other n_rs equal to 0. (Since n_r is the number of times the irreducible representation r appears in the representation we are given, it has to be a non-negative integer.) The given representation $D(g)$ contains the irreducible representation r once and only once. In other words, $D(g)$ is irreducible; in fact, it is just $D^{(r)}(g)$.

In contrast, if the quantity $\sum_r (n_r)^2$ is larger than 1, then the representation we are given is in fact reducible. Indeed, for smallish $\sum_r (n_r)^2 > 1$, we can immediately see the dimensions of the irreducible representations it breaks apart into. For example, suppose $\sum_r (n_r)^2 = 3$; then the only possibility is for three different irreducible representations to each occur once.

We could go further. Take the character of the representation you are being offered and contract it against the character of the representation r:

$$\sum_c n_c \chi^{*(r)}(c) \chi(c) = \sum_c n_c \sum_s n_s \chi^{*(r)}(c) \chi^{(s)}(c) = N(G) n_r \qquad (9)$$

Thus, n_r is determined. Not only can we tell whether the used representation we are being shown will fall apart, we even know how many pieces it will fall apart into.

Notational alert: In this game, many numbers appear. It is easy to confound them. Let's take stock: $N(C)$ is the number of equivalence classes, while n_c is the number of members in the class c; $N(G)$ is the number of elements in the group; $N(R)$ is the number of irreducible representations the group has. These numbers are all properties specific to the given group. In contrast, n_r, the number of times the representation r appears in a possibly reducible representation we are given, merely reflects on that particular representation. It is important to sharply distinguish these different concepts.

The characters of the regular representation are ridiculously easy to compute

In chapter I.2, you learned about Cayley's theorem: any finite group G with $N(G)$ elements is a subgroup of the permutation group $S_{N(G)}$. Since, as was explained in chapter II.1, $S_{N(G)}$ has a defining representation with dimension $N(G)$, any finite group G has an $N(G)$-dimensional representation, known as the regular representation.

Let us apply the result of the previous section to the regular representation of a finite group G. Note that the matrices furnishing the regular representation are rather large, with size given by $N(G)$. (For example, for A_4 they are 12 by 12.) But these matrices are full of zeroes, as was shown for S_4 in chapter II.1.

Dr. Feeling wanders by, muttering, "With that huge factorial size and all those zeroes, I feel that the regular representation has to be massively reducible."

With all those zeroes, the characters for the regular representation are ridiculously easy to compute: except for the identity matrix representing the identity element I, none of the other matrices have any diagonal entries by construction. The characters all vanish, except for the character $\chi(I)$ of the identity (which is in a class by itself, and hence a class with a membership of 1). The character $\chi(I)$ is equal to—as is always the case for any representation—the dimension of the representation, which is $N(G)$ for the regular representation by construction.

Let us apply the reducibility test (8) to the regular representation: $\sum_c n_c \chi^*(c)\chi(c) = (\chi(I))^2 = N(G)^2 = N(G)\sum_r(n_r)^2$. Hence $\sum_r(n_r)^2 = N(G)$, which for sure is larger than 1. We thus learned that the regular representation is reducible. Dr. Feeling's intuition is on the mark.

Also, for the regular representation, (9) gives $\sum_c n_c \chi^{*(r)}(c)\chi(c) = \chi^{*(r)}(I)\chi(I) = d_r N(G) = N(G)n_r$, and thus

$$n_r = d_r \tag{10}$$

This remarkable result says that in the regular representation, each irreducible representation r appears as many times as its dimension d_r. So, the matrices in the regular representation have the form

$$D^{\mathrm{reg}}(g) = \begin{pmatrix} \ddots & 0 & 0 & 0 & 0 \\ 0 & D^{(r)}(g) & 0 & 0 & 0 \\ 0 & 0 & \ddots & 0 & 0 \\ 0 & 0 & 0 & D^{(r)}(g) & 0 \\ 0 & 0 & 0 & 0 & \ddots \end{pmatrix} \tag{11}$$

where the d_r-by-d_r matrix $D^{(r)}(g)$ appears d_r times. Since we know that $D^{\mathrm{reg}}(g)$ is an $N(G)$-by-$N(G)$ matrix, we learn that

$$\sum_r d_r^2 = N(G) \tag{12}$$

The inequality (6) is actually an equality, as promised.

The character table is square

At this point, we can go on and show that $N(C) \leq N(R)$, which together with (5) implies

$$N(C) = N(R) \tag{13}$$

Since the proof of the inequality just cited is a bit involved, I have relegated it to appendix 4.

Knowing the structure of the group practically amounts to knowing how its elements fall into classes, and in particular, $N(C)$, the number of classes. Then we know $N(R)$, the number of irreducible representations, which tells us how many terms there are in the sum $\sum_r d_r^2 = N(G)$. For low $N(G)$, this is such a powerful constraint that the dimensions of the irreducible representations are often uniquely fixed. For example, for A_4 with four equivalence classes, we have $1^2 + d_1^2 + d_2^2 + d_3^2 = 12$. The unique solution is $1^2 + 1^2 + 1^2 + 3^2 = 12$. We conclude that there are three inequivalent 1-dimensional irreducible representations and one 3-dimensional irreducible representation.

Knowing that the character table is square, we could invite ourselves to write $U^c_s = \sqrt{\frac{n_c}{N(G)}} \chi^{(s)}(c)$. Then column orthogonality (3) takes on the suggestive form $\sum_c (U^c_r)^* U^c_s = \sum_c (U^\dagger)^r_c U^c_s = \delta^r_s$ (where we have invented a new notation for the Kronecker delta, as we are free to do). In other words, if we regard the character table suitably normalized as a matrix, that matrix is unitary. But $U^\dagger U = I$ implies $U U^\dagger = I$, and hence $U^\dagger U = I$. But written out, this says that different rows are orthogonal: $\sum_r \chi^{(r)}(c)^* \chi^{(r)}(c') = N(G) \delta^{cc'}/n_c$. Note also that the orthonormality of the first row reproduces (12).

I should caution the reader that this is not a proof* of row orthogonality.[4] In the proof that the character table is square given in appendix 4, we need to show row orthogonality along the way.

Dr. Feeling wanders by again. "What was your intuition regarding how many irreducible representations a group could have?" he asks, and then mutters, "Mine was that the more complicated a group, the more irreducible representations it would have. What (13) shows is that, not only is this feeling correct, but also it is the number of classes that measures how complicated a group is."

Appendix 1: Different irreducible representations are orthogonal

Let us now discuss what happens when $d_r \neq d_s$, the case not treated in the text. As before, construct $A = \sum_g D^{(r)\dagger}(g) X D^{(s)}(g)$, but now A and X are d_r-by-d_s rectangular matrices. As before, we then have $D^{(r)\dagger}(g) A D^{(s)}(g) = A$, so that

$$A D^{(s)}(g) = D^{(r)}(g) A \quad \text{and hence} \quad A^\dagger D^{(r)\dagger}(g) = D^{(s)}(g)^\dagger A^\dagger \qquad (14)$$

Multiply this by A to obtain $A A^\dagger D^{(r)\dagger}(g) = A D^{(s)}(g)^\dagger A^\dagger$, which we can write as

$$A A^\dagger D^{(r)}(g^{-1}) = A D^{(s)}(g^{-1}) A^\dagger \qquad (15)$$

Here we have used the unitarity of the two representations $D^{(r)\dagger}(g) D^{(r)}(g) = I$, which together with $D^{(r)}(g^{-1}) D^{(r)}(g) = I$ imply that $D^{(r)\dagger}(g) = D^{(r)}(g^{-1})$, and similarly for $r \to s$. Using (14), we have $A A^\dagger D^{(r)}(g^{-1}) = D^{(r)}(g^{-1}) A A^\dagger$. Note that with A d_r-by-d_s and $A^\dagger$ d_s-by-d_r the matrix $A A^\dagger$ is actually a d_r-by-d_r square matrix. But since g^{-1} is also "any group element," the equality we just obtained amounts to the input to Schur's lemma, which thus tells us that $A A^\dagger = \lambda I_{d_r}$. (Similarly, we can prove that $A^\dagger A \propto I_{d_s}$.)

We can now show that $\lambda = 0$. It is pedagogically clearer if we take specific values of d_r and d_s, say, 7 and 3 respectively. Then picture the equality $\underbrace{(7 \otimes 3)}_{A} \underbrace{(3 \otimes 7)}_{A^\dagger} = \lambda I_7$.

---

* But it is a useful mnemonic.

Look at this schematic picture:

$$
\begin{pmatrix}
x & x & x \\
x & x & x \\
x & x & x \\
x & x & x \\
x & x & x \\
x & x & x \\
x & x & x
\end{pmatrix}
\begin{pmatrix}
x & x & x & x & x & x & x \\
x & x & x & x & x & x & x \\
x & x & x & x & x & x & x
\end{pmatrix}
= \lambda
\begin{pmatrix}
1 & 0 & 0 & 0 & 0 & 0 & 0 \\
0 & 1 & 0 & 0 & 0 & 0 & 0 \\
0 & 0 & 1 & 0 & 0 & 0 & 0 \\
0 & 0 & 0 & 1 & 0 & 0 & 0 \\
0 & 0 & 0 & 0 & 1 & 0 & 0 \\
0 & 0 & 0 & 0 & 0 & 1 & 0 \\
0 & 0 & 0 & 0 & 0 & 0 & 1
\end{pmatrix}
$$

Here x denotes various generic numbers. Simply construct a new matrix B by adding four extra columns of zeroes to A, that is, define $B^i{}_j = A^i{}_j$, $j \leq 3$, and $B^i{}_j = 0$, $j \geq 4$. It follows trivially that $\underbrace{(7 \otimes 7)}_{B} \underbrace{(7 \otimes 7)}_{B^\dagger} = \lambda I_7$.

Again, the following schematic picture makes it clear:

$$
\begin{pmatrix}
x & x & x & 0 & 0 & 0 & 0 \\
x & x & x & 0 & 0 & 0 & 0 \\
x & x & x & 0 & 0 & 0 & 0 \\
x & x & x & 0 & 0 & 0 & 0 \\
x & x & x & 0 & 0 & 0 & 0 \\
x & x & x & 0 & 0 & 0 & 0 \\
x & x & x & 0 & 0 & 0 & 0
\end{pmatrix}
\begin{pmatrix}
x & x & x & x & x & x & x \\
x & x & x & x & x & x & x \\
x & x & x & x & x & x & x \\
0 & 0 & 0 & 0 & 0 & 0 & 0 \\
0 & 0 & 0 & 0 & 0 & 0 & 0 \\
0 & 0 & 0 & 0 & 0 & 0 & 0 \\
0 & 0 & 0 & 0 & 0 & 0 & 0
\end{pmatrix}
= \lambda
\begin{pmatrix}
1 & 0 & 0 & 0 & 0 & 0 & 0 \\
0 & 1 & 0 & 0 & 0 & 0 & 0 \\
0 & 0 & 1 & 0 & 0 & 0 & 0 \\
0 & 0 & 0 & 1 & 0 & 0 & 0 \\
0 & 0 & 0 & 0 & 1 & 0 & 0 \\
0 & 0 & 0 & 0 & 0 & 1 & 0 \\
0 & 0 & 0 & 0 & 0 & 0 & 1
\end{pmatrix}
$$

Now take the determinant of this equation to obtain* $\lambda = 0$. We have thus proved that $A = \sum_g D^{(r)\dagger}(g) X D^{(s)}(g) = 0$ (for any X), as desired.

We are not quite done. What if the two inequivalent representations r and s happen to have the same dimension?

Since $d_r = d_s$, A is square. Case 1: $\det A = 0 \Longrightarrow \lambda = 0$ and the theorem is proved. Case 2: $\det A \neq 0 \Longrightarrow \lambda \neq 0$. Then A^{-1} exists, and our earlier equation $A D^{(s)}(g) = D^{(r)}(g) A$ implies $D^{(s)}(g) = A^{-1} D^{(r)}(g) A$, which says that the representations r and s are in fact equivalent.

Appendix 2: First hint of row orthogonality

In the text, we showed that the representation r appears in the regular representation d_r times, and so the character of the class c in the regular representation is given by $\chi^{\text{reg}}(c) = \sum_r d_r \chi^{(r)}(c)$.

With an infinitesimal abuse of notation, let us denote by I the class of one that the identity proudly belongs to. Since $\chi^{(r)}(I)$ (the character of the identity in the representation r) is just d_r (the dimension of the representation r), we can also write the result just stated about the regular representation as $\chi^{\text{reg}}(c) = \sum_r \chi^{(r)}(I) \chi^{(r)}(c)$. In particular, if c is not I, then $0 = \chi^{\text{reg}}(c) = \sum_r \chi^{(r)}(I) \chi^{(r)}(c)$. But if c is I, then $N(G) = \chi^{\text{reg}}(I) = \sum_r (\chi^{(r)}(I))^2$, where the first equality simply states the dimension of the regular representation. We obtain (12) again.

Interestingly, the result $\sum_r \chi^{(r)}(I) \chi^{(r)}(c) = 0$ for $c \neq 1$ states that in the character table, the first row is orthogonal to all the other rows. This gives a first hint of row orthogonality, which states that all rows are orthogonal to one another. We will prove this in appendix 3.

---

* We have used two results from the linear algebra review: the determinant of the product is the product of the determinants, and the Laplace expansion.

Appendix 3: Frobenius algebra, group algebra, and class algebra

Groups are defined by their multiplicative structure, and it does not make sense in this structure to add group elements together. What would it mean to add the permutations (12)(34) and (123)? But mathematicians are certainly free to consider an algebra consisting of objects of the form $\sum_i a_i g_i$ where the a_i are a bunch of numbers. In a rather natural fashion, we can add and multiply such objects together: $\sum_i a_i g_i + \sum_i b_i g_i = \sum_i (a_i + b_i) g_i$ and $(\sum_i a_i g_i)(\sum_j b_j g_j) = \sum_{ij} a_i b_j (g_i g_j)$. This algebraic structure is known as the Frobenius algebra (or group algebra) of the group G.

Consider an equivalence class $c = \{g_1^{(c)}, \cdots, g_{n_c}^{(c)}\}$. The superscript c indicates the class c the group element belongs to, while the subscript distinguishes one member of that class from another. Define the class average*

$$K(c) \equiv \frac{1}{n_c} \sum_i g_i^{(c)} \tag{16}$$

We name the various classes as $c, d, e, \cdots$ and multiply the various class averages together: $K(c)K(d) = n_c^{-1} n_d^{-1} \sum_i \sum_j g_i^{(c)} g_j^{(d)}$. Note that the double sum contains $n_c n_d$ terms. For any $g \in G$,

$$g^{-1} \left(\sum_i \sum_j g_i^{(c)} g_j^{(d)} \right) g = \sum_i \sum_j (g^{-1} g_i^{(c)} g)(g^{-1} g_j^{(d)} g) = \sum_i \sum_j g_i^{(c)} g_j^{(d)} \tag{17}$$

Thus, under a similarity transformation by any g, the set of $n_c n_d$ group elements just get permuted among themselves.

This does not mean that they necessarily form one single class; they can in general form a bunch of classes e. Hence, the product $K(c)K(d)$ is equal to a linear combination of class averages. We are licensed to write

$$K(c)K(d) = \sum_e \Gamma(c, d; e) K(e) \tag{18}$$

thus defining the class coefficients $\Gamma(c, d; e)$, which are positive integers by construction. The class coefficients measure how many times the class e appears in the product of classes c and d. This subalgebra of the group algebra is known as the class algebra (see appendix 5).[5] If you are a bit confused here, simply work out the class algebra for your favorite finite group. Exercise!

Appendix 4: The character table is square

Now go to the level of the representation matrices and, for some class c, define[†]

$$\mathcal{D}(c) \equiv \frac{1}{n_c} \sum_{g \in c} D(g) \tag{19}$$

for some representation $D(g)$. (To lessen clutter, we suppress the superscript (r) on $D^{(r)}(g)$ temporarily.) Note that $\mathcal{D}(c)$ is a sum of representation matrices (which as matrices can certainly be added together) but is not itself a representation matrix. It is certainly not to be confused with $D(g)$.

Once again, we are up to our usual trick and notice that, for any g',

$$D(g'^{-1}) \mathcal{D}(c) D(g') = \frac{1}{n_c} \sum_{g \in c} D(g'^{-1} g g') = \mathcal{D}(c) \tag{20}$$

Thus, by Schur's lemma, $\mathcal{D}(c) = \lambda(c) I$, with I the identity matrix with the same dimension as $D(g)$ and $\lambda(c)$ some class-dependent constant. Take the trace to obtain $\chi(c) = \lambda(c)\chi(I)$, thus fixing $\lambda(c) = \chi(c)/\chi(I)$.

---

* The analogous object, defined without the factor n_c^{-1}, is known as the class sum.
[†] Be sure to distinguish $\mathcal{D}$ and D.

Referring to (18), we now have

$$\mathcal{D}(c)\mathcal{D}(d) = \sum_e \Gamma(c, d; e)\mathcal{D}(e) \tag{21}$$

Using $\mathcal{D}(c) = \lambda(c)I$, we obtain $\lambda(c)\lambda(d) = \sum_e \Gamma(c, d; e)\lambda(e)$. Plugging in $\lambda(c) = \chi(c)/\chi(I)$, we finally obtain

$$\chi(c)\chi(d) = \chi(I) \sum_e \Gamma(c, d; e)\chi(e) \tag{22}$$

Note that n_c, n_d, n_e, and $\Gamma(c, d; e)$ are properties of the group, while $\chi(c)$, $\chi(d)$, $\chi(e)$, and $\chi(I)$ are also specific to the representation r (which we had agreed earlier to suppress).

Now we restore the index r and write the preceding equation as $\chi^{(r)}(c)\chi^{(r)}(d) = \chi^{(r)}(I) \sum_e \Gamma(c, d; e)\chi^{(r)}(e)$. Sum over r to obtain

$$\sum_r \chi^{(r)}(c)\chi^{(r)}(d) = \sum_e \Gamma(c, d; e) \sum_r \chi^{(r)}(I)\chi^{(r)}(e) = \Gamma(c, d; I)N(G) \tag{23}$$

The last equality follows from the results obtained in the second paragraph of appendix 2, that $\sum_r \chi^{(r)}(I)\chi^{(r)}(e)$ vanishes for $e \neq I$ and is equal to $N(G)$ for $e = I$.

In chapter I.2, we defined the class $\bar{c}$ to consist of the inverse of all the elements contained in the class c. Since $\chi^{(r)}(d) = \chi^{(r)}(\bar{d})^*$, we can write (23) more conveniently for what will follow as $\sum_r \chi^{(r)}(\bar{d})^*\chi^{(r)}(c) = \Gamma(c, d; I)N(G)$.

When we multiply c and $\bar{c}$ together, by construction the identity appears n_c times. In other words, noting the factor of $1/n_c$ in the definition of K, we have $\Gamma(c, \bar{c}; I) = 1/n_c$ (where once again we denote by I the class to which the identity belongs by itself). In contrast, $\Gamma(c, d; I) = 0$ if $d \neq \bar{c}$, since then the product of the two classes c and d would not contain the identity. Thus, we arrive at

$$\sum_r \chi^{(r)}(c)^*\chi^{(r)}(c') = \frac{N(G)}{n_c}\delta^{cc'} \tag{24}$$

The characters of two different classes c and c' are orthogonal.

We have derived row orthogonality, as promised. This means that we have $N(C)$ orthogonal vectors in an $N(R)$-dimensional vector space. Thus, $N(C) \leq N(R)$. As mentioned in the text this means that $N(C) = N(R)$. The character table is indeed square.

Appendix 5: More on the class algebra

In appendix 3, we introduced the class algebra as a tool to prove row orthogonality. In fact, the class algebra is interesting in itself and can tell us a fair amount about the group. In this appendix, we mention some more properties of the class algebra.

First, it is an abelian algebra, since

$$K(c)K(d) = K(d)K(c) \tag{25}$$

Next,

$$K(c)K(\bar{c}) = \frac{1}{n_c}I + \cdots \tag{26}$$

where we have written $K(I)$ as I (without actual notational abuse, since I is in a class of one).

We now invite ourselves to rewrite (18) as

$$K(c)K(d) = \sum_e n_e f(c, d, e)K(\bar{e}) \tag{27}$$

We have defined $f(c, d, e) \equiv \Gamma(c, d, \bar{e})/n_e$ and have noted that summing over e and summing over $\bar{e}$ amounts to the same thing. You will see presently the convenience of making these (trivial) notational changes.

Multiplying (27) by $K(e)$ from the right gives

$$K(c)K(d)K(e) = \sum_a n_a f(c, d, a) K(\bar{a}) K(e) = f(c, d, e) I + \cdots \tag{28}$$

To obtain the second equality, we note that the product $K(\bar{a})K(e)$ is equal to a sum over a bunch of Ks, but $K(I)$ can only appear in the term with $a = e$ (since then $K(\bar{e})$ contains the inverse of the group elements contained in $K(e)$), and then we use (26) and (27). Thus, the structure constant $f(c, d, e)$ of the class algebra measures the extent to which the identity I appears in the product $K(c)K(d)K(e)$ of three class averages.

But since the algebra is abelian, we have shown that the structure constant $f(c, d, e)$ is totally symmetric in its three arguments; that is, it is symmetric under the exchange of any pair of arguments:

$$f(c, d, e) = f(d, c, e) = f(d, e, c) = f(e, c, d) = f(c, e, d) = f(e, d, c) \tag{29}$$

This symmetry property when expressed in terms of $\Gamma(c, d; e)$ would be somewhat more awkward. This is reflected in our notation: while $\Gamma(c, d; e)$ contains a colon, $f(c, d, e)$ does not. Also, $f(c, d, e) = f(\bar{c}, \bar{d}, \bar{e})$. Exercise! Prove this. Next, use associativity to prove that

$$\sum_e f(c, d, e) f(\bar{e}, a, b) = \sum_e f(d, a, e) f(\bar{e}, c, b) \tag{30}$$

Exercises

1   Show that in the 3-dimensional complex vector space, the three vectors $\begin{pmatrix} 1 \\ 1 \\ 1 \end{pmatrix}$, $\begin{pmatrix} 1 \\ \omega \\ \omega^* \end{pmatrix}$, $\begin{pmatrix} 1 \\ \omega^* \\ \omega \end{pmatrix}$ (where $\omega = e^{i2\pi/3}$) are orthogonal to one another. Furthermore, a vector $\begin{pmatrix} u \\ v \\ w \end{pmatrix}$ orthogonal to all three must vanish. Prove that in d-dimensional complex vector space there can be at most d mutually orthogonal vectors.

2   Determine the multiplication table of the class algebra for $D_5 = C_{5v}$.

3   Show that $f(c, d, I) = \delta_{\bar{c}d}/n_c$.

4   Show that $f(c, d, e) = f(\bar{c}, \bar{d}, \bar{e})$.

5   Prove (30).

6   Use Schur's lemma to prove the almost self-evident fact that all irreducible representations of an abelian group are 1-dimensional.

Notes

1. Issai Schur is not to be confused with Friedrich Schur, another German mathematician.
2. Great in the sense used in connection with symphonies and emperors; see Newton's two superb theorems and Gauss's Theorema Egregium.
3. Let us recall that many of the greats in physics and mathematics were artillery officers: C. Fourier, E. Schrödinger, K. Schwarzschild, R. Thun, and so forth.
4. I have seen this argument given erroneously as a proof of row orthogonality in some books.
5. Not to be confused with the algebra class you no doubt cruised through (ha ha).

II.3 | Character Is a Function of Class

In this chapter* I use the results of the preceding one to construct the character table of various finite groups. I strongly recommend that you do not sit back and read this like a novel with lots of characters; the construction of character tables is a participatory sport. Trust me, for the mathematically inclined, constructing character tables can be quite enjoyable.

Before we start, let us summarize some of the key results we have.

Dimensions of the irreducible representations:

$$\sum_r d_r^2 = N(G) \tag{1}$$

Column orthogonality:

$$\sum_c n_c (\chi^{(r)}(c))^* \chi^{(s)}(c) = N(G) \delta^{rs} \tag{2}$$

Row orthogonality:

$$\sum_r \chi^{(r)}(c)^* \chi^{(r)}(c') = \frac{N(G)}{n_c} \delta^{cc'} \tag{3}$$

The character table is square:

$$N(C) = N(R) \tag{4}$$

These results† impose powerful constraints on the character table.

For good measure, I display the Great Orthogonality theorem from which these results were derived:

$$\sum_g D^{(r)\dagger}(g)^i{}_j D^{(s)}(g)^k{}_l = \frac{N(G)}{d_r} \delta^{rs} \delta^i{}_l \delta^k{}_j \tag{5}$$

---

* I need hardly suggest that the study of group theory will strengthen your mastery of character and enhance your appreciation of class.

† Note that we obtain (1) as a special case of (3) by setting c and c' equal to the class populated solely by the identity.

We also derived from (2) two results giving the number of times n_r that the irreducible representation r appears in a given (possibly) reducible representation:

$$\sum_c n_c \chi^*(c)\chi(c) = N(G)\sum_r n_r^2 \tag{6}$$

and

$$\sum_c n_c \chi^{*(r)}(c)\chi(c) = N(G)n_r \tag{7}$$

As you will see, these two results are much used in this chapter.

The characters of A_3

When I was a student, after I worked my way through the proofs of some powerful theorems, I was often disappointed that those theorems were then applied to some pitifully trivial examples. Well, for pedagogical clarity, my first example here is also rather trivial. Consider A_3, the group of even permutations of three objects. It contains three elements: the identity I, the clockwise permutation $c = (123)$, and the anticlockwise permutation $a = (132)$. The group is abelian, with c and a the inverse of each other.

As was noted in chapter I.2, in an abelian group, every element is a single member of its own equivalence class. There are thus three equivalence classes and hence, according to (4), three irreducible representations. Then (1) can only be satisfied by $1^2 + 1^2 + 1^2 = 3$: there can only be three 1-dimensional irreducible representations; call them 1, 1′, and 1″.

Since $c^3 = I$, and since the representation matrix, being 1-by-1, is just a number, c can only be represented by 1, $\omega \equiv e^{i2\pi/3}$, or $\omega^* = \omega^2$. These three possibilities correspond to the three 1-dimensional irreducible representations, 1, 1′, and 1″. The number representing a is determined by $ca = I$. The character table* is thus fixed to be

A_3	n_c		1	1′	1″
1		I	1	1	1
Z_3	1	$c = (123)$	1	ω	ω^*
Z_3	1	$a = (132)$	1	ω^*	ω

$$(8)$$

Note that column and row orthogonality are satisfied, because $1 + \omega + \omega^* = 0$, namely, the three cube roots of unity sum to 0.

---

* I have already explained in chapter II.2 the convention used in this book. The equivalence classes are listed along the vertical, the irreducible representations along the horizontal. For each class c, the table indicates the number of members n_c, a typical member, and the cyclic subgroup $Z_?$ that a member of that class would generate.

Cyclic groups

Many of the considerations for A_3 apply to any abelian finite group with N elements. The N equivalence classes imply N irreducible representations, all 1-dimensional so that they satisfy $1^2 + 1^2 + \cdots + 1^2 = N$. This certainly makes sense, since 1-by-1 matrices (namely, numbers) all commute. Furthermore, we know that all irreducible representations are unitary, and hence each irreducible representation must be simply some appropriate root of unity.

In fact, since A_3 is isomorphic to Z_3, the discussion of A_3 here is just a special case of how Z_N can be represented. As was already mentioned in chapter II.2, the N 1-dimensional irreducible representations of Z_N are labeled by the integer $k = 0, 1, 2, \cdots, N - 1$. The group element $e^{i2\pi j/N}$ is represented by $D^{(k)}(e^{i2\pi j/N}) = e^{i2\pi kj/N}$ (so that $k = 0$ is the trivial identity representation). As was also remarked there, for Z_N, character orthogonality (2) gives

$$\sum_{j=0}^{N-1} e^{i2\pi(k-k')j/N} = N\delta_{kk'} \tag{9}$$

which is surely one of the most important identities in mathematics, science, and engineering: the identity that motivates Fourier series.

From A_3 to S_3

It is instructive to go from A_3 to S_3 with its $3! = 6$ elements. As we shall see, S_3 is nonabelian. At this point, our friend Confusio comes by and offers, "This game is easy; I could do it. Let me construct the character table of S_3 for you!"

Going from A_3 to S_3, we add the elements (12), (23), and (31), namely, the three transpositions or exchanges, which are the odd permutations not included in A_3. They clearly form an equivalence class by themselves.

"Four equivalence classes, and hence four irreducible representations," Confusio mutters. Since we always have the 1-dimensional trivial representation, the dimensions (call them a, b, and c) of the three nontrivial irreducible representations must satisfy $1^2 + a^2 + b^2 + c^2 = 6$.

But $1^2 + 1^2 + 1^2 + 2^2 = 7 > 6$, and $1^2 + 1^2 + 1^2 + 1^2 = 4 < 6$. "Oops!" exclaims Confusio. Dear reader, can you see what's wrong before reading on?

Confusio has unwittingly made a careless error. You and I gently point out that, since (23) is now in the group, $c = (123)$ and $a = (132)$ become equivalent, and the two distinct classes they belong to in A_3 merge.

Thus, S_3, just like A_3, has only three equivalence classes. There are only three irreducible representations, not four, and now things work out nicely: $1^2 + 1^2 + 2^2 = 6$. Call the irreducible representations 1, $\bar{1}$, and 2, according to their dimensions.

Let us now construct the character table. Note that, as always, the first column and first row are automatically filled in:

S_3	n_c		1	$\bar{1}$	2
	1	I	1	1	2
Z_3	2	(123), (132)	1	1	-1 $_x$
Z_2	3	(12), (23), (31)	1	-1	0 $_y$

(10)

Here $\bar{1}$ denotes the signature representation: it is ± 1, according to whether the permutation is even or odd. This explains why it does not appear in A_3. Indeed, when we restrict S_3 to A_3, $\bar{1}$ becomes 1.

The peculiar notation in the third column after the vertical line, the characters for the representation 2, is because I want to show you how to start with the column $(2, x, y)$ and use the theorems we have learned to determine x and y.

There are many ways to do it, since the various orthogonality theorems actually over-determine a simple character table like this one. One way to start is to observe that the characters for 1 and $\bar{1}$ differ only in the last row. This immediately implies, by sub-tracting two orthogonality relations, that $y = 0$. (In other words, take the difference of $\sum_c n_c (\chi^{(1)}(c))^* \chi^{(2)}(c) = 0$ and $\sum_c n_c (\chi^{(\bar{1})}(c))^* \chi^{(2)}(c) = 0$.)

Next, the orthonormality (weighted by n_c) of $\chi^{(2)}(c)$ (that is, $\sum_c n_c (\chi^{(2)}(c))^* \chi^{(2)}(c) = 6$) gives $1 \cdot 2^2 + 2 \cdot x^2 + 3 \cdot 0^2 = 6 = 4 + 2x^2 \Longrightarrow x^2 = 1$, and so $x = \pm 1$. Choosing the $+$ sign would contradict orthogonality with 1 and $\bar{1}$. So we are forced to choose $x = -1$, and the table is completed. Let us double check that the solution $x = -1$ satisfies orthogonality with the identity representation: $1 \cdot 1 \cdot 2 + 2 \cdot 1 \cdot (-1) + 3 \cdot 1 \cdot 0 = 0$, indeed.

From the character table to the representation matrices

Now that we have constructed the character table, we might want to exhibit the 2-dimensional representation matrices explicitly. In fact, the character table, particularly for smaller groups, contains enough information to determine the actual representation matrices.

The identity I is represented by the 2-by-2 identity matrix, of course. To exhibit repre-sentation matrices explicitly, we have to commit to a particular basis. Let's go to a basis in which (123) and (132) are diagonal. Invoking the theorem that these representation ma-trices must be unitary and the fact that these two elements generate a Z_3 subgroup, we fix that* (123) $\sim \left(\begin{smallmatrix} \omega & 0 \\ 0 & \omega^* \end{smallmatrix} \right)$ and (132) $\sim \left(\begin{smallmatrix} \omega^* & 0 \\ 0 & \omega \end{smallmatrix} \right)$. (Which is which is a matter of convention.) The traces of these matrices are equal to $\omega + \omega^* = -1$, in agreement with the character table.

---

* Henceforth, we will denote "is represented by" by $\sim$.

Next, what about (12), (23), and (31)? Confusio comes along and ventures an educated guess: they must be represented by the three Pauli matrices $\begin{pmatrix} 0 & 1 \\ 1 & 0 \end{pmatrix}$, $\begin{pmatrix} 0 & -i \\ i & 0 \end{pmatrix}$, and $\begin{pmatrix} 1 & 0 \\ 0 & -1 \end{pmatrix}$. Sounds very plausible: they are traceless (in accordance with the character table) and unitary.

But, Confusio, the representation matrices must satisfy the multiplication table. Let (12) be represented by $\begin{pmatrix} 0 & 1 \\ 1 & 0 \end{pmatrix}$. Then since $(12)(123) = (12)(12)(23) = (23)$, (23) is represented by $\begin{pmatrix} 0 & 1 \\ 1 & 0 \end{pmatrix} \begin{pmatrix} \omega & 0 \\ 0 & \omega^* \end{pmatrix} = \begin{pmatrix} 0 & \omega^* \\ \omega & 0 \end{pmatrix}$. So, Confusio guessed wrong. The three transpositions (12), (23), and (31) are represented by $\begin{pmatrix} 0 & 1 \\ 1 & 0 \end{pmatrix}$, $\begin{pmatrix} 0 & \omega^* \\ \omega & 0 \end{pmatrix}$, and $\begin{pmatrix} 0 & \omega \\ \omega^* & 0 \end{pmatrix}$.

Some simple checks: they are unitary, traceless, and generate Z_2, since, for example, $\begin{pmatrix} 0 & \omega^* \\ \omega & 0 \end{pmatrix} \begin{pmatrix} 0 & \omega^* \\ \omega & 0 \end{pmatrix} = \begin{pmatrix} 1 & 0 \\ 0 & 1 \end{pmatrix}$. We could check the multiplication table some more, for example, $(23)(31) = (231) = (123)$: $\begin{pmatrix} 0 & \omega^* \\ \omega & 0 \end{pmatrix} \begin{pmatrix} 0 & \omega \\ \omega^* & 0 \end{pmatrix} = \begin{pmatrix} \omega & 0 \\ 0 & \omega^* \end{pmatrix}$, indeed. Or row orthogonality, between the first two rows, for example: $1 \cdot 1 + 1 \cdot 1 + 2 \cdot (-1) = 1 + 1 - 2 = 0$, as required.

By the very presence of the 2-dimensional irreducible representation, S_3, unlike its subgroup A_3, is not abelian.

The Great Orthogonality theorem is so constraining that there is usually more than one way to arrive at the same conclusion. From (5), we note that any nontrivial irreducible representation r must satisfy

$$\sum_g D^{(r)}(g)^i{}_j = 0 \tag{11}$$

From this we can see that Confusio's guess cannot be correct. We have $D^{(2)}(I) + D^{(2)}(123) + D^{(2)}(132)$ and $D^{(2)}(12) + D^{(2)}(23) + D^{(2)}(31)$ separately summing to 0, so that (11) holds a fortiori, but the three Pauli matrices do not sum to 0. Note that for the correct set of representation matrices, the even and odd permutations must separately sum to 0, because of orthogonality with the $\bar{1}$ as well as with the 1 representation.

Link between group theory and geometry: Fixed points

You may recall from chapter I.1 that S_3 is also the invariance group of the equilateral triangle. The two elements (123) and (132) correspond to rotations through $2\pi/3$ and $4\pi/3$, respectively, and the three transpositions (12), (23), and (31) to reflections across the three medians. If we label the three vertices of the triangle by a, b, c or 1, 2, 3 they can be thought of as the objects being permuted by S_3.

This remark provides a nice link between group theory and geometry.

Confusio asks, "You mean you paint 1, 2, 3 on the three vertices? Then the vertices would be distinguishable, and rotation through $2\pi/3$ would not leave the triangle invariant."

No, Confusio. The labeling is just to help us keep track of which vertex we are talking about. The three vertices are to be treated as identical. The triangle is meant to be a mathematical triangle in your mind's eye, not a physical triangle. For instance, suppose you were to draw a triangle on a piece of lined paper which is blank on the other side. Then an exchange, (12) say, would flip the piece of paper over, and you could tell that this is not quite the same triangle as before.

In chapter II.1, I remarked that S_n has an n-dimensional representation, its defining[1] or fundamental representation. Since we surely do not have room for a 3-dimensional representation for S_3, this defining representation, which we shall refer to as 3, must be reducible.

Now the orthogonality theorem leaps into action again. Let's start by writing down the characters for the 3:
$$\boxed{\begin{array}{c} 3 \\ \hline 0 \\ \hline 1 \end{array}}$$

Think of this as an extra column one could attach to the character table in (10). The top entry is of course 3: the character of the identity is always just the dimension of the representation. To understand the bottom entry, note that in the 3, an element like (12) is represented by $\begin{pmatrix} 0 & 1 & 0 \\ 1 & 0 & 0 \\ 0 & 0 & 1 \end{pmatrix}$. The third basis vector is untouched: it's a "fixed point," to borrow the language of topological maps. Thus, the trace (namely, the character) is just 1. In contrast, (123) is represented by $\begin{pmatrix} 0 & 0 & 1 \\ 1 & 0 & 0 \\ 0 & 1 & 0 \end{pmatrix}$, which has no fixed point and hence has a null character.[2]

OK, any guesses as to what irreducible representations this 3 reduces to? It could be $3 \rightarrow 1 + \bar{1} + \bar{1}$, or $3 \rightarrow 2 + 1$, and so on. For a simple example like this, there are only so many possibilities.

We simply plug in (7). Orthogonality of characters between 2 and 3 gives $1 \cdot 2 \cdot 3 + 2 \cdot (-1) \cdot 0 + 3 \cdot 0 \cdot 1 = 6 = 1(6)$, and between 1 and 3 gives $1 \cdot 1 \cdot 3 + 2 \cdot 1 \cdot 0 + 3 \cdot 1 \cdot 1 = 6 = 1(6)$. Thus, 3 contains 2 once and 1 once; in other words, $3 \rightarrow 2 + 1$. Indeed, the characters add up correctly:
$$\begin{array}{c} 3 \\ \hline 0 \\ \hline 1 \end{array} = \begin{array}{c} 2 \\ \hline -1 \\ \hline 0 \end{array} + \begin{array}{c} 1 \\ \hline 1 \\ \hline 1 \end{array}.$$

The reader might have recognized that this is the same problem as decomposing a vector that we started the review of linear algebra with: let $\begin{pmatrix} 3 \\ 0 \\ 1 \end{pmatrix} = x \begin{pmatrix} 1 \\ 1 \\ 1 \end{pmatrix} + y \begin{pmatrix} 1 \\ 1 \\ -1 \end{pmatrix} + z \begin{pmatrix} 2 \\ -1 \\ 0 \end{pmatrix}$, solve for x, y, z. (For ease of writing we use the "vector" notation instead of the square boxes.) Write this as a matrix equation $C \begin{pmatrix} x \\ y \\ z \end{pmatrix} = \begin{pmatrix} 3 \\ 0 \\ 1 \end{pmatrix}$, where we regard the character table of S_3 in (10) as a matrix $C = \begin{pmatrix} 1 & 1 & 2 \\ 1 & 1 & -1 \\ 1 & -1 & 0 \end{pmatrix}$. Use your knowledge of linear algebra to find C^{-1} and verify that $C^{-1} \begin{pmatrix} 3 \\ 0 \\ 1 \end{pmatrix}$ gives the solution listed above. For groups with large numbers of equivalence classes, this procedure could then be performed by a machine.

We can perform some additional checks, for example, orthogonality between 3 and $\bar{1}$ gives $1 \cdot 1 \cdot 3 + 2 \cdot 1 \cdot 0 + 3 \cdot (-1) \cdot 1 = 0$, as expected (3 does not contain $\bar{1}$). Also, from (6), $1 \cdot 3^2 + 2 \cdot 0^2 + 3 \cdot 1^2 = 12 = 2 \cdot 6$ tells us that 3 contains two irreducible representations.

As you can see, it is difficult to make a careless arithmetical error without catching ourselves while playing this game; the orthogonality theorems provide a web of interlocking checks.

The class of inverses: Real and complex characters

Given the class c, we defined (in chapter I.2) the class $\bar{c}$ to consist of the inverse of all elements belonging to c. Since $D(g)$ is unitary, we have $D(g^{-1}) = (D(g))^{-1} = D(g)^\dagger$. Tracing, we obtain

$$\chi(\bar{c}) = \chi(c)^* \tag{12}$$

One consequence of this is that if a class contains the inverses of its members, then $\bar{c} = c$, and $\chi(c)$ is real. Verify that this holds for the examples given thus far. For example, the characters of S_3 are all real, but not those of its subgroup A_3.

Various approaches for determining the character table

We might summarize the discussion above by enumerating the various approaches for determining the character table, at least in the case of a simple example like A_3.

1. Method of algebraic insight

   Recognize that A_3 is the same as Z_3, and look up the Fourier representations of Z_3.

   $$\begin{array}{ccc} 1 & 1 & 1 \\ 1 & \omega & \omega^* \\ 1 & \omega^* & \omega \end{array} \tag{13}$$

   (Here I am displaying only the heart of the character table.)

2. Method of geometrical insight

   We realize that the three objects being permuted by A_3 can be thought of as the three vertices of an equilateral triangle. It follows that A_3 has a 2-dimensional representation consisting of $\{I, R, R^2\}$, with $R = \begin{pmatrix} c & s \\ -s & c \end{pmatrix}$, $c \equiv \cos(2\pi/3)$, and $s \equiv \sin(2\pi/3)$. But since the group is abelian, we know that there is no irreducible 2-dimensional representation, so it must be reducible.[3] Diagonalize R. Since $\det R = 1$, the diagonalized form of R must have the form $\begin{pmatrix} e^{i\theta} & 0 \\ 0 & e^{-i\theta} \end{pmatrix}$. But $R^3 = I$; hence $e^{i\theta} = e^{2\pi i/3} = \omega$. (Indeed, by inspection we know that R is diagonalized by $S^{-1}RS$, with $S = \begin{pmatrix} 1 & i \\ i & 1 \end{pmatrix}/\sqrt{2}$.)

3. Method of exploiting the orthogonality theorems

   $$\begin{array}{ccc} 1 & 1 & 1 \\ 1 & x & u \\ 1 & y & v \end{array} \tag{14}$$

   Various orthogonality constraints, for example, $1 + x + y = 0$, $1 + |x|^2 + |y|^2 = 3 \Longrightarrow |x| = |y| = 1$, and so on, quickly determine $x = v = \omega$, $y = u = \omega^*$.

The invariance group T of the tetrahedron and A_4

Plato particularly liked the tetrahedron and associated it with fire.[4] Let's find the characters of its invariance group, known as T. We proceed using a mix of geometrical insight, appeal to the orthogonality theorems, and informed guessing.

By now, it doesn't take much brilliance to realize that if we mentally label the four vertices of the tetrahedron* by 1, 2, 3, and 4, we see that T is isomorphic to A_4, with $4!/2 = 12$ elements. Recall that in chapter I.2, we already listed the equivalence classes of A_4. There are four of them, with representative members I, (12)(34), (123), and (132), respectively. Thus, there should be four irreducible representations.

The isomorphism with T immediately tells us that A_4 has a 3-dimensional representation. Indeed, $T = A_4$, being the invariance group of a geometrical object that lives in 3-dimensional Euclidean space, is necessarily a subgroup of the rotation group and hence represented by 3-by-3 matrices.

Indeed, the orthogonality theorems assert that the dimensions of the irreducible representations satisfy $1^2 + d_1^2 + d_2^2 + d_3^2 = 12$. Since $1^2 + 1^2 + 2^2 + 2^2 = 10 < 12$, $1^2 + 2^2 + 2^2 + 2^2 = 13 > 12$, and so on, with the only possible solution $1^2 + 1^2 + 1^2 + 3^2 = 12$, we can proclaim (even if we are not genetically endowed with a modicum of geometrical insight) that there must be a 3-dimensional representation. We also learned that A_4 has two other 1-dimensional representations, traditionally known as $1'$ and $1''$, besides the trivial representation 1.

Playing around a bit[†] with the orthogonality theorems, we soon manage to construct the character table

A_4	n_c		1	1′	1″	3
	1	I	1	1	1	3
Z_2	3	(12)(34)	1	1	1	−1
Z_3	4	(123)	1	ω	ω^*	0
Z_3	4	(132)	1	ω^*	ω	0

(15)

As always, we indicate for each class the number of members, a typical member, and the cyclic subgroup a member of that class would generate.

The first column (all 1s) and the first row (just the dimensions) are immediately fixed. The two 1-dimensional representations $1'$ and $1''$ are easily fixed by virtue of their 1-dimensionality, unitarity, and their having to represent elements generating Z_3. Column and row orthogonality are of course so restrictive as to practically determine everything.

---

* We have in mind a solid tetrahedron. If we also allow reflections, then we would be dealing with S_4.
† And indulging in what pool players call shimmying or body English.

For example, the two bottom entries in the column for the irreducible representation 3 follow immediately from the orthogonality between 3 and 1, and between 1' and 1''.

You might think that the fourth, rather than the cube, root of unity would appear in the character table for A_4. One way of seeing the appearance of ω and ω^* is by restricting A_4 to A_3. We erase the second row (corresponding to (12)(34)) and the fourth column, get rid of a few entries, and necessarily should recover the known character table for A_3.

Beginners are also sometimes surprised that A_4 does not have a 4-dimensional irreducible representation. In exercise 1 you will show that the defining representation of A_4 is in fact reducible: $4 \rightarrow 1 + 3$.

Exhibiting the 3 of A_4

Once again, we are seized by ambition and would like to determine the actual matrices in the 3-dimensional irreducible representation.[5] A tedious but straightforward way would be to center the tetrahedron and work out the rotation matrices that leave it unchanged. Instead, we proceed by inspired and informed guessing. Actually, the character table contains enough hints so that hardly any guesswork is needed.

First, note that the three elements (12)(34), (13)(24), and (14)(23) commute with one another.* Thus, the matrices representing them can be simultaneously diagonalized, according to a theorem discussed in the review of linear algebra. Let's go to a basis in which they are diagonal. The matrices each generate a Z_2, and thus their diagonal entries can only be ± 1. Since A_4 consists of even permutations, these matrices have determinant $= +1$. They can only be

$$r_1 \equiv \begin{pmatrix} 1 & 0 & 0 \\ 0 & -1 & 0 \\ 0 & 0 & -1 \end{pmatrix}, \quad r_2 \equiv \begin{pmatrix} -1 & 0 & 0 \\ 0 & 1 & 0 \\ 0 & 0 & -1 \end{pmatrix}, \quad r_3 \equiv \begin{pmatrix} -1 & 0 & 0 \\ 0 & -1 & 0 \\ 0 & 0 & 1 \end{pmatrix} \tag{16}$$

That there are precisely three such matrices and that they have trace $= -1$, agreeing with the entry of -1 in the character table, provide strong evidence that we are on the right track.

What about (123)? Since it generates Z_3, it is clearly represented by $c \equiv \begin{pmatrix} 0 & 0 & 1 \\ 1 & 0 & 0 \\ 0 & 1 & 0 \end{pmatrix}$, which cyclically permutes the three basis vectors in the clockwise sense. What about the other three guys in the same equivalence class as c? Well, they are $r_1 c r_1$, $r_2 c r_2$, and $r_3 c r_3$.

The inverse of (123), namely, (132), is then represented by the anticlockwise $a \equiv \begin{pmatrix} 0 & 1 & 0 \\ 0 & 0 & 1 \\ 1 & 0 & 0 \end{pmatrix}$, which shares its equivalence class with $r_1 a r_1$, $r_2 a r_2$, and $r_3 a r_3$.

As a check, all six of these matrices have zero trace, in agreement with the character table.

---

* Let's check: $((12)(34))((13)(24)) = \begin{pmatrix} 1 & 2 & 3 & 4 \\ 2 & 1 & 4 & 3 \end{pmatrix} \begin{pmatrix} 1 & 2 & 3 & 4 \\ 3 & 4 & 1 & 2 \end{pmatrix} = \begin{pmatrix} 1 & 2 & 3 & 4 \\ 3 & 4 & 1 & 2 \\ 4 & 3 & 2 & 1 \end{pmatrix} = \begin{pmatrix} 1 & 2 & 3 & 4 \\ 4 & 3 & 2 & 1 \end{pmatrix}$ versus

$((13)(24))((12)(34)) = \begin{pmatrix} 1 & 2 & 3 & 4 \\ 3 & 4 & 1 & 2 \end{pmatrix} \begin{pmatrix} 1 & 2 & 3 & 4 \\ 2 & 1 & 4 & 3 \end{pmatrix} = \begin{pmatrix} 1 & 2 & 3 & 4 \\ 2 & 1 & 4 & 3 \\ 4 & 3 & 2 & 1 \end{pmatrix} = \begin{pmatrix} 1 & 2 & 3 & 4 \\ 4 & 3 & 2 & 1 \end{pmatrix}$.

Figure 1

The tray method

For more practice with representations, I show you one way of constructing a larger (and hence reducible) representation of A_4. To be concrete, let us think of the four objects permuted by A_4 as four balls labeled 1, 2, 3, and 4. Denote by $\langle 12 \rangle$ a tray into which we have put a 1-ball and a 2-ball (using an obvious terminology). With the rule of not allowing the two balls in each tray to be of the same type, we can have $6 = 4 \cdot 3/2$ different types of trays, namely $\langle 12 \rangle$, $\langle 13 \rangle$, $\langle 14 \rangle$, $\langle 23 \rangle$, $\langle 24 \rangle$, and $\langle 34 \rangle$. This furnishes a 6-dimensional representation of A_4, each element of which takes a tray into some other type of tray. See figure 1.

The characters of this representation, referred to as 6, are easy to write down: $\chi = 6$ for I as always (namely, the dimension of the representation); $\chi = 0$ for the class typified by (123), which has no fixed point (so that the 6-by-6 matrix representing (123) vanishes along the diagonal); similarly, $\chi = 0$ for the class typified by (132); and finally, $\chi = 2$ for the class typified by (12)(34), which leaves the tray $\langle 12 \rangle$ and the tray $\langle 34 \rangle$ unchanged (so that the 6-by-6 matrix representing (12)(34) has two 1s along the diagonal).

Thus, the characters for the 6 are given by $\begin{pmatrix} 6 \\ 2 \\ 0 \\ 0 \end{pmatrix}$. First check: $1 \cdot 6^2 + 3 \cdot 2^2 + 0 + 0 = 36 + 12 = 48 = 4(12)$. So the 6 reduces into four irreducible representations. Orthogonality with 1: $1 \cdot 1 \cdot 6 + 3 \cdot 1 \cdot 2 + 0 + 0 = 6 + 6 = 12$, so 6 contains 1 once. Similarly, 6 contains $1'$ once, and $1''$ once. Orthogonality with 3: $1 \cdot 3 \cdot 6 + 3 \cdot (-1) \cdot 2 + 0 + 0 = 18 - 6 = 12$, so 6 also contains 3 once.

We conclude that $6 \rightarrow 1 + 1' + 1'' + 3$ (and of course $6 = 1 + 1 + 1 + 3$ provides another check). Indeed, $\begin{pmatrix} 6 \\ 2 \\ 0 \\ 0 \end{pmatrix} = \begin{pmatrix} 1 \\ 1 \\ 1 \\ 1 \end{pmatrix} + \begin{pmatrix} 1 \\ 1 \\ \omega \\ \omega^* \end{pmatrix} + \begin{pmatrix} 1 \\ 1 \\ \omega^* \\ \omega \end{pmatrix} + \begin{pmatrix} 3 \\ -1 \\ 0 \\ 0 \end{pmatrix}$. Note that $1'$ and $1''$ are complex conjugate of each other, and thus they must occur in 6 with equal weight, since the representation matrices of 6 are manifestly real.

A geometrical fact

Before moving on, I mention one fact that seems strange at first sight. Consider the element (12)(34): can we really simultaneously exchange vertices 1 and 2, and vertices 3 and 4, of a solid tetrahedron by rotation? Group theory assures us that yes, it is possible.

Denote by m_{12} the midpoint on the edge joining 1 and 2, and by m_{34} the midpoint on the edge joining 3 and 4, as shown in figure 2. Imagine a line joining m_{12} and m_{34}. Rotate the

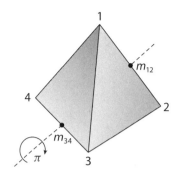

Figure 2

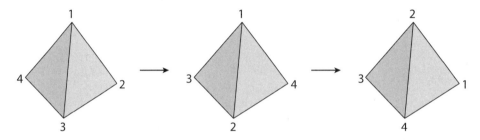

Figure 3

tetrahedron around this line through 180°. You should be able to see that this exchanges vertices 1 and 2, and vertices 3 and 4.

Since there are three such lines joining midpoints to opposite midpoints, there are three 180° rotations, one around each line, corresponding to (12)(34), (13)(24), and (14)(23).

Another way of seeing this is to first rotate the base of the tetrahedron through $2\pi/3$, then rotate again through $2\pi/3$ around a vertex on the base, as shown in figure 3. The net result is indeed the simultaneous exchanges of vertices 1 and 2 and of vertices 3 and 4. Perhaps you realize that this is just $(12)(34) = (12)(23)(23)(34) = (123)(234)$ (recall the rules you learned in chapter I.2).

Fun with character tables

It is instructive to go from A_4 to S_4. Going from four to five equivalence classes (with typical members I, (12)(34), (123), (12), and (1234), respectively), we now should have five irreducible representations, and the only solution for their dimensions is $1^2 + 1^2 + 2^2 + 3^2 + 3^2 = 24$. Next, we determine the number of members in each equivalence class, namely, 1, 3, 8, 6, and 6 which adds up to 24, of course. Note that as remarked earlier, upon $S_4 \to A_4$, the equivalence class to which (123) belongs splits up into two equivalence classes: $8 \to 4 + 4$.

I leave it to you to derive the character table. Here it is:

S_4	n_c		1	$\bar{1}$	2	3	$\bar{3}$
1		I	1	1	2	3	3
Z_2	3	(12)(34)	1	1	2	-1	-1
Z_3	8	(123)	1	1	-1	0	0
Z_2	6	(12)	1	-1	0	1	-1
Z_4	6	(1234)	1	-1	0	-1	1

$$(17)$$

The last two rows pertain to the $6 + 6 = 12$ elements that do not belong to A_4.

Perhaps the appearance of a 2-dimensional irreducible representation is a bit of surprise, now that we understand the presence of the 3-dimensional irreducible representations. On restriction to A_4, 2 breaks into $1' + 1''$.

The dihedral group D_4

By now, you should be able to construct the character table of just about any finite group that comes your way. Personally, I rather enjoy[6] constructing the character tables of (small) finite groups. You might want to try a few.

Let us look at the invariance group of the square, called D_4. As was already mentioned in chapters I.1 and II.2, it consists of rotations R around its center, reflections r across the two medians joining the center of the square to the midpoints of its sides, and reflections d across the two diagonals. See figure 4. The corresponding 2-by-2 transformation matrices are

$$I = \begin{pmatrix} 1 & 0 \\ 0 & 1 \end{pmatrix}$$

$$R = \begin{pmatrix} 0 & -1 \\ 1 & 0 \end{pmatrix}, \quad R^2 = \begin{pmatrix} -1 & 0 \\ 0 & -1 \end{pmatrix} = -I, \quad R^3 = \begin{pmatrix} 0 & 1 \\ -1 & 0 \end{pmatrix}$$

$$r_x = \begin{pmatrix} -1 & 0 \\ 0 & 1 \end{pmatrix}, \quad r_y = \begin{pmatrix} 1 & 0 \\ 0 & -1 \end{pmatrix}$$

$$d_1 = \begin{pmatrix} 0 & 1 \\ 1 & 0 \end{pmatrix}, \quad d_2 = \begin{pmatrix} 0 & -1 \\ -1 & 0 \end{pmatrix} \tag{18}$$

I would like to pause and clear up a potential source of confusion. In a group, minus the identity (that is, $-I$) does not exist and has no more meaning than, say, $3g$ or $-g$ for a group element g. Groups are defined in the abstract by some composition law, which in everyday usage is often referred to as multiplication. It does not make sense,

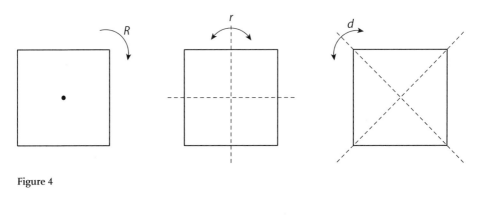

Figure 4

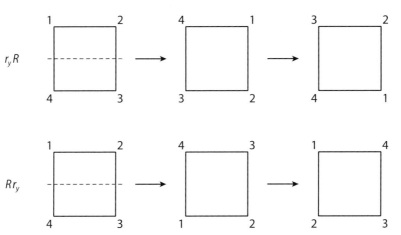

Figure 5

in general, to multiply a group element g by some number. Here, R represents a rotation through $90°$, and R^2, a rotation through $180°$, happens to be represented by a matrix that is numerically (-1) times the identity matrix. If you like, you can think of $-I$ as a convenient symbol for R^2. Or we should simply use some other symbol and not introduce the symbol $-I$ at all. Similarly, it may be confusing for some texts to refer to r_y as $-r_x$.

Incidentally, to show that composition of group elements is noncommutative, the rotational group $SO(3)$ offers the canonical example, as was given in chapter I.1, but D_4 provides an even simpler example. Under $r_y R$, the northwest corner of the square goes to the southeast, but under $R r_y$, the northwest stays put. See figure 5.

The eight elements of D separate into five equivalence classes: $\{I\}$, $\{-I\}$, $\{R, R^3\}$, $\{r_x, r_y\}$, and $\{d_1, d_2\}$. (Verify this!) Thus, there are five irreducible representations. Furthermore, $8 = 1^2 + 1^2 + 1^2 + 1^2 + 2^2$ is the only way to decompose 8. We already know about the 2-dimensional representation, the defining representation. Count them: four 1-dimensional representations!

Thus, we immediately have most of the character table for D_4 filled in already:

D_4	n_c		1	1′	1″	1‴	2
	1	I	1	1	1	1	2
Z_2	1	$-I$	1	1	1	1	-2
Z_4	2	R, R^3	1	a	d	g	0
Z_2	2	r_x, r_y	1	b	e	h	0
Z_2	2	d_1, d_2	1	c	f	i	0

$$(19)$$

Row orthogonality between first and second rows: $1^2 + 1^2 + 1^2 + 1^2 + 2 \cdot (-2) = 0$, check. Column orthogonality between first and fifth columns: $1 \cdot 2 + 1 \cdot (-2) + 0 + 0 + 0 = 0$, check.

I leave it for you to fill in the entries purposely left undetermined in the table. Do exercise 4 now!

Incidentally, in the language of chapter I.2, we can present this group as

$$D_4 : \langle R, r | R^4 = r^2 = I \rangle \qquad (20)$$

Thus, the eight elements are $\{I, R, R^2, R^3, r, rR, rR^2, rR^3\}$.

The group D_4 is a particular example of D_n, known as the dihedral group and defined as the invariance group of the n-sided regular polygon, as already mentioned in chapters I.1 and I.2. It so happens that $D_3 = S_3$, but in general D_n is not isomorphic to some permutation group. (Just count the number of elements: D_4 has 8 elements, but S_4 already has 24 elements. Also, number the vertices of the square by 1, 2, 3, and 4. No way that (12) can be effected by a rotation or a reflection.)

The quarternionic group

In chapter I.2, I introduced the quarternionic group Q with the eight elements 1, -1, i, $-i$, j, $-j$, k, and $-k$, with the multiplication table determined by the rules $i^2 = j^2 = k^2 = -1$ and $ij = -ji = k$, $jk = -kj = i$, and $ki = -ik = j$. As an exercise, construct the character table before reading on.

As always, 1 and -1, since they commute with everybody else, form two exclusive equivalence classes with a single member each. Our intuition tells us that i and $-i$ form one equivalence class. Similarly, j and $-j$. And similarly, k and $-k$. (To check this explicitly, compute $(-j)ij = -jk = -i$, and so indeed, $i \sim -i$. We can then cyclically permute this computation.) Thus, there are five equivalence classes altogether, which means that there are five irreducible representations.

The dimensions of the four nontrivial irreducible representations are determined by $1^2 + a^2 + b^2 + c^2 + d^2 = 8$, which has the unique solution $a = b = c = 1$ and $d = 2$.

Again, I leave it to you to fill in the rest:

Q	n_c		1	1'	1''	1'''	2
	1	1	1	1	1	1	2
Z_2	1	-1	1	1	1	1	-2
Z_4	2	$i, -i$	1	1	-1	-1	0
Z_4	2	$j, -j$	1	-1	1	-1	0
Z_4	2	$k, -k$	1	-1	-1	1	0

$$(21)$$

Here is a sketch of how the construction might proceed. After filling in the first column and the first row automatically, we proceed to the second row. For the 1-dimensional representations 1', 1'', 1''' the entries can only be ± 1, since the corresponding cyclic group is Z_2. Row orthonormality (3) can only work if the entry in the last column is ± 2. Row orthogonality (3) between the first two rows can then only work with the entries shown. Column orthonormality (2) for the last column $2^2 + (-2)^2 + |a|^2 + |b|^2 + |c|^2 = 8$ implies the unique solution $a = b = c = 0$. (The notation here is self-explanatory.) Row orthonormality (3) of the third row $1^2 + |x|^2 + |y|^2 + |z|^2 + 0^2 = 8/2 = 4$ only works if $|x| = |y| = |z| = 1$. The cyclic group associated with the third row also requires $x^4 = y^4 = z^4 = 1$. Row orthogonality (3) between the first and third row then fixes one of the x, y, z to be 1, with the other two equal to -1. (It doesn't matter which is which, since 1', 1'', and 1''' are labeled arbitrarily. Democracy between i, j, and k then fixes the pattern shown.)

Did you notice that the character tables for Q and for D_4 are the same? (Surely you did exercise 4, didn't you?) This answers the question about whether a character table uniquely characterizes a group.

Hardy's criteria

I conclude this chapter with three appendices, two of which involve constructing the character table for S_5 and A_5. After working through these, you will be fully equipped to go out into the world and construct character tables for hire.

I already mentioned G. H. Hardy's criteria for great mathematics in the preface. The theory of characters beautifully satisfies his three criteria: unexpectedness, inevitability, and economy.

Appendix 1: S_5

Here we construct the character table of S_5 and A_5. In exercise (I.2.4) you worked out the seven partitions of 5, which correspond to the seven equivalence classes denoted here by their typical members I, (12)(34), (123), (12345), (12), (1234), and (12)(345). (Note that the last three permutations in the list are odd and therefore do not appear in A_5.) Using the formula you derived in (I.2.4), we determine n_c to be, respectively, 1, 15, 20, 24, 10, 30, and 20, which, as a check, sum up to $N(G) = 5! = 120$. Thus, we know that there must be seven irreducible representations.

Of these we know the identity representation of course, but there is also the 1-dimensional signature representation, being +1 on the even permutations and −1 on the odd permutations. This gives us the first two columns of the character table for free.

To go farther, we again invoke the regular representation, with characters given by $(5, 1, 2, 0, 3, 1, 0)^T$, where, as in our discussion of A_4 in the text, we switch from the square boxes to vector notation. (It is understood that the order of the equivalence classes is as in the character table.) By now, you should be able to write down this column of characters by inspection: the characters are just the number of fixed points. For example, (12) is represented

by $\begin{pmatrix} 0 & 1 & 0 & 0 & 0 \\ 1 & 0 & 0 & 0 & 0 \\ 0 & 0 & 1 & 0 & 0 \\ 0 & 0 & 0 & 1 & 0 \\ 0 & 0 & 0 & 0 & 1 \end{pmatrix}$, with trace $= 3$.

Now use column orthonormality of the 5 with itself: $1 \cdot 5^2 + 15 \cdot 1^2 + 20 \cdot 2^2 + 24 \cdot 0^2 + 10 \cdot 3^2 + 30 \cdot 1^2 + 20 \cdot 0^2 = 2(120)$. Thus, the 5 contains two irreducible representations. But as always (and from exercise 2), we know that the vector with all 1's is invariant when acted upon by the matrices in the regular representation, and thus $5 \rightarrow 4 + 1$. (We could of course check by column orthogonality that the 1 is indeed contained in the 5 once: $1 \cdot 5 + 15 \cdot 1 + 20 \cdot 2 + 24 \cdot 0 + 10 \cdot 3 + 30 \cdot 1 + 20 \cdot 0 = 120$. Note that this can be obtained from the calculation above by erasing the squares.)

The characters of the 4 are given by subtracting 1 from the characters of the 5, and thus equal $(4, 0, 1, -1, 2, 0, -1)^T$. (Check column orthonormality.)

Now we have another trick up our sleeves. Given the irreducible representation 4, we could construct an irreducible representation $\bar{4}$ by simply reversing the signs of the matrices representing the odd permutations, leaving unchanged the matrices representing the even permutations. (This works, since the product of two odd permutations is even, of an odd with an even is odd, and of two even is even.) One way to say this is that the direct product of 4 with $\bar{1}$ is $\bar{4}$.

Thus far, we have gotten 1, $\bar{1}$, 4, and $\bar{4}$. Our next task is to determine the other three irreducible representations. Let us solve the constraint (1) that the sum of the dimensions squared of the irreducible representations is equal to $N(G) = 120$: namely, $1^2 + 1^2 + 4^2 + 4^2 + a^2 + b^2 + c^2 = 120$, that is, $a^2 + b^2 + c^2 = 86$. Interestingly, there are three solutions: $1^2 + 2^2 + 9^2 = 86$, $1^2 + 6^2 + 7^2 = 86$, and $5^2 + 5^2 + 6^2 = 86$. Here is a quick argument that the third solution is the right one, since by the preceding discussion, we might expect the irreducible representations 5 and $\bar{5}$ to appear in a pair. If so, then the 6 and $\bar{6}$ must be equivalent. Also, we might argue that the other two solutions are unlikely on the grounds that we can't think of some way of constructing another 1-dimensional irreducible representation. Indeed, its characters are severely constrained by column and row orthogonality and by the various cyclic subgroups.

To be more sure of ourselves, however, we could invoke the tray argument used in connection with A_4. (For exercise, figure out the dimension of the reducible representation and its characters before reading on.) There are $5 \cdot 4/2 = 10$ trays, namely, $\langle 12 \rangle$, $\langle 13 \rangle$, $\langle 14 \rangle$, $\langle 15 \rangle$, $\langle 23 \rangle$, $\langle 24 \rangle$, $\langle 25 \rangle$, $\langle 34 \rangle$, $\langle 35 \rangle$, and $\langle 45 \rangle$. Once again, the characters of this 10-dimensional reducible representation can be written down by counting fixed points mentally: $(10, 2, 1, 0, 4, 0, 1)^T$. For example, the permutation (12)(34) leaves $\langle 12 \rangle$ and $\langle 34 \rangle$ unchanged, giving $\chi = 2$, while (12) leaves $\langle 12 \rangle$, $\langle 34 \rangle$, $\langle 35 \rangle$, $\langle 45 \rangle$ unchanged. Column orthonormality gives $1 \cdot 10^2 + 15 \cdot 2^2 + 20 \cdot 1^2 + 24 \cdot 0^2 + 10 \cdot 4^2 + 30 \cdot 0^2 + 20 \cdot 1^2 = 3(120)$, telling us that the 10 contains three irreducible representations. A quick calculation $1 \cdot 10 + 15 \cdot 2 + 20 \cdot 1 + 24 \cdot 0 + 10 \cdot 4 + 30 \cdot 0 + 20 \cdot 1 = 120$ and $1 \cdot 10 \cdot 4 + 15 \cdot 2 \cdot 0 + 20 \cdot 1 \cdot 1 + 24 \cdot 0 \cdot (-1) + 10 \cdot 4 \cdot 2 + 30 \cdot 0 \cdot 0 + 20 \cdot 1 \cdot (-1) = 120$, telling us that 10 contains 1 and 4 each once. Therefore, $10 \rightarrow 1 + 4 + 5$.

Indeed, there is a 5-dimensional irreducible representation. Its characters can be determined by subtracting the characters of 1 and 4 from the characters of 10:

$$\begin{pmatrix} 10 \\ 2 \\ 1 \\ 0 \\ 4 \\ 0 \\ 1 \end{pmatrix} - \begin{pmatrix} 1 \\ 1 \\ 1 \\ 1 \\ 1 \\ 1 \\ 1 \end{pmatrix} - \begin{pmatrix} 4 \\ 0 \\ 1 \\ -1 \\ 2 \\ 0 \\ -1 \end{pmatrix} = \begin{pmatrix} 5 \\ 1 \\ -1 \\ 0 \\ 1 \\ -1 \\ 1 \end{pmatrix} \tag{22}$$

By our previous argument, there is also a $\bar{5}$.

Finally, the characters of the remaining 6-dimensional irreducible representation are readily determined by orthogonality. We thus obtain the table shown here:

S_5	n_c		1	$\bar{1}$	4	$\bar{4}$	5	$\bar{5}$	6
	1	I	1	1	4	4	5	5	6
Z_2	15	(12)(34)	1	1	0	0	1	1	-2
Z_3	20	(123)	1	1	1	1	-1	0	0
Z_5	24	(12345)	1	1	-1	-1	0	0	1
Z_2	10	(12)	1	-1	2	-2	1	-1	0
Z_4	30	(1234)	1	-1	0	0	-1	1	0
Z_6	20	(12)(345)	1	-1	-1	1	1	-1	0

(23)

(You could of course also check row orthogonality.) Note that $c = \bar{c}$ for every c in S_5, and thus the character table is real.

Appendix 2: A_5

Go ahead, try your hand at this game. Construct the character table for A_5 before reading on.

Going from S_5 to A_5, we lose three classes, typified by (12), (1234), and (12)(345), namely, those classes on which the irreducible representation $\bar{1}$ is represented by -1. From prior experience, we know to watch out for classes splitting up. Indeed, the class containing (12345) now splits into two; since (45) is no longer around, (12345) and (12354) are no longer equivalent. Some guys don't stay friends when a mutual friend disappears.

Here is an easy mistake to make. You would think that the class containing (123) also splits; (123) shouldn't be equivalent to (124) without (34). But you would be wrong. Behold:

$$(45321)(123)(12354) = (45)(53)(32)(21)(12)(23)(12)(23)(35)(54)$$
$$= (45)(53)(12)(23)(35)(54) = (12)(45)(53)(32)(35)(54)$$
$$= (12)(45)(25)(53)(35)(54) = (12)(45)(25)(54)$$
$$= (124) \tag{24}$$

(The fourth equality uses $(53)(32) = (532) = (253) = (25)(53)$.) There are of course many alternate routes to the same end result. Okay, (34) leaves town, but (123) and (124) are still both friends with (12354).

So, from S_5 to A_5, we lose three classes but have (only) one class splitting into two. The number of equivalence classes goes down from seven to five. Thus, while $N(G)$ halves from 120 to 60, the number of irreducible representations only drops from seven to five.

Now we hunt for these five irreducible representations. There is always the trivial but trusty 1-dimensional representation. As is familiar by now, the regular or defining representation decomposes as $1 + 4$. The dimensions of the other three irreducible representations are constrained by $a^2 + b^2 + c^2 = 60 - 1^2 - 4^2 = 43$, with the unique solution $a = b = 3$, $c = 5$.

The characters for the 4 and for the 5 can be read off from the character table for S_5 by chopping off the last three entries (since those three classes disappear), keeping in mind that one class has split into two. Thus, $(4, 0, 1, -1)^T$ and $(5, 1, -1, 0)^T$ become $(4, 0, 1, -1, -1)^T$ and $(5, 1, -1, 0, 0)^T$, respectively.

Here is an alternative way of getting at the characters of the 4 and of the 5. Suppose we have not yet constructed the character table of S_5, or that we are interested only in A_5. We could follow our usual procedure of starting with the 5-dimensional regular representation $(5, 1, 2, 0, 0)^T$ and subtract off the characters of the trivial 1-dimensional representation to obtain the characters of the 4. (Note that the 5-dimensional irreducible representation is not to be confused with the reducible regular representation.)

Again, we can use the "tray trick" to write down the characters of a $(10 = 5 \cdot 4/2)$-dimensional reducible representation $(10, 2, 1, 0, 0)^T$. By column orthogonality, we find that $10 \to 1 + 4 + 5$. Since we know the characters of the 1 and of the 4, we can obtain by subtraction the characters of the 5 to be $(5, 1, -1, 0, 0)^T$, in agreement with what we had above.

At this stage we have

A_5	n_c		1	3	3	4	5
	1	I	1	3	3	4	5
Z_2	15	(12)(34)	1	x	y	0	1
Z_3	20	(123)	1	z	w	1	-1
Z_5	12	(12345)	1	u	v	-1	0
Z_5	12	(12354)	1	r	s	-1	0

The characters of the two 3-dimensional representations are still to be determined. Using row orthogonality between the first row (the identity I row) and the second row, and row orthonormality of the second row, we obtain $x = y = -1$. Similarly, $z = w = 0$.

Row orthogonality between the first row (the identity I row) and the fourth row gives $1 + 3(u + v) - 4 + 0 = 0 \implies u + v = 1$, while row orthonormality of the fourth row gives $12(1 + u^2 + v^2 + 1 + 0) = 60 \implies u^2 + v^2 = 3$.

Now a small surprise awaits us. We have to solve a quadratic equation $u^2 + (1 - u)^2 = 3$. The solution is $u = (1 + \sqrt{5})/2 \equiv \zeta$, and hence $v = (1 - \sqrt{5})/2 = 1 - \zeta$. With u and v fixed, column orthonormality then fixes r and s.

We finally end up with

A_5	n_c		1	3	3	4	5
	1	I	1	3	3	4	5
Z_2	15	(12)(34)	1	-1	-1	0	1
Z_3	20	(123)	1	0	0	1	-1
Z_5	12	(12345)	1	ζ	$1 - \zeta$	-1	0
Z_5	12	(12354)	1	$1 - \zeta$	ζ	-1	0

(25)

The rabid empiricists among physicists might have been tempted to conclude, from the character tables we have seen until now, that characters have to be either an integer or a root of unity.[7] The appearance of ζ here is like the sighting of a white sheep in our joke. The learned among the readers will recognize[8] ζ as the golden ratio[9] or the divine section (sectio divina) that fascinated the ancients and played an important role in classical architecture and art.[10]

Appendix 3: More about the tetrahedron and A_4

I can't resist saying a few more words about the tetrahedron. If Plato likes it, then so do we.[11] We know from chapter I.1 that the three elements (12)(34), (13)(24), and (14)(23) (together with I, of course) form an invariant subgroup of A_4, namely, Klein's $V = Z_2 \otimes Z_2$. In this chapter, we learned about their representation by the three matrices r_1, r_2, and r_3 given in (16) and about their geometrical significance as 180° rotations about the three median lines.

According to the discussion about invariant subgroups in chapter I.2, we can form the quotient group $Q = A_4/V$. It has $12/4 = 3$ elements. What can they be?

The three left cosets are V, cV, and aV. Although there are $12 - 4 = 8$ elements not in V, they fall into two equivalence classes. For example, $r_1 c r_1 \{I, r_1, r_2, r_3\} = c\{I, r_1, r_2, r_3\}$. Geometrically, if we think of c as a rotation through $2\pi/3$ around the line joining vertex 1 to the center of the triangle formed by vertices 2, 3, and 4, then its fellow members in its equivalence class (namely, $r_1 c r_1$, $r_2 c r_2$, and $r_3 c r_3$) are also rotations through $2\pi/3$ but giving the other vertices 2, 3, and 4 an equal opportunity for a turn.

The quotient group Q, as already noted, has three elements, and so by Lagrange's theorem, can only be Z_3. We can also check this explicitly: $cV \cdot cV = aV$, $cV \cdot aV = V$, and so on.

Group theory provides a bridge between geometry and algebra.

Appendix 4: A_5 is simple: A simple proof

Of the various proofs that A_5 is simple, I rather like this simple (in the everyday sense)"physicist-style" demonstration.

We begin with a self-proving lemma. Let $H \subset G$. If an element is in H, then all the elements in its equivalence class are also in H. Proof: If $h \in H$, then $g^{-1}hg \in H$ by definition. (Have you ever seen something easier to prove than this?)

It follows that H is the union of a bunch of equivalence classes of G. In particular, the bunch contains the "class of one" the identity I belongs to.

Let us now prove that A_5 is simple.[12] Suppose that A_5 is not simple and contains an invariant subgroup H. From the character table in (25), we know that A_5 has five equivalence classes, with $n_c = 1$, 15, 20, 12, and 12, respectively. According to the lemma, $N(H)$, the number of elements in H, can take on only a few possible values, such as $1 + 15 = 16$, $1 + 20 = 21$, $1 + 15 + 12 = 28$, and so on. The important point is that there are a finite number of possibilities that we can list.

But Lagrange's theorem requires that $60 = 5!$ (the number of elements in A_5) divided by $N(H)$ is an integer. Let's try the different possibilities for $N(H)$. OK, $60/16$ is not an integer. Next, $60/21$, not an integer either. Next, and so on. We find the only possibility is $1 + 15 + 20 + 12 + 12 = 60$, which means that the only (nontrivial) invariant subgroup is A_5 itself.

It is instructive to see how A_4 evades this argument: there are four equivalence classes with $n_c = 1$, 3, 4, and 4, and $4!/(1 + 3)$ is an integer. Sure enough, the three elements in the equivalence class $(12)(34)$ together with the identity form $V = Z_2 \otimes Z_2$, the invariant subgroup of A_4.

Appendix 5: The cube: Geometry and algebra

The neat linkage between geometry and algebra, exemplified by the tetrahedron and A_4, suggests that we play the same game for the other regular polyhedra. As you may know, one of the greatest discoveries in the history of mathematics was the realization by the ancient Greeks that there are only five[13] regular polyhedra: the tetrahedron, the cube, the octahedron, the icosahedron, and the dodecahedron. In this brief appendix, I sketch how things go with the cube, offering only physicist-style suggestive "proofs." I also touch on the octahedron in passing.

Align the cube with the x-, y-, and z-axes, and center it at the origin, so that the eight vertices have coordinates (± 1, ± 1, ± 1). Consider G the invariance group of the cube under rotations. How many elements does G have? Let's count, leaving the identity I aside (see figure 6).

Call the center of the top and bottom faces $c_t = (0, 0, 1)$ and $c_b = (0, 0, -1)$, respectively. Picture the line going from c_t to c_b through the center of the cube. Rotations around this line through angles $\pi/2$, π, and $3\pi/2$ leave the cube invariant. Since we have three such lines (for example, going from $(1, 0, 0)$ to $(-1, 0, 0)$) and for each line three rotations, we count $3 \cdot 3 = 9$ elements thus far. Needless to say, you should draw a figure as you read along.

Next, picture the line going from the vertex $(1, 1, 1)$ to the vertex farthest away from it, namely, $(-1, -1, -1)$, through the center of the cube. This is known as a principal diagonal of the cube (of length $2\sqrt{3}$ just to make sure that you are still following). Rotations through $2\pi/3$ and $4\pi/3$ leave the cube invariant. Since there are four principal diagonals, we count $4 \cdot 2 = 8$ elements.

Finally, consider the midpoint $m = (1, 0, 1)$ of one of the four edges on the top face of the cube. Consider the line going from it to the midpoint of one of the four edges on the bottom face, the midpoint farthest away from m, namely $m' = (-1, 0, -1)$, through the center of the cube. Rotation through π leaves the cube invariant. Since there are $4 + 2 = 6$ such edges (don't forget the ones on the side!), this accounts for $6 \cdot 1 = 6$ elements.

Hence G has in all $9 + 8 + 6 + 1 = 24$ elements. (We did not forget the identity!) Cayley's theorem assures us that G is a subgroup of S_{24}, but a very tiny subgroup indeed. A naive first thought might be to generalize the discussion for the tetrahedron and consider A_8, since the cube has 8 vertices. But $8!/2$ is still much larger than 24; clearly, many permutations in A_8 cannot be realized on the eight vertices of the cube. The correct idea is to realize that each of the rotations we described permute the four principal diagonals of the cube. Thus, the physicist would guess that $G = S_4$, which has precisely $4! = 24$ elements.

The octahedron can be constructed readily from the cube as follows. See figure 7. We have named the center of the top and bottom faces of the cube c_t and c_b, respectively. Now name the center of the four side faces of the

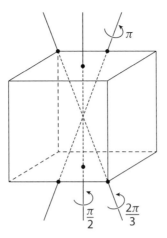

Figure 6

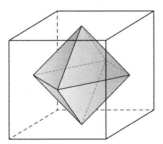

Figure 7

cube c_1, c_2, c_3, and c_4 in order. (There are six faces all together, of course.) Connecting c_1 to c_2, c_2 to c_3, c_3 to c_4, and c_4 to c_1 by straight lines, we obtain a square. Then, connecting c_1, c_2, c_3, and c_4 to c_t by straight lines, we obtain a regular pyramid with a square base. Next, connecting c_1, c_2, c_3, and c_4 to c_b by straight lines gives us another regular pyramid, inverted relative to the pyramid we already have. The two pyramids glued together form an octahedron.

The octahedron is inscribed in the cube. Hence, any rotation that leaves the cube invariant also leaves the octahedron invariant. The invariance group of the octahedron is also S_4.

Note that the octahedron has $V = 6$ vertices, $E = 12$ edges, and $F = 8$ faces, while the cube has $V = 8$ vertices, $E = 12$ edges, and $F = 6$ faces. The number of vertices and the number of faces are interchanged, so that the Euler characteristic[14] $\chi \equiv V - E + F$ equals[15] 2 for both the cube and the octahedron.

The icosahedron with 20 triangular faces and the dodecahedron with 12 pentagonal faces are similarly related and have the same invariance group.[16]

A quick way of counting the number of elements in the rotational invariance group G of the cube is as follows. Imagine putting a cube down on a table. We can choose one of six faces to be the bottom face in contact with the table. For each such choice, we can rotate around the vertical axis connecting the center of the top face and the center of the bottom face through 0, $\pi/2$, π, and $3\pi/2$. Thus, G contains $6 \cdot 4 = 24$ elements, in agreement with what we obtained above.

We can apply the same argument to the icosahedron. Choose one of 20 triangles to be the bottom face, and for each choice, we can rotate through 0, $2\pi/3$, and $4\pi/3$, and thus its rotational invariance group G has $20 \cdot 3 = 60 = 5!/2$ elements. You might guess that $G = A_5$.

The same counting works for the tetrahedron, with its four triangular faces, leading to $4 \cdot 3 = 12 = 4!/2$ elements in the rotational invariance group G, which we know full well is A_4.

Exercises

1   How does the 4-dimensional regular representation of A_4 reduce?

2   Show that the defining representation of S_n, when it breaks up, always contains the identity representation 1.

3   Determine the character table of S_4, and check the various orthogonality theorems.

4   Complete the table in (19).

5   Work out the character table (21) for $\mathcal{Q}$.

6   For $n \geq 5$, any two 3-cycles in A_n are equivalent. Why is the restriction $n \geq 5$ needed?

7   Work out the character table for $D_5 = C_{5v}$, the invariance group of the pentagon. You will find, perhaps surprisingly (perhaps not), that the character table for D_5 is smaller than the character table for D_4.

Notes

1. Not to be confused with its regular representation, which is $n!$-dimensional.
2. Perhaps like some people you know?
3. We discuss this point further when we come to the 2-dimensional rotation group. The issue will be $SO(2)$ versus $U(1)$.
4. For reasons we don't care to know.
5. For reasons quite different from Plato's, some particle physicists are quite fond of A_4. E. Ma first suggested that the 3-dimensional irreducible representation of A_4 may have something to do with the mysterious existence of three families in Nature. See chapter IX.4.
6. Perhaps it could be considered fool's delight, but there is a certain peculiar pleasure to watching the laws of mathematics actually work out consistently.
7. Incidentally, the correct theorem states that the character of a finite group has to be an algebraic integer, defined to be a root of some monic polynomial (a polynomial whose leading coefficient is 1) with coefficients equal to integers. For example, ζ is the solution of $x^2 - x - 1 = 0$.
8. The two numbers here, ζ and $1 - \zeta$, also appear in the solution to the Fibonacci series. Some trivia: According to Nahin (p. 97) the name Fibonacci (= son of Bonacci) was not used until long after his death. The mathematician Leonardo Pisano ("Lenny from Pisa") went by his nickname Bigollo during his lifetime.
9. The golden ratio will appear again in a discussion of crystals in interlude II.i.1. That the golden ratio can be constructed using ruler and compass (an interesting result of classical Greek mathematics) implies that the regular pentagon (unlike most regular n-sided polygons) can be constructed using ruler and compass. See Hahn, Mathematical Excursions to the World's Great Buildings. Here is the relevant mathematics in modern language, given without further explanation: Let $e^{i\theta} = x + iy$. We require $(e^{i\theta})^5 = 1$. The imaginary part of this equation gives $5x^4 - 10x^2y^2 + y^4 = 0$ which upon the substitution $y^2 = 1 - x^2$ becomes $16x^4 - 12x^2 + 1 = 0$ which has a solution $x = (\sqrt{5} - 1)/4$.
10. For a discussion saying that the role of the golden ratio in good design is just a myth, see http://www.fastcodesign.com/3044877/the-golden-ratio-designs-biggest-myth.
11. I am sort of reasoning like a humanities professor here. (Come on, lighten up, that was a joke.)
12. That A_5 is simple is the key to Galois theory, which we do not go into here.
13. Listed by Plato (427–347 BC) in Timaeus.
14. For an elementary discussion of the Euler characteristic and Descartes angular deficit, see G Nut, pp. 725–727.
15. An easy proof, suitable for elementary school children, involves gluing tiny tetrahedrons together to form any object topologically like a blob and showing that $\Delta\chi = \Delta V - \Delta E + \Delta F = 1 - 3 + 2 = 0$.
16. Using the fact that the Euler characteristic $\chi = V - E + F = 2$ for objects with the same topology as the sphere, we can give an easy derivation of the regular polyhedra by making the simplifying assumption that

each polyhedron is made of N regular polygons with s sides each. For example, for the cube, $N = 6$ and $s = 4$; that is, the cube is made of six squares. (By this assumption, we exclude the possibility that the polyhedron is made of different types of polygons.) Let ζ denote the number of polygons that come together at a vertex (for example, the cube has $\zeta = 3$). Then $V = sN/\zeta$, $E = sN/2$, and $F = N$. Plugging this into Euler's formula $V - E + F = 2$, we have

$$N = \frac{4\zeta}{2s - (s - 2)\zeta}$$

For each s, the possible value of ζ is bounded above by positivity, and of course, to be sensible, we must have $\zeta \geq 3$. Thus, we simply list all possible cases.

For $s = 3$, $N = 4\zeta/(6 - \zeta)$. We have for $\zeta = 3$, $N = 4$, the tetrahedron; for $\zeta = 4$, $N = 8$, the octahedron; and for $\zeta = 5$, $N = 20$, the icosahedron.

For $s = 4$, $N = 2\zeta/(4 - \zeta)$. The only possibility is $\zeta = 3$, $N = 6$, the cube.

For $s = 5$, $N = 4\zeta/(10 - 3\zeta)$. The only possibility is $\zeta = 3$, $N = 12$, the dodecahedron.

Incidentally, a neat way of drawing the icosahedron starting with a cube is given in A. Borovik, Mathematics under the Microscope, p. 42. The method allegedly goes back to Piero della Francesca around 1480.

II.4 | Real, Pseudoreal, Complex Representations, and the Number of Square Roots

Complex or not

If someone gives us a number z, we can tell by a glance whether or not it is complex. (Formally, we check if z is equal to its complex conjugate z^*.) But if someone gives us a bunch of matrices $D(g)$ furnishing an irreducible representation* r of a group G, it may not be immediately evident whether or not the irreducible representation is complex.

Even if some or all the entries of $D(g)$ are complex, it may not mean that the representation r is complex. A bunch of apparently complex representation matrices could be equivalent to their complex conjugates. In other words, even if $D(g)^* \neq D(g)$, there might exist an S such that

$$D(g)^* = SD(g)S^{-1} \tag{1}$$

for all $g \in G$. The representation matrices $D(g)$ may not look real, but they and their complex conjugates might be "secretly" related by a similarity transformation and hence not really complex. Looks are not enough.

Conjugate representations

Before going on, let us make sure that $D(g)^*$ indeed forms a representation, which we denote by r^* and is known as the conjugate of r. That's easy: simply complex conjugate $D(g_1)D(g_2) = D(g_1g_2)$ to obtain $D(g_1)^*D(g_2)^* = D(g_1g_2)^*$.

Another almost self-evident comment: the characters of the representation r^* are given by $\chi^{(r^*)}(c) = \text{tr } D(g)^* = (\text{tr } D(g))^* = \chi^{(r)}(c)^*$, the complex conjugate of the characters of r. We denote the class g belongs to by c.

---

* To lessen clutter, I suppress the superscript (r) on $D(g)$; until further notice, we will be focusing on a specific representation.

The issue facing us is whether the two representations r and r^* are equivalent or not.

By now, you know to trace over (1) to get rid of the unknown S. Thus, if r and r^* are equivalent, then $\chi^{(r^*)}(c) = \chi^{(r)}(c)$; in other words, the characters are real.

It follows that if a character $\chi^{(r)}(c)$ is complex, then r^* and r are not equivalent. We say that the representations r and r^* are complex. A rather trivial example: the two 1-dimensional representations of A_4, $1'$ and $1''$, are complex and are conjugate to each other.

In contrast, if the characters $\chi^{(r)}(c)$ are real, then $\chi^{(r^*)}(c) = \chi^{(r)}(c)^* = \chi^{(r)}(c)$; that is, r and r^* have the same characters. This does not, however, on the face of it, imply (1), but is merely strongly suggestive, since the equality of characters holds for all equivalence classes.

At this point, we have a binary classification of irreducible representations into complex and noncomplex. We now show that the noncomplex representations can be subdivided further.

Aside from the intrinsic interest of knowing whether a representation is complex or not, this entire discussion is of great interest to high energy theorists. You have surely heard of antimatter and antiparticles. In quantum field theory, if the particle transforms under some symmetry group according to the representation r, then its antiparticle[1] transforms according to the representation r^*.

A restriction on the similarity transformation S: Real versus pseudoreal

Suppose that a given irreducible representation is not complex. Transposing (1), we have $D(g)^\dagger = D(g)^{*T} = (S^{-1})^T D(g)^T S^T$. Noting that the representation matrix is unitary so that $D(g)^\dagger = D(g^{-1})$, we obtain

$$D(g^{-1}) = (S^{-1})^T D(g)^T S^T \tag{2}$$

This tells us that $D(g^{-1})$ is related to $D(g)$.

Our strategy, clearly, is to use this relationship twice to relate $D(g)$ to itself and hence obtain a condition on S. Substituting g for g^{-1} in this equation, we have $D(g) = (S^{-1})^T D(g^{-1})^T S^T = (S^{-1})^T (S D(g) S^{-1}) S^T = (S^{-1} S^T)^{-1} D(g) (S^{-1} S^T)$. (Here we use (2) and its transpose, and also the elementary identity $(M^{-1})^T = (M^T)^{-1}$.) In other words, $S^{-1} S^T$ commutes with $D(g)$ for all g. Invoking Schur's lemma once again, we conclude that $S^{-1} S^T = \eta I$, that is, $S^T = \eta S$, with η some constant. This implies $S = (S^T)^T = \eta S^T = \eta^2 S$, and hence $\eta = \pm 1$. We conclude that $S^T = \pm S$; in other words, S is either symmetric or antisymmetric.

If S is symmetric, we say that the representation r is real.

If S is antisymmetric, we say that the representation r is pseudoreal.

From the result of exercise 8 in the review of linear algebra, we conclude that a representation r can be pseudoreal only if its dimension is even.

To be sure, many of the representations we have encountered are manifestly real, in the sense that the matrices $D(g)$ have only real entries. In that case, S is just the identity matrix.

We now prove that up to an overall constant not fixed by (1), S is also unitary. From (1), we have $SD(g) = D(g)^*S = (D(g)^{-1})^T S$, and so $S = D(g)^T SD(g)$. Hermitean conjugating, we obtain $S^\dagger = D(g)^\dagger S^\dagger D(g)^*$. Now multiply $S^\dagger$ and S to check unitarity: $S^\dagger S = D(g)^\dagger S^\dagger D(g)^* D(g)^T SD(g) = D(g)^\dagger S^\dagger S D(g)$, and hence $DS^\dagger S = S^\dagger SD$. Thus, $S^\dagger S$ commutes with $D(g)$ for all g and hence must be proportional to the identity. You might have noticed that nothing thus far fixes the scale of S. In other words, we can multiply S by a constant to make $S^\dagger S = I$.

Real representation is really real

Dr. Feeling wanders by, and says, "Don't you feel that, if a representation is real, then it really ought to be real? That is, in some basis, the matrices $D(g)$ should only have real entries, none of this 'it is related to its complex conjugate by a similarity transformation' stuff. Indeed, that's true in all the examples we have seen!"

To prove this, we need a lemma. Given a unitary symmetric matrix U (that is, a unitary matrix that also happens to be symmetric), there exists a unitary symmetric matrix W such that $W^2 = U$. More loosely speaking, the square root of a unitary symmetric matrix is also unitary symmetric. With your permission, I will defer the proof of the lemma until chapter IV.4 as an exercise.

Given the lemma, and given that S is unitary symmetric, let us write $S = W^2$ with W unitary symmetric, which implies that $W^{-1} = W^\dagger = W^*$. Then (1) gives

$$W^2 D(g) W^{-2} = D(g)^*$$
$$\implies WD(g)W^{-1} = W^{-1}D(g)^*W = W^*D(g)^*(W^{-1})^* = (WD(g)W^{-1})^* \tag{3}$$

Thus, the representation matrices $D'(g) \equiv WD(g)W^{-1}$ are real.

An invariant bilinear for a noncomplex representation

Denote by x the set of d_r objects that transform under the d_r-dimensional irreducible representation r; thus, $x \to D(g)x$. Similarly, denote by y some other d_r objects that transform under r; thus, $y \to D(g)y$. (Think of x and y as two column vectors.) As before, suppress a superscript r on $D(g)$ to lessen clutter.

Let us now prove an important[2] theorem.

If the irreducible representation r is real or pseudoreal, then $y^T Sx$ is an invariant bilinear.

To see this, take the inverse of (1): $D^T(g) = SD^\dagger(g)S^{-1}$. Then

$$y^T Sx \to y^T D(g)^T SD(g)x = y^T SD^\dagger(g)S^{-1}SD(g)x = y^T Sx \tag{4}$$

does not change.

Conversely, if $y^T Sx$ is invariant, this implies that $D(g)^T SD(g) = S$, which implies $SD(g) = D(g)^*S$, and hence the equivalence of D and D^*.

To summarize, an invariant bilinear exists if and only if the irreducible representation is real or pseudoreal.

The reality checker

Given an irreducible representation, how can we tell if it is real, pseudoreal, or complex?

We want to build a reality checker, so that, given a representation, we can give it a reality check.

To start the construction, we play a by-now familiar game. Construct

$$S \equiv \sum_{g \in G} D(g)^T X D(g)$$

for an arbitrary X. Then $D(g)^T S D(g) = \sum_{g' \in G} D(g)^T D(g')^T X D(g') D(g) = S$. Since this holds for all g, we find that $y^T S x$ is an invariant bilinear: $y^T S x \to y^T D(g)^T S D(g) x = y^T S x$. But we just showed that the existence of this bilinear would imply that D and D^* are equivalent.

Thus, if the irreducible representation is complex, then this $S = \sum_{g \in G} D(g)^T X D(g)$ must vanish to avoid a contradiction.

As before, we now write this out explicitly. In the context of the present discussion, let us suspend the upstairs-downstairs convention for indices, and treat all indices as superscripts. (The reason for this is that transpose, unlike hermitean conjugate, is not a "natural" operation on unitary matrices.)

Since $S = 0$ holds for any X, let us choose X to have only one nonvanishing entry in the ith row and lth column, equal to 1. Then $(D(g)^T X D(g))^{jk} = (D(g)^T)^{ji}(D(g))^{lk} = D(g)^{ij} D(g)^{lk}$. Thus, the jk-entry of $S = 0$ gives us

$$\sum_{g \in G} D(g)^{ij} D(g)^{lk} = 0 \tag{5}$$

Remarkably, this holds for any i, j, k, and l. Set $j = l$, sum, and use the fact that $D(g)$ is a representation (of course!), so that $D(g)^{ij} D(g)^{jk} = D(g^2)^{ik}$. We thus find $\sum_{g \in G} D(g^2) = 0$, which when traced over gives $\sum_{g \in G} \chi(g^2) = 0$, if the representation is complex.

What if the representation is not complex?

Then $S^T = \eta S$ with $\eta = \pm 1$, as we discovered earlier. Transpose $S = \sum_{g \in G} D(g)^T X D(g)$ to obtain $S^T = \sum_{g \in G} D(g)^T X^T D(g) = \eta \sum_{g \in G} D(g)^T X D(g)$. Once again, set X to be nonzero only in the ith row and lth column. Then the jkth-entry of the second equality becomes $\sum_{g \in G} D(g)^{lj} D(g)^{ik} = \eta \sum_{g \in G} D(g)^{ij} D(g)^{lk}$. Setting $i = j, k = l$, and summing, we obtain

$$\sum_{g \in G} \chi(g^2) = \eta \sum_{g \in G} \chi(g) \chi(g) = \eta N(G) \tag{6}$$

where we used character orthogonality in the last step (recall that χ is not complex by assumption).

We have thus built our trusty reality checker: for an irreducible representation r (here we restore the superscript (r)),

$$\sum_{g \in G} \chi^{(r)}(g^2) = \eta^{(r)} N(G), \quad \text{with } \eta^{(r)} = \begin{cases} 1 & \text{if real,} \\ -1 & \text{if pseudoreal,} \\ 0 & \text{if complex} \end{cases} \tag{7}$$

Thus, to give an irreducible representation r a reality check, use the character table to evaluate the peculiar looking sum $\sum_{g \in G} \chi^{(r)}(g^2)$. It returns $+N(G)$ if the representation is real, $-N(G)$ if pseudoreal, and 0 if complex.*

At this point, we shout in unison "Character is a function of class!" The sum over group elements in $\sum_{g \in G} \chi(g^2)$ reduces to a sum over equivalence classes. Here a tiny lemma is needed. If two elements g_1 and g_2 are equivalent to each other, then their squares g_1^2 and g_2^2 are also equivalent to each other. Exercise!

Checking the reality checker

Let's try out our brand new reality checker. Notice first that for the trivial representation, which is as real as it can get, the sum $\sum_{g \in G} \chi(g^2)$ trivially gives $N(G)$, and thus, $\eta = +1$. The checker tells us that the trivial representation is real. Thank you very much.

Take as our guinea pig the group S_3, whose character table was worked out in chapter II.3, which you should look up now. For each of the three equivalence classes, the typical member squares as follows: $I^2 = I$ (of course), $(123)^2 = (132)$, and $(12)^2 = I$. Thus, examining the character of the $\bar{1}$, we evaluate the sum to be $1 \cdot 1 + 2 \cdot 1 + 3 \cdot 1 = 6$. The $\bar{1}$ is indeed real. (We could also reach this conclusion in another way. The 1-by-1 representation matrices for $\bar{1}$ are manifestly real, and we know that the $\bar{1}$ cannot be pseudoreal, since its dimension is odd.)

On the 2 in S_3, the sum gives $1 \cdot 2 + 2 \cdot (-1) + 3 \cdot 2 = 6$. Thus, we learn that the 2 is real without ever having to write it out explicitly.

Confusio interjects excitedly, "The reality checker does not work! Remember that I guessed the representation matrices of 2 incorrectly? So I made a point of memorizing the correct matrices, as given in chapter II.3. They sure look complex to me!"

We did in fact work out the representation matrices of 2 as follows: $I \approx \begin{pmatrix} 1 & 0 \\ 0 & 1 \end{pmatrix}$, $(123) \approx \begin{pmatrix} \omega & 0 \\ 0 & \omega^* \end{pmatrix}$, $(132) \approx \begin{pmatrix} \omega^* & 0 \\ 0 & \omega \end{pmatrix}$, $(12) \approx \begin{pmatrix} 0 & 1 \\ 1 & 0 \end{pmatrix}$, $(23) \approx \begin{pmatrix} 0 & \omega^* \\ \omega & 0 \end{pmatrix}$, and $(31) \approx \begin{pmatrix} 0 & \omega \\ \omega^* & 0 \end{pmatrix}$. Confusio is right that some of them are indeed complex, but he forgot that we can always make a similarity transformation. Consider the unitary matrix $V = \frac{1}{\sqrt{2}} \begin{pmatrix} 1 & i \\ i & 1 \end{pmatrix}$. Denote the matrices

---

* And if it does not return one of these three possibilities, you better check your arithmetic. So the reality checker is almost foolproof.

listed here by $D^{(2)}(g)$. Verify that $V^\dagger D^{(2)}(g) V$ are real matrices. I did not point this out in chapter II.3 in order to make a pedagogical point here.

Our next example is the $1'$ of A_4, which is manifestly complex. Recall that A_4 has four equivalence classes, whose typical member squares as follows: $((12)(34))^2 = I$, $(123)^2 = (132)$, and $(132)^2 = (123)$. Thus, the sum in (7) returns $1 \cdot 1 + 3 \cdot 1 + 4 \cdot (\omega + \omega^*) = 4(1 + \omega + \omega^*) = 0$.

The number of square roots

In school we learned that a number has not one, but two square roots.[3] Similarly, given an element f of G, let σ_f be the number of square roots of f, in other words, the number of different solutions to the equation $g^2 = f$.

Recall that in chapter I.2, we learned if a group has an even number of elements, then it has at least one element not the identity that squares to the identity. Now that we are "grown up," we want to know more; we want to know how many, that is, what σ_I is.

The key observation is that g^2 appears in (7). Hence, we can write our trusty reality checker as

$$\sum_f \sigma_f \chi^{(r)}(f) = \eta^{(r)} N(G) \tag{8}$$

Note that σ_f is a property of the group and does not depend on the representation r. Now that we have restored r, we use row orthogonality. Multiplying (8) by $\chi^{(r)*}(f')$ and summing, we obtain

$$\sum_r \left(\sum_f \sigma_f \chi^{(r)}(f) \right) \chi^{(r)*}(f') = \sum_r \eta^{(r)} \chi^{(r)*}(f') N(G)$$

$$= \sum_f \sigma_f \left(\sum_r \chi^{(r)}(f) \chi^{(r)*}(f') \right)$$

$$= \sum_f \sigma_f \frac{N(G)}{n_c} \delta_{cc'} = \sigma_{f'} N(G) \tag{9}$$

The first equality is due to (8). The third equality comes from row orthogonality, with c and c' denoting the equivalence classes f and f' belong to, respectively. We committed a minor abuse of notation: by $\delta_{cc'}$ we mean that there is a contribution only if f and f' are in the same class, and hence the fourth equality follows. The factor of n_c cancels out. We thus obtain the interesting result

$$\sigma_f = \sum_r \eta^{(r)} \chi^{(r)}(f) \tag{10}$$

We have dropped the complex conjugation symbol $*$ on $\chi^{(r)}$: since $\eta^{(r)} = 0$ for complex representations, the complex characters do not contribute to the sum anyway.

Remarkably, given the character table, if we know whether each irreducible representation is real, pseudoreal, or complex (that is, if we know $\eta^{(r)}$ for all r), then we can determine the number of square roots of any group element. In particular, the number of square roots of the identity* I is given by

$$\sigma_I = \sum_r \eta^{(r)} d_r = \sum_{r=\text{real}} d_r - \sum_{r=\text{pseudoreal}} d_r \tag{11}$$

In other words, the sum of the dimension of the real irreducible representations minus the sum of the dimension of the pseudoreal irreducible representations.

In particular, for groups without pseudoreal irreducible representations, the number of square roots of the identity is equal to the sum of the dimension of the irreducible representations $\sum_r d_r$.

Let us try these results out on $A_3 = Z_3$. For the rest of this discussion, it will be convenient to have various character tables given in chapter II.3 in front of you, unless you have them memorized.[4] Recall that it has three irreducible representations, the 1, $1'$, and $1''$. We have $\sigma_{(123)} = \sum_r \eta^{(r)} \chi^{(r)}(123) = 1$, because the $1'$ and $1''$ are both complex, leaving only the trivial identity representation to contribute to the sum. Indeed, the query $g^2 = (123)$ has only one response, namely, $g = (132)$.

Similarly, $\sigma_I = 1 \cdot 1 + 0 \cdot 1 + 0 \cdot 1 = 1$, indicating that in Z_3, I has only one square root, namely, I itself.

What about Z_2? It has two real irreducible representations. Thus, $\sigma_I = 2$, and the identity has two square roots, exactly what we learned long ago in school! How about Z_4? It has two real and two complex irreducible representations. Again, the identity has two square roots.

Next, let's try A_4. It has no pseudoreal irreducible representation, two real irreducible representations (namely, the 1 and the 3), and two complex irreducible representations (which do not enter here). Thus, $\sigma_I = 1 + 3 = 4$. Yes indeed, in A_4 the identity I has four square roots $\{I, (12)(34), (23)(41), (13)(24)\}$.

All hail the powers of mathematics!

Let us next count the square roots of $(12)(34)$ in A_4. It should be 0. According to (10), $\sigma_f =$ sum of the characters of the real irreducible representations $-$ sum of the characters of the pseudoreal irreducible representations, but there is no pseudoreal irreducible representation. Thus, $\sigma_{(12)(34)} = 1 + (-1) = 0$, check. I leave it to you to work out how many square roots (123) has.

Next, how many square roots does the identity of S_4 have? Plugging in, we find that it has $\sigma_I = 1 + 1 + 2 + 3 + 3 = 10$ square roots. It is instructive to count them all. I list them by naming a representative member of each equivalence class, followed by the number in each class: $\{I \to 1; (12)(34) \to 3; (12) \to 6\}$, adding up to $1 + 3 + 6 = 10$. These 10 elements do indeed all square to the identity.

---

* The jargon guy tells us that the square roots of the identity are called involutions.

Sum of the representation matrices of squares

Now that we have (7), which gives the sum of the characters of squares, we can be more ambitious and ask what we can say about the sum of the representation matrices of squares.

Given an irreducible representation, define $A \equiv \sum_g D(g^2)$. We play a by-now more-than-familiar game. Consider

$$D^{-1}(g')AD(g') = D^{-1}(g')\left(\sum_g D(g^2)\right)D(g') = \sum_g D(g'^{-1}gg'g'^{-1}gg') = A$$

So by Schur's lemma, $A = cI$, with the constant c to be determined presently by tracing this equation (also, I restore the superscript (r)):

$$\sum_g \text{tr } D^{(r)}(g^2) = cd_r = \sum_g \chi^{(r)}(g^2) = \sum_f \sigma_f \chi^{(r)}(f)$$

$$= \sum_s \eta^{(s)} \sum_f \chi^{(s)*}(f)\chi^{(r)}(f) = \eta^{(r)}N(G) \tag{12}$$

We used (10) in the fourth equality and column orthogonality in the fifth equality. We have thus proved that

$$\sum_g D^{(r)}(g^2) = N(G)(\eta^{(r)}/d_r)I \tag{13}$$

How many ways can a group element be written as a product of two squares?

By now, perhaps you can see how we can have more fun and games along the same lines. Let f be some arbitrary element (with a new discussion comes a new assignment of letters; there are only so many of them) of the group G. Multiply (13) by $D^{(r)}(f^2)$ to obtain $\sum_g D^{(r)}(f^2g^2) = N(G)(\eta^{(r)}/d_r)D^{(r)}(f^2)$. Tracing gives

$$\sum_g \chi^{(r)}(f^2g^2) = N(G)(\eta^{(r)}/d_r)\chi^{(r)}(f^2)$$

Now sum over f:

$$\sum_f \sum_g \chi^{(r)}(f^2g^2) = N(G)(\eta^{(r)}/d_r)\sum_f \chi^{(r)}(f^2) = (N(G)\eta^{(r)})^2/d_r \tag{14}$$

I used (7) to obtain the second equality. To proceed further, use the same "trick" as before. Denote by τ_h the number of solutions of the equation $f^2g^2 = h^2$ for a given element h of G. Then we can write (14) as $\sum_h \tau_h \chi^{(r)}(h) = (N(G)\eta^{(r)})^2/d_r$. Multiplying this by $\chi^{(r)*}(h')$ with h' some arbitrary element of G and summing over r, we obtain

$$\sum_r \left(\sum_h \tau_h \chi^{(r)}(h) \right) \chi^{(r)*}(h') = N(G)^2 \sum_r ((\eta^{(r)})^2/d_r) \chi^{(r)*}(h')$$

$$= \sum_h \tau_h \left(\sum_r \chi^{(r)}(h) \chi^{(r)*}(h') \right)$$

$$= \sum_h \tau_h \frac{N(G)}{n_c} \delta^{cc'} = \tau_{h'} N(G) \tag{15}$$

In the third equality, we used row orthogonality. As is evident (and just as in the earlier discussion), c and c' denote the classes that h and h' belong to respectively, with $\delta^{cc'}$ picking out only the hs that are in the same class as h' (and there are n_c of these). Thus, we obtain

$$\tau_h = N(G) \sum_r (\eta^{(r)})^2 \chi^{(r)}(h)/d_r \tag{16}$$

As before, we are entitled to drop the complex conjugation symbol.

In particular, let us ask, In how many distinct ways can we write the identity as a product of two squares $I = f^2 g^2$? The answer is

$$\tau_I = N(G) \sum_r (\eta^{(r)})^2 \tag{17}$$

Since $(\eta^{(r)})^2 = 1$ for real and pseudoreal irreducible representations and vanishes for complex irreducible representations, the sum here is simply equal to the total number of real and pseudoreal irreducible representations. Remarkably, this number times the order of the group determines the number of solutions to the equation $f^2 g^2 = I$! Compare with (11): now we don't even need to know the dimensions of the irreducible representations.

Let's hasten to check this result on A_3. Write the identity as a product of two squares: $(132)^2(123)^2 = I$, $(123)^2(132)^2 = I$, and $I^2 I^2 = I$. In contrast, our result (17) gives $3 \sum_r 1 = 3 \cdot 1 = 3$, check. (Evidently, in the present context, $\sum_r 1$ ranges over the noncomplex irreducible representations.)

Next, look at S_3. How many ways can we write the identity as a product of two squares? There are the ways just listed for A_3, plus $(23)^2(12)^2 = I$ with all possible exchanges, so these count as $3 \cdot 3 = 9$, plus $I^2(12)^2 = I$, $(12)^2 I^2 = I$, $3 \cdot 2 = 6$ of these. So, altogether $3 + 9 + 6 = 18$, while (17) gives $6 \sum_r 1 = 6 \cdot 3 = 18$, check.

Some readers might have realized that we can keep on trucking, going down the same road, figuring out how many solutions are there to the equation $f^2 g^2 h^2 = k^2$ for a given element k of G.

Exercises

1 Show that the matrices $D(g)^T$ (in contrast to $D(g)^*$) do not form a representation.

2 Show that squares of elements that are equivalent to each other are also equivalent to each other.

3 Verify that the 2-dimensional irreducible representation of D_4, the invariance group of the square, is real, as is almost self-evident geometrically.

4 Give the 3 of A_4 a reality check.

5 Give the 3 of S_4 a reality check.

6 Show that the 2 of the quarternionic group is pseudoreal.

7 How many square roots does (123) in A_4 have?

8 For S_4, evaluate $\sigma_{(12)(34)}$, $\sigma_{(123)}$, $\sigma_{(12)}$, and $\sigma_{(1234)}$, and check your answers against the multiplication table.

9 For A_5, evaluate σ_I, $\sigma_{(12)(34)}$, $\sigma_{(123)}$, $\sigma_{(12345)}$, and $\sigma_{(12354)}$, and check your answers against the multiplication table.

10 Verify the (13) for $\sum_g D^{(r)}(g^2)$ in A_4.

11 For A_4, verify the result for the number of solutions of $f^2 g^2 = I$.

Notes

1. For instance, quarks transform like the 3 of $SU(3)$, while antiquarks transform like the 3*. We will discuss this in parts V and IX in detail.
2. The existence of the invariant bilinears described here is important for neutrino masses. See chapter VII.4.
3. I daresay that for most of us, that was quite an eye opener!
4. Probably not a good use of your disk space.

II.i1 Crystals Are Beautiful

I start with a story. When Eugene Wigner wanted to leave Budapest to go study physics in Germany, his father asked him how many jobs in physics there were in his native Hungary. Wigner said that he thought there was perhaps one. So sure enough, after obtaining his doctorate in physics, Wigner ended up working in his father's leather factory. He soon realized that he did not want to tan his life away, and so he wrote to his professors in Berlin begging for a job. Fortunately, he found one with the crystallographer Weissenberg, who told him to calculate the equilibrium position of atoms in various crystals, giving the young Wigner a book on modern algebra and saying that some of this stuff might be useful.[1]

This started Wigner on the road to applying group theory to physics and to his eventual Nobel Prize.

Crystallography

First, I certainly cannot and will not go into crystallography in any detail; there are enormously thick treatises on the subject. Rather, I am content to discuss a fundamental, and well celebrated, result[2] that turns out to be rather easy to prove.

A crystal is defined to be a lattice of atoms invariant under translations $\vec{T} = n_1 \vec{u}_1 + n_2 \vec{u}_2 + n_3 \vec{u}_3$, with $\vec{u}_i$ three vectors and n_i three integers. The group consisting of these translations, plus rotations, reflections, and possibly inversion ($\vec{x} \to -\vec{x}$) is known as the space group[3] of the crystal. If translations are taken out, the resulting group is known as a point group.[4] All this is just terminology and jargon. The reader should be warned, however, that crystallographers have their own notations.[5]

A great achievement of the subject is the classification of all point groups; there are only a finite number of possibilities.

Crystallography has become a hugely important subject in light of the interest in materials science.

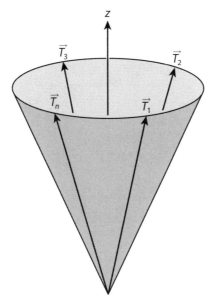

Figure 1

No five-fold symmetry

Here is the famous theorem I would like to introduce you to.

Theorem: Let a crystal be invariant under rotations through $2\pi/n$ around an axis. The only allowed possibilities are $n = 1, 2, 3, 4, 6$. In other words, $n = 5$ and $n > 6$ are not allowed. (Note that $n = 1$ is a trivial special case that we do not need to cover in the following argument.)

Call the axis of rotation the z-axis. Let $\vec{T}_1$ be a translation vector. Under rotation this gets taken into $\vec{T}_2, \cdots, \vec{T}_n$. See figure 1. Because translations form a group, $\vec{T} = \vec{T}_1 - \vec{T}_2$ is also a translation vector. By elementary geometry,[6] $\vec{T}$ is perpendicular to the rotation axis.[7] Thus, we can restrict our attention to translation vectors in the x-y plane.

Denote the difference vectors (such as $\vec{T}$) between T_i and T_j generically by $\vec{t}_{ij}$. As was just noted, these vectors $\vec{t}_{ij}$ live in the x-y plane. Now pick out the shortest such vector (there may be more than one) and call it $\vec{t}$. Choose units so that $\vec{t}$ has length 1. Now rotate around the z-axis through angle $2\pi/n$, and call the vector $\vec{t}$ gets rotated to $\vec{t}'$. Evidently, $\vec{t}'$ also has length 1.

Then $\vec{t} - \vec{t}'$ is also a translation vector. It has length $l_n = 2 \sin \frac{\pi}{n}$, a result obtained readily by either using basic trigonometry or squaring $(\vec{t} - \vec{t}')^2 = 2(1 - \cos \frac{2\pi}{n}) = 4 \sin^2 \frac{\pi}{n}$. But we assumed $\vec{t}$ to be the shortest such vector, and hence we require $4 \sin^2 \frac{\pi}{n} \geq 1$, that is, $\sin \frac{\pi}{n} \geq \frac{1}{2}$.

This condition is violated for $n > 6$. We see from figure 2 that the result is almost trivial: if the opening angle $2\pi/n$ between the vectors $\vec{t}$ and $\vec{t}'$ is too small, then the distance between their tips will be too short.

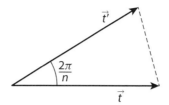

Figure 2

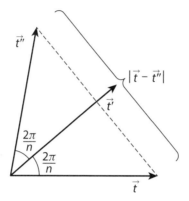

Figure 3

To rule out $n = 5$, rotate $\vec{t}'$ around the z-axis through angle $2\pi/n$ (in the same sense as the earlier rotation), and call the vector $\vec{t}'$ gets rotated to $\vec{t}''$. The angle between $\vec{t}$ and $\vec{t}''$ is thus $2 \cdot 2\pi/n = 4\pi/n$. See figure 3. Then the translation vector $\vec{t} + \vec{t}''$ (note the + sign!) has length $2|\cos \frac{2\pi}{n}|$ (note the 2π rather than π). Again, we obtain this by either using basic trigonometry or squaring $(\vec{t} + \vec{t}'')^2 = 2(1 + \cos \frac{4\pi}{n}) = 4 \cos^2 \frac{2\pi}{n}$. By assumption we require $|\vec{t} + \vec{t}''|$ to be larger than 1, and hence $|\cos \frac{2\pi}{n}| \geq \frac{1}{2}$. (The absolute value is put in because $\cos \frac{2\pi}{n}$ is negative for $n = 2, 3$.) Since $\cos \frac{2\pi}{5} = (\sqrt{5} - 1)/4 \approx 0.309$, this inequality is violated for $n = 5$.[*] The point is that $|\cos \theta|$ has a downward cusp touching 0 at $\theta = \frac{2\pi}{4}$ and $\frac{2\pi}{5}$ is too close to $\frac{2\pi}{4}$.

An alternative and somewhat simpler proof[8] is the following. We proceed as before until we get to the point of choosing units such that $\vec{t}$ has length 1. Let A be a point on a symmetry axis and B be a point A gets translated to under $\vec{t}$. See figure 4. Let B' be the point B gets rotated to under a rotation through $2\pi/n$ around A. Similarly, let A' be the point A gets rotated to under a rotation through $2\pi/n$ around B. The distance between A' and B', equal to $1 + 2 \sin(\frac{2\pi}{n} - \frac{\pi}{2}) = 1 - 2 \cos \frac{2\pi}{n}$, must be, by assumption, some nonnegative integer k. This implies $\cos \frac{2\pi}{n} = \frac{1}{2}(1 - k)$. Since the cosine is bounded by ± 1, the nonnegative integer k is restricted to have the value 0, 1, 2, 3, corresponding to $n = 6, 4, 3, 2$, respectively. (In

---

[*] Recall the endnote about the golden ratio in chapter II.3. You might think that the inequality is also violated for $n = 4$, but in that case $\vec{t} + \vec{t}''$ does not really exist.

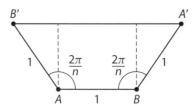

Figure 4

fact, the $n = 2$ case is trivial so that there are actually only 3 cases, for each of which I encourage you to draw the quadrilateral $ABA'B'$.)

One appealing feature of this proof is the immediate exclusion of $n = 5$ without further ado. Indeed, unlike the previous proof, we do not need to know the approximate value of $\cos \frac{2\pi}{5}$. We can see graphically, since $\frac{2\pi}{5} = 75°$ is less than $90°$, that the distance between A' and B' cannot be an integer multiple of 1.

The discovery of quasicrystals leading to a Nobel Prize for materials scientist Dan Shechtman in 2011 makes for a fascinating story,[9] particularly the fact that nontranslation invariant five-fold symmetries exist in Islamic architecture.

I started this interlude talking about Wigner,[10] and I now end with him also. By 1929 Wigner had started wondering about how he could contribute to quantum physics. He said in his recollections, "[I thought that] perhaps the dabbling I had done for Dr. Weissenberg might lead somewhere worthwhile. There was nothing brilliant in this insight—just a bit of good instinct and much good fortune."

Notes

1. Wigner's autobiography cited in chapter I.1.
2. First proved by the French crystallographer Auguste Bravais in 1850. An early hint of this was provided by the great Johannes Kepler; in his treatise (1619) on the harmony of the world, he failed to find a mosaic with five-fold symmetry. See M. Křížek, J. Šolc, and A. Šolcová, Not. AMS 59 (2012), p. 22.
3. In two dimensions, the space group is also called the wallpaper group.
4. Thus, the point group applies to molecules also.
5. The Schönflies notation and the Hermann-Mauguin notation.
6. Or, by analytic geometry, simply observe that $\vec{T}$ has no z-component.
7. For relativistic physicists, $\vec{T}_1, \cdots, \vec{T}_n$, together with $-\vec{T}_1, \cdots, -\vec{T}_n$, form a sort of discrete future and past light cones with z as the time axis.
8. L. Landau and E. Lifshitz, Statistical Physics (second edition, 1969), p. 408.
9. There were also important contributions by the Chinese American mathematician Hao Wang and the English physicist Roger Penrose. See Martin Gardner, Sci. Am. (1976), p. 110.
10. At that time, Weyl's book on group theory was the only one available to physicists, and it was, according to Wigner, "too dense for most people." So Leo Szilard urged Wigner to write his own textbook, which was published in 1931 and is still regarded as a classic.

II.i2 Euler's φ-Function, Fermat's Little Theorem, and Wilson's Theorem

Quick, if you divide 10^{10} by 11, what is the remainder?[1]

I follow this question with a story[2] about the influential[3] and Nobel Prize winning physicist Ken Wilson. During a seminar at Caltech conducted by Feynman for graduate students, Ken Wilson was caught talking to a fellow student. When asked by Feynman what he was talking about, he said that they were discussing Wilson's theorem, at which point Feynman had Wilson up at the blackboard stating and proving Wilson's theorem.

The group G_n

On several occasions, we have seen the close connection between group theory and number theory. Here we explore one connection.[4]

Let n be some positive integer. Denote by G_n the set of integers m, $1 \leq m \leq n-1$, such that the highest common factor of m and n is 1. In other words, m and n have no factor in common except for 1 and hence are relatively prime; for example, neither 9 nor 16 is a prime number, but they are relatively prime.

The claim is that under multiplication modulo n, G_n is a group.

Consider 2 elements m_1 and m_2 of G_n. We want to show closure, that is, $m_1 m_2 \pmod n \in G_n$.

To show this, first note that, since by assumption m_1 and n have no factor in common and since m_2 and n have no factor in common, it follows that $m_1 m_2$ and n have no factor in common except for 1. But what about $m_1 m_2 \pmod n$ and n? Let us prove it by contradiction. Suppose that $m_1 m_2 \pmod n$ and n have a highest common factor, call it $d > 1$. In other words, $n = kd$ and $m_1 m_2 \pmod n = jd$ for some integers k and j. Then $m_1 m_2 = jd + ln = (j + lk)d$ for some integer l. (Henceforth, I will not specify that various quantities are integers when the context makes it clear that they are.) Thus, $m_1 m_2$ and n have a highest common factor $d > 1$, which contradicts what we just said. This proves closure: $m_1 m_2 \pmod n = m_3$. For example, G_9 consists of the integers $m = 1, 2, 4, 5, 7, 8$, and $7 \cdot 8 \pmod 9 = 2$, $2 \cdot 7 \pmod 9 = 5$, and so forth.

Associativity is trivial.

Next, the identity is just 1.

Finally, to prove that the inverse exists, we need a lemma.

Lemma: Given integers a and b with highest common factor $= 1$, then there exist integers x and y such that $xa + yb = 1$.

For example, with $a = 16$ and $b = 9$, we have[5] $4 \cdot 16 - 7 \cdot 9 = 1$. Note that the solution is not unique: we also have $9 \cdot 9 - 5 \cdot 16 = 1$. Try a few cases, and you might see why the lemma holds. A proof is given in appendix 1.

Apply the lemma to m and n: $xm + yn = 1$ means that $xm = 1 \pmod{n}$. Hence x is the inverse of m.

Thus, we have proved that G_n is a group: in fact, an abelian group.

Euler's φ-function

The order of G_n, that is, the number of elements in G_n, is called Euler's φ-function $\varphi(n)$.

We have already seen that $\varphi(9) = 6$. Since there are not many groups with order 6 (or any relatively small integer), we suspect that G_9 is in fact one of the groups we have already encountered. Any intuitive guess here? Exercise! As another example, G_{16} consists of 1, 3, 5, 7, 9, 11, 13, 15, so that $\varphi(16) = 8$. Have you encountered the group G_{16} before? Exercise!

Now Euler[6] proves a theorem. If the highest common factor of a and n is 1, then

$$a^{\varphi(n)} = 1 \pmod{n} \tag{1}$$

To prove this highly nonobvious theorem, all we have to do is to apply just about the most elementary group theory theorem there is, mentioned in chapter I.1. Take any element g of a finite group G and keep multiplying it by itself. Then $g^{N(G)}$ is the identity. Here goes. Divide a by n and take the remainder m, so that $m < n$ (in other words, $a = nk + m$). We claim that the highest common factor of m and n is 1, since otherwise the highest common factor of a and n would not be 1. (To see this, note that if $m = bf$, $n = cf$ with $f > 1$, then $a = (ck + b)f$, and a and n would have f for a common factor.) Thus, $m \in G_n$. The order of G_n is denoted by $\varphi(n)$. Hence, $m^{\varphi(n)} = 1 \pmod{n}$. Then $a^{\varphi(n)} = (nk + m)^{\varphi(n)} = m^{\varphi(n)} + \varphi(n)m^{\varphi(n)-1}nk + \cdots = 1 \pmod{n}$.

Fermat's little theorem (called "little" to distinguish it from his last theorem) is then an immediate corollary. For p prime and a not a multiple of p,

$$a^{p-1} = 1 \pmod{p} \tag{2}$$

Proof: For p prime, $G_p = \{1, 2, \cdots, p - 1\}$ and $\varphi(p) = p - 1$. Apply Euler's theorem.

In particular, $10^{10} = 10^{11-1} = 1 \pmod{11}$, and the question at the beginning of this section is answered.[7] Furthermore, since $a = 1 \pmod{n}$ and $b = 1 \pmod{n}$ implies $ab = 1 \pmod{n}$, we have $10^{10^{10}} = 1 \pmod{11}$, and so forth. Dazzle your friends with such useless facts!

Another way of stating Fermat's little theorem is that, for p prime and a not a multiple of p, $a^{p-1} - 1$ is divisible by p.

I hope that this brief interlude has convinced you that group theory can play an important role in number theory.

Appendix 1: Proof of a lemma

First, some basic facts about modular arithmetic.

If $a = b$ (mod n), then $ka = kb$ (mod n); but the reverse clearly does not hold.

If $a = b$ (mod n) and if $c = d$ (mod n), then $ac = bd$ (mod n).

The proof of the lemma stated in the text is by iterative reduction. With no loss of generality, let $a > b$. The claim that there exist x and y such that $xa + yb = 1$ means that we can solve the equation $xa = 1$ (mod b). Let $a = a'$ (mod b) with $a' < a$. If we could solve $xa' = 1$ (mod b), then we have solved $xa = 1$ (mod b). (Note that $xa' = jb + 1$ and $a = a' + lb$ mean that $xa = (j + xl)b + 1$, that is, $xa = 1$ (mod b).)

Hence we have reduced the problem specified by (a, b) to a problem specified by (b, a') (where we have exchanged the role played by the two integers, since $b > a'$). After another step, this problem is reduced to a problem specified by (a', b') with $a' > b'$. At every step, one integer gets strictly smaller and so eventually one of the two integers reaches 1, and we are done.

Appendix 2: Wilson's theorem

The first theorem we proved in this interlude allows us to prove Wilson's theorem with almost no effort. The theorem, first proved by Lagrange in 1771, was discovered by John Wilson around 1770 but was known earlier to Ibn al-Haytham around 1000. It was announced by Edward Waring in 1770, but neither he nor his student Wilson was able to prove it.

Wilson's theorem: $(n − 1)! + 1$ is divisible by n if and only if n is a prime. (For example, 5 is a prime and $4! + 1 = 25$ is divisible by 5, but 6 is not a prime and $5! + 1 = 121$ is not divisible by 6.) In other words, $(n − 1)! = −1$ (mod n) if and only if n is a prime.

Proof: If n is not a prime, then its factors are to be found among the set of integers $\{1, 2, \cdots, n − 1\}$, and so $(n − 1)!$ is divisible by n, that is, $(n − 1)! = 0$ (mod n).

In contrast, if n is a prime, we have proved that the set of integers $\{1, 2, \cdots, n − 1\}$ form a group G_n under multiplication modulo n. The inverse of 1 is itself, namely 1, and the inverse of $n − 1$ is also itself (since $(n − 1)^2 = n^2 − 2n + 1 = 1$ (mod n)). The inverse of each of the remaining integers $\{2, \cdots, n − 2\}$ is to be found among this subset of $n − 3$ integers. (For example, in G_7, the inverse of 2 is 4, of 3 is 5, and so on.) This sentence implies that $(n − 1)! = (n − 1) \cdot 1 = n − 1$ (mod n), where the first equality follows because in the factorial the integers other than $n − 1$ and 1 cancel in pairs. This implies $(n − 1)! = −1$ (mod n). QED[8]

Appendix 3: A physicist's proof of Fermat's little theorem

Gutfreund and Little[9] have given a nice physicist's proof of Fermat's little theorem. I content myself with giving a baby version of the theorem, and challenge the reader to extend it.[10]

We now prove (2) for $a = 2$. Let p be a prime, and consider a ring of p Ising spins.* Physicists call a sequence of p + and − signs a "spin configuration" (for example, for $p = 5$, $+ + − − +$ is a configuration). We divide the 2^p possible configurations into equivalence classes as follows: define two configurations to be equivalent if one configuration can be mapped into the other configuration by translation around the ring. (For example, for $p = 5$, the 5 configurations $+ + − − +, + − − + +, − − + + +, − + + + −$, and $+ + + − −$ are equivalent. See figure 1.) These equivalence classes each contain p configurations; this, by the way, amounts to the statement proved way back in chapter I.1 that, for p prime, Z_p does not have a subgroup. The configuration with all spins equal to + is special; it is equivalent only to itself. Ditto for the configuration with all spins equal to −.

---

* I assume that the reader knows what an Ising spin is, since, in light of the title of this book, he or she is sort of posing as a physicist. If not, there is always the web.

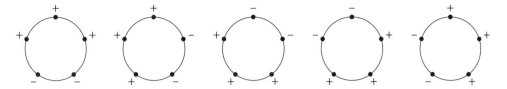

Figure 1

It follows that the $2^p - 2$ configurations that are not these two special configurations are naturally divided into equivalence classes, each with p configurations. Therefore, $2^p - 2$ is divisible by p. But $2^p - 2$ is also manifestly an even integer. Thus, $2^p - 2$ is divisible by $2p$, and hence $2^{p-1} - 1$ is divisible by p. In other words, $2^{p-1} = 1$ (mod p).

Exercises

1 Identify the group G_{10}.

2 Show that $G_9 = Z_6$.

3 Show that $G_{16} = Z_4 \otimes Z_2$.

Notes

1. $10^{10} = 11 \times 909090909 + 1$.
2. P. Ginsparg, arXiv:1407.1855v1, p. 16.
3. In my essay "My Memory of Ken Wilson," I spoke about the impact Wilson had on my career. See Ken Wilson Memorial Volume: Renormalization, Lattice Gauge Theory, the Operator Product Expansion and Quantum Fields, ed. B. E. Baaquie, K. Huang, M. E. Peskin, and K. K. Phua, World Scientific, 2015.
4. I follow the discussion in Armstrong.
5. Think of a horse race in which time is quantized. The two horses run a and b units of distance per unit time, respectively. By adjusting the durations each horse can run, show that we can always arrange to have one horse win by a nose.
6. Leonard Euler had extraordinary powers of concentration. He was extremely productive and averaged over 800 printed pages per year. He maintained his productivity long after he became blind by writing on a slate board by his side, with a team of assistants copying everything down. One evening, while playing with a grandson, he wrote "I die" and fell dead. See Mathematicians Are People, Too, by L. Reimer and W. Reimer, p. 80.
7. At a more elementary level, note that $100/11 = 9 + 1/11$, and so $10^4/11 = 100(9 + 1/11) = 900 + 9 + 1/11$, so on and so forth.
8. For a "physicist's proof," see E. Staring and M. Staring, Am. J. Phys. 51(5) (1983).
9. H. Gutfreund and W. A. Little, Am. J. Phys. 50(3) (1982).
10. Or, failing that, to look up the paper cited in the previous endnote.

II.i3 | Frobenius Groups

Presenting the Frobenius group T_{13}

When you have some spare time, you might want to invent a group or two. Quite a few very smart people did precisely that.

You already know about cyclic groups and direct product groups, and thus the group $Z_{13} \otimes Z_3$. It contains two elements a and b such that

$$a^{13} = I \quad \text{and} \quad b^3 = I \tag{1}$$

with a and b commuting with each other, that is, with

$$bab^{-1} = a \tag{2}$$

which is of course equivalent to $ba = ab$. The group has $13 \cdot 3 = 39$ elements of the form $g = a^m b^n$ with $0 \le m \le 12$ and $0 \le n \le 2$.

Instead of this baby stuff, Frobenius proposed modifying (2) to

$$bab^{-1} = a^r \tag{3}$$

with r an integer to be determined. Equivalently, $ba = a^r b$.

Repeating the transformation in (3) three times, we obtain

$$a = IaI = b^3 a b^{-3} = b^2 a^r b^{-2} = b(bab^{-1})^r b^{-1} = ba^{r^2} b^{-1} = a^{r^3} \tag{4}$$

Consistency thus requires the peculiar number theoretic condition $r^3 = 1 \pmod{13}$.

Consider the solution $r = 3$: indeed, $3^3 = 27 = 2 \cdot 13 + 1$.

The resulting group is known as the Frobenius group $Z_{13} \rtimes Z_3 = T_{13}$. It has the presentation:

$$\langle a, b | a^{13} = I, b^3 = I, bab^{-1} = a^3 \rangle \tag{5}$$

The jargon guy informs us that the funny symbol $\rtimes$ is called a semidirect product. Sounds like a wimpy name to me.

Classes and irreducible representations

Our interest here is to turn the powerful machinery developed in part II on T_{13} and determine its irreducible representations.

The group T_{13} has $13 \cdot 3 = 39$ elements, of the form $g = a^m b^n$ with $0 \leq m \leq 12$ and $0 \leq n \leq 2$, just as in $Z_{13} \otimes Z_3$, but of course the structures of the two groups are quite different. Let us work out the equivalence classes for T_{13}.

First, the usual "class of one" consisting of the identity $\{I\}$. Next, by definition of the group, we have $bab^{-1} = a^3$, and thus the equivalence relations $a \sim a^3 \sim a^9$. We have the class

$$C_3 : \{a, a^3, a^9\} \tag{6}$$

But then also $ba^2b^{-1} = (bab^{-1})^2 = (a^3)^2 = a^6$, and $ba^6b^{-1} = a^{18} = a^5$. Thus, we have found another class

$$C_3' : \{a^2, a^5, a^6\} \tag{7}$$

Similarly, the two classes

$$C_3^* : \{a^4, a^{10}, a^{12}\} \tag{8}$$

and

$$C_3'^* : \{a^7, a^8, a^{11}\} \tag{9}$$

(For example, $ba^7b^{-1} = a^{7 \cdot 3} = a^{21} = a^{13+8} = a^8$.) I hardly have to say that there's a whole lot of number theory going on.

In addition to these five classes, there are elements of the form $a^m b$ and $a^m b^2$, in particular, b. Start with $a^{-1}ba = a^{-1}a^3b = a^2b$, that is, $b \sim a^2b$. Next, $a^{-1}(a^2b)a = a^{-1}a^2a^3b = a^4b$. Note that when we reach $a^{12}b$, we then find $a^{-1}(a^{12}b)a = a^{-1}a^{12}a^3b = a^{14}b = ab$. Proceeding in this way, we obtain the class

$$C_{13}^{(1)} : \{b, ab, a^2b, \cdots, a^{11}b, a^{12}b\}$$

Similarly, we have the class

$$C_{13}^{(2)} : \{b^2, ab^2, a^2b^2, \cdots, a^{11}b^2, a^{12}b^2\}$$

The group T_{13} has seven equivalence classes, and hence, by one of the theorems in chapter II.2, seven irreducible representations. Thus, we have to find six integers d_i such that

$$1 + \sum_{i=1}^{6} d_i^2 = 39 \tag{10}$$

Let us start by seeing whether there are any other 1-dimensional representations besides the trivial representation 1.

If a and b are just numbers, then the condition $bab^{-1} = a^3$ collapses to $a^2 = 1$, but $a = -1$ is not allowed by the condition $a^{13} = 1$. Thus, $a = 1$, and the condition $b^3 = 1$ allows three possibilities $b = 1$, ω, ω^* (as before, $\omega = e^{\frac{2\pi i}{3}}$ denotes the cube root of the identity), which thus correspond to three possible 1-dimensional representations. We will refer to these representations as 1, 1', and 1'*.

After finding the pair of conjugate representations 1' and 1'*, we next have to find four integers whose squares add up to 36. The unique possibility is $3^2 + 3^2 + 3^2 + 3^2 = 36$. (The other possibility, $5^2 + 3^2 + 1^2 + 1^2 = 25 + 9 + 1 + 1 = 36$, is ruled out by the fact that we have already found all the 1-dimensional representations.)

To find the four 3-dimensional irreducible representations, we choose to diagonalize a. Let us start with the condition $b^3 = I$ and represent b by

$$b = \begin{pmatrix} 0 & 1 & 0 \\ 0 & 0 & 1 \\ 1 & 0 & 0 \end{pmatrix} \tag{11}$$

Define $\rho = e^{\frac{2\pi i}{13}}$, the thirteenth root of the identity. Then the diagonal elements of a can be any integral powers of ρ. From the review of linear algebra, we know that the similarity transformation bab^{-1} with (11) moves the diagonal elements around cyclically. Thus, if we choose the first diagonal element of a to be ρ, then the condition $bab^{-1} = a^3$ fixes

$$a = \begin{pmatrix} \rho & 0 & 0 \\ 0 & \rho^3 & 0 \\ 0 & 0 & \rho^9 \end{pmatrix} \tag{12}$$

This is of course consistent, since $(\rho^9)^3 = \rho^{27} = \rho$. Call this irreducible representation 3.

Similarly, the irreducible representation 3' has

$$a = \begin{pmatrix} \rho^2 & 0 & 0 \\ 0 & \rho^6 & 0 \\ 0 & 0 & \rho^5 \end{pmatrix} \tag{13}$$

and b as in (11).

The other two irreducible representations can be obtained by complex conjugation. In 3* and 3'*, a is represented respectively by

$$a = \begin{pmatrix} \rho & 0 & 0 \\ 0 & \rho^3 & 0 \\ 0 & 0 & \rho^9 \end{pmatrix}^* = \begin{pmatrix} \rho^{12} & 0 & 0 \\ 0 & \rho^{10} & 0 \\ 0 & 0 & \rho^4 \end{pmatrix} \tag{14}$$

and

$$a = \begin{pmatrix} \rho^2 & 0 & 0 \\ 0 & \rho^6 & 0 \\ 0 & 0 & \rho^5 \end{pmatrix}^* = \begin{pmatrix} \rho^{11} & 0 & 0 \\ 0 & \rho^7 & 0 \\ 0 & 0 & \rho^8 \end{pmatrix} \tag{15}$$

The 3-dimensional matrices representing a and b all have unit determinant (for example, in (15), det $a = \rho^{11+7+8} = \rho^{26} = 1$). According to a fundamental theorem proved in chapter II.1, these matrices must be unitary, which by inspection is manifestly true. In chapter IV.4 we will learn that the set of all unitary 3-by-3 matrices with unit determinant forms the group $SU(3)$. Thus, the Frobenius group T_{13} is a subgroup[1] of $SU(3)$.

Meanwhile, you could have fun discovering other Frobenius groups, for instance by replacing the number 13 by 7.

Note

1. To determine various properties of T_{13}, such as how the direct product of two of its irreducible representations decompose, it is then easiest to exploit the properties of $SU(3)$ (which we will derive in part V) and figure out how the irreducible representations of $SU(3)$ decompose upon restriction to T_{13}.

Part III | Group Theory in a Quantum World

Group theory has a somewhat limited impact on classical mechanics, but it really blossomed with the coming of the quantum era. Wigner is rightfully honored for introducing group theory into quantum physics. You will learn that symmetry, degeneracy, and representation theory are beautifully intertwined.

Modern physics is almost unthinkable without the Lagrangian and the Hamiltonian, much zippier than the equations of motion. Group theory rules all of them.

III.1 | Quantum Mechanics and Group Theory: Parity, Bloch's Theorem, and the Brillouin Zone

Quantum mechanics is linear

In my study of physics (and of a bit of mathematics), I am often astonished by how the collective mind of the physicists and the collective mind of the mathematicians would converge. Indeed, Eugene Wigner,[1] who won the Nobel Prize for introducing group theory into quantum mechanics, wrote an influential essay[2] on the unreasonable effectiveness[3] of mathematics in physics. He described mathematics as "a wonderful gift we neither understand nor deserve." In physics, representation theory came alive with the advent of quantum mechanics.*

In quantum mechanics, the states of a physical system, be it a molecule or a field or a string, are described by a wave function Ψ, evolving in time according to Schrödinger equation

$$i\hbar \frac{\partial}{\partial t} \Psi = H\Psi \tag{1}$$

Here $\hbar$ denotes Planck's constant,† which we already encountered in chapter I.3. Writing‡ $\Psi(t) = \psi e^{-i\mathcal{E}t/\hbar}$, where ψ depends on the coordinates but not time, we obtain the eigenvalue equation

$$H\psi = \mathcal{E}\psi \tag{2}$$

At the most elementary level, the Hamiltonian H is given by a linear differential operator, for instance, for a particle moving in a potential in 1-dimension, $H = -\frac{1}{2m}\frac{d^2}{dx^2} + V(x)$. In other situations, for a particle with spin in a magnetic field $\vec{B}$, for example, H is given by $H = \vec{B} \cdot \vec{S}$, with $\vec{S}$ the spin operator. At the physicist's level of rigor, and for the purposes

---

* As I said in the Preface, I assume that most readers have at least a nodding acquaintance with quantum mechanics. Evidently, I can only describe the minimum necessary to appreciate the role of group theory in quantum mechanics.

† Later in this text, I will often choose units such that $\hbar = 1$.

‡ I use $\mathcal{E}$ here instead of the more standard E in order to reserve E for later use.

here, we can simply think of any linear operator, including H, as a matrix,[4] albeit often an infinite dimensional one.

Degeneracy was a mystery

The Hamiltonian H has eigenfunctions ψ^α with eigenvalues $\mathcal{E}^\alpha$ and with the index* α possibly taking on an infinite number of values:

$$H\psi^\alpha = \mathcal{E}^\alpha \psi^\alpha \tag{3}$$

(Here we temporarily suspend the repeated index summation convention.)

In general, $\mathcal{E}^\alpha \neq \mathcal{E}^\beta$ for $\alpha \neq \beta$. But often, among all these eigenfunctions ψ^α, there may exist a subset (call them ψ^a) with $a = 1, \cdots, d$, having exactly the same eigenvalue, call it E:

$$H\psi^a = E\psi^a, \quad \text{with } a = 1, \cdots, d \tag{4}$$

Assume that these eigenfunctions are finite in number, so that $d < \infty$. The set of eigenfunctions ψ^a is said to exhibit a d-fold degeneracy. The important point is that E does not depend on a.

In the early days of quantum physics, the existence of degeneracy was enormously puzzling. When you diagonalize a matrix, the eigenvalues will in general be distinct. It became clear that there must be a reason behind the degeneracy. (Also, do not forget the heroic experimental task of unraveling the degeneracy, given that in atomic spectroscopy, experimentalists could only measure the energy difference $|E^\alpha - E^\beta|$, so that degeneracy resulted in many lines superposed on top of one another. The situation was only clarified after tremendous experimental efforts.)

Symmetry implies degeneracy

Eventually, it was realized that degeneracy has a natural explanation. In quantum mechanics, transformations are realized as unitary operators T. Suppose a set of transformations leave H invariant, so that

$$T^\dagger H T = H \tag{5}$$

Then, as explained in chapter I.1, these transformations form a group G: if the transformations T and T' each leave H invariant, then the transformation $T \cdot T'$, which consists of the transformation T' followed by the transformation T, clearly leaves H invariant. The

---

* As in most instances in this textbook, whether I write an index as a superscript or a subscript has no particular significance.

Ts form a symmetry group. For example, G could be the rotation group $SO(3)$. Since T is unitary, $T^\dagger = T^{-1}$, we can write (5) as

$$HT = TH \tag{6}$$

Given that T leaves H unchanged, its action on ψ^a produces an eigenstate of H with the same energy E. More explicitly, if $H\psi^a = E\psi^a$, then, according to (6),

$$H(T\psi^a) = HT\psi^a = TH\psi^a = TE\psi^a = E(T\psi^a) \tag{7}$$

Since $\psi'^a = T\psi^a$ is an eigenstate of H with energy E, it has to be a linear combination of the ψ^as: $\psi^a \to \psi'^a = (D(T))^{ab}\psi^b$. If the transformation T_1 is followed by the transformation T_2, so that we have effectively the transformation $T_2 T_1$, then $D(T_2 T_1) = D(T_2)D(T_1)$.

The crucial fact about quantum mechanics is of course that, unlike classical mechanics, it is linear, and linear superpositions of states are also acceptable states. Consider the linear vector space spanned by the d degenerate eigenstates, that is, the set of all linear combinations of the ψ^as, namely $\sum_a c_a \psi^a$ for c_a a bunch of numbers. On this space, the Hamiltonian H is just a d-by-d identity matrix multiplied by E.

From degeneracy to group representations

Given this entire setup, you can now see that the d degenerate eigenstates ψ^a furnish a d-dimensional irreducible representation of the group G.

If we know, or could guess, what G is, this also determines what the possible values of d are, since group theory fixes the possible irreducible representations of G. This turns out to be of great importance in the development of quantum physics—especially of particle physics.

Conversely, given experimental information on the degeneracy, we can turn the logic around and try to determine what G might be. In particular, G has to have at least one d-dimensional irreducible representation.

Thus, in quantum physics

$G \Longrightarrow$ degeneracy   and $G \Longleftarrow$ degeneracy

---

d = degrees of degeneracy = dimension of irreducible representation

---

The word "degenerate" has a rather negative connotation in everyday usage, but in the history of quantum physics it is regarded very favorably, as the condition can shed some light on the underlying symmetry group.

For example, among the subnuclear particles discovered in the 1950s were eight baryons ("heavy particles," that is, heavy compared to the electron; these include the proton and the neutron) approximately equal in mass. Particle physicists raced each other to find a group with an 8-dimensional irreducible representation (more in chapter V.2).

Connection to Schur's lemma: What group theory can and cannot tell you

The reader can now see that the derivation of Schur's lemma in chapter II.2 parallels exactly the discussion here.[5] The H that commutes with all the d-by-d matrices $D(g)$ in an irreducible representation of a group G corresponds to the Hamiltonian here. Schur tells us that in the d-dimensional subspace, the Hamiltonian H is simply equal to some constant times the identity, which is precisely what we mean when we say that d energy levels are degenerate.

Dr. Feeling wanders by. "In hindsight, it's almost obvious. If a bunch of states get transformed into one another by some symmetry group, which by definition leaves the Hamiltonian unchanged, you would think that these states all have the same energy. In a sense, the Hamiltonian cannot tell these states apart."

Indeed, Schur's lemma is even more powerful if a collection of states form a reducible representation of the symmetry group. In that case, in some basis, the representation matrices are block diagonal:

$$D(g) = \begin{pmatrix} \ddots & 0 & 0 & 0 \\ 0 & D^{(r)}(g) & 0 & 0 \\ 0 & 0 & D^{(s)}(g) & 0 \\ 0 & 0 & 0 & \ddots \end{pmatrix} \quad \text{for all } g \in G \tag{8}$$

Then we know that the Hamiltonian has the form (as discussed in chapter II.2)

$$H = \begin{pmatrix} \ddots & 0 & 0 & 0 \\ 0 & E^{(r)}I & 0 & 0 \\ 0 & 0 & E^{(s)}I & 0 \\ 0 & 0 & 0 & \ddots \end{pmatrix} \tag{9}$$

In many cases, this enables us to diagonalize the Hamiltonian with almost no work. We know that the energy levels break up into a collection of degenerate states ($d^{(r)}$ states with energy $E^{(r)}$, $d^{(s)}$ states with energy $E^{(s)}$, and so on), with group theory telling us what the numbers $d^{(r)}$ and $d^{(s)}$ are.

But of course group theory cannot tell you what the $E^{(r)}$s are.

Think of it this way. Change the Hamiltonian without changing the symmetry group. For example, take the Schrödinger equation with a central force potential $V(r)$. Imagine changing $V(r)$ smoothly. The energy levels would move up and down, but as long as V remains a function of r, rather than of x, y, and z separately, then the degeneracy in the energy spectrum due to rotational symmetry cannot change.

Parity

After this very brief set up, our first example is almost laughably simple, involving a laughably simple group. Consider a particle moving in a potential in one dimension and suppose $V(x) = V(-x)$. Then the Hamiltonian $H = -\frac{1}{2m}\frac{d^2}{dx^2} + V(x)$ is left invariant by the symmetry group Z_2, consisting of the identity and the reflection $r : x \to -x$. Given an eigenfunction $\psi(x)$, the reflection acting on it gives* $r\psi(x) = \psi(-x)$, which, according to the discussion above, has the same energy as $\psi(x)$.

Group theory tells us that Z_2 has only two irreducible representations, in one of which $r = 1$, and in the other $r = -1$ (more precisely, r is represented by 1 and by -1 in the two representations, respectively). In one case, $\psi(x) = \psi(-x)$, and in the other, $\psi(x) = -\psi(-x)$. In conclusion, if $V(x) = V(-x)$, then the solutions of the Schrödinger equation may be chosen to be either even or odd, something you probably knew from day one of your exposure to quantum mechanics. Here we have a classic example of using a sledgehammer to crack open a peanut.

Bloch's theorem and Brillouin zone

Our next application, significantly less trivial, is of fundamental importance to solid state physics. Again, a particle moves in a potential $V(x)$ in one dimension, but now suppose $V(x + a) = V(x)$ is periodic but otherwise unspecified. The symmetry group consists of the elements $\{\cdots, T^{-1}, I, T, T^2, \cdots\}$, where T denotes translation by the lattice spacing $a: x \to Tx = x + a$.

Since the group is abelian, it can only have 1-dimensional irreducible representations. Hence $\psi'(x) \equiv T\psi(x) = \zeta\psi(x)$, so that T is represented by a complex number ζ. But since wave functions are normalized, $\int dx|\psi'(x)|^2 = \int dx|\psi(x)|^2$, we have $|\zeta| = 1$. Thus, we can write $\zeta = e^{ika}$ with k real (and with dimension of an inverse length). Since $e^{i(ka+2\pi)} = e^{ika}$, we may restrict k to the range

$$-\frac{\pi}{a} \leq k \leq \frac{\pi}{a} \tag{10}$$

This range is the simplest example of a Brillouin zone.†

For a 1-dimensional lattice consisting of N sites, we impose periodic boundary conditions and thus require $T^N = I$. The symmetry group is just the familiar cyclic group Z_N. The condition $e^{iNka} = 1$ thus implies that $k = (2\pi/Na)j$ with j an integer. For N macroscopically large, the separation Δk between neighboring values of j, of order $2\pi/Na$, is infinitesimal, and so we might as well treat k as a continuous variable ranging from[6] $-\frac{\pi}{a}$ to $\frac{\pi}{a}$.

---

* The pedant would insist on distinguishing between the transformation acting on x and the transformation acting on $\psi(x)$, introducing more notation, such as r and T_r, and perhaps also inventing some fancy words to scare children.

† Needless to say, this hardly does justice to the subject, but this is not a text on solid state physics for sure.

It is convenient and conventional to write

$$\psi(x) = e^{ikx}u(x) \tag{11}$$

with $u(x + a) = u(x)$. This statement is known as Bloch's theorem. (Of course, any $\psi(x)$ can be written in the form (11); the real content is the condition $u(x + a) = u(x)$.)

Note that this is a general statement completely independent of the detailed form of $V(x)$. Given a specific $V(x)$, the procedure would be to plug (11) into the Schrödinger equation and solve for $u(x)$ for the allowed energy eigenvalues, which of course would depend on k and thus can be written as $E_n(k)$. As k ranges over its allowed range, $E_n(k)$ would vary, sweeping out various energy bands labeled by the index n.

While both of these applications of group theory to quantum mechanics are important, they both involve abelian groups with their relatively simple structure. The most celebrated application of group theory during the early years of quantum mechanics, as is well known, is to atomic spectroscopy and involves the nonabelian rotation group $SO(3)$, but before discussing this, we will have to wait until we work out its irreducible representations in chapters IV.2 and IV.3.

Appendix: Ray or projective representation

One peculiar feature of quantum mechanics is that the wave functions ψ and $e^{i\alpha}\psi$, with $e^{i\alpha}$ an arbitrary phase factor, describe exactly the same state. In other words, since the Schrödinger equation (1) is linear, the wave function can be determined only up to a phase factor. Thus, the group representation condition $D(T_2 T_1) = D(T_2)D(T_1)$ we derived earlier, when we follow a transformation T_1 by another transformation T_2, should be generalized to $D(T_2 T_1) = e^{i\alpha(T_2, T_1)}D(T_2)D(T_1)$, with $\alpha(T_2, T_1)$ depending on T_1 and T_2 as indicated.[7] This generalized type of group representation is sometimes called a ray or projective representation. This freedom to include a phase factor has spawned a subject known as cocycles[8] in group representation theory.

Exercises

1   Let ψ^a be a set of degenerate wave functions that transform under a symmetry group like $\psi^a \rightarrow \psi'^a = (D(T))^{ab}\psi^b$. Show that $D(T_2 T_1) = D(T_2)D(T_1)$.

Notes

1. When I was an undergraduate, Eugene Wigner was the éminence grise of the physics department in Princeton, the Nobel laureate who had witnessed the birth of quantum mechanics and all that. According to a limerick I learned from the chairman's wife,

   There is a clever fellow named Eugene,
   Who invented a wonderful sex machine,
   Concave or convex,
   It fits either sex,
   And besides, is very easy to clean.

2. E. P. Wigner, Comm. Pure Appl. Math. 13 (1960), p. 1.
3. On the thirtieth anniversary of Wigner's essay, a volume of essays was published. See A. Zee, "The Unreasonable Effectiveness of Symmetry in Fundamental Physics," ed. R. S. Mickens, Mathematics and Science, World Scientific, 1990.

4. This statement, now regarded as a vulgarization of profound mathematics by physicists, was of course once a most profound insight in theoretical physics. When Werner Heisenberg showed Pascual Jordan what he had done during his vacation, Jordan allegedly exclaimed, "Hey dude, these are matrices!" This is the version told to me by a postdoc. In more orthodox accounts, Max Born is usually credited as telling Heisenberg about matrices.

5. We even had the foresight to use the same letter, hee hee.

6. See also interlude VII.i2.

7. For a careful discussion, see S. Weinberg, The Quantum Theory of Fields, vol. 1, pp. 50–53.

8. For the reader wishing to learn more, see L. D. Faddeev, Phys. Lett. 145B (1984), p. 81. A considerable literature exists; the paper I am most familiar with is Y-S. Wu and A. Zee, Phys. Lett. 152B (1985), p. 98.

III.2 | Group Theory and Harmonic Motion: Zero Modes

Group theory plays a more important role in quantum mechanics than in classical mechanics

Many beginning students of physics do not realize that, arguably, quantum mechanics is mathematically simpler than classical mechanics: quantum mechanics is linear, while classical mechanics is nonlinear.

Take $F = ma$ for a single particle: $m \frac{d^2 \vec{x}}{dt^2} = -\vec{\nabla} V(\vec{x})$. Let $\vec{x}_1(t)$ and $\vec{x}_2(t)$ be two solutions. There is no sense that the linear combination $\vec{x}(t) = a_1 \vec{x}_1(t) + a_2 \vec{x}_2(t)$ for two numbers $a_{1,2}$ is also a solution: in the central force problem, for example, you can't add two elliptical orbits together to get another orbit. In contrast, if $\psi_1(t)$ and $\psi_2(t)$ are solutions of the Schrödinger equation $i\hbar \frac{\partial}{\partial t} \psi = H\psi$, then $\psi(t) = a_1 \psi_1(t) + a_2 \psi_2(t)$ is also a solution. Thus, in some sense, classical mechanics is significantly more difficult* than quantum mechanics.

The linear superposition principle is why group theory plays a much more important role in quantum mechanics than in classical mechanics, as was explained in chapter III.1. Wave functions furnish a representation of the symmetry transformations that leave H invariant.

Harmonic systems of springs and masses

The exception in classical mechanics is when the force $\vec{\nabla} V(\vec{x})$ is linear in $\vec{x}$, namely, the important case of harmonic oscillation in classical mechanics. Consider the proto-typical case[†] of N particles of equal mass tied together by ideal springs and moving in

---

* Of course, classical mechanics is easier to grasp, because it is closer to our everyday experiences. Furthermore, we can also say that nobody understands quantum mechanics, precisely because it is so remote from our everyday experiences and logic.

[†] Often used as a model for studying molecular vibrations.

D-dimensional space. The deviation of the particles from their equilibrium positions provides the relevant coordinates. Denote the coordinates of the ath particle ($a = 1, \cdots, N$) by x_a^i ($i = 1, 2, \cdots, D$). Absorb the common mass into the spring constant and write Newton's law in the form*

$$\frac{d^2 x_a^i}{dt^2} = -\sum_{b=1}^{N} \sum_{j=1}^{D} H^{ia,jb} x_b^j \tag{1}$$

(We are again temporarily suspending the repeated summation convention.) This system of equations is entirely linear, and thus linear combinations of solutions are also solutions.

Indeed, assemble the coordinates x_a^i into a DN-dimensional vector x^A, $A = 1, 2, \cdots, DN$. Set† $x^A(t) = x^A \sin(\omega t + \phi)$, thus obtaining the eigenvalue equation (reverting to the repeated index summation convention)

$$H^{AB} x^B = \omega^2 x^A \tag{2}$$

As is well known, (2) has the same mathematical form as the eigenvalue problem in quantum mechanics. In fact, as is also well known, harmonic motion (as typified by (2)) of all types, including sound waves, served as the inspiration for quantum mechanics and in particular for Schrödinger's equation.

The real symmetric matrix H will have in general DN eigenvalues ω_α^2 and eigenvectors x_α^A with $\alpha = 1, 2, \cdots, DN$. The vector x_α^A for a given α describes the αth eigenmode or harmonic mode with eigenfrequency squared ω_α^2.

All this will become clear with a couple of examples.

The power of group theory

Even for relatively simple cases, the DN-by-DN matrix H^{AB} can be quite a mess (see the appendix in chapter III.3) to write down and an affront to the eyes. But if the system of springs and masses, or the "molecule," exhibits a symmetry, then H will be invariant under a group of transformations. The awesome power of group theory then manifests itself. Using the character table and a touch of physical intuition, we can often figure out what the harmonic modes are without even writing down H. In favorable cases, we can even learn a lot about the eigenfrequencies.

Let us sketch the approach to be followed here, which amounts to an application of Schur's lemma and its consequences. The action of the symmetry transformation on the masses gives a DN-dimensional representation $D(g)$ of the symmetry group G. Since we know how the coordinates x_a^i change under the transformation g, in principle $D(g)$ is easily written down, although in practice some tedious work would be involved. Fortunately, as we learned, the characters of the representation often suffice. Once we know the characters,

---

* In particular, for a single particle moving in one dimension, we have the familiar $\frac{d^2 x}{dt^2} = -Hx$.

† Here ϕ is an irrelevant phase shift. Or better, set $x^A(t) = x^A e^{i\omega t}$ if you are familiar with the complex formalism for oscillatory phenomena.

we know how $D(g)$ falls apart into the irreducible representations of G, as pictured in (II.2.7), which I reproduce here for your convenience:

$$D(g) = \begin{pmatrix} \ddots & 0 & 0 & 0 & 0 \\ 0 & D^{(r)}(g) & 0 & 0 & 0 \\ 0 & 0 & \ddots & 0 & 0 \\ 0 & 0 & 0 & D^{(s)}(g) & 0 \\ 0 & 0 & 0 & 0 & \ddots \end{pmatrix} \quad \text{for all } g \in G \tag{3}$$

Recall also that the number of times n_r the irreducible representation r appears is determined by

$$\sum_c n_c \chi^*(c)\chi(c) = \sum_c n_c \sum_{r,s} n_r n_s \chi^{*(r)}(c)\chi^{(s)}(c) = N(G)\sum_{r,s} n_r n_s \delta^{rs} = N(G)\sum_r (n_r)^2 \tag{4}$$

and

$$\sum_c n_c \chi^{*(r)}(c)\chi(c) = \sum_c n_c \sum_s n_s \chi^{*(r)}(c)\chi^{(s)}(c) = N(G)n_r \tag{5}$$

Schur's lemma then tells us that in the basis in which (3) is written, H has the form

$$H = \begin{pmatrix} \ddots & 0 & 0 & 0 & 0 \\ 0 & \omega_{(r)}^2 I_{d^r} & 0 & 0 & 0 \\ 0 & 0 & \ddots & 0 & 0 \\ 0 & 0 & 0 & \omega_{(s)}^2 I_{d^s} & 0 \\ 0 & 0 & 0 & 0 & \ddots \end{pmatrix} \tag{6}$$

with I_d the d-by-d identity matrix.

Group theory cannot tell us what the eigenfrequencies $\omega_{(r)}$ are—that clearly has to do with the details of the system and what is sometimes called dynamics in physics—but it does tell us how many modes have the eigenfrequency $\omega_{(r)}$. That is given by the dimension d_r of the irreducible representation r. This is completely analogous to the situation in quantum mechanics: group theory can determine the pattern of degenerate levels in the energy spectrum, but it cannot tell us what the energies of the levels are.

Zero mode

After all this formalism, some examples, as was already promised, would be most welcome. One issue in pedagogy concerns the simplicity of the examples to be given (as I already said in chapter II.3). If the example is too simple, then it could provoke yawns, and worse, obscure or negate the power of the formalism just developed. But if the example is too complicated, it could hide the trees as well as the forest. In any case, we will start with an almost ridiculously simple example, that of two (equal) masses connected by a spring and moving in 1-dimensional space. See figure 1.

Figure 1

We can write down Newton's equations of motion by inspection: $\frac{d^2x_1}{dt^2} = -(x_1 - x_2)$ and $\frac{d^2x_2}{dt^2} = -(x_2 - x_1)$, which we can of course solve immediately. But that's not the point; instead, we would like to see how the group theoretic formalism works.

From the equations of motion we can read off[1]

$$H = \begin{pmatrix} 1 & -1 \\ -1 & 1 \end{pmatrix} \tag{7}$$

(Throughout this chapter, I will absorb all irrelevant constants.)

The symmetry group here is S_2, consisting of the identity I and the element (12) exchanging particles 1 and 2. The character table is trivially constructed; we barely have to use any of the powerful theorems* in chapters II.2 and II.3. But do construct the table for practice before reading on.

Here it is.

S_2	n_c	c	1	$\bar{1}$
	1	I	1	1
Z_2	1	(12)	1	-1

(8)

The group has two elements, separating into two classes each with one element, and two irreducible representations, called the 1 and the $\bar{1}$, also known as even and odd in this context.

The $DN = 2 \cdot 1 = 2$ representation furnished by the two masses is simply given by $D(I) = \begin{pmatrix} 1 & 0 \\ 0 & 1 \end{pmatrix}$ and $D((12)) = \begin{pmatrix} 0 & 1 \\ 1 & 0 \end{pmatrix}$ (that is, in the same basis as used in (7)).

In the basis in which the $D(g)$ are diagonal, consisting of two 1-by-1 blocks, that is, the basis in which $D((12)) = \begin{pmatrix} 1 & 0 \\ 0 & -1 \end{pmatrix}$, Schur's lemma forces H to have the diagonal form in (6), namely,[†] $H = \begin{pmatrix} 0 & 0 \\ 0 & 2 \end{pmatrix}$.

We see that there exists a mode with $\omega = 0$, known as a zero mode.[2] To understand what it means, go back to when we solve the time dependence by setting $x^A(t) = x^A \sin(\omega t + \phi)$. Zero frequency $\omega = 0$ actually does not mean that $x^A(t)$ does not depend on time. Since Newton's equation, unlike the Schrödinger equation, involves the second derivative in time, $x^A(t)$ can actually be a linear function of time for $\omega = 0$. Indeed, we now recognize this zero mode as the center of mass motion of the whole system, moving at some constant velocity. The eigenfrequency is zero precisely because the two masses are moving with the same velocity: the spring is not stretched at all.

---

* For example, $1^2 + 1^2 = 2$.
† The form of H is easily determined by noting that in (7), tr $H = 2$ and det $H = 0$.

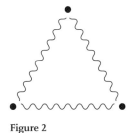

Figure 2

Of course, we do not need fancy schmancy group theory to tell us all this; we could have read off the zero eigenvalue merely by looking at H in (7). Indeed, this follows immediately from $\det H = 0$. The corresponding eigenvector is $x^A = (1, 1)$, which again says that the two masses cruise along together.

As has been repeatedly emphasized, group theory alone cannot tell us what the nonzero eigenfrequency is. However, linear algebra tells us that since H is real symmetric, the other eigenvector must be orthogonal to the eigenvector $(1, 1)$ we already have, and hence can only be $x^A = (1, -1)$. Letting H act on this, we determine by inspection that $\omega^2 = 2$. This is known as a breathing mode for obvious reasons: the two masses are moving in opposite directions, and the whole system expands and contracts as in breathing.

We could have gotten all this by using (4) and (5). Without writing down the $DN = 2 \cdot 1 = 2$ representation furnished by the two masses, we know that the characters are $\chi(I) = 2$ (as always, this is just the dimension of the representation) and $\chi((12)) = 0$ (since the exchange matrix $D((12))$ has zero diagonal elements by definition). From (4) and (5), we deduce easily that $n_1 = 1$ and $n_{\bar{1}} = 1$.

Well, talk about cracking a peanut with a sledgehammer!

The triangular "molecule"

Our next example involves the triangular molecule shown in figure 2, with motion restricted to the 2-dimensional plane, that is, a molecule consisting of three identical hockey pucks connected by identical springs and gliding on frictionless ice. The problem is considerably less trivial. In fact, even writing down the matrix H is quite a chore,[3] but as I said, the whole point is that group theory allows us to avoid doing that.

The invariance group is S_3, whose character table was constructed in chapter II.3:

S_3	n_c		1	$\bar{1}$	2
1		I	1	1	2
Z_3	2	(123), (132)	1	1	−1
Z_2	3	(12), (23), (31)	1	−1	0

(9)

The group consists of six elements divided into three equivalence classes. There are only three irreducible representations, known as 1, $\bar{1}$, and 2, according to their dimensions.

The triangular molecule provides us with a $DN = 3 \cdot 2 = 6$-dimensional representation whose characters we could write down almost immediately as $\chi(c) = \begin{pmatrix} 6 \\ 0 \\ 2 \end{pmatrix}$. For the benefit of some readers, remember how this comes about. The 6 is just the dimension of the representation; $\chi((123)) = 0$, since every mass or "atom" is moved under the transformation, so that all diagonal elements in the representation matrix vanish; and finally $\chi((12)) = 2$, since the x and y coordinates of atom 3 are untouched.

Plugging in (4), we obtain $1 \cdot 6^2 + 2 \cdot 0^2 + 3 \cdot 2^2 = 6 \sum_r n_r^2$, and so $\sum_r n_r^2 = 8$. The possible solutions are (i) $1^2 + 1^2 + 1^2 + 1^2 + 1^2 + 1^2 + 1^2 + 1^2 = 8$, (ii) $1^2 + 1^2 + 1^2 + 1^2 + 2^2 = 8$, and (iii) $2^2 + 2^2 = 8$.

But let's not forget the zero modes. We know of two: translating the whole system in the x direction and in the y direction. This furnishes a 2-dimensional representation under the invariance group, and so we know the irreducible representation 2 occurs at least once, which means that solution (i) is ruled out. (It seems very unlikely anyway, for the given representation to break into eight 1-dimensional pieces.)

As we remarked in chapter II.3, the constraints from group theory are so tightly interlocking that we can afford to forget all sorts of things, such as the physical argument about translation, since we are going to plug in (5) anyway. Write it in the form $n_r = \sum_c n_c \chi^{*(r)}(c) \chi(c) / N(G)$. We obtain

$$n_1 = (1 \cdot 1 \cdot 6 + 2 \cdot 1 \cdot 0 + 3 \cdot 1 \cdot 2)/6 = 2$$
$$n_{\bar{1}} = (1 \cdot 1 \cdot 6 + 2 \cdot 1 \cdot 0 + 3 \cdot (-1) \cdot 2)/6 = 0$$
$$n_2 = (1 \cdot 2 \cdot 6 + 2 \cdot (-1) \cdot 0 + 3 \cdot 0 \cdot 2)/6 = 2 \tag{10}$$

Solution (iii), $2^2 + 2^2 = 8$, is the right one: $n_1 = 2$, $n_{\bar{1}} = 0$, and $n_2 = 2$.

In this example, (6) is realized in the form (remember that $DN = 2 \cdot 3 = 6$, and so H is a 6-by-6 matrix)

$$H = \begin{pmatrix} \omega_{(1)}^2 & 0 & 0 & 0 & 0 & 0 \\ 0 & \omega_{(1)}'^2 & 0 & 0 & 0 & 0 \\ 0 & 0 & \omega_{(2)}^2 & 0 & 0 & 0 \\ 0 & 0 & 0 & \omega_{(2)}^2 & 0 & 0 \\ 0 & 0 & 0 & 0 & \omega_{(2)}'^2 & 0 \\ 0 & 0 & 0 & 0 & 0 & \omega_{(2)}'^2 \end{pmatrix} \tag{11}$$

With a combination of physical intuition and mathematical insight, we can readily figure out these harmonic modes. We already know about the two translational zero modes, as shown in figure 3a. There is also a rotational zero mode, as shown in figure 3b.

Confusio mumbles, "Shouldn't there be two, one clockwise, and one anticlockwise?"

No, Newton's equation involves the second derivative in time and thus is time-reversal invariant. The anticlockwise motion is just the clockwise motion time reversed. Another way of saying this is to notice that (2) determines ω^2, not ω. Also, Confusio can't be right,

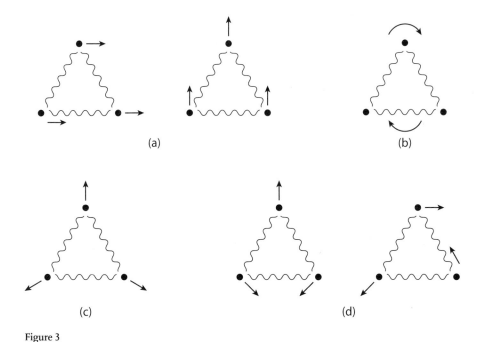

Figure 3

because if he were, then we would also count four, not two, translation zero modes, for translation in the $\pm x$ and $\pm y$ directions.

That takes care of one of the two 1-dimensional representations. Physically, we recognize the breathing mode, being invariant under S_3, furnishes a 1-dimensional representation. See figure 3c.

How about the other 2-dimensional representation? Linear algebra tells us that the eigenvectors of a real symmetric matrix are orthogonal to one another. Since we already know four of the six eigenvectors (two translations, one rotation, and one breather), it is easy to determine the 2-dimensional subspace. These two modes are shown in figure 3d.

If you want to determine the actual frequencies, then you have to find the eigenvalues of a DN-by-DN matrix. For DN large, numerical methods may be the last resort.[4] However, group theory can still be of help. In the example discussed here, we know that in (11), three frequencies vanish, and thus we have to deal with only a 3-by-3 matrix instead of a 6-by-6 matrix. Furthermore, since two of the three eigenvalues are known to be equal by group theory, we know that the equation for the eigenvalues must have the form $(\lambda - a)(\lambda - b)^2 = 0$.

Exercises

1 Let us make the "ridiculously simple" example of two masses connected by a spring and moving in one dimension a bit less simple by having two unequal masses. Without writing down any equations, do you think that there is a zero mode?

2 Use the Great Orthogonality theorem (II.2.2) to derive the projection operators into the desired modes.

Notes

1. It is also instructive to obtain this using the formalism given in the appendix to chapter III.3: for $D = 1$, all vectors are just numbers, in particular, $\hat{X}^0_{12} = 1$ and $V = \frac{1}{2}x^2_{12} = \frac{1}{2}(x_1 - x_2)^2 = \frac{1}{2}\sum_{a,b} x_a H^{ab} x_b$.

2. This has turned out to be an important concept in quantum field theory and in condensed matter physics.

3. It is straightforward (but tedious) to write down H, using the formalism in the appendix to chapter III.3. The relevant matrix is given explicitly in A. Nussbaum, Am. J. Phys. 36 (1968), p. 529, and solved in D. E. O'Connor, Am. J. Phys. 39 (1971), p. 847.

4. As Nussbaum did (see previous endnote).

Physics is where the action is

We talked about the invariance of physical laws as a motivation for group theory back in chapter I.1. At the level of the equation of motion, what we meant is clear: the left and right hand sides have to transform in the same way under the symmetry transformation. In particular, in the case of rotational invariance, force and acceleration both have to transform as a vector for $\vec{F} = m\vec{a}$ to make sense. Otherwise, Newton's force law would depend on how we orient the Cartesian coordinates. More generally, in the equations of motion of physics, every term has to transform in the same way; the equation is said to be covariant.

Here we go over alternative formulations in which the manifestation of symmetry is made even simpler and more transparent. We begin with a lightning review[1] of the Lagrangian, and later of the Hamiltonian, for those readers unfamiliar or rusty with these two concepts from classical mechanics. The Lagrangian will be needed in chapter IV.9 and in some chapters in parts V and VII, while the Hamiltonian has already been used in chapter III.1.

For a particle moving in one dimension in a potential, Newton with his $ma = F$ tells us that

$$m\frac{d^2q}{dt^2} = -V'(q) \tag{1}$$

with q the position of the particle. Lagrange[2] and others discovered later that this could be formulated as an action principle, with the action defined by the integral

$$S(q) \equiv \int dt\, L\left(\frac{dq}{dt}, q\right) = \int dt\left\{\frac{1}{2}m\left(\frac{dq}{dt}\right)^2 - V(q)\right\} \tag{2}$$

In the simplest example of a particle moving in one dimension, the Lagrangian is given by its kinetic energy minus its potential energy

$$L = \frac{1}{2}m\left(\frac{dq}{dt}\right)^2 - V(q) \tag{3}$$

What does the action principle say?

The action $S(q)$ is a functional[3] of the path $q(t)$. With each path, we assign a real number, namely, the action $S(q)$ evaluated for that particular path $q(t)$. We are then instructed to vary $S(q)$ subject to the condition that the initial and final positions $q(t_i) = q_i$ and $q(t_f) = q_f$ are to be held fixed.

The particular path $q(t)$ that extremizes the action $S(q)$ satisfies Newton's law. Often, the more profound a truth is in theoretical physics, the simpler the demonstration is. Here, too, the demonstration is almost laughably simple. We vary the action:

$$\delta S(q) = \int dt \delta L\left(\frac{dq}{dt}, q\right) = \int dt\left(\frac{\delta L}{\delta \frac{dq}{dt}}\delta\frac{dq}{dt} + \frac{\delta L}{\delta q}\delta q\right) \tag{4}$$

Using $\delta\frac{dq}{dt} = \frac{d(q+\delta q)}{dt} - \frac{dq}{dt} = \frac{d\delta q}{dt}$, integrate the first term in (4) by parts to obtain

$$\delta S = \int dt\left(\frac{d}{dt}\left(\frac{\delta L}{\delta \frac{dq}{dt}}\right) - \frac{\delta L}{\delta q}\right)\delta q(t) \tag{5}$$

The boundary terms vanish, because the initial and final positions are held fixed. The action is extremized when δS vanishes, which implies, since $\delta q(t)$ is arbitrary, the Euler-Lagrange equation

$$\frac{d}{dt}\left(\frac{\delta L}{\delta \frac{dq}{dt}}\right) - \frac{\delta L}{\delta q} = 0 \tag{6}$$

To be precise and pedantic, I stress that the notation means the following. We pretend that $L(a, b)$ is an ordinary function of two variables a and b. By $\frac{\delta L}{\delta \frac{dq}{dt}}$ we mean $\frac{\partial L(a,b)}{\partial a}$ with a subsequently set equal to $\frac{dq}{dt}$ and b to $q(t)$. Similarly, by $\frac{\delta L}{\delta q}$ we mean $\frac{\partial L(a,b)}{\partial b}$ with a subsequently set equal to $\frac{dq}{dt}$ and b to $q(t)$.

Switching from Leibniz's notation to Newton's notation $\dot{f}(x) = \frac{df}{dx}$, we can write the Euler-Lagrange equation (5) in the elegantly[4] compact form

$$\frac{\dot{\delta L}}{\delta \dot{q}} = \frac{\delta L}{\delta q} \tag{7}$$

This equation, suitably generalized to quantum fields, underlies all known dynamics in the universe (we will come back to this eventually, in parts VII and VIII).

Applied to the Lagrangian (3), we obtain (1): $\frac{\delta L}{\delta \frac{dq}{dt}} = m\frac{dq}{dt}$ and $\frac{\delta L}{\delta q} = -V'(q)$, and thus (6) states that $\frac{d}{dt}(m\frac{dq}{dt}) + V'(q) = 0$, namely, $ma = F$.

Note that the minus sign in the definition of the Lagrangian is absolutely necessary. For the harmonic oscillator, $V(q) = \frac{1}{2}kq^2$, and the equation of motion (6) reads $m\ddot{q} = -kq$, which is immediately solved by* $q \propto e^{-i\omega t}$ with $\omega = \sqrt{k/m}$.

---

* It is understood, as is standard, that the real part is to be taken.

Local versus global

Newton's equation of motion is described as "local" in time: it tells us what is going to happen in the next instant. In contrast, the action principle is "global": one integrates over various possible trajectories and chooses the best one.[5] While the two formulations are mathematically entirely equivalent, the action principle offers numerous advantages[6] over the equation of motion approach. Perhaps the most important is that the action leads directly to an understanding of quantum mechanics via the so-called Dirac-Feynman path integral, but that is another story for another time.

Historically, the action principle was inspired by Fermat's least time principle for light, which, by the way, goes all the way back to Heron of Alexandria (circa 65 A.D.). The least time principle has a strongly teleological flavor—that light, and particularly daylight, somehow knows how to save time—a flavor totally distasteful to the postrational palate. In contrast, during Fermat's time,[7] there was lots of quasitheological talk about Divine Providence and Harmonious Nature, so no one questioned that light would be guided to follow the most prudent path. After the success of the least time principle for light, physicists naturally wanted to find a similar principle for material particles.

Richard Feynman recalled that when he first learned of the action principle, he was blown away. Indeed, the action principle underlies some of Feynman's deepest contributions to theoretical physics, as was alluded to above.

The laws of physics and symmetry

At the start of this chapter, I mentioned that a law of physics enjoys a symmetry if the different terms in the equation of motion transform in the same way. In the least action formulation, the presence of a symmetry can be stated even more simply: the action S has to be invariant under the symmetry transformation; that is, S does not change. In other words, S must transform like a singlet. Put yet another way, S belongs to the trivial representation of the symmetry group. Again, the reason is clear: otherwise, the action S associated with each path would depend on how we orient the Cartesian coordinates.

All these statements about the action apply, in most* cases, to the Lagrangian as well: L has to be invariant under the symmetry transformation. The simplest example is that of a particle moving in a spherically symmetric potential. The Lagrangian is easy to write down, since we know that out of a vector $\vec{q}$ there is only one invariant we can form, namely, the scalar product $\vec{q} \cdot \vec{q}$. Thus, L must have the form (compare with (3)) $L = \frac{1}{2}m(\frac{d\vec{q}}{dt} \cdot \frac{d\vec{q}}{dt}) - V(|\vec{q}|)$, where $|\vec{q}| = (\vec{q} \cdot \vec{q})^{\frac{1}{2}}$ denotes the length of the vector $\vec{q}$. The rotation group is almost too familiar to be of much pedagogical value in this context.

---

* This caveat covers those cases in which δL does not vanish but is a total time derivative, so that with fixed initial and final conditions, δS still vanishes.

In general, the relevant symmetry group might be much less familiar than the rotation group. For example, we could have four equal masses connected by six identical springs to form a tetrahedron. The symmetry group would then be our old friend A_4 from part II.

The Hamiltonian

After Lagrange invented the Lagrangian, Hamilton invented the Hamiltonian.*

Given a Lagrangian $L(\dot{q}, q)$, define the momentum by $p = \frac{\delta L}{\delta \dot{q}}$ and the Hamiltonian by

$$H(p, q) = p\dot{q} - L(\dot{q}, q) \tag{8}$$

where it is understood that $\dot{q}$ on the right hand side is to be eliminated in favor of p.

Let us illustrate this procedure by a simple example. Given the Lagrangian $L(\dot{q}, q) = \frac{1}{2}m\dot{q}^2 - V(q)$ in (3), we have $p = m\dot{q}$, which is precisely what we normally mean by momentum. The Hamiltonian is then given by

$$H(p, q) = p\dot{q} - L(\dot{q}, q) = p\dot{q} - \frac{1}{2}m\dot{q}^2 + V(q) = \frac{p^2}{2m} + V(q) \tag{9}$$

where in the last step we wrote $\dot{q} = p/m$. You should recognize the final expression as the total energy, namely, the sum of the kinetic energy $\frac{p^2}{2m}$ and the potential energy $V(q)$. The Hamiltonian represents the total energy of the system.

Given a Hamiltonian $H(p, q)$, Hamilton postulated the equations[8]

$$\dot{q} = \frac{\partial H}{\partial p}, \quad \dot{p} = -\frac{\partial H}{\partial q} \tag{10}$$

For the Hamiltonian in (9), these equations read $\dot{q} = \frac{p}{m}$ and $\dot{p} = -\frac{\partial V}{\partial q}$. Plugging the first equation into the second, we obtain $m\ddot{q} = -V'(q)$, thus recovering the correct equation of motion.

The role of formalism in theoretical physics

Practical minded theoretical physicists would question the utility of introducing the Lagrangian and the Hamiltonian. These alternative formulations merely rewrite Newton's equation of motion, they might snort. Indeed, the theoretical physics literature is full of papers that many practitioners dismiss as mere formalism. Most of them do deserve to be consigned to the dustbins of history. But in a few cases, in particular the Lagrangian and the Hamiltonian (otherwise we would hardly be talking about them here), the formalism turns out to be of central importance and points the way toward further developments. In chapter IV.9, we will see that from the Lagrangian, it is but a short skip and hop to quantum field theory.

---

* There is of course a lot more to the Hamiltonian than given here, but this is all we need.

As for the Hamiltonian, it is well known that quantum mechanics was born when Heisenberg promoted q and p, and hence H, to operators. (Indeed, we already invoked this fact in chapter III.1.) We will discuss this further in chapter IV.2.

Take home message: Invariance easier than covariance

Given a Lagrangian, we have to vary it to obtain the equation of motion. For normal everyday use, we might as well stick with the equation of motion. But when symmetry comes to the fore and assumes the leading role, it is significantly easier to render the Lagrangian invariant than to render the equation of motion covariant. For the familiar rotation group, the advantage is minimal; however, when we come to more involved symmetries, finding the correct Lagrangian is usually much easier than finding the correct equation of motion. Indeed, as an example,[9] when Einstein searched for his theory of gravity, he opted for the equation of motion. He could have saved himself a considerable amount of travail if he had determined the relevant Lagrangian instead.

Symmetry and conservation

No free lunch in physics! What I mean to say is that conservation laws abound in physics: energy is conserved, momentum is conserved, angular momentum is conserved, the number of quarks in the universe is conserved, the list goes on and on. Emmy Noether, arguably the deepest woman physicist who ever lived, had the profound insight that each conservation law is associated with a continuous symmetry of the Lagrangian. For example, angular momentum conservation corresponds to invariance of the action under the rotation group $SO(3)$.

Once again, group theory reigns supreme. Indeed, each conserved quantity corresponds to a generator of the Lie algebra. I cannot go into further details here,[10] but mention the Noether relation between conservation and symmetry for future reference.

Appendix: Harmonic motion

Let us consider N equal masses tied together by (ideal, of course) springs and moving in D-dimensional space. As in chapter III.2, denote the small deviation of the ath particle ($a = 1, \cdots, N$) from its equilibrium position by x_a^i ($i = 1, 2, \cdots, D$). Then to leading order in x, the Lagrangian has the form

$$L = \sum_{a=1}^{N} \sum_{i=1}^{D} \frac{1}{2} m \left(\frac{dx_a^i}{dt} \right)^2 - \sum_{a,b=1}^{N} \sum_{i,j=1}^{D} x_a^i H^{ia,jb} x_b^j \tag{11}$$

Since x measures the deviation from equilibrium, in the potential energy the terms linear in x must vanish, leaving us with terms quadratic in x. (Terms that do not depend on x do not enter into the Euler-Lagrange equation.)

Assemble the coordinates x_a^i into a DN-dimensional vector x^A, $A = 1, 2, \cdots, DN$. Then x^A furnishes a representation of the symmetry group G, whatever it might be. Under a symmetry transformation, $x^A \rightarrow W^{AB} x^B$

for some set of matrices W. For the Lagrangian to be invariant, the real symmetric matrix H must satisfy $W^T H W = H$.

It may be instructive to work out the matrix H explicitly. Consider the Lagrangian $L = \frac{1}{2}m \sum_a (\frac{d\vec{X}_a}{dt})^2 - \sum_{a<b} V(\vec{X}_a - \vec{X}_b)$ for N particles of equal mass. Denote the equilibrium positions by $\vec{X}_a^0$, and write $\vec{X}_a(t) = \vec{X}_a^0 + \vec{x}_a(t)$. Introduce the self-evident notation $\vec{X}_{ab} \equiv \vec{X}_a - \vec{X}_b$, $\vec{X}_{ab}^0 \equiv \vec{X}_a^0 - \vec{X}_b^0$, and $\vec{x}_{ab} \equiv \vec{x}_a - \vec{x}_b$. Next, expand $\sum_{a<b} V(\vec{X}_a - \vec{X}_b)$ to quadratic order in the small deviations $\vec{x}_{ab}$. The linear terms sum up to 0 by definition.

To first order in x, we have $|\vec{X}| = |\vec{X}^0 + \vec{x}| = \sqrt{(\vec{X}^0)^2 + 2\vec{X}^0 \cdot \vec{x} + \vec{x}^2} \simeq |\vec{X}^0| + \hat{X}^0 \cdot \vec{x}$, where $\hat{X}^0 \equiv \vec{X}^0/|\vec{X}^0|$ is the unit vector in the direction of $\vec{X}^0$.

Take $V(\vec{X}_{ab})$ to have the form $f(|\vec{X}_{ab}| - l)$, for example, with l the unstretched length of the spring, say. For $f(|\vec{X}| - l) = \frac{1}{2}k(|\vec{X}| - l)^2$ with $l = |\vec{X}^0|$, we obtain, in the harmonic approximation,

$$L = \frac{1}{2}m \sum_a \left(\frac{d\vec{x}_a}{dt}\right)^2 - \sum_{a<b} \frac{1}{2}k(\hat{X}_{ab}^0 \cdot \vec{x}_{ab})^2 \tag{12}$$

To obtain the equation of motion, we plug L into (6); but it is even simpler than that. We merely have to note that the potential energy term in L is quadratic in the collection of xs, and so the $\frac{\delta L}{\delta q}$ term in (6) is linear in the xs. We obtain (III.2.1) in general.

The explicit expression for H can be a bit of a mess, as we could see from (12). For example, for the toy tetrahedral "molecule" alluded to earlier, we are going to have factors like $\sqrt{3}$ all over the place. But, as was explained in chapter III.2, the power of group theory is such that for many purposes, a detailed knowledge of H (which, as given by a toy model, may be unreliable anyway for a real molecule) may not be necessary.

Notes

1. Detailed discussions can be found in all textbooks on classical mechanics. A particularly congenial treatment is given in G Nut, chapter II.3.
2. Starting when he was 18, Joseph Louis, the Comte de Lagrange (1736–1813) (who, by the way, was born Giuseppe Lodovico Lagrangia before the term "Italian" existed), worked on the problem of the tautochrone, which nowadays we would describe as the problem of finding the extremum of functionals. A year or so later, he sent a letter to Leonhard Euler (1707–1783), the leading mathematician of the time, to say that he had solved the isoperimetrical problem: for curves of a given perimeter, find the one that would maximize the area enclosed. Euler had been struggling with the same problem, but he generously gave the teenager full credit. Later, he recommended that Lagrange should succeed him as the director of mathematics at the Prussian Academy of Sciences.
3. A functional is a function of functions; in this case S is a function of $q(t)$. You give S some $q(t)$, and it outputs a number.
4. At Princeton University there is a church-like gothic building with stained glass windows, each of which is inlaid with a fundamental equation of physics. One of the equations is (7).
5. Naturally, the action principle suggests a metaphor for life.
6. See G Nut, p. 141.
7. Pierre Fermat (1601 or 1607/08?–1665). The heavy academic controversy over Fermat's birth year stems from his father marrying twice and naming two sons from two different wives both Pierre. See K. Barner, NTM 9 (4) (2001), p. 209.
8. Hamilton's equations are said to have a symplectic structure. Package q and p into a 2-dimensional vector $Z = \begin{pmatrix} q \\ p \end{pmatrix}$. Then $\frac{dZ_i}{dt} = \varepsilon_{ij} \frac{\partial H}{\partial Z_j}$, where the totally antisymmetric symbol is defined by $\varepsilon_{12} = +1 = -\varepsilon_{21}$. We will study the symplectic group in chapter IV.8, defined in terms of an antisymmetric symbol J, which generalizes ε.
9. See G Nut, chapter VI.1.
10. See, for example, G Nut, chapter II.4. The proof is actually very simple.

Part IV | Tensor, Covering, and Manifold

No need to be tense about tensors at all. Thanks to index democracy, tensors furnish a clever device for generating ever larger representations. The $SO(N)$ groups, which generalize the ordinary rotation group, are used to illustrate this claim.

Another clever trick: introduce raising and lowering operators in the $SO(3)$ algebra. You can then figure out the allowed representations by climbing up and down a ladder and requiring it to end at both ends. Creation and annihilation operators represent another clever trick. You learn how to multiply two representations together and decompose the product.

The method of tensors is extended to $SU(N)$, with the important consequence that the indices now have to live on two floors. From $SU(N)$ we specialize to $SU(2)$ and learn about its many peculiarities. In particular, it double covers $SO(3)$. You will meet the mysterious spinor that turns out to be crucial for modern physics. The behavior of electron spin under time reversal leads to a two-fold degeneracy.

You will learn to integrate over continuous groups, some of which have interesting topologies. The familar sphere appears not as a group manifold, but as a coset manifold.

The symplectic groups and their algebras are studied, with some interesting isomorphisms with more familiar algebras.

You will learn that quantum field theory hardly deserves its frightening reputation among students; it is but a short hop from Lagrangian mechanics.

Finally, in four interludes, we sample some interesting topics, such as crystal field splitting and special functions such as the one belonging to Bessel.

IV.1 | Tensors and Representations of the Rotation Groups SO(N)

Representing the rotation groups

By now, you know how to represent a finite group. How about representing the continuous groups, such as $SO(N)$, the rotation group in N-dimensional Euclidean space?

We defined $SO(N)$, back in chapter I.3, as the group of all N-by-N matrices R satisfying

$$R^T R = I \tag{1}$$

and

$$\det R = 1 \tag{2}$$

We might as well do things more generally in N-dimensional space; it costs us practically nothing except for allowing the relevant matrix indices to run all the way up to N instead of 3. But to help those readers seeing this for the first time focus, I will often specialize to $N = 3$. Henceforth, I will jump back and forth between N and 3 without further ado. We will see later that $SO(3)$ has some special features not shared by $SO(N)$.

So, we already know from (1) and (2) the N-dimensional defining or fundamental representation (as well as the trivial 1-dimensional representation, of course). The elements of the $SO(N)$ are represented, by definition, by the N-by-N matrices transforming the N unit basis vectors $\vec{e}_1, \vec{e}_2, \ldots, \vec{e}_N$ into one another. More precisely, the N-dimensional irreducible representation is furnished by a vector. Let me stress again that a vector is defined* by how it transforms under a rotation:

$$V^i \rightarrow V'^i = R^{ij} V^j \tag{3}$$

with $i, j = 1, 2, \cdots, N$.

---

* In chapter II.1, we referred to a column of numbers in a character table as a vector. In the present context, as in chapter I.3, we revert to a more restrictive definition of a vector, as it appears in physics.

Several questions and some flying guesses

Several questions naturally suggest themselves (as in chapter II.1). How many irreducible representations does the rotation group have? How are they to be characterized? What are their dimensions?

If you are encountering this material for the first time, the answers may be far from self-evident. Indeed, if I were to tell you that the rotation group for good old 3-dimensional Euclidean space has a 5-dimensional irreducible representation, it may very well come as a surprise to you, as it did to me. (Here I am of course not addressing either those readers previously exposed to the rotation group or those born with devastatingly perspicacious mathematical insight.) What are these five beasts living in 3-dimensional space?

Our understanding of the representation theory of finite groups offers some suggestive answers to the questions posed above. As indicated in chapter II.1, the theorems we proved for finite groups depend crucially on the rearrangement identity and on the meaningfulness of the sum over group elements. Our intuition suggests that, with some suitable modifications, they should continue to hold.

Indeed, as proud physicists, let's take a flying guess at the representations of $SO(2)$. Consider the finite group Z_N, which we may think of as generated by the rotation through angle $2\pi/N$. In chapter II.1, we learned that it has N 1-dimensional irreducible representations, labeled by an integer $k = 0, \cdots, N-1$. The group element $e^{i2\pi j/N}$ is represented by $D^{(k)}(e^{i2\pi j/N}) = e^{i2\pi kj/N}$. Well, as $N \to \infty$, the finite group Z_N should turn into the continuous group $SO(2)$. So, $SO(2)$ should have an infinite number of irreducible representations corresponding to $k = 0, \cdots, \infty$, in which a rotation through angle θ is represented by $e^{ik\theta}$.

Ha! The fly-by-night guess works: $D^{(k)}(\theta)D^{(k)}(\theta') = e^{ik\theta}e^{ik\theta'} = e^{ik(\theta+\theta')} = D^{(k)}(\theta + \theta')$ and $D^{(k)}(2\pi) = 1$. (For more on this point, see later in this chapter.) What do you expect? We are self-confessed sloppy physicists in flagrant disregard for mathematical decorum, and we take limits whenever we like.

What about $SO(3)$, a larger group containing $SO(2)$ as a subgroup? Anybody with the absolute minimum amount of mathematical sense can see that the irreducible representations of a group cannot necessarily be "lifted" into the irreducible representations of a larger group containing that group as a subgroup. But still, let's make another flying guess: $SO(3)$ also has an infinite number of irreducible representations labeled by an integer. That also turns out to be true.

Constructing the irreducible representations of $SO(N)$

Both physicists and mathematicians, for reasons mysterious to the unsuspecting guy or gal in the street, would like to construct larger or higher-dimensional irreducible representations.* Back in chapter II.1, in an attempt to construct larger representations of $SO(3)$, we, with laughable naiveté, simply stacked smaller representations together. The resulting representation is, by definition, reducible. No good. To obtain larger irreducible representations, we have to do something slightly more sophisticated: we have to generalize the concept of a vector by considering objects carrying more indices.

Imagine a collection of mathematical entities T^{ij} carrying two indices, with $i, j = 1, 2, \cdots, N$ in N-dimensional space. If the T^{ij}s transform under rotations according to

$$T^{ij} \rightarrow T'^{ij} = R^{ik} R^{jl} T^{kl} \tag{4}$$

then we say that T transforms like a tensor, and hence is a tensor.[1] (Here we are using the Einstein summation convention introduced in the review of linear algebra: The right hand side actually means $\sum_{k=1}^{N} \sum_{l=1}^{N} R^{ik} R^{jl} T^{kl}$, that is, a double sum of N^2 terms.) Indeed, we see that we are just generalizing the transformation law of a vector.

Fear of tensors

Allow me to digress. In my experience teaching, invariably, a couple of students are confused by the notion of tensors. The very word "tensor" apparently makes them tense. Compounded and intertwined with their fear of tensors, the unfortunates mentioned above are also unaccountably afraid of indices. Dear reader, if you are not one of these unfortunates, so much the better for you! You can zip through this chapter. But to allay the nameless fear of the tensorphobe, I will go slow and be specific.

Think of the tensor T^{ij} as a collection of N^2 mathematical entities that transform into linear combinations of one another. Let us list T^{ij} explicitly for $N = 3$. There are $3^2 = 9$ of them: $T^{11}, T^{12}, T^{13}, T^{21}, T^{22}, T^{23}, T^{31}, T^{32}$, and T^{33}. That's it, nine objects that transform into linear combinations of one another. For example, (4) says that $T'^{21} = R^{2k} R^{1l} T^{kl} = R^{21} R^{11} T^{11} + R^{21} R^{12} T^{12} + R^{21} R^{13} T^{13} + R^{22} R^{11} T^{21} + R^{22} R^{12} T^{22} + R^{22} R^{13} T^{23} + R^{23} R^{11} T^{31} + R^{23} R^{12} T^{32} + R^{23} R^{13} T^{33}$. This shows explicitly, as if there were any doubt to begin with, that T'^{21} is given by a particular linear combination of the nine objects. That's all: the tensor T^{ij} consists of nine objects that transform into linear combinations of themselves under rotations.

We could generalize further and define[†] 3-indexed tensors, 4-indexed tensors, and so forth by transformation laws, such as $W^{ijn} \rightarrow W'^{ijn} = R^{ik} R^{jl} R^{nm} W^{klm}$. Here we focus on 2-indexed tensors, and if we say "tensor" without any qualifier we often, but not always, mean a 2-indexed tensor. (With this definition, we might say that a vector is a 1-indexed tensor and a scalar is a 0-indexed tensor, but this usage is not common.) A scalar transforms

---

* Historically, group theory did not develop in the glib way suggested here, of course. Tensors came up naturally in physics and played an important role. Both physicists and mathematicians in the eighteenth and nineteenth centuries tried to solve complicated problems involving rotation, problems with technological importance, such as the precession of the gyroscope.

† Our friend the jargon guy tells us that the number of indices carried by a tensor is known as its rank.

as a tensor with no index at all, namely, $S \to S' = S$; in other words, a scalar does not transform. The scalar furnishes the 1-dimensional trivial representation.

Representation theory

Mentally arrange* the nine objects T^{ij} in a column $\begin{pmatrix} T^{11} \\ T^{12} \\ \vdots \\ T^{33} \end{pmatrix}$. The linear transformation on the nine objects can then be represented by a 9-by-9 matrix $D(R)$ acting on this column.[†]

For every rotation, specified by a 3-by-3 matrix R, we can thus associate a 9-by-9 matrix $D(R)$ transforming the nine objects T^{ij} linearly among themselves. It is fairly clear that $D(R)$ gives a 9-dimensional representation of $SO(3)$, but for the benefit of the abecedarian, let us verify this explicitly. Transform by the rotation R_1 followed by the rotation R_2. Using (4) twice, we obtain

$$T^{ij} \to T'^{\,ij} = R_1^{ik} R_1^{jl} T^{kl}$$

$$\to T''^{\,ij} = R_2^{ik} R_2^{jl} T'^{\,kl} = R_2^{ik} R_1^{km} R_2^{jl} R_1^{ln} T^{mn} = (R_2 R_1)^{im} (R_2 R_1)^{jn} T^{mn} \tag{5}$$

Thus, indeed, $D(R_2)D(R_2) = D(R_2 R_1)$. You see that the key is that each of the two indices on T^{ij} transforms independently, that is, in parallel without interfering with each other. They live in the same household but do not talk to each other.

Another way to think of this (but one that is potentially confusing to some students) is that (4) mandates that a tensor T^{ij} transforms as if it were equal[‡] to the product of two vectors $V^i W^j$, but in general, it isn't.[2] (Indeed, if T^{ij} were actually equal to $V^i W^j$, then $T^{11}/T^{21} = T^{12}/T^{22} = \cdots = T^{1N}/T^{2N} = V^1/V^2$, and the N^2 components of the tensor would not be all independent.)

The tensor T furnishes a 9-dimensional representation of the rotation group $SO(3)$.

Reducible or irreducible?

But is this 9-dimensional representation reducible or not? Of these nine entities T^{ij} that transform into one another, is there a subset among them that only transform into one another? A secret in-club, as it were.

A moment's thought reveals that there is indeed an in-club. Consider $A^{ij} \equiv T^{ij} - T^{ji}$. Under a rotation,

---

* The order of the nine objects does not matter. Changing the order merely corresponds to a similarity transformation of $D(R)$, as indicated in chapter II.1.

[†] Here we are going painfully slowly because of common confusion on this point. In a standard abuse of terminology, some authors refer to this column as a nine-component "vector." In this chapter, as I've already said, the word "vector" is reserved for something that transforms like a vector: $V'^i = R^{ij} V^j$. It is not true that any old collection of stuff arranged in a column transforms like a vector. See exercise I.3.1.

[‡] Compare $V^i W^j \to V'^i W'^j = R^{ik} V^k R^{jl} W^l = R^{ik} R^{jl} V^k W^l$ with (4).

$$A^{ij} \to A'^{ij} \equiv T'^{ij} - T'^{ji} = R^{ik}R^{jl}T^{kl} - R^{jk}R^{il}T^{kl}$$
$$= R^{ik}R^{jl}T^{kl} - R^{jl}R^{ik}T^{lk} = R^{ik}R^{jl}(T^{kl} - T^{lk}) = R^{ik}R^{jl}A^{kl} \tag{6}$$

I have again gone painfully slowly here, but it is obvious, isn't it? We just verified in (6) that A^{ij} transforms like a tensor and is thus a tensor. Furthermore, this tensor changes sign on interchange of its two indices ($A^{ij} = -A^{ji}$) and so is said to be antisymmetric. The transformation law (4) treats the two indices democratically, without favoring one over the other, as was pointed out already, and thus preserves the antisymmetric character of a tensor: if $A^{ij} = -A^{ji}$, then $A'^{ij} = -A'^{ji}$ also. Again, the key is that each of the two indices on T^{ij} transforms independently.

How many objects are here? Let us count. The index i in A^{ij} can take on N values; for each of these values, the index j can take on only $N - 1$ values (since the N diagonal elements $A^{ii} = -A^{ii} = 0$ for $i = 1, 2, \cdots, N$, repeated index summation suspended momentarily here); but to avoid double counting (since $A^{ij} = -A^{ji}$), we should divide by 2. Hence, the number of independent components in A is equal to $\frac{1}{2}N(N - 1)$. For example, for $N = 3$ we have the three objects A^{12}, A^{23}, and A^{31}. The attentive reader will recall that we did the same counting in chapter I.3.

Clearly, the same goes for the symmetric combination $S^{ij} \equiv T^{ij} + T^{ji}$. You can verify as a trivial exercise that $S^{ij} \to S'^{ij} = R^{ik}R^{jl}S^{kl}$. A tensor S^{ij} that does not change sign on interchange of its two indices ($S^{ij} = S^{ji}$) is said to be symmetric. In addition to the components S^{ij} with $i \neq j$, S also has N diagonal components, namely, $S^{11}, S^{22}, \cdots, S^{NN}$. Thus, the number of independent components in S is equal to $\frac{1}{2}N(N - 1) + N = \frac{1}{2}N(N + 1)$. This is a long-winded way of saying that the symmetric tensor S has more components than the antisymmetric tensor A, but I have encountered confusion here among beginning students also.

For $N = 3$, the number of components in A and S are $\frac{1}{2} \cdot 3 \cdot 2 = 3$ and $\frac{1}{2} \cdot 3 \cdot 4 = 6$, respectively. (For $N = 4$, the number of components in A and S are 6 and 10, respectively.) Thus, in a suitable basis, the 9-by-9 matrix referred to above actually breaks up into a 3-by-3 block and a 6-by-6 block. The 9-dimensional representation is reducible.

But we are not done yet. The 6-dimensional representation is also reducible. To see this, note that

$$S^{ii} \to S'^{ii} = R^{ik}R^{il}S^{kl} = (R^T)^{ki}R^{il}S^{kl} = (R^{-1})^{ki}R^{il}S^{kl} = \delta^{kl}S^{kl} = S^{kk} \tag{7}$$

In the third equality, we used the O in $SO(N)$. (Here we are using repeated index summation: the indices i and k are both summed over.) In other words, the linear combination $S^{11} + S^{22} + \cdots + S^{NN}$, the trace of S, transforms into itself; that is, it does not transform at all. It is a loner forming an in-club of one. The 6-by-6 matrix describing the linear transformation of the six objects S^{ij} breaks up into a 1-by-1 block and a 5-by-5 block.

Again, for the sake of the beginning student, let us work out explicitly the five objects that furnish the representation 5 of $SO(3)$. First define a traceless symmetric tensor $\tilde{S}$ for $SO(N)$ by

$$\tilde{S}^{ij} = S^{ij} - \delta^{ij}(S^{kk}/N) \tag{8}$$

(The repeated index k is summed over.) Explicitly, $\tilde{S}^{ii} = S^{ii} - N(S^{kk}/N) = 0$, and $\tilde{S}$ is traceless. Specialize to $N = 3$. Now we have only five objects, namely, $\tilde{S}^{11}, \tilde{S}^{22}, \tilde{S}^{12}, \tilde{S}^{13}$, and $\tilde{S}^{23}$. We do not count $\tilde{S}^{33}$ separately, since it is equal to $-(\tilde{S}^{11} + \tilde{S}^{22})$.

Under an $SO(3)$ rotation, these five objects transform into linear combinations of one another, as just explained. Let us be specific: the object $\tilde{S}^{13}$, for example, transforms into $\tilde{S}'^{13} = R^{1k}R^{3l}\tilde{S}^{kl} = R^{11}R^{31}\tilde{S}^{11} + R^{11}R^{32}\tilde{S}^{12} + R^{11}R^{33}\tilde{S}^{13} + R^{12}R^{31}\tilde{S}^{21} + R^{12}R^{32}\tilde{S}^{22} + R^{12}R^{33}\tilde{S}^{23} + R^{13}R^{31}\tilde{S}^{31} + R^{13}R^{32}\tilde{S}^{32} + R^{13}R^{33}\tilde{S}^{33} = (R^{11}R^{31} - R^{13}R^{33})\tilde{S}^{11} + (R^{11}R^{32} + R^{12}R^{31})\tilde{S}^{12} + (R^{11}R^{33} + R^{13}R^{31})\tilde{S}^{13} + (R^{12}R^{32} - R^{13}R^{33})\tilde{S}^{22} + (R^{12}R^{33} + R^{13}R^{32})\tilde{S}^{23}$, where in the last equality we used $\tilde{S}^{ij} = \tilde{S}^{ji}$ and $\tilde{S}^{33} = -(\tilde{S}^{11} + \tilde{S}^{22})$. Indeed, $\tilde{S}^{13}$ transforms into a linear combination of $\tilde{S}^{11}, \tilde{S}^{22}, \tilde{S}^{12}, \tilde{S}^{13}$, and $\tilde{S}^{23}$.

To summarize, if instead of the basis consisting of the nine entities T^{ij}, we use the basis consisting of the three entities A^{ij}, the single entity S^{kk} (remember repeated index summation!), and the five entities $\tilde{S}^{ij}$, the 9-by-9 matrix $D(R)$ breaks up into a 3-by-3 matrix, a 1-by-1 matrix, and a 5-by-5 matrix "stacked on top of one another." This is represented schematically as

$$S^{-1}D(R)S = \begin{pmatrix} \text{(3-by-3 block)} & 0 & 0 \\ 0 & \text{(1-by-1 block)} & 0 \\ 0 & 0 & \text{(5-by-5 block)} \end{pmatrix} \tag{9}$$

Note that once we choose the new basis, this decomposition holds true for all rotations. In other words, there exists a similarity transformation S that block diagonalizes $D(R)$ for all R.

We say that in $SO(3)$, $9 = 5 \oplus 3 \oplus 1$. More generally, the N^2-dimensional representation furnished by a general 2-indexed tensor decomposes into a $\frac{1}{2}N(N-1)$-dimensional representation, a $(\frac{1}{2}N(N+1) - 1)$-dimensional representation, and a 1-dimensional representation. For example, in $SO(4)$, $16 = 9 \oplus 6 \oplus 1$; in $SO(5)$, $25 = 14 \oplus 10 \oplus 1$.

You might have noticed that in this entire discussion we never had to write out R explicitly in terms of the three rotation angles and how the five objects $\tilde{S}^{11}, \cdots, \tilde{S}^{23}$ transform into one another in terms of these angles. It is only the counting that matters. You might regard that as the difference between mathematics and arithmetic.[3]

An advanced tidbit

Since I am going so slowly in this chapter, I am worried about boring the more advanced readers. Here is a tidbit for them. With some "minor" modifications, both the 6-dimensional and the 10-dimensional representations of $SO(4)$ play a glorious role in theoretical physics. Can you guess what they are? A collection of six objects, and a collection of ten objects.

I offer you a hint. Had Maxwell and Einstein known some group theory, the development of theoretical physics might have been accelerated.[4]

Invariant symbols

At this point, let us formalize what we just did. The discussion, to be given for general N, will end up underlining why $N = 3$ is special.

First, go back to (7) and write it as $S'^{ii} = \delta^{ij} S'^{ij} = (\delta^{ij} R^{ik} R^{jl}) S^{kl} = \delta^{kl} S^{kl} = S^{kk}$. Thus, the crucial property we need to show that the trace does not transform (in other words, that it furnishes the 1-dimensional trivial representation) is embedded in the third equality here, namely

$$\delta^{ij} R^{ik} R^{jl} = \delta^{kl} \tag{10}$$

This is just (1), and so, we have used the O in $SO(N)$.

What about the S in $SO(N)$, namely, the requirement that det $R = 1$?

Recall from the review of linear algebra that the determinant can be written in terms of the antisymmetric symbol $\varepsilon^{ijk\cdots n}$. In N-dimensional space, the antisymmetric symbol carries N indices and is defined by its two properties:

$$\varepsilon^{\cdots l \cdots m \cdots} = -\varepsilon^{\cdots m \cdots l \cdots} \quad \text{and} \quad \varepsilon^{12\cdots N} = 1 \tag{11}$$

In other words, the antisymmetric symbol ε flips sign on the interchange of any pair of indices. It follows that ε vanishes when two indices are equal. (Note that the second property listed is just normalization.) Since each index can take on only values 1, 2, $\cdots$, N, the antisymmetric symbol for N-dimensional space must carry N indices, as already noted. For example, for $N = 2$, $\varepsilon^{12} = -\varepsilon^{21} = 1$, with all other components vanishing. For $N = 3$, $\varepsilon^{123} = \varepsilon^{231} = \varepsilon^{312} = -\varepsilon^{213} = -\varepsilon^{132} = -\varepsilon^{321} = 1$, with all other components vanishing.

The determinant is defined, for any matrix R, by

$$\varepsilon^{ijk\cdots n} R^{ip} R^{jq} R^{kr} \cdots R^{ns} = \varepsilon^{pqr\cdots s} \det R \tag{12}$$

See the review of linear algebra. (Verify this for $N = 2$ and 3.) For R a rotation, det $R = 1$, and hence

$$\varepsilon^{ijk\cdots n} R^{ip} R^{jq} R^{kr} \cdots R^{ns} = \varepsilon^{pqr\cdots s} \tag{13}$$

Colloquially speaking, when a bunch of rotation matrices encounter the antisymmetric symbol, they poof into thin air.

Inspecting (10) and (13) (namely, the O and the S in $SO(N)$, respectively), we see that δ^{ij} and $\varepsilon^{ijk\cdots n}$ can be thought of as invariant symbols: when acted on by rotation matrices in the manner shown, they turn into themselves. (We refer to δ^{ij} and $\varepsilon^{ijk\cdots n}$ as symbols rather than tensors,[5] because they are merely a collection of 0s and 1s.)

We mention in passing that there are various identities involving these symbols; see various exercises in this chapter. One trivial but important identity is that if U^{ij} is a symmetric tensor and V^{ij} an antisymmetric tensor, then $U^{ij} V^{ij} = 0$. To see this, note that $U^{ij} V^{ij} = -U^{ji} V^{ji} = -U^{ij} V^{ij} = 0$, since something equal to its own negative has to vanish. Note that the second equality follows from relabeling the dummy summation indices. In particular, when the antisymmetric symbol is contracted with any symmetric tensor, the result vanishes. For example, $\varepsilon^{ijk\cdots n}U^{ij} = 0$.

Dual tensors

In light of this discussion, given an antisymmetric tensor A^{ij}, we can define another antisymmetric tensor $B^{k\cdots n} = \varepsilon^{ijk\cdots n}A^{ij}$ carrying $N - 2$ indices. Because of (11), the tensor $B^{k\cdots n}$ manifestly flips sign on exchange of any pair of indices. Let us verify that it is in fact a tensor. Under a rotation, $B^{k\cdots n} \rightarrow \varepsilon^{ijk\cdots n}R^{ip}R^{jq}A^{pq}$. But multiplying (13) by a bunch of R^Ts carrying appropriate indices, we have $\varepsilon^{ijk\cdots n}R^{ip}R^{jq} = \varepsilon^{pqr\cdots s}R^{kr}\cdots R^{ns}$. (Derive this!) Thus,

$$B^{k\cdots n} \rightarrow \varepsilon^{ijk\cdots n}R^{ip}R^{jq}A^{pq} = \varepsilon^{pqr\cdots s}R^{kr}\cdots R^{ns}A^{pq} = R^{kr}\cdots R^{ns}B^{r\cdots s} \tag{14}$$

precisely how a tensor carrying $N - 2$ indices should transform. The tensors A and B are said to be dual to each other.

In particular, for $N = 3$, $B^k = \varepsilon^{ijk}A^{ij}$ carries $3 - 2 = 1$ index and transforms like a vector. Thus, in the preceding discussion, when we discovered that the 9-dimensional reducible representation decomposes into $5 \oplus 3 \oplus 1$, the 3 is not a new irreducible representation, but just the good old vector or defining representation. (The 1 is of course just the trivial representation.) The one new irreducible representation that we have discovered is the 5. This result is far from trivial. A priori, if you have never heard of any of this, you might be quite surprised, as I said earlier, that 3-dimensional rotations could transform 5 objects* exclusively into linear combinations of themselves.

For $N = 4$, a 2-indexed antisymmetric tensor is dual to another 2-indexed antisymmetric tensor, since $4 - 2 = 2$. This "peculiar" fact has also played an important role in theoretical physics. Here is a tidbit for the more advanced reader, related to the tidbit given earlier: the electric and magnetic fields are dual to each other. We will come back to this later.

Constructing larger irreducible representations of $SO(N)$

We are now able to construct a large class† of irreducible representations of $SO(N)$, known (not surprisingly) as the tensor representations, each furnished by a tensor $T^{ij\cdots m}$, transforming by definition according to $T^{ij\cdots n} \rightarrow T'^{ij\cdots n} = R^{ik}R^{jl}\cdots R^{nm}T^{kl\cdots m}$. As was explained earlier, since the transformation treats each index on T democratically, T' will have whatever symmetry properties T has. For example, suppose T is symmetric in its first three indices, antisymmetric in its next four indices, symmetric in its next two indices, and so on; then T' will be the same.

---

* In classical mechanics, they appear, for example, in the inertial moment tensor $\int d^3x\rho(x)(x^ix^j - \frac{1}{3}\delta^{ij}x^2)$.
† We will come back to spinor representation in chapter VII.1.

Some people spend some fractions of their lives classifying and studying all possible such patterns, but in the spirit of the Feynman "shut up and calculate" school of physics,[*] we will restrain ourselves from doing so. The language of Young tableaux was invented to keep track of these patterns. Fortunately, Nature is kind to us, and mostly we will have to deal only with smallish groups, such as $SO(3)$ and $SO(4)$. Even in particle physics, where largish groups like $SO(10)$ and $SO(18)$ occasionally pop up, we only have to consider tensors with no more than a few indices. My professor in graduate school told me not to bother learning Young tableaux, and I am giving you the same advice here. If you do run across a tensor that is neither totally symmetric nor totally antisymmetric, such as the example given above, you can usually deal with it on a case-by-case basis. It is rare that you will need a full-blown treatment using Young tableaux. In a book this size, I have to omit some topics, and Young tableaux is one of those that I choose to omit.

Contraction of indices

When we set two indices on a tensor equal and sum, as in (7), we say that we contract the two indices. To see how this works in general, take a general tensor transforming like $T^{ij\cdots np} \to T'^{ij\cdots np} = R^{ik}R^{jl}\cdots R^{nm}R^{pq}T^{kl\cdots mq}$. Take any two indices, say j and n, and contract them. Then

$$T^{ij\cdots jp} \to T'^{ij\cdots jp} = R^{ik}R^{jl}\cdots R^{jm}R^{pq}T^{kl\cdots mq} = R^{ik}\cdots R^{pq}T^{kl\cdots lq} \tag{15}$$

since $R^{jl}R^{jm} = \delta^{lm}$. In other words, $T^{ij\cdots jp}$ transforms like a tensor $T^{i\cdots p}$ with two fewer indices; the contracted indices j and n have disappeared, knocking each other off, so to speak. You see that (7) is just a special case of this: S^{ii} transforms just as if it has no index.

Confusio says, "Yes, I get it. Even though the letter j appears in $T^{ij\cdots jp}$, this actually stands for $T^{i1\cdots 1p} + T^{i2\cdots 2p} + \cdots + T^{iN\cdots Np}$."

Why $SO(3)$ is special

As we saw just now, 3 is special because $3 - 2 = 1$: a pair of antisymmetric indices can always be traded for a single index. For $SO(3)$, we claim that we need do business only with totally symmetric traceless tensors carrying j indices, with j an arbitrary positive integer, that is, a tensor $S^{i_1 i_2 \cdots i_j}$ that remains unchanged on the interchange of any pair of indices and that vanishes when any two indices are contracted.

The claim will be proved inductively in j. We have already seen that the claims hold for $j = 2$: the antisymmetric 2-indexed tensor is equivalent to a 1-indexed tensor.

Let us now move on to a 3-indexed tensor T^{ijk} and ask what new irreducible representation it contains. As before, we could symmetrize and antisymmetrize in the first two

---

[*] Not to mention the theorem that life is short.

indices and decompose the tensor into the symmetric combination $T^{\{ij\}k} \equiv (T^{ijk} + T^{jik})$ and the antisymmetric combination $T^{[ij]k} \equiv (T^{ijk} - T^{jik})$. (The standard notation $\{ij\}$ and $[ij]$ indicates that the tensor is symmetric or antisymmetric, respectively, in the bracketed indices.)

We don't care about the antisymmetric combination $T^{[ij]k}$, because we know that secretly it is just a 2-indexed tensor $B^{lk} \equiv \varepsilon^{ijl} T^{[ij]k}$, and we have already disposed of all* 2-indexed tensors. Our attack is inductive, as I said.

As for the symmetric combination $T^{\{ij\}k}$, we can now proceed to make it symmetric in all three indices by brute force. (Go ahead, do it before reading on.) Explicitly, write $3T^{\{ij\}k} = (T^{\{ij\}k} + T^{\{jk\}i} + T^{\{ki\}j}) + (T^{\{ij\}k} - T^{\{jk\}i}) + (T^{\{ij\}k} - T^{\{ki\}j})$. Verify that the expression in the first parenthesis is completely symmetric in all three indices; indeed, it is just the sum over the $6 = 3!$ permutations of the three indices carried by T^{ijk}. The expressions in the other two round parentheses are antisymmetric in ki and kj, respectively; we can thus multiply them by ε^{kil} and ε^{kjl}, respectively, turning them into 2-indexed tensors, which we drop down the inductive ladder.

Thus, the only thing new is a 3-indexed tensor S^{ijk} totally symmetric in all three of its indices. Furthermore, as in our preceding discussion, we can subtract out its trace, so that the resulting tensor† $\tilde{S}^{ijk}$ is traceless; that is, so that $\delta^{ij} \tilde{S}^{ijk} = 0$. (Note that, due to the complete permutation symmetry in the indices, it does not matter which pair of indices the Kronecker delta is contracted with.) Thus, the new object is a totally symmetric traceless 3-indexed tensor $\tilde{S}^{ijk}$. (We drop the tilde without further ceremony.)

The claim is thus proved.

Dimension of the irreducible representations of $SO(3)$

Let us count the number of independent components contained in $S^{i_1 i_2 \cdots i_j}$, which gives the dimension of the irreducible representation furnished by $S^{i_1 i_2 \cdots i_j}$ and labeled by the integer j.

We will do this in three baby steps. First, suppose that the indices can take on only two values, 1 and 2. Then the independent components are $S^{22 \cdots 2}$, $S^{22 \cdots 21}$, $S^{22 \cdots 211}$, $\ldots$, $S^{11 \cdots 1}$. Since the number of 1s can go from 0 to j, we count $j + 1$ possibilities here. Second, allow the indices to take on the value 3: then the possibilities are $S^{33 \cdots 3xx \cdots x}$ (that is, among the indices are k xs, where x stands for either 1 or 2, with k ranging from 0 to j, and $j - k$ 3s). Thus, the total number is determined by using Gauss's summation formula:

$$\sum_{k=0}^{j} (k+1) = \frac{1}{2}j(j+1) + (j+1) = \frac{1}{2}(j+1)(j+2) \tag{16}$$

(As an interim check, we have $\frac{1}{2} \cdot 2 \cdot 3 = 3$ for $j = 1$, and $\frac{1}{2} \cdot 3 \cdot 4 = 6$ for $j = 2$.)

---

* "All" in the sense of as far as their transformation properties are concerned.
† Explicitly, $\tilde{S}^{ijk} = S^{ijk} - \frac{1}{N+2}(\delta^{ij} S^{hhk} + \delta^{ik} S^{hhj} + \delta^{jk} S^{hhi})$.

But we are not done yet: we have to impose the traceless condition: $\delta^{i_1 i_2} S^{i_1 i_2 \cdots i_j} = 0$. The left hand side here is a totally symmetric tensor carrying $(j - 2)$ indices, which according to (16), has $\frac{1}{2}(j - 2 + 1)(j - 2 + 2) = \frac{1}{2}(j - 1)j$ components. Therefore, setting these to zero amounts to imposing $\frac{1}{2}j(j - 1)$ conditions. So, finally, the dimension of the irreducible representation j is

$$d = \frac{1}{2}(j + 1)(j + 2) - \frac{1}{2}j(j - 1) = \frac{1}{2}(j^2 + 3j + 2 - j^2 + j) = 2j + 1 \tag{17}$$

As a check, note that, for $j = 0, 1, 2, 3, \cdots$, $d = 1, 3, 5, 7, \cdots$, with the first three numbers confirming our earlier discussion. We see that the dimension d goes up only linearly with j. In contrast, an unrestricted tensor carrying j indices, with each index allowed to take on three values, will have 3^j components.

The formula $d = 2j + 1$ is famous in the history of quantum mechanics and atomic physics. In fact, it had already appeared in classical physics in connection with spherical harmonics, to be discussed in chapter IV.2.

The tensors of $SO(2)$

From $N = 3$ let us descend to $N = 2$. Note that the antisymmetric symbol ε^{ij} now carries only two indices. Suppose a tensor $T^{\cdots i \cdots j \cdots}$ carrying m indices is antisymmetric in the pair of indices i and j. We can contract it with ε^{ij} to obtain a tensor $\varepsilon^{ij} T^{\cdots i \cdots j \cdots}$ carrying $m - 2$ indices. Consequently, in our inductive construction, at each step we can immediately proceed to considering only totally symmetric tensors $S^{i_1 i_2 \cdots i_j}$. In the preceding paragraph, we already determined that there are $j + 1$ of these. But we have not yet imposed the traceless condition $\delta^{i_1 i_2} S^{i_1 i_2 \cdots i_j} = 0$. Arguing as before, we see that the left hand side of this condition is a symmetric tensor with $j - 2$ indices, and hence these amount to $j - 2 + 1$ conditions. Hence the dimensions of the irreducible representations are $(j + 1) - (j - 2 + 1) = 2$. All of them are 2-dimensional!

Indeed, a moment's thought reveals what the representation matrices are:

$$D^{(j)}(\theta) = \begin{pmatrix} \cos j\theta & \sin j\theta \\ -\sin j\theta & \cos j\theta \end{pmatrix} \tag{18}$$

In particular, $j = 1$ corresponds to the defining or fundamental representation. You might want to work this out (for $j = 2$, say) as an exercise, or wait until you've read the next section.

Polar decomposition

Now Confusio yells indignantly, "What's going on? You said at the beginning of this chapter that the irreducible representations of $SO(2)$ are 1-dimensional, like the irreducible representations of Z_N. But here the irreducible representations are 2-dimensional!"

Excellent! That Confusio is getting more attentive by the day.

So, what is going on? Well, the representation given in (18) is in fact reducible. Consider the unitary transformation

$$
\begin{aligned}
U^\dagger D^{(j)}(\theta) U &= \frac{1}{\sqrt{2}} \begin{pmatrix} 1 & 1 \\ i & -i \end{pmatrix}^\dagger \begin{pmatrix} \cos j\theta & \sin j\theta \\ -\sin j\theta & \cos j\theta \end{pmatrix} \frac{1}{\sqrt{2}} \begin{pmatrix} 1 & 1 \\ i & -i \end{pmatrix} \\
&= \frac{1}{2} \begin{pmatrix} 1 & -i \\ 1 & i \end{pmatrix} \begin{pmatrix} e^{ij\theta} & e^{-ij\theta} \\ i e^{ij\theta} & -i e^{-ij\theta} \end{pmatrix} \\
&= \begin{pmatrix} e^{ij\theta} & 0 \\ 0 & e^{-ij\theta} \end{pmatrix}
\end{aligned}
\tag{19}
$$

We have shown explicitly that the 2-dimensional representation $D^{(j)}(\theta)$ reduces to two 1-dimensional representations $e^{ij\theta}$ and $e^{-ij\theta}$.

We recognize the two columns $\begin{pmatrix} 1 \\ i \end{pmatrix}$ and $\begin{pmatrix} 1 \\ -i \end{pmatrix}$ in the unitary matrix U as eigenvectors of $D^{(j)}(\theta)$ with eigenvalue equal to $e^{ij\theta}$ and $e^{-ij\theta}$, respectively. This corresponds to going from Cartesian coordinates (x, y) to complex coordinates $(z, z^*) = (x + iy, x - iy)$. In physics, for example in the study of electromagnetic waves, this corresponds to going from transverse to circular polarizations, by writing $\vec{E}_x \pm i \vec{E}_y$, and so on.

It is also instructive to see how this works explicitly with the Cartesian tensors we have been using. Consider $j = 2$, with $S^{11} = -S^{22}$, $S^{12} = S^{21}$. The polar decomposition corresponds to writing $S^{++} = S^{1+i2, 1+i2}$, $S^{--} = S^{1-i2, 1-i2}$. (What happened to S^{+-}?)

In fact, this conversation confirms the isomorphism between $SO(2)$ and $U(1)$, noted way back in chapter I.1. The 1-to-1 map is just $R(\theta) \leftrightarrow e^{i\theta}$.

Rotations in higher-dimensional space

The preceding discussion underlines an important fact. Among the rotation groups $SO(N)$, the two cases $N = 3$ and $N = 2$ that are most familiar to physicists are in fact rather special. Crucially, their antisymmetric symbols carry three and two indices, respectively.

To see what goes "wrong" when we go to higher N, consider a tensor T^{hijkl} that is symmetric in the three indices hij and antisymmetric in the two indices kl. For $N = 3$, contracting with ε^{klm} reduces the number of indices from five to four, a possibility we exploited. We need consider only totally symmetric tensors. In contrast, for $N = 4$, contracting with the antisymmetric symbol gives $\varepsilon^{klmn} T^{hijkl}$, a tensor with the same number of indices, namely, five. Thus, for $N > 3$, we would have to confront, in general, tensors with complicated symmetry patterns on interchanges of indices.

A comment to forestall any potential confusion. The word "symmetry" is necessarily overused in group theory. We saw in the review of linear algebra that symmetry of geometrical figures or of physical laws is associated with a group. In discussing tensors, we often talk about whether a tensor is symmetric or antisymmetric under interchange of its indices. The group involved is some permutation group, which has little or nothing to do with the group that the tensor is furnishing a representation of.

Self-dual and antiself-dual

The rotation group $SO(2n)$ in even-dimensional space enjoys an additional feature, that of self-dual and antiself-dual tensors. Consider the antisymmetric tensor with n indices $A^{i_1 i_2 \cdots i_n}$ with $2n(2n-1) \cdots (n+1)/n! = (2n)!/(n!)^2$ components. From our discussion, you would think that it furnishes an irreducible representation. What could possibly reduce it?

Construct the tensor $B^{i_1 i_2 \cdots i_n} \equiv \frac{1}{n!} \varepsilon^{i_1 i_2 \cdots i_n i_{n+1} i_{n+2} \cdots i_{2n}} A^{i_{n+1} i_{n+2} \cdots i_{2n}}$ dual to A. Then A is dual to B, that is, $A^{i_{n+1} i_{n+2} \cdots i_{2n}} = \frac{1}{n!} \varepsilon^{i_1 i_2 \cdots i_n i_{n+1} i_{n+2} \cdots i_{2n}} B^{i_1 i_2 \cdots i_n}$.

It follows that the two tensors $T_{\pm}^{i_1 i_2 \cdots i_n} \equiv (A^{i_1 i_2 \cdots i_n} \pm B^{i_1 i_2 \cdots i_n})$ are self-dual and antiself-dual, respectively. Schematically, $\varepsilon T_{\pm} \sim \varepsilon(A \pm B) \sim \varepsilon A \pm \varepsilon B \sim B \pm A \sim \pm(A \pm B) \sim \pm T_{\pm}$. Thus, $T_+ \sim \varepsilon T_+$ is dual to itself, while $T_- \sim -\varepsilon T_-$ is dual to minus itself.

Clearly, under an $SO(2n)$ transformation, T_+ transforms into a linear combination of T_+, while T_- transforms into a linear combination of T_-. The two tensors correspond to two irreducible representations with dimension $(2n)!/(2(n!)^2)$, not $(2n)!/(n!)^2$. For example, for $SO(6)$, the dimension of the self-dual and antiself-dual representation is equal to $6 \cdot 5 \cdot 4/(2 \cdot 3 \cdot 2) = 20/2 = 10$.

Restriction to a subgroup

Consider an irreducible representation of some group G. If we restrict ourselves to a subgroup $H \subset G$, that irreducible representation will in general break up into several irreducible representations of H. This makes sense, since we have fewer transformations to take the components of that representation into one another. This is best explained by some examples. Let $G = SO(4)$, with the defining or vector representation consisting of the components of the 4-vector V^i, $i = 1, 2, 3, 4$. Consider the subgroup $H = SO(3)$ consisting of those elements of $SO(4)$ that leave V^4 alone. In other words, the subgroup $SO(3)$ rotates only (V^1, V^2, V^3) into one another. The four objects V^i split into two sets: (V^1, V^2, V^3) and V^4. We write this as $4 \to 3 \oplus 1$: the 4-dimensional vector representation of $SO(4)$ breaks into a 3-dimensional representation and a 1-dimensional representation of $SO(3)$.

In chapter VII.2 we will discuss in detail the Lorentz group, which at this point I simply say is, roughly speaking, just $SO(4)$ suitably modified. The subgroup $SO(3)$ is the good old rotation group. Then the statement $4 \to 3 \oplus 1$ simply states that spacetime breaks up into space plus time in nonrelativistic physics.

How does the 6-dimensional irreducible representation of $SO(4)$ furnished by the antisymmetric tensor A^{ij} break up? We simply enumerate: A^{14}, A^{24}, A^{34} and A^{12}, A^{23}, A^{31}. In other words, $6 \to 3 \oplus 3$. As we shall see in chapter VII.2, this corresponds to the electromagnetic field breaking into the electric and the magnetic fields.

How about the $\frac{1}{2} \cdot 4 \cdot 5 - 1 = 9$-dimensional irreducible representation of $SO(4)$ furnished by the symmetric traceless tensor S^{ij}? Again, we simply list the nine objects. How do they break up? Evidently, it is useful to introduce the indices a, $b = 1$, 2, 3. First, we have S^{44}, which furnishes the 1-dimensional representation of $SO(3)$. Next, S^{a4}, $a = 1$, 2, 3, which furnishes the 3-dimensional representation of $SO(3)$. Finally, we form the symmetric traceless tensor $\tilde{S}^{ab} = S^{ab} - \frac{1}{3}\delta^{ab}S^{cc}$ (with c summed over 1, 2, 3; see (8)), which furnishes the 5-dimensional representation of $SO(3)$. Note that the trace S^{cc} we are taking out here is equal to $-S^{44}$, which has already been counted. Thus, $9 \to 5 \oplus 3 \oplus 1$.

The adjoint representation and the Jacobi identity

As promised, we now discuss in more detail the adjoint representation briefly mentioned in chapter I.3. Perhaps the best way to motivate the adjoint is to go back to the treatment of $SO(3)$ given there. We commute the three matrices

$$J_x = -i\begin{pmatrix} 0 & 0 & 0 \\ 0 & 0 & 1 \\ 0 & -1 & 0 \end{pmatrix}, \qquad J_y = -i\begin{pmatrix} 0 & 0 & -1 \\ 0 & 0 & 0 \\ 1 & 0 & 0 \end{pmatrix}, \qquad J_z = -i\begin{pmatrix} 0 & 1 & 0 \\ -1 & 0 & 0 \\ 0 & 0 & 0 \end{pmatrix} \tag{20}$$

representing the Lie algebra of $SO(3)$ and obtain the commutation relation

$$[J_a, J_b] = i\varepsilon_{abc}J_c \tag{21}$$

with a, b, $c = x$, y, z.

In physics, objects carrying three indices, such as the totally antisymmetric symbol ε_{abc}, are notoriously awkward to handle. In contrast, objects carrying two indices can be regarded as matrices that can be naturally multiplied together. But now let us make a seemingly naive observation. Let us fix a, by setting $a = x$, for example. Then[6] ε_{1bc} carries two indices bc, and we can think of it as a matrix, with its entries in the bth row and cth column given by ε_{abc}.

But the sharp-eyed reader will recognize that this is precisely the first matrix listed in (20). Indeed, the three matrices J_a are precisely equal to $(J_a)_{bc} = -i\varepsilon_{abc}$. (Here $(J_a)_{bc}$ denotes the entry in the bth row and cth column of the matrix J_a.)

Is this an amazing coincidence or what? The three matrices representing the generators in the defining representation are given by the structure constants. Is this a general result?

The answer is more complicated than a simple yes or no.

First, no, it can't be general. A moment's thought reminds us that for $SO(N)$, the defining or fundamental representation is N-dimensional, while the indices on the structure constants f^{abc} range over $N(N-1)/2$ values. It is only for $N = 3$ that $N = N(N-1)/2$.

But then, yes, in fact the structure constants do in general furnish a representation, the adjoint representation.

To understand this, we have to appeal to the Jacobi identity,[7] which states that quite generally, for three matrices or operators A, B, and C,

$$[[A, B], C] + [[B, C], A] + [[C, A], B] = 0 \tag{22}$$

You can prove the Jacobi identity[8] by simply writing out all the terms in (22).

You learned in chapter I.3 that Lie algebras are defined by the commutation relations

$$[T^a, T^b] = i f^{abc} T^c \tag{23}$$

with the indices $a, b, c, \cdots$ ranging over n values, where n denotes the number of generators of the algebra. Now plug this into Jacobi's identity: $[[T^a, T^b], T^c] + [[T^b, T^c], T^a] + [[T^c, T^a], T^b] = 0$. The first term is equal to $if^{abd}[T^d, T^c] = (if^{abd})(if^{dcg})T^g$. Similarly, we can write the other two terms as linear combinations of T^gs. Setting the coefficient of T^g to 0 then gives the remarkable identity involving the structure constants:

$$f^{abd} f^{dcg} + f^{bcd} f^{dag} + f^{cad} f^{dbg} = 0 \tag{24}$$

Now define the matrix T^b by specifying its entry in the cth row and dth column as follows:

$$(T^b)^{cd} = -i f^{bcd} \tag{25}$$

I'd like to alert the reader to a possible notational confusion. In (23), T^b denotes a generator, an abstract mathematical entity, if you like, of the Lie algebra; (23) is a statement about the algebra, independent of any representation we might want to choose. In contrast, (25) defines the specific matrix representing T^b in the adjoint representation.

The claim is that the matrices T^b defined in (25) represent the Lie algebra (23). To prove this, we use (25). Write the first term in (24) as $f^{abd} f^{dcg} = i f^{abd} (T^d)^{cg}$, and the second and third terms as $f^{bcd} f^{dag} + f^{cad} f^{dbg} = -(T^b)^{cd}(-T^a)^{dg} - (-T^a)^{cd}(-T^b)^{dg} = (T^b T^a)^{cg} - (T^a T^b)^{cg} = -([T^a, T^b])^{cg}$. (Note that we used the antisymmetry of the structure constant in the first two indices, for example, $f^{dag} = -f^{adg}$.) Thus, we obtain

$$([T^a, T^b])^{cg} = i f^{abd} (T^d)^{cg} \tag{26}$$

which is just (23) in the adjoint representation.

We have proved that the structure constants of a Lie algebra furnish a representation of the algebra, known as the adjoint, whose dimension is given by the number of generators.

The adjoint of $SO(N)$

We have just proved what we set out to prove for Lie algebras in general, so that's that. No more need be said. Still, it is instructive to see how this works out in the specific case of $SO(N)$. Let's start fresh and go about it using a route slightly different from that followed in the previous section.

Earlier in this chapter, we learned that the antisymmetric 2-indexed tensor T^{ij} in $SO(N)$ furnishes a $\frac{1}{2}N(N-1)$-dimensional irreducible representation.

Recall from chapter I.3 that the number of generators in $SO(N)$ is $\frac{1}{2}N(N-1)$, with the generators represented in the defining N-dimensional representation by the antisymmetric matrices

$$\mathcal{J}^{ij}_{(mn)} = (\delta^{mi}\delta^{nj} - \delta^{mj}\delta^{ni}) \tag{27}$$

(Physicists also define the hermitean matrices $J_{(mn)} = -i\mathcal{J}_{(mn)}$.) As explained in chapter I.3, the matrix indices i and j are not to be confused with the indices m and n, which I put in parenthesis for pedagogical clarity. The indices m and n indicate which of the $\frac{1}{2}N(N-1)$ matrices we are talking about. In particular, note that $\mathcal{J}_{(mn)}$ is an N-by-N matrix, and there are $\frac{1}{2}N(N-1)$ of them.

The poor man might ask, "Is there a connection?" You bet! To see this, first rewrite, for the sake of further clarity, the label (mn), which takes on $\frac{1}{2}N(N-1)$ values, as a, where the symbol a runs from 1 to $\frac{1}{2}N(N-1)$. Then we have $\frac{1}{2}N(N-1)$ matrices $\mathcal{J}_a^{ij}$, each of which is a $\frac{1}{2}N(N-1)$-by-$\frac{1}{2}N(N-1)$ matrix.

Confusio says, "This is why my friends and I get thrown: there are so many $\frac{1}{2}N(N-1)$s all over the place."

Indeed. Furthermore, for the most important case for physics, $N = 3$, both $\frac{1}{2}N(N-1)$ and N are equal to 3!

Here comes the punchline: we can also regard the antisymmetric tensor T^{ij} as an N-by-N matrix, and hence write it as a linear combination of the $\mathcal{J}_a$s, with coefficients denoted by A_a:

$$T^{ij} = \sum_{a=1}^{\frac{1}{2}N(N-1)} A_a \mathcal{J}_a^{ij} = A_a \mathcal{J}_a^{ij} \tag{28}$$

We invoked the repeated index summation convention in the second equality. To forestall confusion, keep in mind the ranges of the indices in $A_a \mathcal{J}_a^{ij}$: $i = 1, \cdots, N$ while $a = 1, \cdots, \frac{1}{2}N(N-1)$.

In other words, we can trade T^{ij} for A_a and vice versa. Check to make sure that we didn't lose anybody: indeed, there are $\frac{1}{2}N(N-1)$ of each. You may think of T^{ij} and A_a as the same thing expressed in two different bases: T^{ij} and A_a are linear combinations of each other.

How do the A_as transform? We go back to (4) and recall $T^{ij} \to T'^{ij} = R^{ik}R^{jl}T^{kl} = R^{ik}T^{kl}(R^T)^{lj}$ and thus

$$T \to T' = RTR^T = RTR^{-1} \tag{29}$$

(If you confuse the T for transpose with the T for tensor, go back to square one.) Regarded as an antisymmetric matrix, the antisymmetric tensor T transforms as if by a similarity transformation in linear algebra. For an infinitesimal rotation, $R \simeq I + \theta_a \mathcal{J}_a$, and so $T' \simeq (I + \theta_a \mathcal{J}_a)T(I + \theta_a \mathcal{J}_a) \simeq T + \theta_a[\mathcal{J}_a, T]$. Thus, the variation of T under the rotation is given by[*] $\delta T = \theta_a[\mathcal{J}_a, T] = \theta_a[\mathcal{J}_a, T_b\mathcal{J}_b] = \theta_a T_b[\mathcal{J}_a, \mathcal{J}_b]$.

In general, a Lie algebra is characterized by the commutation relations between its generators and structure constants. In this context, $[J_a, J_b] = i f_{abc} J_c$ in physicists' notation, or after taking out some trivial factors of i, $[\mathcal{J}_a, \mathcal{J}_b] = f_{abc} \mathcal{J}_c$.

---

[*] Note that here I am treating T as an "abstract" mathematical object and $\mathcal{J}$ as a numerical matrix.

Now we are almost done. Writing $\delta T = \delta A_c \mathcal{J}_c = \theta_a A_b [\mathcal{J}_a, \mathcal{J}_b] = \theta_a A_b f_{abc} \mathcal{J}_c$ and equating coefficients of $\mathcal{J}_c$, we obtain

$$\delta A_c = f_{abc} \theta_a A_b \qquad (30)$$

The $\frac{1}{2} N(N-1)$ objects A_a transform according to (30) and are said to furnish the adjoint representation. See (25).

This conclusion is actually totally familiar in the case of $SO(3)$, for which the structure constant f_{abc} is just the antisymmetric symbol ε_{abc}. Using a vector notation (since for $SO(3)$ the indices a, b, and c take on three values, we recognize A_a as a vector), we can write $\delta \vec{A} = \vec{\theta} \otimes \vec{A}$. With no loss of generality, let $\vec{\theta}$ point along the z-axis, and we obtain (as always, not hesitating to abuse notation if clarity is not compromised) $\delta A_x = -\theta A_y$, $\delta A_y = \theta A_x$. We are back where we started in chapter I.3. It is perhaps more illuminating if we revert to the notation J_x, J_y, J_z. The statement here is simply that the three generators of $SO(3)$ transform like a vector: $\delta \vec{J} = \vec{\theta} \otimes \vec{J}$. Entirely reasonable.

The adjoint representation is clearly of special significance, since it is how the generators transform, so to speak. It is worth emphasizing again that in general the adjoint is not the same as the vector or fundamental representation (certainly $\frac{1}{2} N(N-1) \neq N$ for $N \neq 3$). We will encounter the adjoint representation for other continuous groups later.

Exercises

1 Show that the symmetric tensor S^{ij} is indeed a tensor.

2 Prove the Jacobi identity.

3 Work out $\tilde{S}'^{13}$ for a rotation around the third axis through angle φ.

4 Let T^{ijk} be a totally antisymmetric 3-indexed tensor. Show that T has $\frac{1}{3!} N(N-1)(N-2)$ components. Identify the one component for $N = 3$.

5 Consider for $SO(3)$ the tensor T^{ijk} from the preceding exercise. Show that it transforms as a scalar.

6 In physics, the various abstract entities we talked about—the scalar, the vector, the tensor, and so on—will in general depend on the position $\vec{x}$, and are known as scalar field, vector field, and tensor field, and so on. An example is the temperature at a given point in space: it is a scalar field $T(\vec{x})$. You are of course familiar with the electric field $\vec{E}(\vec{x})$. We can thus take spatial derivatives of these quantities.
   Define $\vec{\nabla} \equiv (\frac{\partial}{\partial x^1}, \frac{\partial}{\partial x^2}, \cdots, \frac{\partial}{\partial x^D})$. Show that if ϕ is a scalar field, then $\vec{\nabla}\phi$ transforms like a vector, while $(\vec{\nabla}\phi)^2 = \vec{\nabla}\phi \cdot \vec{\nabla}\phi = \sum_k (\frac{\partial \phi}{\partial x^k})^2$ and $\nabla^2 \phi$ transform like a scalar. The Laplacian is defined by $\nabla^2 = \vec{\nabla} \cdot \vec{\nabla} = \frac{\partial^2}{\partial (x^1)^2} + \frac{\partial^2}{\partial (x^2)^2} + \cdots + \frac{\partial^2}{\partial (x^D)^2}$.

7 Given a 3-vector $\vec{p} = (p^1, p^2, p^3)$, show that the quantity $p^i p^j$ when averaged over the direction of $\vec{p}$ is given by $\frac{1}{4\pi} \int d\Omega \, p^i p^j = \frac{1}{4\pi} \int d\theta d\varphi \cos\theta \, p^i p^j = \frac{1}{3} \vec{p}^2 \delta^{ij}$. (Here θ and φ are the angles fixing the direction of $\vec{p}$ in spherical coordinates.) In physics, we often encounter integrals of this type. This is the basis for my remark after the orthogonality theorem in chapter II.2.

8 Show that $\varepsilon^{ijk\cdots n} R^{ip} R^{jq} = \varepsilon^{pqr\cdots s} R^{kr} \cdots R^{ns}$.

9 Convince yourself of (18) for $j = 2$.

10 For $SO(3)$, using (13), show that $\varepsilon^{ijk} A^i B^j \equiv C^k$ defines a vector $\vec{C} = \vec{A} \otimes \vec{B}$, the familiar cross product.

11 Prove the identity

$$\varepsilon^{ijk} \varepsilon^{lnk} = \delta^{il} \delta^{jn} - \delta^{in} \delta^{jl} \tag{31}$$

Contracting with A^j, B^l, and C^n, obtain an identity you might recognize: $\vec{A} \otimes (\vec{B} \otimes \vec{C}) = \vec{B}(\vec{A} \cdot \vec{C}) - \vec{C}(\vec{A} \cdot \vec{B})$.

12 List the dimensions of the five smallest irreducible representations of $SO(5)$, and identify the corresponding tensors.

13 Find the dimension of the irreducible representation of $SO(4)$ furnished by the symmetric traceless tensor with h indices.

Notes

1. Long ago, an undergraduate who later became a distinguished condensed matter physicist came to me after a class on group theory and asked me, "What exactly is a tensor?" I told him that a tensor is something that transforms like a tensor. When I ran into him many years later, he regaled me with the following story. At his graduation, his father, perhaps still smarting from the hefty sum he had paid to the prestigious private university his son attended, asked him what was the most memorable piece of knowledge he acquired during his four years in college. He replied, "A tensor is something that transforms like a tensor."
   But this should not perplex us. A duck is something that quacks like a duck. Mathematical objects can also be defined by their behavior.

2. In Newtonian mechanics, tidal forces are described by the quantity $\mathcal{R}^{ij} \equiv \partial^i \partial^j V(r)$, which is manifestly a tensor if $V(r)$ is a scalar. But $\mathcal{R}^{ij}$ certainly does not have the form $V^i W^j$ in general. See, for example, G Nut, p. 58.

3. Those who struggled with Euler angles might recognize the bitterness behind that remark.

4. I am being a bit facetious here. The antisymmetric tensor $F_{\mu\nu}$ and the symmetric tensor $g_{\mu\nu}$ have six and ten components, respectively, if the indices μ and ν can take on four different values. With some fudging involving some signs, these correspond to the electromagnetic field and to the metric of curved spacetime. But I am also being sloppy here: the group relevant for Einstein gravity is actually $GL(4, R)$, not $SO(4)$. I touch on these tensors in later chapters, but a thorough discussion will be far beyond the scope of this book.

5. Many authors, including the author of G Nut, use the potentially confusing terminology of "invariant tensors."

6. As I said in the Preface, I am mixing up x, y, z and 1, 2, 3 on purpose here to show the beginners that these "names" do not matter.

7. This identity plays a crucial role in Riemannian geometry and Einstein gravity.

8. Remarkably, V. I. Arnold pointed out (in an article titled "Sur L'Éducation Mathématique"—I thank H. Orland for sending me this article) that the high school geometry theorem stating that the three altitudes of a triangle intersect at one point follows from Jacobi's identity. (In case you have forgotten, an altitude is the straight line going through a vertex and perpendicular to the opposite side.) Various proofs can be found on the web. I like the one given by Hovik Khudaverdian. By the end of the proof, you would agree with the author that the whole thing is indeed zabavno.

IV.2 | Lie Algebra of $SO(3)$ and Ladder Operators: Creation and Annihilation

The Lie algebra of $SO(3)$ is prototypical of all Lie algebras, and the method used in this chapter to deal with this algebra will give us a strong hint on how to deal with more involved algebras later, in chapters V.3 and VI.3, for example.

You will also learn how to multiply two irreducible representations of $SO(3)$ together, the so-called Clebsch-Gordan decomposition. Then we go on to discuss Casimir invariants, Legendre polynomials, and spherical harmonics.

In appendix 1, I introduce the Heisenberg algebra and the important concept of creation and annihilation operators. In appendix 2, I explain the Jordan-Schwinger construction of the angular momentum algebra in terms of these operators. In appendix 3, I sketch the Dirac construction. So, lots of material here!

Representing the Lie algebra of $SO(3)$

In chapter IV.1, we constructed the irreducible representations of the group $SO(3)$ using the method of tensors. In this chapter we construct the irreducible representations of the Lie algebra* $SO(3)$. We found the algebra already in chapter I.3, namely,

$$[J_x, J_y] = i J_z, \qquad [J_y, J_z] = i J_x, \qquad [J_z, J_x] = i J_y \tag{1}$$

What does it mean to represent the algebra?

It means that we are to find three matrices J_x, J_y, and J_z such that the commutation relations in (1) are satisfied. Since we know that rotations are given by exponentials of linear combinations of the Js, this would lead to matrices representing the rotation group.

We already know by construction the 3-dimensional defining representation of the algebra; the three 3-by-3 matrices J_x, J_y, and J_z were explicitly displayed in chapter I.3.

---

* Some authors carefully distinguish the algebra from the group by writing $so(3)$ for the algebra. We will follow the common usage in the physics literature of denoting the group and the algebra both by $SO(3)$ when no confusion can arise.

However, we know from chapter IV.1 that there exists, for j a non-negative integer, a $(2j + 1)$-dimensional representation of the rotation group. Hence there must exist a corresponding $(2j + 1)$-dimensional representation of the algebra: we simply consider infinitesimal rotations.

We will proceed pretending that we do not know this, so that we will end up confirming this result.

A word of clarification. Strictly speaking, we should distinguish the matrices representing the abstract operators J_x, J_y, and J_z from the operators themselves. But it would only clutter up things if we introduce more notation. Instead, we follow the physicist's sloppy practice of using J_x, J_y, and J_z also to denote the matrices representing the abstract operators J_x, J_y, and J_z.

Climbing up and down on a ladder

Since the three generators J_x, J_y, and J_z do not commute, they cannot be simultaneously diagonalized, as explained in the review of linear algebra. But we can diagonalize one of them. Choose J_z, and work in a basis in which J_z is diagonal.

The move that breaks the problem wide open may be familiar to some students of physics: it is akin to going from the 2-dimensional coordinates x, y to the complex variable* $z = x + iy$, $z^* = x - iy$, and from a transversely polarized electromagnetic wave to a circularly polarized electromagnetic wave. Define $J_\pm \equiv J_x \pm i J_y$. Then we can rewrite (1) as

$$[J_z, J_\pm] = \pm J_\pm, \qquad [J_+, J_-] = 2J_z \tag{2}$$

Instead of working with matrices, we use Dirac's bra and ket notation, also explained in the review of linear algebra. Write the eigenvector of J_z with eigenvalue m as $|m\rangle$; in other words,

$$J_z |m\rangle = m |m\rangle \tag{3}$$

Since J_z is hermitean, m is a real number. What we are doing is going to a basis in which J_z is diagonal; according to (2), $J_\pm$ cannot be diagonal in this basis.

Now consider the state $J_+ |m\rangle$ and act on it with J_z:

$$\begin{aligned} J_z J_+ |m\rangle &= (J_+ J_z + [J_z, J_+]) |m\rangle = (J_+ J_z + J_+) |m\rangle = (m J_+ + J_+) |m\rangle \\ &= (m + 1) J_+ |m\rangle \end{aligned} \tag{4}$$

where the second equality follows from (2). (Henceforth, we will be using (2) repeatedly without bothering to refer to it.)

Thus, $J_+ |m\rangle$ is an eigenvector (or eigenstate; these terms are used interchangeably) of J_z with eigenvalue $m + 1$. Hence, by the definition of $|m\rangle$, the state $J_+ |m\rangle$ must be equal to the state $|m + 1\rangle$ multiplied by some normalization constant; in other words, we have $J_+ |m\rangle = c_{m+1} |m + 1\rangle$ with the complex number c_{m+1} to be determined.

---

* Not to be confused with the third coordinate of course!

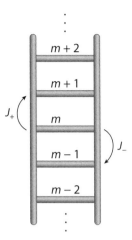

Figure 1

Similarly, $J_z J_- |m\rangle = (J_- J_z + [J_z, J_-]) |m\rangle = (J_- J_z - J_-) |m\rangle = (m-1) J_- |m\rangle$, from which we conclude that $J_- |m\rangle = b_{m-1} |m-1\rangle$ with some other unknown normalization constant.

It is very helpful to think of the states $\cdots, |m-1\rangle, |m\rangle, |m+1\rangle, \cdots$ as corresponding to rungs on a ladder. See figure 1. The result $J_+ |m\rangle = c_{m+1} |m+1\rangle$ tells us that we can think of J_+ as a "raising operator" that enables us to climb up one rung on the ladder, going from $|m\rangle$ to $|m+1\rangle$. Similarly, the result $J_- |m\rangle = b_{m-1} |m-1\rangle$ tells us to think of J_- as a "lowering operator" that enables us to climb down one rung on the ladder. Collectively, $J_\pm$ are referred to as ladder operators.

To relate b_m to c_m, we invoke the hermiticity of J_x, J_y, and J_z, which implies that $(J_+)^\dagger = (J_x + i J_y)^\dagger = J_x - i J_y = J_-$.

Multiplying $J_+ |m\rangle = c_{m+1} |m+1\rangle$ from the left by $\langle m+1|$ and normalizing the states by $\langle m| m\rangle = 1$, we obtain $\langle m+1| J_+ |m\rangle = c_{m+1}$. Complex conjugating this gives us $c^*_{m+1} = \langle m| (J_+)^\dagger |m+1\rangle = \langle m| J_- |m+1\rangle = \langle m| b_m |m\rangle = b_m$, that is, $b_{m-1} = c^*_m$, so that we can write $J_- |m\rangle = c^*_m |m-1\rangle$.

Acting on this with J_+ gives $J_+ J_- |m\rangle = c^*_m J_+ |m-1\rangle = |c_m|^2 |m\rangle$. Similarly, acting with J_- on $J_+ |m\rangle = c_{m+1} |m+1\rangle$ gives $J_- J_+ |m\rangle = c_{m+1} J_- |m+1\rangle = |c_{m+1}|^2 |m\rangle$.

Requiring the ladder to terminate

Since we know that the representation is finite dimensional, the ladder must terminate, that is, there must be a top rung. So, call the maximum value of m by j. At this stage, all we know is that j is a real number. (Note that we have not assumed that m is an integer.) Thus, there is a state $|j\rangle$ such that $J_+ |j\rangle = 0$. It corresponds to the top rung of the ladder.

At this point, we have only used the first part of (2). Now we use the second half: $0 = \langle j| J_- J_+ |j\rangle = \langle j| J_+ J_- - 2 J_z |j\rangle = |c_j|^2 - 2j$, thus determining $|c_j|^2 = 2j$. Furthermore,

$\langle m | [J_+, J_-] | m \rangle = \langle m | (J_+ J_- - J_- J_+) | m \rangle = |c_m|^2 - |c_{m+1}|^2 = \langle m | 2J_z | m \rangle = 2m$. We obtain a recursion relation $|c_m|^2 = |c_{m+1}|^2 + 2m$, which, together with $|c_j|^2 = 2j$, allows us to determine the unknown $|c_m|$.

Here we go: $|c_{j-1}|^2 = |c_j|^2 + 2(j-1) = 2(2j-1)$, then $|c_{j-2}|^2 = |c_{j-1}|^2 + 2(j-2) = 2(3j - 1 - 2)$, then eventually $|c_{j-s}|^2 = 2((s+1)j - \sum_{i=1}^{s} i)$. Recall the Gauss formula $\sum_{i=1}^{s} i = \frac{1}{2}s(s+1)$, and obtain $|c_{j-s}|^2 = 2((s+1)j - \frac{1}{2}s(s+1)) = (s+1)(2j-s)$.

We keep climbing down the ladder, increasing s by 1 at each step. When $s = 2j$, we see that c_{-j} vanishes. We have reached the bottom of the ladder. More explicitly, we have $J_- |-j\rangle = c^*_{-j} |-j-1\rangle = 0$, according to what we just derived. The minimum value of m is $-j$. Since s counts the number of rungs climbed down, it is necessarily an integer, and thus the condition $s = 2j$ that the ladder terminates implies that j is either an integer or a half-integer, depending on whether s is even or odd. If the ladder terminates, then we have the set of states $|-j\rangle$, $|-j+1\rangle$, $\cdots$, $|j-1\rangle$, $|j\rangle$, which totals $(2j+1)$ states. For example, for $j = 2$, these states are $|-2\rangle$, $|-1\rangle$, $|0\rangle$, $|1\rangle$, and $|2\rangle$. Starting from $|2\rangle$, we apply J_- four times to reach $|-2\rangle$. (We will do this explicitly later in this chapter.) To emphasize the dependence on j, we sometimes write the kets $|m\rangle$ as $|j, m\rangle$. Notice that the ladder is symmetric under $|m\rangle \to |-m\rangle$, a symmetry that can be traced to the invariance of the algebra in (1) under $J_x \to J_x$, $J_y \to -J_y$, and $J_z \to -J_z$ (namely, a rotation through π around the x-axis).

In chapter IV.1, we used the method of tensors to show that the representations of the group $SO(3)$ are $(2j+1)$-dimensional with j an integer. So, for j an integer, the method of tensors and the method of Lie algebra agree. Yeah!

Mysterious appearance of the half integers

But what about the representations of the algebra corresponding to $j =$ a half integer? For example, for $j = \frac{1}{2}$, we have a $2 \cdot \frac{1}{2} + 1 = 2$-dimensional representation consisting of the states $\left|-\frac{1}{2}\right\rangle$ and $\left|\frac{1}{2}\right\rangle$. We climb down from $\left|\frac{1}{2}\right\rangle$ to $\left|-\frac{1}{2}\right\rangle$ in one step. Certainly no sight of a 2-dimensional representation in chapter I.3!

The mystery of the $j = \frac{1}{2}$ representation will be resolved in chapter IV.5 when we discuss $SU(2)$, but let's not be coy about it and keep the reader in suspense. I trust that most readers have heard that it describes the electron spin.

We did not go looking for the peculiar number $\frac{1}{2}$; $\frac{1}{2}$ came looking for us.

Ladder operators

It should not escape your notice that as a by-product of requiring the ladder to terminate, we have also determined $|c_m|^2$. Indeed, setting $s = j - m$, we had $|c_m|^2 = (j+m)(j+1-m)$. Recalling the definition of c_m we obtain

$$J_+ |m\rangle = c_{m+1} |m+1\rangle = \sqrt{(j+1+m)(j-m)} \, |m+1\rangle \tag{5}$$

and

$$J_- \, |m\rangle = c_m^* \, |m-1\rangle = \sqrt{(j+1-m)(j+m)} \, |m-1\rangle \tag{6}$$

As a mild check on the arithmetic, indeed $J_+ \, |j\rangle = 0$ and $J_- \, |-j\rangle = 0$. You might also have noticed that, quite rightly, the phase of c_m is not determined, since it is completely up to us to choose* the relative phase of the kets $|m\rangle$ and $|m-1\rangle$. Beware that different authors choose differently. I simply take c_m to be real and positive. Tables of the c_ms for various js are available, but it's easy enough to write them down when needed. Note also that the square roots in (5) and (6) are related by $m \leftrightarrow -m$.

I list the two most common cases needed in physics:

$$j = \frac{1}{2}: \quad J_+ \left|-\frac{1}{2}\right\rangle = \left|\frac{1}{2}\right\rangle, \quad J_- \left|\frac{1}{2}\right\rangle = \left|-\frac{1}{2}\right\rangle \tag{7}$$

$$j = 1: \quad J_+ \, |-1\rangle = \sqrt{2} \, |0\rangle, \quad J_+ \, |0\rangle = \sqrt{2} \, |1\rangle$$

$$J_- \, |1\rangle = \sqrt{2} \, |0\rangle, \quad J_- \, |0\rangle = \sqrt{2} \, |-1\rangle \tag{8}$$

Note that the (nonzero) c_m for these two cases are particularly easy to remember (that is, if for some odd reason you want to): they are all 1 in one case, and $\sqrt{2}$ in the other.

Let us also write down the $j = 2$ case for later use:

$$j = 2:$$

$$J_+ \, |-2\rangle = 2 \, |-1\rangle, \quad J_+ \, |-1\rangle = \sqrt{6} \, |0\rangle, \quad J_+ \, |0\rangle = \sqrt{6} \, |1\rangle, \quad J_+ \, |1\rangle = 2 \, |2\rangle$$

$$J_- \, |2\rangle = 2 \, |1\rangle, \quad J_- \, |1\rangle = \sqrt{6} \, |0\rangle, \quad J_- \, |0\rangle = \sqrt{6} \, |-1\rangle, \quad J_- \, |-1\rangle = 2 \, |-2\rangle \tag{9}$$

Multiplying two $SO(3)$ representations together

We will multiply $SO(3)$ representations together using first the tensor approach of chapter IV.1 and then the ladder approach discussed here.

Tensors are practically begging us to multiply them together. Rather than treat the world's most general case (which will mostly amount to writing a lot of $\cdots\cdots$s), I think it pedagogically much clearer to treat a specific case.

Suppose we are given two $SO(3)$ tensors: a symmetric traceless tensor S^{ij} and a vector T^k. They furnish the 5-dimensional and the 3-dimensional irreducible representations, respectively. The product $P^{ijk} = S^{ij} T^k$ is manifestly a 3-indexed tensor containing $5 \cdot 3 = 15$ components, but P^{ijk} is certainly not a totally symmetric traceless 3-indexed tensor. It is symmetric and traceless in the first two indices, but has no particular symmetry on the interchange of j and k, for example. In chapter IV.1, we learned that the irreducible representations of $SO(3)$ are furnished by symmetric traceless tensors. So, the game is to beat P^{ijk} into a linear combination of symmetric traceless tensors.

---

* The freedom of choice here ends up being a minor irritant for theoretical physicists. For example, it will come back to bite us in (V.1.10), when we attempt to compare results using different formalisms.

First, construct the symmetric tensor $U^{ijk} = S^{ij}T^k + S^{jk}T^i + S^{ki}T^j$. Its trace is given by the vector $U^k = \delta^{ij}U^{ijk} = 2S^{ik}T^i$. (Note that it does not matter which pair of indices on U^{ijk} to contract when calculating the trace, since U^{ijk} is totally symmetric.) Second, take out the trace by defining $\tilde{U}^{ijk} = U^{ijk} - \frac{1}{5}(\delta^{ij}U^k + \delta^{jk}U^i + \delta^{ki}U^j)$. (Check that $\tilde{U}^{ijk}$ is traceless.) There are $(2 \cdot 3 + 1) = 7$ components in $\tilde{U}^{ijk}$: it furnishes a 7-dimensional irreducible representation.

Next we extract the antisymmetric part of the product $S^{ij}T^k$ by contracting it with the antisymmetric symbol: $V^{il} = S^{ij}T^k\varepsilon^{jkl}$. (There is of course no point in contracting the index pair ij with the ε symbol.) The tensor V^{il} is traceless (check this!) and so contains $3 \cdot 3 - 1 = 8$ objects. Extract its symmetric and antisymmetric parts: $W^{il} = V^{il} + V^{li}$ and $X^{il} = V^{il} - V^{li}$.

The antisymmetric part can be written explicitly as a vector: $\frac{1}{2}X^{il}\varepsilon^{mil} = S^{ij}T^k\varepsilon^{jkl}\varepsilon^{mil} = S^{ij}T^k(\delta^{jm}\delta^{ki} - \delta^{ji}\delta^{km}) = S^{im}T^i$, which we recognize as the vector $\frac{1}{2}U^m$. (In the second equality we used a result from exercise IV.1.11.) This furnishes a 3-dimensional irreducible representation.

The symmetric part $W^{il} = S^{ij}T^k\varepsilon^{jkl} + S^{lj}T^k\varepsilon^{jki}$ is manifestly traceless (just set $i = l$ and sum). Thus, it has $\frac{1}{2}(3 \cdot 4) - 1 = 5$ components.

We have thus derived

$$5 \otimes 3 = 7 \oplus 5 \oplus 3 \tag{10}$$

We did not lose anybody: $5 \cdot 3 = 15 = 7 + 5 + 3$.

It should be clear how to treat the general case. Given two totally symmetric traceless tensors, $S^{i_1 \cdots i_j}$ and $T^{k_1 \cdots k_{j'}}$, one with j indices, the other with j' indices, the product is then a tensor with $j + j'$ indices. We first symmetrize this and take out its trace. We get the irreducible representation labeled by $j + j'$.

Next, contract with ε^{ikl}, where i is an index on S and k an index on T. We trade two indices, i and k, for one index l, and hence end up with a tensor with $j + j' - 1$ indices. We get the irreducible representation labeled by $j + j' - 1$. We repeat this process until there is nothing left to work with.

With no loss of generality, take $j \geq j'$. We have nothing left when all the indices on T are gone. We have thus shown that $j \otimes j'$ contains the irreducible representations $(j + j') \oplus (j + j' - 1) \oplus (j + j' - 2) \oplus \cdots \oplus (j - j' + 1) \oplus (j - j')$. Using the absolute value, we can drop the condition $j \geq j'$:

$$j \otimes j' = (j + j') \oplus (j + j' - 1) \oplus (j + j' - 2) \oplus \cdots \oplus (|j - j'| + 1) \oplus |j - j'| \tag{11}$$

For example, (10) can be written in the language of (11) as $2 \otimes 1 = 3 \oplus 2 \oplus 1$.

In chapter IV.3, we will derive this result in another way, but for now, let us count the number of components and check that we did not lose anybody. We start with $(2j + 1)(2j' + 1)$ components. To count the number of components contained in (11), we need $\sum_n^m 1 = (m - n + 1)$ and, using Gauss's trick, $\sum_n^m k = \frac{1}{2}(m - n + 1)(m + n)$, so that $\sum_n^m (2k+1) = (m-n+1)\big((m+n)+1\big) = (m+1)^2 - n^2$. Then the number of components contained in (11) is given by

$$\sum_{|j-j'|}^{j+j'} (2k+1) = (j+j'+1)^2 - (j-j')^2 = (2j+1)(2j'+1) \tag{12}$$

Precisely right.

I trust that you recognize that the very way in which we introduced the notion of a 2-indexed tensor in chapter IV.1 represents a special case of (11). We define the tensor as an object that transforms as if it were composed of the product of two vectors; thus $1 \otimes 1 = 0 \oplus 1 \oplus 2$. Indeed, that was the content of the section "Reducible or irreducible?": we showed that $3 \otimes 3 = 9 = 1 \oplus 3 \oplus 5$.

How do we name irreducible representations?

This is probably as good a place as any to comment on the conventions for naming irreducible representations. Here you have already seen two for $SO(3)$: we can refer to the irreducible representation by the number of indices on the tensor furnishing the representation, as in j, or by its dimension, $2j+1$.

For larger groups, such as $SO(N)$ for $N > 3$, as was explained earlier, a tensor can have rather involved symmetry properties under the interchange of its indices. We need to specify its symmetry properties. For instance, in $SO(N)$, the representation furnished by a totally antisymmetric tensor with n indices is often denoted by $[n]$ and the representation furnished by a totally symmetric traceless tensor with n indices by $\{n\}$. (Obviously, $[1] = \{1\}$. For $SO(3)$, we explained that $[2] = \{1\} = [1]$.)

Different conventions are used in different areas of physics and are favored by different authors. I usually prefer the convention of identifying the irreducible representation by its dimension. (Usually, there is no risk of confusion, but ambiguities arise occasionally: distinct representations may happen to have the same* dimension.) With this convention, you can see instantly that nobody has gone missing when you multiply irreducible representations together, as in (10).

For the group $SO(3)$, with its long standing in physics and its connection with atomic spectroscopy, the confusion over names is almost worse than in reading Russian novels. As you know, the trivial representation 0 or 1 is also called "s-wave"! And the list goes on, 1 or 3 is called p, 2 or 5 is called d, 3 or 7 is called f, and so on.[†]

One glaringly self-evident point is that the letters used in the preceding section, S, T, P, U, V, W, and X, are totally irrelevant; they are merely coatracks on which to hang the

---

* Just as a teaser, I throw out the remark that $SO(8)$, famously or infamously, has three distinct irreducible representations, all with dimension equal to 8. See chapter VII.1.

† For sharp, principal, diffuse, fundamental, and so on, in case you forgot or never learned spectroscopy.

indices. This is reminiscent of one motivation behind the Dirac bra and ket notation: no sense in writing the letter ψ all the time.

Casimir invariant

We saw, in chapter IV.1, that the generators of $SO(3)$, J_x, J_y, and J_z, transform like a vector under rotation: $\delta \vec{J} = \vec{\theta} \otimes \vec{J}$. This implies that rotations do not change $J^2 \equiv \vec{J}^2 = J_x^2 + J_y^2 + J_z^2$, known as the Casimir invariant for $SO(3)$. Indeed, we have* $[J_i, \vec{J}^2] = [J_i, J_j J_j] = i \varepsilon_{ijk}([J_i, J_j]J_j + J_j[J_i, J_j]) = i\varepsilon_{ijk}(J_k J_j + J_j J_k) = 0$. (In general, a Lie algebra may possess several Casimir invariants; one example will be shown in chapter VII.2.)

In particular, $[J_\pm, \vec{J}^2] = 0$. Thus, $\vec{J}^2$ evaluated on the states in an irreducible representation j should give the same value. We can readily verify this statement and determine this value. First, write

$$J^2 = \frac{1}{2}(J_+ J_- + J_- J_+) + J_z^2 \tag{13}$$

Then, $J^2 |j, m\rangle = \left(\frac{1}{2}(J_+ J_- + J_- J_+) + J_z^2\right) |j, m\rangle = \left(\frac{1}{2}(|c_m|^2 + |c_{m+1}|^2) + m^2\right) |j, m\rangle = j(j+1) |j, m\rangle$. As expected, J^2 evaluated on $|j, m\rangle$ depends only on j, but not on m.

This result is sometimes written as

$$J^2 = j(j+1) \tag{14}$$

where it is understood that the equality holds in the irreducible representation j. (Some beginners are confused by this equality, partly because of the standard notation of dropping the arrow on $\vec{J}^2$. Thus, importantly, J^2 is not equal to j^2, because $\vec{J}$ is not a set of numbers, but a set of three noncommuting operators; j is of course just an integer or a half integer.)

Legendre polynomials and spherical harmonics

I trust that most readers have encountered Legendre[1] polynomials and spherical harmonics in their study of physics. It should not come as a surprise that both are intimately connected to the rotation group $SO(3)$, since they pop up in problems with spherical symmetry. To see the connection, let us go back to the remark in chapter I.3 that the generators of $SO(3)$ can be written as differential operators rather than as matrices. Here, to conform to age-old convention, we use L_i instead of J_i to denote the generators. In chapter I.3, I remarked that

$$L_x = i\left(z\frac{\partial}{\partial y} - y\frac{\partial}{\partial z}\right), \qquad L_y = i\left(x\frac{\partial}{\partial z} - z\frac{\partial}{\partial x}\right), \qquad L_z = i\left(y\frac{\partial}{\partial x} - x\frac{\partial}{\partial y}\right) \tag{15}$$

satisfy the commutation relations[†]

---

*The relevant identity is $[A, BC] = [A, B]C + B[A, C]$.

[†] A point of clarification here. The $\vec{L}$s defined here are dimensionless and the commutation relations (16) are just statements about differential calculus. The reader who knows quantum mechanics would recall that

$$[L_x, L_y] = iL_z, \qquad [L_y, L_z] = iL_x, \qquad [L_z, L_x] = iL_y \tag{16}$$

Behold the power of abstract mathematics: by the discussion in this chapter, we know immediately that there are functions of x, y, z, or equivalently r, θ, φ, that represent the Lie algebra (16). These functions satisfy

$$\vec{L}^2 Y_l^m(\theta, \varphi) = l(l+1) Y_l^m(\theta, \varphi) \tag{17}$$

and

$$L_z Y_l^m(\theta, \varphi) = m Y_l^m(\theta, \varphi) \tag{18}$$

Without further ado, we know that $l = 0, 1, 2, \cdots$ is a non-negative integer and that $m = -l, -l+1, \cdots, l-1, l$. (The reason that we can take $Y_l^m(\theta, \varphi)$ to be independent of the radial coordinate r is because $\vec{L}$ acting on r gives nothing. To see this, note $L_z r^2 = i(y\frac{\partial}{\partial x} - x\frac{\partial}{\partial y})(x^2 + y^2 + z^2) = 2i(yx - xy) = 0$.)

The functions Y_l^m are known as spherical harmonics. Note that, while the mathematics of the Lie algebra implies (17) and (18), if we actually want to know what Y_l^m is, we have to slog through what we might refer to as "rather tedious arithmetic." The first step is to write L_x, L_y, and L_z in spherical coordinates. The easiest is $L_z = -i\frac{\partial}{\partial \varphi}$ (which you can see by mentally rotating around the z-axis). This enables us to solve (18) immediately and to isolate the φ dependence of Y_l^m by writing $Y_l^m(\theta, \varphi) = N_l^m e^{im\varphi} P_l^m(\cos\theta)$. Here N_l^m is a pesky normalization constant,* and $P_l^m(\cos\theta)$ is known as the associated Legendre function.

For the record, $L_\pm$ (which are more useful to us than L_x, L_y) are given by $L_\pm = e^{\pm i\varphi}(i\cot\theta\frac{\partial}{\partial\varphi} \pm \frac{\partial}{\partial\theta})$. Using (13), we obtain for the Casimir invariant

$$\vec{L}^2 = \frac{1}{2}(L_+ L_- + L_- L_+) + L_z^2 = -\left(\frac{1}{\sin\theta} \frac{\partial}{\partial\theta} \left(\sin\theta \frac{\partial}{\partial\theta} \right) + \frac{1}{\sin^2\theta} \frac{\partial^2}{\partial\varphi^2} \right) \tag{19}$$

Thus, (17) amounts to the familiar partial differentiation equation satisfied by the spherical harmonics that follows from Laplace's equation after separation of variables.

I refrain from going into further details here, which quite likely may be familiar to you from your study of, say, electrostatics. Instead, I show you a neat connection between the Legendre polynomials $P_l(\cos\theta)$ (which are what $P_l^{m=0}(\cos\theta)$ are called) and tensors. Consider the unit vector V^i. In spherical coordinates, $V^3 = \cos\theta$, which is just the Legendre polynomial $P_1(\cos\theta)$. Indeed, the 3-components of V^i written in the circular basis $V^{1\pm i2}$, V^3 are nothing other than $Y_1^{\pm 1}$, Y_1^0 up to normalization factors. Now consider the symmetric traceless tensor $T^{ij} = V^i V^j - \frac{1}{3}\delta^{ij}\vec{V}^2 = V^i V^j - \frac{1}{3}\delta^{ij}$. Then $T^{33} = \cos^2\theta - 1/3$, which is just $P_2(\cos\theta)$ (as always, up to normalization factors). Next, the symmetric traceless 3-indexed tensor $T^{ijk} = V^i V^j V^k - \frac{1}{5}(\delta^{ij}V^k + \delta^{jk}V^i + \delta^{ki}V^j)$. Then

---

momentum $\vec{p}$ is represented by $-i\hbar\vec{\nabla}$, where the reduced Planck's constant $\hbar$ is required to make the dimension come out right. Thus, in quantum mechanics, the angular momentum operator $\vec{L}_{QM} = \vec{x} \otimes \vec{p} = -i\hbar\vec{x} \otimes \vec{\nabla}$ is equal to $\hbar$ times the $\vec{L}$ defined here.

* These constants can be chalked up to various "historical accidents": the spherical harmonics are normalized by $\int d(\cos\theta)d\varphi |Y_l^m(\theta, \varphi)|^2 = 1$, while $P_l^m(\cos\theta)$ is normalized in some other fashion.

$T^{333} = \cos^3 \theta - \frac{3}{5} \cos \theta$, which is just $P_3(\cos \theta)$. Note that the orthogonality condition $\int_{-1}^{1} d \cos \theta \, P_l(\cos \theta) P_{l'}(\cos \theta) \propto \delta_{ll'}$ also follows. You see the power of the tensor approach.

Perhaps you also see that the spherical harmonics are just the kets $|l, m\rangle$ disguised. In quantum mechanics, the position of a particle is treated as an operator $\hat{x}$, with eigenstates $|x\rangle$ determined by $\hat{x} |x\rangle = x |x\rangle$. In spherical coordinates, $|x\rangle$ would be written as $|r, \theta, \varphi\rangle$. Since we are talking about rotations that do not change r, we suppress the radial coordinate and write $|\theta, \varphi\rangle$. The spherical harmonics are just $Y_l^m(\theta, \varphi) = \langle \theta, \varphi | l, m \rangle$. (Note that while our discussion is motivated by quantum mechanics and uses the bra and ket notation, it actually has nothing to do with quantum mechanics as such; we can perfectly well define $|\theta, \varphi\rangle$ in classical mathematics.)

Appendix 1: Heisenberg algebra and creation and annihilation operators

I should mention that other types of algebras, notably the Heisenberg algebra that undergirds quantum mechanics, are also of interest to physicists, even though, since this is a textbook on group theory, I talk mostly about Lie algebras. Readers who have had some exposure to quantum mechanics know that Heisenberg promoted position q and momentum p to operators satisfying

$$[q, p] = i \tag{20}$$

This algebra can be realized with differential operators: $q = x$ and $p = \frac{1}{i} \frac{d}{dx}$. Acting with the left hand side of (20) on an arbitrary function $f(x)$, we obtain $[q, p]f(x) = \left(x \frac{1}{i} \frac{d}{dx} - \frac{1}{i} \frac{d}{dx} x \right) f(x) = -i \left(x \frac{df}{dx} - \frac{d(xf)}{dx} \right) = i f(x)$, which agrees with $[q, p] = i$.

Following Dirac, we invite ourselves to consider the operators

$$a = \frac{1}{\sqrt{2}}(q + ip) \quad \text{and} \quad a^\dagger = \frac{1}{\sqrt{2}}(q - ip) \tag{21}$$

It follows from (20) that

$$[a, a^\dagger] = \frac{1}{2}[q + ip, q - ip] = -i[q, p] = 1 \tag{22}$$

which we might call the Dirac algebra.

The hermitean operator $N \equiv a^\dagger a$ can be diagonalized, with eigenstate $|n\rangle$ and eigenvalue n. For the moment, n is undetermined but is known to be a real number, since N is hermitean. From (22) we obtain (using an identity mentioned in a footnote in this chapter)

$$[a, N] = [a, a^\dagger a] = [a, a^\dagger]a = a \tag{23}$$

and so

$$Na |n\rangle = (aN + [N, a]) |n\rangle = (n - 1)a |n\rangle \tag{24}$$

Thus, $a |n\rangle$ is an eigenstate of N with eigenvalue equal to $(n - 1)$. Write $a |n\rangle = c_n |n - 1\rangle$ with c_n some normalization factor. We hermitean conjugate to obtain $\langle n| a^\dagger = \langle n - 1| c_n^*$.

Let $|n\rangle$ be normalized: $\langle n | n \rangle = 1$ for all n. Squaring $a |n\rangle$, we obtain $|(a |n\rangle)|^2 = (\langle n| a^\dagger)(a |n\rangle) = \langle n| a^\dagger a |n\rangle = \langle n| N |n\rangle = n = \langle n - 1| c_n^* c_n |n - 1\rangle = c_n^* c_n$ which implies that $c_n = \sqrt{n}$ (after absorbing[2] a phase factor in the definition of $|n\rangle$).

Thus far, we haven't yet shown that n is an integer, in spite of the suggestive notation.

If n were an integer, then, by repeatedly invoking the result $a |n\rangle = \sqrt{n} |n - 1\rangle$, we would have $a |n - 1\rangle = \sqrt{n - 1} |n - 2\rangle$, and so on, and would eventually "climb down" to the state $|0\rangle$ such that

$$a |0\rangle = 0 \tag{25}$$

In other words, $c_0 = 0$. However, if n were not an integer, the sequence $|n\rangle$ would go on indefinitely with n decreasing by 1 at each step, but as soon as n goes negative, the relation $|(a\,|n\rangle)|^2 = \langle n|\,a^\dagger a\,|n\rangle = n$ derived above would be contradicted, since the left hand side is manifestly positive.

Therefore, the sequence $\{n\} = \{0, 1, 2, \cdots, \infty\}$ consists of non-negative integers, and the state $|0\rangle$, known as the vacuum or ground state, must exist.

You can also show (or obtain by hermitean conjugating what we have above) that $[N, a^\dagger] = a^\dagger$, and so $Na^\dagger\,|n\rangle = (n+1)a^\dagger\,|n\rangle$, which then leads to $a^\dagger\,|n\rangle = \sqrt{n+1}\,|n+1\rangle$.

We can conveniently think of the state $|n\rangle$ as containing n quanta or particles; then the relations

$$a\,|n\rangle = \sqrt{n}\,|n-1\rangle \quad \text{and} \quad a^\dagger\,|n\rangle = \sqrt{n+1}\,|n+1\rangle \tag{26}$$

lead to the names creation and annihilation operators for $a^\dagger$ and a, respectively. Note that the Dirac algebra (22) can thus be realized in terms of an infinite-dimensional matrix A representing a with the only nonzero matrix elements $A_{n-1,n} = \langle n-1|\,a\,|n\rangle = \sqrt{n}$ above the diagonal.* It follows that the operators p and q in the Heisenberg algebra (20) can also be realized in terms of infinite-dimensional matrices. Note that in this basis, the number operator $N = a^\dagger a$ is a diagonal matrix with elements equal to $0, 1, 2, \cdots$.

While creation and annihilation operators are usually discussed in quantum mechanics textbooks in connection with the harmonic oscillator, note that the algebraic properties discussed here are logically independent of what the Hamiltonian H is. We haven't even mentioned H, but if H happens to be simply related to the number operator N (for example, by $H = \frac{1}{2}(N+1)$), then the eigenvalues of H are given by $\frac{1}{2}(n+1)$ and so are equally spaced. Let us find out what this H corresponds to:

$$H = \frac{1}{2}(N+1) = \frac{1}{2}a^\dagger a + \frac{1}{2} = \frac{1}{2}\left((q-ip)(q+ip)+1\right) = \frac{1}{2}(p^2+q^2)$$

$$= -\frac{1}{2}\frac{d^2}{dx^2} + \frac{1}{2}x^2 \tag{27}$$

This is precisely the Hamiltonian of the harmonic oscillator (with the mass and the spring constant set to 1 in appropriate units).

I cannot resist remarking here that the additive $+\frac{1}{2}$ in $H = \frac{1}{2}a^\dagger a + \frac{1}{2}$, which implies that $H\,|0\rangle = \frac{1}{2}$, is the mother of all headaches[3] in quantum field theory. The ground state has an irreducible $\frac{1}{2}$ unit of energy, as a direct consequence of Heisenberg's uncertainty principle.

Appendix 2: The Jordan-Schwinger construction of the angular momentum algebra

Jordan and Schwinger showed that, remarkably, the angular momentum algebra in (1) can also be realized in terms of two sets of creation and annihilation operators. Let a and b be mutually commuting operators (that is, $[a, b] = 0$, $[a^\dagger, b] = 0$, and so on) satisfying $[a, a^\dagger] = 1$ and $[b, b^\dagger] = 1$. You can then readily show that

$$[a^\dagger a - b^\dagger b, a^\dagger b] = 2a^\dagger b, \qquad [a^\dagger a - b^\dagger b, b^\dagger a] = -2b^\dagger a, \qquad [a^\dagger b, b^\dagger a] = a^\dagger a - b^\dagger b \tag{28}$$

We recognize that this is precisely the angular momentum algebra:

$$[J_z, J_\pm] = \pm J_\pm, \qquad [J_+, J_-] = 2J_z \tag{29}$$

in (1) if we identify

$$J_z \leftrightarrow \frac{1}{2}(a^\dagger a - b^\dagger b), \qquad J_+ \leftrightarrow a^\dagger b, \qquad J_- \leftrightarrow b^\dagger a \tag{30}$$

The Jordan-Schwinger construction is readily understood physically if we picture $a^\dagger$ creating a spin $\frac{1}{2}$ particle with spin pointing up, and $b^\dagger$ creating a spin $\frac{1}{2}$ particle with spin pointing down. Then $2J_z$ counts the number

---

* The rows and columns of this matrix are numbered from 0 to ∞.

of up particles minus the number of down particles. Thus, we recognize that the state $|j, m\rangle$ discussed in the text corresponds to

$$|j, m\rangle = \frac{1}{\sqrt{(j+m)!(j-m)!}} (a^\dagger)^{j+m} (b^\dagger)^{j-m} |0\rangle \tag{31}$$

You will work out the normalization factor in exercise 4. The raising operator $J_+ \leftrightarrow a^\dagger b$ removes a down particle and puts in an up particle. Similarly J_-. We see also that the number of up particles plus the number of down particles is equal to j.

In this formalism, we can easily recover relations, such as (5):

$$\begin{aligned}
J_+ |j, m\rangle &= a^\dagger b |j, m\rangle = \frac{1}{\sqrt{(j+m)!(j-m)!}} (a^\dagger)^{j+m+1} b (b^\dagger)^{j-m} |0\rangle \\
&= \frac{1}{\sqrt{(j+m)!(j-m)!}} (a^\dagger)^{j+m+1} [b, (b^\dagger)^{j-m}] |0\rangle \\
&= \frac{\sqrt{j+m+1}(j-m)}{\sqrt{(j+m+1)!(j-m)!}} (a^\dagger)^{j+m+1} (b^\dagger)^{j-m-1} |0\rangle \\
&= \sqrt{(j+1+m)(j-m)} \, |m+1\rangle \tag{32}
\end{aligned}$$

As I keep saying, math works.

Appendix 3: The Dirac construction of the angular momentum algebra

The angular momentum generators can also be written in terms of bras and kets. Given two kets $|+\rangle$, $|-\rangle$, and the corresponding bras $\langle +|$, $\langle -|$, normalized so that $\langle +|+\rangle = 1$, $\langle +|-\rangle = 0$, and so on, write

$$J_z = \frac{1}{2}(|+\rangle \langle +| - |-\rangle \langle -|), \qquad J_+ = |+\rangle \langle -|, \qquad J_- = |-\rangle \langle +| \tag{33}$$

Then, for example, $J_+ J_- = |+\rangle \langle -|-\rangle \langle +| = |+\rangle \langle +|$, and similarly, $J_- J_+ = |-\rangle \langle -|$. Thus,

$$[J_+, J_-] = |+\rangle \langle +| - |-\rangle \langle -| = 2J_z \tag{34}$$

As another example, $J_z J_+ = \frac{1}{2}(|+\rangle \langle +| - |-\rangle \langle -|) |+\rangle \langle -| = \frac{1}{2} |+\rangle \langle -|$, and similarly, $J_+ J_z = |+\rangle \langle -| \frac{1}{2}(|+\rangle \langle +| - |-\rangle \langle -|) = -\frac{1}{2} |+\rangle \langle -|$. Thus,

$$[J_z, J_+] = |+\rangle \langle -| = J_+ \tag{35}$$

I think that you can see that, once you get the hang of this game, it offers a very efficient way of calculating stuff.

You might also see, after doing exercise 1, that this amounts to writing the 2-by-2 matrices that represent the Js in the Dirac bra and ket notation.

Exercises

1   Using (7), (8), and (9), write down the matrices representing J_x, J_y, and J_z in the $j = \frac{1}{2}$, 1, and 2 irreducible representations and verify the commutation relations (1).

2   Use the tensor approach to work out $P_4(\cos \theta)$.

3   Show that the orthogonality relation between different Legendre polynomials follows from the fact that $l \otimes l'$ does not contain 0 if $l \neq l'$ and that $\int d \cos \theta \, P_l(\cos \theta) \propto \delta_{l,0}$ (which follows by definition).

4   Show that $|n\rangle = \frac{1}{\sqrt{n!}} (a^\dagger)^n |0\rangle$.

Notes

1. For almost 200 years after his death, Adrien-Marie Legendre (1752–1833) suffered the indignity of having the portrait said to be his being in fact that of somebody else with the same last name.
2. We are allowed to let $|n\rangle \to e^{i\phi_n} |n\rangle$; as a result of this, $c_n \to e^{i(\phi_{n-1}-\phi_n)}c_n$. We can choose the ϕ_ns so that c_n becomes real.
3. See chapter X.7 of G Nut.

IV.3 | Angular Momentum and Clebsch-Gordan Decomposition

Quantized angular momentum

Early quantum theorists confronted by the planetary model of the hydrogen atom faced a serious puzzle. Leaving aside the difficulty of the classical electron radiating electromagnetic waves and spiraling into the proton, they didn't have enough equations. The balance of power between the centrifugal "force" and the electrostatic force, $mv^2/r \sim \alpha/r^2$, provided only one equation for the two unknowns v and r. In classical physics, the initial value of v and r fixes the conserved angular momentum $L = mvr$, which for a given L, allows a solution.

In 1913, Niels Bohr boldly broke the impasse by postulating that angular momentum L is quantized in units of $\hbar$; Planck's constant was known to have the dimension of momentum times distance. With the additional equation $mvr = \hbar$, Bohr was able to solve for r, the Bohr radius, and thence the bound state energy.

When we discussed spherical harmonics in chapter IV.2, we introduced a set of dimensionless generators $\vec{L}_{\text{Lie}}$ of the Lie algebra of $SO(3)$ and remarked in passing that the quantum mechanical angular momentum operators are given by $\vec{L} = \hbar \vec{L}_{\text{Lie}}$. Thus, the angular momentum operators satisfy the commutation relations

$$[L_i, L_j] = i\hbar \varepsilon_{ijk} L_k \tag{1}$$

As $\hbar \to 0$ at fixed[1] $\vec{L}$, angular momentum becomes a classical commuting variable, as we would expect. Also, $\vec{L}^2 |l, m\rangle = \hbar l(l+1) |l, m\rangle$ and $L_z |l, m\rangle = \hbar m |l, m\rangle$. With the operator $\vec{L}$ fixed, we see that for the right hand sides of these two relations not to vanish, the two quantum numbers $l, m \neq 0$ have to become large as $\hbar \to 0$.

It is perhaps worth emphasizing again that although the equation for the spherical harmonics splits out of Schrödinger's equation, it has nothing to do with quantum physics as such.

A note about notation. While the letter J is used to denote angular momentum generically, by convention L is used to denote orbital angular momentum, and S, spin angular momentum. It is understood that, in a group theoretic discussion, they can be used interchangeably.

By the way, this chapter is definitely written in a "learn-as-you-do" style rather than a "learn-as-you-read" style; it contains some conceptually rather involved but in reality quite simple calculations.

Addition of angular momentum

In the prototypical quantum mechanical problem, two particles orbit in a spherically symmetric potential. Particle unprime could be in the state $|l, m\rangle$, and particle prime in the state $|l', m'\rangle$. If the particles do not interact, then the eigenstates of the Hamiltonian could be written using the product states $|l, m\rangle \otimes |l', m'\rangle$. But the particles do interact with each other, and the Hamiltonian H then includes an interaction term H_I (which we take to depend only on the distance between the two particles). To leave H invariant, we would have to rotate both particles, of course. We want to understand what group theory tells us about the wave function of the two particles.

But the mathematical problem involved is precisely the one we did in chapter IV.2 when we multiplied two tensors together. To be specific, suppose particle unprime is in an angular momentum state with $l = 2$; in other words, its wave function is given by the product of a radial wave function that depends only on the radial variable r (and which does not concern us) and an angular wave function that depends on the angular variables θ and φ, and transforms under rotation just like a symmetric traceless tensor S^{ij}. The five components of S^{ij} are various functions of θ and φ, as described in the preceding chapter; they correspond to the $(2l + 1) = 5$ states $|l = 2, m\rangle$. Similarly, suppose particle prime is in an angular momentum state with $l' = 1$; its wave function is given by the product of a radial wave function and an angular wave function that depends on the angular variables θ and φ, and transforms under rotation just like a vector T^k. The three components of T^k correspond to the $(2l' + 1) = 3$ states $|l' = 1, m\rangle$.

How does the product of the two wave functions, one for particle unprime, one for particle prime, transform under rotations? We worked this out and gave the answer in (IV.2.11). For your convenience, I reproduce it with j changed to l:

$$l \otimes l' = (l + l') \oplus (l + l' - 1) \oplus (l + l' - 2) \oplus \cdots \oplus (|l - l'| + 1) \oplus |l - l'| \tag{2}$$

For the specific case described above, $2 \otimes 1 = 3 \oplus 2 \oplus 1$. The combined angular momentum of the two particles can be 3, 2, or 1. Group theory alone tells us that the $5 \cdot 3 = 15$ states we started with split into three sets of degenerate states, with degeneracy of 7, 5, and 3, respectively.

Multiplying two ladders together

In the tensor approach, we produce larger irreducible representations by multiplying tensors together, as described by (2). (Indeed, the attentive reader might have realized that we did something entirely similar with finite groups back in chapter II.3, with that business of "putting balls into a tray.") In the Lie algebraic approach, we should be able to produce larger irreducible representations by multiplying kets together. We shall see presently that that is indeed the case. The following discussion ends up confirming (2). Since the following discussion is entirely general and does not refer to orbital motion in the slightest, I will revert back from l to j.

Suppose we are given two irreducible representations of the Lie algebra of $SO(3)$, labeled by j and j'. We have two sets of kets: $|j, m\rangle$ with $m = -j, -j + 1, \cdots, j - 1, j$, and $|j', m'\rangle$ with $m' = -j', -j' + 1, \cdots, j' - 1, j'$. The $2j + 1$ kets $|j, m\rangle$, when acted on by the generators J_i, transform into linear combinations of one another. Similarly, the $2j' + 1$ kets $|j', m'\rangle$, when acted on by the generators J_i, transform into linear combinations of one another.

Now we write down the product kets $|j, m\rangle \otimes |j', m'\rangle$. There are $(2j + 1)(2j' + 1)$ such states. When acted on by the generators J_i, these kets naturally transform into linear combinations of one another, thus furnishing a $(2j + 1)(2j' + 1)$-dimensional representation of $SO(3)$. We expect this representation to be reducible.

The concept of irreducibility transfers naturally from representations of a Lie group to the representations of a Lie algebra. If the matrices representing the J_is could be block diagonalized, we say that the representation is reducible.

By now, you should be able to see that this construction in reality is not any different, conceptually, from multiplying tensors together. The $2j + 1$ kets $|j, m\rangle$ are in one-to-one correspondence with the $2j + 1$ components of the symmetric traceless j-indexed tensor $T^{i_1 \cdots i_j}$.

When the generators J_i act on the product kets $|j, m\rangle \otimes |j', m'\rangle$, they act on $|j, m\rangle$ and then on $|j', m'\rangle$. We can verify this more-or-less self-evident fact by rotating the product kets. Under an infinitesimal rotation around the z-axis, $R \simeq I + i\theta J_z$, both $|j, m\rangle$ and $|j', m'\rangle$ rotate, of course.* Thus,

$$|j, m\rangle \otimes |j', m'\rangle \to R |j, m\rangle \otimes R |j', m'\rangle$$
$$\simeq (I + i\theta J_z) |j, m\rangle \otimes (I + i\theta J_z) |j', m'\rangle$$
$$= (I + i\theta m) |j, m\rangle \otimes (I + i\theta m') |j', m'\rangle$$
$$\simeq (1 + i\theta(m + m')) |j, m\rangle \otimes |j', m'\rangle + O(\theta^2) \qquad (3)$$

In other words,

$$J_z(|j, m\rangle \otimes |j', m'\rangle) = (J_z |j, m\rangle) \otimes |j', m'\rangle + |j, m\rangle \otimes (J_z |j', m'\rangle)$$
$$= (m + m')(|j, m\rangle \otimes |j', m'\rangle) \qquad (4)$$

The operator J_z acts in turn on $|j, m\rangle$ and $|j', m'\rangle$. Thus, $|j, m\rangle \otimes |j', m'\rangle$ is an eigenstate of J_z with eigenvalue $m + m'$. The eigenvalues of J_z simply add.

---

* As described physically in the example given earlier, to leave physics invariant, we have to rotate both particles. It would hardly make sense to rotate one without touching the other, unless they do not interact at all and hence do not know about each other's presence.

To avoid writing $\otimes$ constantly, we denote $|j, m\rangle \otimes |j', m'\rangle$ by $|j, j', m, m'\rangle$. We just learned that $|j, j', m, m'\rangle$ is an eigenstate of J_z with eigenvalue $m + m'$.

We know that the maximum values m and m' can attain are j and j', respectively, and thus the maximum eigenvalue J_z can have is $j + j'$, attained with the state $|j, j', j, j'\rangle$.

Yes, I know, this is getting hard to follow; I will go to some specific examples presently.

The Clebsch-Gordan decomposition

The plan of attack is to apply the lowering operator J_- repeatedly on $|j, j', j, j'\rangle$. To see what is going on, I will go through three examples. In example (A), we choose $j = \frac{1}{2}$ and $j' = \frac{1}{2}$. Even though we have not yet elucidated the mystery of the $j = \frac{1}{2}$ representation, we know that it does exist as a representation of the Lie algebra. (Somehow it is not a representation of the rotation group, as will be made clear in chapter IV.5.) In example (B), we choose $j = 1$ and $j' = 1$. Finally, in example (C), we choose $j = 2$ and $j' = 1$, so that we can check against our result obtained in the preceding section using tensors.

Example (A): $j = \frac{1}{2}$, $j' = \frac{1}{2}$

There are $(2j + 1)(2j' + 1) = 2 \cdot 2 = 4$ states $|\frac{1}{2}, \frac{1}{2}, m, m'\rangle$ with $m = -\frac{1}{2}, \frac{1}{2}$ and $m' = -\frac{1}{2}, \frac{1}{2}$. Since j and j' are fixed in this discussion, we might as well omit them and simply write $|m, m'\rangle$ instead of $|j, j', m, m'\rangle$. Let's go slow and list the four states $|\frac{1}{2}, \frac{1}{2}\rangle$, $|\frac{1}{2}, -\frac{1}{2}\rangle$, $|-\frac{1}{2}, \frac{1}{2}\rangle$, and $|-\frac{1}{2}, -\frac{1}{2}\rangle$.

As explained above, we expect these four states to furnish a reducible representation and thus to fall apart into a bunch of irreducible representations labeled by J. (We are running out of letters; this J is just a number and definitely not to be confused with any of the three J_is.) Let us denote the states in these irreducible representations by $|J, M\rangle$ with $M = -J, -J + 1, \cdots, J$.

Of these four states, $|\frac{1}{2}, \frac{1}{2}\rangle$ has the maximum eigenvalue J_z can have, namely, $\frac{1}{2} + \frac{1}{2} = +1$. Thus, it can belong only to an irreducible representation labeled by J with $J \geq 1$. In fact, it cannot be that $J > 1$, because then there would have to be states with eigenvalue of J_z greater than 1. So we have

$$|1, 1\rangle = \left|\frac{1}{2}, \frac{1}{2}\right\rangle \tag{5}$$

which to the uninitiated would appear to be an equation forbidden by basic logic.

But you as an initiated recognize that the symbols on the two sides of this equation denote different objects: the left hand side denotes $|J = 1, M = 1\rangle$, while the right hand side denotes $|m = \frac{1}{2}, m' = \frac{1}{2}\rangle$. These two states are each on the top rung of the ladder, having the maximum eigenvalue J_z could have, namely, $M = 1 = m + m'$.

The strategy is to climb down the ladder by applying J_- repeatedly. So, act with J_- on $|1, 1\rangle = \left|\frac{1}{2}, \frac{1}{2}\right\rangle$. But we know from chapter IV.2 how J_- acts on these states.

Using (IV.2.8), we have $J_- |1, 1\rangle = \sqrt{2} |1, 0\rangle$, while using (IV.2.7), we have* $J_- \left|\frac{1}{2}, \frac{1}{2}\right\rangle = \left|-\frac{1}{2}, \frac{1}{2}\right\rangle + \left|\frac{1}{2}, -\frac{1}{2}\right\rangle$. Thus,[†]

$$|1, 0\rangle = \frac{1}{\sqrt{2}} \left(\left|-\frac{1}{2}, \frac{1}{2}\right\rangle + \left|\frac{1}{2}, -\frac{1}{2}\right\rangle \right) \tag{6}$$

Applying J_- again, we obtain $\sqrt{2} |1, -1\rangle = \frac{1}{\sqrt{2}} \left(2 \left|-\frac{1}{2}, -\frac{1}{2}\right\rangle \right)$ and thus $|1, -1\rangle = \left|-\frac{1}{2}, -\frac{1}{2}\right\rangle$ (which we might have expected by applying symmetry to our starting equation, flipping the z-axis).

We have now accounted for three of the four states we started with. The only orthogonal state left is the linear combination $\frac{1}{\sqrt{2}} \left(\left|-\frac{1}{2}, \frac{1}{2}\right\rangle - \left|\frac{1}{2}, -\frac{1}{2}\right\rangle \right)$, which has eigenvalue 0 under J_z; this state, all by its lonesome self, must be $|J = 0, M = 0\rangle$. Alternatively, act with J_- on this state and watch it vanish.

Let me summarize our results, giving $|J, M\rangle$ in terms of $|m, m'\rangle$:

$$|1, 1\rangle = \left|\frac{1}{2}, \frac{1}{2}\right\rangle$$

$$|1, 0\rangle = \frac{1}{\sqrt{2}} \left(\left|-\frac{1}{2}, \frac{1}{2}\right\rangle + \left|\frac{1}{2}, -\frac{1}{2}\right\rangle \right); \qquad |1, 0\rangle = \frac{1}{\sqrt{2}} \left(\left|-\frac{1}{2}, \frac{1}{2}\right\rangle - \left|\frac{1}{2}, -\frac{1}{2}\right\rangle \right)$$

$$|1, -1\rangle = \left|-\frac{1}{2}, -\frac{1}{2}\right\rangle \tag{7}$$

We might say that this has the shape of a python that just swallowed its lunch. The two states with $M = 0$ are linear combinations of the two states $\left|-\frac{1}{2}, \frac{1}{2}\right\rangle$ and $\left|\frac{1}{2}, -\frac{1}{2}\right\rangle$.

We have just showed that

$$\frac{1}{2} \otimes \frac{1}{2} = 1 \oplus 0 \tag{8}$$

Just to whet your appetite, I mention that we are going to apply the result here to proton scattering on proton in chapter V.1.

Example (B): $j = 1$, $j' = 1$

Now that you have gone through example (A), we can practically race through this example. Start with $3 \cdot 3 = 9$ states $|1, 1, m, m'\rangle$ with $m = -1, 0, 1$ and $m' = -1, 0, 1$. Again, we write $|m, m'\rangle$ instead of $|j, j', m, m'\rangle$. These nine states furnish a reducible representation which

---

* Keep in mind that $\left|\frac{1}{2}, \frac{1}{2}\right\rangle$ means $\left|\frac{1}{2}\right\rangle \otimes \left|\frac{1}{2}\right\rangle$.

[†] Alternatively, we could also have argued that the two $J_z = 0$ states, $\left|-\frac{1}{2}, \frac{1}{2}\right\rangle$ and $\left|\frac{1}{2}, -\frac{1}{2}\right\rangle$, must appear with equal weight due to the principle of democracy; the $\frac{1}{\sqrt{2}}$ then follows on normalization. No need to look up (IV.2.7) and (IV.2.8) after all.

decomposes into a bunch of irreducible representations labeled by J. In these irreducible representations, the states are denoted by $|J, M\rangle$ with $M = -J, -J + 1, \cdots, J$.

Of these nine states, the one with the highest value of M is $|1, 1\rangle$, for which $M = 1 + 1 = 2$. So start with $|2, 2\rangle = |1, 1\rangle$ and climb down the ladder. Act with J_-, using (IV.2.8) and (IV.2.9). But as remarked in connection with example (A), we don't even need to look these up. Remembering that $|1, 1\rangle$ means $|1\rangle \otimes |1\rangle$, we lower each of the two kets in turn to $|0\rangle$, so that we end up with a linear combination of $|1, 0\rangle$ and $|0, 1\rangle$. But by the principle of democracy, these two kets must appear with equal weight, and thus $|2, 1\rangle = \frac{1}{\sqrt{2}}(|0, 1\rangle + |1, 0\rangle)$.

Onward! Apply J_- again. Advocating democracy is not enough any more, since this only tells us that we get a state proportional to $|-1, 1\rangle + c\,|0, 0\rangle + |1, -1\rangle$ with an unknown constant c. We have to invoke (IV.2.8) to determine $c = 2$. Thus, $|2, 0\rangle = \frac{1}{\sqrt{6}}(|-1, 1\rangle + 2\,|0, 0\rangle + |1, -1\rangle)$.

At this point we could keep going, but there is no need to even apply J_- any more. By reflection symmetry along the z-axis, we have $|2, -1\rangle = \frac{1}{\sqrt{2}}(|0, -1\rangle + |-1, 0\rangle)$ and $|2, -2\rangle = |-1, -1\rangle$.

These account for five out of the nine states. Of the remaining states, the maximum value M can have is 1, attained by the states $|0, 1\rangle$ and $|1, 0\rangle$. But this state $|J = 1, M = 1\rangle$ has to be orthogonal to the state $|2, 1\rangle = \frac{1}{\sqrt{2}}(|0, 1\rangle + |1, 0\rangle)$ we already have. Thus, with essentially no work, we have found $|1, 1\rangle = \frac{1}{\sqrt{2}}(|0, 1\rangle - |1, 0\rangle)$.

Again, apply J_- on this, and by democracy, we obtain with no work at all $|1, 0\rangle = \frac{1}{\sqrt{2}}(|-1, 1\rangle - |1, -1\rangle)$, and then $|1, -1\rangle = \frac{1}{\sqrt{2}}(|-1, 0\rangle - |0, -1\rangle)$.

So now there is only $9 - 5 - 3 = 1$ state left. This lone state is determined by the fact that it is orthogonal to everybody else. Hence, $|0, 0\rangle = \frac{1}{\sqrt{3}}(|-1, 1\rangle - |0, 0\rangle + |1, -1\rangle)$.

Again, if you line up the nine states as in (7), you will see that there are three states in the middle, and one state at the two ends.

We have just showed that

$$1 \otimes 1 = 2 \oplus 1 \oplus 0 \tag{9}$$

The attentive reader will recall that we obtained exactly the same result, $3 \otimes 3 = 5 \oplus 3 \oplus 1$, multiplying two vectors together back in chapter IV.1. Indeed, that was how the very concept of tensor was invented in the first place.

Example (C): $j = 2$, $j' = 1$

Now that you have worked through examples (A) and (B) thoroughly, you are ready to rip through the multiplication $2 \otimes 1$. I will let you do it, sketching for you how to proceed. We now have $5 \cdot 3 = 15$ states, so things are a bit more involved, but conceptually it is the same game. Your task is to determine the states $|J, M\rangle$ as linear combinations of the 15 states $|m, m'\rangle$.

Start with $|J = 3, M = 3\rangle = |m = 2, m' = 1\rangle$ and climb down with the help of J_-. Refer back to (IV.2.8) and (IV.2.9). First, $J_-|2, 1\rangle = 2\,|1, 1\rangle + \sqrt{2}\,|2, 0\rangle$. (Note that we no longer have democracy as our guiding light; there is no reason for the two kets $|1, 1\rangle$ and $|2, 0\rangle$ to come in with equal weight.) Normalizing* this linear combination, we determine that $|3, 2\rangle = \frac{1}{\sqrt{6}}(2|1, 1\rangle + \sqrt{2}|2, 0\rangle)$. Climb down another rung to get[†] $|3, 1\rangle = \frac{1}{\sqrt{15}}(\sqrt{6}|0, 1\rangle + 2\sqrt{2}|1, 0\rangle + |2, -1\rangle)$. I will let you continue. Climb down rung after rung until you get to $|3, -3\rangle$, referring to (IV.2.8) and (IV.2.9) repeatedly. Then what?

Well, you have accounted for $2 \cdot 3 + 1 = 7$ states. There are $15 - 7 = 8$ states left. The maximum value $m + m'$ can now attain is 2, with the kets $|1, 1\rangle$ and $|2, 0\rangle$. But the linear combination $\frac{1}{\sqrt{6}}(2|1, 1\rangle + \sqrt{2}|2, 0\rangle)$ is already taken, so we are left with the orthogonal linear combination $\frac{1}{\sqrt{6}}(\sqrt{2}|1, 1\rangle - 2|2, 0\rangle)$. Note that $(2\langle1, 1| + \sqrt{2}\langle2, 0|)(\sqrt{2}|1, 1\rangle - 2|2, 0\rangle) = 2\sqrt{2}(\langle1, 1|1, 1\rangle - \langle2, 0|2, 0\rangle) = 0$. Act with J_- to obtain $2 \cdot 2 + 1 = 5$ states.

We are left with $15 - 7 - 5 = 3$ states. The maximum value $m + m'$ can now attain is 1, with the kets $|2, -1\rangle$, $|-1, 2\rangle$, $|1, 0\rangle$, and $|0, 1\rangle$. You have to find the linear combination that is orthogonal to the states that have already appeared, and then apply J_- to obtain $2 \cdot 1 + 1 = 3$ states.

Thus, completing this exercise, you would have shown that

$$2 \otimes 1 = 3 \oplus 2 \oplus 1 \tag{10}$$

After going through these three examples, you see that we have basically proved (2). The general case of $j \otimes j'$ does not involve anything conceptual not contained in these examples, but merely involves more verbiage to recount. Start with $|j + j', j + j'\rangle = |j, j'\rangle$, and climb down. Then start another ladder with the top rung $|j + j' - 1, j + j' - 1\rangle$ given by a linear combination of $|j - 1, j'\rangle$ and $|j, j' - 1\rangle$, and climb down. Repeat until all $(2j + 1)(2j' + 1)$ states are gone. As promised, we have here an alternative derivation of (IV.2.11).

Clebsch-Gordan decomposition and coefficients

This procedure of working out $j \otimes j'$ is known as the Clebsch-Gordan[2] decomposition. The various coefficients that appear are known as Clebsch-Gordan coefficients (for example, the numbers $\frac{1}{\sqrt{6}}$ and $\sqrt{\frac{2}{3}}$ in $|2, 0\rangle = \frac{1}{\sqrt{6}}(|-1, 1\rangle + 2|0, 0\rangle + |1, -1\rangle)$). They are important in various applications in physics. We will see a few examples in chapter V.1.

Write the decomposition of $j \otimes j'$ in the form

$$|J, M\rangle = \sum_{m=-j}^{j} \sum_{m'=-j'}^{j'} |j, j', m, m'\rangle \langle j, j', m, m'|J, M\rangle \tag{11}$$

---

* $\langle3, 2|3, 2\rangle = \frac{1}{\sqrt{6}}\left(2\langle1, 1| + \sqrt{2}\langle2, 0|\right) \frac{1}{\sqrt{6}}\left(2|1, 1\rangle + \sqrt{2}|2, 0\rangle\right) = \frac{1}{6}\left(4\langle1, 1|1, 1\rangle + 2\langle2, 0|2, 0\rangle\right) = 1$.
[†] At the risk of being repetitive, I point out that the numbers inside the ket on the left hand side refer to J, M, while the numbers inside the kets on the right hand side refer to m, m'.

In other words, $|J, M\rangle$ is a linear combination of $|j, j', m, m'\rangle$ with the Clebsch-Gordan coefficients given by the numbers $\langle j, j', m, m'|J, M\rangle$. Since these vanish unless $m + m' = M$, the double sum in (11) reduces to a single sum. Note that the Clebsch-Gordan decomposition is essentially a statement about completeness:

$$\sum_{m=-j}^{j} \sum_{m'=-j'}^{j'} |j, j', m, m'\rangle\langle j, j', m, m'| = I \qquad (12)$$

For the case $j = 1$, $j' = 1$, the results are as follows. (As remarked earlier, since j and j' are fixed once and for all, they will be suppressed.)

$J = 2$:

$$|2, 2\rangle = |1, 1\rangle$$

$$|2, 1\rangle = \frac{1}{\sqrt{2}}(|0, 1\rangle + |1, 0\rangle)$$

$$|2, 0\rangle = \frac{1}{\sqrt{6}}(|-1, 1\rangle + 2|0, 0\rangle + |1, -1\rangle)$$

$$|2, -1\rangle = \frac{1}{\sqrt{2}}(|0, -1\rangle + |-1, 0\rangle)$$

$$|2, -2\rangle = |-1, -1\rangle \qquad (13)$$

$J = 1$:

$$|1, 1\rangle = \frac{1}{\sqrt{2}}(|0, 1\rangle - |1, 0\rangle)$$

$$|1, 0\rangle = \frac{1}{\sqrt{2}}(|-1, 1\rangle - |1, -1\rangle)$$

$$|1, -1\rangle = \frac{1}{\sqrt{2}}(|-1, 0\rangle - |0, -1\rangle) \qquad (14)$$

$J = 0$:

$$|0, 0\rangle = \frac{1}{\sqrt{3}}(|-1, 1\rangle - |0, 0\rangle + |1, -1\rangle) \qquad (15)$$

The Clebsch-Gordan decomposition has so many applications in various areas of physics that for many researchers, calculating Clebsch-Gordan coefficients (or more likely these days, looking them up) is almost a way of life. This is because the conceptual framework behind the decomposition—multiplying two irreducible representations together and separating the resulting reducible representation into irreducible representations—is so basic and natural in group theory. For example, in atomic physics, an electron with orbital angular momentum l will have total angular momentum* j equal to either $j = l + \frac{1}{2}$ or $j = l - \frac{1}{2}$ (excepting the trivial case $l = 0$). Work this out as an exercise for possible later use. The $2(2l + 1)$ states split into $2(l + \frac{1}{2}) + 1$ and $2(l - \frac{1}{2}) + 1$ states.

---

* See the remark early in this chapter regarding the various names for angular momentum; here we are adding orbital and spin angular momentum to form $\vec{J} = \vec{L} + \vec{S}$.

In a course on quantum mechanics, students learn how to combine angular momentum. Here I mention that a common confusion is whether we are multiplying or adding, since both terms are thrown about casually. The answer is both: here we have one section titled "addition of angular momentum" and another titled "multiplying two ladders together." I trust that the detailed analysis given here makes clear what is going on. In a sense, it corresponds to the concept of Lie group versus the concept of Lie algebra: we multiply representations together but add generators.

Wigner-Eckart theorem

The early days of atomic spectroscopy saw a massive mess of confusion, to say the least. Many transition lines with varying intensity were observed, while some expected lines were missing. In quantum mechanics, the probability amplitude for a transition from some initial state $|i\rangle$ to some final state $|f\rangle$ due to some perturbation, such as the electromagnetic field, usually ends up being given by an operator $\mathcal{O}$ evaluated between the two states, that is, $\langle f| \mathcal{O} |i\rangle$.

In atomic spectroscopy, the initial and final states transform like members of some irreducible representations of the rotation group $SO(3)$, respectively, $|i\rangle = |\alpha, j, m\rangle$ and $|f\rangle = |\alpha', j', m'\rangle$. Here α and α' denote generically some other quantum numbers not governed by $SO(3)$, such as the principal quantum number that measures how "high" a given state is in the energy spectrum.

The operator $\mathcal{O}$ also transforms like members of some irreducible representations of $SO(3)$. We write $\mathcal{O}_{JM}$ to indicate that it transforms like the state $|JM\rangle$. For example, in the simplest kind of electromagnetic transition, known as the dipole transition, $\mathcal{O}$ is simply the position operator $\vec{x}$ of the electron. In this case, $\mathcal{O}$ transforms like a vector, and thus, as explained in chapter IV.2, $J = 1$ and $M = -1, 0$, or 1; in other words, $\mathcal{O}_{1M}$ transforms like the spherical harmonic Y_1^M.

The Wigner-Eckart theorem tells us what group theory has to say about the matrix element $\langle \alpha', j', m' | \mathcal{O}_{JM} | \alpha, j, m\rangle$. It states[3] that

$$\langle \alpha', j', m' | \mathcal{O}_{JM} | \alpha, j, m\rangle = \left(\langle j', m'| \left(|JM\rangle \otimes |j, m\rangle \right) \right) \langle \alpha', j'\|O_J\|\alpha, j\rangle$$
$$= \langle j', m' | J, j, M, m\rangle \langle \alpha', j'\|O_J\|\alpha, j\rangle \tag{16}$$

The amplitude factors into a product of two quantities, which we can associate with symmetry and with dynamics. The "thing with the double vertical bars"[4] $\langle \alpha', j'\|O_J\|\alpha, j\rangle$, called the reduced matrix element of the operator $\mathcal{O}$, represents dynamics, about which group theory has nothing to say. Its evaluation requires knowing the Schrödinger wave functions of the initial and final states and doing an integral. One result of the theorem is that the quantity $\langle \alpha', j'\|O_J\|\alpha, j\rangle$ depends on α', α, J, j', and j but does not depend on m' and m.

The group theoretic heart of the theorem resides in the factor $\langle j', m' | J, j, M, m\rangle$, namely, a Clebsch-Gordan coefficient. This makes such perfect sense as to render the theorem almost self-evident, since the quantity $\mathcal{O}_{JM} |\alpha, j, m\rangle$ and the direct product of two states $|JM\rangle \otimes |j, m\rangle = |J, j, M, m\rangle$ transform in exactly the same way. As far as group theory is concerned, they might as well be the same thing.

But we know that $\langle j', m' | J, j, M, m \rangle$ vanishes unless $j' = j + J, \; j + J - 1, \; \cdots,$ $|j - J| + 1, |j - J|$ and $m' = M + m$. Let us write these two conditions more compactly (see exercise 3) in terms of the change $\Delta j = j' - j$ and $\Delta m = m' - m$ in the atomic transition. We have

$$|\Delta j| = |j' - j| \leq J$$
$$\Delta m = m' - m = M \leq J \tag{17}$$

In quantum mechanics, the intensity of a transition line in atomic spectroscopy is given by the absolute square of the probability amplitude $\langle \alpha', j', m' | \mathcal{O}_{JM} | \alpha, j, m \rangle$. Thus, group theory fixes the relative intensity of the various observed lines. Furthermore, the line is forbidden unless the conditions in (17) are satisfied, which thus explains the famous selection rules that played such a pivotal (and mysterious) role in the development of quantum mechanics.

For a given pair of initial and final states (α, j) and (α', j') (and with the type of transition specified by J fixed), $(2j' + 1)(2j + 1)$ transitions are in principle possible, many of which are forbidden by (17). The relative intensities of the observed transition lines are thus entirely fixed by group theory, with the double-vertical-bar thing $\langle \alpha', j' \| O_J \| \alpha, j \rangle$ canceling out.

Imagine yourself back in the early 1920s, before quantum mechanics was developed and before group theory became known to physicists. The puzzle posed by the abundant data in atomic spectroscopy must have seemed awesome.

Exercises

1 Work out example (C) in detail.

2 Work out the Clebsch-Gordan coefficients in $1 \otimes \frac{1}{2} = \frac{3}{2} \oplus \frac{1}{2}$.

3 Derive the selection rules (17).

4 Apply $J_\pm$ to (11) to obtain the following recursion relations for the Clebsch-Gordan coefficients:

$$\sqrt{(J + 1 \pm M)(J \mp M)} \langle m, m' | J, M \pm 1 \rangle$$
$$= \sqrt{(j + 1 \mp m)(j \pm m)} \langle m \mp 1, m' | J, M \rangle$$
$$+ \sqrt{(j' + 1 \mp m')(j' \pm m')} \langle m, m' \mp 1 | J, M \rangle \tag{18}$$

5 Show that the total angular momentum states for an electron with orbital angular momentum $l \neq 0$ are given by

$$\left| j = l + \frac{1}{2}, m \right\rangle = \sqrt{\frac{l + m + \frac{1}{2}}{2l + 1}} \left| m - \frac{1}{2}, \frac{1}{2} \right\rangle + \sqrt{\frac{l - m + \frac{1}{2}}{2l + 1}} \left| m + \frac{1}{2}, -\frac{1}{2} \right\rangle \tag{19}$$

and

$$\left| j = l - \frac{1}{2}, m \right\rangle = -\sqrt{\frac{l - m + \frac{1}{2}}{2l + 1}} \left| m - \frac{1}{2}, \frac{1}{2} \right\rangle + \sqrt{\frac{l + m + \frac{1}{2}}{2l + 1}} \left| m + \frac{1}{2}, -\frac{1}{2} \right\rangle \tag{20}$$

Note that once you obtain $\left| j = l + \frac{1}{2}, m \right\rangle$, then with almost no work, you can determine $\left| j = l - \frac{1}{2}, m \right\rangle$ by appealing to orthogonality.

6 The Clebsch-Gordan decomposition often allows us to discover new identities. Here is an elementary example. Given four vectors, $\vec{u}$, $\vec{v}$, $\vec{w}$, and $\vec{z}$, how many scalars can you form out of them? What about $(\vec{u} \otimes \vec{v}) \cdot (\vec{w} \otimes \vec{z})$?

Notes

1. As always in physics, it is crucial to specify what is being fixed when taking a limit.
2. A piece of trivia: Paul Gordan was Emmy Noether's thesis advisor.
3. The Wigner-Eckart theorem is stated here for $SO(3)$. For $SO(3)$, the product $j \otimes J$ contains j' only once. For other groups, the product of the two given representations (namely, the analogs of j and J) may contain the analog of j' more than once, in which case the Wigner-Eckart theorem involves an additional sum.
4. One of my professors used to speak of this mockingly, mumbling that some people are not satisfied with a single vertical bar.

IV.4 | Tensors and Representations of the Special Unitary Groups *SU(N)*

Since this chapter is fairly involved, I first give you an overview or roadmap. Now that we have worked out the special orthogonal groups $SO(N)$, we simply try to repeat everything for the special unitary groups $SU(N)$. My pedagogical strategy is to work out the tensors for $SU(N)$ by following our noses; soon enough, we will stumble into the fact that the tensors of $SU(N)$, in contrast to the tensors of $SO(N)$, have to carry two kinds of indices, upper and lower. The important message for the beginning reader is that this fact did not capriciously and mysteriously fall from the sky, but merely reflects the presence in $SU(N)$, but not in $SO(N)$, of the basic operation of complex conjugation.

From orthogonal to unitary groups

We worked out the tensors and representations of the special orthogonal groups $SO(N)$ in chapter IV.1. Here we extend that discussion to the tensors and representations of the special unitary groups $SU(N)$. Roughly speaking, the extension corresponds to going from real to complex numbers.

We start with a quick review of the special orthogonal groups $SO(N)$. We learned in chapter I.3 that $SO(N)$ consists of rotations in N-dimensional space,[1] namely, all N-by-N real matrices O that are orthogonal

$$O^T O = 1 \tag{1}$$

and have unit determinant

$$\det O = 1 \tag{2}$$

The condition (1) ensures that the length squared $v^T v$ of the N-component vector v, which transforms according to $v \to v' = Ov$, is left unchanged or invariant. As was remarked in chapter IV.3, the condition that $v^T v$ is left invariant actually implies* the apparently stronger condition that $u^T v$ is left invariant for u and v two arbitrary vectors.

Lie showed us how to solve (1) by writing rotations near the identity as $O \simeq I + A$. To leading order in A, (1) becomes $A^T - A = 0$. The most general A may then be written as a linear combination of $N(N-1)/2$ antisymmetric matrices.

Quantum physics and complex numbers

Classical physics is done with real numbers. With the advent of quantum mechanics, complex numbers came into physics in an essential way. Wave functions, for example, are in general complex.

We wish to generalize the discussion for $SO(N)$ by considering linear transformations on complex vectors ψ. While $v = \{v^j, \ j = 1, \cdots, N\}$ consists of N real numbers, we now take $\psi = \{\psi^j, \ j = 1, \cdots, N\}$ to consist of N complex numbers. We demand that the linear transformations $\psi \to \psi' = U\psi$ leave $\sum_{j=1}^{N} \psi^{j*}\psi^j = \psi^{*T}\psi \equiv \psi^\dagger \psi$ invariant. (The last step defines hermitean conjugation of a complex vector as complex conjugation followed by transposition; see the review of linear algebra.) Just as in the case of $SO(N)$, this implies[†] that for two arbitrary complex vectors ζ and ψ, under the transformations $\zeta \to U\zeta$ and $\psi \to U\psi$, the quadratic forms[‡] $\zeta^\dagger \psi$ and $\psi^\dagger \zeta$ are left invariant. Since ζ and ψ are arbitrary, this condition, that $\zeta^\dagger U^\dagger U \psi = \zeta^\dagger \psi$, leads to the requirement that $U^\dagger U = I$, in other words, that U is unitary.

The group $U(N)$

The group $U(N)$ is defined to consist of all N by N matrices U that are unitary

$$U^\dagger U = I \tag{3}$$

We can readily show that $U(N)$ is indeed a group. (You may want to try it before reading on.)

To prove closure, we need to show that, if U_2 and U_1 are unitary matrices, then the matrix $U_2 U_1$ is also unitary. Simply compute: $(U_2 U_1)^\dagger U_2 U_1 = U_1^\dagger U_2^\dagger U_2 U_1 = U_1^\dagger U_1 = I$. It is crucial that hermitean conjugation reverses the order in a product, that is, $(U_2 U_1)^\dagger = U_1^\dagger U_2^\dagger$, but this follows from the fact that transposition of a product of matrices reverses the order,[§] that is, $(U_2 U_1)^T = U_1^T U_2^T$, a fact which also plays a crucial role in proving closure for the orthogonal groups.

---

* Since $v^T v$ is left invariant for arbitrary v, we could replace v by $v + \lambda u$ with λ an arbitrary real number and conclude that $(v + \lambda u)^T (v + \lambda u) = v^T v + \lambda(u^T v + v^T u) + \lambda^2 u^T u$ is left invariant. Since λ is arbitrary, the coefficient of λ in this expression, namely, $u^T v + v^T u = 2u^T v$, is left invariant.

[†] As before, we replace ψ by $\psi + \lambda \zeta$, with λ now an arbitrary complex number.

[‡] They are of course complex conjugates of each other.

[§] Complex conjugation, in contrast, does not touch the order.

The existence of the identity I and of the inverse $U^\dagger$ for each U holds almost by definition.

$SU(N)$ **as a subgroup of** $U(N)$

For orthogonal matrices, on taking the determinant of the condition (1), $O^T O = 1$, we obtain $(\det O)^2 = 1$, which implies the binary possibilities: $\det O = \pm 1$. Choosing to impose the condition (2), $\det O = 1$, then allows us to eliminate all those transformations that involve reflections. In contrast, for unitary matrices, on taking the determinant of the condition (3), $U^\dagger U = I$, we obtain $\det(U^\dagger U) = (\det U^\dagger)(\det U) = (\det U)^*(\det U) = |\det U|^2 = 1$, which implies a continuum of possibilities: $\det U = e^{i\alpha}$, for $0 \le \alpha < 2\pi$.

The condition

$$\det U = 1 \tag{4}$$

eliminates this dangling phase factor $e^{i\alpha}$.

The two conditions (3) and (4), which are analogous to the two conditions (1) and (2), define the group $SU(N)$, a natural subgroup of $U(N)$.

An alternative way of saying this is as follows.

Among the groups $U(N)$, we have, in particular, the group $U(1)$ consisting of all 1-by-1 unitary matrices, that is, all complex numbers with absolute value equal to 1. In other words, $U(1)$ consists of phase factors $e^{i\varphi}$, $0 \le \varphi < 2\pi$.

The group $U(N)$ contains two sets of group elements.

The first set consists of unitary matrices of the form $e^{i\varphi}I$, where I denotes the N-by-N identity matrix. These matrices clearly form a subgroup of $U(N)$; they satisfy all the group axioms. Indeed, the matrix part (namely, I) is trivial: the different elements are distinguished by the phase factor $e^{i\varphi}$. This is a wordy way of saying that this subgroup is in a one-to-one correspondence with the group $U(1)$ and might as well be called $U(1)$.

The second set of group elements of $U(N)$ consists of N-by-N unitary matrices with determinant equal to 1, namely, the special unitary matrices forming the subgroup $SU(N)$. The matrices in this subgroup trivially commute with matrices of the form $e^{i\varphi}I$.

At this point, we might as well focus on the group $SU(N)$, defined by (3) and (4). (We will return to $U(N)$ late in chapter IV.5.)

The story of $SU(N)$ follows more or less the same plot as the story of $SO(N)$, with the crucial difference that we will be dealing with complex rather than real numbers.

In discussing the orthogonal groups, I started with $SO(2)$ and $SO(3)$, and then proceeded to $SO(N)$. This order of presentation certainly makes sense, given that we live in 3-dimensional Euclidean space. But in discussing the unitary groups, I reverse my pedagogical strategy and start with $SU(N)$, armed as it were with our experience with $SO(N)$.

From a general discussion of $SU(N)$,[2] to be given here, we will then, in subsequent chapters, specialize to $SU(2)$ and $SU(3)$, pointing out the special features that they do not share with $SU(N)$. After our discussion of the orthogonal groups, that the "lower" unitary groups enjoy special features not granted to the "higher" unitary groups is hardly surprising. As in the earlier discussion, this fact has to do with the number of indices the relevant antisymmetric symbols carry.

The Ω^- and counting with our fingers

Let us now merrily repeat what we did back in chapter IV.1 to construct the tensor representations of $SO(N)$. (You might want to do a quick review.) Starting with the fundamental representation* $\psi^i \to \psi'^i = U^{ij}\psi^j$, we consider tensor $\varphi^{i_1 i_2 \cdots i_m}$ carrying m indices and transforming according to $\varphi^{i_1 i_2 \cdots i_m} \to \varphi'^{i_1 i_2 \cdots i_m} = U^{i_1 j_1} U^{i_2 j_2} \cdots U^{i_m j_m} \varphi^{j_1 j_2 \cdots j_m}$. The whole point of tensors, as you will recall, is that their indices transform independently,[†] so that any symmetry of $\varphi^{i_1 i_2 \cdots i_m}$ under permutations of its indices is carried over to $\varphi'^{i_1 i_2 \cdots i_m}$. (For example, if $\varphi^{i_1 i_2 \cdots i_5}$ is totally symmetric in the first three indices and antisymmetric in the last two indices, so too will $\varphi'^{i_1 i_2 \cdots i_5}$ be totally symmetric in the first three indices and antisymmetric in the last two indices.)

Ta dah! We have thus constructed lots of representations of $SU(N)$. Once again, we have to undergo the ordeal of mastering Young tableaux to study the most general symmetry patterns that a tensor of $SU(N)$ carrying m indices can exhibit. And once again, fortunately, for many, perhaps even most, applications (say, in particle physics), we only have to deal with the simplest possible symmetry patterns, such as totally symmetric or totally antisymmetric, and with m small. (See chapter IV.1 for the advice I received long ago and dispense now regarding this point.)

As an example, let us count the dimension of the representation furnished by the 3-indexed totally symmetric tensor φ^{ijk} of $SU(3)$. Indeed, this counting is part of a Nobel prize winning piece of work!

Instead of watching the rich man going through his fancy footwork, let's watch the poor man simply enumerate:[‡] All indices equal to 3: $\{\varphi^{333}\}$; two indices equal to 3: $\{\varphi^{332}, \varphi^{331}\}$; one index equal to 3: $\{\varphi^{322}, \varphi^{321}, \varphi^{311}\}$; and no index equal to 3: $\{\varphi^{222}, \varphi^{221}, \varphi^{211}, \varphi^{111}\}$. We have altogether $1 + 2 + 3 + 4 = 10$ objects. Thus, $SU(3)$ has a 10-dimensional representation.

---

* The repeated index summation convention is enforced here, of course; thus $U^{ij}\psi^j = \sum_j U^{ij}\psi^j$.

† Perhaps a better description is "transform in parallel," as mentioned in chapter IV.1.

‡ In fact, this is by far the most foolproof way to proceed. (In case the student reading this does not realize it, the Swedes in their wisdom tend not to give Nobel prizes to people writing down the general formula for the dimension of tensors with the most general symmetry patterns under permutation of its indices for general N and m.)

In the early 1960s, nine short-lived baryonic* particles with similar properties and masses close to one another were known experimentally. This suggests, as per the argument given in chapter III.1, that these nine particles belong to an irreducible representation of some symmetry group. By "pure thought" (as shown here) and on guessing that the symmetric group of the strong interaction is $SU(3)$, Gell-Mann insisted that there must be ten, not nine, baryonic particles. The dramatic discovery[3] of the tenth baryonic particle, dubbed the Ω^-, confirming Gell-Mann's group theoretic prediction, was not only one of the most stirring episodes[†] in the annals of particle physics, but also ushered in the widespread study of group theory among particle physicists.[4]

Actually, we are only pretending to be poor; you and I can readily work out by induction the general formula for the dimension of $\{m\}$, the irreducible representation furnished by the totally symmetric tensor carrying m indices in $SU(N)$. Exercise! You will realize that the counting thus far is the same as in chapter IV.1. (Our fingers can't tell whether the tensor is for $SO(N)$ or $SU(N)$.)

The trace for $SU(N)$

You ask, what about taking out the trace?

Good question. In chapter IV.1, we had to subtract out the trace $\delta^{ij}\varphi^{ijk}$, so that the irreducible representation of $SO(3)$ is $7 = (10 - 3)$-dimensional,[‡] not 10-dimensional. OK, let's do the same here, but first, let's check how the trace transforms: $\delta^{ij}\varphi^{ijk} \rightarrow \delta^{ij}(U^{if}U^{jg}U^{kh}\varphi^{fgh}) = (\delta^{ij}U^{if}U^{jg})(U^{kh}\varphi^{fgh})$.

But oops! What in the spinning world is $\delta^{ij}U^{if}U^{jg} = (U^T)^{fi}\delta^{ij}U^{jg} = (U^TU)^{fg}$? It certainly is not δ^{fg}, which is what we need if $\delta^{ij}\varphi^{ijk}$ is to transform the way we want, namely, as a vector carrying the index k. We are talking about unitary matrices here, and the combination U^TU has no special meaning. So, this particular "trace" does not make any mathematical sense.

We don't have to take out the trace, contrary to the discussion in the orthogonal case (there, O^TO does have a special meaning, of course—namely, that it is equal to the identity matrix). The representation $\{3\}$ of $SU(3)$ is indeed irreducible, with dimension given by 10. Gell-Mann got it right.

Fine, but this raises a more pressing question: how do we define a trace for $SU(N)$? The clue—no surprise for those readers with mathematical sense!—is that it is the combination $U^\dagger U$, not U^TU, that is equal to the identity matrix. We have to complex conjugate

---

* This merely means that they have masses similar to the proton and neutron. For more, see chapter V.2.

† I describe this in slightly more detail in chapter V.2.

‡ Recall that in $SO(3)$, the dimension of the irreducible representation furnished by the totally symmetric tensor with j indices is given by $2j + 1$. In particular, there does not exist an integer j such that $2j + 1$ is equal to 10.

as well as transpose. (Physically, a quark* is to be combined with an antiquark,† not with another quark!)

Moving downstairs!

Now we come to something that has traditionally confused quite a few beginning students: the sudden appearance of two types of indices. We could go through some fancy schmancy talk about covariant and contravariant, vector spaces and dual vector spaces, et cetera— what Einstein called "more-or-less dispensable erudition"—but we won't; instead, we follow Einstein and adopt a low-brow practical approach, sticking to the nitty-gritty.

In the review of linear algebra, I already mentioned the commonly used notation in which complex vectors carry an upper index, while their complex conjugates carry a lower index. But in view of the confusion alluded to above, it is worthwhile to go back to basics.

As noted at the beginning of this chapter, the quadratic invariant is $\zeta^\dagger \psi = \zeta^{*T} \psi = \sum_{j=1}^{N} \zeta^{j*} \psi^j$, not $\zeta^T \psi = \sum_{j=1}^{N} \zeta^j \psi^j$. Given that $\psi^i \to U^{ij} \psi^j$, we complex conjugate to obtain $\psi^{i*} \to (U^{ij})^* \psi^{j*}$. Next notice that this expression can be written as $\psi^{i*} \to (U^{T*})^{ji} \psi^{j*} = (U^\dagger)^{ji} \psi^{j*} = \psi^{j*} (U^\dagger)^{ji}$; nothing prevents us, in the last step, from interchanging the two complex numbers $(U^\dagger)^{ji}$ and ψ^{j*}. Indeed, we can now verify explicitly that $\psi^{i*} \zeta^i \to \psi^{j*} (U^\dagger)^{ji} U^{ik} \zeta^k = \psi^{j*} \delta^{jk} \zeta^k = \psi^{j*} \zeta^j$ is invariant; after all, that was how we got the condition $U^\dagger U = I$ in the first place.

At this point, people invented a very clever notation: write

$$\psi_i \equiv \psi^{i*} \tag{5}$$

A new object ψ_i appears, but it is literally just ψ^{i*}, nothing mysterious about it at all. At the least, we no longer have to write "*". It may not seem to you like all that much of a notational saving: not having to write "*", we now have to write lower indices. But the nice feature is that the invariant $\zeta^\dagger \psi = \sum_{j=1}^{N} \zeta^{j*} \psi^j$ can be written as $\zeta^\dagger \psi = \sum_{j=1}^{N} \zeta_j \psi^j = \zeta_j \psi^j$ (invoking the repeated index summation convention at the last step). Thus, as this example shows, when we contract and sum over indices, with this new notation, we should impose the rule that we are allowed to contract and sum over an upper index with a lower index, but never an upper index with an upper index, nor a lower index with a lower index. You will see shortly how this works in general.

To summarize, the defining or fundamental representation of $SU(N)$ consists of N objects ψ^j, $j = 1, \cdots, N$, which transform under the action of the group element U according to

$$\psi^i \to \psi'^i = U^i_{\ j} \psi^j \tag{6}$$

---

* If you don't know what this word means, perhaps it is time to read some popular book about particle physics.

† See chapter V.2 on $SU(3)$.

Note that the column index on the matrix U has been typeset to the right of the row index, and lowered to satisfy our convention that the upper index on ψ^j is to be summed with a lower index.

Now we define an object written as ψ_i, which transforms in the same way as ψ^{*i}. The transformation $\psi^{i*} \to \psi^{j*}(U^\dagger)^{ji}$ we had earlier is then translated to $\psi_i \to \psi_j(U^\dagger)^j{}_i$. Thus, the counterpart of (6) is

$$\psi_i \to \psi'_i = \psi_j(U^\dagger)^j{}_i \tag{7}$$

As a check, $\zeta_i\psi^i \to \zeta'_i\psi'^i = \zeta_j(U^\dagger)^j{}_i U^i{}_k\psi^k = \zeta_j(U^\dagger U)^j{}_k\psi^k = \zeta_j\delta^j{}_k\psi^k = \zeta_j\psi^j$. Things work out nicely. Note that the Kronecker delta carries one upper and one lower index. This reinforces the fact (as we shall see shortly in an example) that we are not allowed to contract an upper index with an upper index, nor a lower index with a lower index.

In summary, upper indices transform with U, lower indices with $U^\dagger$.

Tensors with upper and lower indices

We now realize that the discussion earlier in this chapter about tensors in $SU(N)$ was incomplete: in this notation, we must allow the tensors to carry lower indices as well as upper indices; we only had upper indices. In general, we have $\varphi^{i_1 i_2 \cdots i_m}_{j_1 j_2 \cdots j_n}$ carrying m upper indices and n lower indices.*

Instead of obscuring the fairly obvious with a general discussion, let us show how tensors work with a specific example. The tensor φ^{ij}_k transforms as

$$\varphi^{ij}_k \to \varphi'^{ij}_k = U^i{}_l U^j{}_m (U^\dagger)^n{}_k \varphi^{lm}_n = U^i{}_l U^j{}_m \varphi^{lm}_n (U^\dagger)^n{}_k \tag{8}$$

As at the end of the last section, upper indices transform with U, lower indices with $U^\dagger$. In other words, it transforms as if it were equal to the product $\psi^i\psi^j\psi_k$; note that I did not say that φ^{ij}_k is equal to $\psi^i\psi^j\psi_k$, merely that they transform in the same way.

We now know how to take a trace: set an upper index equal to a lower index and sum over them. In our example, consider $\delta^k{}_j\varphi^{ij}_k \equiv \varphi^{ij}_j$. (Note that, as alluded to earlier, the fact that the Kronecker delta carries one upper and one lower index reminds us to contract and sum over an upper index with a lower index.) It transforms as

$$\varphi^{ij}_j \to U^i{}_l U^j{}_m (U^\dagger)^n{}_j \varphi^{lm}_n = U^i{}_l \delta^n{}_m \varphi^{lm}_n = U^i{}_l \varphi^{lm}_m \tag{9}$$

where we have used (3). In other words, the φ^{ij}_j (the trace of φ^{ij}_k), carrying the index $i = 1, 2, \cdots, N$, are N objects that transform into linear combinations of one another in the same way as a vector. Thus, given a tensor, we can always subtract out its trace.

---

* Our friend the jargon guy tells us that the upper indices are known as contravariant and the lower indices as covariant. The reader who has taken a course on Einstein gravity will know that tensors in general relativity also carry upper and lower indices.

Now we see what we did wrong earlier. We erroneously defined a trace by summing over two upper indices, which is not allowed in this formalism.

To summarize, traceless tensors with upper and lower indices furnish representations of $SU(N)$. The discussion proceeds in a way that should by now be familiar. The symmetry properties of a tensor under permutation of its indices are not changed by the group transformation. In other words, given a tensor $\varphi_{j_1 j_2 \cdots j_n}^{i_1 i_2 \cdots i_m}$ we can always require it to have definite symmetry properties under permutation of its upper indices (the is) and under permutation of its lower indices (the js). Note that the upper indices and lower indices are to be permuted separately. It does not make sense to interchange an upper index and a lower index; they are different beasts, so to speak.

In our specific example, we are allowed to take φ_k^{ij} to be either symmetric or antisymmetric under the exchange of i and j and to be traceless. (Here, with one single lower index, we don't have to worry about the symmetry pattern of the lower indices.) Thus, the symmetric traceless tensor $S_k^{ij} = +S_k^{ji}$ furnishes a representation with dimension $\frac{1}{2}N^2(N+1) - N = \frac{1}{2}N(N-1)(N+2)$ and the antisymmetric traceless tensor $A_k^{ij} = -A_k^{ji}$ a representation with dimension $\frac{1}{2}N^2(N-1) - N = \frac{1}{2}N(N-2)(N+1)$.

Thus, in summary, the irreducible representations of $SU(N)$ are realized by traceless tensors with definite symmetry properties under permutation of indices. Convince yourself that for $SU(N)$, the dimensions of the representations defined by the following tensors φ^i, φ^{ij} (antisymmetric), φ^{ij} (symmetric), φ_j^i, and φ_k^{ij} (antisymmetric in the upper indices) are, respectively, N, $N(N-1)/2$, $N(N+1)/2$, N^2-1, and $\frac{1}{2}N(N-2)(N+1)$.

You are now almost ready for the grand unified theories of the strong, weak, and electromagnetic interactions! These are the irreducible representations commonly used in the popular Georgi-Glashow $SU(5)$ theory, with dimensions 5, 10, 15, 24, and 45, respectively.

As in the $SO(N)$ story, representations of $SU(N)$ have many names. For example, we can refer to the representation furnished by a tensor with m upper and n lower indices as (m, n). Alternatively, we can refer to them by their dimensions, with a $*$ to distinguish representations with mostly lower indices from those with mostly upper indices. For example, an alias for $(1, 0)$ is N, and for $(0, 1)$ is N^*. Of course, two entirely different representations could happen to have the same dimension, but this ambiguity does not usually occur for representations with relatively low dimensions. For example, in the popular $SU(5)$ grand unified theory, the known quarks and leptons transform like 5 and 10^*. A square bracket is often used to indicate that the indices are antisymmetric, and a curly bracket that the indices are symmetric. Thus, the 10 of $SU(5)$ is also known as $[2, 0] = [2]$, where as indicated, the 0 (no lower index) is suppressed. Similarly, 10^* is also known as $[0, 2] = [2]^*$.

Moving indices up and down stairs

The astute reader may have noticed that we have not yet used the condition (4), namely, the S in $SU(N)$. It can be written as either

$$\varepsilon_{i_1 i_2 \cdots i_N} U^{i_1}{}_1 U^{i_2}{}_2 \cdots U^{i_N}{}_N = 1 \tag{10}$$

or

$$\varepsilon^{i_1 i_2 \cdots i_N} U^1{}_{i_1} U^2{}_{i_2} \cdots U^N{}_{i_N} = 1 \tag{11}$$

Recall that this parallels what we did with $SO(N)$ in chapter IV.1, except that here we have upper and lower indices. Again, we can immediately generalize (10) to

$$\varepsilon_{i_1 i_2 \cdots i_N} U^{i_1}{}_{j_1} U^{i_2}{}_{j_2} \cdots U^{i_N}{}_{j_N} = \varepsilon_{j_1 j_2 \cdots j_N} \tag{12}$$

Multiplying this identity by $(U^\dagger)^{j_N}_{p_N}$ and summing over j_N, we obtain

$$\varepsilon_{i_1 i_2 \cdots i_{N-1} p_N} U^{i_1}{}_{j_1} U^{i_2}{}_{j_2} \cdots U^{i_{N-1}}{}_{j_{N-1}} = \varepsilon_{j_1 j_2 \cdots j_N} (U^\dagger)^{j_N}_{p_N} \tag{13}$$

Note that the index p_N is "exposed" and not summed over. Clearly, by repeating this process, we can peel off as many Us on the left hand side as we like, and put them back as $U^\dagger$s on the right hand side. We can play a similar game with (11).

In contrast to $SO(N)$, we now have, not one, but two antisymmetric symbols $\varepsilon_{i_1 i_2 \cdots i_N}$ and $\varepsilon^{i_1 i_2 \cdots i_N}$, which we can use to raise and lower indices as our hearts desire.

To avoid drowning in a sea of indices, let me show how we can raise and lower indices in a specific example rather than in general. Consider the tensor φ^{ij}_k in $SU(4)$. We expect that the tensor $\varphi_{kpq} \equiv \varepsilon_{ijpq} \varphi^{ij}_k$ will transform as a tensor with three lower indices. We could trade two upper indices for two lower indices. Let's check that this indeed works: $\varphi_{kpq} \equiv \varepsilon_{ijpq} \varphi^{ij}_k \to \varepsilon_{ijpq} U^i{}_l U^j{}_m (U^\dagger)^n{}_k \varphi^{lm}_n = (U^\dagger)^s{}_p (U^\dagger)^t{}_q ((U^\dagger)^n{}_k \varepsilon_{lmst} \varphi^{lm}_n) = (U^\dagger)^s{}_p (U^\dagger)^t{}_q (U^\dagger)^n{}_k \varphi_{nst}$, as expected. Here we used a version of (13) to trade two Us for two $U^\dagger$s. Similarly, you could verify that $\varphi^{ijlmn} \equiv \varepsilon^{klmn} \varphi^{ij}_k$ transforms like a tensor with five upper indices. Try a few more examples until you catch on.

To summarize, using the two (not one!) antisymmetric symbols, we can move indices on $SU(N)$ tensors up and down stairs.

From group to algebra

As before, Lie instructs us to solve (3) by writing $U \simeq I + iH$, with H some arbitrary "small" complex matrix. (Taking out a factor of i is merely a convenience, as we will see presently.) To leading order in the parameter specifying the smallness of H, $U^\dagger U \simeq (I - iH^\dagger)(I + iH) \simeq I - i(H^\dagger - H) = I$. Thus, (3) leads to $H^\dagger = H$, in other words, to the requirement that H is hermitean. As in the case of $SO(N)$, by multiplying an infinitesimal transformation repeatedly by itself, we can write in general

$$U = e^{iH} \tag{14}$$

The statement is that U is unitary if H is hermitean.

It is worth checking this statement explicitly, although it is already implied by the preceding discussion. As explained in the review of linear algebra, the exponential of a matrix may be defined by a power series: $U = e^{iH} = \sum_{k=0}^{\infty}(iH)^k/k!$. Hermitean conjugating term by term, we have $U^\dagger = \sum_{k=0}^{\infty}(-iH^\dagger)^k/k! = \sum_{k=0}^{\infty}(-iH)^k/k! = e^{-iH}$, and thus $U^\dagger U = e^{-iH}e^{iH} = I$.

Next, to evaluate $\det U$ in order to impose the condition (4), we exploit the fact that H can always be diagonalized and write $H = W^\dagger \Lambda W$, with Λ the diagonal matrix $\mathrm{diag}(\lambda_1, \lambda_2, \cdots, \lambda_N)$ and W a unitary matrix. Then*

$$\det U = \det e^{iH} = \det e^{iW^\dagger \Lambda W} = \det(W^\dagger e^{i\Lambda}W) = \det(WW^\dagger)\det e^{i\Lambda}$$

$$= \det e^{i\Lambda} = \Pi_{j=1}^{N}e^{i\lambda_j} = e^{i\sum_{j=1}^{N}\lambda_j} = e^{i\,\mathrm{tr}\,\Lambda} = e^{i\,\mathrm{tr}\,W^\dagger \Lambda W}$$

$$= e^{i\,\mathrm{tr}\,H} \tag{15}$$

(To obtain the third equality, expand the exponential in a power series and use the unitarity of W.) Thus, the condition $\det U = 1$ implies[†]

$$\mathrm{tr}\,H = 0 \tag{16}$$

This crucial result breaks the problem apart. It says that all we have to do is to write down the most general N-by-N traceless hermitean matrix H. To see how this works, it is easiest to proceed by examples.

For $N = 2$, the hermiticity condition $\begin{pmatrix} u & w \\ z & v \end{pmatrix}^\dagger = \begin{pmatrix} u^* & z^* \\ w^* & v^* \end{pmatrix} = \begin{pmatrix} u & w \\ z & v \end{pmatrix}$ implies that u and v are real and $w = z^*$, while the traceless condition gives $v = -u$. Thus, in general,

$$H = \begin{pmatrix} u & z^* \\ z & -u \end{pmatrix} = \frac{1}{2}\begin{pmatrix} \theta_3 & \theta_1 - i\theta_2 \\ \theta_1 + i\theta_2 & -\theta_3 \end{pmatrix} \tag{17}$$

where θ_1, θ_2, and θ_3 denote three arbitrary real numbers. (The factor of $\frac{1}{2}$ is conventional, due partly to historical reasons. I explain the reason for its inclusion in chapter IV.5.)

It is standard to define the three traceless hermitean matrices, known as Pauli matrices, as

$$\sigma_1 = \begin{pmatrix} 0 & 1 \\ 1 & 0 \end{pmatrix}, \qquad \sigma_2 = \begin{pmatrix} 0 & -i \\ i & 0 \end{pmatrix}, \qquad \sigma_3 = \begin{pmatrix} 1 & 0 \\ 0 & -1 \end{pmatrix} \tag{18}$$

The most general 2-by-2 traceless hermitean matrix H can then be written as a linear combination of the three Pauli matrices: $H = \frac{1}{2}(\theta_1\sigma_1 + \theta_2\sigma_2 + \theta_3\sigma_3) = \sum_{a=1}^{3}\frac{1}{2}\theta_a\sigma_a$. An element of $SU(2)$ can then be written as $U = e^{i\theta_a\sigma_a/2}$ (with the repeated index summation convention).

For $SU(3)$, we can go through analogous steps and define the eight 3-by-3 traceless hermitean matrices, known as Gell-Mann[5] matrices,

---

* Note that this is a special case of the formula $\det M = e^{\mathrm{tr}\,\log M}$ derived in the review of linear algebra.

† We can also obtain this from the infinitesimal form $U \simeq I + iH$, of course. To leading order in H, only the diagonal elements of H contribute to $\det U$. Verify this.

$$\lambda_1 = \begin{pmatrix} 0 & 1 & 0 \\ 1 & 0 & 0 \\ 0 & 0 & 0 \end{pmatrix}, \qquad \lambda_2 = \begin{pmatrix} 0 & -i & 0 \\ i & 0 & 0 \\ 0 & 0 & 0 \end{pmatrix}, \qquad \lambda_3 = \begin{pmatrix} 1 & 0 & 0 \\ 0 & -1 & 0 \\ 0 & 0 & 0 \end{pmatrix},$$

$$\lambda_4 = \begin{pmatrix} 0 & 0 & 1 \\ 0 & 0 & 0 \\ 1 & 0 & 0 \end{pmatrix}, \qquad \lambda_5 = \begin{pmatrix} 0 & 0 & -i \\ 0 & 0 & 0 \\ i & 0 & 0 \end{pmatrix},$$

$$\lambda_6 = \begin{pmatrix} 0 & 0 & 0 \\ 0 & 0 & 1 \\ 0 & 1 & 0 \end{pmatrix}, \qquad \lambda_7 = \begin{pmatrix} 0 & 0 & 0 \\ 0 & 0 & -i \\ 0 & i & 0 \end{pmatrix}, \qquad \lambda_8 = \frac{1}{\sqrt{3}} \begin{pmatrix} 1 & 0 & 0 \\ 0 & 1 & 0 \\ 0 & 0 & -2 \end{pmatrix} \qquad (19)$$

A general 3-by-3 traceless hermitean matrix and an element of $SU(3)$ can then be written, respectively, as $H = \theta_a \frac{\lambda_a}{2}$ and $U = e^{i\theta_a \frac{\lambda_a}{2}}$, where the index a now runs from 1 through 8. Note that the Pauli and Gell-Mann matrices are normalized by tr $\sigma_a \sigma_b = 2\delta_{ab}$ and tr $\lambda_a \lambda_b = 2\delta_{ab}$, respectively.

To make sure that you understand what is going on, you should write down a linearly independent set of 4-by-4 traceless hermitean matrices. (I regret to inform you, however, that physicists no longer have the habit of naming traceless hermitean matrices after anybody, including you.) How many are there?

Indeed, how many are there for a general N? We simply go through mentally the steps we went through for $SU(2)$. The N diagonal elements (the analogs of u and v) are real, and with the traceless condition, there are $N-1$ of these. The $\frac{1}{2}N(N-1)$ entries (each of which is specified by two real numbers) below the diagonal (namely, the analogs of z) are equal to the complex conjugate of the entries above the diagonal (the analogs of w). Thus, the number of real numbers (namely, the analogs of θ_a) required to specify a general N-by-N traceless hermitean matrix is given by $(N-1) + 2 \cdot \frac{1}{2}N(N-1) = N^2 - 1$, which is of course also the number of linearly independent N-by-N traceless hermitean matrices.

The group $SU(N)$ is thus characterized by $N^2 - 1$ real parameters (the analog of the θ_as), namely, 3, 8, 15, 24, 35, . . . for $N = 2, 3, 4, 5, 6, . . .$, respectively. Compare this with the group $SO(N)$ characterized by $\frac{1}{2}N(N-1)$ real parameters, namely, 1, 3, 6, 10, 15, . . . for $N = 2, 3, 4, 5, 6, . . .$, respectively.

Notice that $SU(2)$ and $SO(3)$ are characterized by the same number of parameters. File this fact away in your mind. We will come back to this important observation* later.

As I've said, my pedagogical plan is to continue discussing $SU(N)$ here, describing those aspects that hold for general N. Then we will discuss, in the next two chapters, $SU(2)$ and $SU(3)$ in turn, focusing on features specific to these two groups.

A trivial but potentially confusing point for some students. We just went through a long song and dance about the importance of distinguishing upper and lower indices on tensors. An upper index and a lower index transform quite differently (one might say

---

* Do you observe another potentially interesting numerological equality?

oppositely) under $SU(N)$ transformations. In contrast, the index a on the Pauli matrices is just a label, telling us which one of the three Pauli matrices we are talking about. It is completely arbitrary whether we write σ_a or σ^a. Similarly, the index a on the Gell-Mann matrices λ_a can be written as a subscript or superscript as we please. Correspondingly, the "angles" θ^a, $a = 1, \cdots, N^2 - 1$, can also be written with a superscript or a subscript. I will intentionally not stick to a consistent convention to emphasize this point.*

The structure constants of the Lie algebra

As just explained, any element of $SU(N)$ can be written as $U = e^{iH}$, with H a linear combination of $(N^2 - 1)$ linearly independent N-by-N hermitean traceless matrices T^a $(a = 1, 2, \cdots, N^2 - 1)$, so that

$$U = e^{i\theta^a T^a} \tag{20}$$

where θ^a are real numbers, and the index a is summed over. The T^as are known as the generators of $SU(N)$. The index a, which ranges over $(N^2 - 1)$ values, is clearly not to be confused with the index i, which ranges over N values.

The discussion now parallels that in chapter I.3 for the orthogonal groups. As mentioned in chapter I.1, group elements, in general, do not commute. Lie proposed to capture this essence of group multiplication by focusing on infinitesimal elements. We are thus led to consider two simple unitary matrices, $U_1 \simeq I + A$ and $U_2 \simeq I + B$, near the identity. Just as in chapter I.3, the lack of commutativity is measured by the deviation of the quantity

$$U_2^{-1} U_1 U_2 \simeq (I - B)(I + A)(I + B) \simeq I + A + AB - BA$$
$$= I + A + [A, B] \tag{21}$$

from $U_1 \simeq I + A$, namely, the commutator $[A, B]$. You should notice that this is formally identical to the manipulations in chapter I.3. Indeed, the only difference is that the generators of $SO(N)$ were called J, while the generators of $SU(N)$ are called T. So write $A = i \sum_a \theta^a T^a$ and similarly, $B = i \sum_b \theta'^b T^b$ with θ and θ' small. Hence $[A, B] = i^2 \sum_{ab} \theta^a \theta'^b [T^a, T^b]$, and it suffices to calculate the commutators $[T^a, T^b]$.

Recall from the review of linear algebra that the commutator $[H, K]$ of two hermitean matrices H and K is antihermitean and traceless (since $([H, K])^\dagger = [K^\dagger, H^\dagger] = [K, H] = -[H, K]$, and $\operatorname{tr}[H, K] = \operatorname{tr} HK - \operatorname{tr} KH = 0$).

Since the commutator $[T^a, T^b]$ is antihermitean and traceless, it can also be written as a linear combination of the T^cs multiplied by i:

$$[T^a, T^b] = i f^{abc} T^c \tag{22}$$

(with the index c summed over).

---

* Later, in chapter VI.2, when we discuss Lie algebras from a more mathematical point of view, I will introduce another set of indices, for which it does matter whether the index is up or down.

For an alternative derivation of (22), argue that since $(U_2 U_1)^{-1} U_1 U_2$ is an element of $SU(N)$, it must be equal to $I + i\phi^c T^c + \cdots$ for some ϕ^c.

The commutation relations (22) define the Lie algebra of $SU(N)$ and, as was mentioned way back in chapter I.3, the real numbers f^{abc} are known as the structure constants of the algebra. Note that by construction, $f^{abc} = -f^{bac}$. We will show generally in chapter VI.3 that the structure constant is in fact totally antisymmetric; it changes sign under the interchange of any pair of its three indices.

Given the Pauli matrices (18) and the Gell-Mann matrices (19), you can readily compute the structure constants f^{abc} for $SU(2)$ and $SU(3)$. Exercise!

While physicists fill their brains with specific matrices, mathematicians use their minds to imagine abstract entities T^a satisfying* (22). We started this chapter defining $SU(N)$ as the group formed by unitary matrices with unit determinant. In other words, we thought of the group elements as represented in the fundamental representation by N-by-N matrices. Using tensors, we constructed higher-dimensional representations, in which the group elements U are represented by higher-dimensional matrices. (For example, for $SU(5)$, U can be represented by 10-by-10, or 15-by-15, or 24-by-24 matrices, and so on, as well as by 5-by-5 matrices.) For each of these representations, by considering elements near the identity, $U \simeq I + i\theta^a T^a$, we have a corresponding representation of T^a, for $a = 1, \cdots, N^2 - 1$.

More pedantically and explicitly, consider a d-dimensional representation. The group element $U \simeq I + i\theta^a T^a$ is represented by the d-by-d matrix $D(I + i\theta^a T^a)$. Setting θ to 0, we find that $D(I) = I_d$, the d-by-d identity matrix. Hardly a surprise! Subtracting, or equivalently, differentiating with respect to θ^a, we obtain[†] $D(T^a)$, the d-by-d matrix representing T^a. For example, in the fundamental representation of $SU(2)$, the $D(T^a)$ are just the 2-by-2 Pauli matrices (with the "funny factor" of one half) $\frac{1}{2}\sigma_a$.

Thus, for $SU(5)$, we have a set of 24 10-by-10 matrices T^a constructed to satisfy (22). (Or 15-by-15 matrices, or 24-by-24 matrices, and so on.) In particular, in the fundamental representation, we denote the entry in the ith row and jth column of the matrix T^a by $(T^a)^i_j$.

At the risk of beating a point to death, note that a and i are entirely different beasts. The label a, which ranges over $a = 1, \cdots, N^2 - 1$, tells us which of the $N^2 - 1$ matrices we are talking about, while the indices i and j, which range from 1 to N, are matrix indices. In light of these remarks, it is occasionally convenient, when working with a specific representation, to introduce yet another set of indices $p, q, r, \cdots$ running from 1 to $d =$ the dimension of the representation being discussed.

For example, we might write for the representation 15 of $SU(5)$, 24 15-by-15 matrices $(T^a)^p_q$ (thus, $a = 1, \cdots, 24$, $p, q = 1, \cdots, 15$). Note that it is important to specify the group (of course) as well as the representation. For example, the fundamental representation

---

* In chapter VI.2, we will see how far we can get starting with (22).

[†] This is, strictly speaking, a physicist's abuse of notation: D is supposed to be a matrix function of group elements, not of the generators of the Lie algebra.

15 of $SU(15)$ would consist of 224 15-by-15 matrices $(T^a)^i_j$ (thus, $a = 1, \cdots, 224, i, j = 1, \cdots, 15$). These two sets of 15-by-15 matrices have nothing to do with each other.

In a way, this is neither new nor amazing. For $SO(3)$, for example, we have already learned that the three generators J_x, J_y, and J_z can be represented by three $(2j + 1)$-by-$(2j + 1)$ matrices, for $j = 0, 1, 2, \cdots$.

Henceforth, we will follow the standard usage in the physics literature and will not distinguish between the generators T^a and the matrices $D(T^a)$ representing the generators.

Consider a tensor φ in an arbitrary d-dimensional irreducible representation. Under a group transformation infinitesimally close to the identity, it changes by $\varphi \to (I + i\theta^a T^a)\varphi$, and thus its variation is given by $i\theta^a T^a \varphi$, with T^a as given in that irreducible representation, that is, by d-by-d matrices $(T^a)^p_q$, with $p, q = 1, \cdots, d$. Again, we are at risk of belaboring a point, but some students are invariably confused here. The tensor φ, written out explicitly as $\varphi^{i_1 i_2 \cdots i_m}_{j_1 j_2 \cdots j_n}$, carries m upper indices and n lower indices, each of which runs from 1 through N. But because of various symmetry restrictions and of the traceless condition, these components of the tensor φ are not independent, and thus we often find it convenient to label the components of φ as φ^p, with p running from 1 through d, the dimension of the irreducible representation furnished by φ. Then the change under the transformation may be written as

$$\delta\varphi^p = i\theta^a (T^a)^p_q \varphi^q \tag{23}$$

Note that for N, m, and n equal to some largish numbers, the dimension d can be rather large.

The adjoint representation of $SU(N)$

We've already talked about the adjoint representation in chapters I.3 and IV.1, but let us start our discussion of the adjoint representation of $SU(N)$ without reference to our previous knowledge and take a slightly different tack. We will see, in the end, that the discussions here and in chapter IV.1 are conceptually entirely the same.

For $SU(N)$, the fundamental (or defining) irreducible representation N and its conjugate N^* (defined by φ^i and φ_i, respectively) are of course of, well, fundamental importance. There is another irreducible representation of great importance, namely, the representation defined by the traceless tensor φ^i_j, known as the adjoint representation, or adjoint for short. As explained earlier, it has dimension $N^2 - 1$.

The poor man would make a flying guess here. In light of the earlier discussion, we are to find $(N^2 - 1)(N^2 - 1)$-by-$(N^2 - 1)$ matrices $(T^a)^p_q$ with $p, q = 1, \cdots, N^2 - 1$. The poor man notices that the indices p, q have the same range as the index a. Guessing that they are all the same beasts, the poor man looks around for an object carrying three indices a, b, and c that could be conscripted to serve as $(T^a)^b_c$.

Good guess! The structure constants f^{abc} are practically staring you in the face.

The rich man proceeds more systematically. After all, that was just a wild guess, inspired though it may be. The rich man notes that, by definition, the adjoint transforms according to $\varphi^i_j \rightarrow \varphi'^i_j = U^i{}_l (U^\dagger)^n{}_j \varphi^l_n = U^i{}_l \varphi^l_n (U^\dagger)^n{}_j$. We are thus invited to regard φ^i_j as a matrix transforming according to

$$\varphi \rightarrow \varphi' = U\varphi U^\dagger \tag{24}$$

Note that if φ is hermitean and traceless, it stays hermitean and traceless (since $\varphi'^\dagger = (U\varphi U^\dagger)^\dagger = U\varphi^\dagger U^\dagger = U\varphi U^\dagger = \varphi'$, and tr $\varphi' =$ tr $U\varphi U^\dagger =$ tr $\varphi U^\dagger U =$ tr φ). Thus, with no loss of generality, we can take φ to be a hermitean traceless matrix. (If φ is antihermitean, we can always multiply it by i.) Another way of saying this is that given a hermitean traceless matrix X, then $UXU^\dagger$ is also hermitean and traceless if U is an element of $SU(N)$.

But if φ^i_j is regarded as a hermitean traceless matrix, then we can write it as a linear combination of the complete set of matrices T^a, $a = 1, 2, \cdots, N^2 - 1$:

$$\varphi^i_j = \sum_{a=1}^{N^2-1} A^a (T^a)^i{}_j = A^a (T^a)^i{}_j \tag{25}$$

Here the A^a denote $(N^2 - 1)$ real coefficients. (The last step is simply a reminder that we use the repeated summation convention unless stated otherwise.) For example, for $SU(2)$, this just expresses the 2-by-2 hermitean traceless matrix φ^i_j as a linear combination of the three Pauli matrices.

We are free to think of the $(N^2 - 1)$ objects* A^a as furnishing the $(N^2 - 1)$-dimensional adjoint representation; after all, the A^a are just linear combinations of the $(N^2 - 1)$ objects φ^i_j. (Indeed, all this parallels our discussion for $SO(N)$ in chapter IV.1.)

It is illuminating to work out how A^a transforms. Plugging the infinitesimal transformation $U \simeq 1 + i\theta^a T^a$ into the transformation rule (24), we have

$$\varphi \rightarrow \varphi' \simeq (1 + i\theta^a T^a)\varphi(1 + i\theta^a T^a)^\dagger \simeq \varphi + i\theta^a T^a \varphi - \varphi i\theta^a T^a = \varphi + i\theta^a [T^a, \varphi] \tag{26}$$

In other words, the ath generator acting on the adjoint representation gives $[T^a, \varphi]$. It is convenient to express (26) as giving the infinitesimal change of φ due to an infinitesimal transformation:

$$\delta\varphi = i\theta^a [T^a, \varphi] \tag{27}$$

To find out how A^a transforms, simply plug $\varphi = A^a T^a$ (namely, (25)) into (27) and use (22):

$$(\delta A^b) T^b = \delta(A^b T^b) = i\theta^a [T^a, A^c T^c] = i\theta^a A^c [T^a, T^c] = i\theta^a A^c i f^{acb} T^b \tag{28}$$

Equating the coefficient of T^b, we obtain $\delta A^b = -\theta^a f^{acb} A^c$. Comparing with the definition $\delta A^b = i\theta^a (T^a)^b{}_c A^c$, we obtain

$$(T^a)^{bc} = -if^{abc} \tag{29}$$

---

*The adjoint representation is of particular importance in particle physics. In Yang-Mills theory, the gauge bosons, the analogs of the familiar photon, belong to the adjoint representation. There are 8 gauge bosons for quantum chromodynamics, based on the gauge group $SU(3)$; 4 for the electroweak interaction, based on $SU(2) \otimes U(1)$; and 24 for the grand unified theory, based on $SU(5)$. See part IX.

(As emphasized earlier, there is no particular meaning to whether the indices a, b, and c are written as superscripts or subscripts.) The result here agrees with that in chapter IV.1, of course.

Multiplying representations together

Given two tensors φ and η of $SU(N)$, with m upper and n lower indices and with m' upper and n' lower indices, respectively, we can consider a tensor T with $(m + m')$ upper and $(n + n')$ lower indices that transforms in the same way as the product $\varphi\eta$. We can then reduce T by the various operations described earlier, namely, tracing and contracting with the antisymmetric symbol. The multiplication of two representations is of course of fundamental importance in physics.[6] We will do it in detail for $SU(3)$ in chapter V.2.

$U(1)$ in theoretical physics

I close by mentioning that while the group $U(1)$ is almost trivial, it is of prime importance in theoretical physics. The group consists of elements $e^{i\theta}$ (with $2\pi > \theta \geq 0$) obeying the multiplication rule, $e^{i\theta}e^{i\theta'} = e^{i(\theta+\theta')}$, and hence commuting with one another. The group theory is about as simple as it can get. But electric charge conservation, one of the fundamental facts about the universe, corresponds to the invariance* of physical laws under the multiplication of charged fields by some appropriate powers of $e^{i\theta(x)}$, with the angle θ a function of the spacetime coordinates x. Back in chapter III.3, I mentioned the far-reaching insight of Emmy Noether that conservation laws are associated with symmetries; charge conservation is one of the most celebrated examples.

Exercises

1 Work out the dimension of $\{m\}$ for $SU(3)$.

2 Compute the structure constants f^{abc} for $SU(2)$ and $SU(3)$.

3 In chapter II.4, we needed a lemma stating that given a unitary symmetric matrix U, there exists a unitary symmetric matrix W such that $W^2 = U$. Prove this.

Notes

1. Trivial notational changes: N and O instead of D and R. Emerson: "Consistency is the . . . "
2. While writing this, I came across the following factoid in Wikipedia. "The Nair, also known as Nayar, are a group of Indian castes, described by anthropologist Kathleen Gough as 'not a unitary group but a named category of castes.'"

---

* This statement is explained under the term "gauge invariance" in any textbook on quantum field theory.

3. The Eightfold Way by M. Gell-Mann and Y. Ne'eman.

4. A few years later, while in graduate school, I was told by one of my professors not to waste any time studying such things as Bessel functions. Group theory, in the guise of the kind of counting shown here, was going to be the future. In fact, Bessel functions have a deep connection to group theory. See one of the interludes to this part.

5. Long ago, I heard Murray Gell-Mann tell the story of how he discovered his matrices, a story which, while fun to tell, sounded totally apocryphal to me. According to Gell-Mann, he was visiting France when he realized that he would like to generalize the Pauli matrices. Every morning he would start with $\lambda_{1,2,3}$, add λ_4, commute them, obtain more 3-by-3 matrices, and try to get the algebra to close. Every day, just when he had gotten seven matrices in hand, a French colleague would come to take him to lunch. Keep in mind that the word "lunch" has a different meaning in France than in the rest of the world, especially back in the period we are talking about. Stumbling back into his office in the late afternoon, Gell-Mann would forget what he had done—or so he claimed. The process would start all over again the next morning. Gell-Mann's punchline was that the Frenchman he lunched with was the world's greatest authority on Lie algebra. Even accounting for the fact that everything in physics sounds simple or even trivial in hindsight, I found this story difficult to swallow, given Gell-Mann's intellectual power.

6. In quantum field theory, for example, we multiply fields together to construct the Lagrangian. See part VII of this book, for example.

IV.5 | *SU*(2): Double Covering and the Spinor

From electron spin to internal symmetries

The group $SU(2)$ came into physics with the dramatic discovery of electron spin. Heisenberg, reasoning by analogy, then leaped from the notion of electron spin to the notion of isospin.* This epochal insight opens up a vast vista of internal symmetries, without which none of our modern physical theories could even be conceived. In particular, we would not have been able to write down the nonabelian gauge theories that contemporary physics is entirely based on at the fundamental level.

Just as $SU(2)$ has played an honored role in quantum physics, it is also of great importance in mathematics, due to its many remarkable properties, such as its rather intricate relationship with $SO(3)$. Historically, it is only after physicists had thoroughly mastered the mathematics of $SU(2)$ that they were able to move onto $SU(3)$ and beyond.

Every chapter in this book is important, needless to say, but this chapter is particularly important. I preview some of the highlights to be covered:

1. $SU(2)$ locally isomorphic to $SO(3)$

2. $SU(2)$ covers $SO(3)$ twice

3. irreducible representations of $SU(2)$

4. half angles versus full angles

5. the group elements of $SU(2)$

6. quantum mechanics and double-valued representations

7. $SU(2)$ does not have a downstairs, and its upstairs is all symmetric

8. pseudoreality

9. the precise relation between $U(2)$ and $SU(2)$

---

* We will explore isospin in chapter V.1.

I dare to say that more than a vanishing fraction of theoretical physicists have not totally mastered some of these points, for example, 9.

SU(2) **is locally isomorphic to** *SO*(3)

We start with one striking, I would even say astonishing, property of $SU(2)$: it is locally isomorphic to $SO(3)$. The discussion to follow will make clear what these two words mean.

Any 2-by-2 hermitean traceless matrix X can be written as a linear combination of the three Pauli matrices, as explained in chapter IV.4. For convenience, I list the Pauli matrices again:

$$\sigma_1 = \begin{pmatrix} 0 & 1 \\ 1 & 0 \end{pmatrix}, \qquad \sigma_2 = \begin{pmatrix} 0 & -i \\ i & 0 \end{pmatrix}, \qquad \sigma_3 = \begin{pmatrix} 1 & 0 \\ 0 & -1 \end{pmatrix} \tag{1}$$

The statement is that we can write $X = \vec{x} \cdot \vec{\sigma}$ with three real coefficients, which we can assemble into an array $\vec{x} = (x^1, x^2, x^3)$ and provisionally call a vector. Or, call them the Cartesian coordinates (x, y, z), if you prefer. (I do.) Explicitly,

$$X = x\sigma_1 + y\sigma_2 + z\sigma_3 = \begin{pmatrix} z & x - iy \\ x + iy & -z \end{pmatrix} \tag{2}$$

It is manifestly hermitean and traceless. Conversely, given any hermitean and traceless 2-by-2 matrix, we can write it in the form of (2) with the vector $\vec{x} = (x, y, z)$ uniquely determined.

Calculate the determinant of X: $\det X = -(x^2 + y^2 + z^2) = -\vec{x}^2$. Perhaps not surprisingly, the determinant of X, being an invariant, is given by the length squared[1] of $\vec{x}$. We now make this remark explicit.

Pick an arbitrary element U of $SU(2)$. In other words, U is unitary with determinant equal to 1. Consider $X' \equiv U^\dagger X U$.

I claim that X' is also hermitean and traceless. The proof is by explicit computation: $(X')^\dagger = (U^\dagger X U)^\dagger = U^\dagger X^\dagger U = U^\dagger X U = X'$, and $\text{tr } X' = \text{tr } U^\dagger X U = \text{tr } X U U^\dagger = \text{tr } X = 0$. Since X' is hermitean and traceless, we can write it as $X' = \vec{x}' \cdot \vec{\sigma}$, with three real coefficients (x', y', z'), which we also assemble into a vector $\vec{x}' = (x', y', z')$. The vector $\vec{x}'$ is linearly related to the vector $\vec{x}$: if we scale $\vec{x} \to \lambda\vec{x}$ by multiplying it by a real number λ, then $\vec{x}' \to \lambda\vec{x}'$.

Thus far, we have used the unitarity of U, but not the fact that it has unit determinant. In other words, we have used the U, but not the S, of $SU(2)$. You have done enough math exercises (or read enough detective mysteries) to know that S has to come in somehow somewhere sometime. Here and now in fact! Compute $\det X' = -(\vec{x}')^2 = \det U^\dagger X U = (\det U^\dagger)(\det X)(\det U) = \det X = -\vec{x}^2$. Thus, $\vec{x}'$ and $\vec{x}$ are not only linearly related to each other, but they also have the same length.

By definition, this means that the 3-vector $\vec{x}$ is rotated into the 3-vector $\vec{x}'$. Since X (and hence $\vec{x}$) is arbitrary, this defines a rotation R. In other words, we can associate an element R of $SO(3)$ with any element U of $SU(2)$.

The map $U \to R$ clearly preserves group multiplication: if two elements U_1 and U_2 of $SU(2)$ are mapped to the rotations R_1 and R_2, respectively, then the element $U_1 U_2$ is mapped to the rotation $R_1 R_2$. We can readily verify this: $(U_1 U_2)^\dagger X (U_1 U_2) = U_2^\dagger (U_1^\dagger X U_1) U_2$ $= U_2^\dagger X' U_2 = X''$, so that the 3-vectors associated with X, X', and X'' get rotated in stages, $\vec{x} \to \vec{x}' \to \vec{x}''$, first by R_1 and then by R_2.

$SU(2)$ covers $SO(3)$ twice

Interestingly, this map $f : U \to R$ of $SU(2)$ into $SO(3)$ is actually 2-to-1, since U and $-U$ are mapped into the same R; $f(U) = f(-U)$. (This is not obvious to you? Well, $(-U)^\dagger X (-U) = U^\dagger X U$. In $SU(2)$, U and $-U$ are manifestly not the same: indeed, $U = -U$ would imply, on multiplication by $U^\dagger$, $I = -I$, which is clearly absurd.)

The unitary group $SU(2)$ is said to double cover the orthogonal group $SO(3)$.

Since the map $U \to R$ is 2-to-1, the groups $SU(2)$ and $SO(3)$ are not isomorphic, but only locally isomorphic. In particular, for U near the identity, $-U$ is far from the identity, and so in a neighborhood around the identity,[*] the two groups are isomorphic.

I alluded to this local isomorphism between $SU(2)$ and $SO(3)$ as astonishing, but in fact, we have already encountered several hints of this. For example, in chapter IV.4, I asked you to file away in your mind the apparently coincidental fact that both groups have a 3-dimensional representation. Now we see that this is not a coincidence: identifying the 3 of $SU(2)$ with the vector of $SO(3)$ in fact defines the local isomorphism.

Properties of the Pauli matrices

To proceed further, it is convenient to pause and list some "arithmetical" properties of the three Pauli matrices. You should verify the following statements as you read along. By inspection, we see that each of the Pauli matrices squares to the identity matrix: $(\sigma_a)^2 = I$, for $a = 1, 2, 3$. Distinct Pauli matrices anticommute with each other: $\sigma_a \sigma_b = -\sigma_b \sigma_a$ for $a \neq b$. The product of any two distinct Pauli matrices gives the third with a coefficient given by $\pm i$; thus, $\sigma_1 \sigma_2 = i\sigma_3$, with all other cases given by cyclic permutation and the fact that Pauli matrices anticommute. These facts are summarized by[2] (with the repeated index c summed)

$$\sigma_a \sigma_b = \delta_{ab} I + i \varepsilon_{abc} \sigma_c \tag{3}$$

Here I denotes the 2-by-2 identity matrix; it will often be omitted if there is no risk of confusion.

Interchanging a and b, and then adding and subtracting the resulting equation to and from (3) gives

$$\{\sigma_a, \sigma_b\} = 2\delta_{ab} \tag{4}$$

---

[*] In fact, in a neighborhood around any $SU(2)$ element.

and

$$[\sigma_a, \sigma_b] = 2i\varepsilon_{abc}\sigma_c \tag{5}$$

Note the factors of 2.

In chapter IV.4, we showed that the Lie algebra for $SU(N)$ is given by $[T^a, T^b] = if^{abc}T^c$. Dividing (5) by 2^2, we obtain

$$\left[\frac{\sigma_a}{2}, \frac{\sigma_b}{2}\right] = i\varepsilon_{abc}\frac{\sigma_c}{2} \tag{6}$$

This shows that for the fundamental 2-dimensional representation, the generators T^a are represented by* $\frac{\sigma_a}{2}$, a fact often written somewhat sloppily as $T^a = \frac{\sigma_a}{2}$, and that the structure constant of $SU(2)$ is just the antisymmetric symbol, that is, $f^{abc} = \varepsilon^{abc}$.

As remarked in chapter IV.4, whether the indices $a, b, \cdots$ are written as subscripts or superscripts is of no import. Again, I have capriciously moved the indices $a, b, \cdots$ up- and downstairs to stress this point. Also, I had emphasized that the T^a may be regarded as abstract mathematical entities. In a d-dimensional representation, they are realized as d-by-d matrices. In particular, in the fundamental representation, they are represented by $\frac{\sigma_a}{2}$.

Confusio: "So, T^3 is sometimes a 4-by-4 matrix and sometimes a 17-by-17 matrix?"

Exactly! It depends on the phase of the moon.

The Lie algebra of $SU(2)$ is thus given by $[T^a, T^b] = i\varepsilon^{abc}T^c$; in other words,

$$[T^1, T^2] = iT^3, \qquad [T^2, T^3] = iT^1, \qquad [T^3, T^1] = iT^2 \tag{7}$$

For comparison, the Lie algebra of $SO(3)$ was given in chapter IV.3 as

$$[J_x, J_y] = iJ_z, \qquad [J_y, J_z] = iJ_x, \qquad [J_z, J_x] = iJ_y \tag{8}$$

We see immediately that, with the identification[†] $T^1 \leftrightarrow J_x$, $T^2 \leftrightarrow J_y$, $T^3 \leftrightarrow J_z$, the two Lie algebras are manifestly identical, that is, isomorphic. In fact, who cares which letters[‡] we use to denote the generators of $SU(2)$ and $SO(3)$? It is the same algebra, period.

Representing the Lie algebra of $SU(2)$

Given the preceding remark, we can now find the representations of the Lie algebra of $SU(2)$ without doing a stitch of work. Simply take over what we did in chapter IV.4 for the Lie algebra of $SO(3)$, replacing the letter J by T if you insist. Define $T^{\pm} \equiv T^1 \pm iT^2$ so that

$$[T^3, T^{\pm}] = \pm T^{\pm}, \qquad [T^+, T^-] = 2T^3 \tag{9}$$

and proceed.

---

* Not σ_a! Watch the factors of 2 in this chapter like a hawk.

[†] I again capriciously use 1, 2, and 3 and x, y, and z to denote the same thing to emphasize a point.

[‡] Indeed, different authors use different letters, and in different contexts the same author could use different letters. We will see in later chapters that L, S, and I are also used in physics.

We discover soon enough that the representations of $SU(2)$ are* $(2j+1)$-dimensional, with $j = 0, \frac{1}{2}, 1, \frac{3}{2}, 2, \frac{5}{2}, \cdots$.

In chapter IV.4, I introduced to you the representations of the algebra of $SO(3)$ corresponding to $j =$ half integer as a disturbing puzzle. But now we see why they pop up. They are representations of $SU(2)$! In particular, the 2-dimensional representation with $j = \frac{1}{2}$ consists of the states $\left|-\frac{1}{2}\right\rangle$ and $\left|\frac{1}{2}\right\rangle$. It is the fundamental or defining representation of the group $SU(2)$. More on this shortly.

The group elements of $SU(2)$

Before writing down the group elements of $SU(2)$, we need a useful formula. Given two 3-vectors $\vec{u}$ and $\vec{v}$, contract (3) with u^a and v^b to obtain

$$(\vec{u} \cdot \vec{\sigma})(\vec{v} \cdot \vec{\sigma}) = u^a v^b \sigma_a \sigma_b = u^a v^b (\delta_{ab} I + i \varepsilon_{abc} \sigma_c) = (\vec{u} \cdot \vec{v}) I + i (\vec{u} \otimes \vec{v}) \cdot \vec{\sigma} \tag{10}$$

The dot and cross products (or the scalar and vector products) familiar from elementary vector analysis both appear on the right hand side. This result shows the convenience of thinking of the three Pauli matrices as a vector $\vec{\sigma}$. Also, it follows that $X^2 = x^a x^b \sigma_a \sigma_b = \vec{x}^2$.

A crucial step in proving, in the preceding section, that the 3-vector $\vec{x}$ is rotated into the 3-vector $\vec{x}'$, involves showing that they have the same length. We did this by calculating the determinant of X and the determinant of $X' \equiv U^\dagger X U$. We now see that an alternative method is to note that tr $X^2 =$ tr X'^2. (In fact, the trace is not even necessary, since X^2 is proportional to the identity matrix.)

We already noted in chapter IV.4 that any element of $SU(2)$ can be written as $U = e^{i\varphi_a \sigma_a/2} = e^{i\vec{\varphi}\cdot\vec{\sigma}/2}$ (with the repeated index summation convention). Denote the magnitude and direction of $\vec{\varphi}$ by φ and the unit vector $\hat{\varphi}$, respectively (so that $\vec{\varphi} = \varphi\hat{\varphi}$). Use $(\vec{\varphi} \cdot \vec{\sigma})^2 = \vec{\varphi}^2 = \varphi^2$. We now expand the exponential series into even and odd powers (compare with (I.3.12)) and sum to obtain

$$U = e^{i\vec{\varphi}\cdot\vec{\sigma}/2} = \sum_{n=0}^{\infty} \frac{i^n}{n!} \left(\frac{\vec{\varphi}\cdot\vec{\sigma}}{2}\right)^n$$

$$= \left\{ \sum_{k=0}^{\infty} \frac{(-1)^k}{(2k)!} \left(\frac{\varphi}{2}\right)^{2k} \right\} I + i \left\{ \sum_{k=0}^{\infty} \frac{(-1)^k}{(2k+1)!} \left(\frac{\varphi}{2}\right)^{2k+1} \right\} \hat{\varphi} \cdot \vec{\sigma}$$

$$= \cos\frac{\varphi}{2} I + i\hat{\varphi} \cdot \vec{\sigma} \sin\frac{\varphi}{2} \tag{11}$$

We see that this is structurally the same as Euler's identity $e^{i\varphi/2} = \cos\frac{\varphi}{2} + i \sin\frac{\varphi}{2}$, which can also be derived by splitting the exponential series into even and odd terms.

An easier way to derive (11) is to note that we are free to call the axis of rotation $\hat{\varphi}$ the 3rd or z-axis. Then $\hat{\varphi} \cdot \vec{\sigma} = \sigma_3$, and since σ_3 is diagonal, the exponential can be immediately evaluated by Euler's identity:

---

* You might also write t instead of j if you like.

$$U = e^{i\varphi\sigma_3/2} = \begin{pmatrix} e^{i\frac{\varphi}{2}} & 0 \\ 0 & e^{-i\frac{\varphi}{2}} \end{pmatrix} = \begin{pmatrix} \cos\frac{\varphi}{2} + i\sin\frac{\varphi}{2} & 0 \\ 0 & \cos\frac{\varphi}{2} - i\sin\frac{\varphi}{2} \end{pmatrix}$$

$$= \cos\frac{\varphi}{2} I + i\sigma_3 \sin\frac{\varphi}{2} \tag{12}$$

For the general case, we simply replace σ_3 by $\hat{\varphi} \cdot \vec{\sigma}$ and obtain (11).

Confusio: "For years, I confused $U = \cos\frac{\varphi}{2}I + i\hat{\varphi} \cdot \vec{\sigma}\sin\frac{\varphi}{2}$ and $X = \vec{x} \cdot \vec{\sigma}$ (as defined in (2)) merely because they looked similar."

Indeed, they are quite different: U is unitary and hence an element of $SU(2)$, but it is in general not traceless, while, in contrast, X is hermitean and traceless.

Incidentally, back in the review of linear algebra, I said that the polar decomposition was not particularly attractive. For example, we can write any 2-by-2 complex matrix as $M = (t + \vec{x} \cdot \vec{\sigma})e^{i\theta}e^{i\frac{\varphi}{2}\vec{\sigma}} = (t + \vec{x} \cdot \vec{\sigma})(\cos\frac{\varphi}{2} + i\sin\frac{\varphi}{2}\hat{\varphi} \cdot \vec{\sigma})e^{i\theta}$. A complex matrix is characterized by $2 \cdot 2^2 = 8$ real parameters, namely, t, $\vec{x}$, $\vec{\varphi}$, and θ.

How the half angles become full angles

We had proved earlier that if $X' = \vec{x}' \cdot \vec{\sigma}$ and $X = \vec{x} \cdot \vec{\sigma}$ are related by $X' = U^\dagger X U$, then the two vectors $\vec{x}'$ and $\vec{x}$ are related by a rotation. By construction, it must work. But still, now that we know U explicitly, it is instructive to see how it actually works—how all the wheels and gears fit together, so to speak. In particular, even though we now see how the half angles in (11) come in (to render the Lie algebras of $SU(2)$ and $SO(3)$ the same), it does seem strange at first sight that (11) can lead to the usual expression for rotations (as given in chapter I.3).

So, let us evaluate $U^\dagger X U$ by brute force. As just noted, with no loss of generality, we can take $\hat{\varphi}$ to point along the third axis. So, use (12). For $a = 3$, we obtain trivially $U^\dagger\sigma_3 U = \sigma_3$. For $a = 1, 2$, $U^\dagger\sigma_a U = (\cos\frac{\varphi}{2}I - i\sigma_3\sin\frac{\varphi}{2})\sigma_a(\cos\frac{\varphi}{2}I + i\sigma_3\sin\frac{\varphi}{2})$. To obtain the coefficient of $\sin^2\frac{\varphi}{2}$, we encounter $\sigma_3\sigma_a\sigma_3 = -\sigma_a\sigma_3\sigma_3 = -\sigma_a$ (since σ_3 anticommutes with σ_1 and σ_2 and squares to the identity). Another line of arithmetic (or algebra, whatever you want to call it) gives, for $a = 1, 2$, $U^\dagger\sigma_a U = (\cos^2\frac{\varphi}{2} - \sin^2\frac{\varphi}{2})\sigma_a - i\sin\frac{\varphi}{2}\cos\frac{\varphi}{2}[\sigma_3, \sigma_a]$. From (5), we have $[\sigma_3, \sigma_1] = 2i\sigma_2$ and $[\sigma_3, \sigma_2] = -2i\sigma_1$. If you remember your trigonometric identities (the "double angle formulas") $\cos\varphi = \cos^2\frac{\varphi}{2} - \sin^2\frac{\varphi}{2}$ and $\sin\varphi = 2\sin\frac{\varphi}{2}\cos\frac{\varphi}{2}$, you can practically see the answer emerging:

$$U^\dagger\sigma_1 U = \cos\varphi\,\sigma_1 + \sin\varphi\,\sigma_2, \qquad U^\dagger\sigma_2 U = -\sin\varphi\,\sigma_1 + \cos\varphi\,\sigma_2 \tag{13}$$

We are not quite there yet; we still have to plug in $X' = U^\dagger(x\sigma_1 + y\sigma_2 + z\sigma_3)U = (\cos\varphi\,x - \sin\varphi\,y)\sigma_1 + (\sin\varphi\,x + \cos\varphi\,y)\sigma_2 + z\sigma_3$. In other words, $x' = \cos\varphi\,x - \sin\varphi\,y$, $y' = \sin\varphi\,x + \cos\varphi\,y$, $z' = z$, precisely what we should get* for a rotation around the z-axis through angle φ.

Hey, math works. The half angle $\varphi/2$ becomes the full angle φ.

---

* Compare this with (I.3.1).

Confusio: "I have often read the sloppy and potentially confusing statement that the Pauli matrices transform like a vector."

Indeed, the statement is sloppy, but sometimes useful as a shorthand for (13). Physically, we can talk about two different observers recording the same physical quantity (such as momentum) as $\vec{p}$ and as $\vec{p}'$, but there is no such thing as σ'_a; we simply have a bunch of 2-by-2 matrices with entries like 0, ± 1, and $\pm i$. There exists only one Pauli, so to speak.

Quantum mechanics and the double covering

Now comes a striking fact. From (12) we have $U(\varphi) = e^{i\varphi\sigma_3/2}$, which, as we just checked, leads to a rotation through an angle φ around the z-axis. But note that

$$U(2\pi) = e^{i(2\pi)\sigma_3/2} = e^{i\pi\sigma_3} = \begin{pmatrix} e^{i\pi} & 0 \\ 0 & e^{-i\pi} \end{pmatrix} = -I \tag{14}$$

By the time the rotational angle φ has gone from 0 to 2π, so that the corresponding rotation has gone back to the identity, the $SU(2)$ element U has reached only $-I$. To reach I, the angle φ has to go all the way to 4π:

$$U(4\pi) = I \tag{15}$$

This striking fact, as expressed by (14) and (15), of course just restates what we had learned earlier, that $SU(2)$ double covers $SO(3)$. By the time we get around $SU(2)$ once, the corresponding rotation has gone around $SO(3)$ twice. Another way of saying this is as follows. The 2-dimensional fundamental representation $\psi^i \to U^i{}_j \psi^j$ of $SU(2)$ is strictly speaking not a representation of $SO(3)$. (Some authors call it a double-valued representation of $SO(3)$.) This discussion once again underlines the fact that the groups $SU(2)$ and $SO(3)$ are not globally isomorphic, only locally isomorphic; in other words, only their algebras are isomorphic.

$SU(2)$ does not have a downstairs, and its upstairs is all symmetric

In our discussion of the orthogonal groups, I mentioned that $SO(3)$ has special properties that are not shared by $SO(N)$, a fact that sometimes confuses beginning students. Similarly, the two special unitary groups that most students learn first, namely, $SU(2)$ and $SU(3)$, have special properties that are not shared by $SU(N)$, essentially for the same reasons as in the orthogonal case.

In chapter IV.4, we learned that traceless tensors $T^{i_1 i_2 \cdots i_m}_{j_1 j_2 \cdots j_n}$ with upper and lower indices furnish representations of $SU(N)$. Furthermore, these tensors are to have definite symmetry properties under permutation of their upper indices (the is) and under permutation of their lower indices (the js).

For $SU(2)$, because the antisymmetric symbols ε^{ij} and ε_{ij} carry two indices, we can in fact remove all lower indices. To see how this works, it is clearer not to treat the general case, but to pick a specific example, say, the tensor T^{ijk}_{mn}. (Each of the indices can take on two values, 1 or 2, of course.) Then, following the discussion in chapter IV.4, we can construct $T^{pqijk} = \varepsilon^{pm}\varepsilon^{qn}T^{ijk}_{mn}$, which transforms like a tensor with five upper indices and no lower index. Thus, it suffices to consider only tensors without lower indices.

We have used ε^{ij} but we still have ε_{ij} up our sleeves. We now claim further that it suffices to consider only tensors with upper indices all symmetrized. The argument is inductive (just as in our discussion of the orthogonal groups). Suppose this holds for tensors with less than four upper indices. Now we look at tensors with four indices. Suppose T^{ijkl} has no particular symmetry on the interchange of, say, i and k. We simply write $S^{ijkl} = T^{ijkl} + T^{kjil}$ and $A^{ijkl} = T^{ijkl} - T^{kjil}$. Then $\varepsilon_{ik}A^{ijkl}$ is a tensor with two upper indices, which, by induction, is not something we need to worry about. By construction, S^{ijkl} is of course symmetric on the interchange of i and k.

Repeating this argument leads to the bottom line that we need consider only tensors with upper indices and that do not change sign on the interchange of any pair of indices.

Confusio: "Why doesn't this argument work for $SU(3)$?"

I will let the reader answer Confusio's question.

Ironically, it is harder, in some sense, to understand $SU(2)$ than $SU(5)$, say, precisely because $SU(2)$ is so "simple" and thus has idiosyncratic features.

Dimension of $SU(2)$ irreducible representations

What is the dimension of the representation furnished by $T^{i_1 i_2 \cdots i_m}$? Just count: $T^{11\cdots1}$, $T^{11\cdots12}$, $T^{11\cdots122}$, ..., $T^{22\cdots2}$. The number of 2s ranges from 0 to m, and hence this representation has dimension $m + 1$. To make contact with the earlier discussion, simply write m as $2j$. Since m ranges over the non-negative integers, we have $j = 0, \frac{1}{2}, 1, \frac{3}{2}, 2, \cdots$.

To summarize, an irreducible representation of $SU(2)$ is characterized by j, which can take on integral or half-integral values and has dimension $2j + 1$. Note that $j = \frac{1}{2}$ is the fundamental representation, while $j = 1$ is the vector representation. The existence of the vector representation offers another way of seeing the local isomorphism of $SU(2)$ and $SO(3)$.

Not complex, only pseudoreal

In chapter IV.4, I went over carefully the reason for having both lower and upper indices. You learned that we had to introduce ψ_i as a stand-in, so to speak, for ψ^{i*}. In general, for $SU(N)$, the representations furnished by ψ_i and ψ^i are quite different. But then we just learned that for $SU(2)$, we don't need lower indices, and in particular, ψ_i is not needed. Indeed, we just learned $\varepsilon^{ij}\psi_j$ transforms in exactly the same way as ψ^i.

What is going on? Does this mean that for $SU(2)$, the representations furnished by ψ^{i*} and ψ^i are the same, which would seem to suggest that complex conjugation "doesn't do anything" in some sense? However, we see from (11) that the fundamental representation is explicitly not real.

A clue to what is going on may be found in chapter II.3, where we learned about real, pseudoreal, and complex representations. Recall that if a representation $D(g)$ does not look real, that is, even if $D(g)^* \neq D(g)$, the representation and its complex conjugate could be "secretly" related by a similarity transformation: that is, there may exist an S such that

$$D(g)^* = SD(g)S^{-1} \tag{16}$$

for all $g \in G$. The representation is then said to be pseudoreal.

The fundamental representation of $SU(2)$ is pseudoreal.

We now show this by explicitly finding the similarity transformation S. The fundamental or defining representation $D(g)$ is given by $e^{i\vec{\varphi}\vec{\sigma}/2}$, while its conjugate representation $D(g)^*$ is given by $(e^{i\vec{\varphi}\vec{\sigma}})^* = e^{-i\vec{\varphi}\vec{\sigma}^*}$.

But of the three Pauli matrices (1), σ_1 and σ_3 are manifestly real, while σ_2 is manifestly imaginary. We could say that σ_2 is the "odd man out." Using the facts that different Pauli matrices anticommute and that the Pauli matrices square to the identity, we have $\sigma_2 \sigma_1^* \sigma_2 = \sigma_2 \sigma_1 \sigma_2 = -\sigma_1 \sigma_2 \sigma_2 = -\sigma_1$. Similarly, $\sigma_2 \sigma_3^* \sigma_2 = -\sigma_3$, and $\sigma_2 \sigma_2^* \sigma_2 = -\sigma_2 \sigma_2 \sigma_2 = -\sigma_2$. Thus, we obtain

$$\sigma_2 \sigma_a^* \sigma_2 = -\sigma_a \tag{17}$$

and so

$$\sigma_2 (e^{i\vec{\varphi}\vec{\sigma}})^* \sigma_2 = e^{i\vec{\varphi}\vec{\sigma}} \tag{18}$$

Comparing this with (16), we see that we can choose $S = S^{-1} = \sigma_2$, which is antisymmetric, in accordance with the proof in chapter II.3 that S has to be either symmetric or antisymmetric. Indeed, the fundamental representation of $SU(2)$ is pseudoreal.

We note that (16), together with the condition $S^\dagger S = I$ (also proved in chapter II.3) determines $S = e^{i\alpha}\sigma_2$ only up to a phase factor. It is then convenient to stick in an i and write $S = i\sigma_2 = \begin{pmatrix} 0 & -1 \\ 1 & 0 \end{pmatrix}$.

We suddenly see the connection of this discussion with the tensor approach: the S written here is precisely the ε^{ij} that connects ψ_i to ψ^i.

The groups $U(N)$ and $SU(N)$

In the preceding chapter, we mentioned that the group $U(N)$ contains two sets of group elements. The first set consists of unitary matrices of the form $e^{i\varphi}I$, composing the group $U(1)$. The second set consists of N-by-N unitary matrices with determinant equal to 1, composing the group $SU(N)$. At this point, almost all physicists would say that $U(N)$ is equal to the direct product $SU(N) \otimes U(1)$.

But this is incorrect.

Nobody said that the two sets just mentioned have a nontrivial intersection. The elements of $U(1)$ of the form $e^{i2\pi k/N}I$ with $k = 1, \cdots, N-1$ have determinant equal to 1 and hence also belong to the group $SU(N)$. These elements form the finite group Z_N.

Recall the definition, given in chapter I.1, of the direct product group $H \equiv F \otimes G$, consisting of the elements (f, g), with f and g elements of the groups F and G, respectively. The group F consists of the elements of H of the form (f, I_G), and G of the elements of H of the form (I_F, g). The only element belonging to both F and G is the identity element of H, namely, $I_H = (I_F, I_G)$. (Here we are using the insufferably pedantic language in which we distinguish between the identity elements I_H, I_F, and I_G of the three groups H, F, and G, respectively, and which was alluded to in chapter I.1.)

Since we already included the elements of Z_N in $U(1)$, to avoid overcounting, we have to exclude them from $SU(N)$. Thus, we should actually write

$$U(N) = \left(SU(N)/Z_N\right) \otimes U(1) \tag{19}$$

In the quotient group $SU(N)/Z_N$, the N elements $e^{i2\pi k/N}U$, with U an arbitrary unitary matrix with unit determinant and with $k = 0, 1, \cdots, N-1$, are to be identified as one and the same element. For example, for $N = 2$, the two matrices U and $-U$ are actually the same element, but this is precisely the group $SU(2)/Z_2 = SO(3)$, as discussed earlier in this chapter.

The center of a group was defined way back in exercise I.1.1 as the set of all elements that commute with all the other elements of the group. Thus, Z_N is actually the center of $SU(N)$. We will discuss the physical meaning of the center of $SU(3)$ in chapter V.2.

At the level of Lie algebra, there is in fact no distinction between $SU(N)$ and $SU(N)/Z_N$: the algebra explores only the neighborhood of the identity of the group. Put another way, to determine the algebra of $U(N)$, write an element as $U = e^{iH} \simeq I + iH$. Since U is not required to have unit determinant, H is hermitean but is not required to be traceless. Thus, the generators of the algebra are the $N^2 - 1$ traceless hermitean matrices plus the identity matrix I.

As mentioned back in chapter I.1, the theory of the strong, weak, and electromagnetic interactions is based on $SU(3) \otimes SU(2) \otimes U(1)$. But in typical applications, such as those involving the interaction between gauge bosons, the global properties of the group do not come in at all; only the Lie algebra enters into the field theory Lagrangian.[3] See chapter III.3 and also part IX.

Henceforth, we will follow the sloppy physicists and not distinguish between $U(N) = \left(SU(N)/Z_N\right) \otimes U(1)$ and $SU(N) \otimes U(1)$.

Exercises

1   Show that tr $X^{2k} = 2^{1-k}(\text{tr } X^2)^k$, and that tr $X^{2k+1} = 0$ for $k = 0, 1, \cdots$.

2   Show that the symmetric 2-indexed tensor T^{ij} furnishes the vector representation of $SO(3)$.

Notes

1. Looking at the older literature, I found an 1882 paper by J. J. Sylvester in which he asked (and answered) the question of whether there exists a 2-by-2 matrix linear in (x, y, z) whose determinant is equal to $x^2 + y^2 + z^2$. He found what we might call the Sylvester-Pauli matrices. Sylvester then went on to ask (and answer) the question of whether there exists a 3-by-3 matrix linear in (x, y, z) whose determinant is equal to $x^3 + y^3 + z^3$. Can you match wits with Sylvester? See The Collected Mathematical Papers of James Joseph Sylvester, vol. III, Cambridge University Press, 1909.

2. Thus, the Pauli matrices, together with the identity matrix, represent the quarternions. I read that, when Pauli introduced his matrices in 1927, Jordan had pointed out to him that Cayley had already given a matrix representation of the quarternions. See R. Anderson and G. C. Joshi, arXiv: 9208222 v2 (1992).

3. See, for example, QFT Nut, chapter IV.5.

IV.6 The Electron Spin and Kramer's Degeneracy

The mystique of spin $\frac{1}{2}$

Niels Bohr solved the hydrogen atom in 1913 by boldly postulating that angular momentum is quantized in integral units of $\hbar$. Physicists were shocked, shocked, shocked. But then their minds were really blown in 1925, when George Uhlenbeck and Sam Goudsmit proposed[1] that the electron spins[2] with angular momentum $\frac{1}{2}\hbar$.

A priori, it would seem that any representation in which a rotation through 2π is represented by $-I$ would be worthless for physics, but the incredible mystery of quantum mechanics saved the angular momentum $\frac{1}{2}$ representation from oblivion. The wave function Ψ of a particle can perfectly well change sign upon a 2π rotation, since physical observables are postulated to be bilinears constructed out of $\Psi^{\dagger}$ and Ψ, such as the spin vector $\Psi^{\dagger}\frac{\vec{\sigma}}{2}\Psi$ (see below), which manifestly does not change sign.

Now we know that not only the electron, but also quarks and leptons—the fundamental constituents of matter—all carry spin $\frac{1}{2}$. To many theoretical physicists, spin $\frac{1}{2}$ continues to carry a certain mystique.

As will be explained in chapter VII.4, spin $\frac{1}{2}$ particles are described by the strikingly beautiful Dirac equation.

Several decades later, when Gell-Mann first introduced quarks, many in the theoretical physics community were skeptical, and some were even hostile. For one thing, the fractional electric charges carried by quarks went against preconceived notions. Dirac was among the few eminences who liked quarks. When Gell-Mann asked him why, Dirac replied, "They carry spin $\frac{1}{2}$, don't they?"

The electron spin

The discovery[3] that the electron has spin $\frac{1}{2}$ is surely one of the most stirring episodes in physics. Consider how confused people were in those early days of quantum mechanics, with the mess of quantum numbers. Then they realized that to explain the spectroscopic data, a mysterious additional quantum number had to be introduced. Even after electron spin was proposed, belief would have to hinge on one's philosophical stance. An empirical positivist would demand to know how a spinning particle could carry spin $\frac{1}{2}$ and what that could mean. The mathematical pragmatist would say that a particle whose quantum state transforms like the 2-dimensional representation of the covering group of rotations carries spin $\frac{1}{2}$, period.

In a galaxy far far away, quite possibly physicists knew all about the double covering of rotations and were able to predict the possible occurrence of spin $\frac{1}{2}$. In fact, in our own civilization, there were people who knew, but they were called mathematicians. Possibly, a few physicists, such as Hamilton, might also have known. Communication between mathematicians and physicists, then as now, was blocked by language barriers and any number of other obstacles.[4]

The electron wave function

The proposal that the electron has spin $\frac{1}{2}$ means that the electron wave function ψ has two components and that it transforms like $\psi \to e^{i\vec{\varphi}\cdot\vec{\sigma}/2}\psi$. But, as was already noted in chapter IV.5, for a rotation through 2π around the z-axis,

$$\psi \to e^{i(2\pi)\sigma_3/2}\psi = \begin{pmatrix} e^{i\pi} & 0 \\ 0 & e^{-i\pi} \end{pmatrix} \psi = -\psi \tag{1}$$

But a rotation through 2π is naively no rotation at all, so how can something flip sign?

We have to appeal to the magic of quantum mechanics. The measurable quantities in quantum mechanics are all bilinear in ψ and its complex conjugate $\psi^\dagger$, such as the probability density $\psi^\dagger\psi$. Thus, it is physically acceptable for ψ to flip sign under a rotation through 2π. We now show explicitly that the electron spin does what we expect it to do.

Spin precession

An electron sitting at rest satisfies the Schrödinger equation (set $\hbar = 1$ to lessen clutter)

$$i\frac{\partial}{\partial t}\Psi(t) = H\Psi(t) \tag{2}$$

Since the electron is not moving, $\Psi_\alpha(t)$ does not depend on its position $\vec{x}$ but does depend on an index α indicating that it transforms like the defining representation of $SU(2)$. The spin angular momentum of the electron is then given by $\vec{S} = \Psi^\dagger\frac{\vec{\sigma}}{2}\Psi$. Impose an external magnetic field $\vec{B}$, so that the Hamiltonian is given by

$$H = \mu\vec{B}\cdot\frac{\vec{\sigma}}{2} = \frac{1}{2}\mu(B_1\sigma_1 + B_2\sigma_2 + B_3\sigma_3) \tag{3}$$

with μ a measure of the magnetic moment. The Hamiltonian, effectively a 2-by-2 matrix, breaks rotation invariance explicitly, since $\vec{B}$ picks out a direction. Let $\vec{B}$ be time independent for simplicity; then the Schrödinger equation is immediately solved by

$$\Psi(t) = e^{-i\mu\vec{B}\cdot\frac{\vec{\sigma}}{2}t}\Psi(0) = U(t)\Psi(0) \tag{4}$$

where the unitary time evolution matrix $U(t)$ is precisely the U studied in (IV.5.11) with $\vec{\varphi} = -\mu\vec{B}t$ now linear in time.

Indeed, calling the direction of the magnetic field the third axis, we simply lift (IV.5.12) here, and obtain

$$\Psi(t) = \begin{pmatrix} e^{-i\frac{\mu Bt}{2}} & 0 \\ 0 & e^{i\frac{\mu Bt}{2}} \end{pmatrix} \Psi(0) = \left(\cos\frac{\mu Bt}{2} I - i \sin\frac{\mu Bt}{2}\sigma_3 \right)\Psi(0) \tag{5}$$

The magnetic field simply multiplies the upper and lower components by a relative phase.

Note once again that, when a spin $\frac{1}{2}$ particle is rotated through 2π, after a time $t = 2\pi(\mu B)^{-1}$, its wave function changes sign.

From (IV.5.13), we see that the spin $\vec{S}$ precesses, as would be expected for a classical magnetic moment:

$$\vec{S}_1(t) = \cos\mu Bt\ \vec{S}_1(0) - \sin\mu Bt\ \vec{S}_2(0)$$
$$\vec{S}_2(t) = \sin\mu Bt\ \vec{S}_1(0) + \cos\mu Bt\ \vec{S}_2(0)$$
$$\vec{S}_3(t) = \vec{S}_3(0) \tag{6}$$

After a time $t = 2\pi(\mu B)^{-1}$, the spin $\vec{S}$ indeed returns to its original direction.

Since this is a book on group theory rather than quantum mechanics, I will restrain myself from going into the many interesting phenomena[5] involving the electron spin.

Time reversal and antiunitary operator

In a famous paper in 1932, Wigner showed that time reversal is represented by an anti-unitary operator. Since this subject is potentially confusing, I will go slow and let the physics, as embodied in the equations, lead us.

Take the Schrödinger equation $i\frac{\partial}{\partial t}\Psi(t) = H\Psi(t)$. For definiteness, think of $H = -\frac{1}{2m}\nabla^2 + V(\vec{x})$, just simple one-particle nonrelativistic quantum mechanics. We suppress the dependence of Ψ on $\vec{x}$.

Consider the transformation $t \to t' = -t$. We want to find a $\Psi'(t')$ such that

$$i\frac{\partial}{\partial t'}\Psi'(t') = H\Psi'(t') \tag{7}$$

Write $\Psi'(t') = T\Psi(t)$, where T is some operator to be determined (up to some arbitrary phase factor η). We require Ψ' to satisfy (7) if Ψ satisfies (2).

Plugging in, we have $i\frac{\partial}{\partial(-t)}T\Psi(t) = HT\Psi(t)$. Multiplying by T^{-1}, we obtain $T^{-1}(-i)T\frac{\partial}{\partial t}\Psi(t) = T^{-1}HT\Psi(t)$. Since H does not involve time, we want $HT = TH$, that is, $T^{-1}HT = H$. Then $T^{-1}(-i)T\frac{\partial}{\partial t}\Psi(t) = H\Psi(t)$. If T does what it is supposed to do, then we are forced to conclude, as Wigner was, that

$$T^{-1}(-i)T = i \tag{8}$$

Speaking colloquially, we can say that in quantum physics, time goes with an i, and so flipping time means you have to flip i also.

Let $T = UK$, where K complex conjugates everything to its right. Using $K^2 = I$, we have $T^{-1} = KU^{-1}$, and so (8) holds if $U^{-1}iU = i$, that is, if U^{-1} is just an ordinary (unitary) matrix that does nothing to i. We will determine U as we go along. The presence of K makes T antiunitary.

We check that this works for a spinless particle in a plane wave state $\Psi(t) = e^{i(\vec{p}\cdot\vec{x}-Et)}$. The formalism gives us $\Psi'(t') = T\Psi(t) = UK\Psi(t) = U\Psi^*(t) = Ue^{-i(\vec{p}\cdot\vec{x}-Et)} = \eta e^{-i(\vec{p}\cdot\vec{x}-Et)}$; since Ψ has only one component, U is 1-by-1 and hence is just a phase factor* η, which we can choose to be 1. Rewriting, we have (note the argument of Ψ'!) $\Psi'(t) = e^{-i(\vec{p}\cdot\vec{x}+Et)} = e^{i(-\vec{p}\cdot\vec{x}-Et)}$. Indeed, Ψ' describes a plane wave with wave vector $(-\vec{p})$ that moves in the opposite direction, as we would expect from time reversal. Note crucially that $\Psi'(t) \propto e^{-iEt}$ and thus has positive energy, as it should.[6] Note also that acting on a spinless particle, $T^2 = UKUK = UU^*K^2 = +1$, as we would expect.

Kramer's degeneracy

Next consider a spin $\frac{1}{2}$ nonrelativistic electron, which we can take to be at rest for simplicity. The spin observable is given by $\vec{S} = \psi^\dagger(\frac{1}{2}\vec{\sigma})\psi$, with ψ the 2-component spinor of the electron. Under the time reversal operator $T = UK$, $\vec{S} \to \frac{1}{2}\psi^\dagger(UK)^\dagger\vec{\sigma}UK\psi = \frac{1}{2}\psi^\dagger KU^\dagger\vec{\sigma}UK\psi$. We wish this to be $-\vec{S}$, a wish that would come true if $U = \eta\sigma_2$ with some phase factor η, since

$$KU^\dagger\vec{\sigma}UK = \eta^*\eta K\sigma_2\vec{\sigma}\sigma_2 K = K\begin{pmatrix} -\sigma_1 \\ +\sigma_2 \\ -\sigma_3 \end{pmatrix}K = -\vec{\sigma} \tag{9}$$

Everything fits! The presence of K is crucial. Note that the pseudoreality of the 2-dimensional spinor (and defining) representation of $SU(2)$ is crucial here.

In contrast to the spin 0 case, in the spin $\frac{1}{2}$ case, we need a 2-by-2 matrix $U = \eta\sigma_2$ to flip the spin. The key observation is that, acting on a spin $\frac{1}{2}$ particle,

$$T^2 = \eta\sigma_2 K\eta\sigma_2 K = \eta\sigma_2\eta^*\sigma_2^*KK = -1 \tag{10}$$

Two successive time reversals do not give the identity! Reversing the flow of time twice does not get a spin $\frac{1}{2}$ particle back to itself.

This is the origin of Kramer's degeneracy, which states the following. For an electron moving in an electric field, no matter how complicated the field, each of the energy levels must be two-fold degenerate.

---

* Phase factor rather than an arbitrary complex number, because we require $|\Psi'|^2 = |\Psi|^2$.

The proof is very simple.

Since electric fields do not change under time reversal (think of the electric field as that produced between two charged plates), the Hamiltonian is time reversal invariant ($HT = TH$), and so Ψ and $T\Psi$ have the same energy.

Next, ask whether Ψ and $T\Psi$ represent two distinct states or just one single state.

Suppose they actually represent the same state; this means that $T\Psi$ is proportional to Ψ. Thus, $T\Psi = e^{i\alpha}\Psi$, with $e^{i\alpha}$ some phase factor. But then

$$T^2\Psi = T(T\Psi) = Te^{i\alpha}\Psi = e^{-i\alpha}T\Psi = e^{-i\alpha}e^{i\alpha}\Psi = \Psi \tag{11}$$

But according to (10), $T\Psi$ is supposed to be equal to $-\Psi$. We have reached a contradiction, which implies that Ψ and $T\Psi$ must represent two distinct states! There is a two-fold degeneracy.

The proof of Kramer's theorem is a bit more subtle than most theorems in quantum mechanics, and students[7] often find it difficult to reconstruct from scratch. The key point is that a degeneracy means that two states of the same energy are distinct. To show that, we assume that the two states in question, Ψ and $T\Psi$, are proportional to each other and then reach a contradiction.

Kramer's degeneracy can be generalized immediately to systems with an odd number of electrons[8] in an electric field. The energy levels must be two-fold degenerate. This fact is of far-reaching importance in contemporary condensed matter physics.

The time of mysteries and eeriness

It is difficult to convey to contemporary readers how confusing the half-integral electron spin was at the time; the only way would be to quote[9] the distinguished physicists of the time. For example, C. G. Darwin, who published an early paper on electron spin with Pauli, wrote, "It is rather disconcerting to find that apparently something has slipped through the net, so that physical quantities exist which would be . . . artificial and inconvenient to express as tensors." Paul Ehrenfest in 1932 wrote of "this eerie report that a mysterious tribe by the name of spinor family inhabit isotropic [three-dimensional] space and the Einstein-Minkowski [four-dimensional] world."

Notes

1. Some six months earlier, Ralph de Laer Kronig, then barely 20, had thought of the same idea, but had the misfortune of talking about it to Wolfgang Pauli, who ridiculed the idea. In contrast, Paul Ehrenfest gave Uhlenbeck and Goudsmit the sensible advice that they had no reputation to lose. Many of the objections were apparently quite iron-clad, such as that the surface of the electron would be moving much faster than the speed of light, given the estimated size of the electron at that time. There was also the utter confusion stemming from what Pauli referred to as "classically indescribable two-valuedness." The lesson for the reader might be not to worry about expert opinion.

2. Not only is spin $\frac{1}{2}$ an established fact by now, but it also forms an important part of modern technology.

3. I recommend Sin-itiro Tomonaga, The Story of Spin, University of Chicago Press, 1997.

4. H. Hopf, after whom the Hopf map $S^3 \to S^2$ is named, had an office down the hall from Pauli at the ETH. The Hopf map is intimately related to Pauli matrices.

5. Read about the entire field of spintronics.

6. Note: no nonsense about negative energy states.

7. I am intentionally kind to professors of physics here.

8. For the generalization to relativistic electrons, see, for example, QFT Nut, p. 104.

9. Both quotes are from p. 130 of Tomonaga's The Story of Spin, cited earlier.

IV.7 | Integration over Continuous Groups, Topology, Coset Manifold, and $SO(4)$

In this chapter, we will learn about the following:

1. the character of rotation

2. integration over group manifolds

3. compact versus noncompact groups

4. the topology of $SO(3)$ and of $SU(2)$

5. the sphere as a coset manifold

6. local isomorphism between $SO(4)$ and $SU(2) \otimes SU(2)$

Character orthogonality for compact continuous groups

We derived all those beautiful theorems about character orthogonality back in chapter II.2. As already remarked there, these theorems should work for continuous groups that are compact, that is, if we could simply replace the sum over group elements $\sum_g$ in our discussion of finite groups by an integral $\int d\mu(g)$, with $d\mu(g)$ some kind of integration measure such that the corresponding integrals are finite. In this chapter, we first work out the character of our favorite continuous group, $SO(3)$, and then determine $d\mu(g)$.

The character of rotation

The character of an irreducible representation of $SO(3)$ is easy to compute. Since all rotations through angle ψ are equivalent regardless of the rotation axis, we might as well consider rotations around the z-axis. Take the eigenstate $|jm\rangle$ (for $m = j, j-1, \cdots, -j$) of J_3. Since $J_3 |jm\rangle = m |jm\rangle$, $e^{i\psi J_3} |jm\rangle = e^{im\psi} |jm\rangle$. So, the rotation matrix $R(\psi)$ is diagonal with elements equal to $e^{im\psi}$. Take the trace and use Gauss's summation formula

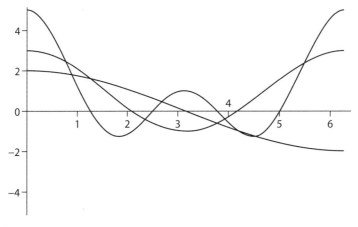

Figure 1

to obtain the character of the irreducible representation j:

$$\chi(j, \psi) = \sum_{m=-j}^{j} e^{im\psi} = e^{ij\psi} + e^{i(j-1)\psi} + \cdots + e^{-ij\psi}$$

$$= e^{-ij\psi}(e^{2ij\psi} + e^{i(2j-1)\psi} + \cdots + 1) = e^{-ij\psi}\frac{e^{i(2j+1)\psi} - 1}{e^{i\psi} - 1}$$

$$= \frac{\sin(j + \frac{1}{2})\psi}{\sin\frac{\psi}{2}} \tag{1}$$

The dimension of the representation j is given by $\chi(j, 0) = 2j + 1$. Note that $\chi(0, \psi) = 1$ (as you would expect), $\chi(1, \psi) = 1 + 2\cos\psi$, and so on.

But wait, didn't we also just calculate the characters of the irreducible representations of $SU(2)$? Indeed. All we have to do is to allow j in (1) to take on half-integral as well as integral values. But the Gauss summation formula doesn't care whether j is integral or half-integral as well as integral. Thus, the expression we just derived applies to $SU(2)$ as well. Thus, $\chi(\frac{1}{2}, \psi) = 2\cos\psi/2$, and so on.

Figure 1 plots $\chi(j, \psi)$ for $j = \frac{1}{2}, 1, 2$. You can figure out which is which using $\chi(j, 0) = 2j + 1$.

To try out the various orthogonal theorems, we need to figure out $d\mu(g)$ next.

Group manifolds

For continuous groups, the notion of a group manifold is completely natural. We also have an intuitive sense of the topology of the manifold. The manifold of the simplest continuous group, $SO(2)$, is clearly a circle. We naturally arrange the group elements, $R(\theta)$ with $0 \leq \theta < 2\pi$ and $R(2\pi) = R(0)$, around a unit circle.

We are invited to introduce a measure of distance on group manifolds, namely, a metric, in the language of differential geometry.[1] In $SO(2)$, the elements $R(\theta)$ and $R(\theta + d\theta)$ are infinitesimally close to each other (with, as usual, $d\theta$ an infinitesimal), and it makes total sense to set $d\mu(g)$ to be $d\theta$.

In general, for a continuous group G, we have a sense that two elements g_1 and g_2 are near each other if $g_1^{-1}g_2$ is close to the identity. (Of course, if $g_1 = g_2$, then $g_1^{-1}g_2 = I$.) Thus, if we understand the metric structure of the manifold near the origin, we naturally understand the structure near an arbitrary element g by simply multiplying what we know by g. This somewhat vague remark will become clearer later in this chapter.

Axis-angle parametrization

Fine, the group manifold of $SO(2)$ is a circle.

The naive guy might guess that the next rotation group in line, $SO(3)$, has the sphere for its group manifold. But that is clearly wrong, since it is characterized by three parameters, while the sphere is 2-dimensional. In fact, the group manifold of $SO(3)$ has a rather unusual topology, as we now discuss.

An easy way to parametrize a rotation is to write $R(\theta_1, \theta_2, \theta_3) = e^{i \sum_k \theta_k J_k}$, as noted in chapter I.3. For many purposes, however, it is more convenient to use the axis-angle parametrization (which is in fact used in everyday life; just try $e^{i \sum_k \theta_k J_k}$ out on your typical car mechanic): identify the rotation by specifying the axis of rotation $\vec{n}$ and the angle ψ through which we rotate around $\vec{n}$. Since $\vec{n}$ is specified* by two angles, θ and φ, in spherical coordinates, a general rotation $R(\vec{n}, \psi)$ is specified by three angles, θ, φ, and ψ, in agreement with the other parametrization just mentioned. So, a rotation is characterized by a vector $\vec{\psi} \equiv \psi \vec{n}$ whose direction and length determine the axis and angle of rotation, respectively.

Thus, the group manifold appears to be the 3-dimensional ball[†] B^3. But this is incorrect, because a rotation by π around $\vec{n}$ is the same as a rotation by π around $-\vec{n}$. (Verify this!) Thus, the group manifold is actually B^3 with the antipodal points on its surface identified. See figure 2, in which the "two points" labeled A are actually a single point. Similarly for the "two points" labeled B. It has no boundary: a traveler approaching $\psi = \pi$ would find himself or herself emerging smoothly at the "other side of the world." The "other side" is of course right in front of him or her. This manifold, while easily defined, is hard to visualize, but that should not perturb us.

In chapter I.2, we already noted that two rotations with different rotation axes but the same rotation angle ψ are equivalent:[‡] $R(\vec{n}, \psi) \sim R(\vec{n}', \psi)$. The similarity transformation

---

* Namely, $\vec{n} = (\sin \theta \cos \varphi, \sin \theta \sin \varphi, \cos \theta)$.

[†] Note that in mathematics, in contrast to everyday life, sphere and ball are totally distinct concepts: S^2 and B^3 are both subsets of Euclidean 3-space, but with the former characterized by points satisfying $x^2 + y^2 + z^2 = 1$, and the latter by $x^2 + y^2 + z^2 \leq 1$.

[‡] Thus, more explicitly, $R(\theta', \varphi', \psi) \sim R(\theta, \varphi, \psi)$, but $R(\theta, \varphi, \psi') \nsim R(\theta, \varphi, \psi)$.

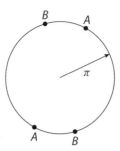

Figure 2

relating the two is simply the rotation that takes $\vec{n}'$ to $\vec{n}$. For a fixed value of ψ, as $\vec{n}$ varies, the rotations $R(\vec{n}, \psi)$ trace out a sphere of radius ψ. Interestingly, the equivalence classes of $SO(3)$ correspond to spheres of different radii. In comparison, the equivalence classes of $SO(2)$ correspond to different points on the circle. (This is consistent with my earlier remark that in an abelian group, every element is in its own equivalence class.)

Note also that in $SO(3)$, the identity corresponds to a sphere of vanishing radius, that is, the point at the origin. As always, the identity is proudly in a class by itself.

Do exercises 1 and 2 now. We will need the results later in this chapter.

Integration measure

Now that we have the notion of a group manifold, we would naturally like to integrate various functions $F(g)$ of the group elements over the group G. So, we need an integration measure $d\mu(g)$ to formulate such integrals as $\int_G d\mu(g) F(g)$.

What properties would we like the integration measure to have? Well, our ultimate goal is to carry over some of the marvelous results obtained about the representations of finite groups to continuous groups. Recall that the crucial step was to form the matrix $A = \sum_g D^\dagger(g) X D(g)$ for some arbitrary matrix X and to observe that $D^\dagger(g) A D(g) = A$. As already mentioned in chapter II.1, we want to replace $\sum_g$ by $\int d\mu(g)$. In particular, for $A = \int d\mu(g) D^\dagger(g) X D(g)$, we would like to have $D^\dagger(g) A D(g) = A$, just as in finite groups.

The left hand side of this condition is

$$D^\dagger(g) \left(\int d\mu(g_1) D^\dagger(g_1) X D(g_1) \right) D(g) = \int d\mu(g_1) D^\dagger(g_1 g) X D(g_1 g)$$

$$= \int d\mu(g_2 g^{-1}) D^\dagger(g_2) X D(g_2) \qquad (2)$$

In the last step, we simply write $g_2 = g_1 g$ and solve for $g_1 = g_2 g^{-1}$. We want this to be equal to $A = \int d\mu(g_2) D^\dagger(g_2) X D(g_2)$. Our little heart's desire would be fulfilled if $d\mu(g_2 g^{-1}) = d\mu(g_2)$ for any g and g_2. But since $g_2 g^{-1}$ and g_2 are just any two group elements, we might as well write

$$d\mu(g) = d\mu(g') \tag{3}$$

which just says that the invariant measure $d\mu(g)$ does not depend on where you are in the group.

Kind of what you might have expected from the beginning, no? In particular, we can write

$$d\mu(g) = d\mu(I) \tag{4}$$

Dr. Feeling drops by. "Tsk, tsk, so much verbiage, but it's obvious," he grumbles. "This is exactly what anybody would do in everyday life if told to measure the area of a house. You would pick a standard measuring stick, start at some convenient place, and transport that stick from room to room, taking care that it is not deformed in any way." The requirement (4) simply forbids us from changing measuring sticks as we move from room to room. A few examples, to be given shortly, will completely clarify these admittedly somewhat vague statements. But first a couple of remarks.

The integral $\int_G d\mu(g)1$, sometimes called the volume of the group $V(G)$, thus offers a measure of how big the group G is. It is the direct analog of $\sum_{g \in G} 1 = N(G)$ for finite groups, namely, the number of group elements. Note that (4) does not tell us how $d\mu(g)$ is to be normalized. There is no particularly natural way to normalize $V(G)$, in contrast to $N(G)$. If you are obsessive compulsive, you could always write $d\mu(g)/(\int_G d\mu(g)1)$ instead of $d\mu(g)$.

Unlike mathematicians, physicists often descend to a specific parametrization of a group manifold. This is similar to discussing the sphere S^2 in the abstract versus actually calculating something using spherical coordinates θ, φ. Let $y^1, y^2, \cdots, y^D$ be a set of coordinates on a D-dimensional manifold. Then we can write $d\mu(g) = dy^1 dy^2 \cdots dy^D \rho(y^1, y^2, \cdots, y^D)$, with ρ sometimes called the density. (We are of course familiar with the infinitesimal area element on a sphere being equal to $d\theta d\varphi \sin\theta$, with $\rho(\theta, \varphi) = \sin\theta$. For fixed $\delta\theta$ and $\delta\varphi$, the vaguely rectangular patch with sides $\delta\theta$ and $\delta\varphi$ covers more area near the equator than it does near the two poles.) In a specific coordinate system, the invariant measure can get a bit involved, as we will see. It is like the difference between writing $d\Omega$ and be done with it, versus writing $d\theta d\varphi \sin\theta$.

Invariant measures on group manifolds: Some examples

As promised, here are some examples of measures on group manifolds. The baby example is $SO(2)$ with the group manifold a circle parametrized by θ, as already mentioned. If we have any sense at all, we would feel that the invariant measure is simply $d\theta$.

Let us derive this more laboriously, like some kind of insufferable pedant. Lay down our ruler, with one end at $\theta = 0$ and the other end at $\delta\theta$. Apply (4). Since $R(\theta')R(\theta) = R(\theta' + \theta)$, left multiplication by $R(\theta')$ transports the two ends of the ruler to θ' and $\theta' + \delta\theta$. So by (4), the distance between θ' and $\theta' + \delta\theta$ is just $\delta\theta$, independent of θ'.

The restricted Lorentz group $SO(1, 1)$ is perhaps more illuminating, since, unlike the circle, it has two common parametrizations. Parametrizing Lorentz transformations with the boost angle φ, we have the multiplication law $L(\varphi')L(\varphi) = L(\varphi' + \varphi)$. Thus, going through the same argument we just went through for $SO(2)$, we find that the invariant measure $SO(1, 1)$ is $d\varphi$. The key, of course, is that the convenient boost angle parametrization maps multiplication of group elements into addition of their respective parameters.

Compact versus noncompact

Now we see the huge difference between $SO(2)$ and $SO(1, 1)$. The volume of $SO(2)$ (namely, $\int_0^{2\pi} d\theta = 2\pi$) is finite; in contrast, the volume of $SO(1, 1)$ (namely, $\int_{-\infty}^{\infty} d\varphi = \infty$) is infinite. As hinted at in chapter II.1, this marks the difference between compact and noncompact groups, a difference with drastic consequences. For example, we saw that compact groups, just like finite groups, have unitary representations, while noncompact groups do not.

The restricted Lorentz group illustrates another important point. Just as we can set up many different coordinates on a curved manifold, we can parametrize a given group in many different ways. Indeed, in physics, the Lorentz group is parametrized more physically by the velocity v:

$$L(v) = \begin{pmatrix} \frac{1}{\sqrt{1-v^2}} & \frac{v}{\sqrt{1-v^2}} \\ \frac{v}{\sqrt{1-v^2}} & \frac{1}{\sqrt{1-v^2}} \end{pmatrix} \tag{5}$$

The readers who have studied some special relativity know that this leads to the addition of velocities[2] as follows: $L(u)L(v) = L(v')$, with $v' = \frac{v+u}{1+vu}$. Readers who do not know this can simply verify this relation by multiplying the 2-by-2 matrices $L(u)$ and $L(v)$ together. Physically, if two observers are moving relative to each other with velocity u, then a particle seen by one observer to move with velocity v would be seen by the other observer to move with velocity v'.

Now we simply follow our noses and lay down our beloved ruler, with one end at v and the other end at $v + \delta v$. Left multiplication by $L(u)$ transports the two ends of the ruler to v' and $v' + \delta v'$. We next compute $v' + \delta v' = \frac{v+\delta v+u}{1+(v+\delta v)u} = \frac{v+u}{1+uv} + \frac{1-u^2}{(1+uv)^2}\delta v$, so that $\delta v' = \frac{1-u^2}{(1+uv)^2}\delta v$.

You may have realized that all we did was to compute the Jacobian of the transformation from v to $v'(v)$ (suppressing u). Indeed, at the level of a specific parametrization, all that (4) is saying is that $dy^1 dy^2 \cdots dy^D \rho(y^1, y^2, \cdots, y^D) = dy'^1 dy'^2 \cdots dy'^D \rho(y'^1, y'^2, \cdots, y'^D)$ $= dy^1 dy^2 \cdots dy^D J(\frac{\partial y'}{\partial y})\rho(y'^1, y'^2, \cdots, y'^D)$, where $J(\frac{\partial y'}{\partial y})$ denotes the Jacobian going from y to y', so that

$$\rho\left(y^1, y^2, \cdots, y^D\right) = J\left(\frac{\partial y'}{\partial y}\right)\rho\left(y'^1, y'^2, \cdots, y'^D\right) \tag{6}$$

Let us return to our simple example in which there is only one parameter, the velocity. This relation then reads $\rho(v)dv = \rho(v')dv' = \rho(v')\frac{1-u^2}{(1+uv)^2}dv$, that is,

$$\rho(v) = \frac{1-u^2}{(1+uv)^2}\rho\left(\frac{v+u}{1+uv}\right) \tag{7}$$

Setting $v = 0$, we solve this instantly to give $\rho(u) = \frac{1}{1-u^2}\rho(0)$. (Alternatively, set $u = -v$ to obtain the same (of course) solution.) As remarked earlier, the measure is determined only up to an overall constant, and so we might as well set $\rho(0) = 1$.

The volume of the group is given by $\int_{-1}^{1}\frac{dv}{1-v^2} = \infty$. Of course, whether or not a group is compact does not depend on the parametrization.

Indeed, if we already know about the boost angle parametrization, then we could go from one parametrization to another by a simple change of variable: from the relation $v = \tanh\varphi$, we have $dv = \frac{d\varphi}{\cosh^2\varphi}$ and $\cosh^2\varphi = \frac{1}{1-v^2}$, and thus, as expected, $d\varphi = \cosh^2\varphi\, dv = \frac{dv}{1-v^2}$.

The measure of the $SO(3)$ group manifold

Mathematicians would leave (4) alone, but physicists, with their insistence on calculating and with their penchant for descending to some specific set of coordinates, would want to work out the measure of the $SO(3)$ group manifold explicitly. So, keep in mind that, when in the rest of the section what some readers would call "math" gets to be quite involved, the mathematicians would say that the physicists are merely making work for themselves.

Write[3] $d\mu(g) = d\theta d\varphi \sin\theta d\psi f(\psi) \equiv d\Omega d\psi f(\psi)$. (We have invoked rotational invariance to say that $f(\psi)$ cannot depend on θ or φ.)

Our task is to determine $f(\psi)$. Were the group manifold a Euclidean 3-space, we would have $f(\psi) \propto \psi^2$, since ψ plays the role of the radial variable. (Here we are just showing off our childhood knowledge that the volume measure in 3-space is $r^2 dr \sin\theta d\theta d\varphi$.)

Dr. Feeling saunters by, mumbling, "In a sufficiently small neighborhood near the identity, to lowest order we don't feel the curvature."[4] The manifold is nearly Euclidean, and so $f(\psi) \propto \psi^2$ should hold for small ψ. However, this can't possibly hold for finite ψ, since the manifold is manifestly not Euclidean. We also sense that $f(\psi)$ can't be something nontrigonometric like ψ^2.

Onward to determine $f(\psi)$! We are going to do it three ways, the hard way first, then an easy way in the next section, and eventually, an even easier way.

Write an infinitesimal rotation as $R(\delta, \varepsilon, \sigma) = I + \begin{pmatrix} 0 & -\delta & \sigma \\ \delta & 0 & -\varepsilon \\ -\sigma & \varepsilon & 0 \end{pmatrix} \equiv I + A$. In a small neighborhood around the identity, δ, ε, and σ can serve as Cartesian coordinates, with the volume measure given by $d\delta d\varepsilon d\sigma$. So much for the right hand side of (4), which we reproduce here for your convenience: $d\mu(g) = d\mu(I)$.

To obtain the left hand side, we now transport this typical infinitesimal rotation to the neighborhood around $R(\vec{n}, \psi)$ by multiplying $R(\delta, \varepsilon, \sigma)$ from the left (or the right, if you like) by $R(\vec{n}, \psi)$. Since what we are after depends only on ψ, we can pick $\vec{n}$ to be pointing along the z-axis to make life easier.

So let's calculate $R(\vec{n}, \psi') = R(\vec{e}_z, \psi)R(\delta, \varepsilon, \sigma)$. With $R(\vec{e}_z, \psi) = \begin{pmatrix} \cos\psi & -\sin\psi & 0 \\ \sin\psi & \cos\psi & 0 \\ 0 & 0 & 1 \end{pmatrix}$, we have $R(\vec{n}, \psi') \simeq R(\vec{e}_z, \psi)(I + A) = R(\vec{e}_z, \psi) + R(\vec{e}_z, \psi)A$.

Evaluating $R(\vec{e}_z, \psi)A$ is easy, but determining $\vec{n}$ and ψ' requires some huffing and puffing. Fortunately, you, that is, you, have done exercise 2. Invoking what you got, we obtain $n_1 = (\sin\psi\, \varepsilon + (1 + \cos\psi)\sigma)/(2\sin\psi)$, $n_2 = (-\sin\psi\, \sigma + (1 + \cos\psi)\varepsilon)(2\sin\psi)$, and $n_3 = 1$. Happily, there is no need to normalize $\vec{n}$, since that would introduce terms of $O(\varepsilon^2)$ (where we regard δ, ε, σ collectively as of order $O(\varepsilon)$), while we are working only up to $O(\varepsilon)$.

Now that we have determined the rotation axis, we have to determine the rotation angle. Without doing any calculations, we intuit that $\psi' = \psi + \delta + O(\varepsilon^2)$, since in the infinitesimal rotation $R(\delta, \varepsilon, \sigma)$, the rotation angle around the z-axis is δ. Of course, to be absolutely sure, we can also use the result of exercise 1, which you have also done. Tracing $R(\vec{e}_z, \psi)R(\delta, \varepsilon, \sigma)$, we obtain $1 + 2\cos\psi' = 1 + 2\cos\psi - 2(\sin\psi)\delta$, which verifies our intuition.

Call the local Cartesian coordinates near ψ' by (x^1, x^2, x^3). Then putting everything together, we have $(x^1, x^2, x^3) = \psi'(n_1, n_2, n_3) = (\psi(\sin\psi\, \varepsilon + (1 + \cos\psi)\sigma)/(2\sin\psi)$, $\psi(-\sin\psi\, \sigma + (1 + \cos\psi)\varepsilon)(2\sin\psi)$, $\psi + \delta) + O(\varepsilon^2)$. The Jacobian $J = \det(\frac{\partial(x^1, x^2, x^3)}{\partial(\varepsilon, \sigma, \delta)})$ for the transformation from the Cartesian coordinates $(\varepsilon, \sigma, \delta)$ near the identity to the Cartesian coordinates (x^1, x^2, x^3) near ψ' is particularly easy to evaluate, since δ appears only in x^3 and hence the determinant is that of a 2-by-2 matrix. We obtain $4J = 2\psi^2(1 + \cos\psi)/\sin\psi^2 = 2\psi^2/(1 - \cos\psi) = \psi^2/\sin^2(\psi/2)$. Dropping an irrelevant overall constant, we find $d\varepsilon d\sigma d\delta = dx^1 dx^2 dx^3/J = \psi^2 d\psi (\sin\theta d\theta d\varphi)\sin^2(\psi/2)/\psi^2 = d\Omega d\psi \sin^2(\psi/2)$. We watch with satisfaction that the "nontrigonometric" factor ψ^2 cancels out. We obtain, up to an arbitrary normalization, $f(\psi) = \sin^2(\psi/2)$. Note also that, by construction, $f(\psi) \propto \psi^2$ for small ψ.

The usual solid angle element $d\Omega$ drops out when we integrate over a class function $F(g)$, that is, a function that depends only on the class the group element belongs to, which, as we have learned, is labeled by ψ. Thus, integrals of class functions over the $SO(3)$ group manifold are given by

$$\int_{SO(3)} d\mu(g) F(g) = \int_0^\pi d\psi \left(\sin^2 \frac{\psi}{2}\right) F(\psi) \tag{8}$$

Character orthogonality

So finally we are in a position to verify character orthogonality. Recall that we have computed the character for the irreducible representation j to be $\chi(j, \psi) = \frac{\sin(j+\frac{1}{2})\psi}{\sin\frac{\psi}{2}}$. Thus,

$$\int_{SO(3)} d\mu(g)\chi(k,\psi)^*\chi(j,\psi) = \int_0^\pi d\psi \sin^2\left(\frac{\psi}{2}\right)\frac{\sin(k+\frac{1}{2})\psi \sin(j+\frac{1}{2})\psi}{\sin^2\frac{\psi}{2}}$$

$$= \int_0^\pi d\psi \sin\left(k+\frac{1}{2}\right)\psi \sin\left(j+\frac{1}{2}\right)\psi$$

$$= \frac{1}{2}\int_0^\pi d\psi (\cos(j-k)\psi - \cos(j+k+1)\psi)$$

$$= \frac{\pi}{2}\delta_{jk} \tag{9}$$

The density $\sin^2(\frac{\psi}{2})$ cancels out rather neatly. Indeed, for $j \neq k$, the characters of the irreducible representations j and k are orthogonal to each other. This calculation also indicates that the normalization factor is $2/\pi$ if we want to normalize the volume of $SO(3)$ to 1.

We can also turn this calculation around and use orthogonality to determine the integration measure, that is, by demanding that $\int_0^\pi d\psi f(\psi)\chi(k,\psi)^*\chi(j,\psi) \propto \delta_{jk}$ to determine $f(\psi)$. This is the promised "easy" way.

Fourier series

At this point, you might have suddenly realized that the Fourier series, perhaps one of the most useful tools in physics and mathematics, is based on character orthogonality for the group $U(1)$.

The element $e^{i\theta}$ (with $0 \leq \theta < 2\pi$) is represented in the irreducible representation j (with $-\infty < j < \infty$) by the complex number $e^{ij\theta}$, which is thus also the character. The invariant measure is just $d\theta$, and thus character orthogonality implies

$$\int_0^{2\pi} d\theta (e^{ik\theta})^* e^{ij\theta} = 2\pi\delta_{jk} \tag{10}$$

A sophisticated way of understanding the Fourier series indeed!

Clebsch-Gordan decomposition

Character orthogonality also affords us a rather nifty way of obtaining the Clebsch-Gordan series for $SO(3)$. Start with the observation (noted in chapter II.1 and used in chapter II.3) that the character of the tensor product of two irreducible representations j and k is the product of the characters of j and k. The method is then to write $\chi(k)\chi(j)$ as a sum $\sum_l \chi(l)$. (To lessen clutter, we temporarily suppress the ψ dependence. The character depends on ψ, of course. Remember? Character is a function of class, and classes are labeled by ψ here.) The characters are given in (1), but for the present purpose, it turns out to be less convenient to use the final form involving the sine function. Instead, we define $\zeta = e^{i\psi}$ and write

$$\chi(j) = (\zeta^j + \zeta^{j-1} + \cdots + \zeta^{-j}) = \zeta^{-j}(\zeta^{2j+1} - 1)/(\zeta - 1) = (\zeta^{j+1} - \zeta^{-j})/(\zeta - 1) \tag{11}$$

Then (assume $j \geq k$ for ease of reading and with no loss of generality),

$$\chi(k)\chi(j) = \left(\zeta^k + \zeta^{k-1} + \cdots + \zeta^{-k}\right)(\zeta^{j+1} - \zeta^{-j})/(\zeta - 1)$$

$$= \left(\zeta^{j+k+1} + \zeta^{j+k} + \cdots + \zeta^{j-k+1} - \zeta^{j-k} - \cdots - \zeta^{-j-k+1} - \zeta^{-j-k}\right)/(\zeta - 1)$$

$$= \chi(j+k) + \chi(j+k-1) + \cdots + \chi(j-k) \tag{12}$$

Note that in the first equality in (12), we use two different forms for χ, and in the third equality the terms in the parenthesis pair off "head and tail."

Thus, we obtain the result, well known to us since chapter IV.2, that

$$k \otimes j = (j+k) \oplus (j+k-1) \oplus \cdots \oplus (j-k) \tag{13}$$

Note that this holds for both $SO(3)$ and $SU(2)$.

Topology of group manifolds

A basic notion in topology is the homotopy of closed curves. I give an extremely brief introduction to the essential concepts for the benefit of the reader totally unfamiliar with the notion of homotopy. Two closed curves in a manifold are said to be homotopic to each other if they can be continuously deformed into each other. In Euclidean space, any two closed curves are homotopic, and in fact, each can be continuously deformed to a point. Similarly, on the sphere, any two closed curves can be continuously deformed into each other and to a point.

The (first) homotopy group $\pi_1(M)$ of a manifold M is defined as follows. Pick a point P on the manifold, and consider closed curves starting at P and ending at P. An element g of the homotopy group corresponds to all the curves that are homotopic to one another. The product $g_1 \cdot g_2$ is naturally defined by the set of closed curves that start at P, follow a curve in g_2, and return to P, and then continue on following a curve in g_1, ending finally at P. See figure 3. I leave it to you to figure out how the identity and the inverse are defined.

You are right of course. The identity corresponds to all the curves that can be shrunk to the point P. The inverse of g corresponds to the curves in g traversed in the opposite

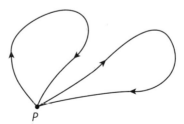

Figure 3

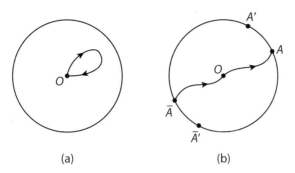

(a)   (b)

Figure 4

direction. For our purposes here, the emphasis would not be on the group aspect of $\pi_1(M)$. Rather, the important point to note is that if two manifolds have different homotopy groups, they are manifestly not topologically equivalent.

Consider, for example,[5] the Euclidean plane with a hole (say, the unit disk centered at the origin) cut out of it. Its homotopy group π_1 is then given by Z. A closed curve starting and ending at a point P could wind around the hole j times, with j an integer.

While this exceedingly brief discussion barely touches the subject of homotopy, it suffices for our purposes.[6]

The group manifold of $SO(3)$

Let us then return to $SO(3)$ with its peculiar group manifold as discussed earlier. A closed curve starting and ending at the identity element (which we will call the origin O) without getting to $\psi = \pm\pi$ can clearly be shrunk to a point. See figure 4a. In contrast, a curve that proceeds from O to a point A with $\psi = \pi$, emerges at the antipodal point $\bar{A}$, and then returns to O (figure 4b) cannot be shrunk. If we deform the curve and move A to A' in a futile attempt to bring A and $\bar{A}$ together, $\bar{A}$ will move to $\bar{A}'$ as shown in the figure. Call the element of $\pi_1(SO(3))$ corresponding to this class of curves g.

Next, consider the class of curves described by $g \cdot g$. A typical member of this class would be a curve that proceeds from O to the point A, emerges at the antipodal point $\bar{A}$, returns to O, proceeds to the point B with $\psi = \pi$, emerges at the antipodal point $\bar{B}$, and finally returns to O. See figure 5a. Remarkably, these curves could be shrunk to a point. Simply bring A and $\bar{B}$ together (so that $\bar{A}$ and B also come together). When A and $\bar{B}$ are on top of each other (so that $\bar{A}$ and B also come together on top of each other), the curve could be "tucked in" away from $\psi = \pi$, and then brought back to O. See figure 5b,c.

Thus, $\pi_1(SO(3)) = Z_2$.

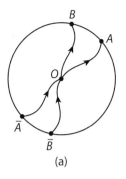 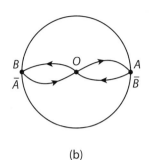 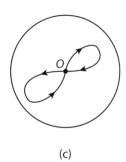

(a)　　　　　　　　　(b)　　　　　　　　　(c)

Figure 5

The group manifold of $SU(2)$

The group manifold of $SO(3)$ has an interestingly nontrivial topology. What about the group manifold of $SU(2)$?

An arbitrary element $U = e^{i\vec{\varphi}\cdot\vec{\sigma}/2}$ of $SU(2)$ is parametrized by three real parameters $\vec{\varphi}$, as we learned in chapter IV.5, and so the desired group manifold is 3-dimensional. Actually, we could already read off from (IV.5.11) what the group manifold is, at least with the benefit of hindsight. But for a clearer presentation, let us introduce a closely related parametrization of $SU(2)$.

Consider the matrix $U = t + i\vec{x}\cdot\vec{\sigma}$ parametrized by four real numbers* (t, x, y, z). The condition for U to be unitary is $U^\dagger U = (t - i\vec{x}\cdot\vec{\sigma})(t + i\vec{x}\cdot\vec{\sigma}) = t^2 + \vec{x}^2 = 1$ (where I have used a special case of (IV.5.10)). By direct computation, we find that the determinant of U is equal to $t^2 + \vec{x}^2$, and the condition $\det U = 1$ leads to the same condition as before. Thus, for U to be an element of $SU(2)$, the four real numbers (t, x, y, z) must satisfy

$$t^2 + \vec{x}^2 = t^2 + x^2 + y^2 + z^2 = 1 \tag{14}$$

If we think of (t, x, y, z) as coordinates of a point in a 4-dimensional Euclidean space, then (14) just says that the distance squared of that point from the origin is equal to 1. Hence the set of points (t, x, y, z) that satisfy (14) is just the 3-dimensional unit sphere S^3 living in a 4-dimensional Euclidean space.

The parametrization used here is just the parametrization in (IV.5.11), $U = \cos\frac{\varphi}{2}I + i\hat{\varphi}\cdot\vec{\sigma}\sin\frac{\varphi}{2}$, in another guise. The group manifold can be characterized by φ and $\hat{\varphi}$, namely, the length and direction of $\vec{\varphi}$ fixed by three real numbers. The correspondence is just $t = \cos\frac{\varphi}{2}, \vec{x} = \sin\frac{\varphi}{2}\hat{\varphi}$.

---

* Evidently, this is much favored by relativistic physicists. See chapter VII.2.

Topology distinguishes $SU(2)$ from $SO(3)$

It is perhaps gratifying that $SU(2)$ turns out to have such a nice group manifold. This also shows that $SU(2)$ and $SO(3)$ cannot possibly be isomorphic: their group manifolds have rather different topologies. In particular, any closed curve on S^3 can be shrunk to a point, so that $\pi_1(SU(2)) = \varnothing$, while we just saw that there are closed curves on the group manifold of $SO(3)$ that cannot be shrunk. This topological consideration confirms what we already know, that $SU(2)$ is locally isomorphic to $SO(3)$, but since $SU(2)$ double covers $SO(3)$, they cannot possibly be globally isomorphic.

The group manifold $SU(2)$ and its integration measure

We now determine the integration measure of $SU(2)$; the group manifold is the sphere S^3, and we know spheres well. There are many ways[7] to determine the integration measure on S^3. Here we follow a particularly elementary approach.

Every school child knows that a line of constant latitude traces out a circle, with radius $\sin\theta$ that approaches 0 as we approach the north pole. With the usual spherical coordinates, we thus determine the measure on S^2 to be $d\Omega = d\theta(d\varphi\sin\theta)$, where we recognize $(d\varphi\sin\theta)$ as the infinitesimal line element on a circle S^1 of radius $\sin\theta$.

Similarly, the surface of constant latitude on S^3 traces out a sphere, with radius $\sin\zeta$ (in other words, set $t = \cos\zeta$, so that $x^2 + y^2 + z^2 = \sin^2\zeta$). We simply generalize the S^2 story and write down the measure on S^3 as $d\zeta(d\Omega\sin^2\zeta) = d\zeta d\theta d\varphi\sin\theta\sin^2\zeta$. Again, we recognize $d\Omega\sin^2\zeta = d\theta d\varphi\sin\theta\sin^2\zeta$ as the infinitesimal area element on a sphere with radius $\sin\zeta$.

To make contact with the notation used here, recall from chapter IV.5 that, in the parametrization used here, $U = \cos\zeta + i\sin\zeta\sigma_z$ actually describes a rotation through angle $\psi = 2\zeta$ around the z-axis. Double covering, remember? Dropping $d\Omega$, we obtain for the integral of a class function $\int_0^\pi d\zeta\sin^2\zeta F(\psi) = \int_0^\pi d\psi\sin^2\frac{\psi}{2}F(\psi)$, in complete agreement with what we obtained earlier in (8). We have invoked the fact that the measure is a local concept and so should be the same for $SO(3)$ and $SU(2)$. As promised, this is the easiest way yet.

How does the sphere appear in group theory?

We learned here that the group manifold of $SO(2)$ is the circle S^1, of $SO(3)$ is some weird topological space, and of $SU(2)$ is the 3-sphere S^3. Where does the ordinary everyday sphere S^2 appear?

The sphere is in fact not a group manifold but an example of a coset manifold.

I introduced the notion of cosets for finite groups back in chapter I.2. First, a quick review. Given a group G with a subgroup H, let us identify two elements g_1 and g_2 of G

if there exists an element h of H such that $g_1 = g_2 h$. Use the notation $g_1 \sim g_2$ if g_1 and g_2 are identified. If there does not exist an h such that $g_1 = g_2 h$, then $g_1 \not\sim g_2$. This procedure thus divides the elements of G into distinct cosets labeled by $g_1 H$. The collection of cosets is denoted by G/H.

Going from finite groups to continuous groups, we have the additional notion that two elements, g_1 and g_2, may be infinitesimally close to each other. In that case, we say that the cosets $g_1 H$ and $g_2 H$ are infinitesimally close to each other. With this notion of distance, the collection of cosets becomes a coset manifold.

Let $G = SO(3)$ and $H = SO(2)$. For definiteness, let H consist of rotations around the z-axis, and $\vec{e}_z$ denote the unit vector pointing along the z-axis. Then $g_1 \sim g_2$ implies $g_1 \vec{e}_z = g_2 h \vec{e}_z = g_2 \vec{e}_z$; in other words, g_1 and g_2 are identical in the sense that they take $\vec{e}_z$ into the same unit vector. Colloquially speaking, the rotation $h \in SO(2)$ did nothing nohow. Somewhat more academically speaking, the rotations g_1 and g_2 map the north pole into the same point. Thus, each coset corresponds to a point on the unit sphere, and different cosets correspond to different points. The ordinary sphere is in fact $S^2 = SO(3)/SO(2)$.

By the way, this discussion is consistent with the remark in chapter I.2 that $SO(2)$ is not an invariant subgroup of $SO(3)$.

$SO(4)$ is locally isomorphic to $SU(2) \otimes SU(2)$

We just learned that the elements of $SU(2)$ are in one-to-one correspondence with the unit vectors in Euclidean 4-space, each defining a point on the 3-dimensional sphere S^3. So, consider an $SU(2)$ element $W = t + i\vec{x} \cdot \vec{\sigma}$ parametrized by four real numbers (t, x, y, z) with $t^2 + \vec{x}^2 = t^2 + x^2 + y^2 + z^2 = 1$. (Note that, in contrast to $X = \vec{x} \cdot \vec{\sigma}$ in chapter IV.5, W is neither traceless nor hermitean.)

Observe that, for U and V two arbitrary elements of $SU(2)$, the matrix $W' = U^\dagger W V$ is also an element of $SU(2)$, for the simple reason that $SU(2)$ is a group!

Since W' is an element of $SU(2)$, it can be written in the form $W' = (t' + i\vec{x}' \cdot \vec{\sigma})$, with $t'^2 + \vec{x}'^2 = 1$. Thus, (t, x, y, z) and (t', x', y', z') are both unit vectors. The transformation $W \to W'$ turns one unit vector into another unit vector. Hence, U and V determine a 4-dimensional rotation, namely, an element of $SO(4)$.

The pair of $SU(2)$ elements (U, V) defines the group $SU(2) \otimes SU(2)$ (it might seem self-evident to you, but do verify the group axioms; for example, $(U_1, V_1)(U_2, V_2) = (U_1 U_2, V_1 V_2)$). We conclude that $SO(4)$ is locally isomorphic[8] to $SU(2) \otimes SU(2)$. (Recall that we had already obtained this result by looking at the Lie algebras way back in chapter I.3.)

We say locally, because the pair of $SU(2)$ elements (U, V) and the pair $(-U, -V)$ correspond to the same $SO(4)$ rotation, since $U^\dagger W V = (-U)^\dagger W(-V)$. Thus, $SO(4)$ is covered by $SU(2) \otimes SU(2)$.

Quite likely, this discussion may have confused our friend Confusio. Admittedly, it sounds eerily similar to the demonstration in chapter IV.5 that $SU(2)$ covers $SO(3)$. But the astute reader should have noticed the differences rather than the similarities. As already mentioned, the matrix X is hermitean and traceless, while W is neither. Here, not one $SU(2)$ element U, but two elements U and V, are involved.

But, indeed, the two discussions are related in one respect. Consider the subgroup of $SU(2) \otimes SU(2)$ defined by the elements (V, V). (Verify that the subgroup is an $SU(2)$, known as the diagonal subgroup of $SU(2) \otimes SU(2)$.) In the transformation $W \to W' = V^\dagger W V = V^\dagger(t + i\vec{x} \cdot \vec{\sigma})V = (t + i\vec{x} \cdot V^\dagger \vec{\sigma} V) = (t' + i\vec{x}' \cdot \vec{\sigma})$, evidently $t' = t$. Only $\vec{x}$ is changed and $\vec{x}'$. Thus, the diagonal $SU(2)$ subgroup maps into the $SO(3)$ subgroup of $SO(4)$. Note that the calculation of $\vec{x}'$ is the same as in chapter IV.5.

Another $SU(2)$ subgroup consists of the elements $(V^\dagger, V)$. Verify that this does change t. What does this subgroup correspond to in $SO(4)$?

Running a reality check on $SO(3)$ and $SU(2)$

In chapter II.3, we derived a reality checker. Evaluate $\eta^{(r)} \equiv \frac{1}{N(G)} \sum_{g \in G} \chi^{(r)}(g^2)$. If $\eta^{(r)} = 1$, the irreducible representation is real; if $= -1$, the irreducible representation is pseudoreal; and if 0, the irreducible representation is complex. We expect that the reality checker would also work for compact groups if we replace the sum $\frac{1}{N(G)} \sum_{g \in G}(\cdots)$ by the integral $\int_G d\mu(g)(\cdots)$ (with the integration measure normalized, so that $\int_G d\mu(g)\, 1 = 1$).

Let us feed the compact groups $SO(3)$ and $SU(2)$ into the reality checker. It would be fun to watch the reality checker churning out the right answer, which in fact we already know. The irreducible representations of $SO(3)$ are real, while half of the irreducible representations of $SU(2)$ are pseudoreal, with the other half real.

So, evaluate (watch the angles and half angles like a hawk!)

$$\int_0^\pi d\mu(g)\chi(j, 2\psi) = \frac{2}{\pi} \int_0^\pi d\psi \sin^2\left(\frac{\psi}{2}\right) \frac{\sin((2j+1)\psi)}{\sin \psi}$$

$$= -\frac{1}{2\pi} \int_0^\pi d\psi (e^{i\psi} - 2 + e^{-i\psi})\left\{e^{ij2\psi} + e^{i(j-1)2\psi} + \cdots + e^{-ij2\psi}\right\} \tag{15}$$

(We made sure to use the normalized measure here.)

This is an interesting integral, whose value depends on whether j is integral or half integral, as it better, if the machine is going to work. For j integral, the curly bracket contains a term equal to 1 but no term equal to either $e^{i\psi}$ or $e^{-i\psi}$, and thus the integral evaluates to $+1$. For j half integral, the curly bracket contains $e^{i\psi} + e^{-i\psi}$, but does not contain a term equal to 1, and thus the integral evaluates to -1. Math triumphs once again!

Exercises

1   Given a rotation matrix R, determine the rotation angle ψ.

2   Given a rotation matrix R, determine the rotation axis $\vec{n}$.

3 What does the $SU(2)$ subgroup of $SO(4)$, consisting of the elements $(V^\dagger, V)$, correspond to? Verify that these transformations do change t.

4 Run the reality checker on $U(1)$.

Notes

1. For an easy introduction, see, for example, chapters I.5–7 in G Nut.
2. See any textbook on special relativity. For example, chapter III.2 in G Nut.
3. If we were really pedantic, we would write $d\mu(R)$.
4. Some readers would know that this remark plays the central role in differential geometry and in Einstein gravity. See, for example, chapters I.6 and V.2 in G Nut.
5. This example underlies the Aharonov-Bohm effect. See QFT Nut, p. 251.
6. There is of course a vast literature on homotopy, with books and papers. My favorite paper is D. Ravenel and A. Zee, Comm. Math. Phys. (1985), p. 239.
7. For instance, if you know the metric on S^3, you could simply evaluate the determinant of the metric. See, for example, p. 75 of G Nut.
8. For an application of this isomorphism to particle physics, see the celebrated paper by M. Gell-Mann and M. Levy, Nuovo Cimento 16 (1960), p. 45.

IV.8 Symplectic Groups and Their Algebras

I give a brief introduction to symplectic groups and their algebras, for the sake of pretending to a modicum of completeness.[1] We encounter the symplectic algebras again in part VI.

Symplectic groups: $Sp(2n, R)$ and $Sp(2n, C)$

Orthogonal matrices are defined as real matrices R that satisfy $R^T R = R^T I R = I$, where with insufferable pedantry, I've inserted the identity matrix between R^T and R.

Now, consider the $2n$-by-$2n$ "canonical" antisymmetric matrix

$$J = \left(\begin{array}{c|c} 0 & I \\ \hline -I & 0 \end{array} \right) \tag{1}$$

with I the n-by-n identity matrix. Define symplectic[2] matrices R as those $2n$-by-$2n$ real matrices satisfying

$$R^T J R = J \tag{2}$$

The symplectic matrices form a group, known as $Sp(2n, R)$. Check that all the group axioms are satisfied.

A potentially confusing point for some students is merely due to the use of the letter R in (2): physicists are used to associating R with rotation. (Here, the letter R stands for real, not rotation.) The matrix R is manifestly not orthogonal.

Recall that in our discussion (back in chapter I.3) of the orthogonal matrices satisfying $R^T R = I$, taking the determinant of this defining relation gives $\det R = \pm 1$. We then impose the condition $\det R = 1$ to exclude reflections.

Here, in contrast, it turns out that (2) already implies that $\det R = 1$, not just $\det R = \pm 1$. We do not have to impose $\det R = 1$ as a separate condition. To get a feel for how this comes about, note that while the diagonal reflection matrix $r \equiv \text{diag}(-1, +1, \cdots, +1)$ satisfies the orthogonality condition $r^T I r = I$, it most certainly does not satisfy the skew-orthogonality or symplectic condition (2): $r^T J r \neq J$. A simple proof that (2) implies $\det R = 1$ will be given in appendix 1. Meanwhile, you can check this explicitly for a few simple cases. For example, for $n = 1$, plugging $R = \left(\begin{smallmatrix} p & q \\ r & s \end{smallmatrix} \right)$ into (2), you find that[3] $\det R = ps - qr = 1$.

The symplecticity condition (2) imposes (since its left hand side is antisymmetric) $2n(2n - 1)/2$ constraints on the $(2n)^2$ real parameters in R. (For $n = 1$, one condition on the four real parameters p, q, r, and s.) Thus $Sp(2n, R)$ is characterized by $4n^2 - n(2n - 1) = n(2n + 1)$ parameters, namely 3 for $n = 1$, 10 for $n = 2$, 21 for $n = 3$, and so on.

So much for real symplectic matrices. Let us move on to complex symplectic matrices, namely, $2n$-by-$2n$ complex matrices C that satisfy

$$C^T J C = J \tag{3}$$

Clearly, they form a group, known as $Sp(2n, C)$.

The same counting goes through; the condition (3) now imposes $2n(2n - 1)/2$ complex constraints on the $(2n)^2$ complex parameters in C. Hence $Sp(2n, C)$ is characterized by $4n^2 - n(2n - 1) = n(2n + 1)$ complex parameters, that is, $2n(2n + 1)$ real parameters, twice as many as $Sp(2n, R)$.

I already mentioned in an endnote in chapter III.3 that classical mechanics, as manifested in Hamilton's equations, enjoys a symplectic structure. For a problem with n coordinates q_i $(i = 1, \cdots, n)$, package q_i and the conjugate momenta p_i into a $2n$-dimensional vector $Z = \left(\begin{smallmatrix} q \\ p \end{smallmatrix} \right)$ with the index a on Z_a running from 1 to $2n$. Then $\frac{dZ_a}{dt} = J_{ab} \frac{\partial H}{\partial Z_b}$. Several applications of symplectic matrices to physics flow from this observation.[4]

$USp(2n)$

Consider the set of $2n$-by-$2n$ unitary matrices satisfying

$$U^T J U = J \tag{4}$$

Verify that these matrices form a group, known as* $USp(2n)$.

Confusio asks: "It's U^T in (4), not $U^\dagger$? You sure?"

Yes, indeed it is U^T, not $U^\dagger$, another point that has often confused students. While the letter R in (2) does not denote an orthogonal matrix, the letter U in (4), in contrast, does denote a unitary matrix. In other words, in addition to (4), we also impose $U^\dagger = U^{-1}$. Or, to put it more mathematically,

$$USp(2n) = U(2n) \cap Sp(2n, C) \tag{5}$$

Confusio was justified to double check; we are not used to seeing the transpose of a unitary matrix.

---

* Unfortunately, the notation is not unified. Some authors call this group $Sp(2n)$ (and some even call it $Sp(n)$), thus causing considerable confusion.

Note that matrices of the form $e^{i\varphi} I_{2n}$, with I_{2n} the $2n$-by-$2n$ identity matrix, manifestly do not satisfy (4). Thus, $USp(2n)$ is a subgroup of $SU(2n)$, not of* $U(2n) = SU(2n) \otimes U(1)$.

To count the number of real parameters in $USp(2n)$, it is perhaps slightly clearer to go to the Lie algebra by writing $U \simeq I + iH$ with H hermitean. Then (4) becomes $H^T J + J H = 0$, which implies

$$H^T = JHJ \tag{6}$$

since $J^2 = -I$. You should figure out how many conditions this imposes on the $(2n)^2$ real parameters in H.

Perhaps the poor man's way is the most foolproof. Plug the general $2n$-by-$2n$ hermitean matrix $H = \left(\begin{array}{c|c} P & W^\dagger \\ \hline W & Q \end{array} \right)$ with P and Q hermitean and W complex into (6). Obtain $Q = -P^T$ and $W = W^T$. Thus, Q is entirely determined by the hermitean P, which contains n^2 real parameters. Nothing says that P is traceless, but, since $\text{tr } P = -\text{tr } Q$, H is traceless (which, by the way, we already know, since $USp(2n)$ is a subgroup of $SU(2n)$, not of $U(2n)$, as noted earlier). Meanwhile, a symmetric complex matrix W has $n + n(n-1)/2$ complex parameters, that is, $n^2 + n$ real parameters. So, altogether H has $2n^2 + n = n(2n+1)$ real parameters.

The group $USp(2n)$ is characterized by $n(2n+1)$ real parameters (the same as $Sp(2n, R)$), which, as noted earlier, is equal to 3 for $n = 1$, 10 for $n = 2$, 21 for $n = 3$, and so on. It is generated by the $2n$-by-$2n$ hermitean matrix of the form

$$H = \left(\begin{array}{c|c} P & W^* \\ \hline W & -P^T \end{array} \right) \tag{7}$$

with P hermitean and W symmetric complex.

Since there are not that many continuous groups with a relatively small number of generators, we might suspect that, with three generators, $USp(2) \simeq SU(2) \simeq SO(3)$, and with ten generators, $USp(4) \simeq SO(5)$. It turns out that both suspicions are correct.

The first local isomorphism is easy to confirm. The general $SU(2)$ group element $U = e^{i\vec{\theta} \cdot \vec{\sigma}}$ (with σ_1, σ_2, and σ_3 the three Pauli matrices) in fact satisfies (6). The second suspected local isomorphism won't be confirmed until chapter* VI.5. See also appendix 2 to this chapter.

Let us also remark how the peculiar condition (4) can arise in physics. For example, in quantum field theory, we might consider a $2n$-component field[†] $\Psi = \left(\begin{array}{c} \chi \\ \xi \end{array} \right)$, composed of two n-component fields χ and ξ. Then $\Psi^T J \Psi = \chi^T \xi - \xi^T \chi$ is invariant under the unitary transformation $\Psi \to U\Psi$ if $U^T J U = J$.

---

* For simplicity, here we do not write the more correct form $U(2n) = (SU(2n)/Z_{2n}) \otimes U(1)$ that is explained in chapter IV.5.

† We will discuss Majorana fields in chapter VII.5.

Symplectic algebras

It is convenient to use the direct product notation introduced in the review of linear algebra. Then $J = I \otimes i\sigma_2$ (with I the n-by-n identity matrix). Consider the set of $2n$-by-$2n$ hermitean traceless matrices

$$iA \otimes I, \qquad S_1 \otimes \sigma_1, \qquad S_2 \otimes \sigma_2, \qquad S_3 \otimes \sigma_3 \tag{8}$$

Here I denotes the 2-by-2 identity matrix; A an arbitrary real n-by-n antisymmetric matrix; and S_1, S_2, and S_3 three arbitrary real n-by-n symmetric matrices. For example,

$$S_1 \otimes \sigma_1 = \left(\begin{array}{c|c} 0 & S_1 \\ \hline S_1 & 0 \end{array} \right)$$

The claim is that these matrices generate the Lie algebra of $USp(2n)$.

A straightforward way to verify this claim is to show that the H obtained earlier can be written as a linear combination of these four sets of matrices. Check this!

Better yet, we could start fresh, and verify that these four sets of matrices form a Lie algebra under commutation and that they each satisfy (6), namely, $JHJ = H^T$.

First, count to make sure that we have the right number of generators. A real n-by-n antisymmetric matrix has $\frac{1}{2}n(n-1)$ real parameters, and a real n-by-n symmetric matrix has $\frac{1}{2}n(n+1)$ real parameters. In other words, there are $\frac{1}{2}n(n-1)$ linearly independent As, $\frac{1}{2}n(n+1)$ linearly independent S_1s, and so on. The total number of generators is $\frac{1}{2}n(n-1) + \frac{3}{2}n(n+1) = n(2n+1)$, which is just right.

Second, let us show that (8) closes. To begin with, note that

$$[iA \otimes I, S_a \otimes \sigma_a] = i[A, S_a] \otimes \sigma_a = iS_a' \otimes \sigma_a \tag{9}$$

since the commutator of an antisymmetric matrix and a symmetric matrix is a symmetric matrix. Next, note that

$$[S \otimes \sigma_1, S' \otimes \sigma_1] = [S, S'] \otimes I = i(-iA) \otimes I \tag{10}$$

(we suppress irrelevant subscripts), since the commutator of two symmetric matrices is an antisymmetric matrix. Then note that

$$[S_1 \otimes \sigma_1, S_2 \otimes \sigma_2] = iS_1 S_2 \otimes \sigma_3 - (-1)iS_2 S_1 \otimes \sigma_3 = i\{S_1, S_2\} \otimes \sigma_3 = iS_3 \otimes \sigma_3 \tag{11}$$

since the anticommutator of two symmetric matrices is a symmetric matrix. I will let you go through the rest.

Finally, we have to verify $JHJ = H^T$:

$$(I \otimes i\sigma_2)(A \otimes I)(I \otimes i\sigma_2) = (A^T \otimes I) \tag{12}$$

and

$$(I \otimes i\sigma_2)(S_a \otimes \sigma_a)(I \otimes i\sigma_2) = (S_a \otimes \sigma_a^T), \quad a = 1, 2, 3 \tag{13}$$

Indeed, the generators in (8) form the symplectic algebra of $USp(2n)$.

Note that an arbitrary linear combination of the generators $I \otimes iA$ and $\sigma_3 \otimes S$ have the matrix form

$$\left(\begin{array}{c|c} S + iA & 0 \\ \hline 0 & -(S - iA) \end{array} \right) \tag{14}$$

Since $S + iA$ is an arbitrary hermitean matrix, these are precisely the generators of $U(n) \simeq SU(n) \otimes U(1)$ in the reducible $n \oplus n^*$ representation. We have identified* the $U(n)$ sub-algebra of the $USp(2n)$ algebra. It corresponds to $W = 0$ in (7). The matrix (for $S = I$ and $A = 0$)

$$\Sigma \equiv \left(\begin{array}{c|c} I & 0 \\ \hline 0 & -I \end{array} \right) \tag{15}$$

generates the $U(1)$.

Appendix 1: The characteristic polynomial of a symplectic matrix is palindromic

We prove here that a $2n$-by-$2n$ matrix R satisfying the symplectic condition $R^T J R = J$ has determinant equal to $+1$.

First, note that the characteristic polynomial satisfies

$$\begin{aligned} P(z) &\equiv \det(R - zI) = z^{2n} \det(z^{-1}R - I) = z^{2n} \det R \det(z^{-1} - R^{-1}) \\ &= z^{2n} \det R \det(z^{-1} + JR^T J) = z^{2n}(\det J)^2 \det R \det(R^T - z^{-1}) \\ &= z^{2n} \det R \det(R^T - z^{-1}) \\ &= (\det R) z^{2n} P(1/z) \end{aligned} \tag{16}$$

The third equality follows on multiplying (2) by J from the left and by R^{-1} from the right: $R^{-1} = -JR^T J$.

Hence, if $P(\lambda) = 0$, then $P(1/\lambda) = 0$ also. The eigenvalues of R come in pairs $\{\lambda_i, 1/\lambda_i\}$, and hence $\det R = 1$. More involved proofs may be found elsewhere.[5]

Amusingly, it follows that the characteristic polynomial of R, namely $P(z) = \det(R - zI) = a_{2n}z^{2n} + a_{2n-1}z^{2n-1} + \cdots + a_1 z + a_0$ (with $a_{2n} = 1$ of course) is palindromic[6] in the sense that $a_j = a_{2n-j}$ (so that the "sentence" $a_{2n}a_{2n-1}\cdots a_1 a_0$ is a palindrome).[†] Using $\det R = 1$, we see that (16) states that

$$P(z) = z^{2n} P(1/z) = a_{2n} + a_{2n-1}z + \cdots + a_1 z^{2n-1} + a_0 z^{2n} \tag{17}$$

The assertion has been proved. (That $a_0 = a_{2n} = 1$ is of course the same as $\det R = 1$.)

---

* Note: $U(n)$, not $U(2n)$.

† A palindrome is a sentence that reads the same forward and backward. Two classic examples are "Madam, I'm Adam" and "Sex at noon taxes."

Appendix 2: Isomorphs of the symplectic algebras

Physicists tend to be quite sloppy and often call $USp(2n)$ simply $Sp(2n)$, at the risk of confounding $Sp(2n, R)$ and $Sp(2n, C)$. Here I make a few remarks about a couple of the smaller symplectic algebras. As already mentioned in the text, $Sp(2) \simeq SU(2)$ with three generators, and $Sp(4) \simeq SO(5)$ with ten generators.

Referring to (7), we see that for $n = 1$, P is a real number and W is a complex number, and thus $H = \begin{pmatrix} p & c-id \\ \hline c+id & -p \end{pmatrix}$, which is just $c\sigma_1 + d\sigma_2 + p\sigma_3$, which indeed generates $SU(2)$.

The $Sp(4)$ algebra has the 10 generates given by the matrices in (8) for $n = 2$:

$$\tau_2 \otimes I, \quad (I, \tau_1, \tau_3)_i \otimes \sigma_i, \quad i = 1, 2, 3 \tag{18}$$

We denote by I the 2-by-2 identity matrix, and by $(I, \tau_1, \tau_3)_i$ an arbitrary linear combination of I, τ_1, and τ_3 (with the subscript i indicating that the linear combination varies according to different values of i). An easy check by counting: $1 + 3 \cdot 3 = 10$. It is manifestly a subgroup of $SU(4)$ which is generated by $4^2 - 1 = 15$ 4-by-4 hermitean traceless matrices. In chapter VII.1, we will show that $SU(4) \simeq SO(6)$, with the defining vector representation 6 of $SO(6)$ corresponding to the antisymmetric 2-indexed tensor A^{ab}, $a, b = 1, 2, 3, 4$ of $SU(4)$ (with dimension $6 = 4 \cdot 3/2$). Under an element U of $SU(4)$, we have[7] $A^{ab} \to A^{cd} U^{ca} U^{db} = (U^T A U)^{ab}$. Referring back to (4), we conclude that under $Sp(4)$, one component of the 6 of $SO(6)$ is left invariant. Thus, $Sp(4)$ is locally isomorphic to $SO(5)$.

Exercises

1   Show that matrices satisfying (2) form a group.

2   Show that the H in (7) can indeed be written in terms of the matrices in (8).

Notes

1. Here is a story about Howard Georgi, one of the founders of grand unified theory, which we discuss in part IX. In the first edition of his well-known book about Lie algebras, he dismissed the symplectic algebras with some statement along the line that the symplectic algebras had never been relevant to particle physics. Some decades later, a seminar speaker at Harvard (where Georgi was, and still is, a professor) talked about some string theory stuff involving a symplectic algebra. In the last slide, he included a quote from Georgi's book knocking the symplectic algebras, with the intention of mocking Georgi. Howard instantly responded, "To the contrary, your work has resoundingly confirmed my statement!" A physicist who witnessed the exchange laughingly told me that this was filed under "asking for it."

2. Some authors use the rather unattractive terms "skew-orthogonal" and "pseudoorthogonal." I am rather adverse in general to jargon terms with the prefix "pseudo," but less so than to those with the prefix "super," "hyper," "duper," and "superduper."

3. Thus showing that $Sp(2, R) \simeq SL(2, R)$.

4. See R. Littlejohn, Phys. Rept. 138 (1986), pp. 193–291, and the references therein. See also D. Holm, T. Schmah, and C. Stoica, Geometric Mechanics and Symmetry: From Finite to Infinite Dimensions, Oxford University Press. Symplectic integrators are important in such areas as molecular dynamics and celestial mechanics.

5. Other proofs, which I don't find particularly illuminating, may be found in M. Hamermesh, Group Theory and Its Applications to Physical Problems, and Z.-Q. Ma, Group Theory for Physicists.

6. I am grateful to Joshua Feinberg for telling me about this and for many helpful discussions.

7. That the notation and convention used here is somewhat different from what we used before is of no importance.

From the Lagrangian to Quantum Field Theory: It Is but a Skip and a Hop

Field theory: The bare bones minimum

Here I attempt the nearly impossible, to convey the essence of quantum field theory in a few pages. All we need here, and for later use,* is merely the rudiments of field theory. I will try to convey the bare bones minimum in the simplest possible terms. Readers at both ends of the spectrum, those who already know field theory and those who are struggling with quantum mechanics, could readily skip this discussion. On the other hand, readers in the middle could, and should, read this impressionistically. It is of course understood that the discussion here is only a caricature of a serious discussion.

To read the rest of this book, it suffices to have a vague impression of the material presented here. For the convenience of some readers, I provide an executive summary at the end of this chapter.

One step at a time. First, we need to review classical mechanics, but we did that already in chapter III.3. Next, onward to quantum mechanics! As described back in chapter IV.2, Heisenberg promoted q and p to operators satisfying

$$[q, p] = i \tag{1}$$

Dirac then introduced creation and annihilation operators, respectively, $a^\dagger = \frac{1}{\sqrt{2}}(q - ip)$ and $a = \frac{1}{\sqrt{2}}(q + ip)$, which according to (1) satisfy $[a, a^\dagger] = \frac{1}{2}[q + ip, q - ip] = -i[q, p] = 1$. For the harmonic oscillator, the Hamiltonian $H = \frac{1}{2}\omega a^\dagger a$ is proportional to the number operator. In Heisenberg's formulation of quantum mechanics, an operator O evolves in time according to $\frac{dO}{dt} = i[H, O]$. In particular, the position operator has the time dependence

$$q(t) = \frac{1}{\sqrt{2}}(ae^{-i\omega t} + a^\dagger e^{i\omega t}) \tag{2}$$

Note that $q(0) = q$.

---

* In chapters V.4, VII.3, and VII.4, for example.

At this breakneck pace, we are already almost at the door of quantum field theory. I assume that the reader is familiar with the concept of a field, the electric field $\vec{E}(t, \vec{x})$, for example. For our purposes, we can define a field as a dynamical variable that also depends on the spatial coordinates $\vec{x}$.

From discrete to continuous labels

The progression from particle to field[1] may be thought of as a two-step process:

$$q(t) \rightarrow q_a(t) \rightarrow \varphi(t, \vec{x}) \tag{3}$$

First, we replace the single particle by N particles labeled by $a = 1, 2, \cdots, N$. Second, we promote the discrete label a to the continuous label $\vec{x}$ (and move it to live inside the parenthesis). By the way, the switch from q to φ for the dynamical variable is merely a bow to tradition.

The important point is to understand that the spatial coordinate $\vec{x}$ is a mere label, not a dynamical variable. Thus, the notation $\vec{E}(t, \vec{x})$ tells us that it is the time dependence of the electric field at $\vec{x}$ we are focusing on: for each $\vec{x}$, $\vec{E}(t, \vec{x})$ is the dynamical variable, not[2] $\vec{x}$.

In the many-particle case, the Lagrangian would contain terms linking the different particles, such as $\sim -\frac{1}{2}k(q_a - q_b)^2$, due to springs connecting them, for example. After the discrete variable a got promoted to the continuous variable $\vec{x}$, these terms take on the form[3] $\sim (\varphi(\vec{x} + \vec{\varepsilon}) - \varphi(\vec{x}))^2 \sim \varepsilon^i (\frac{\partial \varphi}{\partial x^i})^2$; at an intermediate step, we might think of space as a lattice with a microscopic lattice spacing ε, which in the end we will take to zero, rendering space continuous.

In the intermediate stage, the Lagrangian is given as a sum $L = \sum_a (\cdots)$. In the continuum limit, the sum over a becomes an integral over $\vec{x}$. The Lagrangian ends up being given by an integral over space $L = \int d^3x \mathcal{L}$, with $\mathcal{L}$ known as the Lagrangian density. If you have followed the "poetry" in the preceding paragraphs, you could perhaps see that the Lagrangian density assumes the form

$$\mathcal{L} = \frac{1}{2}\left(\frac{\partial \varphi}{\partial t}\right)^2 - \frac{c^2}{2}\sum_{i=1}^{3}\left(\frac{\partial \varphi}{\partial x^i}\right)^2 - V(\varphi) \tag{4}$$

The first term is just the usual kinetic energy. A certain amount of cleaning up has already been performed. For example, a constant multiplying $\frac{1}{2}(\frac{\partial \varphi}{\partial t})^2$ can be absorbed into the definition of φ. The lattice spacing ε has been similarly absorbed. Here c is needed for the dimension to come out right; hence it has the dimension of length over time, that is, that of a speed. In a relativistic theory, c is the speed of light. Units are usually chosen (with length measured in lightseconds rather than in terms of some English king's foot, for example) so that $c = 1$. In other words, we can absorb c by scaling $\vec{x}$.

It is common practice to abuse language somewhat and to refer to $\mathcal{L}$ as the Lagrangian, dropping the word density. Note that the action $S = \int dt L = \int dt d^3x \mathcal{L} = \int d^4x \mathcal{L}$ may be written as an integral of the Lagrangian (density) $\mathcal{L}$ over 4-dimensional spacetime.

In a word, field theory is what you get if you've got an oscillator at each point $\vec{x}$ in space.[4] The field $\varphi(t, \vec{x})$ describes the amplitude of the oscillation at time t of the oscillator located at $\vec{x}$. The term $(\frac{\partial \varphi}{\partial x^i})^2$ links the oscillators at neighboring points in space.

Packaging the time and space coordinates into spacetime coordinates $x^\mu = (t, x^i)$ with $x^0 = t$ and introducing the compact notation $\partial_\mu \equiv \frac{\partial}{\partial x^\mu} = (\frac{\partial}{\partial t}, \frac{\partial}{\partial x^i})$, we can write the Lagrangian density in (4) as

$$\mathcal{L} = \frac{1}{2}(\partial_\mu \varphi)^2 - V(\varphi) \tag{5}$$

Here we are anticipating the discussion of relativistic physics in chapter VII.2, devoted to the Lorentz group. (Not to strain the reader even further, I would say that the simplest tactic here would be to regard[5] $(\partial_\mu \varphi)^2$ as shorthand for the combination $(\frac{\partial \varphi}{\partial t})^2 - \sum_{i=1}^{3} (\frac{\partial \varphi}{\partial x^i})^2$ in (4).)

Recall the Euler-Lagrange equation

$$\frac{d}{dt}\left(\frac{\delta L}{\delta \frac{dq}{dt}}\right) = \frac{\delta L}{\delta q} \tag{6}$$

from chapter III.3. When we jump from mechanics to field theory, $q(t)$ is replaced by $\varphi(t, \vec{x})$. Thus, I presume that you would not be surprised to learn that the Euler-Lagrange equation is naturally extended to include space derivatives as well as a time derivative, thus replacing the left hand side of (6) by $\partial_\mu(\frac{\delta \mathcal{L}}{\delta \partial_\mu \varphi}) \equiv \frac{\partial}{\partial t}(\frac{\delta \mathcal{L}}{\delta \frac{\partial \varphi}{\partial t}}) + \frac{\partial}{\partial x^i}(\frac{\delta \mathcal{L}}{\delta \frac{\partial \varphi}{\partial x^i}})$. For a field theory, we have the Euler-Lagrange equation

$$\partial_\mu\left(\frac{\delta \mathcal{L}}{\delta \partial_\mu \varphi}\right) = \frac{\delta \mathcal{L}}{\delta \varphi} \tag{7}$$

with the corresponding equation of motion

$$\left(\frac{\partial^2}{\partial t^2} - \frac{\partial^2}{\partial \vec{x}^2}\right)\varphi + V'(\varphi) = 0 \tag{8}$$

In the simplest case $V(\varphi) = \frac{1}{2}m^2\varphi^2$ (which we might think of as the harmonic approximation), then the equation of motion becomes $(\frac{\partial^2}{\partial t^2} - \frac{\partial^2}{\partial \vec{x}^2} + m^2)\varphi = 0$, which is solved by $\varphi(t, \vec{x}) \propto e^{-i(\omega_k t - \vec{k}\cdot\vec{x})}$, with $\omega_k^2 = \vec{k}^2 + m^2$.

To quantize this field theory, we "merely" have to follow Heisenberg's prescription. The momentum conjugate to $\varphi(t, \vec{x})$ is $\pi(t, \vec{x}) \equiv \frac{\delta \mathcal{L}}{\delta \frac{\partial \varphi}{\partial t}} = \frac{\partial \varphi}{\partial t}$. (This is almost an immediate generalization of $p(t) = m\frac{dq}{dt}$.) The end result, not surprisingly, is that the field can again be expanded, generalizing (2), in terms of creation and annihilation operators:

$$\varphi(t, \vec{x}) = \int \frac{d^3k}{\sqrt{(2\pi)^3 2\omega_k}}\left(a(\vec{k})e^{-i(\omega_k t - \vec{k}\cdot\vec{x})} + a(\vec{k})^\dagger e^{i(\omega_k t - \vec{k}\cdot\vec{x})}\right) \tag{9}$$

The assumption that φ is real, or more precisely, hermitean ($\varphi^\dagger = \varphi$), fixes the second term in (9) in terms of the first.

A big difference is that the creation and annihilation operators, respectively, $a(\vec{k})^\dagger$ and $a(\vec{k})$, are now functions of $\vec{k}$. Evidently, the dependence of φ on $\vec{x}$ gets translated into a dependence of $a^\dagger$ and a on $\vec{k}$. Physically, $a(\vec{k})^\dagger$ acting on the vacuum state $|0\rangle$ creates a state containing a particle of momentum $\vec{k}$ and energy $\omega_k = +\sqrt{\vec{k}^2 + m^2}$. (Note that we have set $\hbar = 1$ throughout; the momentum and energy are actually $\hbar\vec{k}$ and $\hbar\omega_k$, respectively.) Incidentally, the normalization factor $\sqrt{(2\pi)^3 2\omega_k}$ in (9) is chosen so that this state is normalized conventionally.[6]

Fields and symmetry groups

Since this is a textbook on group theory rather than quantum field theory, we are less interested in the dynamics of fields, about which volumes could be, and have been, written, but more in how symmetries are incorporated into field theory. We have already talked about rotational invariance (in chapter III.3) and relativistic invariance (in this chapter). In chapter IV.5, we mentioned that Heisenberg, by introducing isospin, opened up a vast vista of internal symmetries, so that physicists have many more symmetries to play with beyond these spacetime symmetries. The exploration of these internal symmetries turned out to be a central theme in subatomic physics.* Here we discuss the group theoretic framework in general and come to specific examples in later chapters.

A symmetry group G is postulated, deduced from experimental observation or theoretical considerations, with various fields furnishing various representations $\mathcal{R}_1, \mathcal{R}_2, \cdots$ of G (and of the Lorentz group of special relativity). The task at hand is then to construct the Lagrangian (density). By definition, that G is a symmetry means that the Lagrangian is invariant under the transformations of G.

Thus, the mathematical problem is to simply construct, out of the $\mathcal{R}$s, an object $\mathcal{L}$ that does not change under G.

Let us hasten from this rather airy yak yak yak to a specific example. Suppose that $G = SU(N)$ (as discussed in chapter IV.4) and that our theory contains two fields,[7] φ_{ij} (transforming as a symmetric tensor with two lower indices) and η^k (belonging to the defining representation). Then $\varphi_{ij}\eta^i\eta^j$ is an invariant term we can include in $\mathcal{L}$, describing the interaction of a particle of the φ type with two particles of the η type. Physically, we can see that the Lagrangian, after we expand the fields φ and η into creation and annihilation operators, contains a term that would annihilate a φ particle and create two η particles. This describes the decay of a φ particle into two η particles.

We now elaborate on these remarks.

Lagrangians with internal symmetries

The simplest example is a $U(1)$ invariant theory containing a field φ belonging to the defining representation: under the group element $e^{i\theta}$, $\varphi \to e^{i\theta}\varphi$. Since $\varphi^\dagger \to e^{-i\theta}\varphi^\dagger$, the hermiticity condition $\varphi = \varphi^\dagger$ cannot be maintained. Necessarily, $\varphi \neq \varphi^\dagger$, and φ is a nonhermitean field, commonly referred to as a complex field.

---

* We discuss these developments in parts V and VIII.

An immediate consequence is that we have to write

$$\varphi(t, \vec{x}) = \int \frac{d^3k}{\sqrt{(2\pi)^3 2\omega_k}} \left(a(\vec{k}) e^{-i(\omega_k t - \vec{k}\cdot\vec{x})} + b(\vec{k})^\dagger e^{i(\omega_k t - \vec{k}\cdot\vec{x})} \right) \tag{10}$$

That $\varphi^\dagger \neq \varphi$ compels us to introduce two sets of creation and annihilation operators, $(a, a^\dagger)$ and $(b, b^\dagger)$. In contrast to (9), the second term here is not related to the first term.

You should verify that the two creation operators, $a^\dagger$ and $b^\dagger$, transform oppositely under the $U(1)$. If we say that the state $a^\dagger(\vec{k})\,|0\rangle$ describes a particle of momentum $\vec{k}$, then the state $b^\dagger(\vec{k})\,|0\rangle$ describes an antiparticle[8] of momentum $\vec{k}$.

Electromagnetism in fact exhibits a $U(1)$ symmetry, as explained in any quantum field theory textbook. The two states just described carry equal but opposite electric charges.

Since $U(1)$ is isomorphic to $SO(2)$, a theory of a complex field can also be written in terms of two real fields, just like a complex number can be written in terms of two real numbers $z = x + iy$. Simply set

$$\varphi = \frac{1}{\sqrt{2}}(\varphi_1 + i\varphi_2) \tag{11}$$

Each of the two real fields, φ_1 and φ_2, can be expanded as in (9), thus confirming the need for two sets of creation and annihilation operators, $(a_1, a_1^\dagger)$ and $(a_2, a_2^\dagger)$.

The important factor of $\frac{1}{\sqrt{2}}$ in (11) can be understood as follows. Call the states $a_1^\dagger\,|0\rangle$ and $a_2^\dagger\,|0\rangle$ created by φ_1 and φ_2, respectively, $|1\rangle$ and $|2\rangle$. Let these states be normalized[9] such that $\langle 1|1\rangle = 1$ and $\langle 2|2\rangle = 1$. If φ as defined in (11) is expanded as in (10), then the creation operator $a^\dagger$ in φ creates the state $|s\rangle = \frac{1}{\sqrt{2}}(|1\rangle + |2\rangle)$. The factor $\frac{1}{\sqrt{2}}$ here comes from the $\frac{1}{\sqrt{2}}$ in (11). The state $|s\rangle$ is then normalized correctly: $\langle s|s\rangle = \frac{1}{2}(\langle 1|1\rangle + \langle 2|2\rangle) = \frac{1}{2}(1 + 1) = 1$.

From this discussion it is but a short leap to field theories symmetric[10] under some group G. The various fields appearing in the Lagrangian $\mathcal{L}$ transform under various irreducible representations of G, and they are to be combined in such a way that $\mathcal{L}$ is invariant. In other words, $\mathcal{L}$ transforms as a singlet under G.

All the known internal symmetry groups in physics are compact. Thus, the irreducible representations that the known fields belong to are unitary. In other words, under the group element g, the field φ transforms like $\varphi \to D(g)\varphi$, with $D(g)^\dagger = D(g)^{-1}$. This means that the mass term $m^2 \varphi^\dagger \varphi$, and the "kinetic" terms $\frac{\partial\varphi^\dagger}{\partial t}\frac{\partial\varphi}{\partial t}$ and $\frac{\partial\varphi^\dagger}{\partial x^i}\frac{\partial\varphi}{\partial x^i}$, are always allowed.[11]

Given the symmetry group G and the collection of irreducible representations realized as fields in the theory, the construction of the invariant Lagrangian is thus a group theoretic problem. We multiply the various irreducible representations together to form a singlet. Concrete examples will soon be given in chapters V.1 and V.4.

An executive summary

Given the lightning speed with which I introduced quantum field theory, it would be good to provide an executive summary here:

1. The coordinates $q(t)$ in mechanics are promoted to a collection of fields $\varphi(t, \vec{x})$.

2. Under an internal symmetry group G, the fields $\varphi(t, \vec{x})$ belong to various irreducible representations and transform accordingly.

3. The Lagrangian density $\mathcal{L}$, constructed out of the fields $\varphi(t, \vec{x})$, must be invariant under G. This requirement poses a group theoretic problem.

Notes

1. See QFT Nut, p. 18.
2. This is why I used q instead of x for the position of the particle in the discussion above; this represents a classic source of confusion for students of quantum field theory. See QFT Nut, p. 19.
3. QFT Nut, p. 17.
4. I heard from Kerson Huang that T. D. Lee told him excitedly that Yang and Mills have invented a field theory with a spinning top at each point $\vec{x}$ in space.
5. More precisely, we could define $\partial^\mu \equiv (\frac{\partial}{\partial t}, -\frac{\partial}{\partial x^i})$ so that $(\partial_\mu \varphi)^2 = \partial^\mu \varphi \partial_\mu \varphi = (\frac{\partial \varphi}{\partial t})^2 - \sum_{i=1}^{3} (\frac{\partial \varphi}{\partial x^i})^2$. I would advise against worrying too much about signs at this stage.
6. These somewhat annoying but necessary normalization factors are explained in every quantum field theory textbook. See, for example, QFT Nut, p. 63.
7. We assume implicitly that these fields transform like a scalar under the Lorentz group. See part VII for more details.
8. For further discussion, see QFT Nut, p. 65.
9. For the sake of simplicity, we gloss over some irrelevant details, such as the fact that the states $|1\rangle$ and $|2\rangle$ depend on momentum $\vec{k}$ and that different states with different $\vec{k}$ are orthogonal. Again, see any book on quantum field theory.
10. For more details, see, for example, QFT Nut, chapter I.10.
11. In chapters VII.2–VII.5 we encounter some interesting exceptions to this discussion.

IV.i1 Multiplying Irreducible Representations of Finite Groups: Return to the Tetrahedral Group

Descent from the rotation group to the tetrahedral group

Now that we know how to multiply irreducible representations together and to Clebsch-Gordan decompose, we can go back and do these things for finite groups.

It is particularly easy if the finite group is naturally a subgroup of a continuous group, about which we know a lot. A good example is given by the tetrahedral group $T = A_4$, discussed in chapter II.2. We learned there that T, a subgroup of $SO(3)$, has four irreducible representations, 1, 1′, 1″, and 3. Let us now work out $3 \otimes 3$ in T.

In $SO(3)$, $3 \otimes 3 = 1 \oplus 3 \oplus 5$. But 5 does not exist in T. Restricting $SO(3)$ to T, we expect the 5 to decompose either as $5 \to 3 \oplus 1 \oplus 1$ or as $5 \to 3 \oplus 1' \oplus 1''$. Note that since 5 is real, if its decomposition contains a complex representation, such as 1′, it must also contain the conjugate of that representation.

It is instructive to work out the decomposition explicitly. Given two vectors $\vec{u}$ and $\vec{v}$ of $SO(3)$, the 3 is of course given by the vector cross product $\vec{u} \times \vec{v}$, while the 5, the symmetric traceless tensor, consists of the three symmetric combinations $u_2 v_3 + u_3 v_2$, $u_3 v_1 + u_1 v_3$, and $u_1 v_2 + u_2 v_1$, together with the two diagonal traceless combinations $2u_1 v_1 - u_2 v_2 - u_3 v_3$ and $u_2 v_2 - u_3 v_3$.

The last two guys are clearly the "odd men out." Referring back to the discussion of $A_4 = T$ in chapter II.3, recall that the Z_3 subgroup of A_4 is generated by cyclic permutation on the irreducible representation 3: for example, $u_1 \to u_2 \to u_3 \to u_1$ (or in the reverse order). Under these cyclic permutations, $u_2 v_3 + u_3 v_2$ transforms into his two friends, thus generating the 3 in the product $3 \otimes 3$, while the two odd guys transform into linear combinations of each other, for example, $u_2 v_2 - u_3 v_3 \to u_3 v_3 - u_1 v_1$. This also means that we cannot have $5 \to 3 \oplus 1 \oplus 1$. Thus, we have established that $5 \to 3 \oplus 1' \oplus 1''$.

Therefore, in $T = A_4$,

$$3 \otimes 3 = 1 \oplus 3 \oplus 3 \oplus 1' \oplus 1'' = 1 \oplus 1' \oplus 1'' \oplus 3 \oplus 3 \tag{1}$$

The second equality merely reflects our desire to write things more "symmetrically."

To see how the two odd guys correspond to $1' \oplus 1''$, recall that in our construction of the character table for A_4 in (II.3.15) the cube root of unity $\omega \equiv e^{i2\pi/3} = -\frac{1}{2} + \frac{\sqrt{3}}{2}i$, the number naturally associated with Z_3, plays a prominent role. Some useful identities are

$$1 + \omega + \omega^2 = 0, \qquad \omega^2 = \omega^* \tag{2}$$

so that

$$\begin{pmatrix} 1 \\ \omega \\ \omega^2 \end{pmatrix} + \begin{pmatrix} 1 \\ \omega^2 \\ \omega \end{pmatrix} = \begin{pmatrix} 2 \\ -1 \\ -1 \end{pmatrix} \tag{3}$$

and

$$\begin{pmatrix} 1 \\ \omega \\ \omega^2 \end{pmatrix} - \begin{pmatrix} 1 \\ \omega^2 \\ \omega \end{pmatrix} = \sqrt{3}i \begin{pmatrix} 0 \\ 1 \\ -1 \end{pmatrix} \tag{4}$$

Thus, the $1'$ and $1''$ can be taken to be

$$1' \sim q' = u_1 v_1 + \omega u_2 v_2 + \omega^2 u_3 v_3 \tag{5}$$

and

$$1'' \sim q'' = u_1 v_1 + \omega^2 u_2 v_2 + \omega u_3 v_3 \tag{6}$$

The combinations $2u_1 v_1 - u_2 v_2 - u_3 v_3$ and $u_2 v_2 - u_3 v_3$ contained in the 5 of $SO(3)$ correspond to $q' + q''$ and $q' - q''$.

It is perhaps worth emphasizing that while $1'$ and $1''$ furnish 1-dimensional representations of A_4, they are not invariant under A_4. For example, under the cyclic permutation c, $q' \to \omega q'$ and $q'' \to \omega^2 q''$. Evidently, $1' \otimes 1'' = 1$, $1' \otimes 1' = 1''$, and $1'' \otimes 1'' = 1'$, and also $(1')^* = 1''$.

Note that the two 3s on the right hand side of $3 \otimes 3 = 1 \oplus 1' \oplus 1'' \oplus 3 \oplus 3$ may be taken to be $(u_2 v_3, u_3 v_1, u_1 v_2)$ and $(u_3 v_2, u_1 v_3, u_2 v_1)$. Recall from exercise I.3.1 that neither of these transforms as a vector under $SO(3)$. Under the rotations restricted to belong to T, however, they do transformation properly, like a 3.

Decomposition of the product using characters

We can also derive the decomposition of $3 \otimes 3$ by using the character table of A_4 given in (II.3.15). As explained in chapter II.1, the characters of the product representation $3 \otimes 3$ are given by the square of the characters of the irreducible representation 3, namely, $\begin{pmatrix} 3^2 \\ (-1)^2 \\ 0^2 \\ 0^2 \end{pmatrix} = \begin{pmatrix} 9 \\ 1 \\ 0 \\ 0 \end{pmatrix}$. We now use two results, derived in (II.2.8) and (II.2.9), which I reproduce here for your convenience, telling us about the number of times n_r the irreducible representation r appears in a given representation:

$$\sum_c n_c \chi^*(c)\chi(c) = N(G) \sum_r n_r^2 \tag{7}$$

and

$$\sum_c n_c \chi^{*(r)}(c)\chi(c) = N(G)n_r \tag{8}$$

Applied here, (7) gives $1 \cdot 9^2 + 3 \cdot 1^2 + 0 + 0 = 81 + 3 = 84 = 12 \cdot 7$ and thus $\sum_r n_r^2 = 7$. The two solutions are $1^2 + 1^2 + 1^2 + 1^2 + 1^2 + 1^2 + 1^2 = 7$ and $1^2 + 1^2 + 1^2 + 2^2 = 7$. The former seems highly unlikely, and so we will root for the latter.

To determine n_r, we simply plug into (8). Orthogonality of the product representation with 1 gives $1 \cdot 1 \cdot 9 + 3 \cdot 1 \cdot 1 + 0 + 0 = 12 \cdot 1$, and thus $3 \otimes 3$ contains 1 once. Similarly, $1'$ and $1''$ each occur once. Finally, orthogonality of the product representation with 3 gives $1 \cdot 3 \cdot 9 + 3 \cdot (-1) \cdot 1 + 0 + 0 = 27 - 3 = 24 = 12 \cdot 2$. Thus, $3 \otimes 3$ contains 3 twice.

We obtain $3 \otimes 3 \to 1 \oplus 1' \oplus 1'' \oplus 3 \oplus 3$, in agreement with our earlier discussion. Indeed,

$$\begin{pmatrix} 9 \\ 1 \\ 0 \\ 0 \end{pmatrix} = \begin{pmatrix} 1 \\ 1 \\ 1 \\ 1 \end{pmatrix} + \begin{pmatrix} 1 \\ 1 \\ \omega \\ \omega^* \end{pmatrix} + \begin{pmatrix} 1 \\ 1 \\ \omega^* \\ \omega \end{pmatrix} + \begin{pmatrix} 3 \\ -1 \\ 0 \\ 0 \end{pmatrix} + \begin{pmatrix} 3 \\ -1 \\ 0 \\ 0 \end{pmatrix} \tag{9}$$

Note that this method is of more direct applicability. In most cases, we may not know a continuous group that contains the given finite group, or, if we know the continuous group, we may not be familiar with its Clebsch-Gordan decomposition rules.

IV.i2 | Crystal Field Splitting

An impurity atom in a lattice

In 1929, Hans Bethe explained the phenomenon of crystal field splitting,[1] merely a few years after the birth of quantum mechanics. An atom typically has many degenerate energy levels due to rotation symmetry, as explained in chapter III.1. For instance, those states with angular momentum j have a degeneracy of $(2j + 1)$, as determined in chapter IV.3. But when this atom is introduced ("doped") as an impurity into a crystal lattice, it suddenly finds itself in a less symmetric environment. The degenerate levels then split, in a pattern determined by group theory and discovered by Bethe.

But now that you have mastered some group theory, this situation should be familiar to you: when a group G is restricted to a subgroup H, an irreducible representation of G in general falls apart into a bunch of irreducible representations of H. Here $G = SO(3)$, and H is whatever symmetry group the lattice respects, typically cubic symmetry. We have already discussed this group theoretic phenomenon several times. Indeed, in interlude IV.i1, we saw precisely an example of this: when $SO(3)$ is restricted to the tetrahedral group T, the 5-dimensional irreducible representation of $SO(3)$ decomposes as $5 \rightarrow 3 \oplus 1' \oplus 1''$.

Evidently, crystal field splitting is of tremendous importance in solid state physics and in material science. But since this is a textbook on group theory rather than on solid state physics, we opt to discuss a much simplified toy model for the sake of pedagogical clarity. To achieve a realistic description,[2] we would have to take into account all sorts of effects and discuss a multitude of finite groups characteristic of various lattice structures,[3] a task that lies far outside the scope of this book.

In fact, since we have already discussed the tetrahedral group T in detail, in chapter II.3 and in interlude IV.i1, we will treat a somewhat artificial example of an idealized atom placed into a tetrahedral cage. Imagine ions placed on the vertices of the tetrahedron, so that the atom is acted on by an electric field that respects tetrahedral symmetry. The beauty of group theory is precisely that we do not have to descend to a nitty-gritty calculation of the electric field, as long as it respects tetrahedral symmetry. In fact, this example highlights the group theoretic techniques you need to tackle more realistic problems.

Using characters to find how irreducible representations break up

Actually, you already have all the tools needed to work out crystal field splitting. Knowing the characters of $SO(3)$ (from chapter IV.7) and the character table of $T = A_4$ (from (II.3.15)), we can practically read off the desired splitting patterns. Again, the key results are (II.2.8) and (II.2.9), which we already used in interlude IV.i1; they determine the number of times n_r that the irreducible representation r appears in a given representation. Before reading on, you might try to determine how the 7-dimensional $j = 3$ irreducible representation of $SO(3)$ decomposes.

Given a $(2j + 1)$-dimensional irreducible representation of $SO(3)$, we need to work out its character for each of the four equivalence classes of $T = A_4$ (see chapter II.3). The character $\chi(j, \psi) = \sum_{m=-j}^{j} e^{im\psi} = \frac{\sin(j+\frac{1}{2})\psi}{\sin\frac{\psi}{2}}$ is a function of j and of ψ. As in chapter II.3, we identify the equivalence class by one of its members.

The identity class is the easiest: $\psi = 0$, and hence its character is equal to $\chi(j, 0) = (2j + 1)$, as always just the dimension of the irreducible representation.

The class containing $(12)(34)$ involves a rotation through π, as shown in chapter II.3, and so has character $\chi(j, \pi) = \frac{\sin(j+\frac{1}{2})\pi}{\sin\frac{\pi}{2}} = (-)^j$.

The other two classes, containing (123) and (132), involve a rotation through $2\pi/3$. To determine their characters, it is easiest to go back to the sum defining $\chi(j, 2\pi/3)$ and recognize the iterative relation $\chi(j, 2\pi/3) = \chi(j - 1, 2\pi/3) + \omega^j + \omega^{*j}$, where $\omega \equiv e^{i2\pi/3}$ is the cube root of unity introduced earlier in various chapters. Since $\omega^j + \omega^{*j} = 2, -1, -1$, for $j = 0, 1, 2 \bmod 3$, respectively, we obtain $\chi(j, 2\pi/3) = 0, -1, 1, 0, -1, 1, \cdots$ for $j = 1, 2, 3, 4, 5, 6, \cdots$.

To see what is going on, perhaps it is clearest to simply extend the character table for $T = A_4$, including, to the right of the second vertical line (marking the limit of the character table proper), one column each for the reducible representation $j = 2, 3, \cdots$.

A_4	n_c		1	$1'$	$1''$	3	5	7	9	11	13	15	
	1	I	1	1	1	3	5	7	9	11	13	15	
Z_2	3	$(12)(34)$	1	1	1	-1	1	-1	1	-1	1	-1	(1)
Z_3	4	(123)	1	ω	ω^*	0	-1	1	0	-1	1	0	
Z_3	4	(132)	1	ω^*	ω	0	-1	1	0	-1	1	0	

I presume that you already worked out how the 7-dimensional $j = 3$ of $SO(3)$ decomposes. We simply plug in (II.2.9):

$$1 \cdot 1 \cdot 7 + 3 \cdot 1 \cdot (-1) + 4(1 \cdot 1 + 1 \cdot 1) = 12 \Longrightarrow n_1 = 1$$

$$1 \cdot 1 \cdot 7 + 3 \cdot 1 \cdot (-1) + 4(\omega \cdot 1 + \omega^* \cdot 1) = 0 \Longrightarrow n_{1'} = n_{1''} = 0$$

$$1 \cdot 3 \cdot 7 + 3 \cdot (-1) \cdot (-1) + 4(0 \cdot 1 + 0 \cdot 1) = 24 \Longrightarrow n_3 = 2 \tag{2}$$

Thus, we find $7 \to 3 \oplus 3 \oplus 1$. A somewhat redundant check is provided by (II.2.9):

$$1 \cdot 7^2 + 3 \cdot (-1)^2 + 4(1^2 + 1^2) = 60 \Longrightarrow \sum_r n_r^2 = 5 \tag{3}$$

which has the solution $2^2 + 1^2 = 5$.

Indeed, as noted in the preceding interlude, we could also have decomposed the characters:

$$\begin{pmatrix} 7 \\ -1 \\ 1 \\ 1 \end{pmatrix} = \begin{pmatrix} 3 \\ -1 \\ 0 \\ 0 \end{pmatrix} + \begin{pmatrix} 3 \\ -1 \\ 0 \\ 0 \end{pmatrix} + \begin{pmatrix} 1 \\ 1 \\ 1 \\ 1 \end{pmatrix} \tag{4}$$

As was explained in chapter II.3, we could also obtain this decomposition by regarding the character table of $T = A_4$ as a 4-by-4 matrix $\mathcal{C}$ and inverting it. The decomposition is then determined by $\mathcal{C}^{-1}$, subject to our knowledge that $1'$ and $1''$ are to occur with equal weight.

The decomposition of $j = 4$ was already implied in interlude IV.i1. We obtain $9 \to 3 \oplus 3 \oplus 1 \oplus 1' \oplus 1''$. By the way, the $1'$ and $1''$ should still be degenerate, since they are complex conjugates of each other. Just to check, $9^2 + 3 = 84 = 7(12)$, and $2^2 + 1^2 + 1^2 + 1^2 = 7$. You might want to amuse yourself by doing some other cases.

Symmetry breaking

The significance of this discussion extends far beyond crystal field splitting in solid state physics. The notion of a symmetry group breaking to a subgroup plays a crucial role in many areas of physics. In part V, we explore how the irreducible representations of Gell-Mann's $SU(3)$ decompose when the symmetry breaks to Heisenberg's $SU(2)$.

Exercises

1   Work out the decomposition of $j = 5$.

2   Work out the decomposition of $j = 6$.

Notes

1. H. Bethe, Ann. Physik (Leipzig) 3 (1929), p. 133.
2. For a much more realistic treatment than that given here, see M. Hamermesh, Group Theory and Its Application to Physical Problems, pp. 337 ff.
3. See M. Tinkham, Group Theory and Quantum Mechanics.

| **Group Theory and Special Functions**

You are probably not surprised to discover that group theory and the special functions you have encountered in physics are intimately connected. We have already seen that spherical harmonics, and by specialization the Legendre polynomials, are connected to $SO(3)$. Here I briefly discuss the Euclidean group $E(2)$ and its connection to Bessel functions.[1] The treatment here is exceedingly brief, because entire books, which can get into frightening details, have been written on the subject.[2] I merely want to give you some flavor of how this connection comes about in one particularly simple case.

The group $E(2)$ is also worth studying as a baby example of the Poincaré group we will encounter in chapter VII.2.

The Euclidean group $E(2)$

After years of walking around, we are quite familiar with $E(2)$, the invariance group of the Euclidean plane. Rotation and translation of 2-dimensional vectors

$$\vec{x} \to \vec{x}' = R\vec{x} + \vec{a} \tag{1}$$

clearly form a group. Following one transformation by another gives $\vec{x} \to R_2(R_1\vec{x} + \vec{a}_1) + \vec{a}_2 = R_2R_1\vec{x} + R_2\vec{a}_1 + \vec{a}_2$. Denote the group elements by $g(R, \vec{a})$ characterized by three real parameters, the angle θ implicit in R and the vector $\vec{a}$. The composition law is then

$$g(R_2, \vec{a}_2)g(R_1, \vec{a}_1) = g(R_2R_1, R_2\vec{a}_1 + \vec{a}_2) \tag{2}$$

To avoid clutter, I will abuse notation slightly and do what physicists usually do, confounding group elements and the operators that represent them.

Let us call the pure translation $T(\vec{a}) = g(I, \vec{a})$ and pure rotation $R = g(R, \vec{0})$ (abuse of notation!). Then

$$g(R, \vec{a}) = T(\vec{a})R \tag{3}$$

An element of $E(2)$ can be written as a rotation followed by a translation. Note that the order is important: in fact $RT(\vec{a}) = g(R, R\vec{a}) \neq T(\vec{a})R = g(R, \vec{a})$.

The elements $T(\vec{a})$ form a subgroup, and so do the elements R. In fact, the translations form, not only a subgroup, but an invariant subgroup, since

$$g(R, \vec{b})^{-1}g(I, \vec{a})g(R, \vec{b}) = R^{-1}T(-\vec{b})T(\vec{a})T(\vec{b})R = R^{-1}T(\vec{a})R = g(R^{-1}, 0)g(R, \vec{a})$$
$$= g(I, R^{-1}\vec{a}) \tag{4}$$

Note that $\vec{b}$ drops out.

The Lie algebra of the plane

Following Lie, we expand around the identity $R \simeq I - i\theta J$ and $T(\vec{a}) \simeq I - i\vec{a} \cdot \vec{P}$, and obtain the Lie algebra

$$[J, P_i] = i\varepsilon_{ij}P_j, \quad i = 1, 2, \quad \text{and} \quad [P_1, P_2] = 0 \tag{5}$$

Then $R = e^{-i\theta J}$ and $T(\vec{a}) = e^{-i\vec{a}\cdot\vec{P}}$.

Define $P_{\pm} = P_1 \pm iP_2$. Then $[J, P_{\pm}] = \pm P_{\pm}$. Note that $P^2 \equiv P_1^2 + P_2^2 = P_-P_+ = P_+P_-$ commutes with both J and $P_{\pm}$. Thus, P^2 and J form a maximal commuting subset of operators, and we can diagonalize both of them. Denote the eigenstates by $|pm\rangle$:

$$P^2 |pm\rangle = p^2 |pm\rangle, \quad \text{and} \quad J |pm\rangle = m |pm\rangle \tag{6}$$

Here $p^2 \geq 0$, since

$$\langle pm| P^2 |pm\rangle = p^2 = \langle pm| P_+^\dagger P_+ |pm\rangle = \langle pm| P_-^\dagger P_- |pm\rangle \tag{7}$$

where we have implicitly normalized $\langle pm|pm\rangle = 1$.

Infinite-dimensional representations of the Euclidean algebra

The case $p = 0$ is trivial, since $|0m\rangle$ does not respond to $\vec{P}$. Translations are inert, so to speak, and we simply recover a representation of $SO(2)$. So take $p > 0$. Acting with J on $P_{\pm} |pm\rangle$ and using $[J, P_{\pm}] = \pm P_{\pm}$, we see that $P_{\pm} |pm\rangle$ is equal to $p |p, m \pm 1\rangle$ up to an undetermined phase factor we choose to be $\mp i$ (so as to agree with the standard definition of the Bessel function that will pop up shortly).

Thus, the Lie algebra is represented by

$$\langle pm'| J |pm\rangle = m\delta_{m'm}, \quad \text{and} \quad \langle pm'| P_{\pm} |pm\rangle = \mp ip\delta_{m',m\pm1} \tag{8}$$

The representation is labeled by the real number $p > 0$ and is infinite dimensional, since $m = 0, \pm 1, \pm 2, \cdots$ ranges over the infinite number of integers.

Infinite-dimensional representations of the Euclidean group

Now that we have represented the Lie algebra, it is a mere matter of exponentiation to represent the Lie group $E(2)$. The element $g(R, \vec{a})$ is represented by the (infinite-dimensional) matrix

$$D^{(p)}(\theta, \vec{a})_{m'm} \equiv \langle pm' | \, g(R, \vec{a}) \, | pm \rangle = \langle pm' | \, T(\vec{a}) R \, | pm \rangle = \langle pm' | \, T(\vec{a}) e^{-i\theta J} \, | pm \rangle$$
$$= \langle pm' | \, T(\vec{a}) \, | pm \rangle \, e^{-im\theta} \tag{9}$$

with θ the angle implicit in R. We can always rotate $\vec{a}$ from a "standard" vector $(a, 0)$ pointing along the x-axis; thus $\langle pm' | \, T(\vec{a}) \, | pm \rangle = e^{i(m-m')\varphi} \langle pm' | \, g(I, \vec{a} = (a, 0)) \, | pm \rangle = e^{i(m-m')\varphi} \langle pm' | \, e^{-iaP_1} \, | pm \rangle$, where a and φ denote the length and angle, respectively, of $\vec{a}$ in polar coordinates. Thus, the real juice in $D^{(p)}(\theta, \vec{a})_{m'm}$ is contained in the factor $\langle pm' | \, e^{-iaP_1} \, | pm \rangle$.

The claim is that

$$\langle pm' | \, e^{-iaP_1} \, | pm \rangle = J_{m-m'}(pa) \tag{10}$$

where J_n denotes the nth Bessel function (of the first kind). Students of physics typically first encounter Bessel functions in problems involving cylindrical symmetry.[3] Let the axis of the cylinder be the z-axis. After the z dependence is factored out, what remains is the symmetry of the 2-dimensional plane perpendicular to the z-axis, described by $E(2)$. So it is no surprise that Bessel functions, whether you love 'em or hate 'em, pop up here.

Since the proof of (10) is straightforward and not particularly illuminating, I simply sketch it here. Write e^{-iaP_1} in (10) as $e^{-ia(P_+ + P_-)/2} = e^{-iaP_+/2} e^{-iaP_-/2}$, since P_+ and P_- commute. Expand the two exponentials separately, so that we have two infinite series, summing over n_+ and n_-. We encounter matrix elements of the form $\langle pm' | \, P_+^{n_+} P_-^{n_-} \, | pm \rangle$. Since, according to (8), when acting on $| pm \rangle$, P_+ raises and P_- lowers the "quantum number" m, respectively, these matrix elements vanish unless $n_+ - n_- = m' - m$. Thus, the two infinite series collapse into a single infinite series, which we recognize, by consulting a handbook, as the series that defines the Bessel function.

Induced representation

There is an alternative approach to representing $E(2)$ known as the method of induced representation. In essence, we choose a different maximal set of commuting operators, namely, P_1 and P_2 (instead of the J and P^2 chosen above), which we diagonalize simultaneously. Denote the eigenstates by $| \vec{p} \rangle = | p, \varphi \rangle$ (where to write the second form we have gone to polar coordinates) such that $P_i \, | \vec{p} \rangle = p_i \, | \vec{p} \rangle$, $i = 1, 2$. These states are by construction eigenstates under translations

$$T(\vec{a}) \, | \vec{p} \rangle = e^{-i\vec{a} \cdot \vec{P}} \, | \vec{p} \rangle = e^{-i\vec{a} \cdot \vec{p}} \, | \vec{p} \rangle \tag{11}$$

but not under rotations. Evidently, $R(\theta) \, | p, \varphi \rangle = | p, \varphi + \theta \rangle$.

As indicated in (10), we are interested in the matrix element $\langle pm'| e^{-i\vec{a}\cdot\vec{P}} |pm\rangle$ of the translation operator.

But in the $|\vec{p}\rangle = |p, \varphi\rangle$ basis, the matrix element of the translation operator, according to (11), is just a simple phase factor $e^{-iap\cos\varphi(\vec{a}, \vec{p})}$, with $\varphi(\vec{a}, \vec{p})$ the angle between $\vec{a}$ and $\vec{p}$. So what we need is the transformation from $|pm\rangle$ to $|p, \varphi\rangle$, but Fourier taught us how to do that. Up to a normalization factor c_m,

$$|pm\rangle = c_m \int_0^{2\pi} \frac{d\varphi}{2\pi} e^{im\varphi} |p, \varphi\rangle \tag{12}$$

The matrix element $\langle pm'| e^{-iaP_1} |pm\rangle$ in (10) thus ends up being given by an integral over an angle. The integral turns out to be, as you would expect, the integral representation of the Bessel function, namely, $J_n(z) = \int_0^{2\pi} \frac{d\varphi}{2\pi} e^{i(n\varphi - z\sin\varphi)}$.

It will probably not surprise you that a study of $E(3)$, the invariance group of 3-dimensional Euclidean space, would lead us to the dreaded spherical Bessel functions.

I trust that this brief discussion gives you the essence of how special functions are connected to group theory.

Exercises

1 Show that the 3-by-3 matrices $D(R, \vec{a}) = \left(\begin{array}{c|c} R & \vec{a} \\ \hline 0 & 1 \end{array}\right)$, with R the corresponding 2-by-2 $SO(2)$ matrix, furnish a 3-dimensional representation of $E(2)$. Note that the representation, while finite dimensional, is manifestly not unitary.

2 Determine c_m in (12).

Notes

1. We follow the treatment of W.-K. Tung, Group Theory in Physics, chapter 9.
2. For example, J. D. Talman, Special Functions, Benjamin, 1968.
3. For example, an electromagnetic wave in a cylindrical wave guide. See any textbook on electromagnetism.

IV.i4 Covering the Tetrahedron

The double tetrahedral group T'

Now that we know that $SU(2)$ double covers $SO(3)$, we naturally wonder whether the tetrahedral group $T = A_4$ can also be covered. Recall from chapter IV.5 that two elements U and $-U$ map into the same rotation R. Thus, to cover T, we simply write down all those elements of $SU(2)$ that map into the rotations corresponding to the elements of T.

We know from chapter II.3 that the 12 elements of T fall into four equivalence classes: $I, \{r_1, r_2, r_3\}, \{c, r_1cr_1, r_2cr_2, r_3cr_3\}$, and $\{a, r_1ar_1, r_2ar_2, r_3ar_3\}$. These are to be understood in the present context as 2-by-2 matrices belonging to $SU(2)$. Given that U and $-U$ map into the same rotation R, we double the number of elements to 24:

$$I, \quad r_1, r_2, r_3, c, r_1cr_1, r_2cr_2, r_3cr_3, a, r_1ar_1, r_2ar_2, r_3ar_3,$$
$$-I, \quad -r_1, -r_2, -r_3, -c, -r_1cr_1, -r_2cr_2, -r_3cr_3, -a, -r_1ar_1, -r_2ar_2, -r_3ar_3 \tag{1}$$

The multiplication of these 24 $SU(2)$ matrices defines T'.

Equivalence classes and irreducible representations

Quick, how many equivalence classes does T' have?

It would be easy to say eight equal to four times two, but that would be wrong.[1]

Flip back to chapter II.3 and look at the 3-by-3 diagonal matrices representing r_1, r_2, and r_3; manifestly $r_1^{-1}r_3r_1 = r_3$, for example. We also learned that the r_is correspond to rotations through angle π around three orthogonal axes (namely, the lines joining the median points of two nonadjoining edges of the tetrahedron, in geometric language). From chapter IV.5, we know that a rotation around $\vec{\varphi}$ through angle $\varphi = |\vec{\varphi}|$ can be described by the $SU(2)$ element $U = e^{i\vec{\varphi}\cdot\vec{\sigma}/2} = \cos\frac{\varphi}{2} I + i\hat{\varphi}\cdot\vec{\sigma}\sin\frac{\varphi}{2}$. Thus, a rotation through angle π around the third axis corresponds to $i\sigma_3$. Now, lo and behold,

$$r_1^{-1}r_3r_1 = (-i\sigma_1)(i\sigma_3)(i\sigma_1) = -i\sigma_3 = -r_3 \tag{2}$$

This shows that r_3 and $-r_3$ belong to the same equivalence class. In other words, the six elements $\{r_1, r_2, r_3, -r_1, -r_2, -r_3\}$ form a single equivalence class.

The covering group T' has only seven, not eight, equivalence classes, and hence only seven, not eight, irreducible representations. The requirement $\sum_r d_r^2 = N(G)$ is then satisfied by $1^2 + 1^2 + 1^2 + 3^2 + 2^2 + 2^2 + 2^2 = 24$. The irreducible representations of T' are labeled by 1, 1′, 1″, 3, 2, 2′, and 2″. You can now work out the character table:

T'	n_c		1	1′	1″	3	2	2′	2″
	1	I	1	1	1	3	2	2	2
Z_2	1	$-I$	1	1	1	3	-2	-2	-2
Z_4	6	$r, -r$	1	1	1	-1	0	0	0
Z_6	4	c	1	ω	ω^*	0	1	ω	ω^*
Z_6	4	a	1	ω^*	ω	0	1	ω^*	ω
Z_6	4	$-c$	1	ω	ω^*	0	-1	$-\omega$	$-\omega^*$
Z_6	4	$-a$	1	ω^*	ω	0	-1	$-\omega^*$	$-\omega$

$$(3)$$

The group T' may be relevant to neutrino physics.[2]

Exercises

1 Determine, in the irreducible representation 2, the character of the class that the cyclic permutation c belongs to.

2 Write down explicitly the 2-by-2 matrix representing c in 2.

3 Work out how the irreducible representations 2, 3, 4, 5, 6, and 7 of $SU(2)$ decompose on restriction to T'.

Notes

1. Indeed, a student whom I asked to construct the character table for T' found the task to be impossible. He had quickly assumed the number of equivalence classes to be eight.
2. M. C. Chen and K. T. Mahanthappa, arXiv: 0904.1721v2 [hep-ph], and references therein. For earlier work, see W. M. Feirbairn, T. Fulton, and W. H. Klink, JMP 5 (1964), p. 1038; K. M. Case, R. Karplus, and C. N. Yang, Phys. Rev. 101 (1956), p. 874. The material here is condensed from Y. Bentov and A. Zee, Int. J. Mod. Phys. A 28 (2013), 1350157.

Part V | Group Theory in the Microscopic World

Understanding the groups $SU(2)$ and $SU(3)$ was crucial in our exploration of elementary particles. In particular, the irreducible representations of $SU(3)$ led to the notion of quarks. Furthermore, in working out the $SU(3)$ algebra, we acquire the skill set we need to deal with Lie algebras in general.

Group theory was needed to work out the implications of $SU(2)$ and $SU(3)$ for experiments.

V.1 | Isospin and the Discovery of a Vast Internal Space

A small number in the subnuclear world

In 1932, James Chadwick discovered one of the most important small numbers in the history of physics, the mass difference between the proton and the neutron in units of the neutron mass:

$$(M_n - M_p)/M_n \simeq (939.6 - 938.3)/939.6 \simeq 0.00138 \tag{1}$$

Almost immediately,[1] in the same year, Werner Heisenberg (and others[2]) proposed that the strong interaction is invariant under a symmetry group, which transforms the proton and the neutron into linear combinations of each other. Since the proton is charged while the neutron is not, the electromagnetic interaction was thought to be responsible for the small mass difference. It was postulated that in a world with electromagnetism turned off, the proton p and neutron n would have equal mass,[3] and they would furnish the spinor representation $N \equiv \begin{pmatrix} p \\ n \end{pmatrix}$ of a symmetry group $SU(2)$ that leaves the strong interaction invariant. Since the electromagnetic interaction is much weaker than the strong interaction, we expect that neglecting electromagnetic effects would give a reasonably approximate description of the real world.

I have often been struck by Nature's kindness toward physicists; it is almost as if we were offered a step-by-step instruction manual. Theoretical physicists did not even have to learn more group theory: they already knew about $SU(2)$ and half-integral spin. Hence the name isospin* for this symmetry. The invariance under the strong interaction is the first example of an approximate symmetry[4] in physics, in contrast to rotation invariance, which, as far as we know, is an exact symmetry.

---

* Replacing the antiquated term "isotopic spin," and even more so, "isobaric spin."

Prior to isospin, the symmetries of physics (translation invariance, rotation invariance, Lorentz invariance, and so on) were confined to the spacetime we live and love in. Heisenberg's enormous insight[*] led to the discovery of a vast internal space, the ongoing exploration of which has been a central theme of fundamental physics for close to a hundred years now, as was already alluded to in chapter IV.9.

The pions and the electric charge

The charged pions, π^+ and π^-, were discovered in 1947 after much experimental confusion.[5] Yukawa, with remarkable insight, had postulated in 1935 that the exchange of a then-unknown meson[†] could generate the strong interaction, much like the exchange of the photon generates the electromagnetic interaction. Another example of Nature's kindness to theoretical physicists! Or, you might say that Nature knows only so many readily solvable problems, sort of like a professor making up an exam.

The power of the symmetry argument is that, since the π^+ and π^- interact with the nucleons, they are immediately required to also transform under isospin. Much as charged particles could emit and absorb photons, nucleons could emit and absorb pions via the following fundamental processes: $p \rightarrow n + \pi^+, n \rightarrow p + \pi^-$, and so on.

The initial states in these two processes shown have isospin $I = \frac{1}{2}$. Denote the unknown isospin of the pions by I_π. Then, according to the group theoretic result in chapter IV.3, the final states have isospin $\frac{1}{2} \otimes I_\pi = (I_\pi + \frac{1}{2}) \oplus |I_\pi - \frac{1}{2}|$. For isospin to be a symmetry of the strong interaction, this must contain the $I = \frac{1}{2}$ representation of the initial state; hence I_π can only be either 0 or 1. But $I_\pi = 0$ cannot accommodate both π^+ and π^-, and so we conclude that $I_\pi = 1$, a 3-dimensional irreducible representation.

Isospin thus predicts that π^+ and π^- must have an electrically neutral partner, the π^0. The predicted neutral pion was soon discovered[6] in 1950. Another triumph (small by today's standards perhaps, but nevertheless a triumph) for group theory.

How does the electric charge operator Q fit into the Lie algebra of $SU(2)$?

Isospin is supposed to be exact in an ideal world without electromagnetism. It is important to realize that, even in that world, it is meaningful and legitimate to ask how the electric-charge operator Q transforms under isospin. The following simple observation provides the answer.

Observe that the difference in charge of the proton and of the neutron is given by $\Delta Q = Q_p - Q_n = 1 - 0 = \frac{1}{2} - (-\frac{1}{2}) = I_{3,p} - I_{3,n} = \Delta I_3$, from which we deduce[‡]

$$Q = I_3 + \frac{1}{2}Y \tag{2}$$

---

[*] For a more accurate historical account, see Fearful, pp. 333–334. In particular, Heisenberg's original proposal did not involve $SU(2)$ at all. What I present here is known as textbook pseudohistory. See also the endnote about the Matthew principle.

[†] For the story of how the π meson was so named, based on a multilingual pun involving Chinese and Greek, see Fearful, pp. 168–169 and 335.

[‡] This is a version of the Gell-Mann Nishijima formula.

where Y, known as hypercharge, is an operator lying outside $SU(2)$. Evidently, $Y = 1$ for nucleons and $Y = 0$ for pions.[7] Note that the formula $Q = I_3$ for the pions indicates that the missing $I_3 = 0$ partner of the $\pi^{\pm}$ indeed has $Q = 0$.

Scattering cross sections and isospin

The predictions of isospin have been verified by an almost countless number of measurements. A particularly elegant, and simple, prediction[8] concerns deuteron production in nucleon-nucleon collision, involving the two processes

(A) $p + p \rightarrow d + \pi^+$

and

(B) $p + n \rightarrow d + \pi^0$

The deuteron, as you may know, is a bound state of the proton and the neutron.

We can now put our vast knowledge of group theory to good use. The first step is to establish the isospin of the deuteron. Since $\frac{1}{2} \otimes \frac{1}{2} = 1 \oplus 0$, the deuteron, as a p-n bound state, has isospin equal to either 1 or 0. But if the deuteron's isospin is 1, then it would be a member of a triplet (since $2 \cdot 1 + 1 = 3$) with nearly equal mass. Applying the isospin raising and lowering operators, we see that the other two members are a p-p bound state and an n-n bound state. Since neither of these were seen,* we deduce that the deuteron has isospin 0.

Thus, the final state in both process (A) and process (B), containing a deuteron and a pion, has isospin $0 \otimes 1 = 1$.

On the other hand, as far as isospin is concerned, the initial state is, in process (A), given by $\left| I = \frac{1}{2}, I_3 = \frac{1}{2} \right\rangle \otimes \left| I = \frac{1}{2}, I_3 = \frac{1}{2} \right\rangle = \left| \frac{1}{2}, \frac{1}{2}; \frac{1}{2}, \frac{1}{2} \right\rangle = \left| \frac{1}{2}, \frac{1}{2} \right\rangle$ (to use the notation of chapter IV.3), and in process (B), it is given by $\left| \frac{1}{2}, -\frac{1}{2} \right\rangle$.

But we know about Clebsch-Gordan decomposition! We had obtained in chapter IV.3 that

$$|1, 1\rangle = \left| \frac{1}{2}, \frac{1}{2} \right\rangle$$

$$|1, 0\rangle = \frac{1}{\sqrt{2}} \left(\left| -\frac{1}{2}, \frac{1}{2} \right\rangle + \left| \frac{1}{2}, -\frac{1}{2} \right\rangle \right)$$

$$|0, 0\rangle = \frac{1}{\sqrt{2}} \left(\left| -\frac{1}{2}, \frac{1}{2} \right\rangle - \left| \frac{1}{2}, -\frac{1}{2} \right\rangle \right) \tag{3}$$

The initial proton-proton state in process (A), $|p, p\rangle = \left| \frac{1}{2}, \frac{1}{2} \right\rangle = |1, 1\rangle$, is an $I = 1$ state. In contrast, to obtain the initial proton-neutron state in process (B), $|p, n\rangle = \left| \frac{1}{2}, -\frac{1}{2} \right\rangle$, we have

---

* Without isospin, we would have been hard put to use the absence of the p-p bound state to predict the absence of the n-n bound state. We might have argued that two protons repel electrically, but then it would be puzzling that two neutrons do not bind, given that a proton and a neutron do bind.

to invert (3) using high school algebra. We obtain $|p, n\rangle = \left|\frac{1}{2}, -\frac{1}{2}\right\rangle = \frac{1}{\sqrt{2}}(|1, 0\rangle - |0, 0\rangle)$, a mixture of an $I = 1$ state and an $I = 0$ state. But since the final state is $I = 1$, the $I = 0$ component of the initial state cannot contribute because of isospin conservation. Thus, compared to the amplitude for $p + p \to d + \pi^+$, the amplitude for $p + n \to d + \pi^0$ is down by a factor $\frac{1}{\sqrt{2}}$.

More formally, we can introduce a transition operator $\mathcal{T}$, which takes the initial state to the final state. Then $\langle d\pi^+ | \mathcal{T} | pp \rangle = \langle 1, 1 | \mathcal{T} | 1, 1\rangle$, while

$$\left\langle d\pi^0 \right| \mathcal{T} \left| pn \right\rangle = \langle 1, 0 | \mathcal{T} \left(\frac{1}{\sqrt{2}}(|1, 0\rangle - |0, 0\rangle) \right) = \frac{1}{\sqrt{2}} \langle 1, 0 | \mathcal{T} |1, 0\rangle \tag{4}$$

Since in quantum mechanics, the cross section is determined by the absolute square of the amplitude, we obtain

$$\frac{\sigma(p + p \to d + \pi^+)}{\sigma(p + n \to d + \pi^0)} = 2 \tag{5}$$

The point is that although almost nothing was known about the strong interaction when the prediction was made, isospin allows us to fix the ratio of two cross sections, regardless of how complicated the actual details of the dynamics might be. Knowing that isospin is an approximate symmetry, we might expect this to hold to an accuracy of a few percent. When the experiments were done (in 1951 and 1953), sure enough, the ratio of the two measured cross sections comes out to be about 2 within the error bars.

I have marveled elsewhere[9] about the almost magical power of theoretical physics. Picture the heroic efforts involved in building the necessary accelerator, all the wires, the vacuum pumps, and what not; and designing and setting up the detectors. In the end, the hardworking experimentalists measured the two cross sections, and bam! They confirmed that Nature knew about group theory.

It is worth emphasizing that had Heisenberg merely postulated a Z_2 symmetry under which $p \leftrightarrow n$, we would not get the prediction above. That the strong interaction is invariant under an $SU(2)$ is by now an established fact, and all known strongly interacting particles, known as hadrons, have been classified into $SU(2)$ irreducible representations. To give just one example, the $\pi^+ + p \to \pi^+ + p$ scattering cross section shows an enormous bump[10] when the center of mass energy reaches about 1238 MeV. This resonance phenomenon is interpreted as the formation of $\pi^+ p$ into a short-lived particle known as N^{*++} with mass $\simeq 1238$ MeV, which then quickly decays back into $\pi^+ + p$.

Since the initial and final states contain a pion and nucleon, and since $1 \otimes \frac{1}{2} = \frac{3}{2} \oplus \frac{1}{2}$, the total isospin is either $I = \frac{3}{2}$ or $I = \frac{1}{2}$. But in the initial and final states, we have $I_3 = 1 + \frac{1}{2} = \frac{3}{2}$, and hence I can only be $I = \frac{3}{2}$, not $I = \frac{1}{2}$. We conclude that the particle N^{*++} has isospin $I = \frac{3}{2}$, and is thus a member of a quadruplet (since $2 \cdot \frac{3}{2} + 1 = 4$). Sure enough, three other particles, N^{*+}, N^{*0}, and N^{*-} (with the indicated charges and with masses approximately equal to that of N^{*++}), were eventually found.

A particularly clean prediction is that at resonance, the ratio of the total cross section for $\pi^+ + p$ and the total cross section for $\pi^- + p$ should be equal to 3. You should be able to work this out.

Nucleon nucleon scattering and Feynman diagrams

Historically, measuring and understanding the coupling strength of the three pions to the proton and to the neutron was of great importance in elucidating the strong interaction. Group theoretically, this simply corresponds to picking out the $\frac{1}{2}$ on the right hand side of the multiplication $1 \otimes \frac{1}{2} = \frac{1}{2} \oplus \frac{3}{2}$: we want "pion times nucleon = nucleon," so to speak. Thus, the coupling strengths are proportional to a bunch of Clebsch-Gordan coefficients.

In fact, you already worked out the relevant coefficients in exercise IV.3.5. Group theory does not care whether you are combining the spin angular momentum $S = \frac{1}{2}$ of an electron with orbital angular momentum $L = 1$ or coupling the isospin $I = \frac{1}{2}$ of the nucleon with the isospin $I = 1$ of the pion. Simply plug in the result of the exercise.

$$\left| \frac{1}{2}, m \right\rangle = -\sqrt{\frac{\frac{3}{2} - m}{3}} \left| m - \frac{1}{2}, \frac{1}{2} \right\rangle + \sqrt{\frac{\frac{3}{2} + m}{3}} \left| m + \frac{1}{2}, -\frac{1}{2} \right\rangle \tag{6}$$

to obtain

$$\left| \frac{1}{2}, \frac{1}{2} \right\rangle = -\sqrt{\frac{1}{3}} \left| 0, \frac{1}{2} \right\rangle + \sqrt{\frac{2}{3}} \left| 1, -\frac{1}{2} \right\rangle$$

$$\left| \frac{1}{2}, -\frac{1}{2} \right\rangle = -\sqrt{\frac{2}{3}} \left| -1, \frac{1}{2} \right\rangle + \sqrt{\frac{1}{3}} \left| 0, -\frac{1}{2} \right\rangle \tag{7}$$

Now change the names to protect the innocent and write (7) as

$$|p\rangle \sim -\sqrt{\frac{1}{3}} \left| \pi^0, p \right\rangle + \sqrt{\frac{2}{3}} \left| \pi^+, n \right\rangle$$

$$|n\rangle \sim -\sqrt{\frac{2}{3}} \left| \pi^-, p \right\rangle + \sqrt{\frac{1}{3}} \left| \pi^0, n \right\rangle \tag{8}$$

Of course, this does not mean that the proton and neutron are simply equal to whatever appear on the right hand sides of (8); even Confusio would not think that. The strong interaction is far too strong and complicated for that. Rather, under isospin, the state $|p\rangle$ transforms like a linear combination of $|\pi^0, p\rangle$ and $|\pi^+, n\rangle$. Group theory fixes the coefficients, and tells us, for example, that the amplitude of $|p\rangle$ to be in the quantum state $|\pi^+, n\rangle$ is larger than the amplitude of $|p\rangle$ to be in the quantum state $|\pi^0, p\rangle$ by a factor $(-\sqrt{2})$.

This implies that the relative strength of the various couplings of a pion to a nucleon can be read off from (8):

$$g_{p, \pi^0 p} = g, \qquad g_{p, \pi^+ n} = -\sqrt{2}g, \qquad g_{n, \pi^- p} = \sqrt{2}g, \qquad g_{n, \pi^0 n} = -g \tag{9}$$

Here g is a coupling strength characteristic of the strong interaction, which we are not able to calculate from first principles to this very day. We have chosen the sign of g for later convenience. The notation is such that $g_{p, \pi^+ n}$, for example, is proportional to the probability amplitude of the process $\pi^+ + n \to p$, in which a neutron n absorbs a π^+ to become a proton p. By a basic tenet of quantum field theory, this is the same as the amplitude for $p \to \pi^+ + n$, in which p becomes n by emitting a π^+. (Furthermore, this

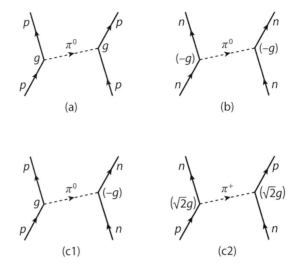

Figure 1

is the same as the amplitude for $n \to \pi^- + p$ and for $\pi^- + p \to n$. In other words, we could remove a particle from one side of a process and put it back on the other side as an antiparticle, for example, turning $p \to \pi^+ + n$ into $\pi^- + p \to n$. See any textbook on quantum field theory.)

The relative signs and factors of $\sqrt{2}$ sure look strange in (9), but in fact are entirely physical. There are three nucleon-nucleon scattering processes that experimentalists can measure, at least in principle, namely, $p + p \to p + p$, $n + n \to n + n$, and $p + n \to p + n$. According to Yukawa, the scattering is caused by the exchange of a pion between the two nucleons, as shown in the little diagrams (surely you've heard of Feynman diagrams!) in Figure 1. For our purposes here, you could think of the diagrams as showing what's happening in spacetime, with the time axis running upward.

For example, in the diagram in (c2), the proton becomes a neutron by emitting a π^+, which is then absorbed by the neutron it is scattering with, turning it into a proton. In other words, we have $p + n \to n + \pi^+ + n \to n + p$. According to the discussion above, the emission and the absorption are both characterized by the coupling strength $g_{p,\pi^+ n} = -\sqrt{2}g$. The various relevant coupling strengths are shown in these Feynman diagrams.

Now comes the fun. Let us write the relative strength of the quantum amplitude for each of these scattering processes. For $p + p \to p + p$, we have $(-g)^2 = g^2$. For $n + n \to n + n$, we have g^2. So far so good: proton-proton scattering and neutron-neutron scattering are the same, as we would expect in an isospin-invariant world. Next, for $p + n \to p + n$, we have from the π^0 exchange in (c1) $g(-g) = -g^2$. Oops, this is not equal to g^2. But let's not forget the contribution due to exchange of the charged pion shown in (c2): this gives $(\sqrt{2}g)(\sqrt{2}g) = 2g^2$. And so we have $-g^2 + 2g^2 = g^2$. Indeed, the three scattering processes come out to be the same. Once again, math works.

Pion nucleon coupling and the tensor approach

It is instructive also to use tensors and spinors to work out the relative strengths of the pion nucleon couplings.

Heisenberg started this story by proposing that the nucleons $N = \begin{pmatrix} p \\ n \end{pmatrix}$ transform like the 2-dimensional $I = \frac{1}{2}$ defining representation N^i of $SU(2)$, written here as a column spinor. It transforms as $N^i \to U^i{}_j N^j$. Recall that the hermitean conjugate of N^i is denoted by* N_i, which can be written as a row spinor[†] $\bar{N} = (\bar{p} \quad \bar{n})$. It transforms as $\bar{N}_i \to \bar{N}_j (U^\dagger)^j{}_i$, so that $N_i N^i$ is invariant. As explained in detail in chapter IV.4, upper indices transform with U, lower indices with $U^\dagger$.

We also learned in chapter IV.4 that the $I = 1$ vector representation of $SU(2)$, which happens to be also the adjoint representation, can be written as a traceless tensor $\phi^i{}_j$, which can also be viewed as a 2-by-2 hermitean matrix. (This is an example of what I alluded to before; the simpler groups are often more confusing to beginners!) In other words, we can package the three pions π as a traceless 2-by-2 hermitean matrix:

$$\phi = \vec{\pi} \cdot \vec{\tau} = \pi_1 \tau_1 + \pi_2 \tau_2 + \pi_3 \tau_3 = \begin{pmatrix} \pi_3 & \pi_1 - i\pi_2 \\ \pi_1 + i\pi_2 & -\pi_3 \end{pmatrix} = \begin{pmatrix} \pi^0 & \sqrt{2}\pi^+ \\ \sqrt{2}\pi^- & -\pi^0 \end{pmatrix} \tag{10}$$

Note that we may pass freely between the Cartesian basis (π_1, π_2, π_3) and the circular basis (π^+, π^0, π^-). The appearance of $\sqrt{2}$ here was explained for (IV.9.11). The pion fields transform according to

$$\phi^i{}_j \to \phi'^i{}_j = U^i{}_l \phi^l{}_n (U^\dagger)^n{}_j \tag{11}$$

As far as group theory is concerned, the representations that the pions and the nucleons belong to under isospin determine their interaction with one another; the interaction term in the Lagrangian can only be, precisely as we learned in chapter IV.4,

$$\mathcal{L} = \cdots + f N_i \phi^i{}_j N^j \tag{12}$$

with f some unknown coupling constant.[‡] The important point is that we do not have to know the details of the strong interaction: group theory rules!

Let us then write out the isospin invariant $N_i \phi^i{}_j N^j$, using (10), as

$$N_i \phi^i{}_j N^j = (\bar{p} \quad \bar{n}) \, \vec{\pi} \cdot \vec{\tau} \begin{pmatrix} p \\ n \end{pmatrix} = (\bar{p}\pi^0 p - \bar{n}\pi^0 n) + \sqrt{2}(\bar{p}\pi^+ n + \bar{n}\pi^- p) \tag{13}$$

---

* Indeed, I mentioned this as far back as the review of linear algebra, and in more detail in chapter IV.4.

[†] The hermitean conjugate is written as a bar here for reasons that do not concern the reader, and which will be explained in chapter VII.3.

[‡] Which we cannot calculate analytically from first principles to this very day.

It is gratifying to see that the relative strength of $\sqrt{2}$ of the coupling of the charged pions to the neutral pion emerges here also, in agreement* with (9).

What we have done here is known as the effective field theory approach.[11] You, the astute reader, will realize that the essential point is that group theory does not care whether the objects being transformed are states or fields or whatever.

Appendix: The Fermi-Yang model and Wigner's $SU(4)$

In the now-obsolete Fermi-Yang model, pions are regarded as bound states of a nucleon and an antinucleon. Regardless of whether this is a useful way to understand the existence of the pions, the group theoretic classification should be valid to the extent that isospin is a fairly good symmetry of the strong interaction. Since the two nucleons furnish the $\frac{1}{2}$ defining representation of $SU(2)$, and similarly the two antinucleons, the nucleon-antinucleon bound states should transform as $\frac{1}{2} \otimes \frac{1}{2} = 1 \oplus 0$. Thus, in addition to the pions, which correspond to the isospin 1 triplet, there might also be an isospin 0 singlet. This particle is commonly called the σ.

In nuclear physics, the interaction between nucleons is, to leading approximation, spin independent. This led Wigner to propose that the isospin symmetry and spin symmetry of nuclear forces, that is, the group $SU(2) \otimes SU(2)$, be extended to $SU(4) \supset SU(2) \otimes SU(2)$. Thus, the four states, consisting of the proton and of the neutron, with spin up and down, furnish the defining representation 4 of $SU(4)$. Explicitly, we write $4 \rightarrow (2, 2)$, indicating that on restricting $SU(4)$ to $SU(2) \otimes SU(2)$, the 4 transforms like a doublet under each of the two $SU(2)$s. The four corresponding antinucleon states evidently furnish the conjugate representation $\bar{4}$.

Now put what you have learned in chapter IV.4 to good use. Consider the 16 bound states of a nucleon and an antinucleon. Under $SU(4)$, they transform like $4 \otimes \bar{4} = 15 \oplus 1$. To identify these purported cousins of the pions and the σ, we resort to our usual tactics and decompose the representations of $SU(4)$ into the irreducible representations of $SU(2) \otimes SU(2)$:

$$4 \otimes \bar{4} \rightarrow (2, 2) \otimes (2, 2) = (3 \oplus 1, 3 \oplus 1)$$
$$= (3, 3) \oplus (3, 1) \oplus (1, 3) \oplus (1, 1) \tag{14}$$

The (3, 1) and the (1, 1) states have spin 0 and correspond to the pions and the σ. The other $9 + 1 = 10$ states describe spin 1 particles, the so-called vector mesons, with an isospin triplet (3, 3) named the ρ and an isospin singlet (1,3) named the ω.

Note the power of group theory! We do not have to understand the detailed strongly interacting dynamics that produces mesons (and indeed, arguably, we do not fully understand the relevant physics to this day). As long as Wigner's proposal makes sense, then regardless, a simple group theoretic calculation yields concrete predictions. In the glory days of particle physics, experimentalists would rush out and, lo and behold, would find these particles, such as the $\rho^{\pm,0}$ and the ω.

Exercises

1   The ^{3}H nucleus (a pnn bound state) and the ^{3}He nucleus (a ppn bound state) are known to form an isospin $\frac{1}{2}$ doublet. Use isospin to predict the ratio of the cross sections for $p + d \rightarrow {}^3$H $+ \pi^+$ and $p + d \rightarrow {}^3$He $+ \pi^0$.

2   Using the Clebsch-Gordan coefficients you found in exercise IV.3.5, work out the theoretical prediction that $\frac{\sigma(\pi^+ + p)}{\sigma(\pi^- + p)} = 3$ at center-of-mass energy $\simeq 1238$ MeV. Hint: The total scattering cross section $\pi^- + p$ should include both the processes $\pi^- + p \rightarrow \pi^- + p$ and $\pi^- + p \rightarrow \pi^0 + n$.

* Some readers may have noticed that the relative sign between $g_{p, \pi^+ n}$ and $g_{n, \pi^- p}$ is missing here. This has to do with the choice of phase factors mentioned in chapter IV.2.

Notes

1. Everything happened so much faster in those days. Chadwick's Nobel Prize came a mere three years later.
2. The Matthew principle, coined by the sociologist R. K. Merton, operates in full force in theoretical physics. For a few examples, see G Nut, pp. 169, 376.
3. The actual story is slightly more complicated. See A. Zee, Phys. Rept., 3C (127) (1972). We now understand that even without electromagnetism, the proton and neutron would not have equal mass, because the up quark and down quark have different masses. Historically, however, Heisenberg was motivated by the near equality of M_p and M_n.
4. I should have said an approximate internal symmetry; Galilean invariance is an example of an approximate spacetime symmetry, for phenomena slow compared to the speed of light. But that was in hindsight; before special relativity, Galilean invariance was thought to be exact.
5. In particular, the pion was confounded with the muon.
6. The reason the π^0 was discovered after $\pi^\pm$ is that it has a fast decay mode $\pi^0 \to \gamma + \gamma$ into two photons, which were much more difficult to detect than charged particles at that time.
7. For a long time, it was not understood why there was a shift between Q and I_3. See any particle physics textbook for the eventual resolution. See also chapter V.2.
8. C. N. Yang, unpublished. See R. H. Hildebrand, Phys. Rev. 89 (1953) p. 1090.
9. QFT Nut, p. 139.
10. S. Gasiorowicz, Elementary Particle Physics, p. 293.
11. Pioneered in the 1960s by Murray Gell-Mann and Maurice Lévy, Feza Gürsey, and many others.

The Eightfold Way of $SU(3)$

While most of our colleagues were put off by the unfamiliar math, [Sidney Coleman and I] became traveling disciples of the Eightfold Way.

—S. L. Glashow[1]

$SU(3)$ in particle physics

Laugh out loud; the math[2] that put off the leading particle theorists in the early 1960s was just the group $SU(3)$, something that students these days are expected to breeze through lickety split. Indeed, you the reader have already breezed through $SU(N)$ in chapter IV.4.

While the fame and glory of the group $SU(2)$ permeate quantum physics (and parts of classical physics as well), the group $SU(3)$ essentially appears only* in particle physics.[3] But what fame and glory it enjoys there! It appears twice, originally as the internal symmetry group that led to the discovery of quarks, and then, later, as the gauge group of quantum chromodynamics.

We will work out the representations of $SU(3)$ by following two lines of attack: first, in this chapter, by using tensors, and second, in chapter V.3, by generalizing the ladder approach used for $SU(2)$.

We now need two floors

Moving from the house of $SU(2)$ to the house of $SU(3)$, we find ourselves now living on two floors.

For $SU(2)$ tensors, recall that we can remove all lower indices using ε^{ij}. In particular, ψ_i is equivalent to ψ^i. This no longer holds for $SU(3)$. The reason is that, for $SU(3)$, in contrast to $SU(2)$, the antisymmetric symbols ε^{ijk} and ε_{ijk} carry three indices. Thus, the irreducible representations furnished by ψ^i and ψ_i are not equivalent to each other: if we try the same trick that worked for us for $SU(2)$ and contract ψ_i with ε^{ijk}, the resulting tensor

---

* See the appendix, however.

$\psi^{jk} \equiv \varepsilon^{ijk}\psi_i$ transforms like a tensor with two upper indices, not just one. As explained in chapter IV.1, the contracted index i has disappeared.

The irreducible representations furnished by ψ^i and ψ_i are commonly denoted by 3 and 3*. They are not equivalent. (Physically, as we will see, the quark corresponds to 3, while the antiquark corresponds to 3*.) Thus, for $SU(3)$, in contrast to $SU(2)$, we have to include tensors with both upper and lower indices.

However, in contrast to $SU(N)$ for $N > 3$, $SU(3)$ does have an important simplifying feature: it suffices to consider only traceless tensors $\varphi^{i_1 i_2 \cdots i_m}_{j_1 j_2 \cdots j_n}$ with all upper indices symmetrized and all lower indices symmetrized. We prove this claim presently. Thus, the representations of $SU(3)$ are uniquely labeled by two integers (m, n), where m and n denote the number of upper and lower indices, respectively.

Recall that for $SO(N)$, it suffices to consider only tensors* with all upper indices symmetrized. To show this, we used an iterative approach. Essentially, the same approach works here. The only catch is that with two integers (m, n) instead of one, we have to move on a 2-dimensional lattice, so to speak. We iterate in $p \equiv m + n$; thus, starting at the lattice point $(m, n) = (0, 0)$, we move on to $(1, 0)$ and $(0, 1)$; then on to $(2, 0)$, $(1, 1)$, and $(0, 2)$; and so on and so forth, increasing p by 1 at each step.

To save verbiage, I assume from the outset that the reader, when presented with a tensor symmetric in its upper indices and symmetric in its lower indices, knows how to render it traceless by subtracting out its trace. For example, given T^{ij}_{kl}, define[4] $\tilde{T}^{ij}_{kl} \equiv T^{ij}_{kl} - \frac{1}{5}(\delta^i_k T^j_l + \delta^j_k T^i_l + \delta^i_l T^j_k + \delta^j_l T^i_k) + \frac{1}{20}(\delta^i_k T^j_l + \delta^i_k \delta^j_l + \delta^i_l \delta^j_k)T$, where $T^j_l \equiv T^{ij}_{il}$ and $T \equiv T^j_j$.

We now prove our claim that, for $SU(3)$, it suffices to consider only traceless tensors with all upper indices symmetrized and all lower indices symmetrized.

The claim holds trivially for the tensors characterized by $(0, 0)$, $(1, 0)$, and $(0, 1)$.

Now onward to $p = 2$. First up is the tensor φ^{ij} with two upper indices. As in chapter IV.1, we break it up into the symmetric combination $\varphi^{\{ij\}} \equiv (\varphi^{ij} + \varphi^{ji})$ and the antisymmetric combination $\varphi^{[ij]} \equiv (\varphi^{ij} - \varphi^{ji})$ (using the notation $\{ij\}$ and $[ij]$ introduced earlier). We can get rid of the antisymmetric combination by writing $\varphi_k \equiv \varepsilon_{ijk}\varphi^{[ij]}$; in other words, the antisymmetric tensor with two upper indices transforms just like a tensor with one lower index, namely, $(0, 1)$, which we have dealt with already. The remaining symmetric tensor $\varphi^{\{ij\}}$, now called $(2, 0)$, satisfies our claim. Similarly, we can take care of $(0, 2)$ by using ε^{ijk}.

Note that, while we have to deal with both upper and lower indices, we also have two ε symbols, which we can use to trade two upper antisymmetric indices for one lower index, and two lower antisymmetric indices for one upper index, respectively.

Finally, we come to the tensor with one upper index and one lower index. If it is not traceless, we simply subtract out its trace and obtain $(1, 1)$. We are done with $p = 2$.

Moving from $p = 2$ to $p = 3$, we first encounter the tensor φ^{ijk}. The analysis is the same as that we gave for $SO(3)$ in chapter IV.1. We break it up into the symmetric combination $\varphi^{\{ij\}k} \equiv (\varphi^{ijk} + \varphi^{jik})$ and the antisymmetric combination $\varphi^{[ij]k} \equiv (\varphi^{ijk} -$

---

* Recall also that for $SO(N)$, tensors do not have lower indices.

φ^{jik}). Once again, we don't care about the antisymmetric combination $\varphi^{[ij]k}$, because it transforms just like a tensor $\varphi^k_l \equiv \varepsilon_{ijl}\varphi^{[ij]k}$ with one upper index and one lower index, namely, $(1, 1)$, which we have already disposed of. As for the symmetric combination $\varphi^{\{ij\}k}$, add and subtract its cyclic permutations and write, as in chapter IV.1, $3\varphi^{\{ij\}k} = (\varphi^{\{ij\}k} + \varphi^{\{jk\}i} + \varphi^{\{ki\}j}) + (\varphi^{\{ij\}k} - \varphi^{\{jk\}i}) + (\varphi^{\{ij\}k} - \varphi^{\{ki\}j})$. Note that the expression in the first set of parentheses is completely symmetric in all three indices, while the expressions in the other two sets of parentheses are antisymmetric in ki and kj, respectively. We can thus multiply the latter two expressions by ε_{kil} and ε_{kjl}, respectively, and turn them into 2-indexed tensors, which we have already taken care of.

Speaking colloquially, 123 is 123, and we don't much care if it is the 123 of $SO(3)$ or of $SU(3)$.

You can carry the ball from here on. For instance, you will next encounter φ^{ij}_k. Its two upper indices can be symmetrized and antisymmetrized. The antisymmetric piece contains $(0, 2)$ and $(1, 0)$.

So, the crucial observation, as noted above, is that we can always trade two antisymmetric indices for one index on a different floor, thus reducing $p = m + n$ by 1.

The dimensions of $SU(3)$ tensors

Let us now compute the dimension of the $SU(3)$ tensor $\varphi^{i_1i_2\cdots i_m}_{j_1j_2\cdots j_n}$ labeled by (m, n). Once again, you should quickly review what we did for $SO(3)$. At the risk of repetition, note that our fingers don't care whether we are talking about $SO(3)$ or $SU(3)$: we have "merely" a counting problem here. The number of distinct configurations of m indices symmetrized, according to a result in chapter IV.1, is* $\frac{1}{2}(m + 1)(m + 2)$. So at this stage, our tensor has $\frac{1}{4}(m + 1)(m + 2)(n + 1)(n + 2)$ components.

You remember, of course, that we have yet to impose the traceless condition $\delta^j_i \varphi^{ii_2\cdots i_m}_{jj_2\cdots j_n} = 0$, and here comes the crucial difference between $SO(3)$ and $SU(3)$: an upper index and a lower index are contracted, rather than two upper indices. The left hand side of this condition transforms like a tensor labeled by $(m - 1, n - 1)$; in other words, $\frac{1}{4}m(m + 1)n(n + 1)$ linear combinations of the tensor $\varphi^{i_1i_2\cdots i_m}_{j_1j_2\cdots j_n}$ are to be set to 0. We conclude that the dimension of the irreducible representation (m, n) of $SU(3)$ is equal to

$$\mathcal{D}(m, n) = \frac{1}{4}(m + 1)(m + 2)(n + 1)(n + 2) - \frac{1}{4}m(m + 1)n(n + 1)$$

$$= \frac{1}{2}(m + 1)(n + 1)(m + n + 2) \tag{1}$$

Keep in mind that the tensor furnishing (m, n) is symmetric in its m upper indices and in its n lower indices.

---

* Our notation has changed from j to m; I am trusting that you are not a divine.

Let us list the dimension of a few low-lying irreducible representations:

$(1, 0)$	$\frac{1}{2} \cdot 2 \cdot 1 \cdot 3 = 3$
$(1, 1)$	$\frac{1}{2} \cdot 2 \cdot 3 \cdot 4 = 8$
$(2, 0)$	$\frac{1}{2} \cdot 3 \cdot 1 \cdot 4 = 6$
$(3, 0)$	$\frac{1}{2} \cdot 4 \cdot 1 \cdot 5 = 10$
$(2, 1)$	$\frac{1}{2} \cdot 3 \cdot 2 \cdot 5 = 15$
$(2, 2)$	$\frac{1}{2} \cdot 3 \cdot 3 \cdot 6 = 27$

The numbers 3, 8, and 10 all have celebrity status in particle physics, and 27 is involved in an interesting Feynman story (see chapter V.4). We have already mentioned 10 in connection with Gell-Mann's stunning prediction of the hitherto unknown Ω^- back in chapter IV.4.

Multiplication of $SU(3)$ irreducible representations

You're almost an old hand by now at multiplying irreducible representations together using the tensor approach. In fact, the most important cases for physics are the simplest.

Consider, for example, $3 \otimes 3^*$. The product of ψ^i and χ_j, namely $\psi^i \chi_j$, transforms like the tensor T^i_j. We render it traceless by subtracting out its trace, which transforms like the trivial representation. We write this important result using two different notations:

$$(1, 0) \otimes (0, 1) = (1, 1) \oplus (0, 0) \tag{2}$$

and

$$3 \otimes 3^* = 8 \oplus 1 \tag{3}$$

To write the second form, we used (1). We need hardly mention that nobody is missing $(3 \cdot 3 = 9 = 8 + 1)$.

The next easiest example is

$$(1, 0) \otimes (1, 0) = (2, 0) \oplus (0, 1) \tag{4}$$

which, according to (1), corresponds to

$$3 \otimes 3 = 6 \oplus 3^* \tag{5}$$

(Now $3 \cdot 3 = 9 = 6 + 3$.)

How about $3 \otimes 6$, that is, $(1, 0) \otimes (2, 0)$? Try to figure it out before reading on.

Write out the tensors ψ^i and φ^{jk} (with $\varphi^{jk} = \varphi^{kj}$) explicitly. The product $T^{ijk} = \psi^i \varphi^{jk}$ is a tensor with three indices upstairs, symmetric in jk but not totally symmetric. So symmetrize in ij; extract the antisymmetric part by contracting with the antisymmetric symbol: $\varepsilon_{mij} \psi^i \varphi^{jk} \equiv \zeta^k_m$. Show that ζ^k_m is traceless, that is, $\zeta^k_k = 0$, and thus it corresponds to

$(1, 1) = 8$. We remove these eight components from T^{ijk}. What is left is a totally symmetric tensor S^{ijk} with three indices, namely, $(3, 0) = 10$.

We thus obtain

$$(1, 0) \otimes (2, 0) = (3, 0) \oplus (1, 1) \tag{6}$$

and

$$3 \otimes 6 = 10 \oplus 8 \tag{7}$$

Putting (5), (3), and (7) together, we have

$$3 \otimes 3 \otimes 3 = (6 \oplus 3^*) \otimes 3 = (6 \otimes 3) \oplus (3^* \otimes 3) = 10 \oplus 8 \oplus 8 \oplus 1 \tag{8}$$

$(3^3 = 27 = 10 + 8 + 8 + 1.)$ As we shall see, this is a highly celebrated result in particle physics.

Our next example, which played an important role historically, is more involved, but it will give us a hint on how to do the general case: $8 \otimes 8 = ?$

Let us go slow and be as explicit as possible. Multiply ψ_j^i and χ_l^k together to form $\tilde{T}_{jl}^{ik} = \psi_j^i \chi_l^k$. Take out the two traces $P_j^k = \psi_j^i \chi_i^k$ and $Q_l^i = \psi_j^i \chi_l^j$. Render these two tensors traceless by taking out the trace $P_j^j = Q_i^i$. Thus far, we have gotten $8 \oplus 8 \oplus 1$ on the right hand side of $8 \otimes 8 = ?$. (The reader should note that, as remarked in an earlier chapter, the letters $\tilde{T}$, ψ, χ, P, and Q are just coatracks on which to hang indices; we are "wasting" letters, so to speak, in the service of pedagogy, even though they come for free.) Next, subtract out all these traces from $\tilde{T}_{jl}^{ik}$ and call the resulting traceless tensor T_{jl}^{ik}.

Antisymmetrize in the upper indices i and k: $A_{jl}^{ik} = T_{jl}^{[ik]}$. You might think that we would have to symmetrize and antisymmetrize in the lower indices j and l next. But thanks to an identity involving the ε symbols, it turns out that, without further massaging, A_{jl}^{ik} is already symmetric in its two lower indices j and l.

To prove this, we first define $B_{mjl} \equiv \varepsilon_{ikm} A_{jl}^{ik}$. The claim is that B_{mjl} is symmetric in j and l. We verify the claim by showing that $B_{mjl} \varepsilon^{jln} = \varepsilon_{ikm} \varepsilon^{jln} A_{jl}^{ik} = 0$, since by construction A_{jl}^{ik} is traceless. (You will show this in exercise 2.) Furthermore, $B_{mjl} \varepsilon^{mjp} = 0$ for a similar reason. Thus, B_{mjl} is totally symmetric and furnishes the irreducible representation $(0, 3) = 10^*$.

Going through similar steps with $T_{[jl]}^{ik}$ gives us the irreducible representation $(3, 0) = 10$.

After taking out $T_{jl}^{[ik]}$ and $T_{[jl]}^{ik}$ from T_{jl}^{ik}, what is left is then symmetric in its upper indices and in its lower indices, and hence furnishes the irreducible representation $(2, 2)$, which from (1) we learn is the 27.

Thus, we obtain

$$(1, 1) \otimes (1, 1) = (2, 2) \oplus (3, 0) \oplus (0, 3) \oplus (1, 1) \oplus (1, 1) \oplus (0, 0) \tag{9}$$

and

$$8 \otimes 8 = 27 \oplus 10 \oplus 10^* \oplus 8 \oplus 8 \oplus 1 \tag{10}$$

Again, we accounted for everybody,[5] of course: $8 \cdot 8 = 64 = 27 + 10 + 10 + 8 + 8 + 1$.

Multiplication rule for $SU(3)$

How do we multiply (m, n) and (m', n')?

After all the examples I went through, I am joking only slightly if I say that the most difficult part of obtaining the general multiplication rule for $SU(3)$ is to invent a suitable notation for the beasts we encounter on our way to the answer. If the following gets hard to follow, simply refer back to the $(1, 1) \otimes (1, 1)$ example we just did.

The product of (m, n) and (m', n') is a tensor with $m + m'$ upper indices and $n + n'$ lower indices.

Let us denote by $(m, n; m', n')$ the traceless tensor with $m + m'$ upper indices, symmetric in the first m indices and symmetric in the second m' indices, and with $n + n'$ lower indices, symmetric in the first n indices and symmetric in the second n' indices. In other words, $(m, n; m', n')$ denotes what remains of the product of (m, n) and (m', n') after all traces have been taken out.

Note that $(m, n; m', n')$, since it is defined to be traceless, is certainly not $(m, n) \otimes (m', n')$, which has no reason to be traceless. Let us record the process of taking out traces:

$$
\begin{aligned}
(m, n) \otimes (m', n') = {}& (m, n; m', n') \\
& \oplus (m - 1, n; m', n' - 1) \oplus (m, n - 1; m' - 1, n') \\
& \oplus (m - 1, n - 1; m' - 1, n' - 1) \oplus (m - 2, n; m', n' - 2) \\
& \oplus (m - 2, n - 1; m' - 1, n' - 2) \\
& \oplus \cdots \|
\end{aligned}
\tag{11}
$$

Note that (11) is merely saying that we first obtain $(m - 1, n; m', n' - 1)$ from $(m, n) \otimes (m', n')$ by contracting an upper index from the first m upper indices with a lower index from the second n' lower indices, and then removing all traces. Since (m, n) is already traceless, there is no point in contracting an upper index from the first m upper indices with a lower index from the first n lower indices. And since (m, n) and (m', n') are both respectively symmetric in their upper indices and in their lower indices, it does not matter which upper index and which lower index we choose to contract. By the way, all this is harder to say than to understand.

Next, we obtain $(m, n - 1; m' - 1, n')$ by contracting an upper index from the second m' upper indices with a lower index from the first n' lower indices, and then removing all traces. We keep repeating this process. The notation $\cdots \|$ in (11) indicates that the process stops when we run out of indices to contract.

After doing all this, we end up with a bunch of reducible representations furnished by objects $(m - p, n - q; m' - q, n' - p)$. Note the word "reducible." As in our $8 \otimes 8$ example, we next take an index from the $(m - p)$ upper indices and an index from the $(m' - q)$ upper indices, and antisymmetrize. The claim, as in the explicit example, is that the result is already symmetric in the lower indices. Similarly, we can exchange the words "upper" and "lower" in the preceding paragraph.[6] It is important to emphasize that $(m, n; m', n')$ (and $(m - p, n - q; m' - q, n' - p)$ in particular) is neither symmetric in all its upper indices nor symmetric in all its lower indices, and hence is certainly not $(m + m', n + n')$. Thus, $(m, n; m', n')$ furnishes a reducible representation, not an irreducible one.

Let us illustrate this procedure with our $8 \otimes 8$ example:

$$(1, 1) \otimes (1, 1) = (1, 1; 1, 1) \oplus (0, 1; 1, 0) \oplus (1, 0; 0, 1) \oplus (0, 0; 0, 0) \tag{12}$$

followed by

$$
\begin{aligned}
(1, 1; 1, 1) &= (2, 2) \oplus (3, 0) \oplus (0, 3) \\
(0, 1; 1, 0) &= (1, 1) \\
(1, 0; 0, 1) &= (1, 1) \\
(0, 0; 0, 0) &= (0, 0)
\end{aligned}
\tag{13}
$$

Putting (12) and (13) together, we recover (9).

From the Λ to the Eightfold Way

At this point, for pedagogical clarity, it would be best to interject some particle physics, which I recount in as abridged and cartoonish a form as possible. The Λ baryon[7] was discovered in 1950 by a team of Australians.[8] Its mass, $\simeq 1115$ MeV, is not far from the mass of the proton and the neutron, $\simeq 939$ MeV, and its other properties (for instance, it also has[9] spin $\frac{1}{2}$) are also similar to that of the proton p and the neutron n. This naturally led S. Sakata in 1956 to generalize Heisenberg's $SU(2)$ to $SU(3)$, proposing that p, n, and Λ furnish the fundamental or defining representation of $SU(3)$.

But Nature was unkind to Sakata. Soon, other baryons (collectively known as hyperons, but never mind) were discovered: Σ^+, Σ^0, and Σ^-, with masses around 1190 MeV, and Ξ^- and Ξ^0, with masses around 1320 MeV. They form an isospin triplet and an isospin doublet, respectively. Also, four K mesons, denoted by K^+, K^0 and $\bar{K}^0$, K^-, with properties similar to the three pions, were discovered.[10] One troubling feature was that their masses, around 495 MeV, were quite different from the pion masses, around 138 MeV.

That there were seven pseudoscalar spin 0 mesons (pseudoscalar because they were odd under parity, that is, under spatial reflection) plus the fact that some eminent theorists presented a series of arguments that the Λ and Σ^0 had opposite parities and hence could not belong together, sent a number of theorists on a search[11] for a symmetry group with a 7-dimensional irreducible representation.

But eventually another pseudoscalar spin 0 meson, the η with mass around 550 MeV, was discovered. Furthermore, better measurements established that the Λ and Σ^0 do in fact have the same parity, namely, the parity of the proton and the neutron.

Gell-Mann,[12] and independently Ne'eman, then proposed that the eight spin 0 mesons[13] and the eight spin $\frac{1}{2}$ baryons furnish the 8-dimensional adjoint representation of $SU(3)$, namely $(1, 1)$ in the tensor notation.

As explained in chapter IV.4, Gell-Mann then made the striking prediction that the nine known baryon resonances (recall the N^{*++} from that chapter) belong to a 10-dimensional irreducible representation (namely, $(3, 0)$) of $SU(3)$. The missing particle, named the

Ω^-, was soon found. Taking his inspiration from Buddhism,[14] Gell-Mann referred to his scheme as the Eightfold Way in honor of the 8-dimensional adjoint representation.

If $SU(2)$ was regarded as an approximate symmetry, then $SU(3)$ is a badly broken symmetry. We would expect the predictions to hold to an accuracy of at most $\sim$20–30%, as measured by, say, the fractional mass difference[15] between the Ξs and the nucleons. We will discuss in chapter V.4 what group theory has to say about baryon and meson masses.

Quarks and triality

You might have noticed that the irreducible representations (m, n) actually observed experimentally, namely, the octet $8 = (1, 1)$ and the decuplet $10 = (3, 0)$, have $m - n = 0 \bmod 3$. It was soon understood that the irreducible representations of $SU(3)$ may be classified according to their triality, which we define to be equal to $(m - n) \bmod 3$.

Recall that, way back in exercise 1 in chapter I.1, we defined the center of a group as the set of all elements that commute with all the other elements. Also, we have already discussed the center of $SU(N)$. Still, it is worthwhile to discuss here the physical meaning of the center of $SU(3)$.

What is the center of $SU(3)$? Consider the set of all unitary 3-by-3 matrices with unit determinant. To commute with all these matrices, a matrix has to be proportional to the identity matrix and to have unit determinant. Thus, the proportionality constant has to be a multiple of $e^{2\pi i/3}$. The center of $SU(3)$ is the group Z_3, consisting of the three elements I, z, and z^2, where

$$z \equiv \begin{pmatrix} e^{2\pi i/3} & 0 & 0 \\ 0 & e^{2\pi i/3} & 0 \\ 0 & 0 & e^{2\pi i/3} \end{pmatrix} = e^{2\pi i/3} \begin{pmatrix} 1 & 0 & 0 \\ 0 & 1 & 0 \\ 0 & 0 & 1 \end{pmatrix} \tag{14}$$

(I wrote this out in two different forms for emphasis.) Acting on the fundamental representation 3, this "triality matrix" z multiplies the representation by the phase $e^{2\pi i/3}$. Acting on the 3*, z multiplies it by the phase $e^{-2\pi i/3}$. More generally, acting on the tensor (m, n), z multiplies it by $e^{2\pi i/3}$ for each upper index and by $e^{-2\pi i/3}$ for each lower index, that is, by $e^{2\pi i(m-n)/3}$. This explains the origin of triality.

Historically, the mysterious fact that the experimentally observed irreducible representations are precisely those with triality 0 provided another important clue to the underlying theory of the strong interaction.

Indeed, at this point, you might well ask,* "Where is the fundamental representation 3?" Gell-Mann soon worked out the properties of the particles furnishing the 3 and named them quarks,† specifically the up quark u, the down quark d, and the strange quark s. The conjugate representation 3* is then furnished by the antiquarks, $\bar{u}$, $\bar{d}$, and $\bar{s}$.

---

* The story goes that Gell-Mann was asked precisely this question during a lunch at Columbia University.
† G. Zweig, at that time Gell-Mann's graduate student, independently proposed the notion of quarks, but unfortunately he named them aces, a name that fortunately did not catch on.

Figure 1 With the Wizard of Quarks

Group theory tells us, very nicely, how the observed strongly interacting particles are built up from quarks and antiquarks. More specifically, each upper index in the tensor furnishing a given irreducible representation corresponds to a quark and each lower index to an antiquark. For example, $3 \otimes 3^*$ contains the 8 according to (3), and the observed octet of pseudoscalar mesons are now known to be bound states of a quark and an antiquark: $\bar{u}u, \bar{d}u, \bar{u}s$, and so on. And $3 \otimes 3 \otimes 3$ also contains the 8 according to (8), and so the protons and neutrons and their cousins are bound states of 3 quarks, uud, udd, uus, and so on. Interestingly, since $3 \otimes 3 \otimes 3$ also contains the 10 according to (8), the celebrated Ω^- is also a 3-quark bound state, namely, sss.

You see that triality is just the number of quarks minus the number of antiquarks mod 3.

Experimentalists have searched for quarks in vain. It is now believed that quarks, and more generally, particles with nonzero triality, are confined; that is, they cannot be produced experimentally. A major challenge for the theory of the strong interaction is to prove confinement.[16]

You shall know the whole by its parts: The analog of crystal field splitting

You learned way back in chapter II.1 that an irreducible representation r of a group G will, in general (when we restrict ourselves to a subgroup H of G), break up into a bunch of irreducible representations of H. As explained there, this makes total sense, because H has fewer transformations than G does. Fewer transformations are available to transform the members of the irreducible representation r into linear combinations of one another, and so these guys can segregate themselves into different cliques.

In everyday life, you might discover how something is put together by taking it apart. Historically, one way that particle physicists became acquainted with $SU(3)$ was to ask how the irreducible representations of $SU(3)$ break up upon restriction of $SU(3)$ to $SU(2)$. By the late 1950s, particle physicists were well acquainted with Heisenberg's isospin and hence with the group $SU(2)$.

Mathematically, the problem of breaking up the irreducible representations of $SU(3)$ is almost trivial, at least in hindsight, using the method of tensors: we simply divide the index set $i = \{1, 2, 3\}$ into two sets $\{1, 2\}$ and $\{3\}$. Introduce indices $a, b, \cdots$ that are allowed to range over $\{1, 2\}$.

Given a tensor, replace the index i by either a or 3. This is more easily explained with the aid of a few examples than with a bunch of words. Consider the fundamental representation 3 of $SU(2)$. We separate ψ^i, $i = 1, 2, 3$ into ψ^a, $a = 1, 2$, and ψ^3. The $SU(2)$ subgroup of $SU(3)$ acts on the index a but leaves the index 3 alone. This is a long-winded way of stating the obvious: upon restriction[17] of $SU(3)$ to $SU(2)$, the irreducible representation 3 of $SU(3)$ decomposes as

$$3 \to 2 \oplus 1 \tag{15}$$

Physically, $SU(3)$ transforms the three quarks, u, d, and s, into linear combinations of one another, while its subgroup $SU(2)$, namely, Heisenberg's isospin, transforms u and d but leaves s alone. Under isospin, the up and down quarks form a doublet, while the strange quark is a singlet.

We can convey more information by going to $SU(2) \otimes U(1)$, the maximal subgroup of $SU(3)$. Here $U(1)$ denotes the group consisting of the $SU(3)$ elements $e^{i\theta Y}$ with the hypercharge matrix

$$Y = \frac{1}{3} \begin{pmatrix} 1 & 0 & 0 \\ 0 & 1 & 0 \\ 0 & 0 & -2 \end{pmatrix} \tag{16}$$

Note that Y is hermitean and traceless, so that $e^{i\theta Y}$ is unitary and has unit determinant and hence belongs to $SU(3)$. In other words, the up and down quarks have hypercharge $\frac{1}{3}$, while the strange quark has hypercharge $-\frac{2}{3}$. Thus, upon restriction of $SU(3)$ to $SU(2) \otimes U(1)$, we have, instead of (15), the more informative $3 \to (2, 1) \oplus (1, -2)$. Here, in $(2, 1)$, for example, the first number indicates the dimension of the isospin representation, while the second number is $3Y$. (I put in the 3 to avoid writing $\frac{1}{3}$ constantly.) The compact notation I_{3Y}, with the hypercharge (multiplied by 3) written as a subscript to the isospin, is often more convenient. For example, we write the decomposition just given as

$$3 \to 2_1 \oplus 1_{-2} \tag{17}$$

The decomposition of the fundamental or defining representation specifies how the subgroup H is embedded in G. Since all representations may be built up as products of the fundamental representation, once we know how the fundamental representation decomposes, we know how all representations decompose.

Conjugating (17), we have $3^* \to 2_{-1} + 1_2$. Notice that it is not necessary to write 2^* (and certainly not necessary to write 1^*). Given these two basic facts, we can then work out how a higher-dimensional irreducible representation of $SU(3)$ decomposes by multiplying representations together.

For example, the multiplication law $3 \otimes 3^* = 8 \oplus 1$ in $SU(3)$ becomes in $SU(2)$ (using (15))

$$
\begin{aligned}
3 \otimes 3^* &= 8 \oplus 1 \\
&\rightarrow (2 \oplus 1) \otimes (2 \oplus 1) \\
&= (2 \otimes 2) \oplus (2 \otimes 1) \oplus (1 \otimes 2) \oplus (1 \otimes 1) \\
&= 3 \oplus 1 \oplus 2 \oplus 2 \oplus 1
\end{aligned}
\tag{18}
$$

Thus, we learn that

$$
8 \rightarrow 3 \oplus 1 \oplus 2 \oplus 2
\tag{19}
$$

Group theory tells us that the octet of mesons should consist of an isospin triplet (the π^+, π^0, π^-), two isospin doublets (K^+, K^0 and $\bar{K}^0$, K^-), and an isospin singlet (the η), exactly as was observed experimentally. Similarly, the octet of baryons consists of an isospin triplet (the Σ^+, Σ^0, Σ^-), two isospin doublets (our beloved proton and neutron, and Ξ^-, Ξ^0), and an isospin singlet (the Λ that we started our discussion with). Note that the decomposition (19) is a property of the irreducible representation 8, regardless of whether we thought of it as being contained in $3 \otimes 3^*$ or in $3 \otimes 3 \otimes 3$.

We can readily include hypercharge. Using (17) and its conjugate $3^* \rightarrow 2_{-1} \oplus 1_2$, and noting that hypercharges simply add (since the group is simply $U(1)$), we obtain a more informative version of (19):

$$
8 \rightarrow 3_0 \oplus 1_0 \oplus 2_3 \oplus 2_{-3}
\tag{20}
$$

(This follows because the first two terms, for example, come from $2_1 \otimes 2_{-1}$.) Recall that the subscript is actually $3Y$. Thus, in the octet 8, the isospin triplet has $Y = 0$; the two isospin doublets $Y = \pm 1$, respectively; and the isosinglet $Y = 0$.

Alternatively, we can simply inspect the tensors involved. For example, the adjoint 8 decomposes like (on splitting the indices i and j into a, b, and 3, as explained earlier)

$$
\varphi^i_j = \{\bar{\varphi}^a_b, \varphi^a_3, \varphi^3_a, \varphi^3_3\}
\tag{21}
$$

where the bar on $\bar{\varphi}^a_b$ reminds us that it is traceless. This corresponds precisely to (19).

In chapter IX.2, we will decompose various representations of $SU(5)$. Everything we do there will simply be somewhat more elaborate versions of what we have done here.

Dear reader, you might realize that what is being done here is the analog of crystal field splitting (interlude IV.i2) for particle physics.

The electric charge

Theorists may think of isospin and hypercharge, but experimentalists identify particles by their electric charges. Back in chapter V.1, we saw that the proton has $I_3 = \frac{1}{2}$ but electric charge $Q = 1$, while the π^+ has both I_3 and Q equal to 1. In the early 1950s, physicists proposed the empirical formula $Q = I_3 + \frac{Y}{2}$, but they were deeply puzzled by this "dislocation" between Q and I_3, and there was certainly no understanding of this mysterious hypercharge.

But within $SU(3)$ and the Eightfold Way, this formula suddenly makes sense: the photon couples to a particular linear combination of two $SU(3)$ generators, but group theory tells us how I_3 and $\frac{Y}{2}$ are correlated. For instance, (20) says that the isodoublet (consisting of the proton and neutron) has $Y = 1$ and hence charge $Q = (\frac{1}{2} + \frac{1}{2}, -\frac{1}{2} + \frac{1}{2}) = (1, 0)$, while the isotriplet (the pions) has $Y = 0$ and hence charge $Q = (1, 0, -1)$.

But this understanding also led to widespread skepticism of the notion of quarks. According to (17), that is, $3 \to 2_1 \oplus 1_{-2}$, the up quark must have electric charge

$$Q = \frac{1}{2} + \frac{1}{2} \cdot \frac{1}{3} = \frac{2}{3} \tag{22}$$

This was quite a shock to physicists used to seeing integral electric charges! The down quark has charge $-\frac{1}{3}$. (One way of seeing this instantly is to note $Q_{\text{up}} - Q_{\text{down}} = 1$.) The strange quark has charge $Q = 0 + \frac{1}{2} \cdot \frac{1}{3} \cdot (-2) = -\frac{1}{3}$. The early searches for quarks often involved looking for these peculiar fractional charges.

We will return to the applications of $SU(3)$ to particle physics[18] in chapter V.4, but not before attending to the Lie algebra of $SU(3)$ in the next chapter.

Appendix: Harmonic oscillator in 3-dimensional space and $SU(3)$

I do not want to give the impression that $SU(3)$ is relevant only to particle physics. Indeed, symmetry groups often pop up rather unexpectedly. A simple example is given by the 3-dimensional harmonic oscillator described by the Hamiltonian

$$H = \frac{\vec{p}^2}{2m} + \frac{1}{2}kr^2 = \sum_{i=1}^{3} \left(\frac{p_i^2}{2m} + \frac{1}{2}kx_i^2 \right) = \sum_{i=1}^{3} \left(a_i^\dagger a_i + \frac{1}{2} \right) \hbar\omega \tag{23}$$

which evidently is the sum of three independent 1-dimensional harmonic oscillators. Note that we have introduced three sets of creation and annihilation operators as per appendix 1 in chapter IV.2. The energy levels are thus given by

$$E_n = \left(n + \frac{3}{2} \right) \hbar\omega = \left(n_1 + n_2 + n_3 + \frac{3}{2} \right) \hbar\omega \tag{24}$$

Note that E_n depends only on the sum $(n_1 + n_2 + n_3)$, but not on $n_1, n_2,$ or n_3 separately.

The spectrum thus exhibits a high degree of degeneracy above and beyond what is mandated by the rotational group $SO(3)$. Indeed, we see that the Hamiltonian actually enjoys a higher symmetry, namely, $SU(3)$, under which $a_i \to U_i^j a_j, a_i^\dagger \to (U_i^j)^* a_j^\dagger$. The eigenstates with energy E_n are given by n creation operators acting on the ground state, $a_{i_1}^\dagger a_{i_2}^\dagger \cdots a_{i_n}^\dagger |0\rangle$, which then manifestly transforms like an $SU(3)$ tensor with n lower indices (or upper indices, depending on your convention). (Note that we can afford to be sloppy here, since we merely want to count the number of degenerate states.)

We now simply look up the dimension of this tensor as given in (1), namely,

$$\mathcal{D}_n = \frac{1}{2}(n+1)(n+2) \tag{25}$$

Let us check the simple case of $\mathcal{D}_3 = 10$. Indeed, we have the states $(3, 0, 0), (0, 3, 0), (0, 0, 3), (2, 1, 0), (2, 0, 1), (1, 2, 0), (0, 2, 1), (1, 0, 2), (0, 1, 2),$ and $(1, 1, 1)$ in a self-evident notation.

Exercises

1 We talked about possible ambiguities when we refer to irreducible representations by their dimensions back in chapter IV.4. Find an $SU(3)$ irreducible representation with the same dimension as $(4, 0)$.

2 Complete the proof that A^{ik}_{jl} is symmetric in j and l.

3 Check $\mathcal{D}_4$ and $\mathcal{D}_5$.

4 Show that the symmetry group of the 3-dimensional harmonic oscillator is actually $U(3) = SU(3) \otimes U(1)$. What does the $U(1)$ describe?

Notes

1. Phys. Today, May 2008, p. 69. Glashow received the Nobel Prize in 1979 for work based to a large extent on group theory.
2. The ghost of the Gruppenpest, which I mentioned in chapter I.1, haunted the physics community for a surprisingly long time. For example, the well-known solid state physicist John Slater (1900–1976) wrote in his autobiography: "As soon as my paper became known, it was obvious that a great many other physicists were as 'disgusted' as I had been with the group theoretical approach to the problem. . . . there were remarks made such as 'Slater has slain the Gruppenpest.' I believe that no other work I have done was so universally popular." I was astonished that the book was published in 1975! This was long after numerous resounding successes of group theory, and a year or so after the standard $SU(3) \otimes SU(2) \otimes U(1)$ theory of the strong, electromagnetic, and weak interactions was established. I would hope that the sentiment Slater was expressing referred to the 1930s, not the 1970s. (See J. C. Slater, Solid State and Molecular Theory, J. Wiley 1975.)
3. There are of course rumblings in other areas, such as the Elliott model in nuclear physics.
4. Recall that we did similar manipulations when discussing Legendre polynomials in chapter IV.2.
5. "Not One Less," http://www.imdb.com/title/tt0209189/?ref_=nv_sr_1.
6. S. Coleman, Aspects of Symmetry, pp. 13ff. Cambridge University Press, 1985. He was my PhD advisor, and so quite naturally my treatment follows his closely.
7. Incidentally, the name refers to the tracks made by the proton and the negatively charged pion in the decay $\Lambda \to p + \pi^-$.
8. V. D. Hopper and S. Biswas of the University of Melbourne.
9. This is easy for me to say now, but in fact a method to determine the spin of the Λ was proposed by T. D. Lee and C. N. Yang only in 1958 (Phys. Rev. 109 (1958), p. 1755). A modern textbook explanation of the method takes more than four pages; see S. Gasiorowicz, Elementary Particle Physics, pp. 214–217. The early history of particle physics was full of inspired guesses and leaps of faith.
10. Again, we gloss over an enormous amount of confusion; it took heroic efforts, both experimental and theoretical, to establish that the K^0 and $\bar{K}^0$ were in fact distinct particles.
11. The search led to the exceptional group G_2. Fortunately, Nature is kind to us; the correct group $SU(3)$ is orders of magnitude simpler to learn.
12. See G. Johnson, Strange Beauty, 1999.
13. It should be noted that while Sakata had put the baryons in the 3, he did assign the mesons correctly to the 8.
14. See http://en.wikipedia.org/wiki/Noble_Eightfold_Path.
15. The large fractional mass difference between the K mesons and the pions was particularly troubling, and led some theorists to reject $SU(3)$. We now understand why the pions have an exceptionally small mass compared to other strongly interacting particles, but that is another story for another time.
16. While some plausible arguments have been advanced, there is as yet no definitive proof.
17. More precisely, this specifies how $SU(2)$ is to be embedded into $SU(3)$. We can specify that on restriction of $SU(3)$ to $SU(2)$, $3 \to 3$. Strictly speaking, this corresponds to specifying how $SO(3)$ is to be embedded into $SU(3)$.
18. I might also mention that $SU(3)$ is also manifest in the Elliott model in nuclear physics. See, for example, A. Arima, J. Phys. G: Nucl. Part. Phys. 25 (1999), p. 581.

The Lie Algebra of $SU(3)$ and Its Root Vectors

The way the Lie algebra of $SU(3)$ works out is not only important for its own sake, but it also indicates to us how to analyze a general Lie algebra, as we will see in part VI.

The eight Gell-Mann matrices

To start with, we can readily write down the eight traceless hermitean matrices that represent the algebra of $SU(3)$, as already explained in chapter IV.4. They generalize the Pauli matrices and are known as Gell-Mann matrices in particle physics:

$$
\lambda_1 = \begin{pmatrix} 0 & 1 & 0 \\ 1 & 0 & 0 \\ 0 & 0 & 0 \end{pmatrix}, \qquad
\lambda_2 = \begin{pmatrix} 0 & -i & 0 \\ i & 0 & 0 \\ 0 & 0 & 0 \end{pmatrix},
$$

$$
\lambda_4 = \begin{pmatrix} 0 & 0 & 1 \\ 0 & 0 & 0 \\ 1 & 0 & 0 \end{pmatrix}, \qquad
\lambda_5 = \begin{pmatrix} 0 & 0 & -i \\ 0 & 0 & 0 \\ i & 0 & 0 \end{pmatrix},
$$

$$
\lambda_6 = \begin{pmatrix} 0 & 0 & 0 \\ 0 & 0 & 1 \\ 0 & 1 & 0 \end{pmatrix}, \qquad
\lambda_7 = \begin{pmatrix} 0 & 0 & 0 \\ 0 & 0 & -i \\ 0 & i & 0 \end{pmatrix},
$$

$$
\lambda_3 = \begin{pmatrix} 1 & 0 & 0 \\ 0 & -1 & 0 \\ 0 & 0 & 0 \end{pmatrix}, \qquad
\lambda_8 = \frac{1}{\sqrt{3}} \begin{pmatrix} 1 & 0 & 0 \\ 0 & 1 & 0 \\ 0 & 0 & -2 \end{pmatrix} \tag{1}
$$

These matrices are normalized by $\mathrm{tr}\,\lambda_a \lambda_b = 2\delta_{ab}$. It would not have escaped your notice that the three Gell-Mann matrices λ_1, λ_2, and λ_3 correspond to the three Pauli matrices σ_1, σ_2, and σ_3, respectively, with a third row and column of 0s added. This evidently specifies how the $SU(2)$ algebra is embedded in the $SU(3)$ algebra, as noted earlier.

You would also have noticed that I have separated the eight matrices into four groups as follows:

(a) λ_3 and λ_8 are the two diagonal guys;

(b) λ_1 and λ_2, together with λ_3, form an $SU(2)$ subalgebra;

(c) λ_4 and λ_5 are basically the same as λ_1 and λ_2, except that they live in the "1-3 sector," while λ_1 and λ_2 live in the "1-2 sector";

(d) λ_6 and λ_7 are also basically the same as λ_1 and λ_2, except that they live in the "2-3 sector".

Next, note that

$$[\lambda_4, \lambda_5] = \left[\begin{pmatrix} 0 & 0 & 1 \\ 0 & 0 & 0 \\ 1 & 0 & 0 \end{pmatrix}, \begin{pmatrix} 0 & 0 & -i \\ 0 & 0 & 0 \\ i & 0 & 0 \end{pmatrix}\right] = 2i \begin{pmatrix} 1 & 0 & 0 \\ 0 & 0 & 0 \\ 0 & 0 & -1 \end{pmatrix} \equiv 2i\lambda_{[4,5]} = i(\lambda_3 + \sqrt{3}\lambda_8) \quad (2)$$

You understand that the matrix I named $\lambda_{[4,5]}$ is the analog of λ_3 living in the 1-3 sector, and together with λ_4 and λ_5, it generates an $SU(2)$ subalgebra.

Similarly, λ_6 and λ_7, together with a matrix we might call $\lambda_{[6,7]}$ (which you should guess and then work out), the analog of λ_3 living in the 2-3 sector, generate an $SU(2)$ subalgebra.

Thus, not surprisingly, $SU(3)$ contains three overlapping $SU(2)$ subalgebras. They overlap because there are only two traceless hermitean matrices in the $SU(3)$ algebra, namely, λ_3 and λ_8. (Of course, the three $SU(2)$ subalgebras have to overlap, since otherwise $SU(3)$ would be $SU(2) \otimes SU(2) \otimes SU(2)$, which it sure isn't.[1]) Three $SU(2)$s have to share two J_3s, so to speak.

In the defining or fundamental representation, the generators are represented by[2] $T^a = \frac{1}{2}\lambda_a$.

As explained in chapter I.3, the generators of any Lie algebra satisfy

$$[T^a, T^b] = if^{abc}T^c \quad (3)$$

with some structure constants f^{abc} characteristic of the algebra. Since (3) holds for any representation of the algebra, we can work out the f^{abc}s for $SU(3)$ by explicitly commuting the Gell-Mann matrices with one another. The result is:

$$f^{123} = 1$$

$$f^{147} = -f^{156} = f^{246} = f^{257} = f^{345} = -f^{367} = \frac{1}{2}$$

$$f^{458} = f^{678} = \frac{\sqrt{3}}{2} \quad (4)$$

I have already worked out f^{453} and f^{458} in (2) for you. You should check the rest, or at least a few of them. Recall that f^{abc} is totally antisymmetric, a fact that we will prove in chapter VI.3. Thus, $f^{453} = f^{345}$, which you should also check explicitly.

It is important to realize that while we determined f^{abc} using the matrices in the fundamental representation, Lie's relation (3) holds for the generators in any representation.*

There is, of course, absolutely no point in memorizing these structure constants; you will see that, even when we actually calculate using $SU(3)$ in chapter V.4, we will not need the explicit values given here. It is, however, essential that you understand what the structure constants embody about the algebra.

From ladders to jungle gyms

For what to do next, we seek inspiration from the ladder approach used for $SU(2)$ and try to generalize it to $SU(3)$. Recall that, of the three generators J_x, J_y, and J_z of $SU(2)$, we diagonalized J_z and defined $J_\pm \equiv J_x \pm i J_y$. This enabled us to write the $SU(2)$ algebra as

$$[J_z, J_\pm] = \pm J_\pm, \qquad [J_+, J_-] = 2J_z \qquad (5)$$

Using Dirac's bra and ket notation and thinking of the matrix J_z as an operator, we wrote the eigenvector of J_z with eigenvalue m as $|m\rangle$: $J_z |m\rangle = m |m\rangle$. Manipulating (5), we deduced that $J_+ |m\rangle = c_{m+1} |m+1\rangle$ and $J_- |m\rangle = c_m^* |m-1\rangle$, with c_m a normalization constant that we determined by a clever use of the algebra (following Dirac!). We thus visualized the states $\cdots, |m-1\rangle, |m\rangle, |m+1\rangle, \cdots$ as rungs on a ladder, and thought of J_+ and J_- as a raising operator and a lowering operator, respectively, that enable us to climb up and down the ladder. The next step was to show that the ladder terminates, and that the maximum and minimum values of m are given by $\pm j$, with j an integer for $SO(3)$, and an integer or half-integer for $SU(2)$.

It is crucial for you to be thoroughly familiar with what we did for $SU(2)$, as we will be simply generalizing the reasoning involved in that case to $SU(3)$.

From the explicit forms of λ_3 and λ_8 in (1), we see that the generators T^3 and T^8 commute: $[T^3, T^8] = 0$, and thus we can simultaneously diagonalize them. At this point, there is a trivial complication: history. We now rename T^3 as the third component of isospin I_3, and the T^8 as hypercharge Y up to the normalization factor $\frac{\sqrt{3}}{2}$. In a given irreducible representation, the states are characterized by the eigenvalue of I_3 and the eigenvalue of Y. In other words, we use the eigenstates of I_3 and Y to represent the algebra $SU(3)$, namely, the kets $|i_3, y\rangle$ such that $I_3 |i_3, y\rangle = i_3 |i_3, y\rangle$ and $Y |i_3, y\rangle = y |i_3, y\rangle$. The states $|i_3, y\rangle$ generalize the states $|m\rangle$ for $SU(2)$.

The states in an irreducible representation of $SU(3)$ are then labeled by two numbers and hence may be arranged in a 2-dimensional lattice rather than on a ladder.

The number of generators that can be simultaneously diagonalized is known as the rank of the Lie algebra. The rank of $SU(2)$ is 1: we can only diagonalize J_3. The rank of $SU(3)$, in contrast, is 2. The states in an irreducible representation of a Lie algebra of rank[†] l are

---

* As mentioned in chapter I.3, we adopt the standard physicist's practice of confounding the generators and the matrices representing the generators.

[†] Physicists mostly denote rank by r, while mathematicians use l. We defer to mathematicians here, so as to make contact with the discussion in part VI.

labeled by l numbers and hence may be arranged in an l-dimensional lattice. We will refer to the 2-dimensional lattice for $SU(3)$ as a jungle gym.

What about the other $6 = 8 - 2$ generators besides I_3 and Y? We use $SU(2)$ as a guide and follow our noses, so to speak. Define

$$I_{\pm} = T_1 \pm i T_2$$
$$U_{\pm} = T_6 \pm i T_7$$
$$V_{\pm} = T_4 \pm i T_5$$
$$I_3 = T_3$$
$$Y = \frac{2}{\sqrt{3}} T_8 \tag{6}$$

Generalizing isospin, particle physicists refer to the $SU(2)$ subalgebra generated by $I_{\pm}$, $U_{\pm}$, and $V_{\pm}$ as I spin, U spin, and V spin, respectively.[*]

Then, generalizing (5), we rewrite the commutation relations (3) as

$$[I_3, I_{\pm}] = \pm I_{\pm}, \qquad [I_3, U_{\pm}] = \mp \frac{1}{2} U_{\pm}, \qquad [I_3, V_{\pm}] = \pm \frac{1}{2} V_{\pm} \tag{7}$$

$$[Y, I_{\pm}] = 0, \qquad [Y, U_{\pm}] = \pm U_{\pm}, \qquad [Y, V_{\pm}] = \pm V_{\pm} \tag{8}$$

and

$$[I_+, I_-] = 2I_3 \tag{9}$$

$$[U_+, U_-] = \frac{3}{2} Y - I_3 = \sqrt{3} T_8 - T_3 \equiv 2U_3 \tag{10}$$

$$[V_+, V_-] = \frac{3}{2} Y + I_3 = \sqrt{3} T_8 + T_3 \equiv 2V_3 \tag{11}$$

$$[I_+, V_-] = -U_- \tag{12}$$

$$[I_+, U_+] = V_+, \tag{13}$$

$$[U_+, V_-] = T_-, \tag{14}$$

$$[I_+, V_+] = 0, \tag{15}$$

$$[I_+, U_-] = 0, \tag{16}$$

$$[U_+, V_+] = 0 \tag{17}$$

Other commutation relations can be obtained from the ones listed here by hermitean conjugation, for example, $([I_+, V_+])^{\dagger}$ gives $[I_-, V_-]$.

Clearly, with eight generators instead of three, we have many more commutation relations in $SU(3)$ than in $SU(2)$, but we will soon make sense of them. In the meantime, you might want to focus on the three commutation relations $[I_+, I_-] = 2I_3$, $[U_+, U_-] = 2U_3$, and $[V_+, V_-] = 2V_3$, with U_3 and V_3 defined above. These commutation relations, as remarked earlier, show that $SU(3)$ may be pictured as three overlapping $SU(2)$s, with $I_{\pm}$, $U_{\pm}$, and $V_{\pm}$ the raising and lowering operators for the three $SU(2)$ algebras, respectively.

---

[*] I don't know who first chose the terminology I, U, V, but as students, I and my friends were taught this stuff singing a variant of the American song "I scream, you scream, we all scream for ice cream!"

Climbing around on a jungle gym

The ladder for $SU(2)$ generalizes to a kind of 2-dimensional jungle gym for $SU(3)$ (and in general, an l-dimensional jungle gym). Instead of each state or ket corresponding to a rung on a ladder, each ket $|i_3, y\rangle$ now corresponds to a node or point on a jungle gym or lattice, laid down on a plane with one axis labeled by i_3 and the other by y. The six generators, $I_\pm$, $U_\pm$, and $V_\pm$, play the role of the raising and lowering operators $J_\pm$. Instead of taking us from one rung of the ladder to another, they now take us from one node on the jungle gym to another.

Since the commutation relations between I_3, $I_\pm$ are exactly the same as those between J_3, $J_\pm$, we have $I_\pm |i_3, y\rangle \propto |i_3 \pm 1, y\rangle$, exactly as in (IV.2.5) and (IV.2.6): $I_\pm$ increases and decreases i_3 by one unit, leaving y untouched. (This time we won't bother to write down and determine the proportionality constants, the analogs of the c_ms. I leave it to you to work it out.)

Next, what does $U_\pm$ do to $|i_3, y\rangle$? Find out by acting with I_3 and Y on $U_\pm |i_3, y\rangle$ and using (7) and (8). We obtain

$$I_3 U_\pm |i_3, y\rangle = \left(U_\pm I_3 \mp \frac{1}{2}U_\pm\right)|i_3, y\rangle = U_\pm\left(I_3 \mp \frac{1}{2}\right)|i_3, y\rangle$$

$$= \left(i_3 \mp \frac{1}{2}\right)U_\pm |i_3, y\rangle \tag{18}$$

and

$$YU_\pm |i_3, y\rangle = (U_\pm Y \pm U_\pm)|i_3, y\rangle = U_\pm(Y \pm 1)|i_3, y\rangle$$

$$= (y \pm 1)U_\pm |i_3, y\rangle \tag{19}$$

We conclude that $U_\pm |i_3, y\rangle \propto |i_3 \mp \frac{1}{2}, y \pm 1\rangle$. (In this discussion, the $\pm$ signs are clearly correlated.)

Going through the analogous manipulations for $V_\pm$, you would conclude that $V_\pm |i_3, y\rangle \propto |i_3 \pm \frac{1}{2}, y \pm 1\rangle$.

Root vectors

Thus, the operators $I_\pm$, $U_\pm$, and $V_\pm$ move us around this lattice. Each of them can be characterized by a vector specifying the change in location $(\Delta i_3, \Delta y)$. From the calculation we just did, we see that

$I_\pm$ moves us by $(\pm 1, 0)$

$U_\pm$ moves us by $\left(\mp\dfrac{1}{2}, \pm 1\right)$

$V_\pm$ moves us by $\left(\pm\dfrac{1}{2}, \pm 1\right)$ \hfill (20)

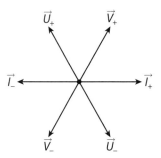

Figure 1

in the (i_3, y) plane. The vectors $(\pm 1, 0)$, $(\mp\frac{1}{2}, \pm 1)$, and $(\pm\frac{1}{2}, \pm 1)$ associated with $I_\pm$, $U_\pm$, and $V_\pm$, respectively, are known as root vectors of the Lie algebra, or roots for short.

What can we say about the lengths of the root vectors and the angles between them?

But wait, the question has no meaning unless the two axes are normalized in the same way.* Earlier, we made sure that the Gell-Mann matrices λ_a were normalized in the same way; referring to (1), we see that, in the fundamental representation, while the third component of isospin is given by $I_3 = \frac{1}{2}\lambda_3$, the hypercharge, given by $Y = \frac{1}{\sqrt{3}}\lambda_8$, is normalized differently for "historical" reasons (namely the Gell-Mann-Nishijima formula mentioned in chapter V.1).

Physicists typically use I_3 and Y, bowing to history, but the great art of mathematics could care less about how physicists fumbled around in the 1950s. Thus, if we hope to discover any universal truths about root vectors, we better go back to $T^3 = I_3$ and $T^8 = \frac{\sqrt{3}}{2}Y$. In other words, we need to multiply the second component of the root vectors listed above by $\frac{\sqrt{3}}{2}$.

Denote the correctly normalized root vectors using a self-evident notation, and write

$$\vec{I}_\pm = (\pm 1, 0)$$

$$\vec{U}_\pm = \left(\mp\frac{1}{2}, \pm\frac{\sqrt{3}}{2}\right)$$

$$\vec{V}_\pm = \left(\pm\frac{1}{2}, \pm\frac{\sqrt{3}}{2}\right) \tag{21}$$

Remembering your equilateral triangle, you see that, interestingly, the root vectors are of equal length (length squared $= 1^2 + 0^2 = (\frac{1}{2})^2 + (\frac{\sqrt{3}}{2})^2 = 1$), and the angles between them are all equal to either 60° or 120°. See figure 1, known as a root diagram. Indeed, the 60° between a neighboring pair of root vectors merely reflects democracy: $\vec{I}_\pm$, $\vec{U}_\pm$, and $\vec{V}_\pm$ have exactly the same status. The root vectors $\vec{I}_\pm$ appears to be more special than $\vec{U}_\pm$ and $\vec{V}_\pm$ merely because of physics history.

---

* Here the notion of a metric creeps into the discussion. Even though the discussion is on a lattice rather than on a continuous space, we can still talk about lengths and angles. We will return to the metric of a general Lie algebra later in chapter VI.3.

In this language, the root vectors of $SU(2)$ live in one dimension. A 1-dimensional vector is of course just a number.

Working with roots

From the root diagram, we can read off various commutation relations. At least we can see which ones ought to vanish. For example, $[I_-, U_+] = 0$.

To show this, let us first introduce some useful notation to save writing. Write the ket $|i_3, y\rangle$ as $|\vec{\omega}\rangle$; that is, define the vector $\vec{\omega} = (i_3, y)$. Then $I_- U_+ |\vec{\omega}\rangle \propto I_- |\vec{\omega} + \vec{U}_+\rangle \propto |\vec{\omega} + \vec{U}_+ + \vec{I}_-\rangle$. Similarly, $I_- U_+ |\vec{\omega}\rangle \propto I_- |\vec{\omega} + \vec{U}_+\rangle \propto |\vec{\omega} + \vec{I}_- + \vec{U}_+\rangle$. Thus, $[I_-, U_+] |\vec{\omega}\rangle \propto |\vec{\omega} + \vec{I}_- + \vec{U}_+\rangle$.

The proportionality factor could well vanish. If it does not vanish, then the root vector associated with $[I_-, U_+]$ would be pointing in the direction corresponding to $\vec{I}_- + \vec{U}_+$. But, referring to figure 1, we see that there isn't a root vector in that direction. Thus, $[I_-, U_+] = 0$.

With some practice, you can see almost instantly from the root diagram that certain commutators must vanish.

In contrast, the vector $\vec{I}_+ + \vec{U}_+$ points in the direction of $\vec{V}_+$, and sure enough, $[I_+, U_+] \propto V_+$. We will be using this sort of reasoning extensively in chapter VI.4.

Some "empirical facts"

From this example of $SU(3)$, we observe several "empirical facts."

All the root vectors have equal length.

The angles between them have "nice" values, that is, not equal to something like $17°$.

The sum of two roots is sometimes equal to another root and sometimes not.

But the sum of a root with itself does not exist; in other words, if $\vec{\alpha}$ is a root, then $2\vec{\alpha}$ (for example, $2\vec{V}_+ = (1, \sqrt{3})$) is not a root.

OK, I won't repeat the black sheep–white sheep joke. But dear reader, try to either prove or disprove these statements for a general Lie algebra. I will come back to them in part VI.

Positive and simple roots

Two notions about roots also naturally suggest themselves. Here we have six roots living in a 2-dimensional space. We ought not to have to specify all six; specifying two roots ought to suffice. To begin with, three of the six roots are just given by (-1) times the other three.

To define the notion of positive roots, we first have to agree to some ordering of the coordinate axes; here we put T^3 before T^8. Then a root is said to be positive if its first nonzero component is positive. Referring to (21), we see that $\vec{I}_+, \vec{U}_-$, and $\vec{V}_+$ are positive.

Colloquially, the roots that point "mostly east" are defined to be positive. Note that which roots are positive depends on which axis we favor. If we had favored T^8 over T^3, then $\vec{I}_+$, $\vec{U}_+$, and $\vec{V}_+$ would be called positive.

Of the three positive roots, $\vec{I}_+$, $\vec{U}_-$, and $\vec{V}_+$, any one of them can be written as a linear combination of the other two. Hence we can choose any two of the three positive roots as basic, but one particular choice recommends itself: speaking colloquially, we see that $\vec{I}_+$ is flanked by $\vec{U}_-$ and $\vec{V}_+$, that is, it lies inside the wedge or cone defined by $\vec{U}_-$ and $\vec{V}_+$. Note that $\vec{I}_+ = \vec{U}_- + \vec{V}_+$. The following more mathematical definition thus emerges naturally.

Given a set of positive roots in l-dimensional space, a subset is said to be simple if any of the positive roots can be written as a linear combination of the simple roots with non-negative coefficients. Thus, $\vec{U}_-$ and $\vec{V}_+$ are the two simple roots of $SU(3)$.

Thus far, I have carefully distinguished an operator, for example U_+, from the root vector $\vec{U}_+$ associated with it. Now that you understand these concepts, I will often drop the arrow.

Weight diagrams

For a given irreducible representation $\mathcal{R}$ of $SU(3)$, we can plot each state in the representation as a point in the $(i_3, i_8) = (i_3, \frac{\sqrt{3}}{2}y)$ plane. The collection of points form a lattice known as the weight diagram for $\mathcal{R}$, the appropriate generalization of the ladder for $SU(2)$. The number of points in the weight diagram corresponds to the dimension of $\mathcal{R}$. (Recall that the ladder for the irreducible representation j of $SU(2)$ has $2j + 1$ rungs.)

The root vectors take us from one point in the weight diagram to another. They generalize the raising and lowering $J_\pm$ that take us from one rung of the ladder to another. Since we are talking about a finite-dimensional representation, a sequence of root vectors acting on a given state either takes us out of the weight diagram (that is, yields 0) or might eventually take us back to our starting point.

It is important to note what should be self-evident for most readers: for a given Lie algebra, there is only one root diagram, but there are many weight diagrams, one for each irreducible representation.

Let's see how this works for a few typical representations. Consider 3, the defining representation, and for convenience, use particle physics labels for the states in the representation. The u quark is located at $(i_3, \frac{\sqrt{3}}{2}y) = (\frac{1}{2}, \frac{1}{2\sqrt{3}})$, the d quark at $(-\frac{1}{2}, \frac{1}{2\sqrt{3}})$, and the s quark at $(0, -\frac{1}{\sqrt{3}})$. The weight diagram consists of three points forming an inverted equilateral triangle. See figure 2.

The roots listed in (21) take us from one point in the weight diagram to another point, just as $J_\pm$ take us from one rung on the ladder to the next in the $SU(2)$ case. For example,

$$\left(0, -\frac{1}{\sqrt{3}}\right) + \vec{V}_+ = \left(0, -\frac{1}{\sqrt{3}}\right) + \left(\frac{1}{2}, \frac{\sqrt{3}}{2}\right) = \left(\frac{1}{2}, \frac{1}{2\sqrt{3}}\right) \tag{22}$$

In other words, $\vec{V}_+$ takes the s quark to the u quark.

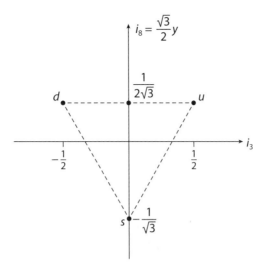

Figure 2

Indeed, you can verify that the requirement that the roots in (21) connect three points in the $(i_3, \frac{\sqrt{3}}{2}y)$ plane fixes the positions of the 3 points and the weight diagram. Analogously, in $SU(2)$, if we require the ladder to terminate, we cannot start at an arbitrary value of m: it has to be integral or half-integral. Indeed, since $SU(3)$ contains three overlapping $SU(2)$ algebras, the argument is basically a direct repeat of the argument we went through for $SU(2)$: I spin seems privileged compared to U spin and V spin merely because we chose the horizontal axis in our plot to describe i_3.

But $U_{\pm}$ and $V_{\pm}$ are in no way second class citizens compared to $I_{\pm}$. The attentive reader might have noticed that I have quietly defined, in (10) and (11), $[U_+, U_-] = 2U_3 \equiv \frac{3}{2}Y - I_3 = \sqrt{3}T_8 - T_3$ and $[V_+, V_-] = 2V_3 \equiv \frac{3}{2}Y + I_3 = \sqrt{3}T_8 + T_3$. You can immediately verify, as you would expect, that the d and s quarks form a U spin doublet.

As an exercise, work out the weight diagram of 3*.

We can do another example. Requiring that the roots connect eight points fixes the weight diagram for the famous octet into which Gell-Mann placed the nucleons and their cousins. Alternatively, start with the known position of the proton $(\frac{1}{2}, \frac{\sqrt{3}}{2})$. We already know that $\vec{I}_-$ takes the proton to the neutron at $(-\frac{1}{2}, \frac{\sqrt{3}}{2})$. Let us act with $\vec{U}_-$ on the proton: $(\frac{1}{2}, \frac{\sqrt{3}}{2}) + (\frac{1}{2}, -\frac{\sqrt{3}}{2}) = (1, 0)$. We get to the Σ^+ hyperon. (See figure 3.)

Suppose we apply $\vec{U}_-$ again, now on the Σ^+ hyperon. We get 0, just like applying the generator J_- of $SU(2)$ to the bottom rung of the ladder produces 0. (Again, see figure 3.) On the other hand, applying $\vec{V}_-$ to Σ^+ gets us to $(1, 0) + (-\frac{1}{2}, -\frac{\sqrt{3}}{2}) = (\frac{1}{2}, -\frac{\sqrt{3}}{2})$, which we recognize as the Ξ^0. In this way, we built the 8 of the baryons proposed by Gell-Mann.

As an exercise, you should now work out the weight diagram of 10 using the root vectors. See figure 4.

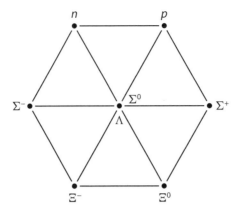

Figure 3

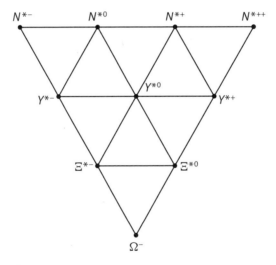

Figure 4

Theorems about weight diagrams

Clearly, due to the hexagonal symmetry of the root vectors, the weight diagrams all exhibit hexagonal symmetry. It is possible to prove various theorems about the shape* of the weight diagrams. For example, the boundary shown schematically in figure 5 is not possible. Roughly speaking, weight diagrams have to be convex outward. To prove this, label the

---

* Note that for $SU(2)$, this notion does not even arise.

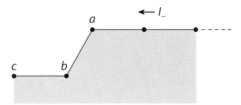

Figure 5

states as shown in the figure. (The shaded region in figure 5 indicates the interior of the weight diagram.)

Along the top of the weight diagram, we keep lowering with I_- until we get to the state $|a\rangle$ when the I spin ladder terminates: $I_- |a\rangle = 0$. As shown, $V_- |a\rangle = \beta |b\rangle$ and $I_- |b\rangle = \gamma |c\rangle$, with β and γ two Clebsch-Gordan coefficients. Applying I_- to $V_- |a\rangle = \beta |b\rangle$ then gives

$$I_- V_- |a\rangle = \beta I_- |b\rangle = \beta\gamma |c\rangle = ([I_-, V_-] + V_- I_-) |a\rangle = [I_-, V_-] |a\rangle = 0 \tag{23}$$

The last equality follows from (15). Hence either β or γ vanishes, or both vanish, which shows that the figure as drawn is not allowed. In the 10, for example, the state $|b\rangle$ does not exist; the boundary continues in the U_- direction. In contrast, in the 8, $|b\rangle$ exists but not $|c\rangle$; at $|b\rangle$ the boundary continues also in the U_- direction.

Clearly, we can also learn a lot about the weight diagrams of $SU(3)$ from the fact that the algebra consists of three overlapping $SU(2)$ algebras. For example, slicing the weight diagram along the U direction, we should see the states lining up neatly into U spin multiplets. Thus, as we have already seen, in the 8 of baryons, the proton and the Σ^+ form a U spin doublet.

We used isospin to relate processes involving different members of isospin multiplets in chapter V.1. In exactly the same way, we can use U spin to relate processes involving the Σ^+ to processes involving the proton. The resulting predictions telling us about the behavior of the Σ^+ were confirmed by experiments, and as a result we have considerable confidence in the correctness of $SU(3)$. In the next chapter, we study further what the group theory of $SU(3)$ can tell us.

Appendix: The d symbols

We mention in passing that sometimes it is useful, for some applications in particle physics, to have the analog of the relation $\{\sigma_a, \sigma_b\} = 2\delta^{ab}$. The Gell-Mann matrices satisfy*

$$\{\lambda_a, \lambda_b\} = \frac{4}{3}\delta^{ab} + 2d^{abc}\lambda_c \tag{24}$$

---

* Again, at this point in the book, whether the indices a, b, and c are written as superscripts or subscripts has no significance.

with the d symbols given by

$$d^{118} = d^{228} = d^{338} = -d^{888} = \frac{1}{\sqrt{3}}$$

$$d^{448} = d^{558} = d^{668} = d^{778} = -\frac{1}{2\sqrt{3}}$$

$$d^{146} = d^{157} = -d^{247} = d^{256} = d^{344} = d^{355} = -d^{366} = -d^{377} = \frac{1}{2}. \tag{25}$$

Check at least a few. Again, it must be emphasized that these anticommutation relations are not part of the Lie algebra but are merely properties of the Pauli and Gell-Mann matrices. The d symbols, in contrast to the f symbols, the structure constants, do not pertain to all representations.

Exercise

1 Work out the weight diagram of 10 using the root vectors.

Notes

1. As I've mentioned in another book, Murph Goldberger, one of my professors, loved to say, "If my aunt had balls, she would be my uncle!" when confronted with faulty logic.
2. As said in the Preface, whether the indices a, b, and c are written as superscripts or subscripts here has no particular significance.

V.4 | Group Theory Guides Us into the Microscopic World

Throw away the 27!
—Richard Feynman
  to George Zweig

To a generation who grew up with superstrings and the multiverse, even quarks and gluons may sound old hat. But not that long ago, they were totally unknown, and utter confusion reigned in the subnuclear world of the hadrons. The strong interaction was gradually unveiled only through decades of assiduous experimental and theoretical work. In this endeavor, group theory was an indispensable guide.

We now know that $SU(3)$ symmetry, as explained in chapters V.2 and V.3, is due to the approximate invariance of the strong interaction under a unitary transformation of the up u, down d, and strange s quarks into one another, approximate because these three quarks have different masses. But none of that was known around 1960.

Here we discuss one particular success of $SU(3)$ and its breaking, namely, the Gell-Mann-Okubo mass formula.[1] To focus on $SU(3)$ breaking, we assume that Heisenberg's $SU(2)$ is not broken. Readers totally unfamiliar with quantum mechanics might wish to skip this chapter.

$SU(3)$ breaking

In a world with exact $SU(3)$ symmetry, the π, K, and η mesons would all have the same mass. In the language of chapter III.1, they would be degenerate. But in fact their masses are quite different. Our world is only roughly $SU(3)$ invariant.

Without a full understanding of the strong interaction, we had (and still have) no way of calculating these meson masses analytically. Instead, what particle physicists accomplished in the 1960s amounted to an impressive mix of science and art, of group theory and inspired guesses.

For readers not about to specialize in particle physics (which may well be most readers), the details of meson mass splitting are not important. The important point to grasp is that what we are discussing here is conceptually similar to other symmetry breaking phenomena, such as crystal field splitting discussed in interlude IV.i2. As another example, for the triangular "molecule" consisting of masses tied together by springs discussed in chapter III.2, by making one spring slightly different from the other two, we break the symmetry group, and the shifts in the eigenfrequencies can be studied using the methods discussed here. As we shall see, Gell-Mann and Okubo, by making a simplifying* assumption about the nature of the symmetry-breaking interaction, were able to go further than just figuring out the pattern of how the 8-dimensional irreducible representation of $SU(3)$ breaks upon restriction to $SU(2)$.

The meson octet

Let us start with the eight mesons, π^+, π^0, π^-, K^+, K^0, K^-, $\bar{K}^0$, and η, whose acquaintance we made in chapter V.2. They were postulated to populate the adjoint representation 8 of $SU(3)$.

Since we are assuming that the $SU(2)$ isospin subgroup is unbroken, π^+, π^0, and π^- have the same mass. Similarly, K^+ and K^0 have the same mass, and K^- and $\bar{K}^0$ also. However, since K^- is the antiparticle of K^+, they have the same mass, and thus also K^0 and $\bar{K}^0$. The data thus consists of three masses, and so if group theory restricts the dynamics sufficiently to reduce the number of unknown constants to less than three, we would have a prediction, or strictly, a postdiction.

We will use the tensor and effective field theory approach to discuss meson masses. By now, you know well that the adjoint is given by a traceless tensor $\Phi^i_{\ j}$, $i, j = 1, 2, 3$ and can thus be written as a traceless 3-by-3 matrix:

$$\Phi = \frac{1}{\sqrt{2}} \sum_{a=1}^{8} \phi_a \lambda_a = \begin{pmatrix} \frac{1}{\sqrt{2}}\pi^0 + \frac{1}{\sqrt{6}}\eta & \pi^+ & K^+ \\ \pi^- & -\frac{1}{\sqrt{2}}\pi^0 + \frac{1}{\sqrt{6}}\eta & K^0 \\ K^- & \bar{K}^0 & -\frac{2}{\sqrt{6}}\eta \end{pmatrix} \tag{1}$$

In the first equality, we merely expand Φ in terms of the eight Gell-Mann matrices λ_a (as given in (V.3.1)). The eight fields ϕ_a may be thought of as the Cartesian components of Φ. In the second equality, the fields are identified by their street names, thus $\pi^0 = \phi_3$, $\eta = \phi_8$, and so on.

Note that the 2-by-2 submatrix contained in the northwest corner of Φ, with η removed, is precisely $\frac{1}{\sqrt{2}}$ times the matrix φ discussed in chapter V.1 when we discussed Heisenberg's $SU(2)$. In particular, the relative factor of $\sqrt{2}$ between π^0 and π^+ was explained there.

Let us warm up by imagining a world with exact $SU(3)$ symmetry and write down the effective Lagrangian

$$\mathcal{L} = \frac{1}{2} \text{tr} \left((\partial_\mu \Phi)^2 - m_0^2 \Phi^2 \right) = \frac{1}{2} \sum_{a=1}^{8} \left((\partial_\mu \phi_a)^2 - m_0^2 \phi_a^2 \right) \tag{2}$$

---

* Refer to the Feynman story at the end of this chapter.

Here we used the normalization tr $\lambda_a \lambda_b = 2\delta_{ab}$ adopted in chapter V.3 so that, for example, tr $\Phi^2 = \text{tr}(\frac{1}{\sqrt{2}} \sum_{a=1}^{8} \phi_a \lambda_a)^2 = \frac{1}{2} \sum_{a=1}^{8} \sum_{b=1}^{8} \phi_a \phi_b (\text{tr } \lambda_a \lambda_b) = \sum_{a=1}^{8} \phi_a^2$. In this imaginary world, the eight mesons would be equally massive, with mass squared $= m_0^2$.

But $SU(3)$ is broken in our world.

We now have to appeal to basic quantum mechanics. Suppose we have found the eigenstates and eigenvalues of a Hamiltonian H_0. Consider another Hamiltonian $H = H_0 + \alpha H_1$. In general, finding the eigenstates and eigenvalues of H would pose an entirely different problem. However, if α is known to be small, then to leading order in α, we would expect the eigenstates and eigenvalues of H to be closely approximated by the eigenstates and eigenvalues of H_0. As some readers may know, this is known as perturbation theory: αH_1 is treated as a small perturbation on H_0. One important result of perturbation theory is that to order α, the eigenvalue of the eigenstate $|s\rangle$ of H_0 is shifted by the amount $\langle s|\alpha H_1|s\rangle$. (To this order, knowing the eigenstate $|s\rangle$ of H_0 suffices; we do not have to worry about how $|s\rangle$ is shifted by the perturbation.)

Indeed, I have already alluded to this perturbative approach to atomic spectrum in chapter IV.3 and mentioned the Wigner-Eckart theorem. If we follow this approach, then we will have to calculate quantities like $\langle 8, a| H_1 |8, b\rangle$, with $|8, b\rangle$ the eigenstates in an $SU(3)$ symmetric world. Even to this day, we are far from being able to calculate such quantities starting from first principles; instead, we rely on group theory to tell us something. As already mentioned, rather than deriving the analog of the Wigner-Eckart theorem for $SU(3)$, we follow the effective Lagrangian approach.

The Gell-Mann-Okubo mass formula for mesons

You might question whether first order perturbation theory is applicable at all: back in chapter V.2, we learned that $SU(3)$ is badly broken, with the corresponding α at least as large as ~ 0.2–0.3. Nevertheless, Gell-Mann (and others at that time) bravely pushed ahead.

To apply group theory, we have to specify how the $SU(3)$ breaking interaction transforms under $SU(3)$. In $\mathcal{L}$, the mass term is quadratic in Φ; it transforms like the product of two Φs. But we have already worked out, in chapter V.2, that $8 \otimes 8 = 27 \oplus 10 \oplus 10^* \oplus 8 \oplus 8 \oplus 1$. However, we had multiplied two different traceless tensors together. Here we are multiplying Φ by itself, and so the antisymmetric part of the product just given drops out. If you go through the multiplication we did, you would see[2] that the symmetric part of $8 \otimes 8$ is given by

$$(8 \otimes 8)_s = 27 \oplus 8 \oplus 1 \tag{3}$$

while the antisymmetric part is given by

$$(8 \otimes 8)_a = 10 \oplus 10^* \oplus 8 \tag{4}$$

One easy check is that the counting works out: $8 \cdot (8 + 1)/2 = 36 = 27 + 8 + 1$, while $8 \cdot (8 - 1)/2 = 28 = 10 + 10 + 8$.

Therefore, there are two possible $SU(3)$-breaking terms, one transforming like the 27 and the other like the adjoint 8. (The term transforming like 1 in (3) does not break $SU(3)$, of course.)

Boldly, Gell-Mann and Okubo (working independently) proposed to throw the 27 out, as Feynman (later) advised Zweig. Thus, to break $SU(3)$, we add the term* tr $\Phi^2\lambda_8$ to the effective Lagrangian $\mathcal{L}$.

It is straightforward to multiply three 3-by-3 matrices together and then take the trace to obtain tr $\Phi^2\lambda_8$. But with some forethought, we can minimize our arithmetical exertions. Observe that $\lambda_8 = \frac{1}{\sqrt{3}} \begin{pmatrix} 1 & 0 & 0 \\ 0 & 1 & 0 \\ 0 & 0 & -2 \end{pmatrix}$ can be written as a linear combination of the 3-by-3 identity matrix and the matrix $\begin{pmatrix} 0 & 0 & 0 \\ 0 & 0 & 0 \\ 0 & 0 & 1 \end{pmatrix}$. The contribution of the identity matrix to tr $\Phi^2\lambda_8$ (namely, tr Φ^2) can be absorbed into the m_0^2 term in $\mathcal{L}$. Thus, the $SU(3)$-breaking term in $\mathcal{L}$ is effectively given by

$$\text{tr } \Phi^2\lambda_8 \to \text{tr } \Phi\Phi \begin{pmatrix} 0 & 0 & 0 \\ 0 & 0 & 0 \\ 0 & 0 & 1 \end{pmatrix} = \text{tr} \begin{pmatrix} X & X & X \\ X & X & X \\ K^- & \bar{K}^0 & -\frac{2}{\sqrt{6}}\eta \end{pmatrix} \begin{pmatrix} 0 & 0 & K^+ \\ 0 & 0 & K^0 \\ 0 & 0 & -\frac{2}{\sqrt{6}}\eta \end{pmatrix}$$

$$= K^-K^+ + K^0\bar{K}^0 + \frac{2}{3}\eta^2 \tag{5}$$

Here X denotes those entries we won't bother to write down, since we intend to take the trace after multiplying the two matrices.

The mass of the K and of the η are shifted, but not that of the π. To get the normalization right, it is easiest to compare with the $SU(3)$ invariant mass term tr $\Phi^2 = (\pi^0)^2 + \eta^2 + 2K^-K^+ + \cdots$. Since we know that this term implies that the K meson and the η meson have equal mass, we have all together $m_\pi^2 = m_0^2$, $m_K^2 = m_0^2 + \frac{1}{2}\kappa$, and $m_\eta^2 = m_0^2 + \frac{2}{3}\kappa$, where κ is an unknown constant.

Happy are we. With three masses and two unknown constants, we obtain one relation, namely, the Gell-Mann Okubo mass formula for mesons,

$$4m_K^2 = 3m_\eta^2 + m_\pi^2 \tag{6}$$

You can plug in the measured masses given in chapter V.2 and verify that this mass formula is satisfied to within the expected accuracy.

Note that if we had included the 27 as well as the 8, we would have an additional unknown constant and thus lose the mass formula (6). In 1961–1962, the proposal that the $SU(3)$-breaking term transforms purely like the 8 was just a guess. As I said earlier, physics at the cutting edge often involves inspired guesses, and is art as well as science.

This guess is now understood in terms of quantum chromodynamics, in which $SU(3)$ breaking is entirely due to the strange quark being much more massive than the up and down quarks. Since these three quarks transform as the defining 3, their mass terms transform as $3^* \otimes 3 = 1 \oplus 8$, as was worked out in chapter V.2.

---

* Note that we do not include tr $\Phi^2\lambda_3$, because we are assuming that isospin is not broken.

Baryon mass splitting

We next turn to the eight baryons, p, n, Σ^+, Σ^0, Σ^-, Λ^0, Ξ^-, and Ξ^0, which we met in chapter V.2 and which form the baryon octet. As in (1), the eight baryon fields[3] can be arranged in a 3-by-3 traceless matrix:

$$
B = \begin{pmatrix} \frac{1}{\sqrt{2}}\Sigma^0 + \frac{1}{\sqrt{6}}\Lambda^0 & \Sigma^+ & p \\ \Sigma^- & -\frac{1}{\sqrt{2}}\Sigma^0 + \frac{1}{\sqrt{6}}\Lambda^0 & n \\ \Xi^- & \Xi^0 & -\frac{2}{\sqrt{6}}\Lambda^0 \end{pmatrix} \tag{7}
$$

Now that we have worked through the derivation of the mass formula for the mesons, we should be able to breeze through a similar derivation for the baryons. However, there are some differences.

One difference is that, unlike the mesons, which carry spin 0, the baryons carry spin $\frac{1}{2}$ and thus transform under the Lorentz group (which we will study in part VII). Thus, as already mentioned in chapter V.1, the hermitean conjugate of B transforms differently from B and is denoted by

$$
\bar{B} = \begin{pmatrix} \frac{1}{\sqrt{2}}\bar{\Sigma}^0 + \frac{1}{\sqrt{6}}\bar{\Lambda}^0 & \bar{\Sigma}^- & \bar{\Xi}^- \\ \bar{\Sigma}^+ & -\frac{1}{\sqrt{2}}\bar{\Sigma}^0 + \frac{1}{\sqrt{6}}\bar{\Lambda}^0 & \bar{\Xi}^0 \\ \bar{p} & \bar{n} & -\frac{2}{\sqrt{6}}\bar{\Lambda}^0 \end{pmatrix} \tag{8}
$$

Consequently, we now have two possible $SU(3)$-breaking terms that transform as 8 (rather than the single term in the meson case), namely,

$$
\text{tr}\, \bar{B}B \begin{pmatrix} 0 & 0 & 0 \\ 0 & 0 & 0 \\ 0 & 0 & 1 \end{pmatrix} = \text{tr}\, \bar{B} \begin{pmatrix} 0 & 0 & p \\ 0 & 0 & n \\ 0 & 0 & -\frac{2}{\sqrt{6}}\Lambda^0 \end{pmatrix} = (\bar{p}p + \bar{n}n + \frac{2}{3}\bar{\Lambda}^0\Lambda^0) \tag{9}
$$

and

$$
\text{tr}\, \bar{B} \begin{pmatrix} 0 & 0 & 0 \\ 0 & 0 & 0 \\ 0 & 0 & 1 \end{pmatrix} B = \text{tr} \begin{pmatrix} 0 & 0 & -\bar{\Xi}^- \\ 0 & 0 & \bar{\Xi}^0 \\ 0 & 0 & -\frac{2}{\sqrt{6}}\bar{\Lambda}^0 \end{pmatrix} B = (\bar{\Xi}^-\Xi^- + \bar{\Xi}^0\Xi^0 + \frac{2}{3}\bar{\Lambda}^0\Lambda^0) \tag{10}
$$

We see that (i) the two terms have different effects and (ii) the symmetry breaking does not affect the Σs. Again, we calculate $\text{tr}\, \bar{B}B$ to make sure that things are properly normalized. To save computational effort, we can set the fields Σ, n, Ξ^0 to zero in light of the preceding remark and since we are not breaking isospin; then $\text{tr}\, \bar{B}B = \bar{p}p + \bar{\Lambda}^0\Lambda^0 + \bar{\Xi}^-\Xi^- + \cdots$.

Denoting the unknown coefficients of the two terms in (9) and (10) by κ and ζ, we thus obtain $M_N = M_0 + \kappa$, $M_\Lambda = M_0 + \frac{2}{3}(\kappa + \zeta)$, $M_\Xi = M_0 + \zeta$, $M_\Sigma = M_0$. This time around there are three unknowns, but we have four masses. So we still get one mass relation, namely,[*]

$$
3M_\Lambda + M_\Sigma = 2(M_N + M_\Xi) \tag{11}
$$

---

[*] Both sides are equal to $4M_0 + 2(\kappa + \zeta)$.

The attentive reader will have noticed that I wrote down masses in (11) but masses squared in (6). You will see in chapter VII.3 that in the field theory Lagrangian for spin $\frac{1}{2}$ particles, mass appears, but for spin 0 particles, mass squared appears (as was written in chapter IV.9 generalizing the harmonic oscillator). Actually, since we are working to linear order in the symmetry breaking, we have no right to quibble about the quadratic versus the linear mass formula. Expanding $m_a^2 = (m_0 + \delta m_a)^2 = m_0^2 + 2m_0 \delta m_a + O((\delta m_a)^2)$, we could just as well have written (6) as $4m_K = 3m_\eta + m_\pi$.

Consequences of $SU(3)$

The consequences[4] of $SU(3)$ were soon worked out in the early 1960s. With the setup given in this chapter, you can join in the fun retroactively. For instance, in chapter V.1, we worked out the pion-nucleon couplings. You can now work out the couplings of the meson octet to the baryon octet. Two terms are allowed by $SU(3)$:

$$g_1 \operatorname{tr} \bar{B} B \Phi + g_2 \operatorname{tr} \bar{B} \Phi B \tag{12}$$

In principle, 17 coupling constants (for example, the $\bar{N} K \Sigma$ coupling) are measurable. That they can be expressed in terms of two unknowns[5] is an impressive achievement of group theory.

Our faith in Nature's simplicity

We close with a story. George Zweig, who discovered quarks independently[6] of Gell-Mann, recalled that he thought that the weak interaction currents of the strongly interacting particles should also be classified in representations of $SU(3)$, and that both the 8- and 27-dimensional representations were to be used. At the time, a decay process requiring the presence of the 27 had been seen by an experimentalist of high reputation[7] working with a strong team using a well understood technique.

Zweig went to talk to Feynman, who liked the idea of applying $SU(3)$ to the decay, but kept saying "Throw away the 27!"

Feynman turned out to be right; the experimentalists were mistaken.

We need to use group theory in physics, but somehow we never have to mess with anything beyond the simplest irreducible representations. Another example verifying the principle of Nature's kindness to theoretical physicists!

Notes

1. M. Gell-Mann, California Institute of Technology Synchrotron Laboratory Report CTSL-20, 1961, and S. Okubo, Prog. Theor. Phys. 27 (5) (1962), 949–966.
2. An easy way to see this is that the 10 corresponds to a symmetric 3-indexed tensor T^{ijk}, one of whose components is T^{111}. To form this out of two adjoints, we try to write $\Phi_i^1 \Phi_j^1 \varepsilon^{ij1}$, which vanishes.

3. Some authors define Ξ^- with an extra minus sign. See S. Gasiorowicz, Elementary Particle Physics, p. 281.

4. As another example, since we know how the electromagnetic interaction transforms under $SU(3)$, we can use the method described here, or equivalently the Wigner-Eckart theorem generalized to $SU(3)$, to relate the various electromagnetic mass splittings between the baryons in an isospin multiplet. The result is known as the Coleman-Glashow formula.

5. This is worked out in S. Gasiorowicz, Elementary Particle Physics, p. 281.

6. See G. Zweig's talk in "Proceedings of the Conference in Honour of Murray Gell-Mann's 80th Birthday," World Scientific.

7. W. H. Barkas at Berkeley; the "Barkas event" was described as exceptionally clean.

Part VI | Roots, Weights, and Classification of Lie Algebras

Learning how to analyze the $SU(2)$ and $SU(3)$ algebras provides us with the concepts and tools needed to deal with Lie algebras in general, an adventure that culminates in the Killing-Cartan classification. The Dynkin diagrams capture the information contained in the root diagrams in a simple pictorial fashion. An ingenious mix of elegant mathematics and powerful logic is on display throughout.

VI.1 | The Poor Man Finds His Roots

Classification of Lie algebras

Before launching into a formal mathematical development, I think that it would be peda-gogical for us to watch over the poor man's shoulders to see how he would grope his way around this problem of classifying Lie algebras. The poor man was highly intrigued by the pattern of the roots of $SU(3)$ and the hexagonal weight diagrams they generated.

The poor man naturally wonders whether this kind of rigid geometrical pattern would continue to hold for $SU(N)$. He first goes back to $SU(2)$. There we diagonalize the Pauli matrix $\sigma_3 = \begin{pmatrix} 1 & 0 \\ 0 & -1 \end{pmatrix}$, and form* the raising matrix $\frac{1}{2}\sigma_{1+i2} = \begin{pmatrix} 0 & 1 \\ 0 & 0 \end{pmatrix}$. Then the commutator $[\frac{1}{2}\sigma_3, \frac{1}{2}\sigma_{1+i2}] = \frac{1}{2}\sigma_{1+i2}$ gives us the 1-dimensional root "vector," namely, the number 1, associated with the raising operator $J_+ = J_{1+i2}$. Fine.

The poor man then moves on to $SU(3)$. It has rank 2, which simply means that we can now diagonalize two matrices, rather than merely one matrix, namely, $\lambda_3 = \begin{pmatrix} 1 & 0 & 0 \\ 0 & -1 & 0 \\ 0 & 0 & 0 \end{pmatrix}$ and $\lambda_8 = \frac{1}{\sqrt{3}} \begin{pmatrix} 1 & 0 & 0 \\ 0 & 1 & 0 \\ 0 & 0 & -2 \end{pmatrix}$. The commutators of $\frac{1}{2}\lambda_3$ and $\frac{1}{2}\lambda_8$ with each of the three raising matrices, $\frac{1}{2}\lambda_{1+i2}$, $\frac{1}{2}\lambda_{4+i5}$, and $\frac{1}{2}\lambda_{6+i7}$, give a 2-dimensional root vector, as discussed in detail in chapter V.3. For example, to determine the root vector of $\frac{1}{2}\lambda_{6+i7}$, we calculate its commutator with $\frac{1}{2}\lambda_3$; the result determines the first component of the root vector. Then we calculate its commutator with $\frac{1}{2}\lambda_8$; the result determines the second component of the root vector. Now is the time to review chapter V.3 if you are a bit shaky about this.

---

* For this heuristic discussion, we are free from historical constraints and will normalize things to keep various expressions as clean as possible.

Determining roots

Let us mechanize this process. Since only the 2-by-2 block of the northwest corner of $\frac{1}{2}\lambda_{1+i2} = \begin{pmatrix} 0 & 1 & 0 \\ 0 & 0 & 0 \\ 0 & 0 & 0 \end{pmatrix}$ is nonzero, and since the corresponding 2-by-2 block of $\frac{1}{2}\lambda_8$ is proportional to the identity matrix, clearly $[\frac{1}{2}\lambda_{1+i2}, \frac{1}{2}\lambda_8] = 0$. We thus obtain for $\frac{1}{2}\lambda_{1+i2}$ the 2-dimensional root vector $(1, 0)$: it is obtained from the 1-dimensional root vector 1 for $SU(2)$ by adding a vanishing second component. The root vector stays in the 1-dimensional subspace, so to speak.

Next, consider $\frac{1}{2}\lambda_{4+i5} = \begin{pmatrix} 0 & 0 & 1 \\ 0 & 0 & 0 \\ 0 & 0 & 0 \end{pmatrix}$. To calculate its commutators with λ_3 and λ_8, we need, effectively, to look at only its 2-by-2 submatrix in the 1-3 sector, namely, $\begin{pmatrix} 0 & 1 \\ 0 & 0 \end{pmatrix}$, and commute it with the corresponding 2-by-2 submatrix in the 1-3 sector of λ_3 and λ_8, namely, $\begin{pmatrix} 1 & 0 \\ 0 & 0 \end{pmatrix}$ and $\begin{pmatrix} 1 & 0 \\ 0 & -2 \end{pmatrix}$.

It turns out that we will be continually commuting 2-by-2 submatrices like these. It is thus convenient to calculate, once and for all, the commutator (for some unspecified n)

$$\left[\begin{pmatrix} 1 & 0 \\ 0 & -n \end{pmatrix}, \begin{pmatrix} 0 & 1 \\ 0 & 0 \end{pmatrix} \right] = \left[\begin{pmatrix} n+1 & 0 \\ 0 & 0 \end{pmatrix}, \begin{pmatrix} 0 & 1 \\ 0 & 0 \end{pmatrix} \right] = (n+1) \begin{pmatrix} 1 & 0 \\ 0 & 0 \end{pmatrix} \begin{pmatrix} 0 & 1 \\ 0 & 0 \end{pmatrix}$$
$$= (n+1) \begin{pmatrix} 0 & 1 \\ 0 & 0 \end{pmatrix} \tag{1}$$

In the first equality, we added n times the identity matrix to the first matrix (which we are free to do, since the identity matrix commutes with everything).

For the commutator of $\frac{1}{2}\lambda_{4+i5}$ with λ_3 we apply this formula (1) for $n = 0$, and with λ_8 this formula for $n = 2$. In other words, we simply write down $(n + 1)$, keeping in mind the relative normalization between λ_3 and λ_8, thus obtaining—zap zap zap—the root vector

$$\frac{1}{2}\left((0+1), \frac{1}{\sqrt{3}}(2+1) \right) = \frac{1}{2}(1, \sqrt{3}) \tag{2}$$

By symmetry, the root vector associated with $\frac{1}{2}\lambda_{6+i7}$ is $\frac{1}{2}(-1, \sqrt{3})$.

Thus, we quickly obtain the three root vectors $(1, 0)$, $\frac{1}{2}(1, \sqrt{3})$, and $\frac{1}{2}(-1, \sqrt{3})$. (The other three root vectors are, of course, the negative of these.) Their lengths squared, $1 = \frac{1}{4}(1 + 3)$, are manifestly equal. The scalar products between them, $(1, 0) \cdot \frac{1}{2}(\pm 1, \sqrt{3}) = \pm\frac{1}{2}$ and $\frac{1}{2}(1, \sqrt{3}) \cdot \frac{1}{2}(-1, \sqrt{3}) = \frac{1}{4}(-1 + 3) = \frac{1}{2}$, are also equal up to a sign. The angles between the root vectors are either $60°$ or $120°$.

Having warmed up, the poor man now presses onward to $SU(4)$ easily. The algebra is now rank 3, with the additional diagonal matrix $\lambda_{15} = \frac{1}{\sqrt{6}} \begin{pmatrix} 1 & 0 & 0 & 0 \\ 0 & 1 & 0 & 0 \\ 0 & 0 & 1 & 0 \\ 0 & 0 & 0 & -3 \end{pmatrix}$, where the $\frac{1}{\sqrt{6}}$ follows from the normalization $\operatorname{tr} \lambda_{15}^2 = 2$. The raising matrices $\begin{pmatrix} 0 & 1 & 0 & 0 \\ 0 & 0 & 0 & 0 \\ 0 & 0 & 0 & 0 \\ 0 & 0 & 0 & 0 \end{pmatrix}$ and $\begin{pmatrix} 0 & 0 & 1 & 0 \\ 0 & 0 & 0 & 0 \\ 0 & 0 & 0 & 0 \\ 0 & 0 & 0 & 0 \end{pmatrix}$, namely, the "descendants" of $\frac{1}{2}\lambda_{1+i2}$ and $\frac{1}{2}\lambda_{4+i5}$, clearly commute with λ_{15} and so lead to root vectors $(1, 0, 0)$ and $\frac{1}{2}(1, \sqrt{3}, 0)$; these are as before except for the addition of a vanishing third component. In other words, they effectively stay in the 2-dimensional root space of $SU(3)$.

To get out of this subspace, we need to look at the raising matrix $\begin{pmatrix} 0 & 0 & 0 & 1 \\ 0 & 0 & 0 & 0 \\ 0 & 0 & 0 & 0 \\ 0 & 0 & 0 & 0 \end{pmatrix}$, which is new to $SU(4)$. As before, only the 2-by-2 submatrix in the 1-4 sector matters. The relevant commutators of this raising matrix with λ_3, λ_8, and λ_{15} are (reading off from (1))

$$\left[\begin{pmatrix} 1 & 0 \\ 0 & 0 \end{pmatrix}, \begin{pmatrix} 0 & 1 \\ 0 & 0 \end{pmatrix} \right] = \begin{pmatrix} 0 & 1 \\ 0 & 0 \end{pmatrix}, \qquad \left[\frac{1}{\sqrt{3}} \begin{pmatrix} 1 & 0 \\ 0 & 0 \end{pmatrix}, \begin{pmatrix} 0 & 1 \\ 0 & 0 \end{pmatrix} \right] = \frac{1}{\sqrt{3}} \begin{pmatrix} 0 & 1 \\ 0 & 0 \end{pmatrix} \tag{3}$$

and the truly new one

$$\left[\frac{1}{\sqrt{6}} \begin{pmatrix} 1 & 0 \\ 0 & -3 \end{pmatrix}, \begin{pmatrix} 0 & 1 \\ 0 & 0 \end{pmatrix} \right] = \frac{4}{\sqrt{6}} \begin{pmatrix} 0 & 1 \\ 0 & 0 \end{pmatrix} = 2\sqrt{\frac{2}{3}} \begin{pmatrix} 0 & 1 \\ 0 & 0 \end{pmatrix} \tag{4}$$

We thus find the root vector $\frac{1}{2}(1, \frac{1}{\sqrt{3}}, 2\sqrt{\frac{2}{3}})$. As a check, the length squared of this root vector is $\frac{1}{4}(1 + \frac{1}{3} + 4 \cdot \frac{2}{3}) = \frac{3+1+8}{12} = 1$, equal to that of the other root vectors.

The scalar dot product between this root vector and the other root vectors is $\frac{1}{2}(1, \frac{1}{\sqrt{3}}, 2\sqrt{\frac{2}{3}}) \cdot (1, 0, 0) = \frac{1}{2}$ and $\frac{1}{2}(1, \frac{1}{\sqrt{3}}, 2\sqrt{\frac{2}{3}}) \cdot \frac{1}{2}(1, \sqrt{3}, 0) = \frac{1}{2}$. So once again the angles between the roots are either 60° or 120°.

The poor man thinks that he may be on to something! For fun, you could work out* $SU(5)$ and see whether the pattern persists.

---

* This happens to be an important group for fundamental physics. See part IX.

VI.2 | Roots and Weights for Orthogonal, Unitary, and Symplectic Algebras

Setting the stage with $SU(2)$ and $SU(3)$

Our work with the $SU(2)$ and $SU(3)$ algebras paves the way to solving more general Lie algebras. Indeed, with $SU(3)$ fresh in our minds, it is almost a cinch to work out the root vectors of $SU(N)$. Already in chapter VI.1, the poor man discovered a definite pattern to the root diagram of $SU(N)$. In this chapter, we adopt a more systematic approach, with an eye toward a general classification of Lie algebras.

Let's start with a quick review of $SU(2)$ and $SU(3)$.

For $SU(2)$, we identify the diagonal matrix σ_3, and the states of the fundamental or defining representation are just labeled by the eigenvalues of $\frac{1}{2}\sigma_3$.

For $SU(3)$, there are two diagonal matrices, λ_3 and λ_8, and so the states of the defining representation are labeled by two numbers, namely, the eigenvalues of $\frac{1}{2}\lambda_3$ and $\frac{1}{2}\lambda_8$. The weights characterizing the states thus consist of two numbers and thus may be thought of as vectors in a 2-dimensional space.

It is almost not worth repeating that you should not read this chapter (and hardly any other chapter) passively. Instead, you should work things out, as we move along, drawing appropriate figures as needed.

Rank and the maximal number of mutually commuting generators

The matrices λ_3 and λ_8 form a maximal subset of generators* that commute with each other. None of the other six generators commutes with both λ_3 and λ_8. As already noted in chapter V.3, the maximal number of mutually commuting generators is known as the rank of the algebra, often denoted by r in physics and by l in mathematics. To make contact with the more mathematical discussion in chapter VI.4, we use l here.

---

* We physicists confound generators and the matrices representing the generators, as usual.

Given a Lie algebra, the first step is thus to identify the maximal set of mutually commuting generators and diagonalize them simultaneously. Instead of the absurd physicist's practice of inventing a new notation for these l matrices for each algebra they encounter and naming them after some über-physicists like Pauli and Gell-Mann, we prefer here the mathematician's notation* of H^i $(i = 1, \cdots, l)$ for them.[†]

Each state in an irreducible representation is then characterized by l numbers, given by the eigenvalues of the H^is. Thus, the weights, and the root vectors that connect them, may be thought of as vectors in an l-dimensional space.

There is certainly more than one method of determining the root vectors of an algebra. The method I favor and use in this chapter is to write down, given the H^is, the weights of an "easy" representation, such as the defining representation, and then to find the vectors that connect the weights (that is, the vectors that take one state in the representation into another). (For example, for $SU(3)$, the three sides of the equilateral triangle with the vertices given by the weights define six root vectors, $6 = 3 \cdot 2$, since the negative of a root vector is also a root vector.)

Review of $SU(3)$

Let us review how this method works with $SU(3)$. We have[‡]

$$H^1 = \text{diag}(1, -1, 0)/\sqrt{2}$$
$$H^2 = \text{diag}(1, 1, -2)/\sqrt{6} \tag{1}$$

(As usual, $\text{diag}(1, -1, 0)$ denotes a 3-by-3 diagonal matrix with diagonal elements equal to $1, -1$, and 0.) The normalization chosen,[§] $\text{tr}\, H^i H^j = \delta^{ij}$, is such that H^1 and H^2 correspond to $\lambda_3/\sqrt{2}$ and $\lambda_8/\sqrt{2}$, respectively.

The weights of the three different states in the fundamental or defining representation live in 2-dimensional space and hence are vectors with two components. We can simply read off these three vectors from (1), scanning vertically. In other words,[‖]

$$w^1 = \left(1, \frac{1}{\sqrt{3}}\right)/\sqrt{2}$$

$$w^2 = \left(-1, \frac{1}{\sqrt{3}}\right)/\sqrt{2}$$

$$w^3 = \left(0, -\frac{2}{\sqrt{3}}\right)/\sqrt{2} \tag{2}$$

---

* This also accords with the notation to be used in chapters VI.4 and VI.5.

[†] The jargon guy tells us that the algebra consisting of the H^is is known as a Cartan subalgebra.

[‡] Concerning the pesky, but trivial, matter of normalization: since the H^is commute, their normalization is more or less arbitrary. We choose whatever normalization is most convenient for the discussion at hand, and hence normalization may vary from chapter to chapter. Hopefully, nobody is a divine around here.

[§] A table of roots for the unitary, orthogonal, and symplectic groups is given later in this chapter. With this normalization, the roots for $SU(N)$ listed in the table will not have a factor of $\sqrt{2}$.

[‖] By rights, we should write w vertically, but for typographical reasons (pace the publisher), we write it horizontally.

which agree with what we had in chapter V.2: w^1, w^2, and w^3 correspond to the up, down, and strange quark, respectively, as explained there. The six roots are then

$$\alpha^1 = w^1 - w^2 = \sqrt{2}(1, 0)$$
$$\alpha^2 = w^2 - w^3 = (-1, \sqrt{3})/\sqrt{2}$$
$$\alpha^3 = w^1 - w^3 = (1, \sqrt{3})/\sqrt{2} \tag{3}$$

corresponding to $\sqrt{2}I_+$, $\sqrt{2}U_+$, and $\sqrt{2}V_+$, respectively, and their negatives (namely, $\alpha^4 = w^2 - w^1$, and so on). I need hardly mention that the labeling of weights and roots is arbitrary. Note that we are not putting arrows on top of the weight and root vectors. It should be understood that they are 2-component vectors.

As another check, the length squared of the weight vectors is equal to $(w^1)^2 = (w^2)^2 = (w^3)^2 = \frac{2}{3}$. In other words, the tips of the three weight vectors w^1, w^2, and w^3 are equidistant from the origin, which confirms what we know, namely, that the three states in the fundamental representation 3 form an equilateral triangle.

Positive and simple roots

Recall from chapter V.2 the notions of positive and negative roots and of simple roots. Order the H^is in some definite, but arbitrary, way. Read the components of the root vectors in this particular order.

For the positive root vectors, the first nonzero values are positive. Here we choose to favor H^2 over H^1, and thus the roots* $w^m - w^n$ for $m < n$ are defined as positive. Again, physicist's versus mathematician's convention! The choice here conforms to the more mathematical discussion to be given in chapters VI.3 and VI.4, and unfortunately differs from the physicist's convention used in chapter V.2. (Again, let the Lord smite the divines!) Roughly speaking, the roots pointing mostly north are called positive here, while the roots pointing mostly east were called positive there.

Next, recall that a simple root is a positive root that cannot be written as a sum of two positive roots with positive coefficients. Here the two simple roots are then $w^1 - w^2$ and $w^2 - w^3$, namely, $\sqrt{2}I_+$ and $\sqrt{2}U_+$. (Clearly, since $w^1 - w^3 = (w^1 - w^2) + (w^2 - w^3)$, it is not simple.)

The notions of positive and simple roots are hardly profound. Pick your favorite direction. The roots that point mostly in that direction are called positive, and out of those, the ones that live on the "outer flanks" are simple.

Onward to the orthogonals: Our friend the square appears

At this point, we could go on and work out the roots of $SU(N)$, but it is pedagogically clearer to work out the roots of $SO(N)$ first, and perhaps, also preferable, so we could enjoy

---

* Note also that the index i on H^i and the index m on w^m have different ranges.

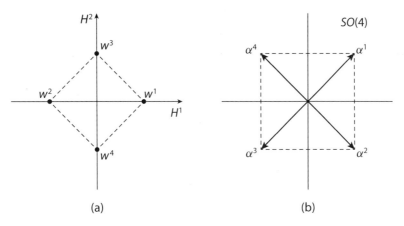

Figure 1

a change of pace. You might want to challenge yourself and figure it out before reading on. You will soon discover that the cases N odd or even have to be treated differently.

Since $SO(3) \simeq SU(2)$ is almost trivial for you by now, let's start with the next less trivial case of $SO(4)$. Of the $6 = 4 \cdot 3/2$ generators,[*] J_{12} and J_{34} form a maximal subset of mutually commuting generators: they do their things in different spaces (namely, the 1-2 plane and the 3-4 plane). Diagonalize them simultaneously, and call them H^1 and H^2, respectively:

$$H^1 = \text{diag}(1, -1, 0, 0)$$
$$H^2 = \text{diag}(0, 0, 1, -1) \tag{4}$$

This means that we are not[†] working in the Cartesian basis (that is, x^1, x^2, x^3, x^4) favored by physicists, but in the circular or polar basis $(x^1 \pm ix^2, x^3 \pm ix^4)$.

Compare and contrast with (1). Trivially, for the orthogonal groups, we normalize by $\text{tr}\, H^i H^j = 2\delta^{ij}$. Note also that we do not have pesky factors like $\sqrt{3}$.

The rank of $SO(4)$ is $l = 2$, so that the weights and roots live in 2-dimensional space. Following the method outlined earlier, we can, again, read off the four weights of the defining representation immediately, scanning (4) vertically:

$$w^1 = (1, 0)$$
$$w^2 = (-1, 0)$$
$$w^3 = (0, 1)$$
$$w^4 = (0, -1) \tag{5}$$

The 4 points defined by w^m, $m = 1, \cdots, 4$ form a square (tilted by $45°$ with respect to the axes defined by H^1 and H^2). After meeting the equilateral triangle, it is rather pleasing to see our other friend the square emerging. See figure 1a.

---

[*] Here we use the notation of chapter I.3.
[†] In the Cartesian basis, H^1 and H^2 would not be diagonal.

The root vectors connect the weights and so are given by

$$\alpha^1 \equiv w^1 - w^4 = (1, 1)$$
$$\alpha^2 \equiv w^1 - w^3 = (1, -1)$$
$$\alpha^3 \equiv w^4 - w^1 = (-1, -1)$$
$$\alpha^4 \equiv w^3 - w^1 = (-1, 1) \tag{6}$$

See figure 1b. (Note that no root takes the weight w^1 into w^2; this is because rotations cannot transform $x^1 \pm i x^2$ into each other. Similarly for $x^3 \pm i x^4$.) The roots also form a square, but one that is now aligned with the axes.

The 45° "misalignment" between the weights and the roots underlies the local isomorphism $SO(4) \simeq SU(2) \otimes SU(2)$ discussed in chapter IV.7. Recall that we defined (with the notation used there) the generators $(J_i \pm K_i)$ and found that the two sets of generators decouple. Interestingly, we arrive at the same result here without* evaluating a single commutator! (The statement that the algebra of $SO(4)$ falls apart into two pieces follows from the fact that $\alpha^1 + \alpha^2 = (2, 0)$ is not a root, and thus the corresponding commutator vanishes, as was explained in chapter V.2.) Note of course that we are free to choose $H^1 \pm H^2$ as our Hs and thus rotate the weight and root diagrams both by 45°.

This example shows clearly once again that it is unnecessary to write down all the roots: half of the roots are negatives of the other half. Let us choose the positive roots to be α^1 and α^2. Denote the Cartesian basis vectors by $e^1 = (1, 0)$ and $e^2 = (0, 1)$. Then we can write the four root vectors as

$$\pm e^1 \pm e^2 \quad \text{(signs uncorrelated)} \tag{7}$$

and the two positive root vectors as

$$e^1 \pm e^2 \tag{8}$$

They are both simple.

Although we can now jump almost immediately to $SO(2l)$, it is instructive to tackle $SO(5)$ first.

$SO(5)$ and a new feature about roots

Again, try to work out the weights and roots of $SO(5)$ before reading on.

First of all, the maximal subset of mutually commuting generators diagonalized now consists of

$$H^1 = \text{diag}(1, -1, 0, 0, 0)$$
$$H^2 = \text{diag}(0, 0, 1, -1, 0) \tag{9}$$

We still have only 2 H's but they are now 5-by-5 matrices. There is no other generator of $SO(5)$ that commutes with both H^1 and H^2. Although the algebra of $SO(5)$ manifestly

---

* We only input (4).

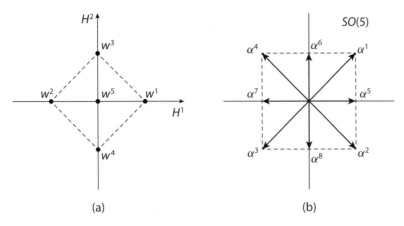

Figure 2

contains the algebra of $SO(4)$ as a subalgebra, they have the same rank $l = 2$. The weights and roots of $SO(5)$ still live in 2-dimensional space. Two guys can have the same rank, although one is bigger than the other.

Again scanning (9) vertically, we can read off the weights of the defining representation 5: besides the weights listed in (5), we have the additional weights $w^5 = (0, 0)$ (from $(H^1)_{55}$ and $(H^2)_{55}$). The weight diagram is now a square with a dot in its center. See figure 2a.

Thirdly, as a result of the appearance of w^5, we have, besides the roots listed in (6), 4 additional root vectors, connecting the weights w^1, w^2, w^3, w^4 to w^5, given by

$$\alpha^5 \equiv w^1 - w^5 = (1, 0)$$
$$\alpha^6 \equiv w^3 - w^5 = (0, 1)$$
$$\alpha^7 \equiv w^2 - w^5 = (-1, 0)$$
$$\alpha^8 \equiv w^4 - w^5 = (0, -1) \tag{10}$$

See figure 2b. The roots form a square, but now with the midpoints of the side included.

We have a bit of freedom in choosing which half of the roots we call positive, as was mentioned before. For $SO(4)$, we chose α^1 and α^2. Now, for $SO(5)$, in addition to these two roots, we also include α^5 and α^6. The algebra $SO(4)$ has two positive roots; $SO(5)$ has four positive roots.

Here comes a new feature that we have not encountered before! Four of the eight roots, α^1, α^2, α^3, and α^4, have length $\sqrt{2}$, while the other four, α^5, α^6, α^7, and α^8, are shorter, having length 1. The former are known as the long roots, the latter the short roots.*

It is worth emphasizing that, while the length of some specific root depends on how we normalize the H^is and so is not particularly meaningful, the ratio of the lengths of two specified roots is, in contrast, an intrinsic property of the algebra.

With the same notation as before, the eight root vectors of $SO(5)$ are given by

$$\pm e^1 \pm e^2 \text{ (signs uncorrelated)}, \quad \pm e^1, \quad \pm e^2 \tag{11}$$

---

* This is the kind of jargon I like, descriptive and to the point.

Compare with (7). The four positive root vectors are (with a particular but natural choice)

$$e^1 \pm e^2, \ e^1, \ e^2 \tag{12}$$

See if you can figure out the simple roots before reading on. They are

$$e^1 - e^2, \ e^2 \tag{13}$$

The other two positive roots, $e^1 = (e^1 - e^2) + e^2$ and $e^1 + e^2 = (e^1 - e^2) + 2e^2$, are not simple: they can be written in terms of $e^1 - e^2$ and e^2 with positive coefficients.

$SO(6)$

Again, instead of jumping immediately to $SO(2l)$ and $SO(2l + 1)$, it is good pedagogy to do $SO(6)$, for which another new feature will emerge. The maximal subset of mutually commuting generators consists of the matrices

$$
\begin{aligned}
H^1 &= \text{diag}(1, -1, 0, 0, 0, 0) \\
H^2 &= \text{diag}(0, 0, 1, -1, 0, 0) \\
H^3 &= \text{diag}(0, 0, 0, 0, 1, -1)
\end{aligned}
\tag{14}
$$

Again, we read off the weights for the defining representation by scanning (14) vertically:

$$
\begin{aligned}
w^1 &= (1, 0, 0) \\
w^2 &= (-1, 0, 0) \\
w^3 &= (0, 1, 0) \\
w^4 &= (0, -1, 0) \\
w^5 &= (0, 0, 1) \\
w^6 &= (0, 0, -1)
\end{aligned}
\tag{15}
$$

The weights live in a 3-dimensional space.

The roots are thus given by

$$\pm e^1 \pm e^2, \ \pm e^2 \pm e^3, \ \pm e^1 \pm e^3 \ \text{(signs uncorrelated)} \tag{16}$$

There are $4 \cdot 3 = 12$ of them, which together with the 3 Hs, give $15 = 6 \cdot 5/2$ generators, as expected. We choose the positive roots to be

$$e^1 \pm e^2, \ e^2 \pm e^3, \ e^1 \pm e^3 \tag{17}$$

The simple roots are

$$e^1 - e^2, \ e^2 - e^3, \ e^2 + e^3 \tag{18}$$

To see that these are simple, show that those not listed here are not simple.

A new feature is the peculiar arrangement of signs: both $e^2 \pm e^3$ are simple.

$SO(2l)$ **versus** $SO(2l+1)$

By now, you should see how this process goes in general. For $SO(2l)$ (with $l \geq 2$), the matrices

$$H^1 = \text{diag}(1, -1, 0, 0, \cdots, 0, 0)$$
$$H^2 = \text{diag}(0, 0, 1, -1, 0, 0, \cdots, 0, 0)$$
$$\vdots$$
$$H^l = \text{diag}(0, 0, \cdots, 0, 0, 1, -1) \tag{19}$$

form the maximal subset of mutually commuting generators, from which we read off the $2l$ weights for the defining representation:

$$w^1 = (1, 0, \cdots, 0)$$
$$w^2 = (-1, 0, \cdots, 0)$$
$$w^3 = (0, 1, 0, \cdots, 0)$$
$$\vdots$$
$$w^{2l-1} = (0, \cdots, 0, 1)$$
$$w^{2l} = (0, \cdots, 0, -1) \tag{20}$$

The weights w^m live in an l-dimensional space.

Let us write the $2l$ weights more compactly as $\pm e^i$ in terms of the l unit vectors e^i, for $i = 1, \cdots, l$. The roots are then given by

$$\pm e^i \pm e^j \text{ (signs uncorrelated) } (i < j) \tag{21}$$

as in (7). As a quick check, there are $4l(l-1)/2$ of these, which together with the l H^is, give $2l(l-1) + l = 2l(2l-1)/2$, as expected.

We take the positive roots to be $e^i \pm e^j$. The simple roots are then (recall the $SO(6)$ example)

$$e^{i-1} - e^i, \quad e^{l-1} + e^l \tag{22}$$

(with the understanding that e^0 does not exist and so effectively $i = 2, \cdots, l$).

For $SO(2l+1)$, the H^is are now $(2l+1)$-by-$(2l+1)$ matrices; we simply add an extra 0 to the matrices in (19). Everything proceeds just as in the discussion going from $SO(4)$ to $SO(5)$. In addition to the weights in (20), we have the weight $w^{2l+1} = (0, 0, \cdots, 0)$, and hence the additional roots $\pm e^i$, taking this extra weight to the weights in (20).

The roots for $SO(2l+1)$ are then given by

$$\pm e^i \pm e^j \text{ (signs uncorrelated) } (i < j), \quad \pm e^i \tag{23}$$

The $2l$ additional roots bring the total number of roots up to $l(2l-1) + 2l = 2l^2 + l = (2l+1)(2l)/2$, as expected.

We take the positive roots to be $e^i \pm e^j$ and e^i. The simple roots are then (recall the $SO(6)$ example)

$$e^{i-1} - e^i, \quad e^l \tag{24}$$

(again, effectively, $i = 2, \cdots, l$).

For the reader's convenience, a table is provided at the end of this chapter.

The roots of $SU(N)$

We now return to $SU(N)$. All that is required of us is a tedious but straightforward generalization of (1). Again, consider the defining representation N. Evidently, there are $l = N - 1$ traceless N-by-N matrices (namely, σ_3 for $N = 2$, and λ_3 and λ_8 for $N = 3$) that commute with one another and hence can be simultaneously diagonalized. They are

$$H^1 = \text{diag}(1, -1, 0, \cdots, 0)/\sqrt{2}$$
$$H^2 = \text{diag}(1, 1, -2, 0, \cdots, 0)/\sqrt{6}$$
$$\vdots$$
$$H^i = \text{diag}(\underbrace{1, 1, 1, 1, \cdots, 1}_{i}, -i, 0, \cdots, 0)/\sqrt{i(i+1)}$$
$$\vdots$$
$$H^{l-1} = \text{diag}(1, 1, 1, 1, \cdots, 1, 1, 1, -(l-1), 0)/\sqrt{(l-1)l}$$
$$H^l = \text{diag}(1, 1, 1, 1, \cdots, 1, -l)/\sqrt{l(l+1)} \tag{25}$$

If we think of the diagonal elements of H^i as a vector, then it is a vector with $N = l + 1$ components. There are l such vectors. In particular, aside from the overall normalization factor $1/\sqrt{i(i+1)}$, the vector H^i contains i ones in a row, followed by a single $(-i)$, which in turn is followed by $(l + 1 - i - 1) = (l - i)$ zeroes. As before with $SU(3)$, we normalize by tr $H^i H^j = \delta^{ij}$: $\text{tr}(H^i)^2 = (i + i^2)/(i(i+1)) = 1$.

The weights of the $N = l + 1$ different states in the fundamental or defining representation live in l-dimensional space and hence are vectors with l components. Once again, we can simply read them off from (25), scanning vertically:

$$w^1 = \sqrt{2}(1/2, 1/(2\sqrt{3}), \cdots, 1/\sqrt{2m(m+1)}, \cdots, 1/\sqrt{2l(l+1)})$$
$$w^2 = \sqrt{2}(-1/2, 1/(2\sqrt{3}), \cdots, 1/\sqrt{2m(m+1)}, \cdots, 1/\sqrt{2l(l+1)})$$
$$w^3 = \sqrt{2}(0, -1/\sqrt{3}, \cdots, 1/\sqrt{2m(m+1)}, \cdots, 1/\sqrt{2l(l+1)})$$
$$\vdots$$
$$w^{m+1} = \sqrt{2}(0, 0, \cdots, 0, -m/\sqrt{2m(m+1)}, 1/\sqrt{2(m+1)(m+2)} \cdots, 1/\sqrt{2l(l+1)})$$
$$\vdots$$
$$w^{l+1} = \sqrt{2}(0, 0, 0, 0, 0 \cdots \cdots \cdots \cdots \cdots \cdots, 0, 0, -l/\sqrt{2l(l+1)}) \tag{26}$$

The ith component of w^j is given by $(H^i)_{jj}$.

This may look complicated, but we are merely copying from (25) without even having to truly engage our brains. If you are at all confused, simply retreat to $N = 2$ and 3.

The root vectors take us from one state to another, and hence they are given by the differences between the weights, namely, $w^m - w^n$, $m, n = 1, \cdots, N = l + 1$. Thus, there are $N(N - 1)$ roots. Together with the $l = N - 1$ H^is, we have all together $N(N - 1) + (N - 1) = (N + 1)(N - 1) = N^2 - 1$ generators, which is indeed the case.

Since the $N(N - 1)/2$ positive roots of $SU(N)$ live in an $(N - 1)$-dimensional space, they can be written as linear combinations of $(N - 1)$ simple roots. In the present context, there is no need to list the $N(N - 1)/2$ positive roots, namely, $w^m - w^n$ for $m < n$; it suffices to list the $(N - 1)$ simple roots $w^m - w^{m+1}$ for $m = 1, 2, \cdots, N - 1$.

From line segment to equilateral triangle to tetrahedron and so on

Pedagogically, instead of the mess in (26), I think that if you work out the $SU(4)$ case, you will understand the general case completely. So do it!

We now have $3 = 4 - 1$ mutually commuting matrices:

$$H^1 = \text{diag}(1, -1, 0, 0)/\sqrt{2}$$
$$H^2 = \text{diag}(1, 1, -2, 0)/\sqrt{6}$$
$$H^3 = \text{diag}(1, 1, 1, -3)/(2\sqrt{3}) \tag{27}$$

Again, scanning (27) vertically, we write down

$$w^1 = \frac{1}{\sqrt{2}}\left(1, \frac{1}{\sqrt{3}}, \frac{1}{\sqrt{6}}\right)$$
$$w^2 = \frac{1}{\sqrt{2}}\left(-1, \frac{1}{\sqrt{3}}, \frac{1}{\sqrt{6}}\right)$$
$$w^3 = \frac{1}{\sqrt{2}}\left(0, -\frac{2}{\sqrt{3}}, \frac{1}{\sqrt{6}}\right)$$
$$w^4 = \frac{1}{\sqrt{2}}\left(0, 0, -\sqrt{\frac{3}{2}}\right) \tag{28}$$

Note that the first two components of w^1, w^2, and w^3 are exactly the same as in (2). In other words, the three weight vectors w^1, w^2, and w^3 form an equilateral triangle. It is the fourth weight vector that takes us into 3-dimensional space, where $SU(4)$, a rank 3 algebra, lives.

You are surely bold enough to guess that the tips of these four weight vectors form the vertices of a tetrahedron. Indeed, $(w^1)^2 = (w^2)^2 = (w^3)^2 = (w^4)^2 = \frac{3}{4}$. The tips of the four weight vectors are equidistant from the origin, and your intuition works.

We see that the sequence proceeding from $SU(2)$ up to $SU(4)$ and beyond describes a line segment, an equilateral triangle, a tetrahedron, and so on. The roots also have an

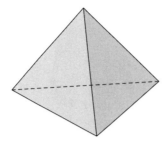

Figure 3 The four vertices of a tetrahedron form the weight diagram of the defining representation of $SU(4)$. The six edges correspond to the $12 = 6 \cdot 2$ roots. The rank is 3, as described by the 3-dimensional space the tetrahedron lives in. So the number of the generators is given by $12 + 3 = 4^2 - 1 = 15$.

appealing geometrical interpretation. For example, for $SU(3)$, they map to the three sides of the equilateral triangle; for $SU(4)$, the six sides of the tetrahedron.

For $SU(4)$, $l = 3$ and the three simple roots (write them down before reading on!) are

$$
\begin{aligned}
\alpha^1 &\equiv w^1 - w^2 = \sqrt{2}(1, 0, 0) \\
\alpha^2 &\equiv w^2 - w^3 = (-1, \sqrt{3}, 0)/\sqrt{2} \\
\alpha^3 &\equiv w^3 - w^4 = (0, -\sqrt{2}, 2)/\sqrt{3}
\end{aligned} \tag{29}
$$

Verify that $(\alpha^1)^2 = (\alpha^2)^2 = (\alpha^3)^2 = 2$, $\alpha^1 \cdot \alpha^2 = -1$, and $\alpha^2 \cdot \alpha^3 = -1$. Indeed, we have a tetrahedron.

In contrast to $\alpha^1 \cdot \alpha^2$ and $\alpha^2 \cdot \alpha^3$, $\alpha^1 \cdot \alpha^3$ vanishes, so that the two roots α^1 and α^3 are orthogonal.

How is your geometric intuition? Can you visualize this result? Yes, precisely, if you label the vertices of the tetrahedron by w^1, w^2, w^3, and w^4, then α^1 and α^3 describe the two edges that do not share a vertex in common. See figure 3.

Generalizing this discussion, we find that the simple roots of $SU(l+1)$ satisfy

$$
(\alpha^i)^2 = 2, i = 1, \cdots l \quad \text{and} \quad \alpha^i \cdot \alpha^{i+1} = -1, i = 1, \cdots, l-1 \tag{30}
$$

The simple roots of $SU(l+1)$ can be written in a more elegant form by going to a space one dimension higher. Let e^i $(i = 1, \cdots, l+1)$ denote unit vectors living in $(l+1)$-dimensional space. Then $(e^i - e^{i+1})^2 = 1 + 1 = 2$, and $(e^i - e^{i+1}) \cdot (e^j - e^{j+1}) = -1$ if $j = i \pm 1$ and 0 otherwise. The l simple roots $SU(l+1)$ are then given by

$$
\alpha^i = e^i - e^{i+1}, i = 1, \cdots l \tag{31}
$$

Note that the simple roots live in the l-dimensional hyperplane perpendicular to the vector $\sum_j e^j$.

$Sp(2l)$

At this point, we could breeze through the algebra of $Sp(2l)$, generated by (as was discussed in chapter IV.7) the set of hermitean traceless matrices

$$iA \otimes I, \quad S_1 \otimes \sigma_1, \quad S_2 \otimes \sigma_2, \quad S_3 \otimes \sigma_3 \tag{32}$$

Recall that I denotes the 2-by-2 identity matrix; A an arbitrary real l-by-l antisymmetric matrix; and S_1, S_2, and S_3, three arbitrary real l-by-l symmetric matrices. The H^i's are practically handed to us on a platter, already diagonalized: namely, for $i = 1, 2, \cdots, l$,

$$H^i = u^i \otimes \sigma_3 = \left(\begin{array}{c|c} u^i & 0 \\ \hline 0 & -u^i \end{array} \right) \tag{33}$$

Here u^i denotes the l-by-l diagonal matrix with a single entry equal to 1 in the ith row and ith column. Verify that these are indeed the correct Hs.

For example, for $Sp(4)$,

$$H^1 = \mathrm{diag}(1, 0, -1, 0)$$
$$H^2 = \mathrm{diag}(0, 1, 0, -1) \tag{34}$$

and so

$$w^1 = (1, 0)$$
$$w^2 = (0, 1)$$
$$w^3 = (-1, 0)$$
$$w^4 = (0, -1) \tag{35}$$

The four points defined by w^m, $m = 1, \cdots, 4$ form a square identical to that appearing in $SO(4)$. Indeed, the weights are given in terms of the unit basis vectors by $w^1 = -w^3 = e^1$ and $w^2 = -w^4 = e^2$.

The eight root vectors connect each of the four weights to the others (note the crucial difference between $SO(4)$ (with $4 \cdot 3/2 = 6$ generators) and the bigger group $Sp(4)$ (with $1 + 3 + 3 + 3 = 10$ generators)). The roots are $\pm e^i \pm e^j$, and $\pm 2e^i$.

As usual, we need only write down the four positive roots, chosen to be

$$\alpha^1 \equiv w^1 - w^3 = (2, 0) = 2e^1$$
$$\alpha^2 \equiv w^1 - w^4 = (1, 1) = e^1 + e^2$$
$$\alpha^3 \equiv w^2 - w^4 = (0, 2) = 2e^2$$
$$\alpha^4 \equiv w^1 - w^2 = (1, -1) = e^1 - e^2 \tag{36}$$

See figure 4.

But this is identical, except for a 45° rotation, to the root digram for $SO(5)$! See figure 2. We have now demonstrated the local isomorphism $Sp(4) \simeq SO(5)$, first suspected in chapter IV.7.

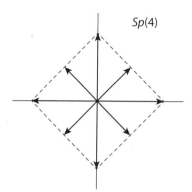

Figure 4

Confusio: "How could that be? The defining representation of the group $Sp(4)$ is 4-dimensional, while that of the group $SO(5)$ is 5-dimensional!"

It is OK, Confusio. Recall that the defining representation of the group $SU(2)$ is 2-dimensional, while that of the group $SO(3)$ is 3-dimensional, yet the two algebras are isomorphic. The root diagrams are the same up to rotation. There is no requirement that the weight diagrams of the two algebras be the same.

For $Sp(2l)$, the roots are $\pm e^i \pm e^j$ and $\pm 2e^i$, with $i, j = 1, \cdots, l$. The positive roots are

$$e^i \pm e^j, i < j, \text{ and } 2e^i \tag{37}$$

with the simple roots (verify this!)

$$e^{i-1} - e^i, i = 2, \cdots, l, \text{ and } 2e^l \tag{38}$$

Let's count. There are $l - 1 + 1 = l$ simple roots, $2l(l-1)/2 + l = l^2$ positive roots, and $2l^2$ roots, plus the l Hs to give a total of $2l^2 + l = l(2l + 1)$ generators, in precise agreement with the result in chapter IV.7.

The roots of the four families

The following table summarizes the four families we have studied.

	Number of generators	Roots	Simple roots
$SU(l)$	$l^2 - 1$	$e^i - e^j$	$e^i - e^{i+1}$
$SO(2l + 1)$	$l(2l + 1)$	$\pm e^i \pm e^j, \pm e^i$	$e^{i-1} - e^i, e^l$
$Sp(2l)$	$l(2l + 1)$	$\pm e^i \pm e^j, \pm 2e^i$	$e^{i-1} - e^i, 2e^l$
$SO(2l)$	$l(2l - 1)$	$\pm e^i \pm e^j$	$e^{i-1} - e^i, e^{l-1} + e^l$

The indices i and j run from 1 to l and the es denote l-dimensional unit vectors. Note that for this statement to hold for all entries in the table, we have to enter the rank $l - 1$ algebra $A_{l-1} = SU(l)$ instead of $SU(l + 1)$. (Mnemonic: $SU(2)$ is rank 1.)

Appendix: The Chevalley basis

You have surely noticed that, in contrast to the orthogonal and symplectic algebras, the H^is, the root vectors, and the weights of the unitary algebras are chock-full of yucky factors like $\sqrt{3}$ (as in (25)). This is partly because for historical reasons (charge, hypercharge, isospin, and so forth) physicists want to keep T_3 and T_8, in the context of $SU(3)$, orthogonal to each other. We can get rid of the nasty square roots in (25), but only at a price. We no longer have tr $H^i H^j$ proportional to δ^{ij}, but why would mathematicians care?

Indeed, mathematicians prefer to use what is known as the Chevalley basis.[1] Pedagogically, the basis is best explained with the example of $SU(3)$. Instead of the Gell-Mann matrices λ_3 and λ_8, we use

$$h^1 = \begin{pmatrix} 1 & 0 & 0 \\ 0 & -1 & 0 \\ 0 & 0 & 0 \end{pmatrix} \quad \text{and} \quad h^2 = \begin{pmatrix} 0 & 0 & 0 \\ 0 & 1 & 0 \\ 0 & 0 & -1 \end{pmatrix} \tag{39}$$

Define the raising matrices*

$$e^1 = \begin{pmatrix} 0 & 1 & 0 \\ 0 & 0 & 0 \\ 0 & 0 & 0 \end{pmatrix} \quad \text{and} \quad e^2 = \begin{pmatrix} 0 & 0 & 0 \\ 0 & 0 & 1 \\ 0 & 0 & 0 \end{pmatrix} \tag{40}$$

Clearly, h^1, e^1, and the lowering matrix $(e^1)^T$ form an $SU(2)$ algebra in the standard form familiar since chapters IV.4 and IV.5: we have $[h^1, e^1] = 2e^1$, and so on and so forth. Similarly, h^2, e^2, and $(e^2)^T$ form another $SU(2)$ algebra. The other raising matrix is given by $e^3 \equiv [e^1, e^2]$. The eight matrices $h^1, h^2, e^1, (e^1)^T, e^2, (e^2)^T, e^3$, and $(e^3)^T$ generate $SU(3)$.

Look, no more pesky square roots, but now tr $h^1 h^2 \neq 0$. We will have slightly more to say about the Chevalley basis in chapter VI.4.

Exercises

1   Show that for $SO(6)$, the positive roots $e^1 - e^3$, $e^1 + e^3$, and $e^1 + e^2$ are not simple.

2   Verify that for $Sp(6)$, the simple roots are $e^1 - e^2$, $e^2 - e^3$, and $2e^3$.

Note

1. There are of course also deeper reasons undergirding Chevalley's work, but we will not go into them in this book.

---

* Unfortunate but standard notation. These es, being matrices, are obviously not to be confused with the root vectors.

VI.3 | Lie Algebras in General

Only four families of Lie algebras, plus a few exceptional algebras

My purpose here is to give an elementary introduction to the Cartan classification of Lie algebras. We already know about the $SO(N)$ and $SU(N)$ families of compact Lie groups and their associated Lie algebras. Offhand, you might think there could exist an infinite variety of Lie algebras.[*] Remarkably, Cartan[†] showed that there are only four[‡] families of Lie algebras, plus a few algebras, known as exceptional algebras, that do not belong to these four families.

How is this remarkable result possible? The root reason lies in our discussion of the roots of $SU(3)$ in chapter V.3. There we saw that the roots, when appropriately normalized, are of equal length, and the angles between them can only take on two possibilities. Thus, the roots of $SU(3)$ form a rigid geometrical pattern. Then our friend the poor man discovered, in chapter VI.1, that only certain patterns might be allowed by the structure of Lie algebras. As we shall see, this heuristic motivation turns out to point us in the right direction.

The poor man is clearly on to something: the geometrical patterns of the root vectors are related for the Lie algebras $SU(N)$. In the rest of this chapter, we shall see that his hunch is in fact valid. The strategy is to derive constraints on the lengths of and the angles between the root vectors, and then to deduce the possible Lie algebras.

A general Lie algebra

Consider a general Lie algebra with n generators defined by

$$[X^a, X^b] = if^{ab}_{\ \ c}X^c \tag{1}$$

---

[*] Here we only discuss the algebras of compact Lie groups.

[†] Our history is admittedly skimpy here. A number of authors, including H. Weyl, B. L. van der Waerden, G. Racah, E. B. Dynkin, and others, all contributed.

[‡] And $SO(N)$ counts as two families. Recall chapter IV.2 showing how $SO(2n)$ and $SO(2n+1)$ differ.

with $a = 1, \cdots, n$. (For example, for $SU(3)$, $n = 3^2 - 1 = 8$.) Here, the index c is summed over.

Note that $f^{ab}_{\ c}$ in (1) carries two superscripts and one subscript. The difference between the two kinds of indices will be explained shortly. For now, we insist that when two indices are set equal and summed over, one has to be an upper index, the other a lower index.

The structure constants $f^{ab}_{\ c} = -f^{ba}_{\ c}$ are antisymmetric in the two upper indices by definition. Hermiticity of X^a implies that the $f^{ab}_{\ c}$ are real.

At this stage, mathematicians would proceed treating X^a as abstract entities, but as physicists, to wrap our heads around something concrete, we can simply think of X^a as matrices in the defining representation, for example, the Pauli matrices or the Gell-Mann matrices.

The adjoint representation

Plug (1) into the Jacobi identity

$$[[X^a, X^b], X^c] + [[X^b, X^c], X^a] + [[X^c, X^a], X^b] = 0 \tag{2}$$

to obtain

$$f^{ab}_{\ d} f^{dc}_{\ g} + f^{bc}_{\ d} f^{da}_{\ g} + f^{ca}_{\ d} f^{db}_{\ g} = 0 \tag{3}$$

Define

$$(T^a)^b_{\ d} \equiv -i f^{ab}_{\ d} \tag{4}$$

In other words, T^a is a matrix labeled by a whose rows and columns are indexed by b and d, respectively. Now rewrite (3) as

$$i^2(-1)(T^a)^b_{\ d}(T^c)^d_{\ g} + i^2(-1)^2(T^c)^b_{\ d}(T^a)^d_{\ g} + i f^{ca}_{\ d}(T^d)^b_{\ g} = 0 \tag{5}$$

We exploited the antisymmetry of f and indicated the factors of (-1) needed to write (3) in this form. (For example, we wrote $f^{dc}_{\ g}$ in the first term in (3) as $-f^{cd}_{\ g}$.)

Recognizing the first term as $+(T^a T^c)^b_{\ g}$ and the second term as $-(T^c T^a)^b_{\ g}$, we see that (5) says

$$[T^a, T^c] = i f^{ac}_{\ d} T^d \tag{6}$$

As is already familiar from earlier discussions of $SU(2)$ and $SU(3)$, for example, and also of $SO(N)$ and $SU(N)$, the matrices T^a, constructed out of the structure constants, represent the entities X^a in the adjoint[1] representation.

The Cartan-Killing metric

Define the Cartan-Killing metric:

$$g^{ab} \equiv \mathrm{Tr}\, T^a T^b = -f^{ac}_{\ d} f^{bd}_{\ c} \tag{7}$$

As physicists, we will, without any apology, immediately restrict ourselves to those Lie algebras for which g^{ab} has an inverse, which we write as g_{bc} with lower indices, such that

$$g^{ab}g_{bc} = \delta^a_c \tag{8}$$

Except in our discussions of $SU(N)$ (for which complex conjugation distinguishes upper and lower indices), whether we write an index as a superscript or a subscript hasn't made a difference in this book until now. Indeed, we were often intentionally sloppy to prove a point, for example, by writing Pauli matrices either as σ^a or σ_a. But now it starts to matter!

For the benefit of some readers, I give an exceedingly brief summary of the notion of a metric in an endnote.[2]

All we need here is that the real symmetric object g_{ab} (and its inverse g^{ab}) provides a natural metric in the present context. We use the metric merely to raise and lower indices and to take scalar products. Indeed, from (7), the metric in the present context is just a real symmetric matrix, given once and for all; there simply isn't even any x for g_{ab} to depend on.

One important consequence of the metric is that when we contract and sum over indices, we can only contract an upper index with a lower index. Note that we have been scrupulously following this rule in this chapter.

In fact, since g^{ab} is manifestly real symmetric, we can, by a similarity transformation, set it to be equal to δ^{ab}, so that the space in question is just good old everyday Euclidean flat space. (Explicitly, by setting $T'^a = S^a_b T^b$, we can diagonalize g^{ab} and by scaling the T'^as set the diagonal elements to be all equal.) Indeed, that is exactly what physicists do implicitly without any handwringing or talk, for example, by normalizing the Pauli matrices for $SU(2)$ and the Gell-Mann matrices for $SU(3)$. With the choice $g^{ab} = \delta^{ab}$, they can afford to be sloppy with upper versus lower indices.

The reason we went through this tedious verbiage is because mathematicians like to keep things general and because for some purposes some choice of bases[3] that makes $g^{ab} \neq \delta^{ab}$ can be more convenient. We will keep it general here until further notice.

Symmetry of structure constants

By construction, the structure constants $f^{ab}_{\ \ c}$ are antisymmetric in ab. Now, it does not even make sense to ask whether they are also antisymmetric on the exchange of b and c, since these two indices are different kinds of beasts. We need to raise the index c for it to be on the same footing as b.

So, let us define structure constants with all upper indices:

$$\begin{aligned}
f^{abc} &\equiv f^{ab}_{\ \ d}g^{dc} = -f^{ab}_{\ \ d}f^{de}_{\ \ g}f^{cg}_{\ \ e} \\
&= (f^{be}_{\ \ d}f^{da}_{\ \ g} + f^{ea}_{\ \ d}f^{db}_{\ \ g})f^{cg}_{\ \ e} \\
&= i^3 \operatorname{tr}(T^b(-T^a)T^c + (-T^a)(-T^b)T^c) \\
&= -i \operatorname{tr}(T^a T^b T^c - T^b T^a T^c) \tag{9}
\end{aligned}$$

where we have used (3) in the third equality (with $c \to e$). The final expression shows that $f^{abc} = -f^{acb}$ and hence is totally antisymmetric.

This verifies what we know about $SU(2)$ and $SU(3)$; for $SU(2)$, the structure constants are just the antisymmetric symbol ε^{abc}.

Cartan subalgebra

Out of the set of n X^as, find the maximal subset of mutually commuting generators H^i, $i = 1, 2, \cdots, l$, so that

$$[H^i, H^j] = 0, \quad i, j = 1, \cdots, l \tag{10}$$

The important number l is known as the rank. In other words, $f^{ij}_{\ c} = 0$ for any c. (For $SU(2)$, $l = 1$ with H corresponding to σ^3. For $SU(3)$, $l = 2$ with H^1 and H^2 corresponding to λ^3 and λ^8, respectively.) The commuting algebra generated by the H^is is known as the Cartan subalgebra, as was already mentioned in chapter VI.2.

Since the H^is mutually commute, they can be simultaneously diagonalized. (For example, the Gell-Mann matrices λ^3 and λ^8 are chosen to be diagonal.)

Call the remaining $(n - l)$ generators Es. The Es will have to carry indices to distinguish themselves from one another, but we will be intentionally vague at this stage. (For $SU(2)$, the Es are known as $J_\pm$, and for $SU(3)$, as $I_\pm$, $U_\pm$, and $V_\pm$.) In the basis in which the H^is are diagonal, we can choose the Es to have all zeroes along the diagonal.[*] Thus, we can choose tr $H^i E = 0$ for all E. (This is clear from the $SU(2)$ and $SU(3)$ examples.)

From the Cartesian basis to the circular basis

We learned earlier in this chapter that the matrices T^a, constructed out of the structure constants, represent the entities X^a in the adjoint representation. Focus on those matrices T^i representing the generators H^i. By definition, they commute with one another, and hence by a theorem given in the review of linear algebra, these l matrices can be simultaneously diagonalized. Denote the diagonal elements of T^i by $-\beta^i(a)$. Here $a, b = 1, \cdots, n$, while $i = 1, \cdots, l$. These l matrices are thus given by

$$(T^i)^a_{\ b} = -\begin{pmatrix} \beta^i(1) & 0 & 0 & 0 \\ 0 & \beta^i(2) & 0 & 0 \\ 0 & 0 & \ddots & 0 \\ 0 & 0 & 0 & \beta^i(n) \end{pmatrix} = -\beta^i(a)\delta^a_{\ b} \tag{11}$$

Indeed, the equality $(T^i)^a_{\ b} = -\beta^i(a)\delta^a_{\ b}$ is just a clumsy way of saying that the diagonal elements of T^i are given by $\beta^i(a)$.

---

[*] Any given E can be written as $N + D$, where N has zeroes along the diagonal while D is diagonal. But then D can be written as a linear combination of the H^is and the identity I, which simply splits off from the Lie algebra as a trivial $U(1)$ piece. Simply subtract D from E to define a new E.

The quantities $\beta^i(a)$ depend on a (or equivalently, b in view of the δ^a_b), of course.* The expression $\beta^i(a)\delta^a_b$ in (11) does not imply summation over a. Note that the hermiticity of T^i implies that the $\beta^i(a)$ are real numbers.

To see more clearly what is going on, let us go back to $SU(2)$ or $SO(3)$. There is only one H^i, namely, the generator J_z of rotation around the third axis. So drop the index i. In the Cartesian basis $\begin{pmatrix} x^1 \\ x^2 \\ x^3 \end{pmatrix}$, J_z is represented by

$$-i \begin{pmatrix} 0 & 1 & 0 \\ -1 & 0 & 0 \\ 0 & 0 & 0 \end{pmatrix} = \begin{pmatrix} 0 & -i & 0 \\ i & 0 & 0 \\ 0 & 0 & 0 \end{pmatrix}. \tag{12}$$

We recognize the second Pauli matrix in the upper left block. This matrix can be diagonalized to $\begin{pmatrix} 1 & 0 & 0 \\ 0 & -1 & 0 \\ 0 & 0 & 0 \end{pmatrix}$ by going from the Cartesian basis to the circular or polar basis. Correspondingly, instead of the generators $\begin{pmatrix} J^1 \\ J^2 \\ J^3 \end{pmatrix}$, we should use

$$\begin{pmatrix} J_+ \\ J_- \\ J^3 \end{pmatrix} \equiv \begin{pmatrix} J^1 + iJ^2 \\ J^1 - iJ^2 \\ J^3 \end{pmatrix} \tag{13}$$

as explained in chapter IV.2. (While we distinguish between upper and lower indices in our general discussion, we don't when we refer to an example; whether we write J^3 or J_3 here is immaterial.) You should realize that, in spite of the unfamiliar notation, we are going over totally familiar stuff.

We want to go through the analogous procedure for a general Lie algebra and find the analogs of $J_\pm$. (To make sure that you follow this discussion, you should work out what $\beta^3(a)$ and $\beta^8(a)$ are for $SU(3)$.)

A better notation

Staring at (11) for a while, recall that we also know, from (4), that these matrices are determined by the structure constants $(T^i)^a_b = -if^{ia}_b$. Thus, in this basis, $if^{ia}_b = \beta^i(a)\delta^a_b$. What an ugly expression! But you know full well what it means.

You should remember that in the original set of generators X^a, $a = 1, \cdots, n$, we have separated out the generators $H^1, H^2, \cdots, H^r$. The remaining $(n - l)$ generators were called Es earlier. Let us now compute the commutator of H^i with one of these $(n - l)$ remaining generators. In our chosen basis, we have

$$[H^i, X^a] = if^{ia}_b X^b = \beta^i(a)\delta^a_b X^b = \beta^i(a)X^a \tag{14}$$

---

* Since otherwise T^i would be proportional to the identity matrix, which we know is not the case.

Note that no summation over a is implied on the right hand side; this says, quite remarkably, that in this basis, H^i commuted with X^a yields X^a again, multiplied by some proportionality constant $\beta^i(a)$. (In the $SU(2)$ example, J^3 commuted with $J_\pm$ yields $\pm J_\pm$.)

When the real numbers $\beta^i(a)$ first popped up in (11), we thought of it as the numbers along the diagonal of each of the l possible T^i matrices. Let us now turn our brains around, and think of

$$\vec{\beta}(a) \equiv (\beta^1(a), \beta^2(a), \cdots, \beta^l(a)) \tag{15}$$

as an l-dimensional vector, known as the root vector* (or root for short), for each a.

Thus, we can associate a vector $\vec{\beta}(a)$ with each a. It is convenient to rename the generator X^a associated with the root $\vec{\beta}(a)$ as $E_{\vec{\beta}}$, or more simply, E_β for ease of writing. In other words, name the generators not in the Cartan subalgebra by their jobs.[4] This makes good sense. What we did for $SU(3)$, inventing names like I, U, and V for different generators, will clearly become unmanageable when we go to larger algebras.

Roots appear

In this revised notation, (14) now reads

$$[H^i, E_\beta] = \beta^i E_\beta \tag{16}$$

To clarify the notation: before, we specified a generator X^a ($a = l + 1, \cdots, n$) and called its associated root vector $\vec{\beta}(a)$. Now we prefer to specify a root vector $\vec{\beta}$ and to label the corresponding generator by E_β. This is simply a matter of convenience and historical precedent.

Again, to see what is going on, think of $SU(2)$ or $SU(3)$, for which the correspondents of the E_βs are $J_\pm$, or $I_\pm$, $U_\pm$, and $V_\pm$.

Indeed, for $SU(2)$ or $SO(3)$, (16) corresponds to $[J^3, J_\pm] = \pm J_\pm$, as has already been mentioned. Since $l = 1$, the root vector $\vec{\beta}(\pm) = \pm 1$ is 1-dimensional, that is, it is just a number; E_+ is J_+ and E_- is J_-. Note that in this basis $\beta^3(+) = +1$ and $\beta^3(-) = -1$ are real, as anticipated.

Hermitean conjugating (16), we obtain $[H^i, E_\beta^\dagger] = -\beta^i E_\beta^\dagger$, where we used the reality of β^i. In other words,

$$E_{-\beta} \equiv E_\beta^\dagger \tag{17}$$

is associated with the root $(-\vec{\beta})$. (Note that while X^a is hermitean, E_β is not; it's just a question of basis: in $SU(2)$, for example, we have the hermitean J_x and J_y versus the nonhermitean $J_\pm$.)

In summary, the original set of generators X^a, $a = 1, 2, \cdots, n$, has now been divided into 2 sets, H^i, $i = 1, 2, \cdots, l$, and E_β, one for each of the $(n - l)$ root vectors. (Again, I remind you that in $SU(3)$, the H^is are I_3 and Y, and the E_βs are $I_\pm$, $U_\pm$, and $V_\pm$.)

---

* As you can tell, we are, slowly but surely, moving toward contact with chapter VI.2.

A theorem about the rank and the number of generators

We just learned that the negative of a root is also a root. It follows that the number of roots[5] is even, and so $(n - l)$ must be an even integer. (For $SU(2)$, $n = 3$, $l = 1$, $n - l = 2$; for $SU(3)$, $n = 8$, $l = 2$, $n - l = 6$.)

Interestingly, when the number of generators is odd, the rank has to be odd, and when the number of generators is even, the rank has to be even.

Let us try out this theorem on the orthogonal algebras: $SO(4)$ has an even number $(4 \cdot 3/2 = 6)$ of generators and has rank 2. Observe that $SO(5)$ still has an even number $(5 \cdot 4/2 = 10)$ of generators and hence its rank must be even, and indeed, it has rank 2, just like $SO(4)$. More generally, $SO(4k)$ and $SO(4k + 1)$ have $4k \cdot (4k - 1)/2$ and $(4k + 1) \cdot 4k/2$ generators, respectively, both even numbers. In contrast, $SO(4k + 2)$ and $SO(4k + 3)$ have $(4k + 2) \cdot (4k + 1)/2$ and $(4k + 2) \cdot (4k + 1)/2$ generators, respectively, both odd numbers. In chapter VII.1, we will see further evidence of different orthogonal algebras behaving quite differently.

Making roots out of roots

Given two roots $\vec{\alpha}$ and $\vec{\beta}$, what is $[E_\alpha, E_\beta]$? To answer this, we ask another question: what do $[E_\alpha, E_\beta]$ do to H^i? We know from (16) what E_α and E_β separately do to H^i.

Exploit the Jacobi identity again, using (16):

$$[H^i, [E_\alpha, E_\beta]] = -[E_\alpha, [E_\beta, H^i]] - [E_\beta, [H^i, E_\alpha]]$$
$$= [E_\alpha, \beta^i E_\beta] - [E_\beta, \alpha^i E_\alpha]$$
$$= (\vec{\alpha} + \vec{\beta})^i [E_\alpha, E_\beta]. \tag{18}$$

Thus, $[E_\alpha, E_\beta]$ is associated with the root $\vec{\alpha} + \vec{\beta}$. We can add two root vectors to get another root vector.

Well, for all we know, $[E_\alpha, E_\beta]$ could vanish, in which case $\vec{\alpha} + \vec{\beta}$ is not a root. More precisely, while you are of course free to add vectors, there does not exist a generator associated with this vector.

A "trivial" corollary tells us that $2\vec{\alpha}$ is not a root, since $[E_\alpha, E_\alpha] = 0$. In other words, given a root $\vec{\alpha}$, if $k\vec{\alpha}$ is also a root, then k can only be equal to 0 or ± 1.

Suppose that $[E_\alpha, E_\beta]$ does not vanish: $\vec{\alpha} + \vec{\beta}$ is a root, and $[E_\alpha, E_\beta]$ is proportional to $E_{\alpha+\beta}$. Giving a name to the proportionality constant, we write

$$[E_\alpha, E_\beta] = N_{\alpha,\beta} E_{\alpha+\beta} \tag{19}$$

We can cover both possibilities by saying that $N_{\alpha,\beta} = 0$ if $\vec{\alpha} + \vec{\beta}$ is not a root. In other words, E_α and E_β commute if $\vec{\alpha} + \vec{\beta}$ is not a root. For example, for $SU(3)$, $[I_+, U_+] = V_+$ while $[U_+, V_+] = 0$.

From (19) we see that the (unknown) normalization factors satisfy $N_{\alpha,\beta} = -N_{\beta,\alpha}$. Hermitean conjugating (19), we obtain, using (17), $[E_{-\beta}, E_{-\alpha}] = N^*_{\alpha,\beta} E_{-\alpha-\beta}$. But by (19), this is also equal to $N_{-\beta,-\alpha} E_{-\alpha-\beta}$. Hence

$$N_{\alpha,\beta} = -N^*_{-\beta,-\alpha} = -N_{-\beta,-\alpha} \tag{20}$$

where in the last step we use the fact that the structure constants are real.

We are not done squeezing the juice out of (18). Set $\vec{\beta} = -\vec{\alpha}$. We learn that $[E_{\alpha}, E_{-\alpha}]$ commutes with H^i, for all i. Since by definition the set $\{H^1, \cdots, H^l\}$ is the maximal set of commuting generators, $[E_{\alpha}, E_{-\alpha}]$ must be a linear combination of the H^is. Just for laughs, call the coefficient in the linear combination α_i, so that

$$[E_{\alpha}, E_{-\alpha}] = \alpha_i H^i \tag{21}$$

At this stage, α_i (note the lower index) are a bunch of coefficients. Our job is to find out whether they are related to α^i, if at all. (Indeed, α_i and α^i are related, as the notation suggests and as we will see presently.)

Space spanned by the H^is and the scalar product between roots

Multiply (21) by H^j and trace:

$$\alpha_i \operatorname{tr} H^i H^j = \operatorname{tr}(E_{\alpha} E_{-\alpha} H^j - E_{-\alpha} E_{\alpha} H^j) = \operatorname{tr} E_{-\alpha}[H^j, E_{\alpha}]$$
$$= \alpha^j \operatorname{tr} E_{\alpha} E_{-\alpha} \tag{22}$$

where we used the cyclicity of the trace. Thus far, E_{α} has not been normalized. We now normalize E_{α} so that

$$\operatorname{tr} E_{\alpha} E_{-\alpha} = \operatorname{tr} E_{\alpha}(E_{\alpha})^{\dagger} = 1 \tag{23}$$

Referring back to (7) we see that

$$\operatorname{tr} H^i H^j = g^{ij} \tag{24}$$

Hence (22) says that

$$\alpha_i g^{ij} = \alpha^j \tag{25}$$

As mentioned earlier, we can choose $\operatorname{tr} H^i E_{\alpha} = 0$, so that the metric is block diagonal. In the space spanned by the H^is (for example, the plane coordinatized by I_3 and Y in $SU(3)$), the metric is then simply g_{ij}, the inverse of g^{ij}, defined by $g_{ij} g^{jk} = \delta^k_i$. Thus, we obtain $\alpha_i = g_{ij}\alpha^j$, and our anticipatory notation is justified.

Again, physicists usually take g^{ij} to be simply the Euclidean metric δ^{ij}. For example, for $SU(3)$, any reasonable physicist would normalize T^3 and T^8 so that $\operatorname{tr} T^3 T^3 = \operatorname{tr} T^8 T^8$ and $\operatorname{tr} T^3 T^8 = 0$. As mentioned in chapter V.3, except for a possible stretching of the axes, physicists much prefer spaces to be Euclidean.

The scalar product between two roots is then defined by

$$(\vec{\alpha}, \vec{\beta}) = g_{ij}\alpha^i\beta^j = g^{ij}\alpha_i\beta_j \tag{26}$$

As you will see in the following, we only need the metric to define these scalar products.

Lie algebra in the Cartan basis

To summarize, in general, a Lie algebra is defined as in (4). First, separate the Xs into two sets, the Hs and the Es, known as the Cartan basis.[6] The commutators are then as follows: H with H gives 0; H with E is just E multiplied by a number; E with another E gives either 0 or another E; however, E with $E^\dagger$ (which is also an E) gives a linear combination of Hs.

More precisely, a general Lie algebra is defined by

$$[H^i, H^j] = 0, \tag{27}$$

$$[H^i, E_\alpha] = \alpha^i E_\alpha, \tag{28}$$

$$[E_\alpha, E_\beta] = N_{\alpha,\beta}E_{\alpha+\beta}, \tag{29}$$

$$[E_\alpha, E_{-\alpha}] = \alpha_i H^i \tag{30}$$

We cannot keep on climbing forever

To see what to do next, we look for inspiration in the two examples that we have worked through: $SU(2) \simeq SO(3)$ and $SU(3)$.

In $SU(2)$, let's start with $[J_+, J_-] \propto J_3$. Then we commute repeatedly with J_+, obtaining $[J_+, [J_+, J_-]] \propto [J_+, J_3] \propto J_+$, and finally $[J_+, [J_+, [J_+, J_-]]] = [J_+, J_+] = 0$. Similarly, commuting $[J_+, J_-]$ repeatedly with J_- eventually also gives 0. The basic idea is that we cannot keep on climbing in the same direction, either up or down, forever. The ladder has a top and a bottom rung.

In $SU(3)$, there is the additional feature of different root vectors pointing in different directions. For example, consider $[U_-, V_+] \propto I_+$, and then $[U_-, [U_-, V_+]] \propto [[U_-, I_+] = 0$.

We are thus inspired, given a general Lie algebra, to consider the sequence of nested commutators $[E_\alpha, E_\alpha, [E_\alpha, \cdots [E_\alpha, E_\beta] \cdots]]$, with two roots $\vec{\alpha} \neq \vec{\beta}$.

According to (29), we encounter $E_{\alpha+\beta}$, $E_{2\alpha+\beta}$, and so on. Eventually, we must reach 0, since the algebra has a finite number of generators. Denote by p the maximum number of E_αs in this chain before it vanishes. In other words, p is determined by requiring that $\vec{\beta} + (p+1)\vec{\alpha}$ not be a root. See figure 1.

Similarly, we can consider the sequence $[E_{-\alpha}, [E_{-\alpha}, [E_{-\alpha}, \cdots [E_{-\alpha}, E_\beta], \cdots]]]$, giving us $E_{-\alpha+\beta}$, $E_{-2\alpha+\beta}$, and so on. Denote by q the maximum number of $E_{-\alpha}$s. In other words, q is determined by requiring that $\vec{\beta} - (q+1)\vec{\alpha}$ not be a root.

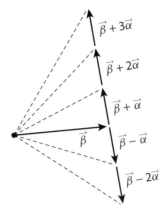

Figure 1 The chain of roots $\vec{\beta} + k\vec{\alpha}$ must terminate on both ends, since the algebra has a finite number of generators. As indicated here, the integers p and q referred to in the text have values $p = 3$ and $q = 2$. We will see later that the figure as drawn is impossible.

The "β chain pointing in the α direction" starts with $\vec{\beta} + p\vec{\alpha}$ and ends with $\vec{\beta} - q\vec{\alpha}$, thus containing all together $p + q + 1$ roots.

For example, in $SU(3)$, the V_+ chain pointing in the U_- direction contains two roots, namely, V_+ and $V_+ + U_- = I_+$. In other words, $p = 1$ and $q = 0$.

Jacobi identity implies a recursion relation

Next consider the Jacobi identity

$$[E_{k\alpha+\beta}, [E_\alpha, E_{-\alpha}]] + [E_\alpha, [E_{-\alpha}, E_{k\alpha+\beta}]] + [E_{-\alpha}, [E_{k\alpha+\beta}, E_\alpha]] = 0. \tag{31}$$

Using (28), (29), and (30), we find that the first term is equal to

$$[E_{k\alpha+\beta}, \alpha_i H^i] = -\alpha_i(k\alpha^i + \beta^i)E_{k\alpha+\beta} \tag{32}$$

while the second and third terms are equal to

$$(N_{\alpha,(k-1)\alpha+\beta}N_{-\alpha,k\alpha+\beta} + N_{-\alpha,(k+1)\alpha+\beta}N_{k\alpha+\beta,\alpha})E_{k\alpha+\beta} \tag{33}$$

Thus, we obtain

$$k(\vec{\alpha}, \vec{\alpha}) + (\vec{\alpha}, \vec{\beta}) = N_{k\alpha+\beta,\alpha}N_{-\alpha,(k+1)\alpha+\beta} + N_{-\alpha,k\alpha+\beta}N_{\alpha,(k-1)\alpha+\beta}. \tag{34}$$

Recall from (19) that $N_{\alpha,\beta} = -N_{\beta,\alpha}$. The first term in (34) can be written as

$$-N_{\alpha,k\alpha+\beta}N_{-\alpha,(k+1)\alpha+\beta} \equiv -M(k, \vec{\alpha}, \vec{\beta}) \tag{35}$$

where, to save writing, we have defined the symbol M. Now we see the second term in (34) is just $+M(k - 1, \vec{\alpha}, \vec{\beta})$. Thus (34) amounts to a recursion relation controlled by the two scalar products $(\vec{\alpha}, \vec{\alpha})$ and $(\vec{\alpha}, \vec{\beta})$:

$$M(k - 1, \vec{\alpha}, \vec{\beta}) = M(k, \vec{\alpha}, \vec{\beta}) + k(\vec{\alpha}, \vec{\alpha}) + (\vec{\alpha}, \vec{\beta}) \tag{36}$$

The initial and final conditions are, as described above,

$$[E_\alpha, E_{p\alpha+\beta}] = 0 \Longrightarrow N_{\alpha, p\alpha+\beta} = 0 \Longrightarrow M(p, \vec{\alpha}, \vec{\beta}) = 0 \tag{37}$$

and

$$[E_{-\alpha}, E_{-q\alpha+\beta}] = 0 \Longrightarrow N_{-\alpha, -q\alpha+\beta} = 0 \Longrightarrow M(-q-1, \vec{\alpha}, \vec{\beta}) = 0 \tag{38}$$

We are now ready to recurse using (36) and the initial condition (37):

$$M(p-1, \vec{\alpha}, \vec{\beta}) = M(p, \vec{\alpha}, \vec{\beta}) + p(\vec{\alpha}, \vec{\alpha}) + (\vec{\alpha}, \vec{\beta}) = cp + d \tag{39}$$

To save writing, we have defined $c \equiv (\vec{\alpha}, \vec{\alpha})$ and $d \equiv (\vec{\alpha}, \vec{\beta})$. Next, from (36), $M(p-2, \vec{\alpha}, \vec{\beta}) = c(p + p - 1) + d + d$, $M(p-3, \vec{\alpha}, \vec{\beta}) = c(p + p - 1 + p - 2) + d + d + d$, and so on, giving us

$$M(p-s, \vec{\alpha}, \vec{\beta}) = c\left(sp - \sum_{j=1}^{s-1} j\right) + sd = s\left(c\left\{p - \frac{1}{2}(s-1)\right\} + d\right) \tag{40}$$

Eventually, we should obtain 0. Indeed, the final condition (38) tells us that $M(-q-1, \vec{\alpha}, \vec{\beta}) = M(p - (p+q+1), \vec{\alpha}, \vec{\beta}) = 0$. Setting $s = p + q + 1$ in (40) and equating the resulting expression to* 0, we obtain $2d/c = 2(\vec{\alpha}, \vec{\beta})/(\vec{\alpha}, \vec{\alpha}) = q - p$.

We conclude that

$$2\frac{(\vec{\alpha}, \vec{\beta})}{(\vec{\alpha}, \vec{\alpha})} = q - p \equiv n \tag{41}$$

with n some integer[†] that can have either sign, since all we know is that p and q are non-negative integers.

Next, we can repeat the same argument with the roles of $\vec{\alpha}$ and $\vec{\beta}$ interchanged; in other words, consider the α chain pointing in the β direction that starts with $\vec{\alpha} + p'\vec{\beta}$ and ends with $\vec{\alpha} - q'\vec{\beta}$. We would obtain

$$2\frac{(\vec{\alpha}, \vec{\beta})}{(\vec{\beta}, \vec{\beta})} = q' - p' \equiv m \tag{42}$$

with m some other integer (that can also have either sign and that may or may not be equal to n). Note that p' and q' are in general not the same as p and q.

Multiplying these two equations, we find that the angle between roots cannot be arbitrary (as we had suspected from our discussion of $SU(3)$):

$$\cos^2 \theta_{\alpha\beta} = \frac{(\vec{\alpha}, \vec{\beta})^2}{(\vec{\alpha}, \vec{\alpha})(\vec{\beta}, \vec{\beta})} = \frac{mn}{4} \tag{43}$$

---

* Recall from our earlier discussion that $p + q + 1$ is equal to the number of roots contained in the chain; it is an integer ≥ 1, since p and q are non-negative integers.

† Not to be confused with the number of generators, clearly. There are only so many letters suitable for integers, and n is standard in this context.

Remarkably, this restricts the unknown integers m and n to only a few possibilities: since $\cos^2 \theta_{\alpha\beta} \leq 1$ we have

$$mn \leq 4 \tag{44}$$

On the other hand, dividing (42) by (41), we find that the ratio of the length squared of the two different roots is required to be a rational number:

$$\frac{(\vec{\alpha}, \vec{\alpha})}{(\vec{\beta}, \vec{\beta})} = \frac{m}{n} \tag{45}$$

These three restrictions, (43), (44), and (45), allowed Cartan to classify all possible Lie algebras.

Our friend the poor man was right; the rather rigid geometrical constraints on the lengths and angles of root vectors restrict root diagrams to only a handful of possible patterns. In chapter VI.4 we will see how this strategy works in detail.

Challenge yourself: see if you can push forward the analysis before reading the next chapter. To make it easier for yourself, first try to classify the rank $l = 2$ algebras; the root vectors then lie in a plane, making them easy to visualize.

Notes

1. In the more mathematical literature used by physicists, for example, in string theory, the adjoint representation is defined as a linear map on the vector space spanned by the Lie algebra. For any two elements X and Y, consider the map $Y \to adj(X)Y \equiv [X, Y]$. Then prove $adj([X, Y]) = [adj(X), adj(Y)]$.

2. In differential geometry, given a decent D-dimensional manifold, with coordinates x^μ (where μ takes on D values; think of the sphere for which $D = 2$), the distance squared between two infinitesimally nearby points with coordinates x^μ and $x^\mu + dx^\mu$ is given by $ds^2 = g_{\mu\nu}dx^\mu dx^\nu$, with μ and ν summed over, of course. Basically, Pythagoras told us that! (For the familiar sphere, $ds^2 = d\theta^2 + \sin^2\theta d\varphi^2$, for example.) Riemann observed that, given the metric $g_{\mu\nu}$, we can obtain the distance between any two points on the manifold by integrating, and by finding the curve giving the shortest distance between these two points, we can define a straight line between these points. This enables us to figure out the geometry of the manifold. The metric $g_{\mu\nu}(x)$ will depend on x (again, think of the sphere coordinatized by the usual θ and φ), and Riemannian geometry is concerned with the study of how $g_{\mu\nu}(x)$ varies from point to point. But we don't need any of that here. The discussion here is necessarily too brief. For further details, see any textbook on general relativity or Einstein gravity. In particular, for a discussion compatible in spirit and style with this book, see G Nut. In particular, see pp. 71 and 183 for detailed arguments on why it is necessary to introduce both upper and lower indices.

3. Such as the Chevalley basis mentioned in the appendix to chapter VI.2.

4. Not so differently from how some people were named in Europe.

5. Strictly speaking, nonvanishing roots.

6. Also known as the Weyl basis.

VI.4 | The Killing-Cartan Classification of Lie Algebras

Constraints on lengths and angles

We learned in chapter VI.3 that the β chain pointing in the α direction, with its two ends at $\vec{\beta} + p\vec{\alpha}$ and $\vec{\beta} - q\vec{\alpha}$ (and thus containing $p + q + 1$ roots), is constrained by*

$$2\frac{(\vec{\alpha}, \vec{\beta})}{(\vec{\alpha}, \vec{\alpha})} = q - p \equiv n \tag{1}$$

Remarkably, performing a series of clever maneuvers starting with this constraint, mathematicians were able to deduce an elegant wealth of information about Lie algebras and to classify them completely.

Note that (1) implies that the α chain pointing in the β direction, with its two ends at $\vec{\alpha} + p'\vec{\beta}$ and $\vec{\alpha} - q'\vec{\beta}$, is similarly constrained by

$$2\frac{(\vec{\alpha}, \vec{\beta})}{(\vec{\beta}, \vec{\beta})} = q' - p' \equiv m \tag{2}$$

Multiplying (1) and (2), we find that the angle between two arbitrary roots is given by

$$\cos^2 \theta_{\alpha\beta} = \frac{(\vec{\alpha}, \vec{\beta})^2}{(\vec{\alpha}, \vec{\alpha})(\vec{\beta}, \vec{\beta})} = \frac{mn}{4} \tag{3}$$

Thus, the integers m and n are limited by

$$mn \leq 4 \tag{4}$$

to only a few possibilities.

Dividing (2) by (1), we see that $\rho_{\alpha\beta} \equiv \frac{(\vec{\alpha}, \vec{\alpha})}{(\vec{\beta}, \vec{\beta})}$ (namely, the ratio of the length squared of the two different roots) is required to be a rational number:

$$\rho_{\alpha\beta} \equiv \frac{(\vec{\alpha}, \vec{\alpha})}{(\vec{\beta}, \vec{\beta})} = \frac{m}{n} \tag{5}$$

---

* Our notation is such that we will often, but not always, drop the vector arrow on the roots.

These restrictions will allow us to classify all possible Lie algebras, as already remarked in chapter VI.3 and as we shall see in detail in this chapter.

Only four possible angles

Note that (3) implies that m and n have to be both positive or both negative.* Furthermore, it says that the angle between two different roots $\vec{\alpha}$ and $\vec{\beta}$ is restricted to be

$$\cos^2 \theta_{\alpha\beta} = \frac{(\vec{\alpha}, \vec{\beta})^2}{(\vec{\alpha}, \vec{\alpha})(\vec{\beta}, \vec{\beta})} = \frac{mn}{4} = 0, \frac{1}{4}, \frac{1}{2}, \frac{3}{4}, \text{ or } 1 \tag{6}$$

Recall that if α is a root, then $-\alpha$ is also a root. Flipping α flips the signs of m and n, and thus we can choose m and n to be both positive. Furthermore, if the two roots α and β are not of equal length, we can always call the longer of the two roots α, so that, with no loss of generality, we can set $m \geq n$. Thus, we can take

$$\cos \theta_{\alpha\beta} = \frac{(\vec{\alpha}, \vec{\beta})}{\sqrt{(\vec{\alpha}, \vec{\alpha})(\vec{\beta}, \vec{\beta})}} \geq 0 \tag{7}$$

and so $0° \leq \theta_{\alpha\beta} \leq 90°$. (In other words, we can always take the angle between α and β to be acute, by flipping α if necessary.) The five cases allowed in (6) correspond to

$$\theta_{\alpha\beta} = 90°, 60°, 45°, 30°, \text{ or } 0° \tag{8}$$

The case $0°$ is not allowed, since this would imply $\vec{\alpha} = \vec{\beta}$, which is not the case by construction. Thus we have only four cases to consider.

The case $90°$ requires special consideration, since $(\vec{\alpha}, \vec{\beta}) = 0$ according to (7). When I divided (2) by (1) to get (5), you should have raised a red flag! I could have divided 0 by 0, something we've been warned against ever since childhood. Indeed, if $(\vec{\alpha}, \vec{\beta}) = 0$, then (1) and (2) tell us that $m = n = 0$, and hence the quantity $\rho_{\alpha\beta}$ defined in (5) is indeterminate.

Let us list the four cases.

Case 1: $\theta_{\alpha\beta} = 90°$ implies that $(\vec{\alpha}, \vec{\beta}) = 0$ and $\rho_{\alpha\beta}$ is indeterminate.

Case 2: $\theta_{\alpha\beta} = 60°$ implies that $\rho_{\alpha\beta} = 1$; the roots $\vec{\alpha}$ and $\vec{\beta}$ have equal length.

Case 3: $\theta_{\alpha\beta} = 45°$ implies that $\rho_{\alpha\beta} = 2$; one root is longer than the other by a factor of $\sqrt{2}$.

Case 4: $\theta_{\alpha\beta} = 30°$ implies that $\rho_{\alpha\beta} = 3$; one root is longer than the other by a factor of $\sqrt{3}$.

We give the allowed values of various quantities for cases[†] 2, 3, and 4 in this table:

m	n	$\frac{(\vec{\alpha}, \vec{\alpha})}{(\vec{\beta}, \vec{\beta})}$	$\cos^2 \theta_{\alpha\beta}$	$\theta_{\alpha\beta}$
1	1	1	$\frac{1}{4}$	$60°$
2	1	2	$\frac{1}{2}$	$45°$
3	1	3	$\frac{3}{4}$	$30°$

---

* Also implied by (5).

† The case $m = n = 2$ gives $\cos^2 \theta_{\alpha\beta} = 1$ and hence $\theta_{\alpha\beta} = 0$.

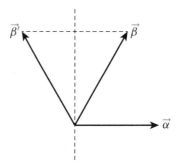

Figure 1

Weyl reflection

Given two roots $\vec{\alpha}$ and $\vec{\beta}$, define

$$\vec{\beta}' \equiv \vec{\beta} - 2\frac{(\vec{\alpha}, \vec{\beta})}{(\vec{\alpha}, \vec{\alpha})}\vec{\alpha} = \vec{\beta} + (p - q)\vec{\alpha} \tag{9}$$

where we invoked (1). Since the β chain pointing in the α direction starts with $\vec{\beta} + p\vec{\alpha}$ and ends with $\vec{\beta} - q\vec{\alpha}$, and since $p \geq p - q \geq -q$, the root $\vec{\beta}'$ belongs to this chain. Denoting the length of $\vec{\alpha}$ and $\vec{\beta}$ by $|\alpha|$ and $|\beta|$, we have

$$\vec{\beta}' = \vec{\beta} - 2\cos\theta_{\alpha\beta}|\beta|\hat{\alpha} \tag{10}$$

where $\hat{\alpha} \equiv \vec{\alpha}/|\alpha|$ is the unit vector pointing in the direction of $\vec{\alpha}$.

The geometrical meaning of $\vec{\beta}'$, known as the Weyl reflection of $\vec{\beta}$, is shown in figure 1. Think of the hyperplane perpendicular to $\vec{\alpha}$ as a mirror. Then $\vec{\beta}'$ is the mirror image of $\vec{\beta}$. Thus, starting with two roots, we can readily generate a whole bunch of other roots by Weyl reflecting repeatedly. For example, in $SU(3)$, starting with $\vec{I}_+$ and $\vec{V}_+$, we generate all six roots.

No more than four roots in a chain

Another big help in constructing root diagrams is the realization that, clearly, chains cannot be arbitrarily long. But we can say more: a chain can contain at most four roots.[1]

To prove this, suppose that a chain contains at least five roots. By calling the root in the middle $\vec{\beta}$, we can always relabel the five roots as $\vec{\beta} - 2\vec{\alpha}$, $\vec{\beta} - \vec{\alpha}$, $\vec{\beta}$, $\vec{\beta} + \vec{\alpha}$, and $\vec{\beta} + 2\vec{\alpha}$.

Now consider the $\beta + 2\alpha$ chain in the β direction. This chain contains only one root, since neither $(\vec{\beta} + 2\vec{\alpha}) - \vec{\beta} = 2\vec{\alpha}$ nor $(\vec{\beta} + 2\vec{\alpha}) + \vec{\beta} = 2(\vec{\beta} + \vec{\alpha})$ is a root.*

---

* We are invoking the "trivial" corollary (mentioned in chapter VI.3) that if $\vec{\alpha}$ is a root, then $2\vec{\alpha}$ is not.

But if the $\beta + 2\alpha$ chain in the β direction contains only one root, this implies $(\vec{\beta} + 2\vec{\alpha}, \vec{\beta}) = 0$ according to (1). Now repeat the argument with the $\beta - 2\alpha$ chain in the β direction to obtain $(\vec{\beta} - 2\vec{\alpha}, \vec{\beta}) = 0$. Adding, we find $(\vec{\beta}, \vec{\beta}) = 0$, which is a contradiction.[2]

Thus, a chain can contain at most four roots. In the notation used earlier, $p + q + 1 \leq 4$.

In particular, the β chain pointing in the $-\beta$ direction contains three roots and has the form $\vec{\beta}, \vec{0}, -\vec{\beta}$, where we are counting $\vec{0}$ as a (null) root. Strictly speaking, $\vec{0}$ represents a linear combination of H^is and should not be called a root; nevertheless, the term "zero root," or "null root," is commonly used.

The by-now familiar $SU(3)$ illustrates all this: for example, the V_+ chain pointing in the U_- direction contains only two roots, as was noted earlier. Indeed, in $SU(3)$, the longest chain contains only three roots, with the form $\vec{\beta}, \vec{0}, -\vec{\beta}$.

All possible rank 2 Lie algebras

In a remarkable achievement, Killing and Cartan[3] were able to classify all possible Lie algebra using these observations.

Let us start with rank 1 and rank 2. We know one rank 1 Lie algebra, namely, $SU(2) \simeq SO(3)$; in fact, the very first one we met. Using the results here, it is easy to prove that this is the only one. For rank 1, the root vector β is 1-dimensional and hence a number, and $\beta, 0, -\beta$ is the only possible chain.

We are now ready to work out all possible rank 2 Lie algebras by using the table just given. Simply examine the possible angles one by one. Once we specify the angle between two roots, their relative lengths are fixed by the table. The root diagrams are easily visualized, since they can be drawn in the plane. In fact, we have already encountered all but one of them:

1. $\theta_{\alpha\beta} = 90°$: The root diagram is shown in figure 2a. Since $[E_\alpha, E_\beta] = 0$, we have two independent $SU(2)$s. (We can normalize the roots to have the same length.) We recognize the algebra of $SU(2) \otimes SU(2) \simeq SO(4)$, with $4 + 2 = 6$ generators (four root vectors, plus the two Hs from the rank). Indeed, we have already seen this in figure VI.2.1b.

2. $\theta_{\alpha\beta} = 60°$: The roots $\vec{\alpha}$ and $\vec{\beta}$ have equal length, and by Weyl reflecting, we obtain all the other roots arranged in a hexagonal pattern. (And of course, if $\vec{\alpha}$ is a root, then so is $-\vec{\alpha}$.) See figure 2b, which we recognize as figure V.3.1. This is the $SU(3)$ beloved by particle physicists, with its $6 + 2 = 8$ generators.

3. $\theta_{\alpha\beta} = 45°$: Put down the first root* $\vec{\beta} = (1, 0)$ and make the second root $\vec{\alpha} = (1, 1)$ a factor of $\sqrt{2}$ longer. Generate the other roots by Weyl reflecting and by multiplying by (-1). We obtain the root diagram in figure 2c, showing an algebra with $8 + 2 = 10$ generators, diamond or tilted square. Note that $10 = 5 \cdot 4/2$. This identifies the algebra as $SO(5)$, whose root diagram

---

* Note that to use the table, we have to adhere to the convention that $\vec{\alpha}$ is longer than $\vec{\beta}$.

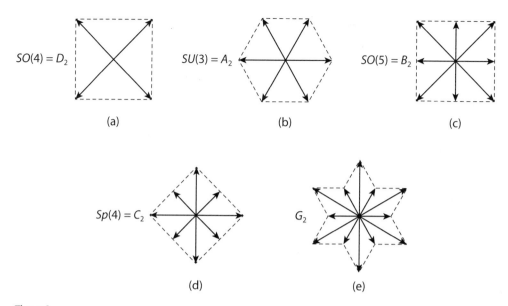

Figure 2

we already saw in figure VI.2.2b. A quick and sloppy way of seeing this is to note that if we took away the four long roots at $(\pm 1, \pm 1)$ (signs uncorrelated) we would get $SO(4)$.

4. $\theta_{\alpha\beta} = 45°$: Put down the first root $\vec{\alpha} = (1, 0)$ as before but now make the second root $\vec{\beta} = \frac{1}{2}(1, 1)$ a factor of $\sqrt{2}$ shorter. Again, generate the other roots by Weyl reflecting and by multiplying by (-1). We obtain the root diagram in figure 2d (which we recognize from figure VI.2.4), showing an algebra with $8 + 2 = 10 = 5 \cdot 4/2$ generators. We obtain the symplectic algebra $Sp(4)$.

But behold, this diagram is just the root diagram for $SO(5)$ tilted by 45°. So, we have discovered the local isomorphism $Sp(4) \simeq SO(5)$.

5. $\theta_{\alpha\beta} = 30°$: Put down the first root $\vec{\beta} = (1, 0)$ and make the second root a factor of $\sqrt{3}$ longer, so that $\vec{\alpha} = \sqrt{3}(\cos 30°, \sin 30°) = \frac{1}{2}(3, \sqrt{3})$. Weyl reflecting and multiplying by (-1) now produces a "star of David" root diagram shown in figure 2e. This is perhaps the most interesting case; it produces a rank 2 algebra we haven't seen before. The algebra has $12 + 2 = 14$ generators and is known as the exceptional algebra G_2.

Dialects

We will go on to classify all possible Lie algebras, but first, we have to mention an awkward notational divergence. While physicists and mathematicians use the same language of logical deduction, they speak different dialects, unfortunately. Mathematicians named the Lie algebras A, B, C, D, E, F, and G. (How imaginative!) The series A, B, C, and D are known to physicists as the algebras of the unitary, orthogonal, and symplectic groups. The translation is as follows:

A_l	$SU(l+1)$
B_l	$SO(2l+1)$
C_l	$Sp(2l)$
D_l	$SO(2l)$

Note that rotations in odd and even dimensional spaces are mathematically distinct.

The algebras E_6, E_7, E_8, and F_4 (which we will mention in the next chapter), and G_2 (which we have already met) are known as exceptional algebras; they don't belong to infinite families.

The mathematical notation is superior in one respect: the integer l corresponds to rank. For example,

$$SU(3) = A_2, \qquad SO(4) = D_2, \qquad SO(5) = B_2, \tag{11}$$

all have the same rank, namely, 2.

Positive and simple roots, and some theorems about roots

At this point, you can readily prove some more theorems[4] about roots, which can be of great help in constructing root diagrams.

Start with an empirical observation: in each of the root diagrams in figure 2, there are only two different lengths. For example, C_2 contains roots of length 1 and of length $\sqrt{2}$ (in some suitable units). So, a "physicist's theorem" suggests itself, backed by five (or is it only four?) "experimental" points: A root diagram contains at most two different lengths.

I leave it as an exercise to prove (or disprove) this theorem. It is in fact true. I already mentioned in chapter VI.2 that the longer root is called the long root, the shorter one the short root.

Next, I remind you of the notions of positive and simple roots, notions we already encountered quite naturally when discussing $SU(3)$, which will serve as a canonical example.

Arrange the Hs in some order, H^1, H^2, $\cdots$, H^r, and set up coordinates accordingly. (For example, we might choose, for $SU(3)$, the first axis to correspond to λ^3, the second to λ^8.) The notion of positive roots can then be defined: A root is said to be positive if the first nonzero component of the root vector is positive. (For example, for $SU(3)$, the roots $(\frac{1}{2}, \frac{\sqrt{3}}{2})$, $(1, 0)$, and $(\frac{1}{2}, -\frac{\sqrt{3}}{2})$, corresponding to V_+, I_+, and U_-, are positive, while $(-\frac{1}{2}, \frac{\sqrt{3}}{2})$, corresponding to U_+, is not.)

The notion of positivity depends on how we choose our coordinates. Of the six roots of $SU(3)$, the three pointing "mostly east" are deemed positive, because we privileged* λ^3.

---

* Notice also that this definition of positive roots does not accord with the physicist's historical choice of which operators (namely, I_+, V_+, and U_+) to call plus or "raising."

Next, we want to codify the fact that not all positive roots are linearly independent. (For example, for $SU(3)$, three positive roots live in 2-dimensional space.) A subset of the positive roots is known as simple roots, if the following holds. When the other positive roots are written as linear combinations of the simple roots, all coefficients are positive. (In the $SU(3)$ example, V_+ and U_- are simple roots, but not I_+, since $V_+ + U_- = I_+$.) Roughly speaking, the simple roots "define," in the root diagram, a privileged sector, somewhat analogous to the upper right quadrant we encountered when we first studied the Cartesian geometry of the plane. While we already introduced both of these notions, positive and simple roots, in chapter V.3, it is worth repeating them to emphasize that they apply to all Lie algebras.

A rank l algebra has l simple roots, which can be used as the basis vectors for the space the root vectors live in. Again, take $SU(3)$: the simple roots V_+ and U_- provide a basis consonant with the hexagonal symmetry of the root diagram, and in this sense are more natural than the basis physicists use for historical reasons. The price we pay is of course that the basis vectors are not orthogonal.

If $\vec{\alpha}$ and $\vec{\beta}$ are two simple roots, then $\vec{\alpha} - \vec{\beta}$ cannot be a root.* (For example, in $SU(3)$, $V_+ - U_- = (0, \sqrt{3})$ is not a root.) Prove this theorem before reading on. (The proof is exceptionally easy!)

Confusio speaks up: "I am not confused for a change. I simply want to mention that some books state this rather confusingly as 'If $\vec{\alpha}$ and $\vec{\beta}$ are two simple roots, then $\vec{\alpha} - \vec{\beta}$ is 0.' Or that '$\vec{\alpha} - \vec{\beta}$ is a 0 root.'"

We agree that it is confusing. The term "0 root" should properly refer to the H^is. Perhaps these authors mean a void or null root, in the sense of a nonexistent root?

Here is the proof. Suppose $\vec{\alpha} - \vec{\beta}$ is a root. With no loss of generality,[†] let it be a positive root. Then a positive root can be written as $\vec{\alpha} - \vec{\beta}$, contradicting the assumption that $\vec{\alpha}$ and $\vec{\beta}$ are simple.

The angle between two simple roots has to be obtuse (or right)

Intuitively, since the simple roots enclose all positive roots, we would expect the angles between the simple roots to be largish. For $SU(3)$, the angle between the roots V_+ and U_- is 120°. This intuitive expectation leads us to the following theorem.

For $\vec{\alpha}$ and $\vec{\beta}$ two simple roots, the angle between them can only be 90°, 120°, 135°, or 150°.

The proof is as follows. Go back to (1). Since we just showed that $\vec{\beta} - \vec{\alpha}$ is not a root, it follows that $q = 0$ in (1), and thus $2\frac{(\vec{\alpha}, \vec{\beta})}{(\vec{\alpha}, \vec{\alpha})} = -p \leq 0$, which implies that $\theta_{\alpha\beta} \geq 90°$.

Confusio mutters, "But we proved earlier that the angle $\theta_{\alpha\beta}$ between two roots could only be 90°, 60°, 45°, or 30°."

---

* Note that if we were merely given two roots $\vec{\alpha}$ and $\vec{\beta}$, then $\vec{\alpha}$ $\vec{\beta}$ may or may not be a root.
[†] Since otherwise interchange $\vec{\alpha}$ and $\vec{\beta}$.

But, Confusio, that was between any two roots $\vec{\alpha}$ and $\vec{\beta}$, and by giving $-\vec{\alpha}$ the name $\vec{\alpha}$ if necessary, we chose $\theta_{\alpha\beta}$ to be less than or equal to $90°$. The statement here is about two simple roots. And if $\vec{\alpha}$ is simple, then $-\vec{\alpha}$ is certainly not simple (it's not even positive).

Similarly, since $\vec{\alpha} - \vec{\beta}$ is not a root, it follows that $q' = 0$ in (2), and thus, $2\frac{(\vec{\alpha},\vec{\beta})}{(\vec{\beta},\vec{\beta})} = -p' \leq 0$. Thus, (3) still holds, that is, $\cos^2 \theta_{\alpha\beta} = \frac{(\vec{\alpha},\vec{\beta})^2}{(\vec{\alpha},\vec{\alpha})(\vec{\beta},\vec{\beta})} = \frac{pp'}{4}$, which together with the information that the $\theta_{\alpha\beta} \geq 90°$, proves the theorem.

Exercise

1 Show that a root diagram contains at most two different lengths.

Notes

1. Confusio says, "The physical example of the four N^* resonances mentioned in connection with Gell-Mann's $SU(3)$ is a living example of a chain with four roots." No, Confusio, that's a weight diagram, not a root diagram!

2. Confusio again. He claims to have a simpler proof. If $\vec{\beta} + 2\vec{\alpha}$ is a root, then subtract $\vec{\beta}$ to get $2\vec{\alpha}$, which we know is not a root if $\vec{\alpha}$ is a root. The fallacy is that $[E_{\beta+2\alpha}, E_\beta^\dagger]$ may vanish. If Confusio's proof is valid, he would have proved that there can be at most three, not four, roots in a chain.

3. When writing this, it occurred to me that since I knew S. S. Chern, I am only separated from a historical figure like Cartan by two degrees. And now, since you know me so well, dear reader, you are only separated from Cartan by three degrees!

4. Some readers might have noticed that I have kept our friend the jargon guy at bay; he loves to talk about simple algebras, semisimple algebras, on and on, instead of the essence of the matter. Whatever. (In fact, he did define semisimple groups for us back in chapter I.2.) You might want to get the names straight; it's up to you. As Dick Feynman said, the name of the bird is not important; it is only important to Murray Gell-Mann. Imagine that one day you would be so fortunate to feel that you are closing in on the ultimate theory of the world. Are you going to pause and worry if the Lie algebra you are using is simple or not?

Lengths of simple roots and the angle between them

For Lie algebras with rank l much larger than 2, the root diagrams are l-dimensional and hence impractical to draw. The Dynkin diagram was invented as one way to capture the relevant information pictorially.[1]

In hindsight at least, it seems clear how one might proceed. First, there is no need to draw the negative as well as the positive roots. Second, the positive roots can be constructed as linear combinations of the simple roots. But we deduced in chapter VI.4 that the angle between two simple roots can only be 90°, 120°, 135°, or 150°. Hence, in the diagram, we need only specify the angle between the simple roots and their relative lengths.

Draw a small circle for each simple root. Connect the two circles* corresponding to two simple roots by one, two, or three lines if the angle between them is 120°, 135°, or 150°, respectively. Do not connect the two circles if the angle between them is 90°.

Recall from chapter VI.4 that for two simple roots α and β,

$$2\frac{(\vec{\alpha}, \vec{\beta})}{(\vec{\alpha}, \vec{\alpha})} = p, \qquad 2\frac{(\vec{\alpha}, \vec{\beta})}{(\vec{\beta}, \vec{\beta})} = p', \qquad \frac{(\vec{\alpha}, \vec{\alpha})}{(\vec{\beta}, \vec{\beta})} = \frac{p'}{p} \tag{1}$$

(Note that the simplicity of the roots implies that the integers q and q' in (VI.4.1) and (VI.4.2) vanish.)

Thus, by construction, the number of lines connecting the two circles corresponding to two simple roots α and β is given by

$$\mathcal{N}_L(\alpha, \beta) = 4\cos^2\theta_{\alpha\beta} = 4\frac{(\vec{\alpha}, \vec{\beta})^2}{(\vec{\alpha}, \vec{\alpha})(\vec{\beta}, \vec{\beta})} = \left(2\frac{(\vec{\alpha}, \vec{\beta})}{(\vec{\alpha}, \vec{\alpha})}\right)\left(2\frac{(\vec{\alpha}, \vec{\beta})}{(\vec{\beta}, \vec{\beta})}\right) = pp' \tag{2}$$

---

* I will drop the adjective "small."

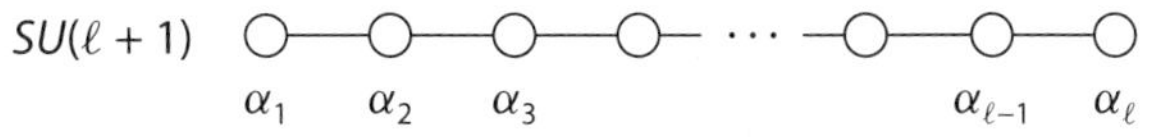

$$SU(\ell+1) \quad \underset{\alpha_1}{\bigcirc}\!-\!\!\underset{\alpha_2}{\bigcirc}\!-\!\!\underset{\alpha_3}{\bigcirc}\!-\!\bigcirc\!-\cdots\!-\bigcirc\!-\!\!\underset{\alpha_{\ell-1}}{\bigcirc}\!-\!\!\underset{\alpha_\ell}{\bigcirc}$$

Figure 1

$$SU(2) \quad \bigcirc \qquad\qquad SU(3) \quad \bigcirc\!-\!\bigcirc$$
$$(a) \qquad\qquad\qquad\qquad (b)$$

Figure 2

It is useful to write this result as[2]

$$\cos\theta_{\alpha\beta} = -\frac{\sqrt{pp'}}{2} = -\frac{1}{2}, \; -\frac{1}{\sqrt{2}}, \; -\frac{\sqrt{3}}{2} \tag{3}$$

The four families

Recall from chapter VI.2 that the simple roots of the Lie algebras we know and love are given by

	Mathematical notation	Simple roots
$SU(l+1)$	A_l	$e^i - e^{i+1}, \; i = 1, \cdots, l$
$SO(2l+1)$	B_l	$e^{i-1} - e^i, e^l, \; i = 2, \cdots, l$
$Sp(2l)$	C_l	$e^{i-1} - e^i, 2e^l, \; i = 2, \cdots, l$
$SO(2l)$	D_l	$e^{i-1} - e^i, e^{l-1} + e^l, \; i = 2, \cdots, l$

Let us see how the Dynkin construction works for $SU(l+1)$. There are l simple roots, $e^i - e^{i+1}, i = 1, \cdots, l$. (For example, for $SU(3)$, the two simple roots are $e^1 - e^2$ and $e^2 - e^3$.) We have $(e^i - e^{i+1})^2 = 1 + 1 = 2$, while $(e^i - e^{i+1})(e^j - e^{j+1}) = 0$ unless $j = i \pm 1$, in which case it is equal to -1. Thus, the angle θ between two neighboring simple roots $e^{i-1} - e^i$ and $e^i - e^{i+1}$ is given by $\cos^2\theta = (-1)^2/(2 \cdot 2) = 1/4$, that is, $\theta = 180° - 60° = 120°$. In contrast, the angle between non-neighboring simple roots is $90°$. The number of circles is equal to the rank of the algebra.

Following the rules stated just now, we obtain the Dynkin diagram for $SU(l+1)$ as shown in figure 1, consisting of l circles, with neighboring circles joined by a single line. A mnemonic: the diagram for our beloved $SU(2)$ consists of a single small circle; for $SU(3)$ it is two circles joined by a line, as shown in figure 2.

Roots of different lengths

Next, consider the $SO(2l+1)$ family. There are now two types of simple roots, $e^{i-1} - e^i, i = 2, \cdots, l$ and e^l, with different lengths. We have to incorporate this information in the diagram.

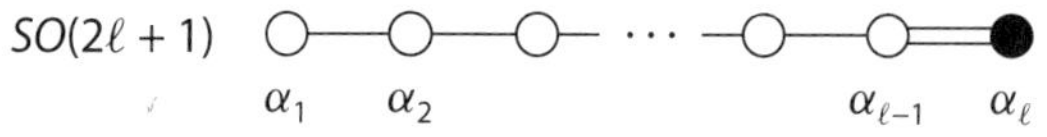

$SO(2\ell + 1)$ $\quad$ O——O——O— $\cdots$ —O——O══●

$\qquad\qquad\qquad\qquad \alpha_1 \quad \alpha_2 \qquad\qquad\qquad \alpha_{\ell-1} \quad \alpha_\ell$

Figure 3

$SO(3)$ $\quad$ O $\qquad\qquad$ $SO(5)$ $\quad$ O══●

$\qquad\qquad$ (a) $\qquad\qquad\qquad\qquad\qquad$ (b)

Figure 4

We impose the additional rule* that we fill the circle (that is, darken it with ink) of the short root. Note that it is crucial, as was proved in an exercise in chapter VI.4, that a root diagram contains at most two different lengths. Mnemonic: short = shaded.

There are $l - 1$ long roots and one short root. The angle between two neighboring long roots is equal to 120° as in the $SU(l + 1)$ case; the open circles for neighboring long roots are connected by a single line. In contrast, the angle between the long root $e^{l-1} - e^l$ and the short root e^l is given by $\cos^2 \theta = (-1)^2/(2 \cdot 1) = 1/2$. Thus, $\theta = 135°$, and we join the circles corresponding to these two roots, one open and one shaded, by two lines. See figure 3.

It is instructive to note two simple cases: $SO(5)$ corresponds to an unfilled circle and a filled circle joined by two lines, while $SO(3)$ consists of a single circle, which there is no point in filling, since there is only one root and so the distinction between long and short roots does not arise. Thus, the algebra $SU(2)$ and $SO(3)$ have the same Dynkin diagram, confirming the isomorphism between them that we have known about for a long time. See figure 4.

Rules for constructing Dynkin diagrams

The rules of the game are surprisingly simple:

1. Each diagram consists of l circles, with l the rank of the Lie algebra. The open or white circles correspond to the long roots, the filled or dark circles to the short roots.

2. Two circles are connected by either zero, one, two, or three lines.

Master these rules, and you could even try to construct the rest of this chapter! Actually, you also need a couple of theorems that we will prove later, such as no more than three lines can come out of a circle.

---

* Caution: some authors reverse the black and white convention used here. Indeed, some authors even have all circles unshaded and superpose an arrow on the lines joining two roots to indicate which of them is shorter, using some convention, such as the short points toward the long. Other authors write the length of the roots inside the circle. Different strokes for different folks.

Figure 5

Symplectics and a black-white symmetry

For $Sp(2l)$, the simple root $2e^l$ is longer than all the other simple roots $e^{i-1} - e^i$, $i = 2, \cdots, l$. The angle between $2e^l$ and $e^{l-1} - e^l$ is now given by $\cos^2 \theta = (-2)^2/(2 \cdot 4) = 1/2$. Once again, $\theta = 135°$, and we join the circles corresponding to the two roots by two lines. The angle between neighboring short roots is given by $120°$ (as for $SU(l + 1)$), and the corresponding circles are connected by one line.

We thus have the Dynkin diagram for $Sp(2l)$ shown in figure 5. Amusingly, it may be obtained from the Dynkin diagram for $SO(2l + 1)$ by interchanging black and white.

Instructively, $Sp(4)$ and $SO(5)$ have the same Dynkin diagram, showing that the two algebras are isomorphic.

Is $SO(2l)$ "better" than $SO(2l + 1)$?

Finally, we come to $SO(2l)$, perhaps the most interesting case of all. The l simple roots, $e^{i-1} - e^i$, $i = 2, \cdots, l$ and $e^{l-1} + e^l$, now have the same length, and so we need not distinguish between open and filled circles, in contrast to the case for $SO(2l + 1)$.

The angle between neighboring simple roots $e^{i-1} - e^i$ and $e^i - e^{i+1}$ is $120°$, so that the corresponding circles are connected by a single line, while the angle between non-neighboring simple roots vanishes. (Thus far, this is reminiscent of our discussion of $SU(l + 1)$.) But interestingly, the angle between the one special simple root $e^{l-1} + e^l$ and the last of this string of simple roots, $e^{l-1} - e^l$, vanishes, and so the corresponding circles are not connected. In contrast, $(e^{l-1} + e^l) \cdot (e^{l-2} - e^{l-1}) = -1$, and the angle between the two roots $e^{l-1} + e^l$ and $e^{l-2} - e^{l-1}$ is $120°$. The corresponding circles are then connected by one line.

Thus, we obtain the Dynkin diagram shown in figure 6 for $SO(2l)$: it has a forked structure at the end. The circle associated with the simple root $e^{l-2} - e^{l-1}$ is connected to both $e^{l-1} - e^l$ and $e^{l-1} + e^l$.

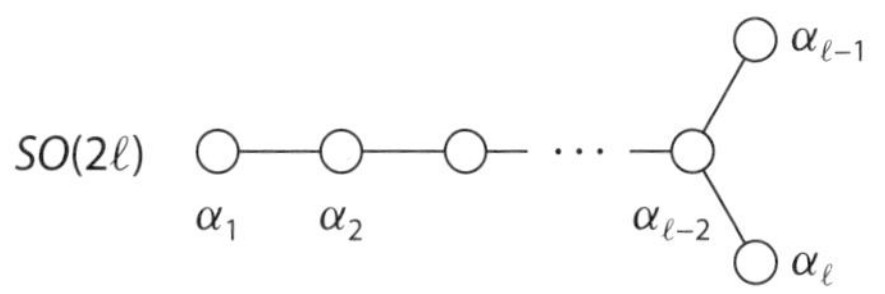

SO(2ℓ) ○──○──○─ ⋯ ─○⟨ $\alpha_{\ell-1}$ / α_ℓ

$\alpha_1 \quad \alpha_2 \qquad\qquad \alpha_{\ell-2}$

Figure 6

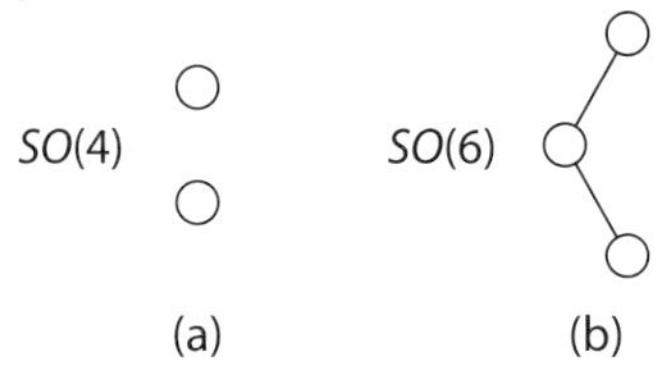

SO(4)

(a)

SO(6)

(b)

Figure 7

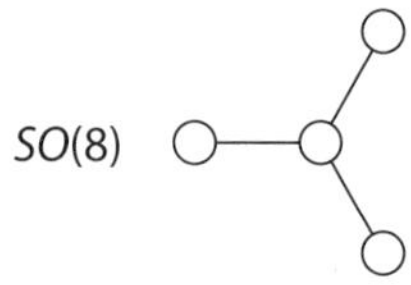

SO(8)

Figure 8

Again, it is instructive to look at what happens for some small values of l. For $l = 2$, the Dynkin diagram breaks up into two unconnected circles, confirming our knowledge that the algebra of $SO(4)$ breaks up into two independent $SU(2)$ algebras. For $l = 3$, the Dynkin diagram consists of three circles; we see with a glance that the algebras of $SO(6)$ and $SU(4)$ are isomorphic. See figure 7a,b. We will discuss this isomorphism in chapter VII.1.

Most interestingly, for $l = 4$, the Dynkin diagram of $SO(8)$ (as shown in figure 8) exhibits a remarkable 3-fold symmetry, sometimes called triality by physicists. People often consider the Lie algebra of $SO(8)$ as the most elegant of them all. We will discuss this further in chapters VII.1 and IX.3.

Dynkin diagrams for low-ranked algebras

It is perhaps amusing to draw all the Dynkin diagrams for the low-ranked algebras. For rank 1, there is only one, consisting of a single circle. All rank 2 algebras, $SO(4)$, $SO(5)$, $SU(3)$, $Sp(4)$, and G_2, are given in figure 9.

We have already encountered the algebra G_2. Simply by inspection of figure VI.4.2e, we see that the simple roots are $(1, 0)$ and $(-3, \sqrt{3})/2$, and the angle between them is $180° - 30° = 150°$. Thus, we draw three lines between an unfilled circle and a filled circle to obtain the Dynkin diagram of G_2, as shown in figure 9. It does not belong to any of the four families and is thus classified as exceptional.

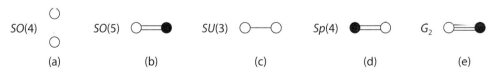

Figure 9

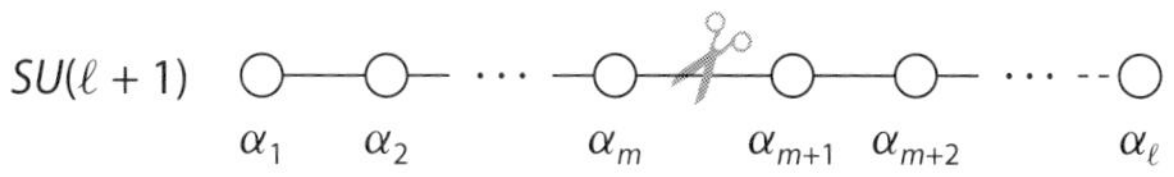

$SU(\ell + 1)$ with nodes α_1, α_2, ..., α_m, α_{m+1}, α_{m+2}, ..., α_ℓ

Figure 10

Cutting Dynkin diagrams

Consider cutting the connection between two circles in a Dynkin diagram. By construction, each of the two pieces that result satisfies the rules for being a Dynkin diagram and thus corresponds to a Lie algebra. Let us see by examples how this works.

The Dynkin diagram of $SU(l + 1)$ consists of l circles connected in a linear chain. Cut the connection between the mth circle and the $(m + 1)$st circle (with $m < l$, of course). See figure 10. The piece consisting of m circles then corresponds to $SU(m + 1)$; the other, consisting of $l - m$ circles, to $SU(l - m + 1)$. Note that $SU(l + 1)$ contains $SU(m + 1)$ and $SU(l - m + 1)$ each as a subalgebra, but not both of them, of course. (Thus, $SU(3)$ contains two overlapping $SU(2)$s, but not $SU(2) \otimes SU(2)$.)

Next, cut the Dynkin diagram of $SO(2l + 1)$ between the $(l - 2)$nd circle and the $(l - 1)$st circle. See figure 11. I chose to cut there simply because this will give the largest simple unitary subalgebra, namely, $SU(l - 1)$. In chapter VII.1, we will see that $SU(l - 1)$ can be embedded naturally into $SO(2l - 2)$, which indeed is a subalgebra of $SO(2l + 1)$. The leftover piece contains one open and one shaded circle, and it corresponds to $SO(5)$.

For $SO(2l)$, cut between the $(l - 3)$rd circle and the $(l - 2)$nd circle. See figure 12a. The two pieces correspond to $SU(l - 2)$ and $SU(4) \simeq SO(6)$, respectively. Alternatively, we could chop off the lth circle from the Dynkin diagram of $SO(2l)$ to obtain the two subalgebras $SU(l)$ and $SU(2)$. See figure 12b.

Finally, for $Sp(2l)$, we can cut between the $(l - 2)$nd circle and the $(l - 1)$st circle. See figure 13a. The two pieces correspond to $SU(l - 1)$ and $SO(5)$, respectively. Alternatively,

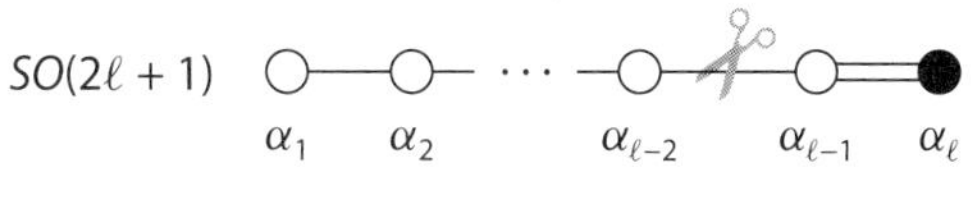

$SO(2\ell + 1)$ with nodes α_1, α_2, ..., $\alpha_{\ell-2}$, $\alpha_{\ell-1}$, α_ℓ

Figure 11

(a) with nodes α_1, α_2, ..., $\alpha_{\ell-3}$, $\alpha_{\ell-2}$, $\alpha_{\ell-1}$, α_ℓ

(b) with nodes α_1, α_2, ..., $\alpha_{\ell-1}$, α_ℓ

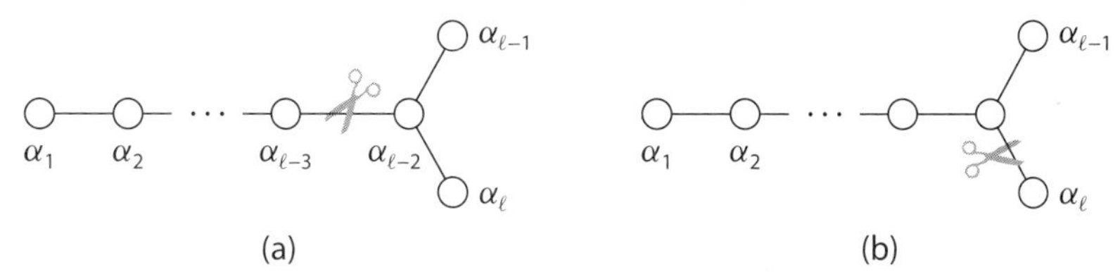

Figure 12

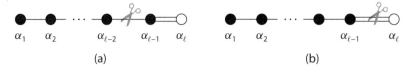

(a)   (b)

Figure 13

we could cut the double lines and obtain an $SU(l)$ and an $SU(2)$ subalgebra. See figure 13b. Interestingly, we showed explicitly in chapter IV.8 that $SU(l)$ is contained in $Sp(2l)$.

Some theorems about Dynkin diagrams

At one level, we can regard Dynkin diagrams as "merely" a compact way of denoting the root diagrams of Lie algebras. There is of course considerably more to Dynkin diagrams than that, with all sorts of theorems one could prove. I mention a few of them here.[3]

First, let us simplify notation, drop the arrows on the roots, and use the standard dot product for the scalar product: $\alpha \cdot \beta = (\alpha, \beta)$, $\alpha^2 = \alpha \cdot \alpha = (\alpha, \alpha) = |\alpha|^2$. Also, henceforth, let α_i, $i = 1, \cdots, l$, denote the l simple roots.* It proves convenient to define $u_i \equiv \alpha_i / |\alpha_i|$ and to work with unit vectors. Then we no longer need to drag along denominators.

By merely rewriting (1) and (3), we have

$$2u_i \cdot u_j = -\sqrt{\zeta_{ij}} \tag{4}$$

Here the integer ζ_{ij} can only take on the values 0, 1, 2, or 3, according to whether the two simple roots are not connected, connected by one line, by two lines, or by three lines, respectively, in the Dynkin diagram. Trivially, if α_i and α_j are connected, then

$$2u_i \cdot u_j \leq -1 \tag{5}$$

$2u_i \cdot u_j$	Lines connecting i and j
-1	1
$-\sqrt{2}$	2
$-\sqrt{3}$	3

We are now ready to prove some theorems.

The no-loop theorem: Loops are not allowed in Dynkin diagrams. (For example, a loop with six roots is shown in figure 14: each root could be connected to some other roots outside the loop, as indicated by dotted lines in the figure.)

---

* In this section, all roots are simple unless stated otherwise. Recall that, by definition, any root appearing in a Dynkin diagram is simple.

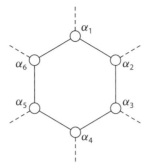

Figure 14

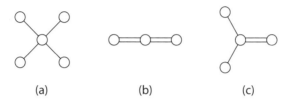

(a)   (b)   (c)

Figure 15

Proof: Consider a loop with k roots. Then α_i is connected to α_{i+1} for $i = 1, 2, \cdots, k - 1$, and α_k is connected to α_1.[*] Hence

$$\left(\sum_{i=1}^{k} u_i\right)^2 = k + \sum_{i \neq j} u_i \cdot u_j = k + 2\sum_{i=1}^{k-1} u_i \cdot u_{i+1} + 2u_k \cdot u_1 \tag{6}$$

The left hand side is strictly positive (that is, > 0), while the right hand side is, according to (5), $\leq k - k$ (that is, ≤ 0). Proof by contradiction.

Note that if we do not have a loop, then the last term $2u_k \cdot u_1$ is absent, and the right hand side is then $\leq k - (k - 1)$, that is, ≤ 1, which is quite consistent[†] with the left hand side being >0. In other words, open strings, but not closed strings, are allowed.

The no-more-than-three lines theorem: The number of lines coming out of a small circle in a Dynkin diagram cannot be more than three.

We conclude immediately that the various situations shown in figure 15 are not allowed. And indeed, we haven't seen anything like these.

Proof: Denote by $w_1, \cdots, w_k$ (which, recall, we are taking to be unit vectors) the k simple roots connected to the given root u. (In figure 15b, for example, four lines come out of the root u to connect to $k = 2$ roots. Recall that all roots in a Dynkin diagram are simple.) The

---

[*] Note that the number of connecting lines must be $\geq k$. Logically, we distinguish between the number of connections and the number of connecting lines. Two roots can be connected by more than one line if they have different lengths. Also, two non-neighboring roots can be connected. (For example, suppose in figure 14 we now additionally connect α_2 to α_4. Then we would actually be dealing with more than one loop.)

[†] In particular, $A_k = SU(k + 1)$ is allowed.

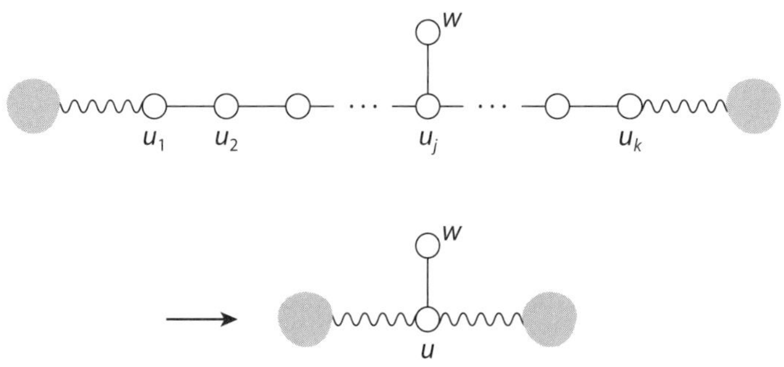

Figure 16

absence of loops means that the ws are not connected to one another, so that $w_i \cdot w_j = 0$. Thus, the k unit vectors w's form an orthonormal set. Note that k cannot be $\geq l$, since the total number of circles in the Dynkin diagram is l. So $k < l$, and the set of vectors w_i, $i = 1, \cdots, k$, cannot form a complete orthonormal basis, and so $\sum_{i=1}^{k}(u \cdot w_i)^2 < 1$. According to (2), the number of lines connecting u to w_i is equal to $4(u \cdot w_i)^2$, and thus the total number of lines coming out of u satisfies $4\sum_i(u \cdot w_i)^2 < 4$. QED.

Given a Dynkin diagram with two circles, this theorem implies that we can join them with no line, one line, two lines, or three lines, but not any more than three lines. These four cases correspond to $SO(4)$, $SU(3)$, $SO(5) \simeq Sp(4)$, and G_2. See figure 9.

The shrinking theorem: Shrinking a linear chain of circles connected to one another by a single line to just one circle leads to a valid Dynkin diagram.

See figure 16 and the proof to understand what this chain of English words means. In the figure, I have indicated the rest of the Dynkin diagram by two shaded blobs, connected to the chain under consideration by two wavy lines. I have also shown a root w connected to a root in this chain (its role will become clear in the proof).

Proof: By assumption, $u_1, \cdots, u_k$ is a chain of simple roots satisfying $u_i \cdot u_{i+1} = -\frac{1}{2}$, $i = 1, \cdots, k - 1$. Consider the vector $u \equiv \sum_i^k u_i$; since $u^2 = \sum_{i=1}^{k} u_i^2 + 2\sum_{i=1}^{k-1} u_i \cdot u_{i+1} = k - 2 \cdot \frac{1}{2}(k - 1) = 1$, it is a unit vector.

Replace the chain by a simple root described by the u. In other words, we have shrunk the chain consisting of $u_1, \cdots, u_k$.

Let w be a unit vector that describes a root in the original Dynkin diagram but not one of the u_is. (In other words, it does not belong to the chain we are shrinking.) It can connect to at most one of the u_is (since otherwise we would have one or more loops); call this root u_j. Then

$$w \cdot u = w \cdot \sum_i^k u_i = w \cdot u_j \tag{7}$$

The unit vector w has the same scalar product with u in the new Dynkin diagram as it did with u_j in the original Dynkin diagram. All of Dynkin's rules are satisfied. This proves the theorem.

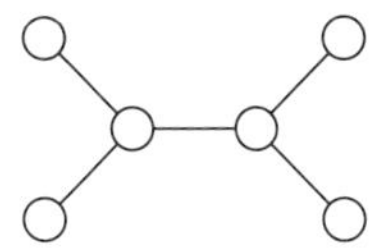

Figure 17

For example, referring to figure 3, we can shrink $SO(2l + 1)$ for $l > 2$ down to $SO(5)$. Note that in figure 16, I show w connecting to u_j by a single line just to be definite, but, as the $SO(2l + 1) \rightarrow SO(5)$ (for $l > 2$) example illustrates, it could be more than one line.

An important corollary follows. I was intentionally vague about the two wavy lines and blobs in figure 16. The theorem implies that the two wavy lines cannot both consist of two connecting lines, since then we could end up with a circle with four lines coming out of it, contradicting a previous theorem. Thus, since the Dynkin diagram of $SO(2l + 1)$ contains a structure at one end, consisting of a double line and a shaded circle, it cannot contain a structure at the other end. The same remark applies to $Sp(2l)$.

This theorem also implies that the diagram in figure 17 is not allowed. While the Dynkin diagram of $SO(2l)$ has a forked structure at one end (figure 6), we cannot have a Dynkin diagram with a forked structure at each of the two ends.

Discovering F_4

Next, consider the double line structure at the end of $SO(2l + 1)$ or $Sp(2l)$. Think of this as a building block in a children's construction set. What can we connect to this block? Call the open circle u_n and shaded circle v_m, and connect to them a chain of n open circles and a chain of m shaded circles, respectively, as shown in figure 18.

Since u_n and v_m are connected by two lines, $u_n \cdot v_m = -1/\sqrt{2}$, according to the table given after (5). By construction, $u_i \cdot u_{i+1} = -\frac{1}{2}, i = 1, 2, \cdots, n - 1$ and $v_j \cdot v_{j-1} = -\frac{1}{2}, j = m, m - 1, \cdots, 2$. Now define $u = \sum_{i=1}^{n} iu_i$ and $v = \sum_{j=1}^{m} jv_j$.

Then

$$u^2 = \left(\sum_{i=1}^{n} iu_i \right)^2 = \sum_{i=1}^{n} i^2 + (-\tfrac{1}{2})2 \sum_{i=1}^{n-1} i(i + 1) = n^2 - \tfrac{1}{2}(n - 1)n = \tfrac{1}{2}n(n + 1) \tag{8}$$

Similarly, $v^2 = \frac{1}{2}m(m + 1)$. Also,

$$u \cdot v = (nu_n) \cdot (mv_m) = -\frac{nm}{\sqrt{2}} \tag{9}$$

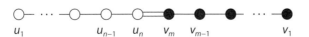

Figure 18

$$F_4 \quad \circ\!\!-\!\!\circ\!\!=\!\!\bullet\!\!-\!\!\bullet$$

Figure 19

Schwartz's inequality insists that $(u \cdot v)^2 \leq u^2 v^2$, which after some algebra (in the high school sense) gives

$$(m-1)(n-1) \leq 2 \tag{10}$$

The strict inequality is satisfied by (i) $m = 1$, n any positive integer; (ii) $n = 1$, m any positive integer; and (iii) $m = 2$, $n = 2$. (In addition, (10) is satisfied as an equality by $m = 3$, $n = 2$ or vice versa. But we can immediately dispose of the case of equality: this holds only if the vectors u and v are equal, which would imply $m(m+1) = n(n+1)$, that is, $m = n$, but $3 \neq 2$.)

Possibility (i) gives an algebra we already know, namely, $SO(2l+1)$, while possibility (ii) gives $Sp(2l)$. See figures 3 and 5.

While it is pleasing to meet old friends again, possibility (iii) is the most interesting, as it leads to a rank 4 exceptional algebra, known as F_4, that we have not seen before, with the Dynkin diagram shown in figure 19.

Looking for more Dynkin diagrams

Let us play the same sort of game as in the preceding section, but this time starting with the Dynkin diagram of $SO(8)$ as the building block (see figure 8). Connect to each of the three ends a chain of open circles, as shown in figure 20a. Note the integers n, m, and p are defined such that $SO(8)$ corresponds to $n = m = p = 1$.

The roots, u_i, v_j, and w_k (with $i = 1, \cdots, n-1$, $j = 1, \cdots, m-1$, and $k = 1, \cdots, p-1$) satisfy $u_i \cdot u_{i+1} = -\frac{1}{2}$, $i = 1, \cdots, n-2$, $v_j \cdot v_{j+1} = -\frac{1}{2}$, $j = 1, \cdots, m-2$, and $w_k \cdot w_{k+1} =$

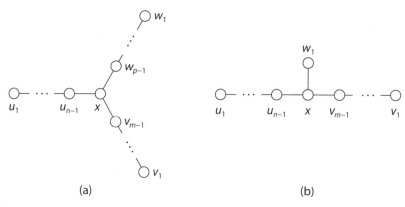

(a)   (b)

Figure 20

$-\frac{1}{2}$, $k = 1, \cdots, p - 2$. Now define $u = \sum_{i=1}^{n-1} i u_i$, $v = \sum_{j=1}^{m-1} j v_j$, and $w = \sum_{k=1}^{p-1} k u_k$. Note the upper limit in each of the sums here. We obtain (by adjusting (8) appropriately)

$$u^2 = \frac{1}{2} n(n - 1), \; v^2 = \frac{1}{2} m(m - 1), \; w^2 = \frac{1}{2} p(p - 1) \tag{11}$$

Furthermore, since u_i, v_j, and w_k (with $i = 1, \cdots, n - 1$, $j = 1, \cdots, m - 1$, and $k = 1, \cdots, p - 1$) are not connected, we have

$$u \cdot v = 0, \qquad v \cdot w = 0, \qquad w \cdot u = 0 \tag{12}$$

Denote by x the root at the center of the Dynkin diagram that connects to all three chains. Since x is connected to u_{n-1}, v_{m-1}, and w_{p-1} each by a single line, we have $x \cdot u_{n-1} = -\frac{1}{2}$, $x \cdot v_{m-1} = -\frac{1}{2}$, and $x \cdot w_{p-1} = -\frac{1}{2}$, and hence

$$x \cdot u = -\frac{1}{2}(n - 1), \qquad x \cdot v = -\frac{1}{2}(m - 1), \qquad x \cdot w = -\frac{1}{2}(p - 1) \tag{13}$$

While x is by definition a unit vector, the vectors u, v, and w are not. Denote their unit vector counterparts by $\hat{u}$, $\hat{v}$, $\hat{w}$, respectively. Thus, the angle between the unit vectors x and $\hat{u}$ is given by (see (11) and (13))

$$\cos^2 \theta_{xu} = (x \cdot \hat{u})^2 = \frac{(x \cdot u)^2}{u \cdot u} = \frac{1}{2}\left(1 - \frac{1}{n}\right) \tag{14}$$

Similarly, $\cos^2 \theta_{xv} = \frac{1}{2}(1 - \frac{1}{m})$ and $\cos^2 \theta_{xw} = \frac{1}{2}(1 - \frac{1}{p})$. Note that the range of the cosine does not impose any restriction, since $n \geq 2$, $m \geq 2$, $p \geq 2$ for the construction to make sense.

We see from (12) that $\hat{u}$, $\hat{v}$, and $\hat{w}$ form an orthonormal basis for a 3-dimensional subspace. Subtract out the component of x in this subspace and define

$$s \equiv x - (x \cdot \hat{u})\hat{u} - (x \cdot \hat{v})\hat{v} - (x \cdot \hat{w})\hat{w} \tag{15}$$

By construction, $s \cdot \hat{u} = 0$, $s \cdot \hat{v} = 0$, and $s \cdot \hat{w} = 0$. Thus, $s^2 = 1 - \left((x \cdot \hat{u})^2 + (x \cdot \hat{v})^2 + (x \cdot \hat{w})^2\right)$. Since s^2 is strictly positive (assume that the rank of the algebra is greater than 3), we have

$$1 > (x \cdot \hat{u})^2 + (x \cdot \hat{v})^2 + (x \cdot \hat{w})^2 \tag{16}$$

Plugging in (14), we obtain the elegant inequality

$$\frac{1}{n} + \frac{1}{m} + \frac{1}{p} > 1 \tag{17}$$

Perhaps you can analyze this inequality before reading on?

Note that if all three of the integers n, m, and p are each ≥ 3, then we have already violated the inequality. Game over. Thus, at least one of these three integers has to be < 3.

As previously noted, they each have to be ≥ 2 for this discussion to even make sense. Let $p = 2$. (In other words, with no loss of generality, take the integer that is < 3 and ≥ 2, and call it p.) Then figure 20a simplifies to figure 20b, and (17) becomes

$$\frac{1}{n} + \frac{1}{m} > \frac{1}{2} \tag{18}$$

Some Platonic solids appear unexpectedly

Now treat the case $m = 2$ and $m > 2$ separately.

For $m = 2$, any integer n satisfies (18). Thus, $(n, m, p) = (n, 2, 2)$ is allowed. Referring to figures 20 and 6, we are pleased to meet our old friends the even orthogonals $SO(2n + 4)$, starting with the celebrated $SO(8)$. (Recall that n must be ≥ 2 for this discussion to make sense.)

Next, for $m > 2$, by high school algebra, write (18) as

$$n\left(1 - \frac{2}{m}\right) < 2 \tag{19}$$

Do you recognize[4] the quantity $(1 - \frac{2}{m})$?

Recall from high school geometry that $(1 - \frac{2}{m})\pi$ is the vertex angle of a regular m-sided polygon (for example, for a triangle, $m = 3$, and the vertex angle is $\pi/3$; for a square, $m = 4$ and $\pi/2$; for a pentagon, $m = 5$ and $3\pi/5$; for a hexagon, $m = 6$ and $2\pi/3$). Consider a regular polyhedron with n regular m-sided polygons that meet at each vertex. Thus, (19), when multiplied by π, "merely" states that at each vertex, the total angle subtended has to be less than 2π.

So, let us now study (18) with $m > 2$. The possible cases are $(n, m) = (3, 3), (3, 4), (3, 5)$. (Note the symmetry of figure 20b under the interchange of n and m. Hence we don't have to worry about $n > 3$.)

Consider these three cases in turn:

1. $(n, m) = (3, 3)$ describes the tetrahedron, with the Dynkin diagram shown in figure 21. The corresponding exceptional algebra is called E_6.

2. $(n, m) = (3, 4)$ describes the cube, with the Dynkin diagram shown in figure 22. The corresponding exceptional algebra is called E_7.

E_6

Figure 21

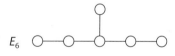

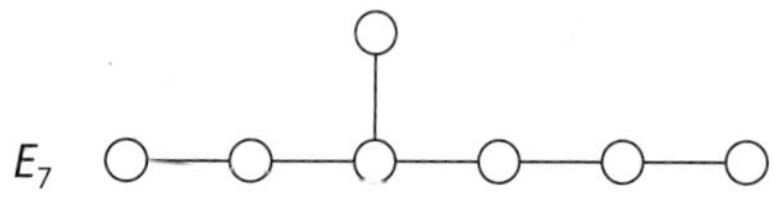

E_7

Figure 22

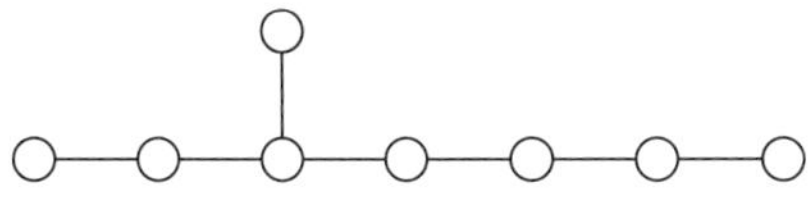

E_8

Figure 23

3. $(n, m) = (3, 5)$ describes the icosahedron, with the Dynkin diagram shown in figure 23. The corresponding exceptional algebra is called E_8.

We might have expected Platonic solids to appear in a discussion of finite groups (such as the tetrahedron for A_4), but who would have anticipated that some of the Platonic solids would appear in a general discussion of Lie algebras? To me, it's the miracle of pure mathematics.

Exceptional algebras

In addition to the four families $A_l = SU(l + 1)$, $B_l = SO(2l + 1)$, $C_l = Sp(2l)$, and $D_l = SO(2l)$, we have encountered the exceptional algebras G_2, F_4, E_6, E_7, and E_8. The beautiful result is that these Lie algebras are all there are!

A lowbrow way to proceed is to simply try to construct more. We end up violating one theorem or another. For example, suppose that, in the Dynkin diagram (figure 3) for $B_l = SO(2l + 1)$, we connect any one of the open circles not at the end to an additional open circle. By shrinking, we would encounter an open circle with four lines coming out of it. Similarly, in the Dynkin diagram for any of the algebra in the E series, we cannot connect any one of the open circles to an additional open circle. Similarly with an additional shaded circle.

Friends, that's all there is!

In summary, starting with the basic definition of a compact Lie algebra in chapter VI.3, mathematicians, by a clever blend of algebra and geometry, have produced a complete classification. As Hardy said in a quote I used earlier in this book, this is mathematics at its best. A complete classification, such as the ancient Greek list of the Platonic solids, is always a monument to the human intellect.

To particle physicists working toward a grand unified gauge theory (whose group theoretic aspects we discuss in part IX) of the strong, weak, and electromagnetic interactions, the Cartan classification is a beacon of light. If the gauge theory is to be based on a compact Lie group, then there are only a finite (in fact, small) number of possibilities.*

---

* As you probably have heard, the leading grand unified theories are built on $SU(5)$, $SO(10)$, and E_6. Furthermore, several possible string theories contain E_8.

Roots of the exceptional algebras

We will not go into how to realize the exceptional algebras.[5] In principle, the procedure is clear. Given the root vectors, we can construct the weight diagrams of various irreducible representations, in particular, the defining or fundamental representation.

How do we determine the root vectors? Simply inspect the Dynkin diagram of each exceptional algebra: the diagram tells us about relative length ratio and angle between the simple roots and should contain enough information to fix the roots. After all, we found the exceptional algebras through their Dynkin diagrams.

Let us write the roots of the exceptional algebras in terms of a set of orthonormal vectors e^i, just as we did in chapter VI.2.

G_2: There are two simple roots, one $\sqrt{3}$ times longer than the other. Let one simple root be $e^1 - e^2$; then the other simple root has to be $-2e^1 + e^2 + e^3$, fixed by the twin requirements of relative length ratio and angle between the two simple roots. (Let's see explicitly how this works. Write the second root as $ae^1 + be^2 + ce^3$. Then $a^2 + b^2 + c^2 = 3 \cdot 2 = 6$ and $a - b = -3$ give $c^2 = -2b^2 + 6b - 3$; the allowed solutions[6] are $b = 1$ and $b = 2$, both leading to essentially the same solution.) The two simple roots of G_2 are

$$e^1 - e^2, -2e^1 + e^2 + e^3 \tag{20}$$

The algebra has rank 2 and so the roots should live in a 2-dimensional space. And indeed they do, since they are orthogonal to $e^1 + e^2 + e^3$.

F_4: We could determine the four simple roots by plugging through as in the preceding example, but an easier way is to note that the Dynkin diagram of F_4 can be constructed from the Dynkin diagram of $B_3 = SO(7)$ (see figures 3 and 19) by adding a short root of the same length as the short root of B_3. From the table in chapter VI.2, the simple roots of B_3 are given by $e^1 - e^2$, $e^2 - e^3$, and e^3. The simple root we are adding has length 1, a dot product of $-\frac{1}{2}$ with e^3, and dot products of 0 with $e^1 - e^2$ and $e^2 - e^3$. The unique solution is $\frac{1}{2}\left(e^4 - (e^1 + e^2 + e^3)\right)$. The four simple roots of F_4 are thus

$$e^1 - e^2, e^2 - e^3, e^3, \frac{1}{2}\left(e^4 - (e^1 + e^2 + e^3)\right) \tag{21}$$

E_8: Construct the Dynkin diagram of E_8 first and then obtain those of E_7 and E_6 by chopping off simple roots. The Dynkin diagram of E_8 is obtained easily by connecting an open circle to a circle at the forked end of $D_7 = SO(14)$. See figures 6 and 23. According to the table in chapter VI.2, the simple roots of D_7 are $e^1 - e^2, \cdots, e^6 - e^7$, and $e^6 + e^7$. See whether you know what to do before reading on.

Well, our first thought is to add $e^8 + e^7$, but to satisfy the requirement of being orthogonal to six of the seven simple roots of D_7, the additional root has to be $\frac{1}{2}\left(e^8 + e^7 - \sum_{i=1}^{6} e^i\right)$. This is orthogonal to $e^1 - e^2, \cdots, e^5 - e^6$, and $e^6 + e^7$. The factor of $\frac{1}{2}$ is to make the length squared equal to 2. The eight simple roots of E_8 are thus[*]

$$e^1 - e^2, \cdots, e^5 - e^6, e^6 - e^7, e^6 + e^7, \frac{1}{2}\left(e^8 + e^7 - \sum_{i=1}^{6} e^i\right) \tag{22}$$

It was not difficult at all!

E_7: Now it is easy to chop off the simple root at the unforked end of the Dynkin diagram of E_8. The seven simple roots of E_7 are

$$e^2 - e^3, \cdots, e^5 - e^6, e^6 - e^7, e^6 + e^7, \frac{1}{2}\left(e^8 + e^7 - \sum_{i=1}^{6} e^i\right) \tag{23}$$

E_6: Chop off another simple root. The six simple roots of E_6 are

$$e^3 - e^4, e^4 - e^5, e^5 - e^6, e^6 - e^7, e^6 + e^7, \frac{1}{2}\left(e^8 + e^7 - \sum_{i=1}^{6} e^i\right) \tag{24}$$

The root $\frac{1}{2}\left(e^8 + e^7 - \sum_{i=1}^{6} e^i\right)$ in E_6, E_7, and E_8 has length squared equal to 2 regardless of the algebra, of course.

Note that as for $A_l = SU(l+1)$, the simple roots of G_2, E_7, and E_6 are expressed in a vector space with dimension higher than the rank of the algebra.

Cartan matrix

Given a rank l Lie algebra with simple roots $\alpha_1, \alpha_2, \cdots, \alpha_l$, the Cartan matrix A is defined by

$$A_{ij} \equiv 2 \frac{(\alpha_i, \alpha_j)}{(\alpha_i, \alpha_i)} \tag{25}$$

Note that A is not necessarily symmetric, since α_i and α_j may have different lengths. The diagonal elements are by definition equal to 2, while the off-diagonal elements can only take on the values 0, -1, -2, and -3 as a result of various theorems we have proved. We can readily write down A using the Dynkin diagram of the algebra.

Knowing the rules for constructing a Dynkin diagram, we can read off the Cartan matrix A from the diagram. For example, for $SU(3)$, we have from figure 2b that $(\alpha_1, \alpha_1) = (\alpha_2, \alpha_3) = 1$ and $(\alpha_1, \alpha_2) = -\frac{1}{2}$ and so

$$A(SU(3)) = \begin{pmatrix} 2 & -1 \\ -1 & 2 \end{pmatrix} \tag{26}$$

We can readily go on. From figure 1, we see that $SU(4)$ involves adding an α_3 orthogonal to α_1 and so forth. Thus,

$$A(SU(4)) = \begin{pmatrix} 2 & -1 & 0 \\ -1 & 2 & -1 \\ 0 & -1 & 2 \end{pmatrix} \tag{27}$$

I trust you to write down $A(SU(l+1))$.

---

[*] Physics is full of surprises! Remarkably, E_8 has emerged in experimental studies of an Ising spin chain.[7]

As another example, from figure 9e for the exceptional algebra G_2, we deduce that $(\alpha_1, \alpha_1) = 3$, $(\alpha_2, \alpha_2) = 1$, and $(\alpha_1, \alpha_2) = \sqrt{3}\cos\theta_{12} = -\sqrt{3}(\sqrt{3}/2) = -3/2$ (with the second equality from (3)) and so

$$A(G_2) = \begin{pmatrix} 2 & -1 \\ -3 & 2 \end{pmatrix} \tag{28}$$

Since the interchange of black and white circles in the Dynkin diagram interchange B_l and C_l, the Cartan matrix does not distinguish B and C.

We end our survey of Lie algebras. Other algebras of interest to physicists lie far beyond the scope of this text.[8]

Appendix 1: Chevalley basis and the Cartan matrix

Let us sketch briefly how the Chevalley basis for $SU(3)$ mentioned in chapter V.2 can be generalized to a general Lie algebra. Denote the simple roots by $\alpha^{(i)}$, $i = 1, \cdots, l$. As we will see, in this discussion, $\alpha_j^{(i)}$ (the jth component of $\alpha^{(i)}$) will appear, and so, for the sake of clarity but at the price of clutter, we introduce a parenthesis to emphasize that i specifies which simple root we are talking about. (Also, we write (i) as a superscript, contrary to the convention used earlier, to avoid too many subscripts.)

Renormalize the raising operator $E_{\alpha^{(i)}}$ associated with the simple root $\alpha^{(i)}$ by $e_i = f_i E_{\alpha^{(i)}}$ (no sum over i); we will choose the numerical factor f_i to our advantage. Denote $(e_i)^\dagger$ by e_{-i}. Referring to the Cartan-Weyl basis given in chapter VI.3, we have $[e_i, e_{-i}] = f_i^2[E_{\alpha^{(i)}}, (E_{\alpha^{(i)}})^\dagger] = f_i^2 \alpha_j^{(i)} H^j \equiv h_i$ (sum over j but no sum over i). Then $[h_i, e_i] = f_i^2 \alpha_j^{(i)} [H^j, E_{\alpha^{(i)}}] f_i = f_i^2 \alpha_j^{(i)} (\alpha^{(i)})^j e_i = f_i^2 (\alpha^{(i)}, \alpha^{(i)}) e_i$. So choose $f_i^2 = 2/(\alpha^{(i)}, \alpha^{(i)})$.

Thus, we obtain

$$[h_i, e_i] = 2e_i, [e_i, e_{-i}] = h_i, \quad i = 1, \cdots, l \tag{29}$$

There are l $SU(2)$ algebras, one for each i, but clearly they cannot all commute with one another; otherwise, the Lie algebra would fall apart into a bunch of $SU(2)$ algebras.

One measure of the noncommutativity is provided by, for $i \neq j$,

$$[h_i, e_j] = f_i^2 \alpha_k^{(i)} [H^k, E_{\alpha^{(j)}}] f_j = f_i^2 \alpha_k^{(i)} (\alpha^j)^k E_{\alpha^{(j)}} f_j$$
$$= 2\frac{(\alpha^{(i)}, \alpha^{(j)})}{(\alpha^{(i)}, \alpha^{(i)})} e_j = A_{ij} e_j \tag{30}$$

Very gratifying! The Cartan matrix A pops right out. We see that its off-diagonal elements measure the linkage between the l "would-be" $SU(2)$ subalgebras.

Appendix 2: Representation theory

I will refrain from working out the irreducible representations of the general Lie algebra.[9] However, given our extensive experience with $SU(2)$ and $SU(3)$, it is fairly clear how to proceed. For a given irreducible representation, find the state with the highest weight, then use the analogs of the lowering operators in $SU(2)$ and $SU(3)$ to act on this state, thus filling in the weight diagram.

Similarly, we can multiply two irreducible representations R_1 and R_2 together by finding the state with the highest weight in $R_1 \otimes R_2$ and then applying the lowering operators.

Exercises

1 Determine the Cartan matrix of F_4 from its Dynkin diagram.

2 Calculate h_i explicitly for $SU(3)$.

Notes

1. Eugene Dynkin invented these diagrams when, as a student in 1944, he was asked by his professor to prepare an overview of the classification of Lie algebras.
2. The three possibilities in (3) correspond to $pp' = 1, 2, 3$, that is, to $\theta_{\alpha\beta} = 120°, 135°, 150°$.
3. One theorem not proved here states that any root can be written as a linear combination of simple roots with integer coefficients. See Z.-Q. Ma, Group Theory for Physicists, p. 291, 7.33 p293 level.
4. In particular, those readers who have read G Nut might. This quantity appears on p. 727 in a discussion of the Descartes theorem on angular deficit.
5. See, for example, R. Slansky, in A. Zee, Unity of Forces in the Universe, volume 1, p. 338.
6. By letting $e^3 \rightarrow -e^3$, we can choose $c > 0$.
7. "Quantum Criticality in an Ising Chain: Experimental Evidence for Emergent E_8 Symmetry," R. Coldea et al. Science 327 (2010), pp. 177–180.
8. In this text, we restrict ourselves to Lie algebras. In recent decades, other algebras have come into physics, notably into string theory, such as Kac-Moody algebra, and so on and so forth. For an overview, see P. West, Introduction to Strings and Branes, Cambridge University Press, 2012.
9. A detailed treatment can be found in more specialized treatments, such as R. Slansky's review article reprinted in Unity.

Part VII | From Galileo to Majorana

The orthogonal groups $SO(N)$ have a more intricate collection of irreducible representations than do the unitary groups $SU(N)$. In particular, the spinor irreducible representations can be complex, real, or pseudoreal, according to whether $N = 4k + 2$, $8k$, or $8k + 4$, respectively, for k an integer. How the spinor irreducible representations of $SO(2n)$ decompose on restriction to the $SU(n)$ subgroup turns out to be crucial in our striving for grand unification.

We then discuss that great chain of algebras of fundamental importance to theoretical physics, namely, the progression through the Galileo, Lorentz, and Poincaré algebras.

In lieu of Dirac's brilliant guess leading to his eponymous equation, group theory leads us by the nose to the Dirac equation via the Weyl equation. A fascinating bit of group theory, lying deep inside the nature of complex numbers, connects Dirac's momentous discovery of antimatter to pseudoreality of the defining representation of $SU(2)$.

In three interludes, we explore the "secret" $SO(4)$ lurking inside the hydrogen atom, the unexpected emergence of the Dirac equation in condensed matter physics, and the even more unexpected emergence of the Majorana equation.

VII.1 | Spinor Representations of Orthogonal Algebras

Spinor representations

That the orthogonal algebras $SO(N)$ have tensor representations is almost self-evident, at least in hindsight. What is much more surprising is the existence of spinor representations. We had a foretaste of this when we saw that the algebra $SO(3)$ is isomorphic to $SU(2)$ and thus has a 2-dimensional spinor representation, and that the algebra $SO(4)$ is isomorphic to $SU(2) \otimes SU(2)$ and thus has not one, but two, 2-dimensional spinor representations.

In this chapter, we show the following.*

1. The algebra $SO(2n)$ has two 2^{n-1}-dimensional spinor irreducible representations.

2. For $SO(4k + 2)$, the spinor representations are complex.

3. For $SO(8m)$, the spinor representations are real.

4. For $SO(8m + 4)$, the spinor representations are pseudoreal.

Don't you find the pattern in points 2, 3, and 4 rather peculiar?

We then show how the spinor in odd-dimensional Euclidean space is related to the spinor in even-dimensional space.

Our discussion will be focused on the Lie algebra of $SO(N)$, rather than on the group.[1]

Clifford algebra

Start with an assertion. For any integer n, we claim that we can find $2n$ hermitean matrices $\gamma_i, i = 1, 2, \cdots, 2n$, which satisfy

$$\{\gamma_i, \gamma_j\} = 2\delta_{ij} I \tag{1}$$

---

* I need hardly say that n, k, and m denote positive integers.

known as a Clifford[2] algebra. As usual in this book, the curly bracket denotes anti-commutation: $\{\gamma_i, \gamma_j\} \equiv \gamma_i \gamma_j + \gamma_j \gamma_i$, and I is the identity matrix.

In other words, to prove our claim, we have to produce $2n$ hermitean matrices γ_i that anticommute with one another ($\gamma_i \gamma_j = -\gamma_j \gamma_i$ for $i \neq j$) and square to the identity matrix ($\gamma_i^2 = I$). We will refer to the γ_is as the γ matrices for $SO(2n)$.

Let us do this by induction.

For $n = 1$, it is a breeze: $\gamma_1 = \tau_1$ and $\gamma_2 = \tau_2$. (Here $\tau_{1,2,3}$ denote the three Pauli matrices.) Check that this satisfies (1).

Next, take a baby step and move on to $n = 2$. We are tasked with producing two more matrices γ_3 and γ_4 such that (1) is satisfied. A moment's thought shows that we are compelled to enlarge the γ matrices. Using the direct product notation explained in the review of linear algebra (and which we already used in chapter IV.8), we write down the four 4-by-4 matrices

$$\gamma_1 = \tau_1 \otimes \tau_3, \; \gamma_2 = \tau_2 \otimes \tau_3, \; \gamma_3 = I \otimes \tau_1, \; \gamma_4 = I \otimes \tau_2 \tag{2}$$

Check that these work. For example, $\{\gamma_1, \gamma_4\} = (\tau_1 \otimes \tau_3)(I \otimes \tau_2) + (I \otimes \tau_2)(\tau_1 \otimes \tau_3) = \tau_1 \otimes \{\tau_3, \tau_2\} = 0$.

Now induct in general! Suppose we have the $2n$ γ matrices for $SO(2n)$, which we denote by $\gamma_j^{(n)}$. We then construct the $2n + 2$ γ matrices for $SO(2n + 2)$ as follows:

$$\gamma_j^{(n+1)} = \gamma_j^{(n)} \otimes \tau_3 = \begin{pmatrix} \gamma_j^{(n)} & 0 \\ 0 & -\gamma_j^{(n)} \end{pmatrix}, \quad j = 1, 2, \cdots, 2n \tag{3}$$

$$\gamma_{2n+1}^{(n+1)} = I \otimes \tau_1 = \begin{pmatrix} 0 & I \\ I & 0 \end{pmatrix} \tag{4}$$

$$\gamma_{2n+2}^{(n+1)} = I \otimes \tau_2 = \begin{pmatrix} 0 & -iI \\ iI & 0 \end{pmatrix} \tag{5}$$

The superscript in parentheses keeps track of which set of γ matrices we are talking about.

Every time we go up in n by 1, we add two γ matrices and double the size of all the γ matrices. Hence, there are $2n$ $\gamma^{(n)}$s, each a 2^n-by-2^n matrix.*

Returning to (3), (4), and (5), we check easily that the $\gamma^{(n+1)}$s satisfy the Clifford algebra if the $\gamma^{(n)}$s do. For example, $\{\gamma_j^{(n+1)}, \gamma_{2n+1}^{(n+1)}\} = (\gamma_j^{(n)} \otimes \tau_3)(1 \otimes \tau_1) + (1 \otimes \tau_1)(\gamma_j^{(n)} \otimes \tau_3) = \gamma_j^{(n)} \otimes \{\tau_3, \tau_1\} = 0$. Do check the rest.

Using the direct product notation, we can write the γ matrices for $SO(2n)$ more compactly as follows, for $k = 1, \cdots n$:

$$\gamma_{2k-1} = 1 \otimes 1 \otimes \cdots \otimes 1 \otimes \tau_1 \otimes \tau_3 \otimes \tau_3 \otimes \cdots \otimes \tau_3 \tag{6}$$

---

* Henceforth, as elsewhere in this book, I will abuse notation slightly and use 1 to denote an identity matrix I of the appropriate size; here I is the same size as $\gamma^{(n)}$. This will allow us to avoid writing awkward-looking objects like iI in (5). The same remarks apply to the symbol 0.

and

$$\gamma_{2k} = 1 \otimes 1 \otimes \cdots \otimes 1 \otimes \tau_2 \otimes \tau_3 \otimes \tau_3 \otimes \cdots \otimes \tau_3 \tag{7}$$

with 1 appearing $k-1$ times and τ_3 appearing $n-k$ times. When and if you feel confused at any point in this discussion, you should of course work things out explicitly for $SO(4)$, $SO(6)$, and so forth.

Representing the generators of $SO(2n)$

But what does this solution or representation of the Clifford algebra (1) have to do with the rotation group $SO(2n)$?

Define $2n(2n-1)/2 = n(2n-1)$ hermitean matrices

$$\sigma_{ij} \equiv -\frac{i}{2}[\gamma_i, \gamma_j] = \begin{cases} -i\gamma_i\gamma_j & \text{for } i \neq j \\ 0 & \text{for } i = j \end{cases}$$

First note that if k is not equal to either i or j, then γ_k clearly commutes with σ_{ij} (since each time we move γ_k past a γ in σ_{ij}, we flip a sign). However, if k is equal to either i or j, then we use $\gamma_k^2 = 1$ to knock off one of the γs in σ_{ij}. In other words,

$$[\sigma_{ij}, \gamma_k] = -i[\gamma_i\gamma_j, \gamma_k] = 2i(\delta_{ik}\gamma_j - \delta_{jk}\gamma_i) \tag{8}$$

Explicitly, $[\sigma_{12}, \gamma_1] = 2i\gamma_2$, $[\sigma_{12}, \gamma_2] = -2i\gamma_2$, and $[\sigma_{12}, \gamma_k] = 0$ for $k > 2$.

Given this, the commutation of the σs with one another is easy to work out. For example,

$$[\sigma_{12}, \sigma_{23}] = (-i)^2[\gamma_1\gamma_2, \gamma_2\gamma_3] = -\gamma_1\gamma_2\gamma_2\gamma_3 + \gamma_2\gamma_3\gamma_1\gamma_2 = 2\gamma_3\gamma_1 = 2i\sigma_{31} \tag{9}$$

Roughly speaking, the γ_2s in σ_{12} and σ_{23} knock each other out, leaving us with σ_{31}. On the other hand, $[\sigma_{12}, \sigma_{34}] = 0$, since the γ_1 and γ_2 in σ_{12} sail right past the γ_3 and γ_4 in σ_{34}.

Thus, the $\frac{1}{2}\sigma_{ij}$s satisfy the same commutation relations satisfied by the generators J^{ij}s of $SO(2n)$, as discussed in chapter I.3. The $\frac{1}{2}\sigma_{ij}$s define a 2^n-dimensional representation of the Lie algebra $SO(2n)$. Note how the factor of $\frac{1}{2}$ has to be included.

We then expect that exponentials of the $\frac{1}{2}\sigma_{ij}$s represent the group $SO(2n)$. We will see in the next section the extent to which this is true.

Explicitly, and for future use, we note that

$$\gamma_1 = \tau_1 \otimes \tau_3 \otimes \tau_3 \otimes \cdots \otimes \tau_3 \tag{10}$$

and

$$\gamma_2 = \tau_2 \otimes \tau_3 \otimes \tau_3 \otimes \cdots \otimes \tau_3 \tag{11}$$

Then

$$\sigma_{12} = \tau_3 \otimes 1 \otimes 1 \otimes \cdots \otimes 1 \tag{12}$$

Thus, σ_{12} is just a diagonal matrix with ± 1 along the diagonal.

From (3), (4), and (5),

$$\sigma_{ij}^{(n+1)} = -i\gamma_i^{(n+1)}\gamma_j^{(n+1)} = \sigma_{ij}^{(n)} \otimes 1$$

$$\sigma_{i,2n+1}^{(n+1)} = \gamma_i^{(n)} \otimes \tau_2$$

$$\sigma_{i,2n+2}^{(n+1)} = -\gamma_i^{(n)} \otimes \tau_1$$

$$\sigma_{2n+1,2n+2}^{(n+1)} = 1 \otimes \tau_3 \tag{13}$$

Spinor and double covering

As 2^n-by-2^n matrices, the σs act on an object ψ with 2^n components, which we will call the spinor ψ. Consider the unitary transformation

$$\psi \to e^{\frac{i}{4}\omega_{ij}\sigma_{ij}}\psi \tag{14}$$

with $\omega_{ij} = -\omega_{ji}$ a set of real numbers. (Note the $\frac{1}{4}$ inserted for later convenience.) Then $\psi^\dagger \to \psi^\dagger e^{-\frac{i}{4}\omega_{ij}\sigma_{ij}}$, since σ_{ij} is hermitean. For ω_{ij} infinitesimal, we have

$$\psi^\dagger \gamma_k \psi \to \psi^\dagger e^{-\frac{i}{4}\omega_{ij}\sigma_{ij}}\gamma_k e^{\frac{i}{4}\omega_{ij}\sigma_{ij}}\psi \simeq \psi^\dagger \gamma_k \psi - \frac{i}{4}\omega_{ij}\psi^\dagger[\sigma_{ij},\gamma_k]\psi \tag{15}$$

Using (8), we see that the set of objects $v_k \equiv \psi^\dagger \gamma_k \psi$, $k = 1, \cdots, 2n$, transforms like a vector* in $2n$-dimensional space:

$$v_k \to v_k - \frac{1}{2}(\omega_{kj}v_j - \omega_{ik}v_i) = v_k - \omega_{kj}v_j \tag{16}$$

To relate ω_{ij} to the rotation angle, let us be explicit and set all the ωs to 0 except for $\omega_{12} = -\omega_{21}$. Then (16) gives $v_1 \to v_1 - \omega_{12}v_2$, $v_2 \to v_2 + \omega_{12}v_1$ with v_k for $k > 2$ untouched. Thus, this corresponds to an infinitesimal rotation in the 12 plane through angle $\varphi = \omega_{12}$.

Under a finite rotation in the 12 plane through angle φ, the spinor $\psi \to e^{\frac{i}{2}\omega_{12}\sigma_{12}}\psi = e^{\pm i\frac{\varphi}{2}}\psi$. (In the last equality, we note by (12) that half of the components of ψ transform by the phase factor $e^{i\frac{\varphi}{2}}$, and the other half by $e^{-i\frac{\varphi}{2}}$.) Thus, under a complete rotation through 2π,

$$\psi \to -\psi \tag{17}$$

As we might have expected from our experience with the spin $\frac{1}{2}$ representation of $SO(3)$, while the spinor ψ furnishes a representation of the Lie algebra $SO(2n)$, it does not furnish a representation of the group $SO(2n)$. Instead, it furnishes a representation of the double covering of the group $SO(2n)$.

This discussion also indicates that in (13) the matrices $\sigma_{ij}^{(n+1)}$, $\sigma_{i,2n+1}^{(n+1)}$, $\sigma_{i,2n+2}^{(n+1)}$, and $\sigma_{2n+1,2n+2}^{(n+1)}$ generate rotations in the (i, j)-plane, the $(i, 2n + 1)$-plane, the $(i, 2n + 2)$-plane, and the $(2n + 1, 2n + 2)$-plane, respectively.

---

* As mentioned in chapter IV.5, in connection with the Pauli matrices, people often speak loosely and say that the γs transform like a vector.

Strictly speaking, the double cover of $SO(N)$ should be called $Spin(N)$, but physicists are typically lax in distinguishing the two groups. At the level of the Lie algebra (which is what we generally focus on rather than the corresponding group), they are the same, of course. We have already seen[*] that $Spin(3) \simeq SU(2)$ and $Spin(4) \simeq SU(2) \otimes SU(2)$.

Left and right handed spinors

Is the 2^n-dimensional spinor representation reducible or irreducible?

Define the hermitean[†] matrix

$$\gamma_F = (-i)^n \gamma_1 \gamma_2 \cdots \gamma_{2n} \tag{18}$$

It anticommutes with all the gamma matrices:

$$\gamma_F \gamma_i = -\gamma_i \gamma_F, \quad \text{for all } i \tag{19}$$

To see this, note that, as γ_i moves through the string of γs in (18), each time it jumps past a γ, according to (1), a minus sign pops up, except when it jumps past itself (so to speak), so that the net sign produced is $(-1)^{2n-1} = -1$.

By explicit computation, using (1), show that γ_F squares to the identity matrix I. Thus, $P_\pm \equiv \frac{1}{2}(I \pm \gamma_F)$ are projection operators in the sense that

$$P_+^2 = P_+, \ P_-^2 = P_-, \text{ and } P_+ + P_- = I \tag{20}$$

Thus, we could decompose the spinor into a left handed spinor $\psi_L \equiv \frac{1}{2}(1 - \gamma_F)\psi$ and a right handed spinor $\psi_R \equiv \frac{1}{2}(1 + \gamma_F)\psi$, such that $\gamma_F \psi_L = -\psi_L$ and $\gamma_F \psi_R = \psi_R$, and $\psi = \psi_L + \psi_R$.

From (19), it follows that γ_F commutes with σ_{ij}. Thus, acting with $P_\pm$ on the transformation $\psi \to e^{\frac{i}{4}\omega_{ij}\sigma_{ij}}\psi$, we obtain $P_\pm \psi \to P_\pm e^{\frac{i}{4}\omega_{ij}\sigma_{ij}}\psi = e^{\frac{i}{4}\omega_{ij}\sigma_{ij}} P_\pm \psi = e^{\frac{i}{4}\omega_{ij}\sigma_{ij}} P_\pm (P_\pm \psi)$. We deduce that

$$\psi_L \to (e^{\frac{i}{4}\omega_{ij}\sigma_{ij}} P_-)\psi_L = e^{\frac{i}{4}\omega_{ij}\sigma_{ij}}\psi_L \quad \text{and} \quad \psi_R \to (e^{\frac{i}{4}\omega_{ij}\sigma_{ij}} P_+)\psi_R = e^{\frac{i}{4}\omega_{ij}\sigma_{ij}}\psi_R \tag{21}$$

In other words, ψ_L and ψ_R transform separately[‡] and just like ψ.

The projection into left and right handed spinors cuts the number of components by two. We arrive at the important conclusion that the irreducible spinor representation of $SO(2n)$ has dimension $2^n/2 = 2^{n-1}$, not 2^n. The representation matrices are given by $(e^{\frac{i}{4}\omega_{ij}\sigma_{ij}} P_\pm)$.

We will often refer to the two spinor representations as S^+ and S^-, or by their dimensions as 2^{n-1}_+ and 2^{n-1}_-. For example, for $SO(10)$, a group of great relevance for grand unified theories (see chapter IX.3), we have 16_+ and 16_-.

It is also useful to note that, in the basis we are using, γ_F is real; explicitly,

$$\gamma_F = \tau_3 \otimes \tau_3 \otimes \cdots \otimes \tau_3 \tag{22}$$

---

[*] We note also that $Spin(5) \simeq Sp(4)$ and $Spin(3) \simeq SU(4)$.

[†] Verify this using the hermiticity of the gamma matrices and (1).

[‡] The peculiar notation γ_F, ψ_L, and ψ_R has its origin in the treatment of the Lorentz group in particle physics, as we will see in chapter VII.3.

with τ_3 appearing n times. (This can be seen from the iterative relation $\gamma_F^{(n+1)} = \gamma_F^{(n)} \otimes \tau_3$, which follows from the definition of γ_F and (3), (4), and (5).) Note that with this explicit form, the key property (19) clearly holds.

A binary code

Given the direct product form of the γ matrices in (6) and (7), and hence of σ_{ij}, we can write the states of the spinor representations as (using the Dirac bra and ket notation already introduced in the review of linear algebra)

$$|\varepsilon_1 \varepsilon_2 \cdots \varepsilon_n\rangle \qquad (23)$$

where each of the εs takes on the values ± 1. From (22) we see that

$$\gamma_F |\varepsilon_1 \varepsilon_2 \cdots \varepsilon_n\rangle = (\Pi_{j=1}^n \varepsilon_j) |\varepsilon_1 \varepsilon_2 \cdots \varepsilon_n\rangle \qquad (24)$$

The right handed spinor S^+ consists of those states $|\varepsilon_1 \varepsilon_2 \cdots \varepsilon_n\rangle$ with $(\Pi_{j=1}^n \varepsilon_j) = +1$, and the left handed spinor S^- those states with $(\Pi_{j=1}^n \varepsilon_j) = -1$. Indeed, the spinor representations have dimension 2^{n-1}.

Some explicit examples: $SO(2)$ and $SO(4)$

The binary notation is very helpful if we want to see explicitly what is going on. Try the simplest possible example, $SO(2)$. We have $\gamma_1 = \tau_1$ and $\gamma_2 = \tau_2$, and hence $\sigma_{12} = \tau_3$. The two irreducible spinor representations S^+ and S^- for $n = 1$ are $2^{n-1} = 2^0 = 1$-dimensional. Since $\gamma_F = \tau_3$, they correspond to $|+\rangle$ and $|-\rangle$, with $SO(2)$ rotations represented by phase factors $e^{i\varphi}$ and $e^{-i\varphi}$, respectively. The spinor representations S^+ and S^- are manifestly complex. The reader will of course recognize that this is just the isomorphism between $SO(2)$ and $U(1)$ discussed back in chapter I.3.

Next, it is also instructive to work out $SO(4)$, the $n = 2$ case. The γ matrices were given in (2); for your convenience, I list them here again:

$$\gamma_1 = \tau_1 \otimes \tau_3, \ \gamma_2 = \tau_2 \otimes \tau_3, \ \gamma_3 = 1 \otimes \tau_1, \ \gamma_4 = 1 \otimes \tau_2 \qquad (25)$$

Then, by either direct computation or by (13),

$$\sigma_{12} = \tau_3 \otimes 1,$$
$$\sigma_{31} = -\tau_1 \otimes \tau_2, \ \sigma_{23} = \tau_2 \otimes \tau_2,$$
$$\sigma_{14} = -\tau_1 \otimes \tau_1, \ \sigma_{24} = -\tau_2 \otimes \tau_1,$$
$$\sigma_{34} = 1 \otimes \tau_3 \qquad (26)$$

You can readily verify the commutation relations befitting $SO(4)$. (For example, $[\sigma_{12}, \sigma_{34}] = 0$, while $[\sigma_{12}, \sigma_{23}] = [\tau_3, \tau_2] \otimes \tau_2 = 2i\sigma_{31}$.) Also, $\gamma_F = \tau_3 \otimes \tau_3$.

While the commutation relations already confirm that the σs represent the Lie algebra of $SO(4)$, it is instructive to verify this using the binary code. Go ahead and do it before reading on!

According to (22), the spinor S^+ consists of $|++\rangle$ and $|--\rangle$, while the spinor S^- consists of $|+-\rangle$ and $|-+\rangle$.

Acting with the σs on $|++\rangle$ and $|--\rangle$, we have

$$
\begin{aligned}
\sigma_{12}|++\rangle &= +|++\rangle \,, \; \sigma_{12}|--\rangle = -|--\rangle \,, \\
\sigma_{23}|++\rangle &= -|--\rangle \,, \; \sigma_{23}|--\rangle = -|++\rangle \\
\sigma_{31}|++\rangle &= -i\,|--\rangle \,, \; \sigma_{31}|--\rangle = i\,|++\rangle \\
\sigma_{14}|++\rangle &= -|--\rangle \,, \; \sigma_{14}|--\rangle = -|++\rangle \\
\sigma_{24}|++\rangle &= -i\,|--\rangle \,, \; \sigma_{24}|--\rangle = i\,|++\rangle \\
\sigma_{34}|++\rangle &= |++\rangle \,, \; \sigma_{34}|--\rangle = -|--\rangle
\end{aligned}
\tag{27}
$$

Thus, in the 2-dimensional space spanned by $|++\rangle$ and $|--\rangle$, we have $\sigma_{12} \sim \tau_3, \sigma_{23} \sim -\tau_1$, $\sigma_{31} \sim -\tau_2$, and $\sigma_{14} \sim -\tau_1, \sigma_{24} \sim -\tau_2, \sigma_{34} \sim \tau_3$.

Similarly, you can work out the σ matrices in the 2-dimensional space spanned by $|+-\rangle$ and $|-+\rangle$.

As was discussed in chapters II.3 and IV.7, $SO(4)$ is locally isomorphic to $SU(2) \otimes SU(2)$. By inspection, we see that $\frac{1}{2}(\sigma_{12} + \sigma_{34})$, $\frac{1}{2}(\sigma_{23} + \sigma_{14})$, and $\frac{1}{2}(\sigma_{31} + \sigma_{24})$ acting on $|++\rangle$ and $|--\rangle$ represent* $SU(2)$, while $(\sigma_{12} - \sigma_{34})$ and its cousins acting on $|++\rangle$ and $|--\rangle$ give 0.

In contrast, you can (and should) check that $\frac{1}{2}(\sigma_{12} - \sigma_{34})$ and its cousins acting on $|+-\rangle$ and $|-+\rangle$ represent the other $SU(2)$, while $(\sigma_{12} + \sigma_{34})$ and its cousins acting on $|+-\rangle$ and $|-+\rangle$ give 0. Furthermore, the two $SU(2)$s commute. Once again, math works, but what did you expect?

To summarize, we can write the spinor representations of $SO(4)$ as

$$
\begin{aligned}
S^+ &\sim \{|++\rangle \text{ and } |--\rangle\} \\
S^+ &\sim \{|+-\rangle \text{ and } |-+\rangle\}
\end{aligned}
\tag{28}
$$

The spinor S^+ transforms like a doublet under one $SU(2)$ and like a singlet under the other $SU(2)$. In other words, it corresponds to the irreducible representation $(2, 1)$ of $SU(2) \otimes SU(2)$. In contrast, S^- corresponds to $(1, 2)$. Explicitly, under one $SU(2)$,

$$
S^+ \to 2 \quad \text{and} \quad S^- \to 1 + 1
\tag{29}
$$

Under the other $SU(2)$, the vice versa holds.

We can also write out the 4-by-4 matrices explicitly projected by $P_\pm$. From (22), $\gamma_F = $

$$
\tau_3 \otimes \tau_3 = \begin{pmatrix} 1 & 0 & 0 & 0 \\ 0 & -1 & 0 & 0 \\ 0 & 0 & -1 & 0 \\ 0 & 0 & 0 & 1 \end{pmatrix}. \text{ Thus, for example,}
$$

$$
\sigma_{12} P_+ = \begin{pmatrix} 1 & 0 & 0 & 0 \\ 0 & -1 & 0 & 0 \\ 0 & 0 & 1 & 0 \\ 0 & 0 & 0 & -1 \end{pmatrix} \begin{pmatrix} 1 & 0 & 0 & 0 \\ 0 & 0 & 0 & 0 \\ 0 & 0 & 0 & 0 \\ 0 & 0 & 0 & 1 \end{pmatrix} = \begin{pmatrix} 1 & 0 & 0 & 0 \\ 0 & 0 & 0 & 0 \\ 0 & 0 & 0 & 0 \\ 0 & 0 & 0 & -1 \end{pmatrix} \text{"="} \begin{pmatrix} 1 & 0 \\ 0 & -1 \end{pmatrix} = \tau_3
\tag{30}
$$

---

* We might have called this $SU(2) \, SU(2)_+$, and the other $SU(2) \, SU(2)_-$. At the risk of potential confusion, we will see in chapter VII.3 that it makes sense, in another context, to write $SU(2) \otimes SU(2)$ as $SU(2)_L \otimes SU(2)_R$.

(the next-to-last equality is an effective equality). This agrees with what we have written above, namely, that, acting on S^+, $\sigma_{12} \sim \tau_3$. Similarly,

$$\sigma_{12} P_- = \begin{pmatrix} 1 & 0 & 0 & 0 \\ 0 & -1 & 0 & 0 \\ 0 & 0 & 1 & 0 \\ 0 & 0 & 0 & -1 \end{pmatrix} \begin{pmatrix} 0 & 0 & 0 & 0 \\ 0 & 1 & 0 & 0 \\ 0 & 0 & 1 & 0 \\ 0 & 0 & 0 & 0 \end{pmatrix} \text{"="} \begin{pmatrix} -1 & 0 \\ 0 & 1 \end{pmatrix} = -\tau_3 \tag{31}$$

Indeed, the whole point of the direct product notation is that we don't have to write out 4-by-4 matrices, such as those written here.

Complex conjugation: The $SO(2n)$s are not created equal

Are the spinor representations real, pseudoreal, or complex?

Let us motivate the discussion with a question raised in chapter II.4. Suppose we have two spinors ψ and ζ, so that $\psi \to D(g)\psi$ and $\zeta \to D(g)\zeta$, where $D(g) = e^{\frac{i}{4}\omega_{ij}\sigma_{ij}}$ (with g an element of $SO(2n)$). We ask whether there exists a matrix C such that $\zeta C \psi$ is invariant.

Since $\zeta^T C \psi \to \zeta^T e^{\frac{i}{4}\omega\sigma^T} C e^{\frac{i}{4}\omega\sigma} \psi \simeq \zeta^T C \psi + \frac{i}{4}\omega\zeta^T (\sigma^T C + C\sigma)\psi$, we see that $\zeta^T C \psi$ would be invariant if C satisfies

$$\sigma_{ij}^T C = -C\sigma_{ij} \tag{32}$$

In other words,

$$C^{-1}\sigma_{ij}^T C = -\sigma_{ij} = C^{-1}\sigma_{ij}^* C \tag{33}$$

where the second equality follows from the hermiticity of σ_{ij}. It is also handy to write

$$e^{\frac{i}{4}\omega\sigma^T} C = C e^{-\frac{i}{4}\omega\sigma} \tag{34}$$

for later use. (Expand the exponential; every time we move C past a σ^T, we pick up a minus sign.)

Note that (32) does not determine C uniquely. For example, given a C that satisfies (32), then $\gamma_F C$ would too. (For some students, this "variability" adds to the confusion surrounding C: it seems that every text presents a different C.)

In summary, we define C by requiring $\zeta^T C \psi$ to be invariant.

To see how this works, let us be specific and look at the σ matrices for $SO(4)$ listed in (26). For what follows, it is helpful to keep in mind that $\tau_2 \tau_a^T \tau_2 = -\tau_a$ for $a = 1, 2, 3$. Note that σ_{13} and σ_{24} are antisymmetric, while the rest are symmetric. We see that C equal to either $i\tau_2 \otimes \tau_1$ or $\tau_1 \otimes i\tau_2$ would work* (but not $\tau_1 \otimes \tau_1$ or $i\tau_2 \otimes i\tau_2$). The two possible choices are related (according to (IV.5.13)) by a similarity transformation $U = e^{\frac{i}{2}(\frac{\pi}{4})\tau_3}$. We choose

$$C = i\tau_2 \otimes \tau_1 \tag{35}$$

---

* Since $\gamma_F = \tau_3 \otimes \tau_3$, this is precisely the issue of C versus $\gamma_F C$ just mentioned.

(Note that we have exploited the freedom allowed by (32) to multiply C by a constant and used $i\tau_2 = \begin{pmatrix} 0 & 1 \\ -1 & 0 \end{pmatrix}$ instead of τ_2.)

Now that we have C for $SO(4)$, we induct* to determine C for $SO(2n)$. Suppose we have C for $SO(2n)$; call it C_n. We want to construct C_{n+1}, starting with C_2 given by (35). We are invited by (13) to try $C_{n+1} = C_n \otimes \kappa$.

First, looking at $\sigma_{2n+1,2n+2}^{(n+1)} = 1 \otimes \tau_3$ in (13), we see that we can choose κ to be either τ_1 or $i\tau_2$. Next, we take care of $\sigma_{i,2n+1}^{(n+1)}$ and $\sigma_{i,2n+2}^{(n+1)}$. For your reading convenience, I repeat them here: $\sigma_{i,2n+1}^{(n+1)} = \gamma_i^{(n)} \otimes \tau_2$, and $\sigma_{i,2n+2}^{(n+1)} = -\gamma_i^{(n)} \otimes \tau_1$. Impose (33) to determine κ:

$$C_{n+1}^{-1}\sigma_{i,2n+1}^{(n+1)\,T}C_{n+1} = C_n^{-1}\gamma_i^{(n)\,T}C_n \otimes (\kappa^{-1}\tau_2\kappa) = -\gamma_i^{(n)} \otimes \tau_2$$

$$C_{n+1}^{-1}\sigma_{i,2n+2}^{(n+1)\,T}C_{n+1} = -C_n^{-1}\gamma_i^{(n)\,T}C_n \otimes (\kappa^{-1}\tau_1\kappa) = \gamma_i^{(n)} \otimes \tau_1 \tag{36}$$

(The second equality here is the condition (33).) This could be satisfied if we alternate choosing κ to be $i\tau_2$ or τ_1 and if

$$C_n^{-1}\gamma_i^{(n)\,T}C_n = (-1)^n\gamma_i^{(n)} \tag{37}$$

Note that, in contrast to the basic requirement on C given in (32), this condition depends explicitly on n.

Putting it all together, we have

$$C_{n+1} = \begin{pmatrix} 0 & C_n \\ (-1)^{n+1}C_n & 0 \end{pmatrix} = \begin{cases} C_n \otimes \tau_1 & \text{if } n \text{ is odd} \\ C_n \otimes i\tau_2 & \text{if } n \text{ is even} \end{cases} \tag{38}$$

One final (trivial) check is that this also works for $\sigma_{ij}^{(n+1)} = \sigma_{ij}^{(n)} \otimes 1$. In particular, this choice agrees with our earlier choice $C_2 = i\tau_2 \otimes \tau_1$ in (35).

From (38), we can write C_n as the direct product of n τ matrices, alternating between $i\tau_2$ and τ_1:

$$C_n = i\tau_2 \otimes \tau_1 \otimes i\tau_2 \otimes \tau_1 \otimes i\tau_2 \otimes \tau_1 \cdots \tag{39}$$

Incidentally, from (38) and (39), we see that $C_1 = i\tau_2$, which works, since σ_{12} (the lone generator of $SO(2)$) is simply τ_3.

From this and (22) we see immediately that

$$C_n^{-1}\gamma_F C_n = (-1)^n\gamma_F \tag{40}$$

This peculiar dependence on n forced on us will lead to some interesting behavior of the orthogonal algebras.

We can determine from (39) whether C_n is symmetric or antisymmetric. The poor man simply looks at the first few C_ns:

$$C_2 = i\tau_2 \otimes \tau_1, \qquad C_3 = i\tau_2 \otimes \tau_1 \otimes i\tau_2,$$
$$C_4 = i\tau_2 \otimes \tau_1 \otimes i\tau_2 \otimes \tau_1, \qquad C_5 = i\tau_2 \otimes \tau_1 \otimes i\tau_2 \otimes \tau_1 \otimes i\tau_2$$
$$C_6 = C_5 \otimes \tau_1, \qquad C_7 = C_6 \otimes i\tau_2, \qquad C_8 = C_7 \otimes \tau_1, \qquad C_9 = C_8 \otimes i\tau_2 \tag{41}$$

---

* I strongly advise you to work through this. You will miss all the fun if you read the following passively.

Thus, $C_n^T = -C_n$ for $n = 2, 5, 6, 9$, while $C_n^T = +C_n$ for $n = 3, 4, 7, 8$. Do you see the pattern?

The rich man, being more systematic, defines the integers a_n by $C_n^T = (-1)^{a_n} C_n$, and plugs this into (38) to obtain the recursion relation $(-1)^{a_{n+1}} = (-1)^{a_n + n + 1}$. Solving this (using Gauss's summation formula, for example), he arrives at the somewhat peculiar-looking result

$$C_n^T = (-1)^{\frac{1}{2}n(n+1)} C_n \tag{42}$$

We remark in passing that (6) and (7) imply $\gamma_i^T = (-1)^{i+1} \gamma_i$ and that (37) gives $\gamma_i^{(n) T} C_n = (-1)^n C_n \gamma_i^{(n)}$. Thus,

$$\gamma_i^{(n)} C_n = (-1)^{n+i+1} C_n \gamma_i^{(n)} \tag{43}$$

In particular, for $SO(10)$, $n = 5$, and γ_i commutes with C for i even and anticommutes with C for i odd.

The observation that $\gamma_i^T = (-1)^{i+1} \gamma_i$ (namely, that γ_i is symmetric for i odd and anti-symmetric for i even) offers us another way to determine C.

We go back to the defining condition (32): $\sigma_{ij}^T C = -C \sigma_{ij}$. Note that C is determined only up to an overall constant. Recall that $\sigma_{ij} = -i\gamma_i\gamma_j$ for $i \neq j$ (and of course 0 for $i = j$). If one, and only one, of the two integers i, j is even, then $(\gamma_i\gamma_j)^T = \gamma_j^T \gamma_i^T = -\gamma_j\gamma_i = \gamma_i\gamma_j$. We demand that $\gamma_i\gamma_j C = -C\gamma_i\gamma_j$. On the other hand, if neither i nor j is even, or if both i and j are even, then $(\gamma_i\gamma_j)^T = -\gamma_i\gamma_j$. We demand that $\gamma_i\gamma_j C = C\gamma_i\gamma_j$.

We claim that C equals the product of the "even gamma matrices" solves these two conditions. In other words, for $n = 1$, $C = \gamma_2$; for $n = 2$, $C = \gamma_2\gamma_4$; for $n = 3$, $C = \gamma_2\gamma_4\gamma_6$; and so on. It would be best if you verify this for a few cases. As a check, for $n = 3$ for example, $C = \gamma_2\gamma_4\gamma_6 = (\tau_2 \otimes \tau_3 \otimes \tau_3)(1 \otimes \tau_2 \otimes \tau_3)(1 \otimes 1 \otimes \tau_2) = -i\tau_2 \otimes \tau_1 \otimes \tau_2$, in agreement with (41). Note that C is determined only up to an overall constant, as remarked just now.

With this form of C, you can readily verify (37), (40), and (42). For example, since γ_F anticommutes with each of the γ matrices, then γ_F passing through C has to jump across n γ matrices, thus producing a factor of $(-1)^n$, in agreement with (40). Similarly, with $C = \gamma_2\gamma_4 \cdots \gamma_{2n}$, we have $C^T = \gamma_{2n}^T \cdots \gamma_4^T \gamma_2^T = (-1)^n \gamma_{2n} \cdots \gamma_4\gamma_2 = (-1)^n(-1)^{n-1}\gamma_2\gamma_{2n} \cdots \gamma_4$. I will leave it to you to complete the process of bringing the γ matrices into the right order and checking that the resulting sign agrees with (42).

Conjugate spinor

Now that we have determined C_n with its peculiar dependence on n (see (38) and (42)), we could go back to the question asked earlier. Are the spinor representations complex, real, or pseudoreal?

It is important to ask this question of the irreducible 2^{n-1}-dimensional representations rather than of the reducible 2^n-dimensional representation. Given that $\psi \to e^{\frac{i}{4}\omega_{ij}\sigma_{ij}} P_\pm \psi$, then its complex conjugate $\psi^* \to e^{-\frac{i}{4}\omega_{ij}\sigma_{ij}^*} P_\pm^* \psi^*$. If there exists a matrix C such that

$$C^{-1} e^{-\frac{i}{4}\omega_{ij}\sigma_{ij}^*} P_\pm^* = e^{\frac{i}{4}\omega_{ij}\sigma_{ij}} P_\pm C^{-1} \tag{44}$$

then the conjugate spinor

$$\psi_c \equiv C^{-1}\psi^* \to C^{-1}e^{-\frac{i}{4}\omega_{ij}\sigma_{ij}^*}P_\pm^*\psi^* = e^{\frac{i}{4}\omega_{ij}\sigma_{ij}}P_\pm C^{-1}\psi^* \tag{45}$$

would transform like ψ. The condition (44) works out to be

$$C^{-1}\sigma_{ij}^*\frac{1}{2}(I \pm \gamma_F)^*C = -\sigma_{ij}\frac{1}{2}(I \pm \gamma_F) \tag{46}$$

From (33) and (40), we have, with the conjugation matrix C that we found,

$$C^{-1}\sigma_{ij}^*\frac{1}{2}(1 \pm \gamma_F)^*C = C^{-1}\sigma_{ij}^T\frac{1}{2}(1 \pm \gamma_F)C = -\sigma_{ij}\frac{1}{2}\left(1 \pm (-1)^n\gamma_F\right) \tag{47}$$

The factor $\frac{1}{2}(1 \pm (-1)^n\gamma_F)$ on the right hand side equals $P_\pm$ or $P_\mp$ for n even or odd, respectively.

Thus, for n even, S^+ and S^- are the complex conjugates of themselves. The irreducible spinor representations are not complex for $SO(4k)$, in particular, $SO(4)$.

In contrast, for n odd, complex conjugation takes S^+ and S^- into each other. In other words, S^+ and S^- are conjugates of each other. The irreducible spinor representations are complex for $SO(4k+2)$, in particular, $SO(10)$.

$SO(4k)$, $SO(4k+2)$, $SO(8m)$, $SO(8m+4)$: **Complex, real, or pseudoreal?**

Actually, we do not even need the precise form of (38). All we need is that C is a direct product of an alternating sequence of τ_1 and τ_2, and thus, acting on $|\varepsilon_1\varepsilon_2 \cdots \varepsilon_n\rangle$, C flips the sign of all the εs. Thus, C changes the sign of $(\Pi_{j=1}^n\varepsilon_j)$ for n odd and does not for n even. In other words, for n odd, C changes one kind of spinor into the other kind, but for n even, it does not.

For n even, the conclusion that S^+ and S^- are complex conjugates of themselves, and hence not complex, then leads to the question of whether they are real or pseudoreal. According to Wigner's celebrated analysis described in chapter II.4, this corresponds to C being symmetric or antisymmetric.

But we already have the answer in (42): for $n = 2k$, we have $C^T = (-1)^{k(2k+1)}C = (-1)^kC$. Thus, for $k = 2m$, $C^T = +C$. The spinors S^+ and S^- are real for $SO(8m)$, in particular, for $SO(8)$. In contrast, for $k = 2m+1$, $C^T = -C$. The spinors S^+ and S^- are pseudoreal for $SO(8m+4)$, in particular for $SO(4)$.

The complexity of the spinor irreducible representations of the orthogonal algebras are summarized in this table:

$SO(4k+2)$	complex
$SO(8m)$	real
$SO(8m+4)$	pseudoreal

Multiplying spinor representations together

We learned how to multiply tensor representations of $SO(N)$ together back in chapter IV.1. Now we want to multiply spinor representations together.

For pedagogical clarity, let us focus on $SO(10)$, for which $n = 5$. There are ten γ_i matrices, each 32-by-32 (note $32 = 2^5$). Let us take a lowbrow physicist approach and write the spinors 16^+ as ψ_α and 16^- as ξ_β, with the spinor indices α, $\beta = 1, 2, \cdots, 32$. (Half of the spinor components are projected out by $P_\pm$, of course.)

So, what is $16^+ \otimes 16^+$? We want to have something like $\psi_\beta(\cdots)\psi_\alpha$ with a matrix denoted by $(\cdots)$ to tie the spinor indices α and β together. Well, the γ matrices $\gamma_{i\alpha\beta}$ carry two spinor indices (and one vector index i), exactly what we need. Indeed, any product of γ matrices would work. So, try objects like $\psi_\alpha(\gamma_i\gamma_j\gamma_k)_{\alpha\beta}\psi_\beta = \psi^T(\gamma_i\gamma_j\gamma_k)\psi$.

We want to know how these objects transform under $SO(10)$. Read off from (21) that $\psi_R \to e^{\frac{i}{4}\omega_{ij}\sigma_{ij}}P_+\psi_R$ and thus, $\psi_R^T(\gamma_i\gamma_j\gamma_k)\psi_R \to \psi^T P_+ e^{\frac{i}{4}\omega\sigma^T}(\gamma_i\gamma_j\gamma_k)e^{\frac{i}{4}\omega\sigma}P_+\psi$. Note that P_+ is symmetric.

For $e^{\frac{i}{4}\omega\sigma^T}$ and $e^{\frac{i}{4}\omega\sigma}$ to knock each other off, we have to ask the conjugation matrix C for help, to do what it did in (34). We are thus forced to insert a C and write $\psi_R^T C\gamma_i\gamma_j\gamma_k\psi_R \to \psi^T P_+\{e^{-\frac{i}{4}\omega\sigma}(\gamma_i\gamma_j\gamma_k)e^{\frac{i}{4}\omega\sigma}\}CP_+\psi$. But the result in (15) allows us to evaluate the quantity in the curly bracket: we thereby learn that the object $T_{ijk} = \psi^T C\gamma_i\gamma_j\gamma_k\psi$ transforms like an $SO(10)$ tensor with three indices, i, j, and k.

Still following thus far? Next, thanks to the Clifford algebra (1) that started this whole business, we can decompose this tensor into its totally antisymmetric part and the rest. For instance, write $\gamma_i\gamma_j = \frac{1}{2}\{\gamma_i, \gamma_j\} + \frac{1}{2}[\gamma_i, \gamma_j] = \delta_{ij}I + i\sigma_{ij}$. The Kronecker delta leads to a tensor T_k with only one index, which we can deal with separately. Thus, inductively, we could in effect take the tensor T_{ijk} to be totally antisymmetric. (Note that these remarks apply to any pair of indices, not just an adjacent pair like ij, because we can use (1) to bring γ_k next to γ_i if we wish.)

We deduce that the product $16^+ \otimes 16^+$ transforms like a sum of antisymmetric tensors.

Good. Now we generalize to $SO(2n)$. The product $S^+ \otimes S^+$ is equal to a sum of objects like $\psi^T C\Gamma_\kappa\psi$, where Γ_κ denotes schematically an antisymmetric product of κ γ matrices (for example, $\gamma_i\gamma_j\gamma_k$ for $\kappa = 3$).

But wait: we still have not paid any attention to P_+ in the transformation

$$\psi_R^T C\Gamma_\kappa\psi_R \to \psi^T P_+ Ce^{-\frac{i}{4}\omega\sigma}\Gamma_\kappa e^{\frac{i}{4}\omega\sigma}P_+\psi \tag{48}$$

The plan is to move the P_+ on the left past all the intervening stuff, C and the various γ matrices, to meet up with the P_+ on the right. We will see that sometimes we get a big fat zero and sometimes not.

Recall that (40) tells us that $\gamma_F C = (-1)^n C\gamma_F$. Thus, we have to treat n odd and n even separately.

For n odd, $P_+ C = CP_-$. Since γ_F anticommutes with γ matrices, if κ is even, this P_- would run into the P_+ in (48), and they would annihilate each other. But if κ is odd, the

P_- would arrive as a P_+ to meet P_+, and $P_+^2 = P_+$. This is summarized as follows:

For n odd, $\quad \psi^T P_+ C \Gamma_\kappa P_+ \psi = \psi^T C \Gamma_\kappa P_+ P_+ \psi = \psi^T C \Gamma_\kappa P_+ \psi \quad$ for κ odd

$$\psi^T P_+ C \Gamma_\kappa P_+ \psi = \psi^T C \Gamma_\kappa P_- P_+ \psi = 0 \quad \text{for } \kappa \text{ even} \tag{49}$$

In particular, for $SO(10)$, $n = 5$, and so κ has to be odd; thus, we obtain

$$16^+ \otimes 16^+ = [1] \oplus [3] \oplus [5] = 10 \oplus 120 \oplus 126 \tag{50}$$

The usual check to make sure that we did not lose anybody: $16 \cdot 16 = 256 = 10 + 120 + 126$. In calculating dimensions, did you remember that the irreducible representation [5] is self-dual? Thus the factor of $\frac{1}{2}$ in $\frac{1}{2}(\frac{10 \cdot 9 \cdot 8 \cdot 7 \cdot 6}{5 \cdot 4 \cdot 3 \cdot 2 \cdot 1}) = 126$.

For n even, $P_+ C = C P_+$, and the situation is reversed. If κ is odd, we would get zero, but if κ is even, the P_+ would arrive as a P_+ to meet P_+. Thus, we simply interchange the words odd and even in (49).

Fine. We have now evaluated the product $S^+ \otimes S^+$. What about the product $S^+ \otimes S^-$?

Simple. Instead of having P_+ and P_+ in (48), we now have P_+ and P_-. We simply flip our conclusions (49). In the product $S^+ \otimes S^-$, for n odd, κ has to be even, while for n even, κ has to be odd. In other words,

For n odd, $\quad \psi^T P_+ C \Gamma_\kappa P_- \psi = \psi^T C \Gamma_\kappa P_+ P_- \psi = 0 \quad$ for κ odd

$$\psi^T P_+ C \Gamma_\kappa P_- \psi = \psi^T C \Gamma_\kappa P_- P_- \psi = \psi^T C \Gamma_\kappa P_- \psi \quad \text{for } \kappa \text{ even} \tag{51}$$

For n even, $P_+ C = C P_+$, and the situation is reversed.

In particular, for $SO(10)$, $n = 5$, and so κ has to be even:

$$16^+ \otimes 16^- = [0] \oplus [2] \oplus [4] = 1 \oplus 45 \oplus 210 \tag{52}$$

In contrast to (50), now it is those Γ_κ with κ even that live. Note that 45 is the adjoint representation.

Let us summarize the results obtained here:

$S^+ \otimes S^+, S^- \otimes S^-$	n	κ
	odd	odd
	even	even

and

$S^+ \otimes S^-$	n	κ
	odd	even
	even	odd

It is instructive to work out a few other cases. For $SO(4)$, $n = 2$ is even, and so

$$2^+ \otimes 2^+ = [0] \oplus [2] = 1 \oplus 3$$
$$2^+ \otimes 2^- = [1] = 4 \tag{53}$$

But we know that $SO(4)$ is locally isomorphic to $SU(2) \otimes SU(2)$, and that $2^+ = (2, 1)$, $2^- = (1, 2)$. The multiplications above correspond to $(2, 1) \otimes (2, 1) = (1, 1) \oplus (3, 1)$ and $(2, 1) \otimes (1, 2) = (2, 2)$, which confirms what we knew about multiplying irreducible representations in $SU(2)$. Once again, we have learned that the 4-dimensional representation $(2, 2)$ describes the 4-vector of $SO(4)$.

We postpone working out the multiplication of spinors in $SO(6)$ and $SO(8)$ until the end of this chapter.

The rotation group in odd-dimensional spaces: Coming down versus going up

All this stuff about $SO(2n)$ is fine, but what about $SO(2n - 1)$? Pardon me, but I can't help being prejudiced in favor of $SO(3)$.

There are two routes to $SO(2n - 1)$. We could move down from $SO(2n)$, or we could move up from $SO(2n - 2)$. In particular, we could get to $SO(3)$ either by moving down from $SO(4)$ or moving up from $SO(2)$.

Suppose we have the γ matrices for $SO(2n)$, namely, $2n$ 2^n-by-2^n matrices γ_i for $i = 1, 2, \cdots, 2n$ satisfying the Clifford algebra (1). How do we find the $2n - 1$ γ matrices for $SO(2n - 1)$?

Simply throw γ_{2n} out.

We still have the 2^n-by-2^n matrices σ_{ij} for $i = 1, 2, \cdots, 2n - 1$, representing the Lie algebra of $SO(2n - 1)$, and hence constituting a spinor representation.

But what about γ_F as defined in (18)? Without γ_{2n}, it looks like we won't have a γ_F any more; but no, we can perfectly well use the matrix called γ_F that was constructed for $SO(2n)$. After all, it is just a 2^n-by-2^n matrix that anticommutes with all the γ matrices for $SO(2n - 1)$, namely, γ_i for $i = 1, 2, \cdots, 2n - 1$. So this spinor representation still splits into two, with each piece 2^{n-1}-dimensional. This shows that the σ_{ij}s for $SO(2n - 1)$ are effectively 2^{n-1}-by-2^{n-1} matrices. In appendix 2, we will see explicitly how this works for $SO(3)$. In particular, for $SO(3)$, $n = 2$, $2^{n-1} = 2$, and the σ_{ij}s are just the 2-by-2 Pauli matrices we have known since chapter IV.5.

As remarked just now, we can also reach $SO(2n + 1)$ by adding a γ matrix to $SO(2n)$. Let us illustrate with $SO(3)$. The Clifford algebra for $SO(2)$ has two γ matrices: $\gamma_1 = \tau_1$ and $\gamma_2 = \tau_2$. Simply add $\gamma_3 = \tau_3$. These three matrices form the Clifford algebra for $SO(3)$. In appendix 2, we show that this is equivalent to what we get starting with $SO(4)$.

In general, since the single γ matrix we are adding is merely required to anticommute with the existing γ matrix of $SO(2n)$ and to square to the identity, we have considerable freedom of choice. Referring back to (6) and (7), we have $\gamma_{2n-1} = 1 \otimes 1 \otimes \cdots \otimes 1 \otimes \tau_1$ and $\gamma_{2n} = 1 \otimes 1 \otimes \cdots \otimes 1 \otimes \tau_2$. Thus, we could simply add

$$\gamma_{2n+1} = 1 \otimes 1 \otimes \cdots \otimes 1 \otimes \tau_3 \tag{54}$$

Alternatively, we could add the matrix known as γ_F in $SO(2n)$ and given in (22), namely, add

$$\gamma_{2n+1} = \tau_3 \otimes \tau_3 \otimes \cdots \otimes \tau_3 \tag{55}$$

How about the charge conjugation matrix C? Again, we could either come down or go up.

Consider coming down from $SO(2n)$ to $SO(2n-1)$. Since C is required to satisfy $\sigma_{ij}^T C = -C\sigma_{ij}$, and since coming down we throw out some of the σs, we could a fortiori use the C for $SO(2n)$. But again, we must remember that γ_F cuts the size of C in half, rendering it a 2^{n-1}-by-2^{n-1} matrix. For example, the C for $SO(4)$ is 4-by-4, but for $SO(3)$ is 2-by-2.

To explain what happens when we go up, it may be clearer to focus on the most important example for physics. As explained earlier, going up from $SO(2)$ to $SO(3)$, we add $\gamma_3 = \tau_3$ to $\gamma_1 = \tau_1$ and $\gamma_2 = \tau_2$. Thus, $C = \tau_2$. In fact, going back to the motivation for C, we see that if ψ and χ are two $SO(3)$ spinors, then $\psi^T C\chi = \psi^T \tau_2 \chi$ is manifestly invariant under $SO(3)$. Since C is antisymmetric, the spinors for $SO(3)$ are pseudoreal, as we have known for a very long time.

Rotation groups inside rotation groups

A natural subgroup of $SO(2n+2m)$ is $SO(2n) \otimes SO(2m)$. The decomposition of the spinors can be read off using the binary code notation $|\varepsilon_1 \varepsilon_2 \cdots \varepsilon_{n+m}\rangle$ by inserting a semi-colon: $|\varepsilon_1 \varepsilon_2 \cdots \varepsilon_n; \varepsilon_{n+1} \cdots \varepsilon_{n+m}\rangle$.

Since γ_F is given by the product of the εs, this immediately gives us

$$2_+^{n+m-1} \to (2_+^{n-1}, 2_+^{m-1}) \oplus (2_-^{n-1}, 2_-^{m-1})$$

$$2_-^{n+m-1} \to (2_+^{n-1}, 2_-^{m-1}) \oplus (2_-^{n-1}, 2_+^{m-1}) \tag{56}$$

For example, on restriction of $SO(10)$ is $SO(4) \otimes SO(6)$, the spinor 16^+ breaks up into $(2^+, 4^+) \oplus (2^-, 4^-)$.

Embedding unitary groups into orthogonal groups

The unitary group $U(n)$ can be naturally embedded into the orthogonal group $SO(2n)$. In fact, I now show you that the embedding is as easy as $z = x + iy$.

Consider the $2n$-dimensional real vectors $x = (x_1, \cdots, x_n, y_1, \cdots, y_n)$ and $x' = (x_1', \cdots, x_n', y_1', \cdots, y_n')$. By definition, $SO(2n)$ consists of linear transformations[3] on these two real vectors, leaving invariant their scalar product $x'x = \sum_{j=1}^n (x_j' x_j + y_j' y_j)$.

Now out of these two real vectors we can construct two n-dimensional complex vectors $z = (x_1 + iy_1, \cdots, x_n + iy_n)$ and $z' = (x_1' + iy_1', \cdots, x_n' + iy_n')$. The group $U(n)$ consists of transformations on the two n-dimensional complex vectors z and z', leaving invariant their scalar product $(z')^*z = \sum_{j=1}^n (x'_j + iy'_j)^*(x_j + iy_j) = \sum_{j=1}^n (x'_j x_j + y'_j y_j) + i \sum_{j=1}^n (x'_j y_j - y'_j x_j)$.

In other words, $SO(2n)$ leaves $\sum_{j=1}^n (x'_j x_j + y'_j y_j)$ invariant, but $U(n)$ consists of the subset of those transformations in $SO(2n)$ that leave not only $\sum_{j=1}^n (x'_j x_j + y'_j y_j)$ invariant, but also leave $\sum_{j=1}^n (x'_j y_j - y'_j x_j)$ invariant as well.

In particular, the $U(1)$ inside $U(n)$ simply multiplies z by a phase $e^{i\xi}$ and z^* by the opposite phase $e^{-i\xi}$. As a result, x_j and y_j, for $j = 1, 2, \cdots, n$, rotate into each other: $x_j \to \cos\xi\, x_j + \sin\xi\, y_j$, $y_j \to -\sin\xi\, x_j + \cos\xi\, y_j$. This certainly is a rotation in the $2n$-dimensional space with coordinates $(x_1, \cdots, x_n, y_1, \cdots, y_n)$; the rotation just described leaves $(x'_j x_j + y'_j y_j)$ and $(x'_j y_j - y'_j x_j)$ invariant for each j, and thus a fortiori leaves $\sum_{j=1}^n (x'_j x_j + y'_j y_j)$ (and also $\sum_{j=1}^n (x'_j y_j - y'_j x_j)$) invariant.

Now that we understand this natural embedding of $U(n)$ into $SO(2n)$, we see that the defining or vector representation of $SO(2n)$, which we will call simply $2n$, decomposes on restriction to[4] $SU(n)$ into the two defining representations of $SU(n)$, n and n^*, thus

$$2n \to n \oplus n^* \tag{57}$$

In other words, $(x_1, \cdots, x_n, y_1, \cdots, y_n)$ can be written as $(x_1 + iy_1, \cdots, x_n + iy_n)$ and $(x_1 - iy_1, \cdots, x_n - iy_n)$.

Given the decomposition law (57), we can now figure out how other representations of $SO(2n)$ decompose when we restrict it to its natural subgroup $SU(n)$. The tensor representations of $SO(2n)$ are easy, since they are constructed out of the vector representation. For example, the adjoint representation of $SO(2n)$, which has dimension $2n(2n-1)/2 = n(2n-1)$, transforms like an antisymmetric two-indexed tensor $2n \otimes_A 2n$ and so decomposes into

$$2n \otimes_A 2n \to (n \oplus n^*) \otimes_A (n \oplus n^*) \tag{58}$$

according to (57). The antisymmetric product $\otimes_A$ on the right hand side is to be evaluated within $SU(n)$, of course. For instance, $n \otimes_A n$ is the $n(n-1)/2$ representation of $SU(n)$. In this way, we see that

$$\text{adjoint of } SO(2n) = n(2n-1) \to (n^2 - 1) \oplus 1 \oplus n(n-1)/2 \oplus (n(n-1)/2)^* \tag{59}$$

In the direct sum on the right hand side, we recognize the adjoint and the singlet of $SU(N)$. As a check, the total dimension of the representations of $SU(n)$ on the right hand side adds up to $(n^2 - 1) + 1 + 2(n(n-1)/2) = n(2n-1)$. In particular, for $SO(10) \to SU(5)$, we have

$$45 \to 24 \oplus 1 \oplus 10 \oplus 10^* \tag{60}$$

and of course, $24 + 1 + 10 + 10 = 45$.

Decomposing the spinor

Now that we know how the generators of $SO(2n)$ decompose, we want to figure out how the spinors of $SO(2n)$ decompose on restriction to $SU(n)$. We will do it intuitively, giving a heuristic argument that satisfies most physicists but certainly not mathematicians. In appendix 3, a more precise (that is, more formal) treatment is given.

For ease of writing, I will just do $SO(10) \to SU(5)$, and let you convince yourself that the discussion generalizes. The question is how the spinor 16 falls apart. Just from numerology and from knowing the dimensions of the smaller representations of $SU(5)$ (namely 1, 5, 10, and 15), we see that there are only so many possibilities, some of them rather unlikely (for example, the 16 falling apart into 16 1s).

Picture the spinor 16 of $SO(10)$ breaking up into a bunch of representations of $SU(5)$. By definition, the 45 generators of $SO(10)$ transform the components of the 16 into each other. Hence the 45 scramble all these representations of $SU(5)$ together. Let us ask what the various pieces of 45 (namely $24 \oplus 1 \oplus 10 \oplus 10^*$, given in (60)) do to these representations.

The 24 generators transform each of the representations of $SU(5)$ into itself, of course, because they are the 24 generators of $SU(5)$, and that is what generators do in life. The singlet 1 of $SU(5)$, since it does not even carry an $SU(5)$ index, can only multiply each of these representations by a real number. (In other words, the corresponding group element multiplies each of these representations by a phase factor: it generates the $U(1)$ subgroup of $U(5)$.) As a challenge to yourself, figure out what these real numbers are. We will figure them out in chapters IX.2 and IX.3.

What does the 10, also known as [2], do to these representations? Suppose the bunch of representations that S breaks up into contains the singlet $[0] = 1$ of $SU(5)$. The $10 = [2]$ acting on [0] gives the $[2] = 10$. (Almost too obvious for words: an antisymmetric tensor of two indices combined with a tensor with no indices is an antisymmetric tensor of two indices.)

What about $10 = [2]$ acting on [2]? It certainly contains the [4], which is equivalent to $[1]^* = 5^*$. But look, $1 \oplus 10 \oplus 5^*$ already adds up to 16. Thus, we have accounted for everybody. There can't be more. So we conclude

$$16^+ \to [0] \oplus [2] \oplus [4] = 1 \oplus 10 \oplus 5^* \qquad (61)$$

We learned earlier that $SO(10)$ contains two spinors 16^+ and 16^- that are conjugate to each other. Indeed, you may have noticed that I snuck in a superscript + in (61). The conjugate spinor 16^- breaks up into the conjugate of the representations in (61):

$$16^- \to [1] \oplus [3] \oplus [5] = 5 \oplus 10^* \oplus 1^* \qquad (62)$$

Note that [0], [2], and [4] are the conjugates of [5], [3], and [1], respectively, in $SU(5)$.

We could now write down how the spinor of $SO(2n)$ decomposes on restriction to $SU(n)$ for arbitrary n, but it is more instructive to see how it works for some specific cases.

For $SO(8) \to SU(4)$,

$$8^+ \to [0] \oplus [2] \oplus [4] = 1 \oplus 6 \oplus 1$$

$$8^- \to [1] \oplus [3] = 4 \oplus 4^* \qquad (63)$$

This example confirms our earlier conclusion that the spinors of $SO(8m)$ are real. Note that in $SU(4)$, $6 \sim 6^*$, because of the 4-indexed antisymmetric symbols.

For $SO(12) \to SU(6)$,

$$32^+ \to [0] \oplus [2] \oplus [4] \oplus [6] = 1 \oplus 15 \oplus 15^* \oplus 1$$

$$32^- \to [1] \oplus [3] \oplus [5] = 6 \oplus 20 \oplus 6^* \qquad (64)$$

Note that a pseudoreal spinor of $SO(8m + 4)$ breaks up into a real reducible representation of $SU(4m + 2)$.

Just for fun, let us do another example of $SO(4k + 2)$. For $SO(6) \to SU(3)$,

$$4^+ \to [0] \oplus [2] = 1 \oplus 3^*$$

$$4^- \to [1] \oplus [3] = 3 \oplus 1 \qquad (65)$$

Indeed, here the $+$ and $-$ spinors are complex and are the conjugates of each other.

In general, we have for n odd,

$$S^+ \to [0] \oplus [2] \oplus \cdots \oplus [n-1]$$

$$S^- \to [1] \oplus [3] \oplus \cdots \oplus [n]$$

and for n even,

$$S^+ \to [0] \oplus [2] \oplus \cdots \oplus [n]$$

$$S^- \to [1] \oplus [3] \oplus \cdots \oplus [n-1]$$

$SO(6)$ and $SO(8)$

The group $SO(6)$ has $6(6 - 1)/2 = 15$ generators. Notice that the group $SU(4)$ also has $4^2 - 1 = 15$ generators. In fact, $SO(6)$ and $SU(4)$ are locally isomorphic, as we have already seen in chapter VI.5 by studying their Dynkin diagrams. An explicit way to see this isomorphism is to note that the spinors $S^\pm$ of $SO(6)$ are $2^{\frac{6}{2}-1} = 4$-dimensional and that the generators of $SO(6)$ acting on them (namely, $\sigma_{ij} P_\pm$) are represented by traceless hermitean 4-by-4 matrices. Thus, the two spinor irreducible representations correspond to the 4-dimensional defining representations 4 of $SU(4)$ and its conjugate 4^*.

It is amusing to identify some low-dimensional representations of $SO(6)$ and $SU(4)$. Try it before reading on! We just noted that the spinors of $SO(6)$ are the fundamental or defining representations of $SU(4)$. The vector* V^i of $SO(6)$ is the antisymmetric[†] $A^{[\alpha\beta]}$ of $SU(4)$ (with dimension $6 = \frac{1}{2}(4 \cdot 3)$); the adjoint J^{ij} is the adjoint T^α_β (with dimension $15 = \frac{1}{2}(6 \cdot 5) = 4^2 - 1$); the symmetric traceless tensor S^{ij} is the antisymmetric traceless tensor $A^{[\alpha\beta]}_\gamma$ (with dimension $20 = \frac{1}{2}(6 \cdot 7) - 1 = 4 \cdot 6 - 4$); and so on. The various

---

* The notation here should be self-explanatory.
† Recall that we already used this correspondence in an appendix to chapter IV.8.

dimensions in $SO(6)$ and in $SU(4)$ are given by different, but numerically identical, arithmetical expressions.

The self-dual and antiself-dual n-indexed tensors of $SO(2n)$ were discussed in chapter IV.1. For $SO(6)$ the 3-indexed self-dual tensor D_+^{ijk} has dimension $10 = \frac{1}{2}(\frac{6 \cdot 5 \cdot 4}{3 \cdot 2})$. What could it correspond to in $SU(4)$?

The symmetric tensor $S^{\alpha\beta}$ of $SU(4)$ indeed has dimension $10 = \frac{1}{2}(4 \cdot 5)$ and corresponds to the self-dual 3-indexed tensor of $SO(6)$.

The antiself-dual 3-indexed tensor of $SO(6)$ evidently corresponds to the conjugate symmetric tensor $S_{\alpha\beta}$ of $SU(4)$:

$SO(6)$	$SU(4)$	Dimension
V^i	$A^{[\alpha\beta]}$	$6 = 6 = \frac{1}{2}(4 \cdot 3)$
J^{ij}	T^α_β	$15 = \frac{1}{2}(6 \cdot 5) = 4^2 - 1$
S^{ij}	$A^{[\alpha\beta]}_\gamma$	$20 = \frac{1}{2}(6 \cdot 7) - 1 = 4 \cdot 6 - 4$
D_+^{ijk}	$S^{\alpha\beta}$	$10 = \frac{1}{2}(\frac{6 \cdot 5 \cdot 4}{3 \cdot 2}) = \frac{1}{2}(4 \cdot 5)$

(66)

Math works, as we have noted many times already in this book.

We noticed, in chapter VI.6, that the Dynkin diagram of $SO(8)$ exhibits a striking 3-fold symmetry sometimes called triality. We now understand what this signifies; namely, that the two irreducible spinor representations 8^+ and 8^- of $SO(8)$ happen to have the same dimension as the defining vector representation 8_v. Check also that the spinors are real according to the table given earlier. Thus, there exists an automorphism of the Lie algebra of $SO(8)$ that maps 8^+, 8^-, and 8_v cyclically into one another.

This is underlined by the following observation. In general, a subgroup H of a group G can be embedded in G in more than one way. As an example, the subgroup $SO(6)$ of $SO(8)$ can be embedded in $SO(8)$ by saying that the vector 8_v decomposes into $6 \oplus 1 \oplus 1$. A physicist would say that this defines the natural embedding. Indeed, this is the embedding specified in (56): on the restriction $SO(8) \to SO(6) \otimes SO(2) \to SO(6)$, the two spinors decompose as $8^+ \to (4_+, 1_+) \oplus (4_-, 1_-) \to 4_+ \oplus 4_-$ and $8^- \to (4_+, 1_-) \oplus (4_-, 1_+) \to 4_+ \oplus 4_-$. But evidently this is quite different from the embedding specified in (63): on the restriction $SO(8) \to SU(4) \simeq SO(6)$, we have $8^+ \to 1 \oplus 6 \oplus 1$ and $8^- \to 4 \oplus 4^*$, and also, $8_v \to 4 \oplus 4^*$, as in (57). Evidently, these two $SO(6)$ subgroups of $SO(8)$ are different, but related to each other by the triality automorphism.

Earlier in the chapter, I promised to work through the multiplication of spinors in $SO(6)$ and $SO(8)$, which we will now do.

The situation in $SO(6)$ is interesting in light of its local isomorphism with $SU(4)$, as discussed above. Since $n = 3$ is odd, our earlier analysis leads to

$$4^+ \otimes 4^+ = [1] \oplus [3] = 6 \oplus 10$$
$$4^+ \otimes 4^- = [0] \oplus [2] = 1 \oplus 15$$

(67)

In light of the table (66) and the remark above that the spinors are just the fundamental 4 and 4* representations of $SU(4)$, these multiplication results are also readily understood within $SU(4)$. Thus, $4 \otimes 4 = 6 \oplus 10$ is just the familiar decomposition of the 2-indexed tensor into its antisymmetric and symmetric parts, and $4 \otimes 4^* = 1 \oplus 15$ corresponds to separating out the trace to obtain the adjoint representation.

For $SO(8)$, $n = 4$ is even, and so, again, according to our earlier analysis

$$8^+ \otimes 8^+ = [0] \oplus [2] \oplus [4] = 1 \oplus 28 \oplus 35$$
$$8^+ \otimes 8^- = [1] \oplus [3] = 8_v \oplus 56 \tag{68}$$

We have indicated that the 8 appearing in the product of 8^+ with 8^- is the vector 8_v. And of course, $8^2 = 64 = 1 + 28 + 35 = 8 + 56$.

More fun with local isomorphism

The local isomorphism $SU(4) \simeq SO(6)$ leads to another proof of the local isomorphism $Sp(4) \simeq SO(5)$, which we already proved by looking at Dynkin diagrams, for example.

Let us consider the antisymmetric tensor $A^{[\alpha\beta]}$ of $SU(4)$ furnishing the $4 \cdot 3/2 = 6$-dimensional irreducible representation. Denote an element of $SU(4)$ by the simple unitary 4-by-4 matrix U. Then $A^{[\alpha\beta]}$ transforms* as $A^{[\alpha\beta]} \to A^{[\gamma\delta]}U^{\gamma\alpha}U^{\delta\beta} = (U^T)^{\alpha\gamma}A^{[\gamma\delta]}U^{\delta\beta}$.

But now we see that the symplectic condition (IV.8.4) $U^T J U = J$ that defines the symplectic algebra $USp(4)$ just says that one particular component of $A^{[\alpha\beta]}$ is invariant under $SU(4)$. Since $SU(4) \simeq SO(6)$, and $A^{[\alpha\beta]}$ corresponds to the vector representation 6 of $SO(6)$, we conclude that $USp(4)$ (more commonly known to physicists as $Sp(4)$) is isomorphic to the subalgebra of $SO(6)$ that leaves one component of the vector 6 invariant, namely, $SO(5)$. This amounts to another proof that $Sp(4) \simeq SO(5)$ locally, as promised earlier.

Appendix 1: $SO(2)$

Confusio speaks up. "I noticed that you started the induction in (38) with $SO(4)$, rather than $SO(2)$."

Yes, very astute of Confusio! As is often the case in group theory, the smaller groups are sometimes the most confusing. For $SO(2)$, $\gamma_1 = \tau_1$, $\gamma_2 = \tau_2$, and $\sigma_{12} = \tau_3$, and so (33) can be solved with either $C = \tau_1$ or $C = i\tau_2$. We seem to have a choice. But this should already raise a red flag, since the discussion in chapter II.4 indicates that whether an irreducible representation is real or pseudoreal is not open to choice. If you go back to that discussion, you will see that to apply Schur's lemma it was crucial for the representation in question to be irreducible. The point is that on the irreducible representations, the generator of $SO(2)$ is represented not by σ_{12}, but by $\sigma_{12}P_\pm = \tau_3 \frac{1}{2}(1 \pm \tau_3) = \frac{1}{2}(\tau_3 \pm 1)$. Thus, the irreducible representations of $SO(2)$ are neither real nor pseudoreal, but complex; that is, the group elements are represented by $e^{i\varphi}$ and $e^{-i\varphi}$ in S^+ and S^-, respectively.

Even more explicitly, since $\gamma_F = -i\gamma_1\gamma_2 = \tau_3$, the two spinors are $\psi_+ = \begin{pmatrix} 1 \\ 0 \end{pmatrix}$ and $\psi_- = \begin{pmatrix} 0 \\ 1 \end{pmatrix}$. The $SO(2)$ group element $e^{i\varphi\tau_3}$ acting on ψ_+ and ψ_- is effectively equal to $e^{i\varphi}$ and $e^{-i\varphi}$, respectively.

This agrees with a table in the text; for $SO(4k + 2)$, the spinor representations are complex, in particular for $k = 0$.

Let us also see how the multiplication rules in (49) and (51) work for $SO(2)$. Since $n = 1$ is odd, we have $1^+ \otimes 1^+ = [1]$ and $1^+ \otimes 1^- = [0]$. At first sight, since [1] is the vector of $SO(2)$, the first result appears puzzling.

---

* I put U on the right to make contact with chapter IV.8.

The resolution is that we should not forget that $1 = \frac{1}{2} \cdot 2$, and the irreducible representation is self-dual or antiself-dual. Indeed, $\psi_+ C \tau_1 \psi_+$ and $\psi_+ C \tau_2 \psi_+$ are proportional to each other.

Appendix 2: $SO(3)$

It is instructive to see how the spinor formalism given in this chapter works for our beloved $SO(3)$. As mentioned earlier, we can get to $SO(3)$ by descending from $SO(4)$.

For $SO(4)$, $n = 2$, $2^{n-1} = 2$, and thus we have a 2-dimensional spinor, that is, 2-component spinors, precisely those that Pauli (and others) discovered. More explicitly, remove γ_4 from (2) and keep $\sigma_{12} = \tau_3 \otimes 1$, $\sigma_{23} = \tau_2 \otimes \tau_2$, and $\sigma_{31} = -\tau_1 \otimes \tau_2$ from (26). We still have $\gamma_F = \tau_3 \otimes \tau_3$, and so the three rotation generators are represented by

$$\frac{1}{2}\sigma_{12}(1 \pm \gamma_F) = \frac{1}{2}(\tau_3 \otimes 1 \pm 1 \otimes \tau_3),$$

$$\frac{1}{2}\sigma_{23}(1 \pm \gamma_F) = \frac{1}{2}(\tau_2 \otimes \tau_2 \mp \tau_1 \otimes \tau_1),$$

$$\frac{1}{2}\sigma_{31}(1 \pm \gamma_F) = -\frac{1}{2}(\tau_1 \otimes \tau_2 \pm \tau_2 \otimes \tau_1) \tag{69}$$

It is more transparent to write out the 4-by-4 matrices. For example, $\sigma_{12} = \tau_3 \otimes 1 = \begin{pmatrix} 1 & 0 & 0 & 0 \\ 0 & -1 & 0 & 0 \\ 0 & 0 & 1 & 0 \\ 0 & 0 & 0 & -1 \end{pmatrix}$. Since

$P_+ = \frac{1}{2}(I + \tau_3 \otimes \tau_3) = \begin{pmatrix} 1 & 0 & 0 & 0 \\ 0 & 0 & 0 & 0 \\ 0 & 0 & 0 & 0 \\ 0 & 0 & 0 & 1 \end{pmatrix}$, to calculate $\sigma_{12} P_+$, we simply cross out the second and third rows and the second and third columns, and thus obtain effectively a 2-by-2 matrix that we recognize as one of the Pauli matrices: $\sigma_{12} P_+$ "$=$" τ_3. Similarly, $\sigma_{23} = \tau_2 \otimes \tau_2 = -\begin{pmatrix} 0 & 0 & 0 & 1 \\ 0 & 0 & -1 & 0 \\ 0 & -1 & 0 & 0 \\ 1 & 0 & 0 & 0 \end{pmatrix}$ and thus, $\sigma_{23} P_+$ "$=$" $-\tau_1$. I leave you to verify that $\sigma_{31} P_+$ "$=$" $-\tau_2$. Thus,

$$\sigma_{12} P_+ \sim \tau_3, \ \sigma_{23} P_+ \sim -\tau_1, \ \sigma_{31} P_+ \sim -\tau_2 \tag{70}$$

(with $\sim$ denoting "represented by"). Indeed, what else could the three generators of the $SO(3)$ algebra be represented by but the three Pauli matrices?

In contrast, to calculate $\sigma_{ij} P_-$, we simply cross out the first and fourth rows and the first and fourth columns of σ_{ij}, since $P_- = \begin{pmatrix} 0 & 0 & 0 & 0 \\ 0 & 1 & 0 & 0 \\ 0 & 0 & 1 & 0 \\ 0 & 0 & 0 & 0 \end{pmatrix}$. Given that there are only three traceless hermitean 2-by-2 matrices, we know that the $\sigma_{ij} P_-$ also have to be Pauli matrices, up to possibly some signs. A quick calculation shows that $\sigma_{12} P_-$ "$=$" $-\tau_3$, $\sigma_{23} P_+$ "$=$" τ_1, and $\sigma_{31} P_+$ "$=$" $-\tau_2$.

The two representations are in fact equivalent (as expected) and are related by a unitary transformation with $U = i\tau_2$.

Appendix 3: Using fermions to describe the spinors of $SO(2n)$

Here we introduce a fermionic formalism to describe the spinor representations of $SO(2n)$. I assume that the reader has heard of the Pauli exclusion principle, stating that you cannot put more than two particles of the type known as fermions into the same quantum state. The electron is perhaps the most famous fermion of them all. If this is totally foreign to you, please skip this appendix. Here we merely do more formally what we did heuristically in the text.

Thus, in our binary code, we could interpret $|+\rangle$ as a state containing one fermion and $|-\rangle$ as a state containing no fermion. It is now convenient to use the creation and annihilation operators introduced back in appendix 1 of chapter IV.2, but, instead of a and $a^\dagger$, we write f and $f^\dagger$ in honor of Enrico Fermi.

You might also want to change notation slightly to fix on the creation and annihilation language used (but it is not necessary). Write the state with no fermion as the vacuum state $|0\rangle$ instead of $|-\rangle$, and the state with one fermion as $|1\rangle \equiv f^{\dagger}|0\rangle$ instead of $|+\rangle$. The Pauli condition stating that states with two or more fermions do not exist can then be written as $f^{\dagger}f^{\dagger}|0\rangle = 0$ or as the operator statement $f^{\dagger}f^{\dagger} = 0$. The correct quantization condition then turns out to involve anticommutation, rather than commutation (see chapter IV.2), relations:

$$\{f, f^{\dagger}\} = ff^{\dagger} + f^{\dagger}f = 1 \tag{71}$$

For your convenience, I list the correspondence here:

$$|0\rangle \leftrightarrow |-\rangle \quad \text{and} \quad |1\rangle \equiv f^{\dagger}|0\rangle \leftrightarrow |+\rangle \tag{72}$$

The operator $N \equiv f^{\dagger}f$ counts the number of fermions. Indeed, using (71), we have

$$N|0\rangle = f^{\dagger}f|0\rangle = 0$$
$$N|1\rangle = f^{\dagger}f|1\rangle = f^{\dagger}ff^{\dagger}|0\rangle = f^{\dagger}\{f, f^{\dagger}\}|0\rangle = f^{\dagger}|0\rangle = |1\rangle \tag{73}$$

When regarded as operators on the states $|\pm\rangle$, the Pauli matrices can then be written as

$$\tau_1 = f + f^{\dagger}, \qquad \tau_2 = i(f - f^{\dagger}), \qquad \tau_3 = 2N - 1 = (-1)^{N+1} \tag{74}$$

(Henceforth, somewhat sloppily, we confound the $\leftrightarrow$ sign and the $=$ sign.) For example, $\tau_1|-\rangle = (f + f^{\dagger})|0\rangle = f^{\dagger}|0\rangle = |1\rangle = |+\rangle$. Note that the last equality in (74) holds because N can take on only the values 0 or 1.

To discuss the spinors of $SO(2n)$ with the states $|\varepsilon_1\varepsilon_2\cdots\varepsilon_n\rangle$ in the binary code notation, we need to generalize to n sets of fermion creation and annihilation operators $f_i^{\dagger}$, f_i with $i = 1, \cdots, n$ satisfying

$$\{f_i, f_j^{\dagger}\} = \delta_{ij}, \quad \text{and} \quad \{f_i, f_j\} = 0 \tag{75}$$

The number operator for fermions of species i is given by $N_i = f_i^{\dagger}f_i$. (For example, the spinor state $|+ - + + -\rangle$ of $SO(2n)$ would be written as $f_1^{\dagger}f_3^{\dagger}f_4^{\dagger}|0\rangle$, with $|0\rangle$ short for the more pedantic $|0, 0, 0, 0, 0\rangle$. In other words, we describe $|+ - + + -\rangle$ as the state containing a fermion of species 1, a fermion of species 3, and a fermion of species 4.)

The spinors $S^{\pm}$ are distinguished by having an even or odd number of fermions present. More formally, this follows from (22) and (74):

$$\gamma_F = \tau_3 \otimes \tau_3 \otimes \cdots \otimes \tau_3 = \Pi_i(-1)^{(N_i+1)} = (-1)^{(N+n)} \tag{76}$$

It is instructive to work out the generators of $SO(2n)$ in this language. From (74) and from what we have learned in the text, we see that the generators are given by all possible bilinear operators of the form* $f_i f_j$, $f_i^{\dagger}f_j^{\dagger}$, and $f_i^{\dagger}f_j$, with $i, j = 1, \cdots, n$. Due to the anticommutation relations (75), the number of these operators are, respectively, $n(n-1)/2$, $n(n-1)/2$, and n^2, giving a total of $2n^2 - n$, which is precisely equal to $2n(2n-1)/2$, the number of generators of $SO(2n)$.

The generators of the $U(n)$ subgroup are then the operators that conserve the number of fermions, namely, $f_i^{\dagger}f_j$. The diagonal $U(1)$ is just the total fermion number $\Sigma_{i=1}^n f_i^{\dagger}f_i$. Given an irreducible spinor representation, with its 2^{n-1} states, those states with the same number of fermions form an $SU(n)$ representation. The generators of $SO(2n)$ that are not in $U(n)$ (namely, $f_i f_j$ and $f_i^{\dagger}f_j^{\dagger}$) then either create or annihilate two fermions at a time. This completely accords with the discussion on how spinors decompose given in the text.

We can also readily make contact with the tensor formalism given in chapter IV.4. Denote by $(\lambda_a)^i{}_j$, with $a = 1, \cdots, n^2 - 1$ and $i, j = 1, \cdots, n$, the $n^2 - 1$ hermitean traceless n-by-n matrices that generate $SU(n)$ (namely, for particle physicists, the generalized Gell-Mann matrices). Then we can write the $n^2 - 1$ generators of $SU(n)$ as $T_a = (\lambda_a)^i{}_j f_i^{\dagger}f^j$, where we have raised the index on f to accord with the notation in chapter IV.4. (Summation over i and j is of course implied.)

---

* Note that due to (75), there is no need to include $f_j f_i^{\dagger} = \delta_{ij} - f_i^{\dagger}f_j$.

Exercise

1   Why doesn't $C = \tau_1 \otimes \tau_1 \otimes \tau_1 \otimes \cdots \otimes \tau_1$ work?

Notes

1. This chapter is based on F. Wilczek and A. Zee, Phys. Rev. D 25 (1982), p. 553. See this paper for detailed references to the relevant physics literature.
2. W. K. Clifford, after translating Riemann's work into English, proposed a theory of gravity based on curved space that anticipated, albeit only in broad outline, Einstein gravity by some 40 years.
3. As usual, we know we can take care of reflections if somebody insists.
4. Recall our earlier remark that, rigorously speaking, $U(n) = \left(SU(n)/Z_n \right) \otimes U(1)$ rather than $U(n) = SU(n) \otimes U(1)$.

The additive group of real numbers

I showed you, way back in chapter II.1, a peculiar 2-dimensional representation

$$D(u) = \begin{pmatrix} 1 & 0 \\ u & 1 \end{pmatrix} \tag{1}$$

of the group of addition, such that

$$D(u)D(v) = D(u + v) \tag{2}$$

I then asked you whether the group described by (1) ever appeared in physics. Where in physics have you encountered* matrices that effectively add two real numbers?

If you didn't know back then, I reveal to you now that $D(u)$ represents the Galilean group of nonrelativistic physics. Denote the 2-dimensional array (call it a vector for short) that $D(u)$ acts on (rather suggestively) as $\begin{pmatrix} t \\ x \end{pmatrix}$. Then $D(u)$ transforms this vector into $\begin{pmatrix} t' \\ x' \end{pmatrix}$, with

$$\begin{aligned} t' &= t \\ x' &= ut + x \end{aligned} \tag{3}$$

This Galilean transformation relates the space and time coordinates t, x and t', x' of two observers in uniform motion with velocity u relative to each other.[†] The sophisticated representation of addition in (1) merely reflects the addition of relative velocity in elementary physics.

---

* Yes, in $SO(2)$ the rotation angles add. But we already know all the irreducible representations of $SO(2)$, and they do not include (1). However, as we shall see, the addition of angles is at some level related to the subject of this chapter.

† The convention here agrees with that used in G Nut (p. 159) and is chosen to minimize the number of signs in the Lorentz transformation to be given below in (6). In Einstein's Gedanken experiment with the train moving to "the right of the page," Ms. Unprime is on the train, while Mr. Prime is on the ground. Thus, if the position of Ms. Unprime is given by $x = 0$ then $x' = ut = ut'$, and the train is moving in the $+x'$ direction.

Relative motion

Observers in motion relative to one another provide one of the most gripping foundational dramas in physics. The issue is how the laws of physics as codified by one observer are related to the laws of physics as codified by another observer in motion relative to the first. Denote, somewhat abstractly, by $T(1 \to 2)$ the transformation that takes the laws of physics seen by observer 1 to the laws of physics seen by observer 2. Similarly, the physics of observer 2 is transformed into the physics of observer 3 by $T(2 \to 3)$. A fundamental postulate of physics asserts that

$$T(2 \to 3)T(1 \to 2) = T(1 \to 3) \tag{4}$$

More precisely, it asserts that the relativity of motion defines a group.

Note that this statement does not specify the precise form of T. As most readers know, one of the two great revolutions of twentieth-century physics involves a disagreement over the form of T, whether it is Galilean or Lorentzian.

It could well be that this fundamental postulate will eventually fail, like so many other truths previously held to be sacred by physicists, but so far, there is no evidence whatsoever to contradict the postulate that motion forms a group.

The fall of absolute time

The form of T encapsulates what we understand about space and time. Thus, the first equation in (3) asserts that time is absolute in nonrelativistic physics. But as we now know, the "self-evident" equality $t' = t$ is actually incorrect.

The fall of absolute time was surely one of the most devastating falls in the history of physics.

Physics in the nineteenth century culminated in the realization that electromagnetism is not invariant under the Galilean transformation, a realization that led Einstein to formulate special relativity. As you probably know,[1] Einstein showed that the Galileo transformation is but an approximation of the Lorentz transformation, which we will derive shortly.

Given that this is a book on group theory in physics, rather than physics per se, our discussion will emphasize the group theoretic aspects, rather than the relevant (and profound) physics, of special relativity.

It is conceivable that, sometime in the nineteenth century, a bright young person could have tried to improve[2] the "lopsided" matrix $\begin{pmatrix} 1 & 0 \\ u & 1 \end{pmatrix}$ in (1) by filling in the 0, that is, by replacing the first equation in (3) for small u by something like $t + \xi^{-1}ux$. She would have seen immediately that ξ would have to have dimension of $(L/T)^2$, that is, of a universal speed[3] squared.

In the absence of an experimentally observed universal speed, this potential star physicist would have been stymied. We now know, of course, that the speed of light c furnishes this universal speed. Once we recognize that the speed of light c provides a universal constant with dimension of L/T, then the door is wide open for modifying[4] the Galilean transformation.

Three derivations of the Lorentz transformation: Each easier than the preceding

Here we offer three derivations of the Lorentz transformation, each one "easier" than the preceding one. All derivations of course must input the astonishing fact that c does not depend on the observer. By an elegant thought experiment in which light is bounced between two mirrors moving in a direction perpendicular to the separation between the mirrors, Einstein showed[5] that this fact implies $(c\Delta t')^2 - (\Delta x')^2 = (c\Delta t)^2 - (\Delta x)^2$, or better,

$$(c\, dt')^2 - (dx')^2 = (c\, dt)^2 - (dx)^2 \tag{5}$$

in the infinitesimal limit.

Now we are ready to derive the Lorentz transformation. For pedagogical clarity, we will continue to stay with the case of one spatial dimension, that is, with $(1+1)$-dimensional spacetime. (In any case, as the reader probably knows, the other two coordinates, y and z, orthogonal to the direction of relative motion, merely go along for the ride: the Lorentz transformation leaves them untouched. See below.)

The first derivation is by brute force. Replace (3) by $t' = w(t + \zeta ux/c^2)$, $x' = \tilde{w}(ut + x)$, with w, $\tilde{w}$, and ζ three unknown dimensionless functions of u/c. Plugging this into (5), we readily determine these functions and obtain

$$ct' = \frac{ct + ux/c}{\sqrt{1 - \frac{u^2}{c^2}}}$$

$$x' = \frac{ut + x}{\sqrt{1 - \frac{u^2}{c^2}}} \tag{6}$$

In all likelihood, students of physics first encounter the Lorentz transformation written in this elementary form, with the infamous $\sqrt{1 - \frac{u^2}{c^2}}$ factor.

Note that, for $c = \infty$, the Lorentz transformation (6) reduces to the Galileo transformation (3), as expected.

It is clearly convenient to use ct instead of t in (6), so that time and space have the same dimension of length. Henceforth, we will set $c = 1$.

For our second derivation, write $(dt)^2 - (dx)^2$ as $dx^T \eta dx$, where dx denotes* the column vector $\begin{pmatrix} dt \\ dx \end{pmatrix}$ and η the 2-by-2 matrix $\eta = \begin{pmatrix} 1 & 0 \\ 0 & -1 \end{pmatrix} = \sigma_3$, known as the Minkowski metric.[†]

---

* Here we follow the standard notation (which can be confusing to the rank beginner) of using x for both the spacetime coordinate and the spatial coordinate along the x-axis. In any case, the level of the presentation here presupposes that the reader has heard of special relativity before.

† We refrain from going into a long detailed discussion of the metric here. See G Nut, chapter III.3.

For convenience, we also identify η as the third Pauli matrix. Finally, denote the 2-by-2 matrix representing the Lorentz transformation by L, so that $\begin{pmatrix} dt' \\ dx' \end{pmatrix} = L \begin{pmatrix} dt \\ dx \end{pmatrix}$. Then (5) may be written as $dx'^T \eta dx' = dx^T L^T \eta L dx = dx^T \eta dx$. Since dx is arbitrary, we can extract the requirement

$$L^T \eta L = \eta \tag{7}$$

Perhaps you recognize that the reasoning is essentially the same as that used regarding rotational invariance back in chapter I.3, where we obtained $R^T R = I$ as the definition of rotation. Indeed, if we write the left hand side of this requirement as $R^T I R$, the resemblance with (7) becomes even more apparent. The correspondence between rotation and Lorentz transformation is $R \to L, I \to \eta$.

Lie taught us that to solve (7) and to obtain the Lorentz group, it suffices to consider infinitesimal transformations and write $L \simeq I + i\varphi K$, with φ some infinitesimal real parameter. (Note that the generator K is defined with a factor of i, in analogy with how the generators J of rotation were defined in chapter I.3 by $R \simeq I + i\vec{\theta}\vec{J}$.) We find $K^T \eta + \eta K = 0$, and hence $K^T \eta = -\eta K$. The solution is immediate:

$$iK = \begin{pmatrix} 0 & 1 \\ 1 & 0 \end{pmatrix} \tag{8}$$

If we recall that $\eta = \sigma_3$ and that Pauli matrices anticommute, the solution of $K^T \eta = -\eta K$ is even more immediate: $iK = \sigma_1$. Note that, while iJ is real and antisymmetric, iK is real and symmetric (and hence a fortiori hermitean).

Finally, I show you an even easier way to derive the Lorentz transformation. Write

$$dt^2 - dx^2 = (dt + dx)(dt - dx) \equiv dx^+ dx^- \tag{9}$$

In the last step, I defined the "light cone" coordinates $x^\pm = t \pm x$. Thus, $dt^2 - dx^2$ is left invariant if we multiply x^+ by an arbitrary quantity and divide x^- by the same quantity. Call this quantity e^φ. We immediately obtain the Lorentz transformation. For example, $t = \frac{1}{2}(x^+ + x^-) \to \frac{1}{2}(e^\varphi x^+ + e^{-\varphi} x^-) = \frac{1}{2}(e^\varphi(t + x) + e^{-\varphi}(t - x)) = \cosh \varphi t + \sinh \varphi x$.

By the way, the analogous step for $SO(2)$ rotation would be to replace x and y by the complex coordinates $z \equiv x + iy$ and $z^* \equiv x - iy$. Then under a rotation, $z \to e^{i\theta}z$, and $z^* \to e^{-i\theta}z^*$.

The Lorentz transformation stretches and compresses the light cone coordinates $x^\pm$ by compensating amounts.

Lorentz is more natural than Galileo

I started this chapter with the Galilean transformation to show you that mathematically, the Lorentz algebra is more natural and aesthetically appealing than the Galilean algebra. Compare $iK = \begin{pmatrix} 0 & 1 \\ 1 & 0 \end{pmatrix}$ with the Galilean generator $\begin{pmatrix} 0 & 0 \\ 1 & 0 \end{pmatrix}$. (Since we have already set $c = 1$, it is no longer manifest that one matrix reduces to the other.)

In summary, to leading order, the Lorentz transformation is gloriously simple:

$$\begin{pmatrix} t' \\ x' \end{pmatrix} \simeq (I + i\varphi K) \begin{pmatrix} t \\ x \end{pmatrix} = \begin{pmatrix} 1 & \varphi \\ \varphi & 1 \end{pmatrix} \begin{pmatrix} t \\ x \end{pmatrix} \tag{10}$$

Could the mystery of spacetime be expressed in a simpler mathematical form than this? Time rotates into space, and space rotates into time.

Note that $\begin{pmatrix} 1 & \varphi \\ \varphi & 1 \end{pmatrix} \begin{pmatrix} 1 & \chi \\ \chi & 1 \end{pmatrix} = \begin{pmatrix} 1+\varphi\chi & \varphi+\chi \\ \varphi+\chi & 1+\varphi\chi \end{pmatrix}$. Thus, if we agree to ignore the quadratic terms $\varphi\chi$, the matrix in (10) also approximately represents the additive group.

Building up to finite relative velocities

The reader might wonder about the square roots in (6) so characteristic of elementary treatments of special relativity. Lie assures us, however, as he did back in chapter I.3, that we do not lose anything by going to infinitesimal transformations. We can always reconstruct the finite transformations using the multiplicative structure of the group. Indeed, from the discussion in chapter I.3, we can simply exponentiate and promote $L \simeq (I + i\varphi K)$ to $L = e^{i\varphi K}$. Expanding this exponential as a Taylor series just as in chapter I.3, separating the series into even and odd terms, and using $(iK)^2 = I$, we obtain the finite Lorentz transformation (keep in mind that iK and hence L are real):

$$L(\varphi) = e^{i\varphi K} = \sum_{n=0}^{\infty} \varphi^n (iK)^n / n! = \left(\sum_{k=0}^{\infty} \varphi^{2k}/(2k)! \right) I + \left(\sum_{k=0}^{\infty} \varphi^{2k+1}/(2k+1)! \right) iK$$

$$= \cosh\varphi I + \sinh\varphi\, iK$$

$$= \begin{pmatrix} \cosh\varphi & \sinh\varphi \\ \sinh\varphi & \cosh\varphi \end{pmatrix} \tag{11}$$

Hence,

$$\begin{pmatrix} t' \\ x' \end{pmatrix} = \begin{pmatrix} \cosh\varphi & \sinh\varphi \\ \sinh\varphi & \cosh\varphi \end{pmatrix} \begin{pmatrix} t \\ x \end{pmatrix} \tag{12}$$

Comparing the result $x' = \sinh\varphi t + \cosh\varphi x$ with (6), we determine

$$u = \frac{\sinh\varphi}{\cosh\varphi} = \tanh\varphi \tag{13}$$

Solving, we obtain $\cosh\varphi = \frac{1}{\sqrt{1-u^2}}$, $\sinh\varphi = \frac{u}{\sqrt{1-u^2}}$, and recover the square roots in (6) (with c set to 1, of course), as expected.

We note that the matrix in (12) now represents the additive group exactly, either by using various identities for the hyperbolic sine and cosine, or by merely writing $e^{i\varphi_1 K} e^{i\varphi_2 K} = e^{i(\varphi_1 + \varphi_2)K}$. The matrix in (10) is of course the small φ approximation of the matrix in (12).

$SO(m, n)$

It probably did not escape your notice that our discussion of the Lorentz transformation in $(1 + 1)$-dimensional spacetime is totally reminiscent of our discussion of rotation in 2-dimensional space. (Indeed, I already mentioned it.) This is hardly an accident.

We defined rotations back in chapter I.3 as those linear transformations $d\vec{x}' = R d\vec{x}$ (with $d\vec{x} = (dx^1, dx^2, \cdots, dx^N)$) that leave the distance squared between two nearby points $ds^2 = \sum_{i=1}^{N} (dx^i)^2 = (dx^1)^2 + (dx^2)^2 + \cdots + (dx^N)^2$ unchanged. We are now invited to generalize and consider linear transformations $d\vec{x}' = L d\vec{x}$, with $d\vec{x} = (dx^1, dx^2, \cdots, dx^{(m+n)})$, which leave the generalized distance squared

$$ds^2 = \sum_{i=1}^{m} (dx^i)^2 - \sum_{i=m+1}^{m+n} (dx^i)^2 \tag{14}$$

unchanged. This set of transformations defines the group[6] $SO(m, n)$. (By now, you should be able to verify the group axioms in your head.)

Thus, the $(1 + 1)$-dimensional Lorentz group we discussed is $SO(1, 1)$. It should hardly surprise you that $SO(1, 1)$ may be analytically continued* from $SO(2)$. Write the time coordinate as $t = iy$ and continue y to a real variable: then $-dt^2 + dx^2 = dy^2 + dx^2$. Setting $\varphi = i\theta$, we continue the Lorentz transformation $t' = \cosh \varphi t + \sinh \varphi x$, $x' = \sinh \varphi t + \cosh \varphi x$ to the rotation $y' = \cos \theta y + \sin \theta x$, $x' = -\sin \theta y + \cos \theta x$. Indeed, in the older literature, the fourth coordinate $x^4 = ict$ is often used, but by now it has mostly been replaced by x^0.

I close with a "pregnant" and "mystic" quote from Minkowski. The words he chose probably won't fly in a contemporary scholarly journal, but I like them all the same.

> The essence of this postulate may be clothed mathematically in a very pregnant manner in the mystic formula $3 \cdot 10^5 \text{ km} = \sqrt{-1}$ secs.
> —H. Minkowski

The Lorentz group $SO(3, 1)$

It is time to remember that we live in $(3 + 1)$-dimensional spacetime rather than $(1 + 1)$-dimensional spacetime, and so we have to deal with the group $SO(3, 1)$. We will focus on the Lie algebra rather than on the group itself. With three spatial coordinates, we can boost in any of three directions, and, of course, we can also rotate. Thus, the Lie algebra $SO(3, 1)$ consists of six generators: J_x, J_y, J_z and K_x, K_y, K_z. It is straightforward to generalize the 2-by-2 matrices in the preceding discussion to 4-by-4 matrices:

---

* Thus, the addition of angles is indeed related to the subject of this chapter, as mentioned in the first footnote in the chapter.

$$
iK_x = \begin{pmatrix} 0 & 1 & 0 & 0 \\ 1 & 0 & 0 & 0 \\ 0 & 0 & 0 & 0 \\ 0 & 0 & 0 & 0 \end{pmatrix}, \quad iK_y = \begin{pmatrix} 0 & 0 & 1 & 0 \\ 0 & 0 & 0 & 0 \\ 1 & 0 & 0 & 0 \\ 0 & 0 & 0 & 0 \end{pmatrix}, \quad iK_z = \begin{pmatrix} 0 & 0 & 0 & 1 \\ 0 & 0 & 0 & 0 \\ 0 & 0 & 0 & 0 \\ 1 & 0 & 0 & 0 \end{pmatrix} \tag{15}
$$

I leave it to you to write the 4-by-4 matrices representing J_x, J_y, and J_z.

To repeat, in our convention, iK_j is real symmetric and hence hermitean. Thus, K_j is imaginary symmetric and hence antihermitean. Therefore, as was already pointed out in the very first chapter, $L = e^{i\varphi K} = e^{\varphi(iK)}$ is manifestly not unitary. Explicitly (see the matrix in (12)), $L^\dagger L = L^T L \neq I$ (the first equality holds since iK and hence L are real in the defining representation); rather, $L^T \eta L = \eta$, according to (7).

Instead of matrices, we will use, for a change of pace, the differential operators introduced and discussed in chapter I.3. In particular, $J_z = i(y\frac{\partial}{\partial x} - x\frac{\partial}{\partial y})$. Similarly,

$$
iK_x = t\frac{\partial}{\partial x} + x\frac{\partial}{\partial t} \tag{16}
$$

Note the plus sign. (Let us check: under an infinitesimal transformation, $\delta t = \varphi(iK_x t) = \varphi x$, $\delta x = \varphi(iK_x x) = \varphi t$, $\delta y = 0$, and $\delta z = 0$.) As noted earlier, y and z merely go along for the ride; nevertheless, the point is that the presence of two other boost operators, $iK_y = t\frac{\partial}{\partial y} + y\frac{\partial}{\partial t}$ and $iK_z = t\frac{\partial}{\partial z} + z\frac{\partial}{\partial t}$, make the group theory much richer.

The rotation subgroup is generated by J_x, J_y, and J_z, as was discussed in chapter I.3. The commutators of the Js with one another are given, as in chapter I.3, by

$$
[J_i, J_j] = i\varepsilon_{ijk} J_k \tag{17}
$$

While it should be clear that the Js rotate the three Ks into one another, it is somewhat instructive to verify this explicitly. For example,

$$
\begin{aligned}
[J_z, iK_x] &= i\left[y\frac{\partial}{\partial x} - x\frac{\partial}{\partial y}, t\frac{\partial}{\partial x} + x\frac{\partial}{\partial t} \right] \\
&= i\left(\left[y\frac{\partial}{\partial x}, x\frac{\partial}{\partial t} \right] - \left[x\frac{\partial}{\partial y}, t\frac{\partial}{\partial x} \right] \right) \\
&= i\left(y\frac{\partial}{\partial t} + t\frac{\partial}{\partial y} \right) = i(iK_y)
\end{aligned} \tag{18}
$$

Thus,

$$
[J_i, K_j] = i\varepsilon_{ijk} K_k \tag{19}
$$

Finally, since the algebra closes, we expect the commutators between the Ks should produce the Js:

$$
\begin{aligned}
[K_x, K_y] &= (-i)^2 \left[t\frac{\partial}{\partial x} + x\frac{\partial}{\partial t}, t\frac{\partial}{\partial y} + y\frac{\partial}{\partial t} \right] \\
&= \left(\left[y\frac{\partial}{\partial t}, t\frac{\partial}{\partial x} \right] - \left[x\frac{\partial}{\partial t}, t\frac{\partial}{\partial y} \right] \right) \\
&= y\frac{\partial}{\partial x} - x\frac{\partial}{\partial y} = -iJ_z
\end{aligned} \tag{20}
$$

Perhaps one small surprise is that $[K_x, K_y]$ does not yield iJ_z, but rather $-iJ_z$. Thus,

$$[K_i, K_j] = -i\varepsilon_{ijk}J_k \tag{21}$$

It goes without saying that you can also obtain (20) and (21) by commuting 4-by-4 matrices, as you should verify.

Note that under parity, that is, spatial reflection* $\vec{x} \to -\vec{x}$, $t \to t$, the generators transform as $J_i \to J_i$, $K_i \to -K_i$. From this, we know[7] that K_k cannot appear in the right hand side of (21). We can also arrive at the same conclusion by appealing to time reversal $\vec{x} \to \vec{x}$, $t \to -t$.

Wait! Haven't we seen something like this before?

The Lorentz algebra falls apart into two pieces

The reader with a long memory will recall that, way back in chapter I.3, when we spoke of rotations in 4-dimensional Euclidean space, we denoted the six generators of $SO(4)$, $J_{mn} = -J_{nm}$ for $m, n = 1, \cdots, 4$, by $J_3 = J_{12}$, $J_1 = J_{23}$, $J_2 = J_{31}$ and $K_1 = J_{14}$, $K_2 = J_{24}$, $K_3 = J_{34}$. Then the commutation relations between these generators (which we wrote down in general for $SO(N)$) took the form

$$[J_i, J_j] = i\varepsilon_{ijk}J_k \tag{22}$$

$$[J_i, K_j] = i\varepsilon_{ijk}K_k \tag{23}$$

$$[K_i, K_j] = i\varepsilon_{ijk}J_k \tag{24}$$

I have copied (I.3.25), (I.3.26), and (I.3.27) here. Evidently, the Ks are analogs of the boosts for the Lorentz group, if we think of the coordinate x^4 as "Euclid's time," so to speak.

For your convenience, I collect the commutation relations for $SO(3, 1)$:

$$[J_i, J_j] = i\varepsilon_{ijk}J_k \tag{25}$$

$$[J_i, K_j] = i\varepsilon_{ijk}K_k \tag{26}$$

$$[K_i, K_j] = -i\varepsilon_{ijk}J_k \tag{27}$$

Comparing the commutation relations (25), (26), and (27) for $SO(3, 1)$ with the commutation relations (22), (23), and (24) for $SO(4)$, we see that there is a subtle (or perhaps not so subtle) minus sign in (27). Evidently, we can formally obtain the commutation relations for $SO(3, 1)$ from the commutation relations for $SO(4)$ by letting $K \to iK$.

We learned in chapter I.3 that $SO(4)$ is locally isomorphic to $SU(2) \otimes SU(2)$: specifically, if we define $J_{\pm, i} = \frac{1}{2}(J_i \pm K_i)$, then the Lie algebra of $SO(4)$ falls apart into two pieces.

---

* With the increasing use of Lorentz invariance, Dirac equation, Majorana particle, and related concepts in condensed matter physics, it is important to note here that while spatial reflection and parity are the same in $3 + 1$ dimensions, they are not the same in odd-dimensional spacetimes. For instance, in $2 + 1$ dimensions, spatial reflection $\vec{x} \to -\vec{x}$ is actually a rotation, since it has determinant $+1$, in contrast to parity, defined to be $x_1 \to -x_1$, $x_2 \to +x_2$ (or, equivalently, $x_1 \to +x_1$, $x_2 \to -x_2$).

Later, in chapter IV.7, we constructed the isomorphism explicitly, showing that in fact $SU(2) \otimes SU(2)$ double covers $SO(4)$.

Since $SO(3, 1)$ can be obtained by analytically continuing from $SO(4)$, we expect something similar to happen here. We only need to take care of the one difference due to the minus sign in (27) by defining

$$J_{\pm,i} = \frac{1}{2}(J_i \pm i K_i) \tag{28}$$

with an i. Note that, since J_i and $i K_i$ are both hermitean operators, $J_{\pm,i}$ are manifestly hermitean.

Let us now verify that the six generators $J_{\pm,i}$ divide into two sets of three generators each, the J_+s and the J_-s, with each set of generators commuting right past the other set. Using (25), (26), and (27), we have indeed

$$[J_{+,i}, J_{-,j}] = \left(\frac{1}{2}\right)^2 [J_i + i K_i, J_j - i K_j]$$

$$= \left(\frac{1}{2}\right)^2 \left([J_i, J_j] - i[J_i, K_j] - i[J_j, K_i] + [K_i, K_j]\right)$$

$$= \left(\frac{1}{2}\right)^2 i\varepsilon_{ijk}(J_k - i K_k + i K_k - J_k) = 0 \tag{29}$$

To obtain the other commutation relations, we simply flip a few signs in (29):

$$[J_{+,i}, J_{+,j}] = \left(\frac{1}{2}\right)^2 [J_i + i K_i, J_j + i K_j]$$

$$= \left(\frac{1}{2}\right)^2 \left([J_i, J_j] + i[J_i, K_j] - i[J_j, K_i] - [K_i, K_j]\right)$$

$$= \left(\frac{1}{2}\right)^2 i\varepsilon_{ijk}(J_k + i K_k + i K_k + J_k) = i\varepsilon_{ijk}J_{+,k} \tag{30}$$

Similarly,

$$[J_{-,i}, J_{-,j}] = i\varepsilon_{ijk}J_{-,k} \tag{31}$$

That the Lorentz algebra falls apart into two pieces will be central to our discussion of the Dirac equation in chapter VII.4.

It is worth noting again that the algebra (25), (26), and (27) allows us to flip the sign of K_i: $K_i \to -K_i$, but not the sign of J_i. This corresponds to the interchange $J_{+i} \leftrightarrow J_{-i}$. Physically, this describes spatial reflection.

Space and time become spacetime

Space and time become spacetime, and we can package the three spatial coordinates and the time coordinate into four spacetime coordinates $dx^\mu = (dx^0, dx^1, dx^2, dx^3) = (dt, dx, dy, dz)$. By definition, a vector p^μ in spacetime is defined as a set of four numbers $p^\mu = (p^0, p^1, p^2, p^3)$, which transform in the same way as the ur vector dx^μ under the

Lorentz transformation. It is sometimes called a 4-vector to distinguish it from ordinary 3-vectors* in space. By convention, we write the index on a 4-vector as a superscript.

The square of the length of the 4-vector p is defined as $p^2 \equiv p \cdot p \equiv \eta_{\mu\nu}p^\mu p^\nu$. (The dot will be often omitted henceforth.) Here we generalize the $(1+1)$-dimensional Minkowski metric to the $(3+1)$-dimensional Minkowski metric given by the diagonal matrix with $\eta_{00} = +1$, $\eta_{11} = -1$, $\eta_{22} = -1$, and $\eta_{33} = -1$. As already noted, Lorentz transformations leave p^2 unchanged. Repeating an argument used in chapter I.3, we deduce[†] that they also leave the scalar dot product

$$p \cdot q \equiv \eta_{\mu\nu}p^\mu q^\nu = p^0 q^0 - (p^1 q^1 + p^2 q^2 + p^3 q^3) = p^0 q^0 - \vec{p} \cdot \vec{q} \tag{32}$$

between two 4-vectors p and q unchanged.

The attentive reader might have noticed that I snuck in lower indices by writing $\eta_{\mu\nu}$ as an object carrying subscripts.

Thus far, $\eta_{\mu\nu}$ is the only object with lower indices. Whenever we want to sum over two indices μ and ν, the rule is that we multiply by $\eta_{\mu\nu}$ and invoke Einstein's repeated summation convention. We say that we have contracted the two indices. For example, given two vectors p^μ and q^μ, we might want to contract the indices μ and ν in $p^\mu q^\nu$ and obtain $p \cdot q \equiv \eta_{\mu\nu}p^\mu q^\nu$. Another example: given $p^\mu q^\nu r^\rho s^\sigma$, suppose we want to contract μ with σ and ν with ρ. Easy, just write $\eta_{\mu\sigma}\eta_{\nu\rho}p^\mu q^\nu r^\rho s^\sigma = (p \cdot s)(q \cdot r)$. Savvy readers will recognize that I am going painfully slowly here for the sake of those who have never seen this material before.

So far so good. All vectors carry upper indices, and the only object that carries lower indices is η.

The next step is purely for the sake of notational brevity. To save ourselves from constantly writing the Minkowski metric $\eta_{\mu\nu}$, we define, for any vector p^μ, a vector with a lower index

$$p_\nu \equiv \eta_{\mu\nu}p^\mu \tag{33}$$

In other words, if $p^\mu = (p^0, \vec{p})$, then $p_\mu = (p^0, -\vec{p})$. Thus,[‡] $p \cdot q = p_\mu q^\mu = p^0 q^0 - \vec{p} \cdot \vec{q}$. With this notation, we can write $p \cdot q = p_\nu q^\nu = p^\nu q_\nu$. Similarly, an expression $\eta_{\mu\nu}p^\mu q^\nu \eta_{\rho\sigma}r^\rho s^\sigma$ can be written more simply as $p_\nu q^\nu r_\sigma s^\sigma$. The Minkowski metric has been folded into the indices, so to speak.

Unaccountably, some students are twisted out of shape by this trivial act of notational sloth. "What?" they say, "there are two kinds of vectors?" Yes, fancy people speak of contravariant vectors (p^μ, for example) and covariant vectors (p_μ, for example), but let

---

* Evidently, the 4-vector p^μ contains the 3-vector $p^i = (p^1, p^2, p^3)$.

† For two arbitrary 4-vectors p and q, consider the vector $p + \alpha q$ (for α an arbitrary real number). Its length squared is equal to $p^2 + 2\alpha p \cdot q + \alpha^2 q^2$. Since Lorentz transformations leave lengths unchanged and since α is arbitrary, $p \cdot q$ cannot change.

‡ Notice that the same dot in this equation carries two different meanings, the scalar product between two 4-vectors on the left hand side and between two 3-vectors on the right hand side, but there should be no confusion.

me assure the beginners that there is nothing terribly profound* going on here. Just a convenient notation.[8]

The next question might be: given p_μ, how do we get back to p^μ?

Here is where I think beginners can get a bit confused. If you have any math sense at all, you would expect that we use the inverse of η, and you would be absolutely right. If you use η to move indices downstairs, surely you would use the inverse of η to move them upstairs. But the inverse of the matrix η is itself. So traditionally, the inverse of η is denoted by the same symbol, but with two upper indices, like this: $\eta^{\mu\nu}$. We define $\eta^{\mu\nu}$ by the diagonal matrix with $\eta^{00} = +1$ and $\eta^{11} = \eta^{22} = \eta^{33} = -1$.

Indeed, $\eta^{\mu\nu}$ is the inverse of $\eta_{\mu\nu}$ regarded as a matrix: $\eta^{\mu\nu}\eta_{\nu\lambda} = \delta^\mu_\lambda$, where the Kronecker delta δ^μ_λ is defined, as usual, to be 1 if $\mu = \lambda$ and 0 otherwise. It is worth emphasizing that while $\eta^{\mu\nu}$ and $\eta_{\mu\nu}$ are numerically the same matrix, they should be distinguished conceptually. Let us check the obvious, that the inverse metric $\eta^{\mu\nu}$ raises lower indices: $\eta^{\mu\nu}p_\nu = \eta^{\mu\nu}\eta_{\nu\lambda}p^\lambda = \delta^\mu_\lambda p^\lambda = p^\mu$. Yes, indeed.

Confusio: "Ah, I get it. The same symbol η is used to denote a matrix and its inverse, distinguished by whether η carries lower or upper indices."

From this we see that the Kronecker delta δ^μ_λ has to be written with one upper and one lower index. In contrast, η^μ_ν does not exist. On the other hand, there is no such thing as $\delta^{\mu\nu}$ or $\delta_{\mu\nu}$. Also, note that the Kronecker delta δ does not contain any minus signs, unlike the Minkowski metric η.

It follows that the shorthand ∂_μ for $\frac{\partial}{\partial x^\mu}$ has to carry a lower index, because

$$\partial_\mu x^\nu = \frac{\partial x^\nu}{\partial x^\mu} = \delta^\nu_\mu \tag{34}$$

In other words, for the indices to match, ∂_μ must be written with a lower index. This makes sense, since the coordinates x^μ carry an upper index but in $\frac{\partial}{\partial x^\mu}$ it appears in the denominator, so to speak. We will use this important fact later. It follows that

$$\partial_\mu x_\nu = \eta_{\mu\nu} \tag{35}$$

in contrast to (34). Again, the difference merely reflects that in the convention used here, $x^\mu = (t, \vec{x})$, while $x_\mu = (t, -\vec{x})$.

The Lorentz transformation revisited

We derived the Lorentz transformation earlier in this chapter, but it is convenient, now that we have upper and lower indices,[9] to write its defining feature again. A Lorentz transformation is a linear transformation on the spacetime coordinates

$$x^\mu \to x'^\mu = L^\mu_{\ \nu} x^\nu \tag{36}$$

---

* Of course, if you woke up one day and discovered that you were a mathematician or a mathematician-want-to-be, you should and could read more profound books.

which leaves unchanged the proper time interval $d\tau$, defined by

$$d\tau^2 = \eta_{\mu\nu}dx^\mu dx^\nu = dt^2 - d\vec{x}^2 \tag{37}$$

This is of course the same invariant in (5) and (9).

More generally, let p and q be two arbitrary 4-vectors. Consider the linear transformation $p'^\mu = L^\mu_{\ \sigma}p^\sigma$ and $q'^\mu = L^\mu_{\ \sigma}q^\sigma$. Notice the upper-lower summation convention. For L to be a Lorentz transformation, we require $p' \cdot q' = p \cdot q$; that is, $p' \cdot q' = \eta_{\mu\nu}p'^\mu q'^\nu = \eta_{\mu\nu}L^\mu_{\ \sigma}p^\sigma L^\nu_{\ \rho}q^\rho = p \cdot q = \eta_{\sigma\rho}p^\sigma q^\rho$. Since p^σ and q^ρ are arbitrary, L must satisfy

$$\eta_{\mu\nu}L^\mu_{\ \sigma}L^\nu_{\ \rho} = \eta_{\sigma\rho} \tag{38}$$

Let us define the transpose by $(L^T)_\sigma^{\ \mu} = L^\mu_{\ \sigma}$. (Note that when the indices are interchanged, the guy who was downstairs stays downstairs, and similarly for the guy who was upstairs.) We may then write (38) as $(L^T)_\sigma^{\ \mu}\eta_{\mu\nu}L^\nu_{\ \rho} = \eta_{\sigma\rho}$, or more succinctly, as $L^T\eta L = \eta$. This is of course the same as (7).

Let me emphasize again an extremely useful feature of this notational device. In the Einstein convention, a lower index is always contracted with an upper index (that is, summed over), and vice versa. Never never sum[10] two lower indices together, or two upper indices together!

Energy and momentum become momentum

I have already apologized for the lack of emphasis on physics, but this is after all a book on group theory. There is of course an enormous amount of profound physics connected with the Lorentz transformation. I limit myself to one particularly important implication, which we will need for later use.

In Newtonian mechanics, the conserved momentum $p^i_{\text{Newton}} = m\frac{dx^i}{dt}$ plays a central role, but this expression does not transform in any mathematically sensible way, considering that it is given by the components of a 3-vector dx^i divided by the time component dt of the 4-vector dx^μ. In Einstein's special relativity, $p^i_{\text{Newton}} = m\frac{dx^i}{dt}$ is promoted to a 4-vector $p^\mu = m\frac{dx^\mu}{d\tau}$, with $d\tau$ defined in (37). Since $d\tau$ is invariant under Lorentz transformations, the momentum transforms just like dx^μ, that is, as a 4-vector (as it should). The time component $p^0 = E$ is identified as the energy of the particle. Henceforth, energy and momentum become 4-momentum, or momentum for short.

Using the expression for $d\tau$ in (37), we obtain $p^2 = \eta_{\mu\nu}p^\mu p^\nu = m^2\eta_{\mu\nu}\frac{dx^\mu}{d\tau}\frac{dx^\nu}{d\tau} = m^2$. Written out in components, $p^2 = m^2$ says that $E^2 - \vec{p}^2 = m^2$, namely, Einstein's relation $E^2 = \vec{p}^2 + m^2$ between the energy and momentum of a particle (that is, the layperson's $E = mc^2$).

Lorentz tensors

Back in chapter IV.1, we worked out the tensor representations of $SO(N)$ in general and $SO(4)$ in particular. Evidently, the discussion of the tensors of the Lorentz group $SO(3, 1)$

would proceed in the same way, with the additional feature of having indices on two floors, as mandated by the Minkowskian metric η. As we have learned, out of the two basic vector representations (namely, the Lorentz vector with an upper index and the Lorentz vector with a lower index), we can construct the general tensor representation.

In particular, the ur vector dx^μ transforms like a vector with an upper index, while ∂_μ transforms like a vector with a lower index.

The multiplication of tensors also works out along a by-now familiar line. As an important example, multiplying two vectors A_μ and B_ν together, we have $4 \otimes 4 = 6 \oplus 10$, corresponding to an antisymmetric tensor and a symmetric tensor, respectively, each with two lower indices. Indeed, in your study of physics, you might already have encountered the Lorentz tensor with two lower indices. Given a 4-vector field* $A_\mu(x)$, we can form the antisymmetric tensor $F_{\mu\nu} \equiv \partial_\mu A_\nu - \partial_\nu A_\mu$. This is of course Maxwell's electromagnetic field, which started the entire relativistic story told here.

You can also readily work out how Lorentz tensors break up on restriction to the rotation subgroup $SO(3)$, starting with how the Lorentz vector decomposes: $4 \to 3 + 1$. In particular, $4 \otimes 4 \to (3+1) \otimes (3+1) = 5 + 3 + 1 + 1 + 3 + 3$. From this we deduce that $6 \to 3 + 3$, which of course we could also work out directly from inspecting [0, 2]. The decomposition $6 \to 3 + 3$ in mathematics corresponds to, in physics, the antihistorical process of the electromagnetic field breaking up into the electric $\vec{E}$ and the magnetic $\vec{B}$ fields. Indeed, I might remind you that, already in chapter IV.1, we have discussed how the 6-dimensional representation of $SO(4)$ breaks up as $6 \to 3 + 3$ on restriction to the natural subgroup $SO(3)$. Evidently, much of what was said about $SO(4)$ carries over to $SO(3)$ with minor or no modification.

Self-dual and antiself-dual tensors and the Lorentz generators

Let us spend a few more moments on the 2-indexed antisymmetric tensor furnishing the 6-dimensional representation of the Lorentz algebra. Recall the discussion of dual tensors in chapter IV.1. For $SO(2k)$, the antisymmetric symbol $\varepsilon_{i_1 i_2 \ldots i_{2k}}$ carries $2k$ indices and the antisymmetric tensor with n indices is dual to a tensor with $2k - n$ indices. In particular, for $SO(4)$, given an antisymmetric tensor A_{ij} with two indices, the dual tensor $B_{ij} = \frac{1}{2}\varepsilon_{ijkl}A_{kl}$ also carries two indices. Thus, we have the possibility of forming the self-dual and antiself-dual tensors $A_{ij} \pm B_{ij}$. This indicates that the 6-dimensional representation furnished by A_{ij} is actually reducible and decomposes into two irreducible representations as $6 \to 3 + 3$. The discussion carries over to $SO(3, 1)$. Indeed, back in chapter IV.1, I promised to come back to the "peculiar" fact $4 - 2 = 2$, which I do now.

Well, dear readers, where oh where have you seen six objects in our discussion of the Lorentz group?

Yes, indeed, the generators $\vec{J}$ and $\vec{K}$ may be packaged into the Lorentz tensor $J_{\mu\nu}$. Indeed, J_{ij} correspond to $\varepsilon_{ijk}J_k$, and J_{0i} to K_i, thus verifying explicitly the decomposition

---

* The word "field" in this context merely means a 4-vector that depends on the spacetime coordinates.

$6 \to 3 + 3$ on restricting the Lorentz group to its rotation subgroup. This also accords with the general statement that the generators of a Lie algebra furnish the adjoint representation. For $SO(3, 1)$, the adjoint is the two-indexed antisymmetric tensor.

But now the statement that duality implies that the adjoint representation is not irreducible just informs us, once again, that the combinations $J_i \pm i K_i$ form two irreducible representations.[11]

The relativistic spacetime notation allows us to package the differential operators $J_z = i(y \frac{\partial}{\partial x} - x \frac{\partial}{\partial y})$, $i K_x = t \frac{\partial}{\partial x} + x \frac{\partial}{\partial t}$, and so forth into

$$J_{\mu\nu} = i(x_\mu \partial_\nu - x_\nu \partial_\mu) \tag{39}$$

Note that since $J_{\mu\nu}$ is defined with two lower indices, our index convention requires $x_\mu = \eta_{\mu\nu} x^\nu$, rather than x^μ, to appear on the right hand side. This leads to various minus signs in the following identification:

$$J_{12} = i(x_1 \partial_2 - x_2 \partial_1) = -i\left(x \frac{\partial}{\partial y} - y \frac{\partial}{\partial x}\right) = J_z$$

$$J_{01} = i(x_0 \partial_1 - x_1 \partial_0) = i(x^0 \partial_1 + x^1 \partial_0) = i\left(t \frac{\partial}{\partial x} + x \frac{\partial}{\partial t}\right) = -K_x \tag{40}$$

Thus, the formalism offers the attractive feature of automatically taking care of the relative signs inside $\vec{J}$ and inside $\vec{K}$.

Now the commutation relations (25), (26), and (27) of the Lorentz algebra can be derived much more compactly. Using (39), we readily calculate* the commutation relations

$$[J_{\mu\nu}, J_{\rho\sigma}] = -i(\eta_{\mu\rho} J_{\nu\sigma} + \eta_{\nu\sigma} J_{\mu\rho} - \eta_{\nu\rho} J_{\mu\sigma} - \eta_{\mu\sigma} J_{\nu\rho}) \tag{41}$$

(Indeed, we can even lift this from the commutation relations between the generators $J_{(mn)}$ given in (I.3.23).) As already remarked in chapter I.3, while the right hand side of (41) looks rather involved, it actually is very simple. From the general definition of a Lie algebra, the right hand side has to be linear in J, which carries two indices. Since there are four indices μ, ν, ρ, and σ, we need an η to soak up two of these indices. Given the first term on the right hand side, the other three terms can be obtained by symmetry considerations (for example, the left hand side is antisymmetric in $\mu\nu$). Only the overall sign has to be fixed, but this can be determined by the rotation subalgebra: $[J_x, J_y] = [J_{23}, J_{31}] = i\eta_{33} J_{21} = +i J_{12} = i J_z$.

Indeed, (41) holds for $SO(m, n)$ with the appropriate $\eta_{\mu\nu}$. Specialize to the Lorentz group $SO(3, 1)$. We simply read off

$$[K_x, K_y] = (-1)^2 [J_{01}, J_{02}] = -i\eta_{00} J_{12} = -i J_z \tag{42}$$

We have thus "recovered" the minus sign in (20)!

In this notation, we can then write a Lorentz transformation as

$$L = e^{-\frac{i}{2} \omega^{\mu\nu} J_{\mu\nu}} \tag{43}$$

* Be sure to keep (35) in mind.

where $J_{\mu\nu}$ is represented by the 4-by-4 matrices in (15), in the same way as our writing a rotation as $R = e^{i\vec{\theta}\cdot\vec{J}}$ in chapter I.3. The six real parameters are $\omega^{\mu\nu} = -\omega^{\nu\mu}$.

Of course, if you prefer, you can always deal with $J^{\mu\nu} = \eta^{\mu\rho}\eta^{\nu\sigma}J_{\rho\sigma} = i(x^\mu\partial^\nu - x^\nu\partial^\mu)$ instead of $J_{\mu\nu}$.

From the Lorentz algebra to the Poincaré algebra

A Lorentz transformation (36) supplemented by a translation is known as a Poincaré transformation:

$$x^\mu \to x'^\mu = L^\mu{}_\nu x^\nu + a^\mu \tag{44}$$

Following one Poincaré transformation by another gives $x \to x' = L_2(L_1 x + a_1) + a_2 = L_2 L_1 x + (L_2 a_1 + a_2)$. Denote the group elements by $g(L, a)$ characterized by $6 + 4 = 10$ real parameters. The composition law is then

$$g(L_2, a_2)g(L_1, a_1) = g(L_2 L_1, L_2 a_1 + a_2) \tag{45}$$

The Lie algebra of the Poincaré group is obtained as usual by expanding the group elements near the identity. The set of generators $J_{\mu\nu}$ is supplemented by the generators of translation $P_\mu = i\partial_\mu$. (We see that P_μ generates translation by acting with it: $(I - ia^\mu P_\mu)x^\lambda = (I + a^\mu\partial_\mu)x^\lambda = x^\lambda + a^\lambda$. This is of course the same way we see that $J_{\mu\nu}$ in (39) generates rotations and boosts.)

The Lorentz algebra defined by (41) is thus extended to the Poincaré algebra, generated by $J_{\mu\nu}$ and P_μ. The commutation relations (41) are supplemented by

$$[J_{\mu\nu}, P_\rho] = -i(\eta_{\mu\rho}P_\nu - \eta_{\nu\rho}P_\mu) \tag{46}$$

and

$$[P_\mu, P_\nu] = 0 \tag{47}$$

In chapter I.3, we distinguished between the purely geometrical generators of rotation and the physical angular momentum operators of quantum mechanics. As you would learn (or perhaps have learned) in a course on quantum physics, angular momentum operators are given by the generators of rotation multiplied by Planck's constant $\hbar$, which have dimensions of angular momentum, namely, $[ML^2/T]$. (We have subsequently touched on this connection on several occasions, for instance, in chapter IV.3.)

Similarly, multiplying by $\hbar$ allows us to promote the purely geometrical (and dimensionless) translation operator $P_\mu = i\partial_\mu$ to the momentum operator $P_\mu = i\hbar\partial_\mu$ of quantum mechanics.[12]

Particles as representations of the Poincaré algebra

The reader with a long memory will have noticed a striking resemblance of the Poincaré group to the Euclidean group $E(2)$ studied in Interlude IV.i3. With the replacement of

Euclid by Minkowski, the Poincaré algebra has the same structure as the Euclidean algebra, and its representations can be studied by the method of induced representation discussed earlier. In particular, the momentum operators P^μ form an invariant subalgebra and can be simultaneously diagonalized. Denote their eigenstates by $|p\rangle$, with $P^\mu |p\rangle = p^\mu |p\rangle$. Under a Lorentz transformation L, $|p\rangle \to |Lp\rangle$.

Evidently, the operator $P^2 = P_\mu P^\mu$ is a Casimir invariant, that is, it commutes with all the generators of the algebra. (That P^2 commutes with $J_{\mu\nu}$ is of course just the statement that it is a Lorentz scalar.) We have $P^2 |p\rangle = p^2 |p\rangle = m^2 |p\rangle$. The single particle state $|p\rangle$ describes an elementary particle with mass m moving with 4-momentum p. Mathematically, the mass m labels a representation and is not changed by Poincaré transformations.

Elementary particles also carry spin, and thus the Poincaré algebra should contain another Casimir invariant that tells us about spin. From our work with $SO(3)$, with its Casimir invariant $\vec{J}^2$, we might have guessed $J^2 \equiv J_{\mu\nu}J^{\mu\nu} = \vec{J}^2 - \vec{K}^2 = J_+^2 + J_-^2$, but this is incorrect, as it also contains the boost operator.[13] The invariant, as we will see presently, should reduce to $\vec{J}^2$ in the rest frame of the particle.

The correct invariant is constructed out of the Pauli-Lubanski vector

$$W_\sigma \equiv -\frac{1}{2}\varepsilon_{\mu\nu\rho\sigma}J^{\mu\nu}P^\rho \tag{48}$$

(Incidentally, if we had defined the dual generator $\tilde{J}_{\rho\sigma} = -\frac{1}{2}\varepsilon_{\mu\nu\rho\sigma}J^{\mu\nu}$, then we could write, somewhat more compactly, $W^\mu = \tilde{J}_{\rho\sigma}P^\rho$.)

To see how this works, note that acting on $|p\rangle$, the Pauli-Lubanski vector is effectively equal to $-\frac{1}{2}\varepsilon_{\mu\nu\rho\sigma}J^{\mu\nu}p^\rho$. In the rest frame of the particle, with momentum $p_*^\mu = (m, \vec{0})$, W_σ reduces to $W_0 = 0$, $W_i = \frac{1}{2}\varepsilon_{ijk}mJ^{jk} = mJ_i$. (The 4-dimensional antisymmetric symbol reduces to the 3-dimensional antisymmetric symbol, and J_i denotes the generators of $SO(3)$.) Incidentally, that $W_0 = 0$ also follows from $W_\mu P^\mu = 0$.

Thus, the correct second Casimir invariant of the Poincaré algebra is $W_\mu W^\mu$, and

$$m^{-2}W_\mu W^\mu |p, j\rangle = \vec{J}^2 |p, j\rangle = j(j+1)|p, j\rangle \tag{49}$$

Note that our states are now characterized by spin j as well as by momentum p.

More formally, we readily* derive

$$[W_\mu, W_\nu] = i\varepsilon_{\mu\nu\rho\sigma}W^\rho P^\sigma \tag{50}$$

With P^σ replaced by p_*^σ, this effectively becomes the $SO(3)$ algebra.

The little group

Physically, what we did was to pick out a special momentum p_* and ask for the subgroup of the Lorentz group that leaves p_* invariant (namely, $SO(3)$) and classify the states according to the irreducible representation of this group, which physicists refer to as the little group.

---

* Schematically, $[W, W] \sim [JP, JP] \sim [J, J]PP + [J, P]JP \sim JPP \sim WP$.

(This is reminiscent of our treatment of $E(2)$, in which we also pick out a special $\vec{p}$, pointing along the x-axis, say.)

What if the particle is massless, with no rest frame for us to go to?

Pick the special momentum $p_* = (p, 0, 0, p)$. (Simply call the direction of motion the z-axis.) We want to find the subgroup that leaves p_* invariant.[14] The condition $L^\mu{}_\nu p^\nu_* = p^\mu_*$, since it amounts to three equations, defines a $6 - 3 = 3$-parameter subgroup. Rotations around the z-axis clearly belong: the subgroup $O(2)$ leaves p^μ_* invariant.

The spin states of a massless particle around its direction of motion are known as helicity states. For a particle of spin j, the helicities, $\pm j$, are transformed into each other by parity and time reversal, and thus both helicities must be present if the interactions the particle participates in respect these discrete symmetries, as is the case with the photon and the graviton.* In particular, the photon, as we have seen repeatedly, has only two polarization degrees of freedom, instead of three, since we no longer have the full rotation group $SO(3)$. (You already learned in classical electrodynamics that an electromagnetic wave has two transverse degrees of freedom.) For more on this, see appendix 3.

To find the other transformations, it suffices, as Lie taught us, to look in the neighborhood of the identity, at Lorentz transformations of the form $L(\alpha, \beta) = I + \alpha A + \beta B + \cdots$. By inspection, we see that

$$
A = \begin{pmatrix} 0 & 1 & 0 & 0 \\ 1 & 0 & 0 & -1 \\ 0 & 0 & 0 & 0 \\ 0 & 1 & 0 & 0 \end{pmatrix} = i(K_1 + J_2), \qquad B = \begin{pmatrix} 0 & 0 & 1 & 0 \\ 0 & 0 & 0 & 0 \\ 1 & 0 & 0 & -1 \\ 0 & 0 & 1 & 0 \end{pmatrix} = i(K_2 - J_1) \tag{51}
$$

acting on p^μ_* gives zero. Referring to (15), we have identified A and B in terms of J and K. Note that A and B are to a large extent determined by the fact that J and K are symmetric and antisymmetric, respectively.

By direct computation or by invoking the celebrated minus sign in (27), we find that $[A, B] = 0$. Also, $[J_3, A] = B$ and $[J_3, B] = -A$. As expected, (A, B) forms a 2-component vector under $O(2)$ rotations around the third axis.

We see that A, B, and J_3 generate the group $E(2)$, the invariance group of the Euclidean 2-plane, consisting of two translations and one rotation. (Perhaps you are surprised, but then again, as was mentioned in Interlude IV.i3, cylindrical symmetry also comes in here.) You may well wonder what the translations do; it turns out that they generate gauge transformations,[15] but this will take us far outside the scope of a book on group theory.

As might be expected, we can also express the preceding discussion in terms of the Pauli-Lubanski vector. Acting with $W_\mu P^\mu = 0$ on $|p_*\rangle$ gives $W_0 |p_*\rangle = W_3 |p_*\rangle$. The relation (50) gives effectively

$$
[W_1, W_2] |p_*\rangle = 0 \tag{52}
$$

---

* But not with the neutrino, as we will discuss in chapter VII.3.

and

$$[W_3, W_1]\,|p_*\rangle = i p_* W_2\,|p_*\rangle\,, \qquad [W_3, W_2]\,|p_*\rangle = -i p_* W_1\,|p_*\rangle \tag{53}$$

which is effectively isomorphic to the $E(2)$ algebra after a trivial normalization. Consistent with the somewhat cryptic (field theoretic in origin and thus outside the scope of this book) remark about gauge transformations just made, $W_{1,2}$ then leave physical states unchanged: $W_{1,2}\,|p_*\rangle = 0$. These results imply that W^μ and P^μ are effectively parallel in spacetime:

$$W^\mu\,|p_*\rangle = -h P^\mu\,|p_*\rangle \tag{54}$$

The invariant proportionality factor h defines the helicity.

More on the Lorentz group and its double cover

In the next chapter, I will have more to say about the Lorentz group $SO(3, 1)$ and its double cover $SL(2, C)$. We will also talk about $SO(4)$, $SO(2, 2)$, and $SO(3, C)$. It is simply more convenient to discuss these topics after we have studied the Weyl spinor, to be introduced shortly.

Appendix 1: From the Galilean algebra to the Poincaré algebra: Missing halves

It is instructive to regress and study the supposedly familiar but in fact somewhat exotic Galilean algebra. We will be mostly restating some of what we already said in the text but in the language of differential operators. This will also be useful in chapter VIII.1, when we discuss contractions.

Let us start by writing down the generators of translation* and rotation as differential operators, as was done for the Lorentz algebra in the text:

$$\text{Translation:} \quad P_x = i\frac{\partial}{\partial x}, \text{ and so forth}$$

$$\text{Rotation:} \quad J_z = \frac{1}{i}\left(x\frac{\partial}{\partial y} - y\frac{\partial}{\partial x}\right) = -x P_y + y P_x, \text{ and so forth} \tag{55}$$

We verify easily the Euclidean algebra $E(3)$, which tells us about the geometry of Euclidean space:

$$[J_i, J_j] = i\epsilon_{ijk} J_k, \qquad [J_i, P_j] = i\epsilon_{ijk} P_k, \qquad [P_i, P_j] = 0 \tag{56}$$

For example, $[J_z, P_x] = [x\frac{\partial}{\partial y}, \frac{\partial}{\partial x}] = -\frac{\partial}{\partial y} = iP_y$, and $[J_z, P_y] = [-y\frac{\partial}{\partial x}, \frac{\partial}{\partial y}] = +\frac{\partial}{\partial x} = -iP_x$.

Thus far, it is all geometry. But humans cannot live by geometry alone; we need to introduce physics, in the person of Newton and his profound laws of motion, namely, $\frac{dp^i}{dt} = m\frac{d^2x^i}{dt^2} = 0$ for a free particle.

Generalizing (3) slightly, we have the Galilean transformation $t' = t$, $x' = x + ut$, $y' = y$, and $z' = z$ relating the space and time coordinates t, x, y, z and t', x', y', z' of two observers in uniform motion relative to each other with velocity u along the x direction. Evidently, this transformation leaves Newton's laws invariant. We can immediately generalize to many interacting particles, provided that the potential between pairs of particle a and b is given by $V(|\vec{x}_a - \vec{x}_b|)$, which is clearly invariant under $\vec{x}'_{a,b} = \vec{x}_{a,b} + \vec{u}t$, since the universal Newtonian time t does not know about a and b, being by definition universal.

---

* To determine the sign here, note that $P_\mu = i\partial_\mu = i\frac{\partial}{\partial x^\mu}$ implies $P_i = i\frac{\partial}{\partial x^i}$.

In a beautiful passage,[16] Galileo imagines a sailing ship moving smoothly in a calm sea. In a cabin below deck, we would hardly notice that the ship is moving; in particular, a butterfly flying around would not be affected by the forward velocity of the ship.

The generators of Galilean transformations are given by

$$\text{Boost:} \quad K_x = t\frac{1}{i}\frac{\partial}{\partial x} = -tP_x, \text{ and so forth} \tag{57}$$

Acting with the boost operator $B_x(v) \equiv (I + ivK_x) = (1 + vt\frac{\partial}{\partial x})$, for example, on the coordinates, we obtain

$$x' = B_x(v)x = x + vt, \quad y' = B_x(v)y = y, \quad z' = B_x(v)z = z, \quad t' = B_x(v)t = t \tag{58}$$

The boost operator B_x translates x by an amount linear in t and leaves y, z, t untouched, as it should. Note that K_x is proportional to the generator of translation P_x with the time coordinate t as the proportionality factor. (This in itself is not particularly strange, since J_z also involves xP_y and yP_x.)

We readily verify that

$$[K_i, K_j] = 0, \qquad [J_i, K_j] = i\epsilon_{ijk}K_k \tag{59}$$

and

$$[K_i, P_j] = 0 \tag{60}$$

In analogy with the generator for translation in space P_i, we introduce the generator for translation in time which we call[17] H

$$\text{Dynamics:} \quad H = i\frac{\partial}{\partial t} \tag{61}$$

Evidently,

$$[P_i, H] = 0, \qquad [J_i, H] = 0 \tag{62}$$

How does H behave under a boost? Simply calculate: $[K_x, H] = [t\frac{1}{i}\frac{\partial}{\partial x}, i\frac{\partial}{\partial t}] = -\frac{\partial}{\partial x} = iP_x$.
In summary, the Galilean algebra is given by

$$[J_i, J_j] = i\epsilon_{ijk}J_k, \quad [J_i, P_j] = i\epsilon_{ijk}P_k, \quad [J_i, K_j] = i\epsilon_{ijk}K_k$$
$$[P_i, H] = 0 \quad [J_i, H] = 0, \quad [P_i, P_j] = 0$$
$$[K_i, H] = iP_i, \quad [K_i, P_j] = 0, \quad [K_i, K_j] = 0 \tag{63}$$

We have already talked about, in this chapter, the extension of the Galilean algebra to the Poincaré algebra. But it is worthwhile looking at this all-important step in physics again.

The ancient Greeks theorized that human beings originally had four legs and four arms. But some god came around and sliced them into two halves. So now everybody is wandering around looking for his or her missing half. This explains why everybody needs to find another person. Very charming theory. When I read this long ago, I thought that it was also relevant for physics, where some expressions looked like they had half of themselves missing.

The generator of Galilean boost $K_x = \frac{1}{i}(t\frac{\partial}{\partial x})$ is exactly like one of those people who have been cut into two halves looking around for the other half. The obvious other half, $\frac{1}{i}(x\frac{\partial}{\partial t})$, seems such a natural match for $\frac{1}{i}(t\frac{\partial}{\partial x})$ that we are immediately tempted to marry them. The strange thing is that K_x has dimension of time over length, that is (T/L), while its purported other half has dimension (L/T). We need a fundamental constant with dimension (L/T), namely, that of a speed, to get the dimensions to come out right.

You all know that this story has a happy ending. Nature obligingly gives us the speed of light c. So if you divide one term by c^2, then you can join together whom men had put asunder. We thus extend the Galilean K_x to

$$K_x = \frac{1}{i}\left(t\frac{\partial}{\partial x} + \frac{1}{c^2}x\frac{\partial}{\partial t}\right) \tag{64}$$

Note that as c goes to infinity, we go back to the Galilean boost.

To extend the Galilean algebra represents of course the great discovery of Lorentz, Einstein, Poincaré, and their colleagues. Adding this extra term also makes the algebra look better, by making K_x more closely resemble $J_z = \frac{1}{i}(x\frac{\partial}{\partial y} - y\frac{\partial}{\partial x})$.

The crucial question is: What does the commutator $[K_x, K_y]$ give us? Rather nicely, it gives us back J_z:

$$[J_i, J_j] = i\epsilon_{ijk}J_k, \qquad [J_i, K_j] = i\epsilon_{ijk}K_k, \qquad [K_i, K_j] = -\frac{1}{c^2}i\epsilon_{ijk}J_k \qquad (65)$$

Next, we should see how the rest of the Galilean algebra changes. We have

$$[K_i, H] = \left[\frac{1}{i}\left(t\frac{\partial}{\partial x^i} + \frac{1}{c^2}x^i\frac{\partial}{\partial t}\right), i\frac{\partial}{\partial t}\right] = iP_i$$

$$[K_i, P_j] = \left[\frac{1}{i}\left(t\frac{\partial}{\partial x^i} + \frac{1}{c^2}x^i\frac{\partial}{\partial t}\right), i\frac{\partial}{\partial x^j}\right] = i\delta_{ij}\frac{1}{c^2}H \qquad (66)$$

(You might also check that (66) is consistent with (46) with $K_i = -J_{0i}$.) The commutator $[K_i, H]$ is the same as before, unchanged by the extra term in K_i, but the commutator $[K_i, P_j]$ no longer vanishes, as in (60).

It is rather pleasing that K_i now turns P_i and H into each other. Note, however, that P_i and H have different dimensions. This is readily fixed by defining $P_0 = \frac{1}{c}H$. Then (66) can be written as

$$c[K_i, P_0] = iP_i, \qquad c[K_i, P_j] = iP_0\delta_{ij} \qquad (67)$$

So now we are quite happy that our algebra looks more symmetrical. Indeed, a bright young person named Albert Einstein now comes along and remarks that these two equations imply

$$[K_i, P_0^2 - \vec{P}^2] = 0 \qquad (68)$$

Furthermore, $[J_i, P_0^2 - \vec{P}^2] = 0$ since both P_0^2 and $\vec{P}^2$ are rotational scalars. Thus, $P_0^2 - \vec{P}^2$ is a Casimir invariant, namely, a quantity that commutes with all the generators in the algebra. It follows that the so-called d'Alembertian, defined by $\partial^2 = \Box = \frac{1}{c^2}\frac{\partial^2}{\partial t^2} - \nabla^2$, and which generalizes the Laplacian ∇^2, is a relativistic invariant.

For more on the Galilean algebra, see chapter VIII.1.

Appendix 2: Connected components of the Lorentz group

In our discussion, we implicitly define the Lorentz group to consist of those Lorentz transformations continuously connected to the identity. A looser definition would include all those transformations satisfying (38), thus admitting time reversal and spatial reflections. To get at the main point here, it suffices to lower the dimension of spacetime by 2 and descend to $SO(1, 1)$. Define time reversal by $\mathcal{T} \equiv \begin{pmatrix} -1 & 0 \\ 0 & 1 \end{pmatrix}$ and spatial reflection by $\mathcal{P} \equiv \begin{pmatrix} 1 & 0 \\ 0 & -1 \end{pmatrix}$.

In chapter I.3, we noted that, topologically, the group $O(2)$ contains two connected components, one continuously connected to the identity, the other including reflection and hence not. In contrast, the subgroup $SO(2)$ has only one component, consisting of $R(\theta) = \begin{pmatrix} \cos\theta & \sin\theta \\ -\sin\theta & \cos\theta \end{pmatrix}$.

In contrast to $O(2)$, the group $O(1, 1)$ contains four connected components, not two, described as follows: (i) the set $L(\varphi) = \begin{pmatrix} \cosh\varphi & \sinh\varphi \\ \sinh\varphi & \cosh\varphi \end{pmatrix}$, (ii) the set $\mathcal{T}L(\varphi)$, (iii) the set $\mathcal{P}L(\varphi)$, and (iv) the set $\mathcal{T}\mathcal{P}L(\varphi) = -L(\varphi)$. Interestingly, $SO(1, 1)$ then contains two connected components, namely, $L(\varphi)$ and $-L(\varphi)$.

The key point is that in $L(\varphi)$, the quantity $\cosh\varphi \geq 1$, while in $R(\theta)$, the quantity $\cos\theta$ can take on either sign. Thus, $(-I)$ is continuously connected to the identity I in $SO(2)$ but not in $SO(1, 1)$.

By the same analysis as given here, we conclude that $O(3, 1)$ also contains four connected components.

Appendix 3: Topological quantization of helicity

I mention and resolve a puzzle here.[18] You learned that the nonlinear algebraic structure of the Lie algebra $SO(3)$ enforces quantization of angular momentum. But the little group for a massless particle is merely $O(2)$. How do we get the helicity of the photon and the graviton quantized?

Perhaps surprisingly, the answer is to be found in chapter IV.7. Recall that the group manifold of $SO(3)$ is $SU(2)/Z_2 = S^3/Z_2$, that is, the 3-sphere with antipodal points identified. Consider closed paths in $SO(3)$.

Starting at some point P on S^3, wander off a bit and come back to P. The path you traced can evidently be continuously shrunk to a point. But suppose you go off to the other side of the world and arrive at $-P$, the antipodal point of P. You also trace a closed path in $SO(3)$, but this closed path obviously cannot be shrunk to a point. However, if after arriving at $-P$ you keep going and eventually return to P, then the entire path you traced can be continuously shrunk to a point. Using the language of homotopy groups mentioned in chapter IV.7, we say that $\Pi_1(SO(3)) = Z_2$: there are two topologically inequivalent classes of paths in the 3-dimensional rotation group.

The resolution of our puzzle is that we have to invoke topological, rather than algebraic, quantization. A rotation through 4π is represented by $e^{i4\pi h}$ on the helicity h state of the massless particle, but the path traced out by this rotation in $SO(3)$ can be continuously shrunk to a point. Hence, we must have $e^{i4\pi h} = 1$ and so $h = 0, \pm\frac{1}{2}, \pm 1, \cdots$.

Exercises

1   Follow a boost in the x direction with a boost in the y direction, and then boost back in the x direction. Take the infinitesimal limit, and verify (20).

2   Work out explicitly how the components F_{0i} and F_{ij} of the antisymmetric tensor $F_{\mu\nu}$ transform under a Lorentz transformation. This tells us how the electric and magnetic fields transform. Compare with standard textbooks on electromagnetism.

3   Work out the commutation relations of $SO(2, 2)$ and verify (41).

4   Show that the defining relation (38) for the Lorentz group implies that $L^0{}_0$ either ≥ 1 or ≤ -1. Lorentz transformations with $L^0{}_0 \geq 1$ are known as orthochronous.

5   Prove the practically self-evident fact that the product of two orthochronous Lorentz transformations is also orthochronous. Hint: You need to use the Cauchy-Schwarz inequality $\vec{a}^2\vec{b}^2 \geq (\vec{a} \cdot \vec{b})^2$.

Notes

1. If you don't, see part III of G Nut.
2. Dyson has indeed conjectured that somebody, proceeding on aesthetics alone, could have proposed this long before Einstein. The point is that the group theory associated with the Lorentz transformation is considerably more appealing than the group theory associated with the Galileo transformation.
3. For a somewhat fishy argument for the existence of a universal speed limit and hence the failure of the Galilean transformation, see G Nut, p. 172.
4. Indeed, the first attempt to modify the Galilean transformation, by W. Voigt in 1887, almost succeeded. Later, several authors obtained the correct Lorentz transformation. See G Nut, p. 169.
5. See G Nut, pp. 166–167. Here is a sketch of Einstein's reasoning without the benefit of a figure (which may be found on p. 167 of G Nut). Let the separation between the two mirrors be L along the y-axis. In the rest frame of the mirrors, light leaving a mirror and returning to that mirror is described by $\Delta x = 0$, $\Delta t = 2L/c$. Thus, $(c\Delta t)^2 - (\Delta x)^2 = 4L^2$. Now let an observer see the whole setup moving at velocity u along the x-axis. To him, $\Delta t' = 2\sqrt{(\Delta x'/2)^2 + L^2}/c$, and $\Delta x' = u\Delta t'$. Thus, $(c\Delta t')^2 - (\Delta x')^2 = 4L^2$, which is indeed equal to $(c\Delta t)^2 - (\Delta x)^2$. A very neat argument indeed!
6. The couplet (m, n) is known as the signature of the group.
7. A similar argument was given in chapter I.3 for $SO(4)$.
8. Let me clear up some potential questions about the notation. Some students ask why there isn't a distinction between upper and lower indices for ordinary vectors. The answer is that we could have, if we wanted to, written the Euclidean metric δ_{ij} with lower indices back in chapter I.3 and risked confusing the reader at

that early stage. But there is no strong incentive for doing that: the Euclidean metric does not contain any minus signs, while the Minkowskian metric necessarily has one positive sign and three negative signs to distinguish time from space. The upper and lower index notation serves to keep track of the minus signs. In the Euclidean case, if we define $p_i = \delta_{ij} p^j$, the vector p_i would be numerically the same as the vector p^i. In Minkowski space, with our convention, $p_0 = p^0$, but $p_1 = -p^1$, $p_2 = -p^2$, and $p_3 = -p^3$.

9. Earlier we saw that $SU(N)$ tensors also require two floors for its indices, but, as is worth emphasizing, for different reasons (namely, the presence of complex conjugation). Similarly, as was discussed in the review of linear algebra, unitary matrices are most conveniently written with upper and lower indices.

10. Mathematically, this amounts to the statement that $\sum_\mu p^\mu q^\mu \to \sum_\mu L^\mu{}_\sigma L^\mu{}_\rho p^\sigma q^\rho$ does not transform into anything nice.

11. Thus, more accurately, the electromagnetic field does not break up into $\vec{B}$ and $\vec{E}$, but into $\vec{B} \pm i\vec{E}$.

12. Note that the sign choice here is such that the Hamiltonian $H = P^0 = P_0 = i\hbar\partial_0 = i\hbar\frac{\partial}{\partial x^0}$, consistent with standard convention used in quantum mechanics $i\frac{\partial}{\partial t}\psi = H\psi$. This will become relevant in chapter VIII.1.

13. Furthermore, it is a Casimir invariant only of the Lorentz algebra, but not of the Poincaré algebra.

14. We follow the treatment given in, for example, QFT Nut, second edition, p. 186.

15. For the curious reader, see chapter III.4 of QFT Nut, second edition.

16. See G Nut, pp. 17–19.

17. Let us not be coy about it. Eventually H will be identified with the energy or Hamiltonian, as was already alluded to in an earlier endnote.

18. See QFT Nut, second edition, p. 532.

VII.3 | *SL*(2,*C*) Double Covers *SO*(3,1): Group Theory Leads Us to the Weyl Equation

Group theoretic understanding versus a brilliant guess

The Dirac equation is surely the most translucently brilliant equation of theoretical physics. Dirac derived his elegant* equation for the relativistic electron by a brilliantly idiosyncratic guess. Arguably, it may be pedagogically the best approach for a first pass at the Dirac equation; indeed, many textbooks[1] introduce this celebrated equation in essentially this way (as we also will in chapter VII.4).

In this book, we would like to develop a group theoretic, and deeper, understanding of the Dirac equation.[2] Here we eschew the "brilliant guess" derivation, and instead, simply allow group theory to lead us by the nose. We will first discover the Weyl equation. In chapter VII.4, we will stack two 2-component Weyl spinors together to form the 4-component Dirac spinor. The Dirac equation appears naturally.

One of the great advances in theoretical physics over the past half-century or so is the understanding that the Weyl spinors are more fundamental than Dirac spinors; the world appears to be constructed out of them (see part IX).

We will also learn that the Lorentz group $SO(3, 1)$ is double covered by $SL(2, C)$ just as the rotation group $SO(3)$ is double covered by $SU(2)$.

Equations of motion in physics and the symmetries of space and time

The quest for equations of motion has been a central theme in physics. Writing down the first equation of motion demanded such a staggering depth of insight that it still awes me and most theoretical physicists. The full brilliance of a true genius like Newton was needed

---

* "Elegance is for tailors," according to Wolfgang Pauli. Most theoretical physicists disagree. Pauli's own work, such as that on the hydrogen atom described in interlude 1, shows that he probably said this just for the fun of saying it.

for the jump from the first[3] to the second derivative in time. The glories of nineteenth-century physics culminated in Maxwell's equations of motion for the electromagnetic field, slowly pieced together through a long series of ingenious experiments involving twitching frogs and such.

But after the discovery of the symmetric group of spacetime, the search for equations of motion was reduced to, with only slight exaggeration, merely an application of group theory.

The point is that once we know how to describe a particle sitting at rest, then a Lorentz boost to a moving frame of reference immediately tells us about a moving particle. (For a massless particle, if we know it in one frame, then Lorentz tells us about it in all frames. In fact, we will deal with this case in this chapter.) More generally, once we know the behavior of a field (say, the electric or the magnetic field) in some frame of reference, and if we know, or can guess, how the field transforms, then we can invoke the power of group theory to determine the behavior of the field in any other frame. If we know Coulomb's law $\vec{\nabla} \cdot \vec{E} = \rho$, then we can determine the full set of Maxwell's equations.*

Historically, things happened the other way around: Lorentz invariance was deduced from the full set of Maxwell's equations.

In truth, group theory alone cannot tell us which representation of the Lorentz group physical quantities like the electromagnetic field and charge density belong to. But typically, it can list the possibilities, and invites us to try a few, starting with the simplest. Ultimately, the consequences of our guesses would have to be compared with experiments, of course.

As a simple example, suppose we somehow discovered, experimentally perhaps, that a static rotationally invariant field $\phi(\vec{x})$ satisfies $\nabla^2 \phi = m^2 \phi$. Then Lorentz invariance immediately tells us that[†] the time-dependent field $\phi(t, \vec{x})$ satisfies the so-called Klein-Gordon equation[4]

$$\left(\frac{\partial^2}{\partial t^2} - \vec{\nabla}^2 + m^2 \right) \phi(x) = (\partial^2 + m^2) \phi(x) = 0 \tag{1}$$

Fourier transforming to energy-momentum space (momentum space for short) by setting[‡] $\phi(t, \vec{x}) = \int d^4 p \, e^{-i(Et - \vec{p} \cdot \vec{x})} \phi(E, \vec{p})$, we obtain

$$(E^2 - \vec{p}^2 - m^2) \phi(E, \vec{p}) = (p^2 - m^2) \phi(p) = 0 \tag{2}$$

The Klein-Gordon equation merely tells us that $\phi(E, \vec{p})$ vanishes unless Einstein's relation $E^2 = \vec{p}^2 + m^2$ (which was mentioned in chapter VII.2) is satisfied. Here we have assumed that the rotational scalar ϕ is also a Lorentz scalar; a priori, it could belong to

---

* Indeed, starting from the Newton-Poisson equation for the gravitational potential $\nabla^2 \phi = G\rho$ (with ρ the mass density), we can deduce the full set of Einstein's equations. See, for example, chapter IX.5 of G Nut.

† Similarly, if we were given $\left(\frac{\partial^2}{\partial x^2} + \frac{\partial^2}{\partial z^2} \right) \phi = m^2 \phi$, rotational invariance would require us to add $\frac{\partial^2}{\partial y^2}$ and promote the differential operator in this equation to the Laplacian.

‡ I will often use the same symbol to denote a field and its Fourier transform, with essentially no risk of confusion.

a more complicated representation. For example, ϕ could be the time component of a Lorentz vector.

The crucial point is of course that Lorentz invariance is a symmetry of spacetime, and so anybody moving in spacetime, be it an electron or a quark, better obey Lorentz invariance. We thus expect the Lorentz group to "give" us the Dirac equation.

From brilliant guess to group theoretic understanding

Now that we have mastered the Lorentz group, we can adopt the following strategy to get to the Dirac equation, as explained above. Write down the wave function of the electron in its rest frame, in which it is completely described* by the rotation group. Then boost the electron to any momentum we like. In other words, the Lorentz group should lead us to the Dirac equation.

We have already treated the spin $\frac{1}{2}$ electron in chapter III.1 by the simple expedient of attaching an index α to the wave function in Schrödinger's equation, $i\hbar \frac{\partial}{\partial t}\psi_\alpha(\vec{x}, t) = -\frac{\hbar^2}{2m}\nabla^2\psi_\alpha(\vec{x}, t)$, so that the wave function ψ_α furnishes the 2-dimensional representation of the rotation group.

In a sense, the index α merely functions as a label; once we choose the angular momentum quantization axis, we might as well be describing two kinds of electrons, called the "up electron" and the "down electron." But, as the discussion in chapter VII.2 indicates, this nonrelativistic treatment is not only aesthetically unappealing but also physically inadequate for relativistic electrons, since boosts and rotations are intimately related: successive boosts can result in a rotation.[5]

Group theoretically, this means that the wave function ψ has to furnish an irreducible representation of the Lorentz group, not merely the rotation group.

For a proper treatment, we need of course to replace the electron wave function by the electron field, which we then have to quantize.[6] Henceforth, we will use the term "wave function" and field interchangeably. See chapter IV.9 for a discussion of the scalar field. Since this is a textbook on group theory, not quantum field theory, we will emphasize the group theoretic aspects, rather than the quantum field theoretic aspects, of the Dirac equation.

$SO(4)$, $SO(3, 1)$, and $SL(2, C)$

I have already touched on some of the group theory we need here; nevertheless, at the cost of repeating a bit, I think it worthwhile to review and reexamine the salient features. We learned in chapter IV.5 that $SU(2)$ and $SO(3)$ are locally, but not globally, isomorphic. The

---

* We deal with the free electron first. Its interaction with, for example, the electromagnetic field can then be added later, and in fact, should properly be treated by the methods of quantum field theory.

group $SU(2)$ double covers the group $SO(3)$: two distinct elements of $SU(2)$ map into the same element of $SO(3)$.

Let us recall the discussion of $SO(4)$ in chapter IV.7, with a slightly different notation. We began by writing down the 2-by-2 matrix

$$X_E = x^4 I + i\vec{x} \cdot \vec{\sigma} \tag{3}$$

with the Euclidean length squared of the 4-dimensional vector $(\vec{x}, x^4)$ restricted by

$$(x^4)^2 + \vec{x}^2 = 1 \tag{4}$$

Then X_E defines an element of $SU(2)$. (Simply verify that X_E is unitary and that det $X_E = (x^4)^2 + \vec{x}^2 = 1$. To do this, it is easiest to invoke rotational invariance and choose $\vec{x}$ to point along the third axis, so that X_E is a diagonal matrix with elements $x^4 \pm ix^3$.)

For any two elements U and V of $SU(2)$, consider the map $X_E \to X_E' = U^\dagger X_E V$. By construction, X_E' is also an element of $SU(2)$: it is unitary with unit determinant. Writing $X_E' = x'^4 I + i\vec{x}' \cdot \vec{\sigma}$, we see that $(x'^4)^2 + \vec{x}'^2 = (x^4)^2 + \vec{x}^2 = 1$, and so the map $(\vec{x}, x^4) \to (\vec{x}', x'^4)$ defines a 4-dimensional rotation and an element of $SO(4)$. This shows explicitly that $SO(4)$ is isomorphic to $SU(2) \otimes SU(2)$, but as explained in chapter IV.5, the isomorphism is local, rather than global, since the pair $(-U, -V)$ and (U, V) correspond to the same $SO(4)$ rotation. In other words,

$$(SU(2) \otimes SU(2))/Z_2 = SO(4) \tag{5}$$

If $U = V$ we have an $SO(3)$ rotation, and if $U^\dagger = V$, the Euclidean analog of a boost.

Now we are ready to look at the group $SO(3, 1)$. Simply strip off an i from (3) and thus write down the most general 2-by-2 hermitean matrix

$$X_M = x^0 I + \vec{x} \cdot \vec{\sigma} = \begin{pmatrix} x^0 + x^3 & x^1 - ix^2 \\ x^1 + ix^2 & x^0 - x^3 \end{pmatrix} \tag{6}$$

By explicit computation,

$$\det X_M = (x^0)^2 - \vec{x}^2 \tag{7}$$

(To see this instantly, again choose $\vec{x}$ to point along the third axis and invoke rotational invariance.) Compare and contrast with (4).

Consider the group $SL(2, C)$, consisting, as the notation indicates, of all 2-by-2 matrices with complex entries and unit determinant. With L an element of $SL(2, C)$, let $X_M \to X_M' = L^\dagger X_M L$. Writing $X_M' = x'^0 I - \vec{x}' \cdot \vec{\sigma}$, we see that this defines a map of the 4-vector $(x^0, \vec{x})$ to the 4-vector $(x'^0, \vec{x}')$. Manifestly, det $X_M' = $ det X_M and so $(x'^0)^2 - \vec{x}'^2 = (x^0)^2 - \vec{x}^2$. Thus, the transformation preserves the Minkowski metric and is in fact a Lorentz transformation. This defines a map from $SL(2, C)$ to $SO(3, 1)$.

Since L and $-L$ define the same transformation $X_M \to X_M'$, and hence the same $x \to x'$, we see that $SL(2, C)$ double covers $SO(3, 1)$. In other words,

$$SL(2, C)/Z_2 = SO(3, 1) \tag{8}$$

(Strictly speaking, $SL(2, C)/Z_2$ gives the component of $SO(3, 1)$ connected to the identity. Show this.)

If L is also unitary, that is, if it belongs to the $SU(2)$ subgroup of $SL(2, C)$, then the transformation is a rotation: $L^\dagger I L = I$, and x^0 is left untouched. The $SU(2)$ subgroup of $SL(2, C)$ double covers the rotation subgroup $SO(3)$ of the Lorentz group $SO(3, 1)$, something we have known for quite a while.

While we already know that everything works, let us, as a simple check, count the number of generators of this $SL(2, C)$. Two conditions on the determinant (real part $= 1$, imaginary part $= 0$) cuts the four complex entries containing eight real numbers down to six, which accounts for the six generators of the Lorentz group $SO(3, 1)$.

Note that while the rotation group $SO(3) \simeq SU(2)$ is compact, the Lorentz group $SO(3, 1)$ is not.[*] In contrast, the group $SO(4)$ is compact and thus can be covered by a compact group, namely, $SU(2) \otimes SU(2)$, but the noncompact group $SO(3, 1)$ could not possibly be. It is double covered by the noncompact group $SL(2, C)$. It is worth mentioning that the group theoretic facts discussed here and the possibility of writing a Lorentz 4-vector as a 2-by-2 hermitean matrix as in (6) are highly relevant for the exciting development of using twistors to understand spacetime.[7]

In going from (3) to (6) by stripping off an i, we are echoing Minkowski's pregnant and mystic expression: $3 \cdot 10^5 \, \text{km} = \sqrt{-1} \, \text{sec}$.

From algebra to representation

The crucial observation about the Lorentz algebra, as explained in chapter VII.2, is that it falls apart into two pieces defined by the combinations $J_{\pm i} = \frac{1}{2}(J_i \pm i K_i)$. (Recall, also, that $J_{\pm i}$ are hermitean.) We obtained

$$[J_{+i}, J_{+j}] = i\epsilon_{ijk} J_{+k}$$
$$[J_{-i}, J_{-j}] = i\epsilon_{ijk} J_{-k} \tag{9}$$

and most remarkably,

$$[J_{+i}, J_{-j}] = 0 \tag{10}$$

This last commutation relation tells us that J_+ and J_- form two separate $SU(2)$ algebras.[†]

Nature, as She has been time and time again, is kind to theoretical physicists!

She first taught them about the representations of $SU(2)$. Now they don't have to learn any more math. To determine the finite-dimensional irreducible representations of $SO(3, 1)$, we simply apply what we already know.

The representations of $SU(2)$ are labeled by $j = 0, \frac{1}{2}, 1, \frac{3}{2}, \cdots$, as we learned in chapter IV.2. We can think of each representation as consisting of $(2j + 1)$ objects ϕ_m with $m = -j, -j + 1, \cdots, j - 1, j$, which transform into one another under $SU(2)$. It follows

---

[*] As already mentioned in chapter I.1.
[†] Strictly speaking, as analytically continued from $SO(4) \simeq SU(2) \otimes SU(2)$.

immediately that the representations of the $SO(3, 1)$ algebra are labeled by (j^+, j^-), with j^+ and j^- each taking on the values $0, \frac{1}{2}, 1, \frac{3}{2}, \cdots$. Each representation consists of $(2j^+ + 1)(2j^- + 1)$ objects $\phi_{m^+m^-}$ with $m^+ = -j^+, -j^+ + 1, \cdots, j^+ - 1, j^+$ and $m^- = -j^-, -j^- + 1, \cdots, j^- - 1, j^-$. The generators J_+ act on the index m^+, while the J_- act on m^-.

Thus, we can list the irreducible representations* of the $SO(3, 1)$ algebra as

$$(0, 0), \; \left(\frac{1}{2}, 0\right), \; \left(0, \frac{1}{2}\right), \; (1, 0), \; (0, 1), \; \left(\frac{1}{2}, \frac{1}{2}\right), \; \left(\frac{3}{2}, 0\right), \; \left(0, \frac{3}{2}\right), \; \left(\frac{1}{2}, 1\right), \; \left(1, \frac{1}{2}\right), \cdots \tag{11}$$

with dimension equal to $1, 2, 2, 3, 3, 4, 4, 4, 6, 6, \cdots$, respectively. The 1-dimensional representation $(0, 0)$ is clearly the trivial representation, realized by the Lorentz scalar ϕ, for which we derived an equation of motion just now. We expect the 4-dimensional irreducible representation $(\frac{1}{2}, \frac{1}{2})$ to be the Lorentz vector, the defining representation of the Lorentz group.

Restriction to the rotation subgroup

Ever since chapter II.1, we have learned that it is useful to ask how irreducible representations decompose on restriction to a subgroup. Here the group of rotations $SO(3)$ form a natural subgroup of $SO(3, 1)$.

Since the generators of rotations are given by $J_i = J_{+i} + J_{-i}$, we already know how to solve the problem of figuring out what a given irreducible representation of the Lorentz group decomposes into on restriction to $SO(3)$. If we think of J_{+i} and J_{-i} as the angular momentum operators of two particles called $+$ and $-$, the problem at hand is just the addition of angular momentum, which we solved back in chapter IV.3.

To determine how (j^+, j^-) decomposes, we simply calculate $j^+ \otimes j^-$, as in chapter IV.3. Thus, with no further work, we deduce that

$$(j^+, j^-) \to (j^+ + j^-) \oplus (j^+ + j^- - 1) \oplus (j^+ + j^- - 2) \oplus \cdots \oplus |j^+ - j^-| \tag{12}$$

on restricting $SO(3, 1)$ to $SO(3)$.

In particular, $(\frac{1}{2}, \frac{1}{2}) \to 1 \oplus 0$; in other words, this 4-dimensional representation decomposes[†] into a 3-dimensional vector plus a 1-dimensional scalar under rotation.[‡] This fact further bolsters our expectation that $(\frac{1}{2}, \frac{1}{2})$ furnishes the defining representation of the Lorentz algebra. The 1-dimensional scalar in the decomposition is the time component of the Lorentz 4-vector.

---

* A heads-up potential notational confusion: the reader should of course not confuse the (j^+, j^-) notation used here with the (m, n) notation used in chapter VII.2 to describe a tensor with m upper and n lower indices. Two different types of animals entirely!

[†] Don't be confused by the coexistence of two commonly used notations! Thus, $(\frac{1}{2}, \frac{1}{2}) \to 1 \oplus 0$ is also written as $4 \to 3 \oplus 1$.

[‡] Note that the other 4-dimensional representations, $(\frac{3}{2}, 0)$ and $(0, \frac{3}{2})$, are disqualified by their contents under rotation, for example, $(\frac{3}{2}, 0) \to \frac{3}{2}$, or $4 \to 4$.

What about the representation $(\frac{1}{2}, 0)$? We have $(\frac{1}{2}, 0) \to \frac{1}{2}$. This 2-dimensional irreducible representation of the Lorentz algebra decomposes into the 2-dimensional spinor of the rotation algebra, as we might we guessed. What else could it be? Similarly, $(0, \frac{1}{2}) \to \frac{1}{2}$.

We now study these two spinor representations.

Spinor representations

Examine the representation $(\frac{1}{2}, 0)$ in more detail. Write the two objects as u_α, with $\alpha = 1, 2$. Well, what does the notation $(\frac{1}{2}, 0)$ mean?

It says that $J_{+i} = \frac{1}{2}(J_i + iK_i)$ acting on u is represented by $\frac{1}{2}\sigma_i$ while $J_{-i} = \frac{1}{2}(J_i - iK_i)$ acting on u is represented by 0. By adding and subtracting, we find that

$$J_i = \frac{1}{2}\sigma_i \tag{13}$$

and

$$iK_i = \frac{1}{2}\sigma_i \tag{14}$$

The equal sign means "represented by" in this context. (By convention, we do not distinguish between upper and lower indices on the 3-dimensional quantities J_i, K_i, and σ_i.)

Repeating the same steps as above, you will find that on the representation $(0, \frac{1}{2})$, which we denote by v,

$$J_i = \frac{1}{2}\sigma_i \tag{15}$$

and

$$iK_i = -\frac{1}{2}\sigma_i \tag{16}$$

To keep the signs straight, we need to say a word about notation here. Regard the three Pauli matrices σ_1, σ_2, and σ_3 as matrices introduced in nonrelativistic physics and fixed once and for all. As a rule, we will always write σ_i with a lower index, and if we were to write on occasion σ^i, we will set $\sigma^i = \sigma_i$. For example, σ_3, and also $\sigma^3 = \sigma_3$, will always be $\left(\begin{smallmatrix} 1 & 0 \\ 0 & -1 \end{smallmatrix}\right)$. The symbol $\vec{\sigma}$ denotes $\vec{\sigma} = (\sigma_1, \sigma_2, \sigma_3)$.

The single sign difference between the set of equations (13) and (14), on the one hand, and the set (15) and (16) on the other, is crucial!

The 2-component spinors[8] u and v are called Weyl spinors* and each furnishes a perfectly good representation of the Lorentz group. As expected, under rotations $e^{i\vec{\theta}\vec{J}}$, both

---

* To distinguish between the two Weyl spinors, van der Waerden invented a notation known informally to particle physicists as "dotted and undotted" notation. One Weyl spinor carries an index with a dot written over it, the other not. It has become important, because the notation is used in supersymmetric physics and in superstring theory. Incidentally, Dirac allegedly said that he wished he invented the dotted and undotted notation. For more details, see QFT Nut, p. 541.

$u \to e^{i\vec{\theta}\cdot\frac{\vec{\sigma}}{2}}u$ and $v \to e^{i\vec{\theta}\cdot\frac{\vec{\sigma}}{2}}v$ transform the same way, like a Pauli spinor. In contrast, under boosts $e^{i\vec{\varphi}\vec{K}}$, $u \to e^{\vec{\varphi}\cdot\frac{\vec{\sigma}}{2}}u$ and $v \to e^{-\vec{\varphi}\cdot\frac{\vec{\sigma}}{2}}v$ transform oppositely. This last remark will be key in what follows.

Transformation of Weyl spinors

First, consider a Weyl spinor u furnishing a representation $(\frac{1}{2}, 0)$ of the Lorentz group. According to (13) and (14), under rotations $e^{i\vec{\theta}\vec{J}}$, the spinor transforms as

$$u \to e^{i\vec{\theta}\cdot\frac{\vec{\sigma}}{2}}u \tag{17}$$

while under boosts $e^{i\vec{\varphi}\vec{K}}$, it transforms as

$$u \to e^{\vec{\varphi}\cdot\frac{\vec{\sigma}}{2}}u \tag{18}$$

More compactly,

$$u \to e^{i(\vec{\theta}-i\vec{\varphi})\cdot\frac{\vec{\sigma}}{2}}u = e^{i\vec{\xi}\cdot\frac{\vec{\sigma}}{2}}u \tag{19}$$

transforms as if in $SU(2)$ but with the angles $\vec{\theta}$ continued to complex angles $\vec{\xi} = \vec{\theta} - i\vec{\varphi}$.

While $e^{i\vec{\theta}\frac{\vec{\sigma}}{2}}$ is special unitary, the 2-by-2 matrix $e^{\vec{\varphi}\frac{\vec{\sigma}}{2}}$, bereft of the i, is merely special but not unitary. (Incidentally, and as I have already mentioned, to verify these and subsequent statements, since you understand rotation thoroughly, you could, without loss of generality, choose $\vec{\varphi}$ to point along the third axis, in which case $e^{\vec{\varphi}\frac{\vec{\sigma}}{2}}$ is diagonal with elements $e^{\frac{\varphi}{2}}$ and $e^{-\frac{\varphi}{2}}$. Thus, while the matrix is not unitary* its determinant is manifestly equal to 1.)

By now, you are hardly surprised that $e^{i\vec{\xi}\cdot\frac{\vec{\sigma}}{2}}$ pops up as an element $SL(2, C)$.

We remark in passing that this discussion also shows that the Lorentz group is locally isomorphic to $SO(3, C)$, known as the complexification of $SO(3)$, obtained by promoting the three angles in the rotation group element $e^{i\vec{\theta}\vec{J}}$ from $\vec{\theta}$ to $\vec{\theta} - i\vec{\varphi}$. In this textbook, I have consciously stayed away from complexification.

By hermitean conjugation, under rotations and boosts, the conjugate spinor $u^\dagger \to u^\dagger e^{-i\vec{\theta}\frac{\vec{\sigma}}{2}}$ and $u^\dagger \to u^\dagger e^{\vec{\varphi}\frac{\vec{\sigma}}{2}}$, respectively. Watch the signs!

Notice that since $i\vec{K}$, rather than $\vec{K}$, is hermitean, the matrix representing the boost $e^{i\vec{\varphi}\vec{K}}$ is not unitary for sure.

You can readily repeat this discussion for a Weyl spinor v furnishing a representation $(0, \frac{1}{2})$ of the Lorentz group. We now have

$$v \to e^{i(\vec{\theta}+i\vec{\varphi})\cdot\frac{\vec{\sigma}}{2}}v = e^{i\vec{\xi}^*\cdot\frac{\vec{\sigma}}{2}}v \tag{20}$$

Compare and contrast with (19).

---

* We already had a hint of this in chapter II.1: the finite representations of noncompact groups are not necessarily unitary.

Pseudoreality of the doublet representation of $SU(2)$: A caution about terminology

Now the pseudoreality of the defining representation of $SU(2)$ discussed in chapter IV.5 becomes highly relevant. Recall that

$$\sigma_2 \sigma_i^* \sigma_2 = -\sigma_i \tag{21}$$

Define $\tilde{u} \equiv i\sigma_2 u^*$. Complex conjugating (19), we find

$$\tilde{u} \to i\sigma_2 e^{-i(\vec{\theta}+i\vec{\varphi})\cdot\frac{\vec{\sigma}^*}{2}} u^* = i\sigma_2 e^{-i(\vec{\theta}+i\vec{\varphi})\cdot\frac{\vec{\sigma}^*}{2}} \sigma_2 \sigma_2 u^* = e^{i(\vec{\theta}+i\vec{\varphi})\cdot\frac{\vec{\sigma}}{2}} i\sigma_2 u^* = e^{i(\vec{\theta}+i\vec{\varphi})\cdot\frac{\vec{\sigma}}{2}} \tilde{u} \tag{22}$$

Referring to (20), we see that $\tilde{u}$ transforms just like v, that is, $\tilde{u}$ furnishes the representation $(0, \frac{1}{2})$.

Note that if we write $u = \begin{pmatrix} u_1 \\ u_2 \end{pmatrix}$, then $\tilde{u} = \begin{pmatrix} u_2^* \\ -u_1^* \end{pmatrix}$ is just u with its two components conjugated and then interchanged, with a minus sign thrown in.

Now comes a cautionary note about terminology. People differ according to whether they regard u and $\tilde{u}$ as independent or not. For the purpose of this book at least, I regard $\tilde{u}$ as derived from u, rather than as an independent entity. In what follows, I will say things like "consider a world with a single Weyl spinor transforming like $(\frac{1}{2}, 0)$." Otherwise excellent theoretical physicists would say that this is impossible, this Weyl spinor is always accompanied by a Weyl spinor transforming like $(0, \frac{1}{2})$.

What is physically relevant is the number of degrees of freedom, and on this everybody can only agree that there are two complex degrees of freedom or fields, namely, u_1 and u_2.

Again, you can repeat the discussion for a Weyl spinor v transforming like $(0, \frac{1}{2})$. The spinor $\tilde{v} \equiv i\sigma_2 v^*$ transforms like $(\frac{1}{2}, 0)$.

Parity forces us to stack two Weyl spinors together to form a Dirac spinor

In introducing this chapter, I mentioned that the electron is described by a 4-component Dirac spinor. Why then is the electron not described by a Weyl spinor?

The reason is parity. Under parity,[9] $\vec{x} \to -\vec{x}$ and $\vec{p} \to -\vec{p}$, and thus $\vec{J} \to \vec{J}$ and $\vec{K} \to -\vec{K}$, as already mentioned in chapter VII.2. Hence $\vec{J}_+ \leftrightarrow \vec{J}_-$. In other words, under parity, the representations* $(\frac{1}{2}, 0) \leftrightarrow (0, \frac{1}{2})$.

To describe the electron, we must use both of these 2-dimensional irreducible representations, $(\frac{1}{2}, 0)$ and $(0, \frac{1}{2})$, to form the 4-dimensional reducible representation $(\frac{1}{2}, 0) \oplus (0, \frac{1}{2})$, known to physicists as a Dirac spinor. We are invited to stack the two 2-component Weyl spinors u and v together to form a 4-component Dirac spinor, as mentioned at the start of this chapter. (In light of the remark about terminology in the preceding section,

---

* In general, under parity the representations $(j^+, j^-) \leftrightarrow (j^-, j^+)$. Yet another reason that $(\frac{1}{2}, \frac{1}{2})$, which is its own parity partner, should be identified as the 4-vector, rather than, say, $(\frac{3}{2}, 0)$.

some authors would insist that we stack four Weyl spinors together, u, v, $\tilde{u}$, and $\tilde{v}$.) We will do this in the chapter VII.4.

Finite dimensional nonunitary representations of the Lorentz group

It is worth recalling the theorem we proved back in chapter II.1 stating that the representations of finite groups, and by extension, compact groups, are necessarily unitary. We had pointed out that, in contrast, the finite-dimensional representations of noncompact groups, in particular the Lorentz group, need not be unitary. In particular, the corresponding generators are not necessarily represented by hermitean matrices. This is explicitly realized here.

The generators J_{+i} and J_{-i}, being generators of (separate) $SU(2)$ algebras, are necessarily represented by hermitean matrices. This implies that the generators $J_i = J_{+i} + J_{-i}$ are represented by hermitean matrices, as they should be, since they generate the compact rotation group. In contrast, the generators $K_i = (J_{+i} - J_{-i})/i$ are necessarily represented by antihermitean matrices. We see that this is indeed the case in (13), (14), (15), and (16).

Indeed, looking back at chapter VII.2, we see that in the defining representation, namely the $(\frac{1}{2}, \frac{1}{2})$, the J_is are represented by hermitean matrices, but the K_is are represented* by antihermitean matrices. Mathematically, it would be best to treat[10] $SO(3, 1)$ as the analytic continuation of $SO(4)$, letting the fourth coordinate $x^4 \to -ix^0 = -it$.

Some aspects of the representations of the Lorentz algebra are developed further in the exercises.

Group theoretic approach to the Weyl equation

For now, let us focus on the Weyl spinors u and v separately. In a world with a single Weyl spinor (say, u) transforming like $(\frac{1}{2}, 0)$, the equation it satisfies is known as a Weyl equation.[†]

Historically, since the Weyl equation did not respect parity, it was rejected in favor of the Dirac equation. But after the proposal of parity violation by Lee and Yang in 1956, the Weyl equation was resurrected to describe the neutrino. In an ironic[‡] twist of history, it is now understood that particle physics at the fundamental level does not respect parity, and that the Weyl equation describes all spin $\frac{1}{2}$ fields, quarks and leptons, before they acquire mass via the Higgs mechanism.

Under a rotation, $u^\dagger u$ is invariant, while $u^\dagger \vec{\sigma} u$ transforms like a 3-vector, as should be known to you by now. Under a boost, in contrast, $u^\dagger u$ is not invariant; rather,

$$u^\dagger u \to u^\dagger e^{\vec{\varphi} \frac{\vec{\sigma}}{2}} e^{\vec{\varphi} \frac{\vec{\sigma}}{2}} u = u^\dagger e^{\vec{\varphi} \vec{\sigma}} u \simeq u^\dagger (I + \vec{\varphi} \vec{\sigma}) u = u^\dagger u + \vec{\varphi} u^\dagger \vec{\sigma} u \qquad (23)$$

---

* As remarked earlier, we follow the common physicist's practice of not introducing different notations for the abstract mathematical entities that appear in the Lie algebra and for the matrices that represent them.

  † Similarly, the spinor $(0, \frac{1}{2})$ also satisfies a Weyl equation, as discussed below.

  ‡ A graduate student even told me that "nobody" bothered with the Dirac equation any more, but he was exaggerating.

for φ infinitesimal. This indicates that the 4-objects $\omega^\mu \equiv (u^\dagger u, u^\dagger \vec{\sigma} u)$ transform like a 4-vector.

To be more explicit, it is convenient to work with infinitesimal boosts and to specify the boost to be in the z direction (with no loss of generality, of course). We have $u \to (I + \varphi \frac{\sigma_z}{2})u$, which we write as $\delta u = \frac{\sigma_z}{2} u$, where to lessen clutter I have simply absorbed the infinitesimal φ into the definition of δu. Hermitean conjugation gives $\delta u^\dagger = u^\dagger \frac{\sigma_z}{2}$. First, we have (watch how the $\frac{1}{2}$ disappears)

$$\delta(u^\dagger u) = (\delta u^\dagger)u + u^\dagger \delta u = u^\dagger \sigma_z u \tag{24}$$

which is of course just a rewrite of (23). Next,

$$\delta(u^\dagger \sigma_i u) = (\delta u^\dagger)\sigma_i u + u^\dagger \sigma_i \delta u = u^\dagger \frac{1}{2}\{\sigma_i, \sigma_z\}u \tag{25}$$

which is equal to $u^\dagger u$ for $i = z$ and to 0 for $i = x$, y. In other words, $u^\dagger u$ and $u^\dagger \sigma_z u$ transform into each other under a boost in the z direction, with $u^\dagger \sigma_x u$ and $u^\dagger \sigma_y u$ unaffected.

As claimed, $\omega^\mu \equiv (u^\dagger u, u^\dagger \vec{\sigma} u)$ transforms like a 4-vector with an upper index. We are now cordially invited by group theory to introduce a fourth Pauli matrix, defined by $\sigma^0 \equiv I$, with I the 2-by-2 identity matrix. Then we can write[11]

$$\sigma^\mu = (I, \vec{\sigma}) = (I, \sigma_1, \sigma_2, \sigma_3) \tag{26}$$

(Note, as remarked earlier, that $\sigma^i = \sigma_i$ by fiat.)

Pause for a moment to recall that in our convention, the coordinates $x^\mu = (t, x, y, z)$ transform like $\delta t = z$, $\delta z = t$, $\delta x = 0$, and $\delta y = 0$ for a boost along the z direction, using the same abbreviated notation employed above. Here, $\delta\omega^0 = \omega^z$, $\delta\omega^z = \omega^0$, $\delta\omega^x = 0$, and $\delta\omega^y = 0$. Indeed, ω^μ transforms just like the spacetime coordinates x^μ. Note for use presently that this implies that $\partial_\mu \equiv \frac{\partial}{\partial x^\mu}$ transforms like a 4-vector with a lower index (see exercise 1). In particular, $\partial_\mu \omega^\mu$ is a Lorentz scalar.

Group theoretically, what we are doing here is easily understood. Since $u \sim (\frac{1}{2}, 0)$, it follows that $u^\dagger \sim (0, \frac{1}{2})$ (because under conjugation, $J_+ \leftrightarrow J_-$). Thus, $u^\dagger u$ transforms like a member of $(\frac{1}{2}, 0) \otimes (0, \frac{1}{2}) = (\frac{1}{2}, \frac{1}{2})$, which is precisely the Lorentz 4-vector.

Here $u(x) = u(t, \vec{x})$ is, of course, a function of the spacetime coordinates x^μ, which we can always Fourier transform to momentum space: $u(p) = \int d^4x e^{-ipx} u(x)$. Henceforth, unless otherwise noted, we will use relativistic notation $x = (t, \vec{x})$, $p = (E, \vec{p})$ and so forth. Also, we prefer not to clutter up the notation by distinguishing between u in position space and u in momentum space.

The Weyl Lagrangian

I gave you a rudimentary grasp of what a Lagrangian is in chapter III.3 and then some vague notion of what a quantum field theory is in chapter IV.9. (If you feel uncomfortable with the Lagrangian, you could skip to the equation of motion below.) At this point, all I require of you, dear reader, is to know that (i) the physics described by a Lagrangian enjoys a symmetry if the Lagrangian is invariant under the transformations of the symmetry, and that (ii) varying the Lagrangian with respect to the dynamical variable and setting the result to 0 gives the equation of motion for that variable.

Thus, our first task is to construct, out of u and $u^\dagger$, a Lagrangian density $\mathcal{L}$ invariant under Lorentz transformations. Since we already have the 4-vector ω^μ, our first thought would be to try $\mathcal{L} = \partial_\mu \omega^\mu$. But it is a total divergence, and so cannot affect the equation of motion.[*]

The solution is to pull ω^μ apart and sandwich ∂_μ in between $u^\dagger$ and u, that is, to construct

$$\mathcal{L} = i u^\dagger \sigma^\mu \partial_\mu u \tag{27}$$

The important point, in this context, is that we only care about how $\mathcal{L}$ transforms. Where we put ∂_μ does not affect the transformation of $u^\dagger$ and u.

I postpone making a couple of technical remarks about $\mathcal{L}$ until the end of this section. Instead, we want to get at the physics immediately. Varying with respect to $u^\dagger$ yields the Weyl equation of motion

$$\sigma^\mu \partial_\mu u = 0 \tag{28}$$

Fourier transforming to momentum space by setting

$$u(t, \vec{x}) = e^{-ip \cdot x} u(p) = e^{-i(Et - \vec{p} \cdot \vec{x})} u(E, \vec{p}),$$

we obtain in momentum space

$$\sigma^\mu p_\mu u = 0 \tag{29}$$

For the readers a bit shaky with the elegantly compact relativistic notation, let us unpack the Lagrangian in (27) and write it as

$$\mathcal{L} = i u^\dagger \left(\frac{\partial}{\partial t} + \vec{\sigma} \cdot \vec{\nabla} \right) u \tag{30}$$

It is evidently invariant under rotations, and by construction, is invariant under boosts (a fact you are invited to check). Varying (30) with respect to $u^\dagger$ and then Fourier transforming, we obtain the Weyl equation (28) and (29) in the less relativistic-looking forms

$$\left(\frac{\partial}{\partial t} + \vec{\sigma} \cdot \vec{\nabla} \right) u = 0 \tag{31}$$

and

$$(E - \vec{\sigma} \cdot \vec{p}) u = 0 \tag{32}$$

Acting with $(E + \vec{\sigma} \cdot \vec{p})$ on this equation from the left, we obtain $E^2 = \vec{p}^2$: the particle the Weyl equation describes is massless, with energy[12] $E = |\vec{p}|$.

If you skipped the construction of the Lagrangian, you could regard (32) (or (31)) as a reasonable guess for the equation of motion (for example, it makes sense dimensionally) and simply verify that it respects Lorentz invariance. Do exercise 4 now.

---

[*] Recall chapter IV.9. In other words, the would-be action $S \equiv \int d^4x \mathcal{L} = \int d^4x \partial_\mu \omega^\mu$ reduces to a surface term at spacetime infinity on integration by parts.

As promised, a couple of technical remarks. First, the asymmetry between $u^\dagger$ and u in (30), with the derivatives acting on u but not on $u^\dagger$, is only apparent. Integrating the action $S \equiv \int d^4x \mathcal{L}$ by parts,[13] we can write, equivalently, $\mathcal{L} = -i((\frac{\partial}{\partial t} + \vec{\sigma} \cdot \vec{\nabla})u^\dagger)u$. Next, this form also shows that the factor of i is needed for the action to be hermitean.[14]

Note that the combination $(E - \vec{\sigma} \cdot \vec{p})$ in (32) indeed corresponds to $\sigma^\mu p_\mu = p_0 I + p_i \sigma^i = p^0 I - p^i \sigma^i = E - \vec{\sigma} \cdot \vec{p}$ in (29). This brings up a question: What does the combination $(E + \vec{\sigma} \cdot \vec{p})$ used above correspond to in the relativistic notation? I might call this a self-answering question. Clearly, in analogy to (26), we should define

$$\bar{\sigma}^\mu = (I, -\vec{\sigma}) = (I, -\sigma_1, -\sigma_2, -\sigma_3) \tag{33}$$

so that the desired combination $\bar{\sigma}^\mu p_\mu = (E + \vec{\sigma} \cdot \vec{p})$. Hah! We just flip some signs. Interestingly, as you will see shortly, the combination $\bar{\sigma}^\mu$ will come looking for us, even if we do not look for it here.

Let these technical points not obscure the central message here. Group theory uniquely determines the Weyl Lagrangian $\mathcal{L}$ and hence the physics (which we explore further in the next section).

A right handed spinor

In our (standard) discussion of angular momentum in chapter IV.3, we chose the z-axis as the "quantization axis," and the states are labeled by the eigenvalues of J_z. For a massless particle, it is natural to choose the direction of its motion as the "quantization axis" for its spin angular momentum. The helicity of a particle is thus defined as the eigenvalue of $h \equiv \vec{\sigma} \cdot \hat{p}$, with $\hat{p} \equiv \vec{p}/|\vec{p}|$ the unit vector pointing in the direction of the particle's motion. Since $E = |\vec{p}|$, we see that the Weyl equation (32) tells us that the helicity of the particle is equal to $+1$.

We say that the spinor u is right handed.

We already mentioned in chapter VII.2 that under spatial reflection, $\vec{p} \to -\vec{p}$ and $\vec{J} \to \vec{J}$. Thus, helicity flips sign. The reader should draw a picture of a particle with a definite helicity and then reflect the given situation in mirrors placed perpendicular and parallel to the direction of the particle's motion to see that helicity always flips sign on reflection. See figure 1. A Weyl particle manifestly violates parity, as was noted historically.

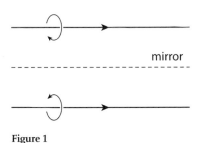

Figure 1

Indeed, by choosing $\vec{p} = (0, 0, p)$ with $p > 0$, we could have extracted the physical content of the Weyl equation (32) almost trivially. The equation tells us that $\begin{pmatrix} E-p & 0 \\ 0 & E+p \end{pmatrix} \begin{pmatrix} u_1 \\ u_2 \end{pmatrix}$ $= 0$, which implies $u_2 = 0$, since E and p are both positive. To have a nontrivial solution $u \neq 0$, we must require $E = p$.

The Weyl equation may be interpreted as a projection. Define the projection operator*
$\mathcal{P} \equiv \frac{1}{2}(1 - \vec{\sigma} \cdot \hat{p}) = \frac{1}{2}(1 - h)$; then the Weyl equation simply says that

$$\mathcal{P}u = 0 \tag{34}$$

It sets the $h = -1$ component of the 2-component spinor u to 0.

The other Weyl equation

The preceding discussion started with the representation $(\frac{1}{2}, 0)$. We could have equally well envisioned a world with a 2-component spinor v furnishing the representation $(0, \frac{1}{2})$. According to (15) and (16), under rotations, v transforms in the same way as u, but under boosts, v transforms with the opposite sign as u: with the same choice of conventions and so forth as before, under boosts $\delta v = -\frac{\sigma_z}{2} v$, in contrast to $\delta u = \frac{\sigma_z}{2} u$.

Convince yourself that as a result, now it is $\rho^\mu \equiv (v^\dagger v, -v^\dagger \vec{\sigma} v)$, with a crucial minus sign, that transforms like a 4-vector with an upper index. Write ρ^μ as $v^\dagger \bar{\sigma}^\mu v$. Thus, as was foretold, the matrices $\bar{\sigma}^\mu = (I, -\vec{\sigma})$ defined in (33) come looking for us.

By now, you can write down the invariant Lagrangian for v as well as anybody else, namely,

$$\mathcal{L} = i v^\dagger \bar{\sigma}^\mu \partial_\mu v \tag{35}$$

Varying with respect to $v^\dagger$ we find the other Weyl equation of motion

$$\bar{\sigma}^\mu \partial_\mu v = 0 \tag{36}$$

and in momentum space

$$\bar{\sigma}^\mu p_\mu v = 0 \tag{37}$$

Again, unpacking into a less relativistic form, we have

$$\mathcal{L} = i v^\dagger \left(\frac{\partial}{\partial t} - \vec{\sigma} \cdot \vec{\nabla} \right) v \tag{38}$$

and the corresponding equations of motion

$$\left(\frac{\partial}{\partial t} - \vec{\sigma} \cdot \vec{\nabla} \right) v = 0 \tag{39}$$

and

$$(E + \vec{\sigma} \cdot \vec{p}) v = 0 \tag{40}$$

---

* By definition, a projection operator satisfies $\mathcal{P}^2 = \mathcal{P}$: if $\mathcal{P}$ projects us into a subspace, then projecting again does not do anything more. Acting with $\mathcal{P}$ on a spinor in the orthogonal subspace gives 0 by the very definition of projection.

A left handed spinor

Compare these equations for the Weyl spinor v with the corresponding equations for the Weyl spinor u and notice the crucial sign flips. Similarly to what we did before, we again conclude, on acting with $(E - \vec{\sigma} \cdot \vec{p})$ on (40) from the left, that the Weyl particle described by (40) is also massless. It also violates parity. Indeed, if we define the projection operator $Q \equiv \frac{1}{2}(1 + \vec{\sigma} \cdot \hat{p}) = \frac{1}{2}(1 + h)$, then we have $Qv = 0$, that is, $h = -1$. The Weyl equation sets the $h = +1$ component of the 2-component spinor v to 0. The spinor v is said to be left handed.

Conjugate Weyl spinors

Let us multiply the complex conjugate of the Weyl equation $\sigma^\mu p_\mu u = 0$ in (29) by σ_2 on the left: $\sigma_2(\sigma^{\mu*} p_\mu) u^* = 0$. Note that

$$\sigma_2 \sigma^{\mu*} \sigma_2 = \bar{\sigma}^\mu \tag{41}$$

Hence, we obtain

$$\bar{\sigma}^\mu p_\mu \tilde{u} = 0 \tag{42}$$

The spinor $\tilde{u} \equiv i\sigma_2 u^*$ not only transforms like v, it satisfies the same Weyl equation as v, entirely as you would expect.

Similarly, $\tilde{v}$ satisfies the same Weyl equation as u, as you would also expect.

The Weyl equation came roaring back

I already mentioned that the Weyl equation, rejected in favor of the Dirac equation, came roaring back after the discovery of parity violation in 1956. It was realized that the neutrino is left handed and massless.*

Since the particle the Weyl equation describes is massless, there is no rest frame, and Lorentz invariance connects one moving frame to another. It is instructive to work out what group theory has to say. With no loss of generality, let the particle be moving along the z direction with some reference 4-momentum $p_* = E_*(1, 0, 0, \pm 1)$. The subgroup of $SO(3)$ that leaves p_* invariant is $SO(2)$, known as the little group.[15] As already discussed in chapter VII.2, we have to specify that J_z is either $+\frac{1}{2}$ or $-\frac{1}{2}$. Since J_z is represented by $\frac{\sigma_z}{2}$, the Weyl spinor u is either $\propto \begin{pmatrix} 1 \\ 0 \end{pmatrix}$ or $\propto \begin{pmatrix} 0 \\ 1 \end{pmatrix}$, respectively.

---

* The more recent discovery that the neutrino has a tiny but nonvanishing mass only made the story even more intriguing.

Majorana mass term

In a world with only u, is there another invariant term we could add to the Lagrangian given in (27) for u? Think for a moment before reading on.

Well, with all this talk about pseudoreality, you might have written down $u^T \sigma_2 u$ (or more sloppily, $u \sigma_2 u$). Let us check: transposing $u \to e^{i \vec{\zeta} \cdot \vec{\sigma}} u$ (with $\vec{\zeta}$ being three complex angles) gives $u^T \to u^T e^{i \vec{\zeta} \cdot \vec{\sigma}^T}$, but since $\vec{\sigma}^T \sigma_2 = -\sigma_2 \vec{\sigma}$ (since the Pauli matrices are hermitean, this is equivalent to (21)), we have $u^T \sigma_2 u \to u^T e^{i \vec{\zeta} \cdot \vec{\sigma}^T} \sigma_2 e^{i \vec{\zeta} \cdot \vec{\sigma}} u = u^T \sigma_2 u$. Thus, this term is indeed invariant.

But now you might object, noting that the quantity $u^T \sigma_2 u$ vanishes identically, since σ_2 is an antisymmetric 2-by-2 matrix. That would indeed be the case had the two components u_1 and u_2 of u been ordinary complex numbers. But here is a subtlety that goes well beyond the scope of this book: for u to describe a spin $\frac{1}{2}$ particle in quantum field theory, u_1 and u_2 have to be quantized as anticommuting Grassman numbers,[16] so that $u_1 u_2 = -u_2 u_1$.

Let us thus add $m u^T \sigma_2 u$ to the Lagrangian with m some constant. To make the Lagrangian hermitean (as is required in quantum field theory), we add the hermitean conjugate of $m u^T \sigma_2 u$: $(u^T \sigma_2 u)^\dagger = u^\dagger \sigma_2 u^* = (u^*)^T \sigma_2 u^*$. Thus, we can extend the Lagrangian to

$$\mathcal{L} = i u^\dagger \left(\frac{\partial}{\partial t} + \vec{\sigma} \cdot \vec{\nabla} \right) u - \frac{1}{2} (m u^T \sigma_2 u + m^* u^\dagger \sigma_2 u^*) \qquad (43)$$

Varying with respect to $u^\dagger$, we obtain

$$i \left(\frac{\partial}{\partial t} + \vec{\sigma} \cdot \vec{\nabla} \right) u = m^* \sigma_2 u^* \qquad (44)$$

Compare and contrast with (31).

Notice that there is no v here! Once again, use our old trick of charge conjugating this to get $-i(\frac{\partial}{\partial t} + \vec{\sigma}^* \cdot \vec{\nabla}) u^* = -m \sigma_2 u$ and then multiply by σ_2 from the left, which gives $(\frac{\partial}{\partial t} - \vec{\sigma} \cdot \vec{\nabla}) \sigma_2 u^* = -i m u$. Acting with $(\frac{\partial}{\partial t} - \vec{\sigma} \cdot \vec{\nabla})$ on (44) from the left gives us finally $\partial^2 u = (\frac{\partial^2}{\partial t^2} - \vec{\nabla}^2) u = -|m|^2 u$. (By redefining $u \to e^{i\alpha} u$ in the Lagrangian, we can make m real with a suitable choice of α.)

The Weyl particle has become massive, acquiring what is known as a Majorana mass m. We will return to this in more detail in chapter VII.5, when we discuss the Majorana equation.

We need hardly mention that, similarly, we could add a Majorana mass term to the Lagrangian given in the preceding section for v.

Appendix 1: About $SO(2, 2)$

Having done $SO(4)$ and $SO(3, 1)$ earlier in this chapter, I might as well throw in the group $SO(2, 2)$ for no extra charge. Let us strip the Pauli matrix σ^2 (kind of a troublemaker, or at least an odd man out) of his i and define (just for this paragraph) $\sigma^2 \equiv \left(\begin{smallmatrix} 0 & -1 \\ 1 & 0 \end{smallmatrix} \right)$. Any real 2-by-2 matrix X_H can be decomposed as

$$X_H = x^4 I + \vec{x} \cdot \vec{\sigma} \qquad (45)$$

Compare and contrast with (3) and (6). Now

$$\det X_H = (x^4)^2 + (x^2)^2 - (x^3)^2 - (x^1)^2 \tag{46}$$

The set of all linear transformations (with unit determinant) on (x^1, x^2, x^3, x^4) that preserve this quadratic form defines the group $SO(2, 2)$ (see chapter VII.2).

Introduce the multiplicative group $SL(2, R)$ consisting of all 2-by-2 real-valued matrices with unit determinant. For any two elements L_l and L_r of this group, consider the transformation $X_H \to X'_H = L_l X_H L_r$. Evidently, $\det X'_H = \det X_H$. This shows explicitly that the group $SO(2, 2)$ is locally isomorphic to $SL(2, R) \otimes SL(2, R)$.

The metric of a world described by $ds^2 = (dx^4)^2 + (dx^2)^2 - (dx^3)^2 - (dx^1)^2$ would have two time coordinates, x^4 and x^2. Although we do not know how to make sense of a world with two "times," scattering amplitudes in quantum field theory are analytic in various kinematic variables, and theorists could, and do, exploit the group $SO(2, 2)$ formally in studying[17] amplitudes in contemporary research.

Appendix 2: $SO(2, 1)$ and $SL(2, R)$

We have shown that $SO(3, 1)$ is locally isomorphic to $SL(2, C)$ by studying the most general 2-by-2 hermitean matrix $X_M = x^0 I + \vec{x} \cdot \vec{\sigma}$ shown in (6). Now set $x^2 = 0$, so that X_M becomes the most general 2-by-2 real symmetric matrix

$$X_R = x^0 I + \vec{x} \cdot \vec{\sigma} = \begin{pmatrix} x^0 + x^3 & x^1 \\ x^1 & x^0 - x^3 \end{pmatrix} \tag{47}$$

By explicit computation, $\det X_R = (x^0)^2 - (x^1)^2 - (x^3)^2$.

Now consider the group $SL(2, R)$, consisting of all 2-by-2 real matrices with unit determinant.[18] With L an element of $SL(2, R)$, let $X_R \to X'_R = L^T X_R L$, which is manifestly real symmetric. Write X'_R as in (47) in terms of (x'^0, x'^1, x'^3). Manifestly, $\det X'_M = \det X_M$, and so $(x'^0)^2 - x'^{12} - x'^{32} = (x^0)^2 - x^{12} - x^{32}$. Thus, the transformation $(x^0, x^1, x^3) \to (x'^0, x'^1, x'^3)$ preserves the Minkowski metric in $(2 + 1)$-dimensional spacetime. The construction here thus defines a map from $SL(2, R)$ to $SO(2, 1)$. Again, it is a double cover, since L and $-L$ define the same map.

As a challenge, show that $SL(2, H)$ and $SO(5, 1)$ are locally isomorphic.

Exercises

1 Explicitly verify that ∂_μ transforms like a 4-vector with a lower index.

2 Multiplying h 4-vectors together, we generate the tensor representations of the Lorentz algebra. Show how this works.

3 Show that the irreducible representation $(\frac{1}{2}h, \frac{1}{2}h)$ corresponds to the traceless symmetric tensor representation of $SO(3, 1)$. Count the number of components. Hint: Recall an exercise you did in chapter IV.1. By the way, this result will also show up in another guise in interlude 1.

4 Decompose $(\frac{1}{2}, 0) \otimes (\frac{1}{2}, \frac{1}{2})$.

5 Verify that the Weyl equation $(E - \vec{\sigma} \cdot \vec{p})u = 0$ in (32) is covariant under the Lorentz group.

Notes

1. See, for example, chapter II.3 in QFT Nut.
2. A deeper understanding of the Dirac spinor not only gives us a certain satisfaction, but is also indispensable for studying supersymmetry, one of the foundational concepts of superstring theory.

3. As still believed in by the proverbial guy in the street.

4. First discovered by Schrödinger. See, for example, QFT Nut, p. 21.

5. This fact underlies Thomas precession.

6. See any reputable textbooks on quantum field theory.

7. For an easy introduction to this frontier subject, see chapter X.6 in G Nut.

8. Our use of u and v runs a slight risk of confusing some readers who have studied quantum field theory. In many quantum field theory textbooks, solutions of the 4-component Dirac equation are written as $\psi(x) = u(p)e^{-ip \cdot x}$ and $\psi(x) = v(p)e^{+ip \cdot x}$. But the risk of confusion is truly slight, since u and v denote 2-component spinors and are initially introduced as functions of the spacetime coordinates $x = (t, \vec{x})$. See the Weyl equations to be derived shortly. In contrast, in the solutions of the Dirac equation, $u(p)$ and $v(p)$ have 4-components and are functions of the 4-momentum.

9. For a pedagogical discussion of parity and other discrete symmetries in particle physics, see T. D. Lee, Particle Physics and Introduction to Field Theory, Harwood, 1981.

10. Known as the Weyl unitarian trick. This is done routinely in quantum field theory; see, for example, QFT Nut, pp. 538–539.

11. See QFT Nut, p. 99, for example.

12. See QFT Nut, p. 113, for a remark about poetic metaphors. Energy is a positive quantity.

13. And dropping a surface term at spacetime infinity.

14. See, for example, QFT Nut.

15. For further discussion, see QFT Nut, pp. 186–187.

16. For Grassman numbers, see, for example, chapter II.5 in QFT Nut. Also, $u_1 u_1 = -u_1 u_1 = 0$, and similarly for u_2, but this is not relevant here, since the diagonal elements of σ_2 vanish.

17. See, for example, chapter N.3 in QFT Nut.

18. According to chapter IV.8, this is isomorphic to $Sp(2, R)$.

VII.4 | From the Weyl Equation to the Dirac Equation

Two Weyls equal one Dirac

I will give you, in this chapter, two "modern" derivations of the Dirac equation in addition to Dirac's original derivation.

We understood that to describe the electron, we need the 4-dimensional reducible representation $(\frac{1}{2}, 0) \oplus (0, \frac{1}{2})$, known to physicists as a Dirac spinor. In other words, we need both the right handed and the left handed Weyl spinors u and v.

So imagine a world with two 2-component Weyl spinors, transforming as $(\frac{1}{2}, 0)$ and $(0, \frac{1}{2})$, respectively. We add the Lagrangian for u we had in (VII.3.27) and the Lagrangian for v in (VII.3.35) to obtain $\mathcal{L} = i u^\dagger \sigma^\mu \partial_\mu u + i v^\dagger \bar{\sigma}^\mu \partial_\mu v$. With this Lagrangian, the two Weyl spinors u and v live independently of each other.

The possibility of mass

But now a new possibility arises. Recall that under rotations and boosts, $u \to e^{(i\vec{\theta} + \vec{\varphi}) \cdot \frac{\vec{\sigma}}{2}} u$, while $v \to e^{(i\vec{\theta} - \vec{\varphi}) \cdot \frac{\vec{\sigma}}{2}} v$. In other words, u and v transform oppositely under boosts. This implies that, with both u and v on hand, we could construct the invariant terms $u^\dagger v$ and $v^\dagger u$ (verify that they are invariant!) and include them in the Lagrangian. Hermiticity requires that we add them in the combination $m(u^\dagger v + v^\dagger u)$, with some real parameter m whose physical meaning will become clear presently. Group theory thus leads us to

$$\mathcal{L} - i u^\dagger \sigma^\mu \partial_\mu u + i v^\dagger \bar{\sigma}^\mu \partial_\mu v - m(u^\dagger v + v^\dagger u) \tag{1}$$

Now varying $\mathcal{L}$ with respect to $u^\dagger$, we obtain, instead of (VII.3.28), $i\sigma^\mu \partial_\mu u = mv$, and thus in momentum space

$$\sigma^\mu p_\mu u = (E - \vec{\sigma} \cdot \vec{p})u = mv \tag{2}$$

Similarly, varying with respect to $v^\dagger$, we obtain $\bar{\sigma}^\mu \partial_\mu v = mu$ and thus

$$\bar{\sigma}^\mu p_\mu v = (E + \vec{\sigma} \cdot \vec{p})v = mu \tag{3}$$

Acting with $(E + \vec{\sigma} \cdot \vec{p})$ on (2) from the left and invoking (3), we find $(E^2 - \vec{p}^2)u = m^2 u$. In other words, u vanishes unless

$$E^2 = \vec{p}^2 + m^2 \tag{4}$$

The term $m(u^\dagger v + v^\dagger u)$ allowed by group theory has endowed the particle with mass m. Furthermore, under parity, $u \leftrightarrow v$, and so physics now respects parity.

The particle possesses two helicity states with $h = \pm 1$, is massive, and respects parity: it could perfectly well be the electron! The two Weyl spinors u and v may be thought of as the right and left handed components of the electron, respectively.

A somewhat highbrow way of defining the mass of a spin $\frac{1}{2}$ Dirac particle is to say that mass is the quantity that connects its left hand and its right hand.

Physically, if a spin $\frac{1}{2}$ particle has mass, then we could go to its rest frame, where rotation invariance requires it to have two spin states. Boosting implies that a massive spin $\frac{1}{2}$ particle, such as the electron, must have both helicity states.

The Dirac equation pops up

The two equations (2) and (3) are practically begging us to stack the two[1] 2-component Weyl spinors together to form a 4-component Dirac spinor[2] $\psi = \begin{pmatrix} u \\ v \end{pmatrix}$ (in both position and momentum space). Then

$$\begin{pmatrix} \bar{\sigma}^\mu p_\mu v(p) \\ \sigma^\mu p_\mu u(p) \end{pmatrix} = \begin{pmatrix} 0 & \bar{\sigma}^\mu p_\mu \\ \sigma^\mu p_\mu & 0 \end{pmatrix} \begin{pmatrix} u(p) \\ v(p) \end{pmatrix} \tag{5}$$

The formalism is inviting us to define the four 4-by-4 matrices

$$\gamma^\mu \equiv \begin{pmatrix} 0 & \bar{\sigma}^\mu \\ \sigma^\mu & 0 \end{pmatrix} \tag{6}$$

so that the left hand side of (5) can be written as $\gamma^\mu p_\mu \psi(p)$.

The matrices in (6) are known as gamma matrices. In this group theoretic approach, we did not go looking for the gamma matrices, the gamma matrices came looking for us.

We can now package (2) and (3) to read

$$(\gamma^\mu p_\mu - m)\psi(p) = \begin{pmatrix} \bar{\sigma}^\mu p_\mu v(p) - mu(p) \\ \sigma^\mu p_\mu u(p) - mv(p) \end{pmatrix} = 0 \tag{7}$$

In position space, this is the famed Dirac equation

$$(i\gamma^\mu \partial_\mu - m)\psi(x) = 0 \tag{8}$$

We did not go looking for the Dirac equation, the Dirac equation came looking for us!

The Clifford algebra and gamma matrices

You might recognize the gamma matrices γ^μ, which popped up here unbidden as analytic continuations of the gamma matrices for $SO(4)$, and more generally for $SO(2n)$, as discussed in chapter VII.1; hence the arrangement of chapters. Indeed, the various spinors discussed in this chapter represent specializations and continuations of the various spinors of $SO(2n)$.

From the discussion in chapter VII.1, we expect that γ^μ here would satisfy the anticommuting Clifford algebra, suitably continued, and indeed, they do.

Let us calculate

$$\gamma^\mu \gamma^\nu = \begin{pmatrix} 0 & \overline{\sigma}^\mu \\ \sigma^\mu & 0 \end{pmatrix} \begin{pmatrix} 0 & \overline{\sigma}^\nu \\ \sigma^\nu & 0 \end{pmatrix} = \begin{pmatrix} \overline{\sigma}^\mu \sigma^\nu & 0 \\ 0 & \sigma^\mu \overline{\sigma}^\nu \end{pmatrix} \tag{9}$$

Recall that $\sigma^\mu = (I, \sigma_1, \sigma_2, \sigma_3)$ and $\overline{\sigma}^\mu = (I, -\sigma_1, -\sigma_2, -\sigma_3)$. We want to add $\gamma^\nu \gamma^\mu$ to this to evaluate $\{\gamma^\mu, \gamma^\nu\}$. Let us calculate $\overline{\sigma}^\mu \sigma^\nu + \overline{\sigma}^\nu \sigma^\mu$ (notice that this is not an anticommutator) term by term: for $\mu = \nu = 0$, $2I$; for $\mu = 0$, $\nu = i$, $I\sigma^i + (-\sigma^i)I = 0$; and for $\mu = i$, $\nu = j$, $(-\sigma^i)\sigma^j + (-\sigma^j)\sigma^i = -2\delta^{ij}I$. Thus,

$$\overline{\sigma}^\mu \sigma^\nu + \overline{\sigma}^\nu \sigma^\mu = 2\eta^{\mu\nu}I \tag{10}$$

with $\eta^{\mu\nu} = \text{diag}(1, -1, -1, -1)$, that is, the Minkowski metric introduced in chapter VII.2. Similarly, you could check that

$$\sigma^\mu \overline{\sigma}^\nu + \sigma^\nu \overline{\sigma}^\mu = 2\eta^{\mu\nu}I \tag{11}$$

Hence

$$\{\gamma^\mu, \gamma^\nu\} = 2\eta^{\mu\nu} \tag{12}$$

as expected. Recall that the Clifford algebra (12) merely says that each γ^μ squares to* either I or $-I$, and that the γ^μs anticommute with one another.

It follows that for any two 4-vectors p and q, we have

$$(p_\mu \gamma^\mu)(q_\nu \gamma^\nu) = \frac{1}{2}\{\gamma^\mu, \gamma^\nu\}p_\mu q_\nu = \eta^{\mu\nu}p_\mu q_\nu = p \cdot q = p^0 q^0 - \vec{p} \cdot \vec{q} \tag{13}$$

For convenience, introduce the Feynman slash notation, which is now standard.[3] For any 4-vector a_μ,

$$\not{a} \equiv \gamma^\mu a_\mu = a_0 \gamma^0 + a_i \gamma^i = a^0 \gamma^0 - a^i \gamma^i = a^0 \gamma^0 - \vec{a} \cdot \vec{\gamma} \tag{14}$$

Then (13) says $\not{p}\not{q} = p \cdot q$, in particular $\not{p}\not{p} = p \cdot p = p^2$. The Dirac equation (8) may be written as

$$(i\not{\partial} - m)\psi = 0 \tag{15}$$

or $(\not{p} - m)\psi = 0$ in momentum space.

---

* Note that, throughout this book, the identity matrix I is often suppressed, for example, in (7), (8), and (12).

Acting with $(\not{p} + m)$ on the Dirac equation, we obtain $(\not{p} + m)(\not{p} - m)\psi = (\not{p}\not{p} - m^2)\psi = (p^2 - m^2)\psi = 0$, using (13). In other words, $\psi(p)$ vanishes unless $p^2 = E^2 - \vec{p}^2 = m^2$, but this is precisely what we deduced in (4). We have merely packaged the previous information more compactly.

For calculations, it is more efficient to write the γ^μ in (6) using the direct-product notation,[4] as we learned in chapter VII.1:

$$\gamma^0 = \begin{pmatrix} 0 & I \\ I & 0 \end{pmatrix} = I \otimes \tau_1, \qquad \gamma^i = \begin{pmatrix} 0 & -\sigma_i \\ \sigma_i & 0 \end{pmatrix} = -\sigma_i \otimes i\tau_2 \tag{16}$$

We will see presently that we are free to change basis. The basis used here is known as the Weyl basis.

Different bases for different folks

If we use the gamma matrices γ^μ, somebody else is certainly free to use $\tilde{\gamma}^\mu = W\gamma^\mu W^{-1}$ instead, with W any 4-by-4 matrix with an inverse. Evidently, $\tilde{\gamma}^\mu$ also satisfies the Clifford algebra. (Verify this.) If ψ satisfies the Dirac equation $(i\not{\partial} - m)\psi = 0$, then $\tilde{\psi} = W\psi$ satisfies $(i\tilde{\gamma}^\mu \partial_\mu - m)\tilde{\psi} = (iW\not{\partial}W^{-1} - m)W\psi = W(i\not{\partial} - m)\psi = 0$.

This freedom of choosing the γ matrices corresponds to a simple change of basis; going from ψ to $\tilde{\psi}$, we scramble the four components of ψ.

Hermitean conjugating the equation $\tilde{\gamma}^\mu = W\gamma^\mu W^{-1}$ and inserting the fact that γ^0 is hermitean while γ^i is antihermitean (see appendix 1), we deduce that $W^{-1} = W^\dagger$, that is, W has to be unitary.

An alternative derivation of the Dirac equation: Too many degrees of freedom

A profound truth should be honored with more than one derivation. We have derived the Dirac equation by detouring through the Weyl equation. Suppose we didn't do that and proceeded directly to the representation $(\frac{1}{2}, 0) \oplus (0, \frac{1}{2})$ as mandated by parity.

Parity forces us to have a 4-component spinor ψ, but we know for a fact that a spin $\frac{1}{2}$ particle has only two physical degrees of freedom. What are we to do? We have too many degrees of freedom.

Clearly, we have to set two of the components of ψ to 0, but this has to be done in such a way as to respect Lorentz invariance.

Let us go to the rest frame, in which the electron has momentum $p_r \equiv (m, \vec{0})$, with m the electron mass. We are to project out two of the four components contained in $\psi(p_r)$, that is, set $P\psi(p_r) = 0$ with P a 4-by-4 matrix satisfying $P^2 = P$.

Parity suggests that we treat u and v on the same footing, since $u \leftrightarrow v$ under parity. With the benefit of hindsight, let us set $u = v$ in the rest frame. In other words, we choose[5] the projection operator to be $P = \frac{1}{2} \begin{pmatrix} I & -I \\ -I & I \end{pmatrix}$, so that $P\psi(p_r) = 0$ corresponds to $u - v = 0$.

Next, write $P = \frac{1}{2}(I - \gamma^0)$, where γ^0 is, at this point, merely a particularly bizarre notation for some 4-by-4 matrix.* (Remember, we are deriving the Dirac equation from scratch once again; we have never heard of the gamma matrices, which are waiting to be discovered, ha ha.) From the form above, we have

$$\gamma^0 = \begin{pmatrix} 0 & I \\ I & 0 \end{pmatrix} \tag{17}$$

with I the 2-by-2 identity matrix.[6]

Write the projection to 2 degrees of freedom, $P\psi(p_r) = 0$, as

$$(\gamma^0 - I)\psi(p_r) = 0 \tag{18}$$

The Dirac equation in disguise

We have derived the Dirac equation, but heavily disguised!

The wave function $\psi(p_r)$ in the rest frame satisfies (18). Since our derivation is based on a step-by-step study of the spinor representation of the Lorentz group, we know how to obtain the equation satisfied by $\psi(p)$ for any p: we simply boost, following the strategy described in chapter VII.3.

With no loss of generality, let us boost in the z direction. With boost "angle" φ, the electron acquires energy $E = m \cosh \varphi$ and 3-momentum $p = m \sinh \varphi$, with m the electron mass (of course). Note for use below that $me^{\varphi\sigma_3} = m(\cosh \varphi + \sigma_3 \sinh \varphi) = E + \sigma_3 p$.

Group theory tells us that, to obtain the wave function $\psi(p)$ satisfied by the moving particle, we simply apply $e^{i\varphi K}$ to $\psi(p_r)$ (with K short for K_z). So act with $e^{i\varphi K}$ on (18) conveniently multiplied by m:

$$e^{i\varphi K} m(\gamma^0 - I)\psi(p_r) = m(e^{i\varphi K}\gamma^0 e^{-i\varphi K} - I)\psi(p) = 0 \tag{19}$$

But we know from chapter VII.3 that $e^{i\varphi K} = \begin{pmatrix} e^{\frac{1}{2}\varphi\sigma_3} & 0 \\ 0 & e^{-\frac{1}{2}\varphi\sigma_3} \end{pmatrix}$, and thus

$$me^{i\varphi K}\gamma^0 e^{-i\varphi K} = \begin{pmatrix} e^{\frac{1}{2}\varphi\sigma_3} & 0 \\ 0 & e^{-\frac{1}{2}\varphi\sigma_3} \end{pmatrix} \begin{pmatrix} 0 & I \\ I & 0 \end{pmatrix} \begin{pmatrix} e^{-\frac{1}{2}\varphi\sigma_3} & 0 \\ 0 & e^{\frac{1}{2}\varphi\sigma_3} \end{pmatrix}$$

$$= m \begin{pmatrix} 0 & e^{\varphi\sigma_3} \\ e^{-\varphi\sigma_3} & 0 \end{pmatrix} = \begin{pmatrix} 0 & E + \sigma_3 p \\ E - \sigma_3 p & 0 \end{pmatrix} \tag{20}$$

Note that the factors of $\frac{1}{2}$ have disappeared. Thus (19) becomes

$$\begin{pmatrix} -m & E + \vec{\sigma} \cdot \vec{p} \\ E - \vec{\sigma} \cdot \vec{p} & -m \end{pmatrix} \begin{pmatrix} u \\ v \end{pmatrix} = 0 \tag{21}$$

---

* Intentionally, we have chosen P so that the γ^0 here agrees with the γ^0 in (6).

Referring to (17), we see that the matrix in (21) can be written as $\gamma^0 p^0 - \gamma^i p^i - m$ if we define

$$\gamma^i = \begin{pmatrix} 0 & -\sigma_i \\ \sigma_i & 0 \end{pmatrix} \tag{22}$$

We can thus write (21) as $(\gamma^\mu p_\mu - m)\psi(p) = 0$. In position space, we have the Dirac equation $(i\gamma^\mu \partial_\mu - m)\psi(x) = 0$.

Projection boosted into an arbitrary frame

We recognize the two coupled equations $(E - \vec{\sigma} \cdot \vec{p})u = mv$ and $(E + \vec{\sigma} \cdot \vec{p})v = mu$ contained in (21) as precisely the two coupled equations (2) and (3) obtained earlier. The two roads of our derivations have come together, so to speak.

The derivation here represents a deep group theoretic way of looking at the Dirac equation: it is "merely" a projection boosted into an arbitrary frame.

This exemplifies the power of symmetry and group theory that pervades modern physics: our knowledge of how the electron field transforms under the rotation group (namely, that it has spin $\frac{1}{2}$) allows us to know how it transforms under the Lorentz group. Symmetry rules!

With group theory, we don't have to be as brilliant as Dirac, not even a small fraction really, to find the Dirac equation. Once we solemnly intone $(\frac{1}{2}, 0) \oplus (0, \frac{1}{2})$, group theory leads us by the nose to the promised land.

From Dirac back to Weyl

Suppose that we are given the reducible representation $(\frac{1}{2}, 0) \oplus (0, \frac{1}{2})$. How do we extract the two irreducible representations $(\frac{1}{2}, 0)$ and $(0, \frac{1}{2})$? In other words, we want to separate out the two Weyl spinors contained in the Dirac spinor.

Define* $\gamma_5 \equiv i\gamma^0\gamma^1\gamma^2\gamma^3$. This product of gamma matrices is so important that it has its own name! (The peculiar name comes about because in some old-fashioned notation, the time coordinate was called x^4 with a corresponding γ^4.) In the Weyl basis used here,

$$\gamma_5 \equiv i\gamma^0\gamma^1\gamma^2\gamma^3 = i(I \otimes \tau_1)(\sigma^1\sigma^2\sigma^3 \otimes (-i)^3\tau_2) = I \otimes \tau_3 = \begin{pmatrix} I & 0 \\ 0 & -I \end{pmatrix} \tag{23}$$

is diagonal. Verify that γ_5 anticommutes with γ^μ.

Since $(\gamma_5)^2 = 1$, we can form two projection operators, $P_L \equiv \frac{1}{2}(1 - \gamma_5)$ and $P_R \equiv \frac{1}{2}(1 + \gamma_5)$ satisfying $P_L^2 = P_L$, $P_R^2 = P_R$, and $P_L P_R = 0$. Note that we have already introduced the

---

* Incidentally, we do not distinguish between γ_5 and γ^5.

analogs* of γ_5 and of these two projections in chapter VII.1, not surprisingly, since many properties of the $SO(N)$ spinors can be analytically continued.

We have now solved the problem just posed. Given a 4-component spinor ψ, the two combinations $\psi_L = \frac{1}{2}(1 - \gamma_5)\psi$ and $\psi_R = \frac{1}{2}(1 + \gamma_5)\psi$ correspond to precisely the two Weyl spinors contained in ψ. Note that $\psi = \psi_L + \psi_R$ (of course). With the convention in (23), $\psi_R = u$ and $\psi_L = v$.

We can readily recover what we learned about the Weyl equation in chapter VII.3. Acting with P_R on the Dirac equation $(\not{p} - m)\psi = 0$ from the left and noting that γ_5 anticommutes with γ^μ (as you have just shown), we obtain

$$\not{p}\psi_L = m\psi_R \tag{24}$$

Acting with P_L, you would have obtained $\not{p}\psi_R = m\psi_L$. Once again, we see that the mass m connects the left and the right.

Now imagine a particle (let's say the neutrino for definiteness) described by a single Weyl field, not two Weyl fields like the electron, that is, a particle with ψ_L but not ψ_R. In other words, $\psi_R = 0$. Then ψ_L satisfies the elegant Weyl equation now written in 4-component form:

$$\not{p}\psi_L = 0 \tag{25}$$

Multiplying the Weyl equation by $\not{p}$ from the left, we obtain $\not{p}\not{p} = p^2 = 0$. Thus, not only does a particle described by the Weyl equation violate parity, but it must also be massless.

One physicist told me that as a student he was confused by a triviality. He saw the Weyl spinor written in the Dirac notation as $\psi_L = \frac{1}{2}(1 - \gamma_5)\psi$, an apparently 4-component object, but then referred to as a 2-component spinor. I trust that the reader understands clearly that projecting two components out of a 4-component Dirac spinor would leave a 2-component spinor. I hope that this book—which introduces the Weyl spinor first as a 2-component spinor, then forms the Dirac spinor by stacking two Weyl spinors together, and finally extracts the left and right handed spinor from the 4-component Dirac spinor—avoids what this physicist call "notational hiccups."

Dirac's brilliant guess

Now that we have gone through the group theoretic derivation of the Dirac equation, we mention Dirac's original path[7] to "his[8] equation." He started with the Klein-Gordon equation $(\partial^2 + m^2)\psi = 0$ (which was actually known before the Schrödinger equation, since it merely states the Einstein relation $p^2 = m^2$ in position space). For misguided reasons now no longer relevant, Dirac looked for an equation with only one power of spacetime derivative[9] rather than the two powers in the Klein-Gordon equation, and wrote down $(c^\mu \partial_\mu - b)\psi = 0$. Note that in the group theoretic treatment, the fact that the Dirac equation contains one, rather than two, power of spacetime derivative was forced on us.

---

* Some readers might have guessed that γ_F in chapter VII.1 is actually short for γ_{Five}.

If c^μ were four numbers, then c would define a 4-vector and thus pick out a privileged direction, which immediately implies that Lorentz invariance is broken. A lesser physicist might have given up immediately at this point. But not Dirac. He multiplied this equation from the left by $(c^\mu \partial_\mu + b)$, thus obtaining $(\frac{1}{2}\{c^\mu, c^\nu\}\partial_\mu\partial_\nu - b^2)\psi = 0$. He then noticed that, if $\{c^\mu, c^\nu\} = -2\eta^{\mu\nu}$ and $b = m$, he would have obtained the Klein-Gordon equation.

Dirac realized that to satisfy the required anticommutation relation the c^μs would have to be matrices rather than numbers. By trial and error, he found that the smallest matrices that worked would have to be 4-by-4. Consequently, ψ must have four components. Write $c^\mu = i\gamma^\mu$, with γ^μ four 4-by-4 matrices satisfying the algebra $\{\gamma^\mu, \gamma^\nu\} = 2\eta^{\mu\nu}$.

Lorentz transformation of the Dirac spinor

This lightning-bolt-out-of-nowhere derivation segues nicely into the following question. How could the Dirac equation $(i\gamma^\mu\partial_\mu - m)\psi = 0$ possibly be Lorentz invariant, while the equation $(c^\mu\partial_\mu - m)\psi = 0$, with c^μ being four numbers, is clearly not? The magic of matrices, as we will see presently.

The group theoretic derivation of the Dirac equation, in contrast, leaves no doubt that it is Lorentz invariant. Indeed, the moment we mutter "$(\frac{1}{2}, 0) \oplus (0, \frac{1}{2})$", we fix completely how the 4-component Dirac spinor ψ transforms. Lorentz invariance is guaranteed by construction. Case closed.

Nevertheless, we (along with the poor man) would like to see, in the context of Dirac's original derivation, the nuts and bolts behind the Lorentz invariance. Let us state the question more formally. Under a Lorentz transformation $x^\mu \to x'^\mu = L^\mu_{\ \nu}x^\nu$, how do the four components of ψ_α transform, and how does the Dirac equation respond under this transformation?

Let us ask Confusio not to confuse the Lorentz index $\mu = 0, 1, 2, 3$ with the spinor or Dirac index $\alpha = 1, 2, 3, 4$. He says that he is not confused, even though both indices take on four different values.

So, let $\psi(x) \to \psi'(x') \equiv S(L)\psi(x)$ and try to determine the 4-by-4 matrix $S(L)$.

A quick review of what we learned in chapter VII.2: following Lie, write $L = e^{\frac{i}{2}\omega_{\mu\nu}J^{\mu\nu}} \simeq (I + \frac{i}{2}\omega_{\mu\nu}J^{\mu\nu})$ near the identity, with the six $(= 4 \cdot 3/2)$ 4-by-4 matrices $J^{\mu\nu} = -J^{\nu\mu}$ generating the three rotations and the three boosts. (Here the indices μ and ν are to be summed over.) The six parameters $\omega_{\mu\nu} = -\omega_{\nu\mu}$ correspond to the three rotation angles $\vec{\theta}$ and the three boost angles $\vec{\varphi}$.

Let $S(L) \simeq (I - \frac{i}{4}\omega_{\mu\nu}\sigma^{\mu\nu})$ under this infinitesimal transformation, with the six 4-by-4 matrices $\sigma^{\mu\nu}$ to be determined. In other words, the issue is how the generators $J^{\mu\nu}$ are represented in the spinor representation. How hard can this be? After all, there are only 16 linearly independent 4-by-4 matrices. We already know six of these matrices: the γ^μs, the identity matrix I, and $\gamma_5 = i\gamma^0\gamma^1\gamma^2\gamma^3$, corresponding to products of k γ matrices, with $k = 1, 0$, and 4, respectively.[10]

What about the products of two or three γ matrices? The products of three γ matrices are in fact equal to $\gamma^\mu\gamma^5$. (For example, $i\gamma^1\gamma^2\gamma^3 = \gamma^0(i\gamma^0\gamma^1\gamma^2\gamma^3) = \gamma^0\gamma_5$.) Next, thanks to the Clifford algebra (12), products of two γ matrices reduce to $\gamma^\mu \gamma^\nu = \frac{1}{2}\{\gamma^\mu, \gamma^\nu\} + \frac{1}{2}[\gamma^\mu, \gamma^\nu] = \eta^{\mu\nu} - i\sigma^{\mu\nu}$, where we define

$$\sigma^{\mu\nu} \equiv \frac{i}{2}[\gamma^\mu, \gamma^\nu] \tag{26}$$

There are $4 \cdot 3/2 = 6$ of these $\sigma^{\mu\nu}$ matrices.

Count them, we got all 16.* The set of $(1 + 4 + 6 + 4 + 1) = 16$ matrices $\{1, \gamma^\mu, \sigma^{\mu\nu}, \gamma^\mu \gamma_5, \gamma_5\}$ form a complete basis of the space of all 4-by-4 matrices; that is, any 4-by-4 matrix can be written as a linear combination of these 16 matrices.

Group theoretically, the question posed earlier is how the generators $J^{\mu\nu}$ are represented on $(\frac{1}{2}, 0) \oplus (0, \frac{1}{2})$. Given the preceding discussion and the fact that there exist six matrices $\sigma^{\mu\nu}$, we suspect that up to an overall numerical factor, the $\sigma^{\mu\nu}$s must represent the six generators $J^{\mu\nu}$ of the Lorentz group acting on a spinor. In fact, since I already knew the answer, I even wrote $S(L) \simeq (I - \frac{i}{4}\omega_{\mu\nu}\sigma^{\mu\nu})$.

To confirm our suspicion and to fix the numerical factor, it is easiest to work out what a rotation $e^{\frac{i}{2}\omega_{ij}J^{ij}}$ does to the Dirac spinor. Choose $\gamma^\mu = \begin{pmatrix} 0 & \bar{\sigma}^\mu \\ \sigma^\mu & 0 \end{pmatrix}$ as in (6), namely, the Weyl basis. (I remind you that $\sigma^\mu = (I, \vec{\sigma})$ and $\bar{\sigma}^\mu = (I, -\vec{\sigma})$.) Then, from (9), we have $\sigma^{\mu\nu} = \frac{i}{2}\begin{pmatrix} \bar{\sigma}^\mu \sigma^\nu - \bar{\sigma}^\nu \sigma^\mu & 0 \\ 0 & \sigma^\mu \bar{\sigma}^\nu - \sigma^\nu \bar{\sigma}^\mu \end{pmatrix}$, that is,

$$\sigma^{0i} = i\begin{pmatrix} \sigma_i & 0 \\ 0 & -\sigma_i \end{pmatrix} \qquad \sigma^{ij} = \varepsilon^{ijk}\begin{pmatrix} \sigma_k & 0 \\ 0 & \sigma_k \end{pmatrix} \tag{27}$$

We see that σ^{ij} are just the Pauli matrices doubly stacked. Thus, for a rotation around the z-axis, $e^{-\frac{i}{2}\omega_{ij}\frac{1}{2}\sigma^{ij}} = e^{-i\omega_{12}\frac{\sigma_3}{2}}$, with ω_{12} the rotation angle. This corresponds to exactly how a spin $\frac{1}{2}$ particle transforms.

We have figured out that a Lorentz transformation L acting on ψ is represented by $S(L) = e^{-\frac{i}{4}\omega_{\mu\nu}\sigma^{\mu\nu}}$.

It is instructive to verify that if $\psi(x)$ satisfies the Dirac equation $(i\gamma^\mu \partial_\mu - m)\psi = 0$, then $\psi'(x') \equiv S(L)\psi(x)$ satisfies the Dirac equation $(i\gamma^\mu \partial'_\mu - m)\psi'(x') = 0$ in the primed frame, where $\partial'_\mu \equiv \frac{\partial}{\partial x'^\mu}$. To show this, note that[†] $[\sigma^{\mu\nu}, \gamma^\lambda] = 2i(\gamma^\mu \eta^{\nu\lambda} - \gamma^\nu \eta^{\mu\lambda})$, and hence for ω infinitesimal, $S\gamma^\lambda S^{-1} = \gamma^\lambda - \frac{i}{4}\omega_{\mu\nu}[\sigma^{\mu\nu}, \gamma^\lambda] = \gamma^\lambda - \omega^\lambda_{\ \mu}\gamma^\mu$. Building up a finite Lorentz transformation by compounding infinitesimal transformations, as instructed by Lie, we have $S\gamma^\lambda S^{-1} = L^\lambda_{\ \mu}\gamma^\mu$. Math works, of course.

The Lorentz generators and the adjoint representation

Everything is coming together. Now that we understand the representation $(\frac{1}{2}, 0) \oplus (0, \frac{1}{2})$ used in the Dirac equation, we might ask about the pair of 3-dimensional representations

---

* If the reader is worried about factors of i, go through the same counting for $SO(4)$, where all these matrices are hermitean thanks to Euclid.

† In fact, the astute reader would have realized that we already did a Euclidean version of this calculation in chapter VII.1.

(1, 0) and (0, 1) in the list of irreducible representations. We instantly understand that parity maps one into the other, and together, we have a 6-dimensional reducible representation (1, 0) ⊕ (0, 1) of the Lorentz group. What could it be?

Or, I could ask you another question. We know that every Lie group has an adjoint representation furnished by the generators of the Lie algebra. Well, the Lorentz group $SO(3, 1)$ has six generators, $\vec{J}$ and $i\vec{K}$. How do they transform?

You answered right. Yes, (1, 0) ⊕ (0, 1) is the adjoint representation. The reducibility of (1, 0) ⊕ (0, 1) reflects the splitting of the set J_i, K_i into $J_{+,i}$, $J_{-,i}$, corresponding to (1, 0) and (0, 1), respectively.

Referring back to the discussion in chapter VII.3, we see that acting on the 4-component spinor ψ, we have the generators of rotation

$$\vec{J} = \begin{pmatrix} \frac{1}{2}\vec{\sigma} & 0 \\ 0 & \frac{1}{2}\vec{\sigma} \end{pmatrix} \tag{28}$$

and the generators of boost

$$i\vec{K} = \begin{pmatrix} \frac{1}{2}\vec{\sigma} & 0 \\ 0 & -\frac{1}{2}\vec{\sigma} \end{pmatrix} \tag{29}$$

As in the preceding chapter, the equality means "represented by." Note once again the all-important minus sign. By the way, we can now see, almost instantly, the commutation relations (VII.2.23)–(VII.2.25): $[J, J] \sim J$, $[J, K] \sim K$, $[K, K] \sim -J$ (as realized on the Dirac spinor).

The Dirac Lagrangian

Go back to the Lagrangian $\mathcal{L} = iu^\dagger(\frac{\partial}{\partial t} + \vec{\sigma} \cdot \vec{\nabla})u + iv^\dagger(\frac{\partial}{\partial t} - \vec{\sigma} \cdot \vec{\nabla})v - m(u^\dagger v + v^\dagger u)$ in (1). Let us now write it in terms of the Dirac spinor ψ.

From (17), we have $(u^\dagger v + v^\dagger u) = \psi^\dagger \gamma^0 \psi$, which we can write as $\bar{\psi}\psi$ if we define

$$\bar{\psi} \equiv \psi^\dagger \gamma^0 \tag{30}$$

The two time derivative terms then come together as $u^\dagger \frac{\partial}{\partial t}u + v^\dagger \frac{\partial}{\partial t}v = \psi^\dagger \frac{\partial}{\partial t}\psi = \bar{\psi}\gamma^0 \frac{\partial}{\partial t}\psi$ $= \bar{\psi}\gamma^0 \partial_0 \psi$. I leave it to you to verify that the two spatial derivative terms combine into $\bar{\psi}\gamma^i \partial_i \psi$. Writing $\displaystyle{\not{\partial}} = \gamma^\mu \partial_\mu$, we obtain the Dirac Lagrangian in 4-component form:

$$\mathcal{L} = \bar{\psi}(i{\not{\partial}} - m)\psi \tag{31}$$

In nonrelativistic quantum mechanics, you are used to writing $\psi^\dagger \psi$. In relativistic physics you have to get used to writing $\bar{\psi}\psi$. Show that $\bar{\psi}\psi$, but not $\psi^\dagger \psi$, transforms like a Lorentz scalar. Hint: Show that $(\sigma^{\mu\nu})^\dagger = \gamma^0 \sigma^{\mu\nu}\gamma^0$, and hence, $S(L)^\dagger = \gamma^0 e^{\frac{i}{4}\omega_{\mu\nu}\sigma^{\mu\nu}}\gamma^0$. (Incidentally, this shows clearly that S is not unitary, which we knew, since σ_{0i} is not hermitean.)

Dirac bilinears

Denote the 16 linearly independent 4-by-4 matrices[11] $\{I, \gamma^\mu, \sigma^{\mu\nu}, \gamma^\mu\gamma_5, \gamma_5\}$ generically by Γ. The 16 objects $\bar{\psi}\Gamma\psi$, known as Dirac bilinears, were tremendously important in the development of particle physics.[12] In the exercises, you will work out how they transform under Lorentz transformations and under spatial reflection. In fact, the notation already suggests how they transform. Our discussion of the Dirac Lagrangian in (31) indicates that $\bar{\psi}\psi\,(=\bar{\psi}I\psi)$ must transform like a Lorentz scalar, since it appears as the mass term. Similarly, $\bar{\psi}\gamma^\mu\psi$ must transform like a 4-vector, since, when contracted with ∂_μ, it appears as the "kinetic energy" term $\bar{\psi}\,\partial\!\!\!/\,\psi$.

You will not be surprised, then, that $\bar{\psi}\sigma^{\mu\nu}\psi$ transforms like an antisymmetric tensor. From exercises 4 and 7, you will learn that $\bar{\psi}\gamma^\mu\gamma_5\psi$ and $\bar{\psi}\gamma_5\psi$ transform like an axial vector and a pseudoscalar, respectively, that is, quantities odd under spatial reflection.

The various Dirac bilinears transform according to how they look like they should transform, as indicated by the Lorentz indices they carry.

Group theoretically, the existence of the 16 Dirac bilinears corresponds to the Clebsch-Gordon decomposition

$$
\left(\left(\tfrac{1}{2},0\right)\oplus\left(0,\tfrac{1}{2}\right)\right)\otimes\left(\left(\tfrac{1}{2},0\right)\oplus\left(0,\tfrac{1}{2}\right)\right)
$$

$$
=\left(\tfrac{1}{2}\otimes\tfrac{1}{2},0\otimes 0\right)\oplus\left(\tfrac{1}{2}\otimes 0,0\otimes\tfrac{1}{2}\right)\oplus\left(0\otimes\tfrac{1}{2},\tfrac{1}{2}\otimes 0\right)\oplus\left(0\otimes 0,\tfrac{1}{2}\otimes\tfrac{1}{2}\right)
$$

$$
=(0\oplus 1,0)\oplus\left(\tfrac{1}{2},\tfrac{1}{2}\right)\oplus\left(\tfrac{1}{2},\tfrac{1}{2}\right)\oplus(0,0\oplus 1)
$$

$$
=(0,0)\oplus\left(\tfrac{1}{2},\tfrac{1}{2}\right)\oplus(1,0)\oplus(0,1)\oplus\left(\tfrac{1}{2},\tfrac{1}{2}\right)\oplus(0,0) \tag{32}
$$

As always, count to make sure that we did not lose anybody: $4\cdot 4=16=1+4+6+4+1$. While the 4s on the left hand side are actually $2+2$, the 4s on the right hand side are in fact 4. (Got that?) The 6 is $3+3$.

Three for the price of one! Here we have gone through two different derivations of the Dirac equation, plus Dirac's original derivation. The take home message is that group theory rules.

Appendix 1: The Dirac Hamiltonian

The Dirac equation (8) can be cast in the Hamiltonian form $i\frac{\partial\psi}{\partial t}=H\psi$ given in chapter III.1 by moving all terms other than the time derivative term to the right hand side and multiplying by γ^0:

$$
i\partial_0\psi=(-i\gamma^0\gamma^i\partial_i+\gamma^0 m)\psi \tag{33}
$$

Indeed, that was how Dirac originally wrote[13] his equation. Thus, we identify the Dirac Hamiltonian as $H=-i\gamma^0\gamma^i\partial_i+\gamma^0 m$.

The requirement that the Hamiltonian is hermitean implies that $(\gamma^0)^\dagger = \gamma^0$ and $(\gamma^0 \gamma^i)^\dagger = \gamma^0 \gamma^i$. The latter leads to $(\gamma^i)^\dagger = -\gamma^i$, since γ^0 and γ^i anticommute. Thus, we deduce that γ^0 is hermitean while γ^i is anti-hermitean, a fact conveniently written as $(\gamma^\mu)^\dagger = \gamma^0 \gamma^\mu \gamma^0$. By inspection, we can verify explicitly that the γ matrices in the Weyl basis in (16) obey these hermiticity requirements, as they must.

Appendix 2: Slow and fast electrons

We learned that we are free to choose the γ matrices up to a similarity transformation. Physics determines which choice is the most convenient. For example, suppose we want to study a slowly moving electron, as would be the case in atomic physics.*

In the discussion leading up to (7) and (18), we chose $\gamma^0 = I \otimes \tau_1$, as in (6). Let us now choose

$$\gamma^0 = \begin{pmatrix} I & 0 \\ 0 & -I \end{pmatrix} = I \otimes \tau_3 \tag{34}$$

instead of (17). We keep the γ^i's the same as in (22): $\gamma^i = \begin{pmatrix} 0 & -\sigma^i \\ \sigma^i & 0 \end{pmatrix}$. This is known as the Dirac basis. Verify that the Clifford algebra (12) is satisfied.

Since in the Dirac basis γ^0 is diagonal by design, the projection $(\gamma^0 - I)\psi = 0$ in the rest frame is trivially solved. Write the 4-component spinor as $\psi = \begin{pmatrix} u \\ v \end{pmatrix}$. Inserting (34), we obtain $v(p_r) = 0$, that is, v vanishes for an electron at rest.

For a slowly moving electron, we thus expect $v(p)$ to be much smaller than $u(p)$. We are invited to develop an approximation scheme treating $v \ll u$.

In the Dirac basis, the Dirac equation works out to be

$$(\not{p} - m)\psi = \begin{pmatrix} E - m & -\vec{\sigma} \cdot \vec{p} \\ \vec{\sigma} \cdot \vec{p} & -E - m \end{pmatrix} \begin{pmatrix} u \\ v \end{pmatrix} = 0 \tag{35}$$

(compare with the Dirac equation (21) in the Weyl basis). This equation implies

$$v = \frac{\vec{\sigma} \cdot \vec{p}}{E + m} u \simeq \frac{\vec{\sigma} \cdot \vec{p}}{2m} u \tag{36}$$

since in the nonrelativistic limit, the kinetic energy K in $E = m + K$ is much less than the rest energy m. This expression for v then allows us to calculate the leading relativistic corrections to various physical quantities.[14]

Notice that in going from the Weyl basis to the Dirac basis, γ^0 and γ_5 trade places (up to a sign). As already mentioned, physics dictates which basis to use: we prefer to have γ^0 diagonal when we deal with slowly moving spin $\frac{1}{2}$ particles, while we prefer to have γ_5 diagonal when we deal with fast moving spin $\frac{1}{2}$ particles.

Appendix 3: An alternative formalism for the Dirac Lagrangian

Let us start with the Lagrangian in (1). Write $v = i\sigma_2 w^*$. As explained in chapter VII.3, this means that the Weyl spinor w transforms like $(\frac{1}{2}, 0)$. Hermitean conjugating, we obtain $v^\dagger = w^T(-i\sigma_2)$. Thus,

$$v^\dagger \bar{\sigma}^\mu \partial_\mu v = w^T \sigma_2 \bar{\sigma}^\mu \sigma_2 \partial_\mu w^* \text{ “} = \text{” } - (\partial_\mu w^T) \sigma_2 \bar{\sigma}^\mu \sigma_2 w^*$$
$$= (-)^2 w^\dagger (\sigma_2 \bar{\sigma}^\mu \sigma_2)^T \partial_\mu w = w^\dagger \sigma^\mu \partial_\mu w \tag{37}$$

* For a treatment of the relativistic hydrogen-like atoms, see J. J. Sakurai and J. Napolitano, Modern Quantum Mechanics, p. 506.

The effective equality results because in the action, the Lagrangian is integrated over spacetime, and thus (disregarding possible surface terms) we are allowed to integrate by parts. In the third equality, the extra minus sign is because the spinor w is to be thought of as a Grassman number, as mentioned in chapter VII.3. Finally, the fourth equality follows from

$$(\sigma_2 \bar{\sigma}^\mu \sigma_2)^T = \sigma_2 \bar{\sigma}^{\mu T} \sigma_2 = \sigma_2(I, -\sigma_1, \sigma_2, -\sigma_3)\sigma_2 = \sigma^\mu \tag{38}$$

With two Weyl spinors transforming like $(\frac{1}{2}, 0)$, we are invited to write W_A with $A = 1, 2$ such that $W_1 = u$ and $W_2 = w$. The Lagrangian allowed by Lorentz invariance is then

$$\mathcal{L} = iW^\dagger \sigma^\mu \partial_\mu W - \frac{1}{2}(m_{AB} W_A^T i\sigma_2 W_B + h.c.) \tag{39}$$

where $h.c.$ stands for hermitean conjugation. Without the mass term, the Lagrangian enjoys a $U(2)$ symmetry acting on the index A, which (depending on the form of m_{AB}) may be broken down to $U(1)$.

Exercises

1   In chapter VII.2 you worked out how the six components of the electric and magnetic fields $\vec{E}$ and $\vec{B}$ transform under the Lorentz group. In fact, the electromagnetic field transforms as $(1, 0) \oplus (0, 1)$. Show that it is parity that once again forces us to use a reducible representation.

2   Show that we don't have to list products of k γ matrices with $k \geq 5$.

3   Find the W that takes the Weyl basis into the Dirac basis, and verify that it is unitary.

4   Work out how the Dirac bilinears transform under the Lorentz group.

5   Show that the bilinears in the preceding exercise are all hermitean.

6   Consider space reflection or parity: $x^\mu \to x'^\mu = (x^0, -\vec{x})$. Show that $\psi'(x') \equiv \gamma^0 \psi(x)$ satisfies the Dirac equation in the space-reflected world.

7   Under a Lorentz transformation, $\bar{\psi}(x)\gamma^5\psi(x)$ and $\bar{\psi}(x)\psi(x)$ transform in the same way. Show that under space reflection, they transform oppositely; in other words, $\bar{\psi}(x)\gamma^5\psi(x)$ transforms like a pseudoscalar.

8   Express $v^\dagger u$ in terms of v and w.

9   Show that for $m_{AB} = m\delta_{AB}$, the Lagrangian in (39) exhibits an $SO(2) \simeq U(1)$ symmetry. Rewrite the Lagrangian in terms of $\psi_\pm = (W_1 + iW_2)/\sqrt{2}$.

Notes

1. In light of the cautionary note about terminology given in chapter VII.3, some authors would say that we need four Weyl spinors.
2. Evidently then, u and v denote 2-component spinors. Hence they are not to be confused with the 4-component spinors u and v that appear in the solution of the Dirac equation.
3. For a story about the Feynman slash notation, see QFT Nut, p. 105.
4. We have changed the notation from chapter VII.1 slightly, using σ to denote one of the two sets of Pauli matrices to emphasize that spin is involved.
5. Different choices of P correspond to different choices of basis for the γ matrices, as was just explained.

6. Alternatively, we could impose the requirement that P is a projection, which implies that $(I - \gamma^0)^2 = 2(I - \gamma^0)$, and hence $(\gamma^0)^2 = I$. The matrix γ^0 squares to the identity matrix; its eigenvalues can only be ± 1.

7. According to a possibly apocryphal story, Dirac was staring into a fire when the equation occurred to him.

8. According to another likely apocryphal legend, when the young Feynman introduced himself to Dirac, the latter said quietly, after a long silence, "I have an equation, do you?"

9. This would be somewhat akin to a confused beginning student writing down $(\frac{\partial}{\partial x} + \frac{\partial}{\partial y} + \frac{\partial}{\partial z})\phi = 0$ in an exam rather than Laplace's equation. The professor would not be amused.

10. I have intentionally changed the superscript on γ^5 to γ_5 just to show that it doesn't matter, to uphold Emerson and to irk the divines.

11. The corresponding 16 matrices for $SO(4)$ generate an $SU(4)$ algebra.

12. Particularly in the theory of the weak interaction, as is detailed in any book on particle physics.

13. With the somewhat antiquated notation $\alpha_i = \gamma^0\gamma^i$ and $\beta = \gamma^0$. This was how the letter γ came into the modern form of the Dirac equation.

14. See, for example, J. J. Sakurai, Invariance Principles and Elementary Particles, Princeton University Press, p. 27.

VII.5 | Dirac and Majorana Spinors: Antimatter and Pseudoreality

We finally come to the shocking prediction of antimatter. Interestingly, the existence of antimatter is related, in the language of group theory, to the existence of pseudoreal representations, which we discussed in chapter II.4.

Here we adopt a more traditional and historical approach, using the 4-component Dirac formalism.

The Majorana mass term has already been mentioned in chapter VII.3. In this chapter, the notion of the charge conjugate field will offer us another approach (which of course does not differ in essence) to the fascinating possibility of a Majorana particle.

Antimatter and pseudoreality

The prediction and discovery of antimatter is surely one of the most astonishing and momentous developments in twentieth-century physics. Here I will show you that antimatter is intimately tied up with the concept of pseudoreal representation.

For the discussion here to proceed, you need to know that a quick way of introducing the electromagnetic field is to replace the spacetime derivative ∂_μ by the so-called covariant derivative $\partial_\mu - ieA_\mu(x)$, a concept already used in nonrelativistic quantum mechanics.[*,1] (Here $A_\mu(x)$ denotes the electromagnetic potential mentioned in chapter VII.2.) If you have never heard of this, you may wish to skip this section.

In the presence of an electromagnetic field, the Dirac equation is thus modified to

$$\left(i\gamma^\mu(\partial_\mu - ieA_\mu) - m \right)\psi = 0 \tag{1}$$

The charge e of the electron measures the strength of the interaction of ψ with A_μ.

---

* You might have seen it in the following form. The Schrödinger equation $i\partial_t\psi = -\frac{1}{2m}\vec{\nabla}^2\psi$ is changed to $i(\partial_t - ieA_t)\psi = -\frac{1}{2m}(\vec{\nabla} - ie\vec{A})^2\psi$, that is, $i\partial_t\psi = (-\frac{1}{2m}(\vec{\nabla} - ie\vec{A})^2 + V)\psi$ if we define the potential $V \equiv -eA_t$. In particular, the case with $\vec{A} = 0$ appears in an early chapter of practically every quantum mechanics textbook.

Our strategy for "finding" antimatter is to show that, given (1), we can find a ψ_c that satisfies the same equation but with the sign of the charge e flipped.

First, complex conjugate (1): $(-i\gamma^{\mu*}(\partial_\mu + ieA_\mu) - m)\psi^* = 0$. This looks like (1) with $e \to -e$, precisely what we want, but with $(-\gamma^{\mu*})$ appearing in place of γ^μ.

Second, complex conjugating the Clifford algebra $\{\gamma^\mu, \gamma^\nu\} = 2\eta^{\mu\nu}$, we see that the four matrices $(-\gamma^{\mu*})$ also satisfy the Clifford algebra. Thus, $(-\gamma^{\mu*})$ must be γ^μ expressed in a different basis. In other words, there exists a matrix $C\gamma^0$ (the notation with an explicit factor of γ^0 is standard; see below) such that

$$-\gamma^{\mu*} = (C\gamma^0)^{-1}\gamma^\mu(C\gamma^0) \tag{2}$$

Third, plugging this into $(-i\gamma^{\mu*}(\partial_\mu + ieA_\mu) - m)\psi^* = 0$ and defining

$$\psi_c \equiv C\gamma^0\psi^* \tag{3}$$

we find

$$\left(i\gamma^\mu(\partial_\mu + ieA_\mu) - m\right)\psi_c = 0 \tag{4}$$

If ψ is the field of the electron, ψ_c is the field of a particle with a charge opposite to that of the electron but with the same mass,[2] namely, the antielectron, now known as the positron.*

So easy[3] in hindsight!

Charge conjugation in the Dirac and Weyl bases

We could continue the discussion in a general basis, but for the sake of definiteness and ease of exposition, let us restrict ourselves to the Dirac basis and the Weyl basis henceforth. To fix our minds, we will keep the explicit form of the γ matrices in front of us.

$$\gamma^0 = I \otimes \tau_1 = \begin{pmatrix} 0 & I \\ I & 0 \end{pmatrix} \quad \text{Weyl,} \quad \text{or} \quad \gamma^0 = I \otimes \tau_3 = \begin{pmatrix} I & 0 \\ 0 & -I \end{pmatrix} \quad \text{Dirac} \tag{5}$$

$$\gamma^i = -\sigma_i \otimes i\tau_2 = \begin{pmatrix} 0 & -\sigma_i \\ \sigma_i & 0 \end{pmatrix} \quad \text{Weyl and Dirac} \tag{6}$$

Also, $\gamma_5 \equiv i\gamma^0\gamma^1\gamma^2\gamma^3$ is given by

$$\gamma_5 = I \otimes \tau_3 = \begin{pmatrix} I & 0 \\ 0 & -I \end{pmatrix} \quad \text{Weyl,} \quad \text{or} \quad \gamma_5 = -I \otimes \tau_1 = -\begin{pmatrix} 0 & I \\ I & 0 \end{pmatrix} \quad \text{Dirac} \tag{7}$$

---

* The physical approach here shows unequivocally that ψ_c describes an antielectron. In a more mathematical treatment, we could write ψ in terms of the Weyl spinors u and v and pass to their conjugates (called $\tilde{u}$ and $\tilde{v}$ in chapter VII.3). But it would require a more sophisticated understanding of electric charge (namely, as determined by transformation under a $U(1)$ gauge group, which we will get to in part IX) to see that they carry charges opposite to those carried by u and v. It may be worthwhile to emphasize that the conjugate field ψ_c is conceptually independent of the existence of the electromagnetic field A_μ.

Note that in going from one basis to the other, we merely have to interchange γ^0 and γ_5, perhaps throwing in a minus sign.

In both bases, γ^2 is the only imaginary gamma matrix. Therefore, the defining equation (2) for C just says

$$(C\gamma^0)\gamma^\mu = -\gamma^\mu(C\gamma^0) \quad \text{for } \mu \neq 2, \quad \text{while} \quad (C\gamma^0)\gamma^2 = \gamma^2(C\gamma^0) \tag{8}$$

We have to find a matrix $C\gamma^0$ that commutes with γ^2 but anticommutes with the other three γ matrices. Well, $C\gamma^0 = \alpha\gamma^2$ (with α some constant) would work; the requirement (derived in chapter VII.4) that the change of basis matrix $C\gamma^0$ has to be unitary gives $|\alpha|^2 = 1$. Let us choose[4] $\alpha = 1$. Then

$$C = \gamma^2\gamma^0 \tag{9}$$

Referring back to (3), we end up with the remarkably simple (and satisfying) relation

$$\psi_c = \gamma^2\psi^* \left(= \gamma^2(\psi^\dagger)^T \right) \tag{10}$$

With the choice (9), $C = -C^*$ is imaginary. Note that γ^0 and γ^2 are both symmetric, and so $C^T = (\gamma^2\gamma^0)^T = \gamma^0\gamma^2 = -C$. Hence C is antisymmetric imaginary and therefore hermitean: $C^\dagger = C$. The unitarity of $C\gamma^0$ implies the unitarity of C: $I = C\gamma^0(C\gamma^0)^\dagger = C\gamma^0\gamma^0 C^\dagger = CC^\dagger$. Hence, $C^2 = I$ (which we can also check directly: $C^2 = \gamma^2\gamma^0\gamma^2\gamma^0 = -\gamma^2\gamma^2\gamma^0\gamma^0 = I$), so that $C^{-1} = C$.

We can summarize all this as

$$C = C^{-1} = -C^* = -C^T = C^\dagger \tag{11}$$

Explicitly,

$$C = -\sigma_2 \otimes \tau_3 = \begin{pmatrix} -\sigma_2 & 0 \\ 0 & \sigma_2 \end{pmatrix} \quad \text{Weyl}, \quad \text{or} \quad C = \sigma_2 \otimes \tau_1 = \begin{pmatrix} 0 & \sigma_2 \\ \sigma_2 & 0 \end{pmatrix} \quad \text{Dirac} \tag{12}$$

The properties of C mentioned above hold true by inspection. Note that $C\gamma^0 = \gamma^2$ is the same in the Weyl basis and in the Dirac basis.

Alternatively, to determine C, let us start by complex conjugating the hermiticity equation $(\gamma^\mu)^\dagger = \gamma^0\gamma^\mu\gamma^0$ derived in chapter VII.4. We obtain $(\gamma^\mu)^T = \gamma^0\gamma^{\mu*}\gamma^0$, since γ^0 is real in the bases we are working with. Plugging $\gamma^{\mu*} = \gamma^0(\gamma^\mu)^T\gamma^0$ into the defining equation (2), we obtain

$$C^{-1}\gamma^\mu C = -(\gamma^\mu)^T \tag{13}$$

Since γ^0 and γ^2 are symmetric (as already noted), while γ^1 and γ^3 are antisymmetric, we see that (13) is satisfied. (For example, for $\mu = 1$, $\gamma^2\gamma^0\gamma^1\gamma^2\gamma^0 = -\gamma^2\gamma^0\gamma^2\gamma^0\gamma^1 = \gamma^1$, while for $\mu = 0$, $\gamma^2\gamma^0\gamma^0\gamma^2\gamma^0 = -\gamma^2\gamma^0\gamma^2\gamma^0\gamma^0 = -\gamma^0$.)

With γ^0 symmetric and C antisymmetric, we can also write (3) as

$$\psi_c \equiv C\gamma^0\psi^* = (\psi^*)^T(C\gamma^0)^T = \psi^\dagger\gamma^0 C^T = -\bar{\psi}C \tag{14}$$

The charge conjugate of a spinor transforms as a spinor

Physically, if the positron is described by ψ_c, then ψ_c better transforms as a spinor. Just from the fact that the antiparticle also lives in spacetime, it has to obey Lorentz. Let us verify this. Under a Lorentz transformation $\psi \to e^{-\frac{i}{4}\omega_{\mu\nu}\sigma^{\mu\nu}}\psi$, and complex conjugating, we have $\psi^* \to e^{+\frac{i}{4}\omega_{\mu\nu}(\sigma^{\mu\nu})^*}\psi^*$, hence

$$\psi_c = C\gamma^0\psi^* \to C\gamma^0 e^{+\frac{i}{4}\omega_{\mu\nu}(\sigma^{\mu\nu})^*}\psi^* = e^{-\frac{i}{4}\omega_{\mu\nu}\sigma^{\mu\nu}}C\gamma^0\psi^* = e^{-\frac{i}{4}\omega_{\mu\nu}\sigma^{\mu\nu}}\psi_c \tag{15}$$

precisely as expected. (The second equality here follows from the defining equation (2) for C, namely, $(C\gamma^0)\gamma^{\mu*} = -\gamma^\mu(C\gamma^0)$, which implies $C\gamma^0(\sigma^{\mu\nu})^* = C\gamma^0(-i/2)[\gamma^{\mu*}, \gamma^{\nu*}] = (-i/2)[\gamma^\mu, \gamma^\nu]C\gamma^0 = -\sigma^{\mu\nu}C\gamma^0$. Note that $\sigma^{\mu\nu}$ is defined with an explicit i.)

As we said, when charge conjugating, it is often more transparent and convenient to go to a specific basis. But notice that the calculation in (15) is done without committing to any specific basis. Clearly, that the antiparticle lives in spacetime is a physical statement that cannot possibly be basis dependent.

Perhaps it is worth repeating that ψ_c transforms correctly already follows from the equation of motion (4), since it would not make sense otherwise.

Charge conjugation and pseudoreality

To exhibit the connection with pseudoreality, let us go to the 2-component notation. Writing $\psi = \begin{pmatrix} u \\ v \end{pmatrix}$ and $\psi_c = \begin{pmatrix} u_c \\ v_c \end{pmatrix} = \gamma^2\psi^*$, we obtain $u_c = \sigma_2 v^*$ and $v_c = -\sigma_2 u^*$. In other words,

$$\psi = \begin{pmatrix} u \\ v \end{pmatrix} \quad \text{and} \quad \psi_c = \begin{pmatrix} -\sigma_2 v^* \\ \sigma_2 u^* \end{pmatrix} = i\begin{pmatrix} \tilde{v} \\ -\tilde{u} \end{pmatrix} \tag{16}$$

in both the Weyl and Dirac bases (since γ^2 is the same in these bases).

Note in (16) the Weyl spinors $\tilde{u}$ and $\tilde{v}$ introduced in chapter VII.3. There we already worked out that $\tilde{u}$ and $\tilde{v}$ transform like v and u, respectively, but it would be good for you to repeat the exercise here. Using the familiar identity

$$\sigma_2\sigma_i^*\sigma_2 = \sigma_2\sigma_i^T\sigma_2 = -\sigma_i, \quad \text{that is, } \sigma_2\sigma_i^* = \sigma_2\sigma_i^T = -\sigma_i\sigma_2 \tag{17}$$

from chapter IV.5, you can show that u and u_c transform in the same way, and v and v_c transform in the same way, which of course must be the case for ψ and ψ_c to transform in the same way. Again, the pseudoreality of the defining representation of $SU(2)$ plays a crucial role.

Evidently, we also have $(u_c)_c = \sigma_2 v_c^* = \sigma_2(-\sigma_2 u^*)^* = u$. Similarly, $(v_c)_c = v$. And of course, we also have

$$(\psi_c)_c = \gamma^2(\gamma^2\psi^*)^* = \gamma^2(-\gamma^2)\psi = \psi \tag{18}$$

The antiparticle of the positron is the electron, as might be expected.

The conjugate of the left hand is the right hand

The charge conjugate of a left handed field is right handed and vice versa. This fact turns out to be crucial in the construction of grand unified theory, as we shall see in chapter IX.2. See whether you can prove this important fact before reading on.

Let us do it together. Suppose ψ is left handed: $(1 - \gamma_5)\psi = 0$. We want to show that $(1 + \gamma_5)\psi_c = 0$. Complex conjugating the definition $\gamma_5 \equiv i\gamma^0\gamma^1\gamma^2\gamma^3$, we obtain, using (2), $\gamma_5^* = -i\gamma^{0*}\gamma^{1*}\gamma^{2*}\gamma^{3*} = -i(C\gamma^0)^{-1}\gamma^0\gamma^1\gamma^2\gamma^3(C\gamma^0) = -(C\gamma^0)^{-1}\gamma_5(C\gamma^0)$. Thus,

$$(1 + \gamma_5)\psi_c = (1 + \gamma_5)C\gamma^0\psi^* = C\gamma^0(1 - \gamma_5^*)\psi^* = C\gamma^0\left((1 - \gamma_5)\psi\right)^* = 0 \tag{19}$$

which is what was to be demonstrated. Again, note that this result is basis independent.

The Majorana equation and the neutrino

Can you possibly construct another relativistic equation for a spin $\frac{1}{2}$ particle besides the Dirac equation? Think for a moment.

Since the Lorentz group transforms ψ and ψ_c in the same way, Majorana* had the brilliant insight that Lorentz invariance allows not only the Dirac equation $i\,\partial\!\!\!/\,\psi = m\psi$ but also the Majorana equation

$$i\,\partial\!\!\!/\,\psi = m\psi_c \tag{20}$$

As noted earlier, γ^2 is the only imaginary gamma matrix, and thus $\gamma^2(\gamma^\mu)^* = -\gamma^\mu\gamma^2$. Complex conjugating (20), recalling that $\psi_c = \gamma^2\psi^*$, and multiplying by γ^2, we obtain $\gamma^2(-i\gamma^{\mu*})\partial_\mu\psi^* = \gamma^2 m(-\gamma^2\psi)$, that is,

$$i\,\partial\!\!\!/\,\psi_c = m\psi \tag{21}$$

Thus, $-\partial^2\psi = i\,\partial\!\!\!/(i\,\partial\!\!\!/\,\psi) = i\,\partial\!\!\!/\,m\psi_c = m^2\psi$. As we might have anticipated, m is indeed the mass, known as a Majorana mass,[5] of the particle associated with ψ.

Note that, since ψ_c is right handed if ψ is left handed, the Majorana equation (20), unlike the Dirac equation, preserves handedness. In other words, a left handed field ψ, satisfying $(1 - \gamma_5)\psi = 0$ and (20), describes a 2-component massive neutrino. The formalism is almost tailor made for the neutrino.

Now, suppose that our theory enjoys another symmetry such that the action is invariant[6] under the $U(1)$ transformation

$$\psi \to e^{i\omega}\psi \tag{22}$$

---

* Ettore Majorana had a shining but tragically short career. While still in his twenties, he disappeared off the coast of Sicily during a boat trip. The precise cause of his death has long been a mystery, although various theories have been advanced.[7] Some time ago, a photo surfaced, showing Majorana living in a small town in Venezuela in the 1950s. In 2015, the authorities in Rome announced that the case was officially closed. See the entry in Wikipedia for further details.

with $0 \leq \omega < 2\pi$. One fairly well-known example is the symmetry associated with the conservation of electric charge. Another example is the symmetry associated with the conservation of "lepton number" in particle physics. The electron and the neutrino each carry one unit of lepton number.

Since

$$\psi_c \rightarrow e^{-i\omega}\psi_c \qquad (23)$$

(remember the complex conjugation in charge conjugation!) transforms oppositely from ψ under $U(1)$ (compare with (22)), this symmetry would forbid the Majorana equation (20). Conversely, if (20) holds, then this symmetry must fail. To the extent that electric charge conservation is considered absolute, the Majorana equation can thus only apply to electrically neutral spin $\frac{1}{2}$ particles. The only such particle known is the mysterious neutrino.[8]

From its conception the neutrino was assumed to be massless. Thus, for decades the Majorana equation remained a mathematical curiosity not particularly relevant for physics. People were not clamoring for a description of an electrically neutral massive spin $\frac{1}{2}$ particle. But in the late 1990s, experiments established that the neutrino has a small but nonvanishing mass. At present, it is not known whether the neutrino's mass is Dirac or Majorana.[9] If the neutrino turns out to have a Majorana mass, lepton number would then necessarily be violated (since the neutrino carries lepton number).

The Majorana Lagrangian

We have already constructed the Lagrangian containing a Majorana mass term in chapter VII.3. Let us now write it using the 4-component formalism, as is more commonly seen in the particle physics literature. From (10), we have $\bar{\psi}_c = \psi^T \gamma^2 \gamma^0$ and $\bar{\psi}_c \psi = \psi^T \gamma^2 \gamma^0 \psi = \psi^T C \psi$. With $v = 0$, $\psi = \begin{pmatrix} u \\ 0 \end{pmatrix}$. Let us go to the Weyl basis. Referring to (12), we have $\psi^T C \psi = u^T \sigma_2 u$. Thus, the Lagrangian is

$$\mathcal{L} = \bar{\psi} i \gamma^\mu \partial_\mu \psi - \frac{1}{2} m (\psi^T C \psi + h.c.) \qquad (24)$$

Here $h.c.$ denotes the hermitean conjugate of the term explicitly displayed.

Again, in quantum field theory, ψ has to be treated as an anticommuting Grassman object, since C is manifestly antisymmetric.

At present, it is known that there are three different neutrinos,* described by three Majorana fields ψ_a, $a = 1, 2, 3$. The Lagrangian is then generalized to $\mathcal{L} = \bar{\psi}_a i \gamma^\mu \partial_\mu \psi_a - \frac{1}{2}(\psi_a^T M_{ab} C \psi_b + h.c.)$, with the indices a and b summed over. The discussion in this

---

* We will discuss this mysterious triplication further in chapter IX.1 and what group theory might have to say about it.

chapter thus requires the 3-by-3 mass matrix M to be symmetric and complex. That it is symmetric is important in the theory of neutrino masses and oscillation.*

For use in part IX, we note here that the bilinear for the Dirac mass term can be written (using (11) and (14)) as

$$-\psi^c C \psi = \bar{\psi} C C \psi = \bar{\psi} \psi \qquad (25)$$

Contrast this with the bilinear $\psi^T C \psi$ for the Majorana mass term.

Summary of spin $\frac{1}{2}$ fields and particles

In the last three chapters, we studied how the Lorentz group deals with the spin $\frac{1}{2}$ field. Since we have covered a lot of territory, it would be helpful to summarize. Given that this is a book on group theory rather than quantum field theory, we often blur the distinction between particle and field, but still, keep in mind that a field is a mathematical construct that creates and annihilates particles in the minds of theorists, while a particle is a physical entity that experimentalists can observe, at least in principle.

In the following discussion, the three fundamental discrete Z_2 symmetries—parity or spatial reflection P, charge conjugation[†] C, and time reversal T—will be important. The physics governing the neutrino, namely the weak interaction, was discovered in 1956 to be neither invariant under P nor under C. It is, however, invariant under the combined transformation CP to an excellent approximation. In fact, it was believed to be exact until 1964, when a slight violation of CP was observed. But for ease of exposition, we simply assume that CP is still an exact symmetry. Alternatively, we could invoke CPT, an exact symmetry thought to be sacred in relativistic local quantum field theory.[10] Note that under T, momentum and spin are both reversed, and hence helicity remains unchanged.

Another cautionary note: the terminology I use is common in the particle physics community; others might prefer a slightly different terminology. A Weyl field is by definition a 2-component field without a Majorana mass.

Here are some important points.

1. A single Weyl field is by definition massless.[‡] It has two components and is either right or left handed.

2. Let us go back to the days when the neutrino was believed to be massless. It was established experimentally that the neutrino is left handed. If parity P were a good symmetry, this would mean that there is also a right handed massless neutrino. That there is no right handed massless neutrino in the world implies that parity is broken.

---

* This was alluded to in the review of linear algebra, where we showed that for a symmetric complex matrix M, there always exists a unitary matrix U such that $U^T M U = D$, with D a positive real diagonal matrix. The unitary transformation U describes neutrino mixing and oscillation.

† I trust the reader not to confound the discrete symmetry transformation C and the matrix C.

‡ Strictly speaking, the particle associated with the Weyl field is massless.

3. To be definite, consider a Weyl field that creates a left handed massless neutrino. Under the CP operation, the C turns the neutrino into an antineutrino, while the P turns the left hand into the right hand. Thus, CP invariance means that the CP transform of this field would create a right handed massless antineutrino, even if P is broken. Note that the word "massless" is redundant here. (The operational definition* of a neutrino is that it produces an electron when scattered off a nucleon; an antineutrino would produce a positron.)

4. A Weyl field that acquires a Majorana mass is called a Majorana field. It creates a Majorana particle which is, by definition, massive.

5. Since a Majorana particle is massive, we can always go to its rest frame, where its spin states can be analyzed using the rotation group $SO(3)$. By boosting in the appropriate direction, we can certainly realize both helicities. Another way of saying this is that since the particle is massive, it cannot be traveling at the speed of light. Thus, we can overrun it and go to a frame in which it is moving in the opposite direction and thus has opposite helicity.

6. By combining (3) and (5), we see that a Majorana particle is necessarily its own antiparticle.

7. While it is meaningless to ask whether a Majorana particle is left or right handed (since by definition it is massive), we can still ask, when given a Majorana field ψ, whether $(1 - \gamma_5)\psi = 0$ or $(1 + \gamma_5)\psi = 0$.

Appendix 1: Majorana spinor

We mention in passing that it is possible for $\psi = \psi_c$, in which case ψ is known as a Majorana spinor.[†] Referring to (16), we see that this implies

$$u = \sigma_2 v^*, \qquad v = -\sigma_2 u^* \tag{26}$$

The two Weyl spinors u and v contained in a Majorana spinor are not independent of each other:

$$\psi = \begin{pmatrix} u \\ -\sigma_2 u^* \end{pmatrix} \tag{27}$$

Note that given a Dirac spinor (call it ξ), we can always construct a Majorana spinor, namely,

$$\psi = \frac{1}{2}(\xi + \xi_c) \tag{28}$$

Recall (18).

Referring back to (7), we see explicitly that this means that in $(3 + 1)$-dimensional spacetime, a spinor cannot be both Majorana and Weyl.

We can also readily prove this fact directly. A 4-component spinor ψ is Majorana if $\psi = \psi_c = \gamma_2 \psi^*$. It is Weyl if either $(1 - \gamma_5)\psi = 0$ or $(1 + \gamma_5)\psi = 0$. Let's say the former is true. Then we have $0 = (1 - \gamma_5)\psi = (1 - \gamma_5)\gamma_2 \psi^* = \gamma_2(1 + \gamma_5)\psi^* = \gamma_2((1 + \gamma_5)\psi)^*$, since γ_5 is real in both the Dirac and the Weyl bases. This implies $(1 + \gamma_5)\psi = 0$, and hence $\psi = 0$.

Note that in this text, we carefully distinguish between Majorana field, Majorana particle, and Majorana spinor.

---

* Consult any textbook on particle physics. For a bit more detail, see part IX.
† Such fields are useful in modern condensed matter theory.

Appendix 2: Momentum space

While going to momentum space, we have to watch the signs a bit because of complex conjugation. The Fourier transform $\psi(x) = \int \frac{d^4p}{(2\pi)^4} e^{-ipx} \psi(p)$ implies $\psi_c(x) = \int \frac{d^4p}{(2\pi)^4} e^{+ipx} \psi_c(p) = \int \frac{d^4p}{(2\pi)^4} e^{-ipx} \psi_c(-p)$. Thus, we have

$$\not{p}\psi(p) = m\psi_c(-p) \quad \text{and} \quad \not{p}\psi_c(-p) = m\psi(p) \tag{29}$$

from which $p^2\psi(p) = m^2\psi(p)$ (and hence $p^2 = m^2$) follows almost immediately.

Exercises

1   Show explicitly that the ψ constructed in (28) has the form given in (27).

2   Show that u and u_c transform in the same way.

Notes

1. Indeed, it already appears in classical mechanics: in the presence of an electromagnetic potential, the canonical momentum is modified $\vec{p} \to \vec{p} - e\vec{A}$.
2. Dirac first thought that ψ_c could describe the proton, but the difference in mass proved to be a stumbling block.
3. I can't resist remarking that when theoretical physicists are shown the Dirac equation, only a small fraction ε would think of conjugating it, and out of those, only a tiny fraction (of order ε^2?) would be so bold as to predict antimatter. Among mathematical physicists, perhaps a substantial fraction, of order unity, might have recognized that γ^* also satisfies the Clifford algebra if γ does.
4. Another popular choice is $\alpha = i$ to knock out the is in γ^2 and thus make C real.
5. For a particularly clear discussion of the Dirac mass and the Majorana mass, see T. P. Cheng and L. F. Li, pp. 412–414.
6. Here the symmetry can be either global or gauged, for readers who know what this means.
7. See F. Guerra and N. Robotti, "Ettore Majorana: Aspects of His Scientific and Academic Activity," 2008.
8. See, for example, Fearful, pp. 34–39. The life of the neutrino is associated with many stories; for instance, how its "discoverer" C. D. Ellis learned physics.
9. Tremendous efforts, not to mention zillions, are being spent to determine this experimentally. Most theorists favor a Majorana mass.
10. See, for example, QFT Nut.

VII.i1 | A Hidden $SO(4)$ Algebra in the Hydrogen Atom

An unexpected degeneracy

In a course on quantum mechanics, students are taught to solve the Schrödinger equation with the potential $V(r) = -\frac{\kappa}{r}$, with the result to be applied to the hydrogen atom. One goes through various standard steps for dealing with partial differential equations in physics: set up spherical coordinates, separate variables using spherical harmonics, solve the resulting differential equation in the radial variable, and so on and so forth.[1] Generically, for a potential with spherical symmetry, the energy levels are specified by three quantum numbers, n, l, and m, where $m = -l, -l+1, \cdots, l-1, l$ runs over a range determined in chapter IV.2 by group theory. For this particular problem, $l = 0, 1, \cdots, n-1$ runs over n values. Astonishingly, after the dust settles, the energy eigenvalues for a given n turn out not to depend on the angular quantum number l.

Please realize that the degeneracy here goes much beyond the degeneracy guaranteed by rotation invariance as explained in chapter III.1. The symmetry group $SO(3)$ only says that the $(2l+1)$ states belonging to a given l have the same energy independent of m. For a given n, if the energy does not depend on l, the degeneracy is increased to

$$\sum_{l=0}^{n-1}(2l+1) = n^2 \tag{1}$$

This additional degeneracy is known as accidental or dynamical degeneracy; but surely, dear reader, you are sophisticated enough to suspect that this remarkable degeneracy did not occur by accident.

Pauli was the first to understand this additional degeneracy as having a group theoretic origin.[2] One clue is that the $(2l+1)$-fold degeneracy holds as long as the potential $V(r)$ depends only on r, while this dynamical degeneracy is specific to the inverse square law.

A clue from the closing of orbits

Another clue is that, in Newtonian mechanics, orbits in a $1/r$ potential close. Beginning students often take this for granted, without realizing that a closed orbit is nothing less than an apparent miracle.

Think of it this way: to determine an elliptical orbit, we extract from Newton's laws expressions for $\frac{dr}{dt}$ and $\frac{d\theta}{dt}$ and hence an expression for $\frac{d\theta}{dr}$. Over a period, r goes from its minimum value (when the planet reaches perihelion), to its maximum value, and back to its minimum value again. Meanwhile, the angle θ changes by $\Delta\theta = 2 \int_{r_{\min}}^{r_{\max}} dr \frac{d\theta}{dr}$. For an arbitrary $V(r)$, there is no reason for this to be equal to 2π (or multiples thereof), the necessary condition for the orbit to close.

But remarkably,[3] for $V(r) = -\frac{\kappa}{r}$, $\Delta\theta = 2\pi$. (In fact, in Einstein gravity, the effective potential is not $\propto 1/r$; planetary orbits do not close, and the precession of the perihelion of Mercury provides a classic test of general relativity.)

Laplace was the first to understand that this apparent mystery is due to existence of a conserved vector $\vec{\mathcal{L}} \equiv \frac{1}{m}\vec{L} \times \vec{p} + \kappa\frac{\vec{r}}{r}$, now known as the Laplace-Runge-Lenz vector. (Here $\vec{L} = \vec{r} \times \vec{p}$ denotes the conserved angular momentum vector, and m the mass of the planet.) Computing the time derivative $\dot{\vec{\mathcal{L}}}$, you can verify that $\vec{\mathcal{L}}$ is conserved[4] for an inverse square central force. Again, we emphasize that, in contrast, $\vec{L}$ is conserved for any $V(r)$. Since $\vec{L}$ is perpendicular to both $\vec{r}$ and to $\vec{p}$, this implies that the motion takes place entirely in a plane orthogonal to $\vec{L}$. In contrast, $\vec{\mathcal{L}}$ lies in this orbital plane.

When $\vec{p}$ is perpendicular to $\vec{r}$, which occurs at perihelion and aphelion, the vector $\vec{\mathcal{L}}$ points in the direction of $\vec{r}$. We can take the constant vector $\vec{\mathcal{L}}$ to point toward the perihelion, and thus, the position of the perihelion does not change. Hence the conservation of $\vec{\mathcal{L}}$ implies that the orbit closes.

Promotion to operators

Everything said thus far pertains to classical mechanics. When we move on to the quantum mechanics of the hydrogen atom, we trivially replace the planet by the electron, the sun by the nucleus, and set $\kappa = e^2$ (or Ze^2 for hydrogenic atoms). More significantly, various physical quantities are promoted to hermitean operators satisfying

$$[r_i, p_j] = i\hbar\delta_{ij} \tag{2}$$

and

$$[L_i, r_j] = i\hbar\varepsilon_{ijk}r_k, \quad [L_i, p_j] = i\hbar\varepsilon_{ijk}p_k \tag{3}$$

Also, $\vec{L}$ generates the $SO(3)$ algebra:

$$[L_i, L_j] = i\hbar\varepsilon_{ijk}L_k \tag{4}$$

The commutation relations in (3) and (4) merely say that $\vec{r}$, $\vec{p}$, and $\vec{L}$ transform as vectors.[5]

Since $\vec{L}$ and $\vec{p}$ do not commute, we have to worry about the order of $\vec{L}$ and $\vec{p}$ in the Laplace-Runge-Lenz vector. The prescription, as is standard in quantum mechanics, is to modify the classical expression for $\vec{\mathcal{L}}$ to

$$\vec{\mathcal{L}} \equiv \frac{1}{2m}(\vec{L} \times \vec{p} - \vec{p} \times \vec{L}) + \kappa \frac{\vec{r}}{r} \tag{5}$$

so that it becomes a hermitean operator and hence observable.

The classical statement that $\frac{d\vec{\mathcal{L}}}{dt} = 0$ with the Hamiltonian

$$H = \frac{p^2}{2m} - \frac{\kappa}{r} \tag{6}$$

is now replaced by

$$[H, \vec{\mathcal{L}}] = 0 \tag{7}$$

(which you can readily verify).

The Laplace-Runge-Lenz vector is evidently a vector, and so we have to add to (3) and (4) the commutation relation

$$[L_i, \mathcal{L}_j] = i\hbar \varepsilon_{ijk} \mathcal{L}_k \tag{8}$$

The emergence of $SO(4)$

Our next question is naturally to ask about the commutation between the $\mathcal{L}$s. A straightforward but rather tedious calculation[6] yields

$$[\mathcal{L}_i, \mathcal{L}_j] = i\hbar \varepsilon_{ijk} \left(\frac{-2H}{m} \right) L_k \tag{9}$$

Let us note several remarkable features of this crucial result.

1. The Hamiltonian appears in the algebra, but in a rather different role from the one it had in chapter III.1. There the generators of the algebra commute with H. Here H pops up in the algebra.

2. Define

$$M_i \equiv \sqrt{-\frac{m}{2H}} \mathcal{L}_i \tag{10}$$

so that we can replace (9) by

$$[M_i, M_j] = i\hbar \varepsilon_{ijk} L_k \tag{11}$$

Note that this makes sense only because H commutes with both L and $\mathcal{L}$.

3. Acting on the eigenstates $|E\rangle$ of H with eigenvalue equal to E (that is, states such that $H|E\rangle = E|E\rangle$), we can replace H effectively by E. The important point is that if we restrict ourselves to bound states so that $E < 0$, then $\sqrt{-\frac{m}{2H}}$ is a real number and M_i is hermitean just like $\mathcal{L}_i$. We also trivially replace (8) by

$$[L_i, M_j] = i\hbar \varepsilon_{ijk} M_k \tag{12}$$

Lo and behold! The commutations (4), (12), and (11) characterize the Lie algebra of $SO(4)$, as was first given in chapter I.3. The dynamics of the hydrogen atom has hidden in it an $SO(4)$ algebra.

We proceed just as in chapter I.3 (and also chapters VII.2 and VII.3, dropping a few is here and there). Define $A_{\pm i} = (L_i \pm M_i)/2$. Then the $SO(4)$ algebra breaks up into two $SU(2)$s, as is familiar by now:

$$[A_{+i}, A_{+j}] = i\hbar\varepsilon_{ijk}A_{+k} \tag{13}$$

$$[A_{-i}, A_{-j}] = i\hbar\varepsilon_{ijk}A_{-k} \tag{14}$$

and

$$[A_{+i}, A_{-j}] = 0 \tag{15}$$

The irreducible representations of $SO(4)$ are then labeled by (a_+, a_-), with $a_+ = 0, \frac{1}{2}, 1, \frac{3}{2}, \cdots$ and $a_- = 0, \frac{1}{2}, 1, \frac{3}{2}, \cdots$. The states in the irreducible representation (a_+, a_-) can be listed as $|b_+, b_-\rangle$, with $b_+ = -a_+, -a_+ + 1, \cdots, a_+ - 1, a_+$ and with $b_- = -a_-, -a_- + 1, \cdots, a_- - 1, a_-$. The irreducible representation (a_+, a_-) thus has dimension $(2a_+ + 1)(2a_- + 1)$.

Finding the energy spectrum

For the problem at hand, an additional constraint comes from the fact that classically, $\vec{L} \cdot \vec{\mathcal{L}} = 0$, since $\vec{L}$ is perpendicular to the orbital plane while $\vec{\mathcal{L}}$ lies in it. You can check that this persists into quantum mechanics. Therefore,

$$\vec{A}_+^2 - \vec{A}_-^2 = \vec{L} \cdot \vec{M} = 0 \tag{16}$$

Thus, only irreducible representations with $a_+ = a_- = a$ are allowed.

In contrast,

$$\vec{A}_+^2 + \vec{A}_-^2 = \frac{1}{2}(\vec{L}^2 + \vec{M}^2) = \frac{1}{2}(\vec{L}^2 - \frac{m}{2E}\vec{\mathcal{L}}^2) \tag{17}$$

Now we need to do another straightforward but tedious calculation[7] to obtain

$$\vec{\mathcal{L}}^2 = \frac{2H}{m}(\vec{L}^2 + \hbar^2) + \kappa^2 \tag{18}$$

Plugging this into (17), we simplify the right hand side to $-\frac{1}{2}(\hbar^2 + \frac{m}{2E}\kappa^2)$. Then, evaluating (17) on a state in the irreducible representation (a, a), we obtain $2 \cdot a(a+1)\hbar^2 = -\frac{1}{2}(\hbar^2 + \frac{m}{2E}\kappa^2)$, which immediately fixes the bound-state energies to be

$$E = -\frac{m\kappa^2}{2\hbar^2}\frac{1}{n^2} \tag{19}$$

where we define $n = 2a + 1$. Since a could be half-integral or integral, n ranges over the positive integers $n = 1, 2, \cdots$. This is the famous Balmer series[8] for the spectrum of the hydrogen atom. The degeneracy is just the dimension of the irreducible representation (a, a), namely, $(2a + 1)^2 = n^2$, in agreement with (1).

As you can see, this algebraic approach to the spectrum (19) is not any easier than the standard ("just solve the partial differential equation!") approach given in standard textbooks. The advantage is that it reveals the accidental degeneracy to be not accidental at all.[9]

Appendix 1: Symmetric traceless tensors of $SO(4)$

Another way to count the degeneracy is instructive from the group theoretic point of view. It consists of noting that the irreducible representation (a, a) corresponds to the symmetric traceless tensor with $h = 2a$ indices of $SO(4)$. For example, $(\frac{1}{2}, \frac{1}{2})$ is the vector, $h = 1$; $(1, 1)$, with dimension equal to $(2 \cdot 1 + 1)^2 = 9$, is the symmetric traceless tensor with $h = 2$; and so on. Note that under the $SO(3)$ subgroup, (a, a) decomposes as $0 \oplus 1 \cdots \oplus (2a - 1) \oplus 2a$, and does not contain any half-integral representations; in other words, it may be described purely as a tensor with no spinorial content.

The diligent reader with a good memory will recall the result of exercise IV.1.13 that the symmetric traceless tensor with h indices of $SO(4)$ has $(h + 1)^2$ components. We thus obtain the degeneracy given in (1). For example, for the principal quantum number $n = 3$, the corresponding $SO(4)$ tensor has $h = n - 1 = 2$ indices, and thus has $\frac{1}{2}(4 \cdot 5) - 1 = 9$ components. On the other hand, $l = 0, 1, 2$, with $2l + 1 = 1, 3, 5$ states, respectively. In other words, $\frac{1}{2}(4 \cdot 5) - 1 = 1 + 3 + 5$.

Appendix 2: The periodic table and "accidental" degeneracy

The pioneers of quantum mechanics realized that this "accidental" degeneracy could explain the periodic table and, in broad outline, the chemical properties of the elements* contained therein. With Z electrons moving around, we would in principle have to solve a rather complicated problem in quantum mechanics, but early on, in 1928, Hartree suggested that we can focus on one electron and think of it as moving in the average electric field generated by the other $Z - 1$ electrons plus the nucleus. The single electron in the Hartree "mean field" approximation thus finds itself in a potential $\sim -Ze^2/r$ near the nucleus and $\sim -e^2/r$ far away. Since the potential is still spherically symmetric, we can still use the principal quantum number n and angular momentum $l = 0, 1, 2, \cdots, n - 1$ to organize the states. For each n and l, there are $2(2l + 1)$ states. To make contact with high school chemistry,[10] it is convenient to use the spectroscopic notation $s, p, d, f, \cdots$ corresponding respectively to $l = 0, 1, 2, 3, \cdots$ with $2 \cdot 1 = 1$, $2 \cdot 3 = 6$, $2 \cdot 5 = 10$, $2 \cdot 7 = 14$, $\cdots$ states.

But since the potential is not $\propto 1/r$, for each n, the states with different l no longer have precisely the same energy.[11] We expect states with larger l to have higher energy, since our electron is spending less time near the nucleus, where the attractive potential is the deepest. Call this the "large l effect." The states in order of increasing energy would thus start out as

$$1s; 2s, 2p; 3s, 3p; \tag{20}$$

giving 2; $2 + 6 = 8$; $2 + 6 = 8$ elements, respectively. The first two are of course hydrogen and helium; the two 8s are the second and third rows in the periodic table, taking us from lithium to neon, and sodium to argon.

But now the funny business starts. Due to the "large l effect," $3d$ has higher energy than $4s$. Similarly for $4d$ versus $5s$. The next set of states are thus

$$4s, 3d, 4p; 5s, 4d, 5p; \tag{21}$$

giving $2 + 10 + 6 = 18$; $2 + 10 + 6 = 18$ elements, respectively. That explains why in the fourth and fifth row of the periodic table, giving ten "extra" elements in each row (including iron, copper, and zinc in the fourth row, and niobium and silver in the fifth row), and causing the characteristic "valley" in the shape of the table.[12]

As is well known from high school chemistry, this shell structure, due to Fermi statistics obeyed by the electrons and to the way these states are organized energetically, accounts for quite a lot. When a shell is completed, such as $4s$, $3d$, and $4p$, it takes a relatively large amount of energy to step up to $5s$, the next set of states. Hence the elements in the last column of the table, with their complete shells, are inert noble gases, so noble that they don't want to interact with the hoi polloi. We can read off their atomic numbers from (20) and

---

* Different use of the word than in the rest of this book!

(21): helium $Z = 2$, neon $Z = 2 + 8 = 10$, argon $Z = 2 + 8 + 8 = 18$, krypton[13] $Z = 2 + 8 + 8 + 18 = 36$, xenon $Z = 2 + 8 + 8 + 18 + 18 = 54$, and so on. Without physics and mathematics, these numbers would pose a total mystery. Interestingly, the two groups $SO(3)$ and $SO(4)$ both play important roles.

In contrast, the elements in the first column of the table have an "extra" electron and are eager to get rid of it. They are the alkali metals: lithium $Z = 2 + 1 = 3$, sodium $Z = 2 + 8 + 1 = 11$, potassium $Z = 2 + 8 + 8 + 1 = 19$, and so on. But the halogens in the next-to-last column are eager to grab an electron to complete their shells: fluorine $Z = 2 + 8 - 1 = 9$, chlorine $Z = 2 + 8 + 8 - 1 = 17$, bromine $Z = 2 + 8 + 8 + 18 - 1 = 35$, and so on.

Beyond the list in (21), the "large l effect" continues to wreak havoc on the periodic table, causing it to have the two "fold-out strips" bedeviling high school students. Now the f states come in, giving us

$$6s, 4f, 5d, 6p; 7s, 5f, \cdots, \tag{22}$$

$2(2 \cdot 3 + 1) = 14$ "additional" elements in the sixth and seventh row, the rare earth or lanthanides, and the actinides.

Without group theory (plus a bit of mean-field dynamics), these "magic numbers" of chemistry—2, 10, 18, 36, 54, and so on—would have been a total mystery.

Notes

1. The hydrogen atom is treated in every book on quantum mechanics. See, for example, J.J. Sakurai and J. Napolitano, Modern Quantum Mechanics, p. 216.
2. W. Pauli, Z. Phys. 36 (1926), p. 336; V. A. Fock, Z. Phys. 98 (1935), p. 145; V. Bargman, Z. Phys. 99 (1936), p. 576. Note that Pauli's group theoretic solution came mere months after the invention of quantum mechanics as we know it. This is also discussed in a number of quantum mechanics textbooks, such as Baym, Lectures on Quantum Mechanics; J.J. Sakurai and J. Napolitano, Modern Quantum Mechanics; and S. Weinberg, Quantum Mechanics.
3. See, for example, chapter I.1 in G Nut.
4. See, for example, G Nut, p. 794.
5. Strictly speaking, we should call $\vec{r}$ and $\vec{p}$ vectors, and $\vec{L}$ an axial vector or a pseudovector: under spatial reflection, $\vec{L}$ transforms oppositely to $\vec{r}$ and $\vec{p}$.
6. I urge the reader to build strength of character: just do it. Since this result is crucial, it is worthwhile to see even schematically how the Hamiltonian pops up. Roughly, the basic commutator of quantum mechanics $[x, p] = i$ means that an x and a p would knock each other out in a back-of-the-envelope type calculation. So here goes: $[\mathcal{L}, \mathcal{L}] \sim [\frac{1}{m}xpp + \kappa \frac{x}{r}, \frac{1}{m}xpp + \kappa \frac{x}{r}] \sim \frac{1}{m^2}[xpp, xpp] + \frac{1}{m}[xpp, \kappa \frac{x}{r}] \sim \frac{1}{m}(\frac{1}{m}xp^3 + \kappa x p \frac{1}{r}) \sim \frac{xp}{m}(\frac{p^2}{m} + \frac{\kappa}{r}) \sim LH/m$.
7. The hard part is to get the $\hbar^2$ in this equation. The rest of the terms just require vector gymnastics in classical mechanics:
$$\vec{\mathcal{L}}^2 = \left(\frac{1}{m}\vec{L} \times \vec{p} + \kappa \frac{\vec{r}}{r}\right)^2$$
$$= \frac{1}{m^2}(\vec{L} \times \vec{p})^2 + \frac{2\kappa}{m}(\vec{L} \times \vec{p}) \cdot \frac{\vec{r}}{r} + \kappa^2 = \frac{1}{m^2}\vec{L}^2\vec{p}^2 - \frac{2\kappa}{mr}\vec{L}^2 + \kappa^2$$
$$= \frac{2H}{m}\vec{L}^2 + \kappa^2 \tag{23}$$
8. The Swedish spectroscopist A. J. Ångström (1814–1874) had published four lines in the visible spectrum of light emitted from hydrogen at wavelengths of 6562.74, 4860.74, 4340.10, and 4101.2 $\times 10^{-8}$ cm. In 1885, Johann Balmer, a 60-year-old mathematician teaching at a girls school, noticed that these measured numbers divided by 3645.6 gave four numbers close to the fractions 9/5, 4/3, 25/21, and 9/8, which he guessed were in fact $3^2/(3^2 - 4)$, $4^2/(4^2 - 4)$, $5^2/(5^2 - 4)$, and $6^2/(6^2 - 4)$. We now understand that when the hydrogen atom makes a transition from level n to level m, the energy of the emitted photon is equal to a constant times $\frac{1}{m^2} - \frac{1}{n^2}$ whose inverse equals $n^2m^2/(n^2 - m^2)$. Luckily for Balmer, all four lines involve quantum jumps to the same level $m = 2$.
9. Quite amazingly, the $SO(4)$ can be extended to $SO(4, 2)$. A single irreducible representation of $SO(4, 2)$ contains all the states of the hydrogen atom.
10. A personal note: I didn't get to take physics in high school; I had chemistry instead.
11. The discussion here is adapted from S. Weinberg, Lectures on Quantum Mechanics, pp. 125 ff.
12. For example, see http://en.wikipedia.org/wiki/Periodictable.
13. Known to every little kid who is into superheroes.

VII.i2 | The Unexpected Emergence of the Dirac Equation in Condensed Matter Physics

From our description of the Lorentz group leading us by the nose to the Dirac equation, you would think that this equation would be the exclusive province of relativistic physics. In fact, the realization that the Dirac equation naturally emerges in a solid was one of the interesting surprises in theoretical physics.

Noninteracting hopping electrons

Consider the quantum Hamiltonian

$$H = -t \sum_j (f_{j+1}^\dagger f_j + f_j^\dagger f_{j+1}) \tag{1}$$

describing noninteracting electrons on a 1-dimensional lattice (figure 1) with sites labeled by j. Here $f_j^\dagger$ and f_j create and annihilate an electron on site j, respectively. Thus, the first term in H removes an electron from site j and places it on site $j + 1$, effectively describing an electron hopping from site j to site $j + 1$ with amplitude t. The second term, required by the hermiticity of H, effectively describes an electron hopping in the opposite direction, from site $j + 1$ to site j. We have suppressed the spin label on the electron; since the electrons are assumed to be noninteracting, we can deal with the spin up electrons and the spin down electrons separately. This is just about the simplest model in solid state physics; a good place to read about it is in Feynman's "Freshman lectures."

To proceed, we Fourier transform* $f_j = \frac{1}{\sqrt{N}} \sum_k e^{ikaj} f(k)$, where a is the spacing between sites. We also impose a periodic boundary condition on a lattice with N sites, so that the lattice effectively forms a ring, and site $N + 1$ is actually site 1. Since j and $N + j$ are merely different labels for the same site, the wave number k must satisfy $e^{ikaj} = e^{ika(N+j)}$, that is, $e^{ikNa} = 1$, which implies that $k = (\frac{2\pi}{Na})n$ with n an integer ranging from $-\frac{1}{2}N$ to

---

* I do not introduce an additional symbol for the Fourier transform such as $\tilde{f}(k)$ and instead trust you to discern whether we are in position or momentum space.

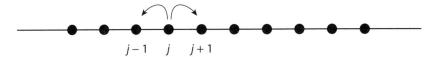

$j-1$ j $j+1$

Figure 1

$\frac{1}{2}N$. The allowed range for k is then $-\frac{\pi}{a} < k \le \frac{\pi}{a}$, which defines the Brillouin zone. The spacing between successive values of k is given by $\Delta k = \frac{2\pi}{Na}$, which vanishes as $N \to \infty$; k becomes a continuous rather than a discrete variable. All this is basic solid state physics, reviewed here for your convenience.

In quantum mechanics, the wave e^{ikaj} describes a state of momentum $p = \hbar k$. In the continuum limit, with $a \to 0$ and $aj \to x$, this becomes the familiar de Broglie wave $e^{ipx/\hbar}$. We will not distinguish the wave number k and the momentum p. In what follows, to lessen clutter, we will choose units with $a = 1$ and $\hbar = 1$. You can always restore them when desired by appealing to dimensional analysis.

To determine the energy spectrum, we plug $f_j = \frac{1}{\sqrt{N}}\sum_k e^{ikj} f(k)$ into the Hamiltonian, written more conveniently as $H = -t\sum_j (f^\dagger_{j+1} + f^\dagger_{j-1})f_j$, and use the Fourier identity $\sum_j e^{ikj} = N\delta_{k,0}$. We obtain

$$H = -t \sum_k \sum_q (e^{-iq} + e^{iq})\left(\frac{1}{N}\sum_j e^{-iqj}e^{ikj}\right)f^\dagger(q)f(k) = -2t \sum_k \cos k \; f^\dagger(k)f(k) \tag{2}$$

Thus, when we create an electron with momentum k, its energy is equal to

$$\varepsilon(k) = -2t \cos k \tag{3}$$

See figure 2.

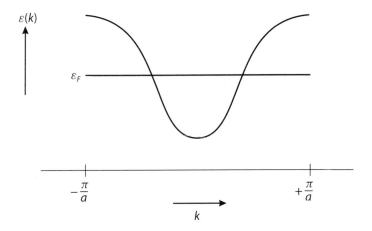

Figure 2

The Fermi sea

Where is the Dirac equation? No trace of it anywhere! There is absolutely nothing relativistic thus far. Indeed, at the bottom of the spectrum, the energy (up to an irrelevant additive constant) goes like $\varepsilon(k) \simeq 2t\left(\frac{1}{2}k^2\right) \equiv \frac{k^2}{2m_{\text{eff}}}$. The electron has a nonrelativistic Newtonian energy with an effective mass $m_{\text{eff}} \propto t^{-1}$.

Yet in a metal we are not dealing with a single electron, but rather a macroscopically large number of electrons. Since electrons obey the Pauli exclusion principle, we can put only one electron in a single quantum state. We thus fill the system with electrons up to some Fermi energy* $\varepsilon_F = -2t \cos k_F$, determined by the density of electrons (see figure 2). This is known as a Fermi sea.

Suppose we now add an electron with energy just above the Fermi sea. Measure its energy from ε_F and momentum from $+k_F$. We are interested in energies small compared to ε_F, that is, $E \equiv \varepsilon - \varepsilon_F \ll \varepsilon_F$, and momenta small compared to k_F, that is, $p \equiv k - k_F \ll k_F$. This electron thus obeys a linear energy momentum relation

$$E = \left.\frac{\partial \varepsilon}{\partial k}\right|_{k=k_F} p = v_F p \tag{4}$$

where we have defined the Fermi velocity v_F.

These electrons with momentum $k \simeq k_F + p$ are known as "right movers" for the obvious reason that they are moving to the "right" (more strictly speaking, moving clockwise, say, around the ring). Fourier transforming (4), we can write down the equation of motion for the wave function ψ_R describing a right moving electron, namely,

$$\left(\frac{\partial}{\partial t} + v_F \frac{\partial}{\partial x}\right)\psi_R = 0 \tag{5}$$

The solution is $\psi_R \propto e^{-i(Et-px)}$, with E and p related as in (4).

Similarly, the electrons with momentum $k \simeq -k_F + p$ obey the energy momentum relation $E = -v_F p$ (with $p < 0$). These electrons are known as "left movers," and the wave function describing them satisfies

$$\left(\frac{\partial}{\partial t} - v_F \frac{\partial}{\partial x}\right)\psi_L = 0 \tag{6}$$

Emergence of the Dirac equation

By introducing a two-component $\psi = \begin{pmatrix} \psi_L \\ \psi_R \end{pmatrix}$, we can package these two equations (5) and (6) into a single equation:

$$\left(\frac{\partial}{\partial t} - v_F \sigma_3 \frac{\partial}{\partial x}\right)\psi = 0 \tag{7}$$

---

* If this is totally unfamiliar to you, you would need to consult a textbook on solid state physics.

Let us choose units for space and time such that $v_F = 1$. Then multiplying from the left by $i\sigma_1$ we can write (7) in a more familiar form as

$$i \left(\sigma_1 \frac{\partial}{\partial t} + i\sigma_2 \frac{\partial}{\partial x} \right) \psi = i\gamma^\mu \partial_\mu \psi = 0 \tag{8}$$

with $\gamma^0 = \sigma_1$ and $\gamma^1 = i\sigma_2$ satisfying the Clifford algebra $\{\gamma^\mu, \gamma^\nu\} = 2\eta^{\mu\nu}$.

Amazingly enough, the massless Dirac equation $i\not\partial\psi = 0$ for $(1+1)$-dimensional spacetime emerges in a totally nonrelativistic situation! Evidently, the key observation is that the presence of the Fermi sea forces a linear relation between energy and momentum.

From the massless to the massive Dirac equation

You might wonder whether it would be possible to go from the massless Dirac equation to the massive Dirac equation $(i\not\partial - m)\psi = 0$.

First of all, you would have noticed that in the massless Dirac equation (8), the right moving and the left moving components ψ_R and ψ_L are merely packaged together. In fact, they do not talk to each other. In other words, they each separately obey a Weyl equation. This is completely in accord with the general discussion in chapters VII.3 and VII.4, but here in the context of $(1+1)$-dimensional spacetime. In contrast, the massive Dirac equation

$$(i\gamma^\mu \partial_\mu - m)\psi = \left(i\sigma_1 \frac{\partial}{\partial t} + i\sigma_2 \frac{\partial}{\partial x} - m \right) \psi = 0 \tag{9}$$

links ψ_R and ψ_L together.

Thus, physics mandates that the massive Dirac equation can emerge only if there is some physics on the lattice scale that would connect the right and left moving electrons. In fact, it turns out that when the "energy band" is half-filled (that is, when $k_F = \pi/2$), there is a lattice distortion known as the Peierls instability that does precisely this. We would wander off much too far beyond the scope of this book to go into this in any detail.[1]

Note

1. See, for example, chapter V.5 in QFT Nut.

VII.i3   The Even More Unexpected Emergence of the Majorana Equation in Condensed Matter Physics

The realization that the Dirac equation naturally emerges in a solid, as described in interlude 2, was one of the interesting surprises in theoretical physics. Perhaps even more surprising is the emergence of the Majorana equation. It almost goes without saying that here we can merely touch on this rapidly evolving subject.

The Kitaev chain

Consider a set of hermitean (or more loosely, real) operators $\mu_j = \mu_j^\dagger$ satisfying the anti-commutation relation[1]

$$\{\mu_j, \mu_l\} = \frac{1}{2}\delta_{jl} \tag{1}$$

(The factor of $\frac{1}{2}$ is merely a convenient normalization.) In particular, $\mu_j^2 = \frac{1}{4}$. Condensed matter physicists call such operators Majorana operators.

Consider a 1-dimensional chain of $2N$ sites indexed by $j = 1, 2, \cdots, 2N$. On each site is a Majorana operator μ_j. For our purposes here, we need not inquire where they came from.

One possibility is that the μ_js are actually constructed as follows out of electron creation and annihilation operators $f_n^\dagger$ and f_n living on a chain of N sites indexed by $n = 1, 2, \cdots, N$:

$$\mu_{2n-1} = \frac{1}{2}(f_n^\dagger + f_n) \quad \text{and} \quad \mu_{2n} = \frac{1}{2i}(f_n^\dagger - f_n) \tag{2}$$

In other words, the complex operator f_n at each site of an N-sited chain is split into its two real components μ_{2n-1} and μ_{2n} living on two sites of a $2N$-sited chain obtained by duplicating each site of the N-sited chain. Using the fermionic commutation relations $\{f_n, f_m\} = 0$ and $\{f_n^\dagger, f_m\} = 1$, we readily check that (1) holds:

$$\mu_{2n-1}^2 = \frac{1}{4}(f_n^\dagger + f_n)^2 = \frac{1}{4} \quad \text{and} \quad \mu_{2n}^2 = -\frac{1}{4}(f_n^\dagger - f_n)^2 = \frac{1}{4}$$

$$\{\mu_{2n-1}, \mu_{2n}\} = \frac{1}{4i}\{f_n^\dagger + f_n, f_n^\dagger - f_n\} = 0 \tag{3}$$

But in fact, as you will see, this level of detail is not necessary for much of what follows.

Consider the Hamiltonian[2]

$$H = -iJ_2(\mu_1\mu_2 + \mu_3\mu_4 + \cdots) - iJ_1(\mu_2\mu_3 + \mu_4\mu_5 + \cdots) \tag{4}$$

(Note that, with J_1 and J_2 real, the overall $(-i)$ is needed for the hermiticity of H.) In other words, the coupling between the Majorana operators on neighboring sites alternates between J_1 and J_2, which we write more conveniently as $J_1 = \frac{1}{2}(t + m)$ and $J_2 = \frac{1}{2}(t - m)$. Then we can rewrite the Hamiltonian as

$$H = -\frac{1}{2}i \sum_{j=1}^{2N-1} \left(t + (-1)^j m\right)\mu_j\mu_{j+1} \tag{5}$$

This is known as a Kitaev chain,[3] proposed in the context of error correction in quantum computing.

To determine the energy spectrum, we proceed as in interlude 2 and Fourier transform to momentum space, $\mu_j = \frac{1}{\sqrt{2N}}\sum_k e^{ikaj}\mu(k)$, where a is the spacing between sites. Again, the periodic boundary condition implies $e^{ikaj} = e^{ika(2N+j)}$, that is, $e^{ik(2N)a} = 1$, in other words, $k = (\frac{2\pi}{2Na})n$ with n an integer ranging from $-N$ to N. The allowed range for k is then $-\frac{\pi}{a} < k \leq \frac{\pi}{a}$, which defines the Brillouin zone. As in interlude 2, we choose units such that $a = 1$ and $\hbar = 1$, so that the momentum $p = \hbar k = k$ ranges between $-\pi$ and π.

Using the anticommutation relation, we write the two terms in the Hamiltonian (5) more symmetrically:

$$\sum_j \mu_j\mu_{j+1} = \frac{1}{2}\sum_j \mu_j(\mu_{j+1} - \mu_{j-1})$$

$$\sum_j (-1)^j \mu_j\mu_{j+1} = \frac{1}{2}\sum_j e^{i\pi j}\mu_j(\mu_{j+1} + \mu_{j-1}) \tag{6}$$

Plugging in $\mu_j = \frac{1}{\sqrt{2N}}\sum_p e^{ipj}\mu(p)$, we see that the first term in (6) becomes

$$\frac{1}{2}\sum_j \frac{1}{2N}\sum_{p'}\sum_p e^{ip'j}e^{ipj}(e^{ip} - e^{-ip})\mu(p')\mu(p) = i\sum_p \sin p\, \mu(-p)\mu(p) \tag{7}$$

That was a warm-up exercise before dealing with the second term in (6). Going through the same steps, we see that the $+$ sign now produces a $\cos p$ rather than a $\sin p$; more interestingly, the factor $(-1)^j = e^{i\pi j}$ kicks in a momentum of π. We obtain

$$\frac{1}{2}\sum_j \frac{1}{2N}\sum_{p'}\sum_p e^{ip'j}e^{ipj}e^{i\pi j}(e^{ip} + e^{-ip})\mu(p')\mu(p)$$

$$= \sum_p \cos p\, \mu(-p - \pi)\mu(p)$$

$$= \frac{1}{2}\sum_p \cos p\left(\mu(-p - \pi)\mu(p) - \mu(-p)\mu(p + \pi)\right) \tag{8}$$

In the last step we used $\cos(p + \pi) = -\cos p$ to write the result more symmetrically. In parallel with this, we can use $\sin(p + \pi) = -\sin p$ to write the sum in (7) as

$$\frac{i}{2} \sum_p \sin p \left(\mu(-p)\mu(p) - \mu(-p - \pi)\mu(p + \pi) \right).$$

That the sum in (8) connects $\mu(-p - \pi)$ with $\mu(p)$ in (8) suggests introducing a 2-component notation $\psi(p) = \begin{pmatrix} \mu(p) \\ \mu(p+\pi) \end{pmatrix}$. We can then express the Hamiltonian $H = \frac{1}{2} \sum_p \psi(-p)h(p)\psi(p)$ in terms of the 2-by-2 matrix Hamiltonian

$$h(p) = t \sin p \begin{pmatrix} 1 & 0 \\ 0 & -1 \end{pmatrix} + m \cos p \begin{pmatrix} 0 & -i \\ i & 0 \end{pmatrix} = t \sin p \, \sigma_3 + m \cos p \, \sigma_2 \tag{9}$$

Interestingly, due to the alternating couplings in (4), a 2-component spinor effectively appears. Note also that the Pauli matrices naturally pop up. (How could they not, given the hermiticity of the Hamiltonian?)

Note that the condition $\mu_j = \mu_j^\dagger$ became $\mu(p)^\dagger = \mu(-p)$; hence, $\psi(p)^\dagger = \psi(-p)$ (since $\mu(-p - \pi) = \mu(-p + \pi)$). Thus, the hermiticity of H translates to $h(p)^\dagger = h(p)$.

The eigenvalues $\pm E(p)$ of $h(p)$ are then easily determined by squaring $h(p)$: we have two bands given by $E(p) = \pm\sqrt{t^2 \sin^2 p + m^2 \cos^2 p} = \pm\sqrt{m^2 + (t^2 - m^2) \sin^2 p}$ as p runs from $-\pi$ to $+\pi$.

Let the bottom band be filled. Then, with $t^2 > m^2$, the top band bottoms out at $p = 0$ and at $p = \pi$. A particle with momentum $p \simeq 0$ is then governed by the effective Hamiltonian $h(p) = t\sigma_3 p + m\sigma_2$ with terms of order $O(p^2)$ dropped.[4] Since $p \simeq 0$, the scale of p is no longer set by π and we are free to absorb t into p.

On Fourier transforming back to space, we have $E\psi(x) = (i\sigma_3 \frac{\partial}{\partial x} + \sigma_2 m)\psi(x)$.

A notational alert: some students are justifiably confused by the distinction between the symbols $\dagger$ and $*$ for hermitean conjugation and complex conjugation, respectively. In linear algebra, hermitean conjugation is defined as complex conjugation followed by transposition. In quantum mechanics, however, hermitean conjugation, also known as the adjoint, carries another layer of meaning associated with operators in Hilbert space. Since this is a text on group theory for physicists, not a text on quantum field theory, we gloss over this point. We regard $\psi(x)$ as a two-component wave function rather than as a quantum field. Then, strictly speaking, we should write the relation $\psi(p)^\dagger = \psi(-p)$ stated earlier as $\psi(p)^* = \psi(-p)$, namely, as an equality between two column spinors. We do not transpose.

After a Fourier transform, the wave function in position space $\psi(x) = \psi(x)^*$ is consequently real. Introduce time by writing $\psi(t, x) = e^{-iEt}\psi(x)$. We thus obtain the wave equation* $i\frac{\partial}{\partial t}\psi(t, x) = (i\sigma_3 \frac{\partial}{\partial x} + \sigma_2 m)\psi(t, x)$. Multiplying this by σ_2, we finally arrive at

$$(i\gamma^\mu \partial_\mu - m)\psi(x) = 0 \tag{10}$$

---

* Needless to say, time, denoted as usual by t, is not to be confused with the parameter $t \equiv J_1 + J_2$ that appears in the lattice Hamiltonian.

where $\gamma^0 = \sigma_2$, $\gamma^1 = -i\sigma_1$ and where x now denotes the spacetime coordinates $x^\mu = (x^0, x^1) = (t, x)$.

Majorana and Weyl in $(1 + 1)$-dimensional spacetime

We can now simply carry over the discussion on charge conjugation and the Majorana equation given in chapter VII.5. You learned that the charge conjugate field is defined by $\psi_c = C\gamma^0\psi^*$, with C a matrix that satisfies $-\gamma^{\mu*} = (C\gamma^0)^{-1}\gamma^\mu(C\gamma^0)$. (Evidently, you can think of $C\gamma^0$ as one unit here.[5])

Interestingly, in the basis used here (and commonly in the condensed matter literature), both $\gamma^0 = \sigma_2$ and $\gamma^1 = -i\sigma_1$ are imaginary. The defining condition given above[6] for C becomes $\gamma^\mu = (C\gamma^0)^{-1}\gamma^\mu(C\gamma^0)$. Thus, we can choose $C\gamma^0 = I$ so that $\psi_c(x) = \psi^*(x)$. Hence, in the condensed matter literature, the term Majorana is often taken to mean

$$\psi_c(x) = \psi(x) = \psi^*(x) \tag{11}$$

In the basis used here, we define $\gamma_5 = -\gamma^0\gamma^1 = \sigma_3$. Thus, the Weyl condition (either $\psi = \frac{1}{2}(I + \gamma_5)\psi$ or $\psi = \frac{1}{2}(I - \gamma_5)\psi$) can coexist peacefully with the Majorana condition. In contrast, recall from chapter VII.5 that a spinor cannot be both Weyl and Majorana in $(3 + 1)$-dimensional spacetime.

Notes

1. I use the letter μ for Majorana.
2. The more advanced reader will have noticed that when written out in terms of the electron operator, this Hamiltonian contains not only $f^\dagger f$ terms, but also $f^\dagger f^\dagger$ and ff terms, and hence describes a system of electrons in the presence of superconducting order.
3. A. Kitaev, arXiv: cond-mat/0010440v2 (2000), Phys.-Usp. 44 (2001) 131. I follow closely the treatment given by Y. BenTov, arXiv:1412.0154v3 (2015).
4. Particles with momentum $p \simeq \pi + q$ can be treated in a similar fashion by expanding in q.
5. Indeed, some authors absorb γ^0 into C, but only to have it pop up somewhere else.
6. Notice that the condition is not basis invariant.

Part VIII | The Expanding Universe

In physics, we often let a parameter, such as the speed of light, tend to infinity. The group or its associated algebra is said to contract. The opposite of contracting an algebra is known as extending an algebra. While contracting an algebra merely requires the turning of a crank, extending an algebra in physics requires revolutionary insights. Indeed, the two great developments of early twentieth-century physics, special relativity and quantum mechanics, can both be thought of as extending the Galilean algebra. Here we discuss how to go from the Galilean algebra to the quantum Galilean algebra, also known as the Schrödinger algebra.

Almost from day one of school, we learned about mass (as in $ma = F$, for instance) but few of us know mass as a central charge that we can add to the Galilean algebra.

By adding the generators of dilation and conformal transformations, we extend the Poincaré algebra to the conformal algebra. As "Jerry the merchant" pointed out, when you are lost, it matters more to you to know that you are going in the right direction than to know how far you are from your destination.

The universe is expanding exponentially. What does group theory have to do with it? A lot, it turns out. The d-dimensional de Sitter spacetime that for $d = 4$ describes our universe to leading approximation is a coset manifold of two analytically continued orthogonal groups: $dS^d \equiv SO(d, 1)/SO(d - 1, 1)$.

VIII.1 | Contraction and Extension

The mathematical notion of contracting a group arises naturally in physics. When a parameter contained in a group, implicitly or explicitly, tends to infinity (or if you prefer, zero), the group or its associated algebra is said to contract.* Indeed, the Lorentz group could be chosen as the poster child of this phenomenon: when $c \to \infty$, it contracts to the Galilean group.

Reverting to the flat earth

For the simplest example, let us contract the rotation algebra $SO(3)$. Set $z = L\zeta$, with L some length and ζ some dimensionless number of order 1. If we let L tend to infinity, then $\frac{\partial}{\partial z} = L^{-1}\frac{\partial}{\partial \zeta} \to 0$. Referring back to the differential representation of the rotation generators given in chapter I.3, we have

$$J_z = -i\left(x\frac{\partial}{\partial y} - y\frac{\partial}{\partial x}\right)$$

$$J_x = -i\left(y\frac{\partial}{\partial z} - z\frac{\partial}{\partial y}\right) \to i(L\zeta)\frac{\partial}{\partial y} = L\zeta P_y$$

$$J_y = -i\left(z\frac{\partial}{\partial x} - x\frac{\partial}{\partial z}\right) \to -i(L\zeta)\frac{\partial}{\partial x} = -L\zeta P_x \tag{1}$$

We have written, as usual, the translation operators $P_a \equiv i\frac{\partial}{\partial x^a}, a = x, y$. Note that a rotation around the x-axis amounts to a large translation along the negative y direction, and around the y-axis to a translation along the positive x direction.

The commutation relation $[J_x, J_y] = iJ_z = -L^2\zeta^2[P_y, P_x]$ becomes $[P_x, P_y] = iL^{-2}\zeta^{-2}J_z \to 0$. The other two commutation relations, $[J_z, J_x] = iJ_y$ and so forth, become $[J_z, P_a] = i\varepsilon_{ab}P_b$, with $\varepsilon_{xy} = +1$ the 2-dimensional antisymmetric symbol.

---

* Our friend the jargon guy informs us that this is sometimes known as the Wigner-Inönü contraction.

The Lie algebra generated by P_x, P_y, and J_z (which we might as well call simply J) is known as $E(2)$, the invariance algebra of the Euclidean plane.*

The group $SO(3)$ contracts[†] to the group $E(2)$. This of course merely represents the regression from the round earth to the flat earth, which we humans thought of as home for millennia. You might have recognized the contraction parameter L as the radius of the earth. This discussion expresses Riemann's essential idea[1] that a (sufficiently smooth) curved space is locally flat. Over an everyday region of size much less than L, we can certainly take our world to be flat, described by $E(2)$.

Indeed, we can readily contract the algebra of $SO(N)$, given way back[2] in chapter I.3: $[J_{mn}, J_{pq}] = i(\delta_{mp}J_{nq} + \delta_{nq}J_{mp} - \delta_{np}J_{mq} - \delta_{mq}J_{np})$, with $m, n, p, q = 1, 2, \cdots, N$. Simply split these indices into two sets, with $i, j, k, l = 1, 2, \cdots, N-1$, and the index N. Set $x^N = L\zeta$, and let $L \to \infty$. Then $J_{iN} = -i(x^i \frac{\partial}{\partial x^N} - x^N \frac{\partial}{\partial x^i}) \to i(L\zeta)\frac{\partial}{\partial x^i} = L\zeta P_i$ is proportional to translation in the ith direction. Meanwhile, the J_{ij}s are unchanged and generate $SO(N-1)$. In other words, $SO(N)$ contracts to[‡] $E(N-1)$, consisting of the rotations $SO(N-1)$ and the $(N-1)$ translations.

From Lorentz to Galileo

Of more physical relevance is the contraction of the Lorentz algebra to the Galilean algebra. Start with the generators $J_{\mu\nu} = i(x_\mu \partial_\nu - x_\nu \partial_\mu)$ and the commutation relations

$$[J_{\mu\nu}, J_{\rho\sigma}] = -i(\eta_{\mu\rho}J_{\nu\sigma} + \eta_{\nu\sigma}J_{\mu\rho} - \eta_{\nu\rho}J_{\mu\sigma} - \eta_{\mu\sigma}J_{\nu\rho}) \tag{2}$$

given in (VII.2.41). Here $\eta_{\mu\nu}$ denotes the Minkowski metric, with the convention $\eta_{00} = +1$, $\eta_{ij} = -\delta_{ij}$.

In light of the preceding example, set $x^0 = ct$, and let the speed of light $c \to \infty$. Then

$$J_{0i} = i(x_0 \partial_i - x_i \partial_0) = i\left(ct\frac{\partial}{\partial x^i} + x^i\frac{1}{c}\frac{\partial}{\partial t}\right) \to ict\frac{\partial}{\partial x^i} = -ctP_i \tag{3}$$

In contrast, $J_{ij} = -i(x^i \frac{\partial}{\partial x^j} - x^j \frac{\partial}{\partial x^i})$ (which does not depend on c, of course) is unchanged. Note that the Galilean boost operator, which we define by $c^{-1}J_{0i}$, is proportional to spatial translation, with the proportionality factor given by the time coordinate t. Analogous to the flat earth example, we recover from $[J_{0i}, J_{0j}] = -iJ_{ij}$ the fact that translations commute: $[P_i, P_j] = O(1/c^2) \to 0$.

The commutation relations $[J_{0i}, J_{kl}] = i(\delta_{il}J_{0k} - \delta_{ik}J_{0l})$, after dividing out the common factor $(-ct)$, simply tell us how translations transform under rotations: $[J_{kl}, P_i] = i(\delta_{ki}P_l - \delta_{li}P_k)$, or in a more familiar notation, $[J_m, P_i] = [\frac{1}{2}\varepsilon_{mkl}J_{kl}, P_i] = \varepsilon_{mik}P_k$, as ex-

---

* As we had studied in interlude IV.13.
  [†] In fact, one motivation of the Wigner-Inönü paper was to study how spherical harmonics become Bessel functions.
  [‡] The jargon guy tells us that $E(N)$ is also known as $ISO(N)$.

pected. Finally, for completeness, we mention that the remaining commutations, $[J_{ij}, J_{kl}]$ (namely, those describing $SO(3)$), are evidently not affected by $c \to \infty$ at all.

Indeed, if we trust the various signs stemming from the metric to take care of themselves, contracting the Lorentz algebra corresponds to contracting $SO(4)$ to $E(3)$, as discussed in the preceding section. Our two examples are thus essentially the same.

Beyond Lorentz invariance

From these two examples, we see that it is algorithmic, and hence straightforward,[3] to contract an algebra. Simply introduce some parameter and let that parameter tend to infinity (or zero). But going in the opposite direction is not so easy. Indeed, going from the flat earth to the round earth, and going from Galilean spacetime to Minkowskian spacetime, mark two enormous, even epoch-changing, steps in physics.

We might invite ourselves to think what the next step might be, going beyond Minkowskian spacetime.

It would seem entirely natural to think that we would go from $SO(3, 1)$ to $SO(4, 1)$. Indeed, by now, it is almost self-evident how $SO(4, 1)$ would contract to the Poincaré algebra (which was mentioned in chapter VII.2 and already discussed in an earlier section modulo some signs). The algebra of $SO(4, 1)$ is given as in (2), where the indices run over 0, 1, 2, 3, and 4. To emphasize this, let us rename these indices pertaining to the 5-dimensional spacetime of $SO(4, 1)$ as A and B, so that A ranges over the Minkowski index $\mu = 0, 1, 2, 3$, and 4. Thus, the 10 generators J_{AB} divide into two sets, $J_{\mu\nu}$, the familiar generators of the Lorentz algebra, and $J_{\mu 4}$, generating translations in spacetime.

Most intriguingly, $SO(4, 1)$ is the invariance group[4] of an expanding universe (see chapter VIII.3) driven entirely by the vacuum energy known as the de Sitter universe.[5] It is in fact a good first approximation to the universe we inhabit. The length scale being sent to infinity in the contraction is none other than the size of the universe.

I am even tempted to speculate[6] that in a civilization in a galaxy far far away a brilliant person could be motivated by the attractive group theory of the 4-sphere S^4 to analyze the de Sitter group[7] $SO(4, 1)$ and thus discover the expanding universe "by pure thought," as Einstein put it.

Two ways of extending the Galilean algebra

The opposite of contracting an algebra is known as extending an algebra. Some authors refer to the process as deformation, but in general, I prefer a more descriptive terminology.[8]

As I just remarked, while contracting an algebra merely requires the turning of a crank, extending an algebra requires revolutionary insights. In chapter VII.2, we discussed how the Galilean algebra can be extended to the Poincaré algebra, corresponding to the development of special relativity. In a sense, we simply work backward. Here we discuss how to extend the Galilean algebra in another way, corresponding to the other great development of early twentieth-century physics: quantum mechanics. This is discussed in appendix 1.

Appendix 1: From the Galilean algebra to the quantum Galilean algebra: Mass as a central charge

Let us first review the Galilean algebra, derived in appendix 1 in chapter VII.2. I copy the algebra here for your convenience:

$$[J_i, J_j] = i\epsilon_{ijk}J_k, \qquad [J_i, P_j] = i\epsilon_{ijk}P_k, \qquad [J_i, K_j] = i\epsilon_{ijk}K_k$$

$$[P_i, H] = 0, \qquad [J_i, H] = 0, \qquad [P_i, P_j] = 0$$

$$[K_i, H] = iP_i, \qquad [K_i, P_j] = 0, \qquad [K_i, K_j] = 0 \tag{4}$$

Recall the generators of translation in space and in time:

$$P_i = i\frac{\partial}{\partial x^i} \quad \text{and} \quad H = i\frac{\partial}{\partial t} \tag{5}$$

We will have to be careful about signs here.[9]

As promised, we now extend (4) to accommodate quantum mechanics. We will see that we have to amend the commutator $[K_i, P_j] = 0$.

The astute reader would have noticed that thus far, we still do not have physics, only rotations and translations in space and in time. Indeed, H and P_i have dimensions of $1/T$ and $1/L$, respectively. To turn these into energy and momentum (as the notation suggests), we need a previously unknown fundamental constant.

Happily, Nature obliges and provides us with Planck's constant $\hbar$. From Planck's relation $E = \hbar\omega$, which turns a circular frequency into an energy, we see that $\hbar$ has dimensions of $M(L/T)^2/(1/T) = ML^2/T$. Alternatively, from the Heisenberg commutation relation $[q, p] = i\hbar$, we can also deduce the dimension of $\hbar$ to be equal to $L(ML/T) = ML^2/T$. Hence, if we multiply H and P by $\hbar$, then they would have dimensions $M(L/T)^2$ and $M(L/T)$, respectively (namely, the dimensions of energy and momentum, respectively).

To avoid introducing more symbols, let us agree to continue to denote by H and P the H and P written down earlier multiplied by $\hbar$. In other words, the generators of translation in time and in space have been promoted to energy (the Hamiltonian) and to momentum, respectively. For instance, $P_i = i\hbar\frac{\partial}{\partial x^i}$.

We do not touch J and K; they continue to be the generator of rotation and of boost, respectively.

Look at the Galilean algebra (4). Let us focus on the zero in the commutation $[K_i, P_j] = 0$. To a mathematician, this zero means that the commutator $[K_i, P_j]$ commutes with all other generators. For example, use the Jacobi identity to write $[H, [K_i, P_j]] = -[K_i, [P_j, H]] - [P_j, [H, K_i]]$. We can check that algebra (4) implies the vanishing of the right hand side, which is consistent with $[K_i, P_j]$ commuting with H on the left hand side.

Therefore, we can modify the algebra by setting $[K_i, P_j]$ equal to a number, known as a c-number (defined in this context as something that commutes with all the generators of the algebra), which we denote by* M:

$$[K_i, P_j] = iM\delta_{ij} \tag{6}$$

The δ_{ij} follows from rotational invariance. In mathematics, the modification of an algebra by changing the 0 in a commutation relation to a c-number is known as a central extension.

Since K has the dimension of T/L, and P the dimension of ML/T, the c-number M has dimension of mass. Hermitean conjugating (6), we deduce that M is real.

We have discovered something called mass, known to mathematicians in this context as a central charge. By the way, I learned this basically the first day in graduate school[10] from Julian Schwinger's course on quantum field theory. So I thought to myself, "Wow, it's a really good idea to come to graduate school, because when I was an undergraduate, they told me that mass was the proportionality factor in $F = ma$. But in graduate school they

---

* An M which, pace Confusio, is not to be confused with the symbol M we used when doing dimensional analysis.

now tell me something much more interesting, that mass is the central charge* in some algebra. So graduate school is definitely worthwhile."

To show that M is indeed the mass we first encountered in our study of physics, consider this particular object $\frac{P^2}{2M}$. Under a boost,

$$\left[K_i, \frac{P^2}{2M}\right] = \frac{1}{2M}\left([K_i, P_j]P_j + P_j[K_i, P_j]\right)$$
$$= \frac{1}{2M}(2iM\delta_{ij})P_j = iP_i \tag{7}$$

But we also have $[K_i, H] = iP_i$ from (4). Thus, $[K_i, H - \frac{P^2}{2M}] = 0$. Furthermore, $(H - \frac{P^2}{2M})$ also commutes with J_i, H, and P. Therefore, it is consistent to write

$$H = \frac{P^2}{2M} \tag{8}$$

with the trivial freedom of adding an arbitrary c-number. Identifying H and P_i with the generators of translation in time and space (multiplied by $\hbar$), we obtain Schrödinger's equation:

$$i\hbar\frac{\partial}{\partial t}\psi = -\frac{\hbar^2\nabla^2}{2M}\psi \tag{9}$$

The classical Galilean algebra (4) is modified by (6) to become the quantum Galilean algebra.[†]

Equivalently, we can also say that we modify the boost generator $K_i = -tP_i$ in such a way as to reproduce (6). Let

$$K_i = -tP_i - Mx^i \tag{10}$$

To avoid any confusion about signs, let us also write this without indices:

$$K_x = -it\frac{\partial}{\partial x} - Mx \tag{11}$$

Thus,

$$[K_x, P_x] = \left[-it\frac{\partial}{\partial x} - Mx, i\frac{\partial}{\partial x}\right] = (-)^2\left[i\frac{\partial}{\partial x}, Mx\right] = +iM \tag{12}$$

This agrees with (6) (clearly, K_i and P_j commute if $i \neq j$).

Another way to understand the central charge is to take a detour through the Poincaré algebra, which tells us that under a boost, energy and momentum transform into each other. Indeed, this is one way to verify that our signs here are correct, since Lorentz invariance interlocks various terms that are decoupled in nonrelativistic physics, so to speak. Pass to the relativistic regime:

$$K_x = -it\frac{\partial}{\partial x} \rightarrow K_x = -i\left(t\frac{\partial}{\partial x} + x\frac{\partial}{\partial t}\right) \tag{13}$$

The relative $+$ sign between the $\frac{\partial}{\partial x}$ and $\frac{\partial}{\partial t}$ was explained in chapter VII.2. Since $P_x = i\frac{\partial}{\partial x}$, we have

$$[K_x, P_x] = \left[-ix\frac{\partial}{\partial t}, i\frac{\partial}{\partial x}\right] = (-i)^2\left[\frac{\partial}{\partial x}, x\frac{\partial}{\partial t}\right] = -\frac{\partial}{\partial t} = +iH \tag{14}$$

---

* Although my memory is vague, I am almost certain that Schwinger did not use this term.

[†] The jargon guy says that this is also known as the centrally extended Galilean algebra and as the Bargmann algebra. I take the liberty to remark here that as an undergraduate, I once took a course from Valentine Bargmann, mostly about the Lorentz group, but I don't remember him mentioning any of this.

Let us restore c momentarily and then return to the nonrelativistic regime:

$$[K_x, P_x] = i\frac{1}{c^2}H = i\frac{1}{c^2}\left(Mc^2 + \frac{P^2}{2M} + \cdots\right) = iM + \cdots \tag{15}$$

In the limit $c \to \infty$, the central charge[11] emerges in front of our very eyes.

Indeed, to check that the signs in (11) are correct, take the non-relativistic limit of (13):

$$K_x = -i\left(t\frac{\partial}{\partial x} + x\frac{\partial}{\partial t}\right) = -i\left(t\frac{\partial}{\partial x} - ixH\right) \to -it\frac{\partial}{\partial x} - Mx \tag{16}$$

We have simply replaced H by M.

Appendix 2: Central charge and de Broglie's relation

I now mention a pedagogically illuminating episode[12] in physics that serves to highlight the crucial role played by the central charge (which for our purpose here, I am using as a shorthand for (6)) in nonrelativistic quantum mechanics.

In 1975, Landé pointed out[13] that de Broglie's relation between the momentum p of a particle and its wavelength λ, namely, $p = \hbar k = h/\lambda$ (with the wave number $k = 2\pi/\lambda$, as usual), is inconsistent with Galilean invariance. This turned out to be a pseudo-paradox, since we know perfectly well that nonrelativistic quantum mechanics should respect Galilean invariance. Let us first go over Landé's remark more explicitly and then seek a resolution.

From $x' = x + ut$, we see that the distance between two points (say, two successive crests of a wave) is unchanged[14] under a Galilean transformation at a given instant of time (that is, $\Delta x' = \Delta x$ since $t' = t$). In other words, the wavelength λ, and hence the wave number k, is unchanged.

In contrast, we all know that momentum of the particle allegedly described by the de Broglie wave changes. Indeed, the everyday law of addition of velocity, alluded to in chapter VII.2, states that $\frac{dx'}{dt'} = \frac{d(x+ut)}{dt} = \frac{dx}{dt} + u$. Thus,

$$p' = M\frac{dx'}{dt'} = M\frac{dx}{dt} + Mu = p + Mu \tag{17}$$

Hence $p' \neq p$ but $k' = k$, which would appear to contradict $p = \hbar k$. So, apparently de Broglie clashes with Galileo.

But . . . but, we just derived Schrödinger's equation (9) from the Galilean algebra, and Schrödinger contains de Broglie. So what's going on?

Landé's argument can be rendered more explicit by writing the expression for a wave: $\psi(x, t) = A\sin(kx - \omega t)$. In a moving frame, since the amplitude of the wave is unchanged, being perpendicular to the direction of relative motion (I'm not bothering to draw a figure here; you can supply one), the wave is given by $\psi'(x', t') = \psi(x, t)$, and thus

$$A\sin(k'x' - \omega't') = A\sin(kx - \omega t) = A\sin(k(x' - ut') - \omega t') = A\sin(kx' - (\omega + uk)t') \tag{18}$$

where we have plugged in $t = t'$ and $x = x' - ut'$.

This simple exercise shows $k' = k$, which is just what we said, that the wavelengths seen by the two observers are the same.

In fact, we obtain more—namely, that $\omega' = \omega + uk$, which we recognize as the Doppler effect. This shows, by the way, that if Landé's argument holds, the Planck-Einstein relation $E = \hbar\omega$ is in just as much trouble as de Broglie's relation, since under a Galilean transformation, energy transforms by

$$E \to E' = \frac{p'^2}{2M} = \frac{(p + Mu)^2}{2M} = E + up + \frac{1}{2}Mu^2 \tag{19}$$

which is quite different from the way frequency transforms.

Do you see the resolution to this apparent paradox?

The resolution is that in quantum mechanics, the wave $\psi(x, t)$ is complex, and Galilean invariance only requires that $|\psi'(x', t')|^2 = |\psi(x, t)|^2$, not $\psi'(x', t') = \psi(x, t)$. In other words, possibly $\psi'(x', t') = e^{i\varphi(x,t)}\psi(x, t)$ with some phase factor.

You can also work out explicitly what the phase factor is for a finite Galilean transformation. Consider a plane wave solution $\psi(x, t) = e^{\frac{i}{\hbar}(px-Et)}$ of the Schrödinger equation $i\hbar\frac{\partial}{\partial t}\psi = -\frac{\hbar^2\nabla^2}{2M}\psi$ in (9). (We restore $\hbar$ temporarily to make a point below.) Then the plane wave solution of the transformed Schrödinger's equation $i\hbar\frac{\partial}{\partial t'}\psi' = -\frac{\hbar^2\nabla'^2}{2M}\psi'$ is given by

$$\psi'(x', t') = e^{\frac{i}{\hbar}(p'x'-E't')} = e^{\frac{i}{\hbar}(p'(x+ut)-E't)} = e^{\frac{i}{\hbar}(p'x-(E'-up')t)}$$

$$= e^{\frac{i}{\hbar}\left((p+Mu)x-(E-\frac{1}{2}Mu^2)t\right)}$$

$$= e^{\frac{iM}{\hbar}(ux+\frac{1}{2}u^2t)}\psi(x, t) \tag{20}$$

where we used (17) and (19): in particular, $E' - up' = E + up + \frac{1}{2}Mu^2 - u(p + Mu) = E - \frac{1}{2}Mu^2$.

We remark that there is no sense in which the phase factor that saves us would disappear in the limit $\hbar \to 0$. The phase factor[15] $e^{i\varphi(x,t)}$ in (20) reflects the presence of the central charge in (10). To see this, take the infinitesimal limit $u \to 0$ to get at the generator of a Galilean boost. To linear order in u,

$$\psi'(x, t) = e^{i\varphi(x-ut,t)}\psi(x - ut, t) \simeq \left(1 + i\frac{M}{\hbar}(ux + \cdots)\right)\left(1 - ut\frac{\partial}{\partial x} + \cdots\right)\psi(x, t)$$

$$= \left(1 + iu\left(\frac{M}{\hbar}x + it\frac{\partial}{\partial x} + \cdots\right)\right)\psi(x, t) = (1 - iuK_x)\psi(x, t) \tag{21}$$

This shows that in the quantum world, the generator K_x has to be modified. By the way, this also verifies the relative sign in (11).

Math and physics work, as always.

Appendix 3: Contraction and the Jacobi identity

Another way of obtaining a subalgebra takes advantage of the Jacobi identity.[16] Given an algebra consisting of the generators $\{A_1, A_2, \cdots\}$, pick one of these generators and call it C. Consider all those elements that commute with C and call them B_n (which of course includes C itself). Then the B_ns form a subalgebra. To see this, observe that $[[B_m, B_n], C] = -[[B_n, C], B_m] - [[C, B_m], B_n] = 0$. Thus, $[B_m, B_n]$ belongs to the subalgebra. Indeed, C is then a central charge of this algebra.

As a simple example, take $SU(3)$ and let $C = \lambda_8$. Then the B_ns form an $SU(2) \otimes U(1)$ algebra.

Notes

1. See, for example, G Nut, p. 82.
2. We are now sophisticated enough to drop the parentheses in $J_{(mn)}$.
3. That is, if we ignore various issues real mathematicians would worry about, such as the existence of the limit, and so on and so forth.
4. This invariance group is known as the isometry group of the spacetime. See chapter IX.6 in G Nut.
5. See G Nut, p. 359; also see chapter IX.10 in that book.
6. Which I did elsewhere. See G Nut, p. 645.
7. For the group theory of $SO(4, 1)$, see chapter VIII.3 and p. 644 in G Nut.
8. Besides, deformation sounds derogatory.
9. We go over some tricky signs here for the benefit of some readers. This note is written in a somewhat telegraphic style, addressed to those with some nodding familiarity with basic special relativity and quantum mechanics. We use the so-called Bjorken-Drell convention (see J. D. Bjorken and S. Drell, Relativistic Quantum Mechanics; the same convention is used in QFT Nut) in which the coordinates are $x^\mu = (x^0, x^i) = (t, \vec{x})$, and the diagonal elements of the metric $\eta_{\mu\nu}$ are given by $(+1, -1, -1, -1)$. We start with an indisputable sign, namely, $H = +i\frac{\partial}{\partial t}$, because in everybody's convention the Schrödinger equation reads $i\hbar\frac{\partial}{\partial t}\psi = H\psi$. (We set $\hbar = 1$ henceforth.) Since $H = P^0 = P_0$, this implies $P_\mu = i\frac{\partial}{\partial x^\mu} = i\partial_\mu$, and thus, $P_i = i\frac{\partial}{\partial x^i}$, as given in the text. In the Bjorken-Drell convention, $P^i = -P_i$, and thus the issue is whether P_i or P^i corresponds to what I will

call the football player's momentum $\vec{P} = m\vec{v}$. This is resolved by the relativistic equation $P^\mu = mdx^\mu/d\tau$, with $P^i = mdx^i/d\tau \rightarrow mdx^i/dt$ in the nonrelativistic limit. Just as $\vec{x}$ has components x^i, the football player's momentum $\vec{P}$ has components P^i. Thus, $P^i = \frac{1}{i}\frac{\partial}{\partial x^i} = -i\frac{\partial}{\partial x^i}$. (Note that, with this sign, the signs in (1) are consistent with the elementary formula for angular momentum $\vec{J} = \vec{x} \times \vec{P}$. Note also that this is consistent with Heisenberg's $[p, q] = -i$.)

10. In fact, Schwinger probably went three times faster than my presentation, and I understood little of what he said. Consequently, as I will explain, I learned quantum field theory and Mandarin during my first year in graduate school. Both turn out to be very important in my life.

   The serious theory students at Harvard in those days attended Schwinger's course year after year in hopes of gleaning some insights. The older students impressed upon the first year students that if you raised your hand, Schwinger might think that you were an idiot and kick you out of his class. So nobody asked questions. Schwinger would already be orating the first sentence of his lecture as he came in the door of the lecture room, and would utter the last sentence as he sailed out the door at the end of the class.

   At that time, the number one physics undergraduate in Taiwan was sent to Harvard every year, and all of them wanted to work with the legendary Nobel Prize winner. Part of the legend was that Schwinger liked to work at night and get up late, and so his lecture was always held at eleven. After the lecture, it was noon and I would try to go to lunch with the students from Taiwan and ask them questions about quantum field theory. But they all spoke Mandarin, and said, "You come from Brazil and only speak Shanghainese and Cantonese; you really should learn Mandarin." So I learned quantum field theory and Mandarin at the same time, which probably squared the difficulty.

   By the way, Schwinger was very fond of the students from Taiwan, and one of them, for his wedding, asked Schwinger to give away the bride (her father could not come; in those days international travel was far less common and affordable). Since Schwinger and his wife did not have children of their own, they agreed enthusiastically and beamed throughout the ceremony. I discovered on this social occasion that Schwinger was in fact extremely approachable and friendly.

11. Incidentally, the material discussed here is very much used on the forefront of physics. See, for example, D. Son et al., arXiv:1407.7540, and the references therein.

12. I am grateful to C. K. Lee for telling me about the Landé pseudo-paradox.

13. I follow the treatment given by J.-M. Lévy-Leblond, Am. J. Phys. 44 (1976), p. 1130. Various references, including that of A. Landé's paper, can be found in this paper.

14. To paraphrase Landé, a photo of ocean waves taken from a ship and a photo of ocean waves taken from a low flying airplane look the same.

15. The jargon guy tells us that this is related to the cocycles discussed in the mathematical physics literature.

16. See D. T. Son, arXiv:0804.3972.

VIII.2 | The Conformal Algebra

Conformal transformations

Besides rotation and translation, we can apply other transformations to space. In particular, one familiar transformation often practiced in everyday life is a change of scale, magnifying or shrinking a picture (now doable with the flick of two fingers). And from space it is but a hop to spacetime by flipping a sign between us friends, so that rotation gets generalized to Lorentz transformation.

As we shall see, the set of transformations that preserve* the angle between any two line segments generates the conformal algebra. Historically, after special relativity was established, people studied the invariances of the free Maxwell equations (that is, the equations without any charges and currents) and found that they generated the conformal algebra.[1]

Here we consider only flat Euclidean space or Minkowskian spacetime, but as is often the case in math and in physics, to set the stage, it is actually better to be more general and consider a curved space (or spacetime). Curved spaces are defined in general by the metric $g_{\mu\nu}(x)$, such that the distance squared between two neighboring points with coordinates x^μ and $x^\mu + dx^\mu$ is given by

$$ds^2 = g_{\mu\nu}(x)dx^\mu dx^\nu \tag{1}$$

(This is in fact the starting point of Riemann's development of differential geometry[2] used in Einstein's theory of gravity.) A familiar example is the sphere, on which $ds^2 = d\theta^2 + \sin^2\theta d\varphi^2$, with θ and φ the familiar angular coordinates. We have already touched on curved spaces in this text on several occasions, for example, when discussing integration over group manifolds in chapter IV.7. The notion of a metric was mentioned peripherally in our study of the general Lie algebra in chapter VI.3.

---

* Clearly, changes of scale do not preserve the length of a line segment.

Thus, we will set up the framework with a general metric $g_{\mu\nu}(x)$, but promptly set $g_{\mu\nu}(x)$ to be equal to $\delta_{\mu\nu}$ for flat space or to $\eta_{\mu\nu}$ for flat spacetime.

Under a change of coordinates $x \to x'(x)$, we have, according to the rules of differential calculus, $dx^\mu = \frac{\partial x^\mu}{\partial x'^\rho} dx'^\rho$. The distance between two nearby points ds is of course independent of our coordinate system. Thus, plugging into (1), we obtain

$$ds^2 = g_{\mu\nu}(x) \frac{\partial x^\mu}{\partial x'^\rho} \frac{\partial x^\nu}{\partial x'^\sigma} dx'^\rho dx'^\sigma = g'_{\rho\sigma}(x') dx'^\rho dx'^\sigma \tag{2}$$

Here the second equality is merely the definition of the metric $g'_{\rho\sigma}(x')$ in the new coordinate system. Thus, under a change of coordinates, the metric transforms according to

$$g'_{\rho\sigma}(x') = g_{\mu\nu}(x) \frac{\partial x^\mu}{\partial x'^\rho} \frac{\partial x^\nu}{\partial x'^\sigma} \tag{3}$$

For example, under a scale transformation $x^\mu \to x'^\mu = \lambda x^\mu$ (for λ a real number), $g'_{\rho\sigma}(x') = g'_{\rho\sigma}(\lambda x) = \lambda^2 g_{\mu\nu}(x)$.

Suppose we now impose the condition

$$g'_{\rho\sigma}(x') = \Omega^2(x') g_{\rho\sigma}(x') \tag{4}$$

for some unknown function* Ω. Transformations $x \to x'(x)$ that satisfy (4) are known as[†] conformal transformations. In fact, we just saw an example, scale transformation, which corresponds to $\Omega(x) = \lambda$ being a constant.

Note, by the way, that in (4), x' is a dummy variable, and so we could, for example, simply erase the prime on x'. But of course we are not allowed to erase the prime on $g'_{\rho\sigma}$; (4) tells us that the new metric is given by the old metric multiplied by the function Ω^2.

Conformal transformations clearly form a group. After a conformal transformation associated with Ω_1, perform another one associated with Ω_2. We end up with a conformal transformation associated with $\Omega_2\Omega_1$:

$$g''_{\rho\sigma}(x) = \Omega_2^2(x) g'_{\rho\sigma}(x) = \Omega_2^2(x)\Omega_1^2(x) g_{\rho\sigma}(x) = (\Omega_2(x)\Omega_1(x))^2 g_{\rho\sigma}(x) \tag{5}$$

Following a long line of luminaries from Newton to Lie, we now go to the infinitesimal limit. Consider the infinitesimal coordinate transformation $x'^\mu = x^\mu + \varepsilon\xi^\mu(x)$. For $\varepsilon \to 0$, we expand $\Omega^2(x') \simeq 1 + \varepsilon\kappa(x') = 1 + \varepsilon\kappa(x) + O(\varepsilon^2)$ with some unknown function $\kappa(x)$. Plug this into the condition (4). Collecting terms of order ε, we obtain what is known as the conformal Killing condition:

$$g_{\mu\sigma}\partial_\rho\xi^\mu + g_{\rho\nu}\partial_\sigma\xi^\nu + \xi^\lambda\partial_\lambda g_{\rho\sigma} + \kappa g_{\rho\sigma} = 0 \tag{6}$$

(For $\kappa = 0$, the conformal Killing condition reduces to what is called the isometry condition. Indeed, setting $\Omega = 1$, we see that (4) just says that the two metrics are the same.)

---

* The square on Ω^2 is conventional.
[†] The important special case with $\Omega(x) = 1$ is known as an isometry (= equal measure). Our friend the jargon guy is overcome with joy.

We could eliminate $\kappa(x)$ in (6) by contracting with $g^{\rho\sigma}$ (the inverse of $g_{\mu\nu}$), so that this amounts to a condition on the metric $g_{\mu\nu}$ and the vector field ξ, known as a conformal Killing vector field. See below.

To the lost, angles are more important than distances

Some readers may be aware of the many motivations—historical, mathematical, and physical—for studying conformal transformations. Here I mention one familiar from everyday life. The key property is of course that while conformal transformations do not preserve the length between neighboring points, they do preserve the angle between two line segments.

When you are lost, it matters more to you to know that you are going in the right direction than to know how far you are from your destination. To the lost, angles are more important than distances. Gerardus Mercator (1512–1594), or "Jerry the merchant," fully appreciated this. The familiar Mercator map of the world is obtained by a conformal transformation of the spherical coordinates θ and φ on the globe, at the price of stretching lengths near the two poles.[3]

Retreat to flat spacetime

If someone hands us a metric $g_{\mu\nu}(x)$, we could in principle find its conformal Killing vectors by solving (6).

The simplest metric to deal with is the Minkowski metric, of course. (By the way, spacetime with the metric given by $\Omega^2(x)\eta_{\mu\nu}$, as in (4), is known as conformally flat.) Here we content ourselves by studying this easy case, for which (6) simplifies to (with $\xi_\sigma \equiv \eta_{\sigma\mu}\xi^\mu$, as usual) $\partial_\rho\xi_\sigma + \partial_\sigma\xi_\rho + \kappa\eta_{\rho\sigma} = 0$. Contracting this with $\eta^{\rho\sigma}$, we obtain $\kappa = -2\partial\cdot\xi/d$ in d-dimensional spacetime. Hence the condition (6) becomes

$$\partial_\rho\xi_\sigma + \partial_\sigma\xi_\rho = \frac{2}{d}\eta_{\rho\sigma}\partial\cdot\xi \tag{7}$$

Infinitesimal transformations $x'^\mu = x^\mu + \varepsilon\xi^\mu(x)$ that satisfy (7) are said to generate the conformal algebra for Minkowski spacetime. Clearly, with the substitution of $\delta_{\rho\sigma}$ for $\eta_{\rho\sigma}$, this entire discussion applies to flat space as well as to flat spacetime. As promised, we are now done with $g_{\mu\nu}$ and can forget about him.

As mentioned at the start of this chapter, we already know that translations and Lorentz transformations solve (7), namely, $\xi^\mu = a^\mu + b^\mu_{\nu}x^\nu$, with $b^{\mu\nu} = b^\mu_{\lambda}\eta^{\lambda\nu} = -b^{\nu\mu}$ required to be antisymmetric (see chapter VII.2). Note that the right hand side of (7) vanishes for this class of solutions.

In search of conformal generators

At this point, you should look for more solutions of (7). Go ahead. I already gave you a huge hint.

We could also wing it like the poor man. Stare at Minkowski spacetime: $ds^2 = \eta_{\mu\nu}dx^\mu dx^\nu$. What transformations on x would change the metric conformally? Well, I already mentioned earlier that the scale transformation, or more academically, dilation,[4] $x^\mu \to \lambda x^\mu$ (for λ a real number), that is, stretching or shrinking spacetime by a constant factor, is a solution. To identify the corresponding ξ^μ, consider an infinitesimal transformation with $\lambda = 1 + \varepsilon c$; then $\xi^\mu = cx^\mu$, with some (irrelevant) constant c. Sure enough, this satisfies (7), of course.

Now, can you find another transformation? Think for a minute before reading on.

The clever poor man notices that inversion* $x^\mu = e^2 y^\mu / y^2$ would solve (4). Plug in $dx^\mu = e^2(\delta^\mu_\lambda y^2 - 2y_\lambda y^\mu)dy^\lambda/(y^2)^2$. We obtain $ds^2 = \eta_{\mu\nu}dx^\mu dx^\nu = (e^4/(y^2)^2)\eta_{\mu\nu}dy^\mu dy^\nu$, which indeed is conformally flat. I introduced e to avoid confusing you, but now that its job is done, we will set it to 1 and define inversion as the transformation (for $x^2 \neq 0$)

$$x^\mu \to \frac{x^\mu}{x^2} \tag{8}$$

Schematically, x goes into $1/x$: inversion does what the word suggests.

You object, saying that the entire discussion has been couched in terms of infinitesimal transformations. The inversion is a discrete transformation, and is in no way no how infinitesimal.

That is a perfectly valid objection. There is no parameter in (8) that we can send to 0. How then can we identify the corresponding ξ^μ?

Invert, translate, then invert

Now the poor man makes another clever move: invert, translate by some vector a^μ, and then invert back. The composition of these three transformations is surely a conformal transformation, since inversion and translation are both conformal transformations. For $a^\mu = 0$, the two inversions knock each other out, and we end up with the identity transformation. Thus, as $a^\mu \to 0$, the composition of inversion, translation, and inversion would indeed be a conformal transformation infinitesimally close to the identity.

Let's work out what I just said in words:

$$x^\mu \to \frac{x^\mu}{x^2} \to \frac{x^\mu}{x^2} + a^\mu \to \left(\frac{x^\mu}{x^2} + a^\mu\right) \bigg/ \left(\eta_{\rho\sigma}\left(\frac{x^\rho}{x^2} + a^\rho\right)\left(\frac{x^\sigma}{x^2} + a^\sigma\right)\right)$$

$$= \left(\frac{x^\mu}{x^2} + a^\mu\right) \bigg/ \left(\frac{1}{x^2} + \frac{2a \cdot x}{x^2} + a^2\right)$$

$$= (x^\mu + a^\mu x^2)/(1 + 2a \cdot x + a^2 x^2) \simeq (x^\mu + a^\mu x^2)(1 - 2a \cdot x) + O(a^2)$$

$$= x^\mu + a_\lambda(\eta^{\mu\lambda}x^2 - 2x^\mu x^\lambda) + O(a^2) \tag{9}$$

* An irrelevant constant e, with dimension of length, is introduced here to ensure that x and y both have dimensions of length.

The transformation $x^\mu \to x^\mu + a_\lambda(\eta^{\mu\lambda}x^2 - 2x^\mu x^\lambda)$ is sometimes known, specifically, as a conformal transformation. You can verify that $\xi^\mu = a_\lambda(\eta^{\mu\lambda}x^2 - 2x^\mu x^\lambda)$ satisfies (7), of course.

The corresponding generators are denoted by[*]

$$K^\mu = (\eta^{\mu\nu}x^2 - 2x^\mu x^\nu)\partial_\nu \tag{10}$$

As I said, you could have also simply solved (7) by brute force, and I am counting on you to have already done so. It is also instructive to act with $\partial^\rho \equiv \eta^{\rho\sigma}\partial_\sigma$ on (7); we obtain

$$d\partial^2\xi_\sigma = (2-d)\partial_\sigma(\partial \cdot \xi) \tag{11}$$

Applying ∂^σ, we obtain further $\partial^2(\partial \cdot \xi) = 0$ (all for $d \neq 1$).

We now draw two important conclusions.

1. The case $d = 2$ is special. We learn from (11) that any solutions of the generalized Laplace equation $\partial^2\xi_\nu = 0$ yield a conformal transformation. Indeed, for $d = 2$, either go to light cone coordinates for Minkowski spacetime, or to complex coordinates for Euclidean space. With complex coordinates $z = x + iy$, we have $(\partial_x^2 + \partial_y^2)\xi_\sigma = (\partial_x + i\partial_y)(\partial_x - i\partial_y)\xi_\sigma = \frac{\partial}{\partial z^*}\frac{\partial}{\partial z}\xi_\sigma = 0$, and hence we can exploit the full power of complex analysis.[5] For $d = 2$, there exist an infinite number of solutions of (7) for ξ^μ.

2. For $d \neq 2$, these equations tell us that ξ^μ can depend on x at most quadratically. Thus, we have in fact found all the solutions of (7) for $d \neq 2$, namely,

$$\xi^\mu = a^\mu + b^\mu_{\ \nu}x^\nu + cx^\mu + d_\nu(\eta^{\mu\nu}x^2 - 2x^\mu x^\nu) \tag{12}$$

with $b^{\mu\nu}$ antisymmetric. We had noted all these terms already. Pleasingly, in (12), the constant term corresponds to translation, the linear terms to Lorentz transformation and to dilation, and the quadratic term to conformal transformation.

Generators of conformal algebra

Associated with each of these terms, we have a generator of the Minkowskian conformal algebra. As in chapters I.3 and VII.2, it is convenient to use a differential operator representation. Recall that in chapter VII.2, by adding the generators[6] of translation P_μ to those of Lorentz transformation $J_{\mu\nu}$, we extended the Lorentz algebra to the Poincaré algebra, defined by commuting

$$P_\mu = \partial_\mu \quad \text{and} \quad J_{\mu\nu} = (x_\mu\partial_\nu - x_\nu\partial_\mu) \tag{13}$$

By adding the dilation generator D and the conformal generator K^μ,

$$D = x^\mu\partial_\mu \quad \text{and} \quad K^\mu = (\eta^{\mu\nu}x^2 - 2x^\mu x^\nu)\partial_\nu \tag{14}$$

---

[*] Even Confusio would not confuse the conformal generators K^μ with the generators of Lorentz boosts discussed in chapters VII.2 and VII.4. In the present context, the latter are contained in $J_{\mu\nu}$, the generators of Lorentz transformations. There are only so many letters in the alphabet, and the use of K for both is more or less standard.

we can now, in turn, extend the Poincaré algebra to the conformal algebra, defined by commuting P_μ, $J_{\mu\nu}$, D, and K_μ.

In other words, the commutators between P, J, D, and K generate an algebra that contains the Poincaré algebra.

The commutators involving D are easy to compute: $[D, x^\nu] = [x^\mu \partial_\mu, x^\nu] = x^\mu[\partial_\mu, x^\nu] = x^\nu$, and $[D, \partial_\nu] = [x^\mu \partial_\mu, \partial_\nu] = [x^\mu, \partial_\nu]\partial_\mu = -\partial_\nu$. (To work out various commutators, keep in mind the identity $[A, BC] = [A, B]C + B[A, C]$.) Evidently, D (as is sensible for a dilation generator) simply counts the length dimension, $+1$ for x^ν and -1 for ∂_ν. Thus, $[D, J_{\mu\nu}] = 0$, since $J \sim x\partial$ has zero length dimension. Interestingly, another way of reading this is to write it as $[J_{\mu\nu}, D] = 0$, which says that D is a Lorentz scalar. Next, we can read off $[D, P^\mu] = -P^\mu$ and $[D, K^\mu] = +K^\mu$ just by counting powers of length dimension ($P \sim \partial$, $K \sim xx\partial$).

The commutators involving K^μ are not much harder to work out. First, $[J^{\mu\nu}, K^\lambda] = -\eta^{\mu\lambda}K^\nu + \eta^{\nu\lambda}K^\mu$ just tells us that K^μ transforms like a vector, as expected. The nontrivial commutator is $[K^\mu, P^\lambda] = -[\partial^\lambda, (\eta^{\mu\nu}x^2 - 2x^\mu x^\nu)\partial_\nu] = -2(\eta^{\mu\nu}x^\lambda - \eta^{\lambda\mu}x^\nu - x^\mu \eta^{\lambda\nu})\partial_\nu = 2(J^{\mu\lambda} + \eta^{\lambda\mu}D)$. Finally, verify that $[K^\mu, K^\nu] = 0$. Can you see why? (Recall that we constructed the conformal transformation as an inversion followed by a translation and then followed by another inversion.)

Collecting our results, we have the conformal algebra

$$[P^\mu, P^\nu] = 0, \qquad [K^\mu, K^\nu] = 0$$

$$[D, P^\mu] = -P^\mu, \qquad [D, J_{\mu\nu}] = 0, \qquad [D, K^\mu] = +K^\mu$$

$$[J^{\mu\nu}, P^\lambda] = -\eta^{\mu\lambda}P^\nu + \eta^{\nu\lambda}P^\mu, \qquad [J^{\mu\nu}, K^\lambda] = -\eta^{\mu\lambda}K^\nu + \eta^{\nu\lambda}K^\mu$$

$$[J^{\mu\nu}, J^{\lambda\rho}] = -\eta^{\mu\lambda}J^{\nu\rho} - \eta^{\nu\rho}J^{\mu\lambda} + \eta^{\nu\lambda}J^{\mu\rho} + \eta^{\mu\rho}J^{\nu\lambda}$$

$$[K^\mu, P^\nu] = 2(J^{\mu\nu} + \eta^{\mu\nu}D) \tag{15}$$

We see that, in some sense, K acts like the dual of P.

As we saw just now, the dilation generator D counts the power-of-length dimension: so that $P \sim \partial$ counts as -1, $D \sim x\partial$ and $J \sim x\partial$ both count as 0, and $K \sim xx\partial$ counts as $+1$. When we commute these generators, we "annihilate" ∂ against x, but this process preserves the power-of-length dimension. Thus, whatever $[K, P]$ is, it counts as 0, and thus must be a linear combination of D and J. This counting also serves as a mnemonic for $[K, K] \sim 0$, since $[K, K]$ counts as $+2$ and there aren't any generators around that count as $+2$. (Another way of remembering this is that K is sort of the dual of P and $[P, P] \sim 0$.)

In light of this counting scheme, the apparent "mess" in (15) is actually easy to understand. We just explained the first and last lines. The second line tells us about the length dimension of D, P, and K. The third line says that P^μ and K^μ are Lorentz vectors. The fourth line is of course just the Lorentz algebra, which in some sense we have already encountered way back in chapter I.3.

Identifying the conformal algebra

Now that we have used our eyeballs and brains, let's use our fingers. Count the number of generators (P, K, D, J): $d + d + 1 + \frac{1}{2}d(d-1) = \frac{1}{2}(d+2)(d+1)$. Do you know an algebra with this many generators? Hint: It has to contain the Lorentz algebra $SO(d-1, 1)$ of d-dimensional Minkowski spacetime.

Yes, $SO(d, 2)$. Good guess!

Remarkably, the conformal algebra of d-dimensional Minkowski spacetime, with the invariance group $SO(d-1, 1)$, is the Lie algebra of $SO(d, 2)$, the Lorentz algebra of $(d+2)$-dimensional Minkowski spacetime. The two algebras are isomorphic. Let that sink in for a minute.

The rule is that, given $SO(d-1, 1)$, the conformal algebra is $SO(d-1+1, 1+1) = SO(d, 2)$: we "go up by $(1, 1)$", so to speak. We can prove this assertion by the "what else could it be" argument. We could of course verify the assertion by direct computation and thus also ascertain the signature. Instead, we will argue our way through.

Denote the generators of $SO(d, 2)$ by J^{MN}, with $M, N = 0, 1, 2, \cdots, d-1, d, d+1$ (and $\mu, \nu = 0, 1, 2, \cdots, d-1$) satisfying

$$[J^{MN}, J^{PQ}] = -\eta^{MP} J^{NQ} - \eta^{NQ} J^{MP} + \eta^{NP} J^{MQ} + \eta^{MQ} J^{NP} \tag{16}$$

with $\eta^{00} = +1$, $\eta^{ij} = -\delta^{ij}$, $\eta^{dd} = -1$, and $\eta^{d+1, d+1} = +1$. The coordinates x^0 and x^{d+1} are both like time coordinates.

The isomorphism between the conformal algebra of d-dimensional Minkowski spacetime and the $SO(d, 2)$ algebra is almost fixed by symmetry considerations.

We already have the generators of the Lorentz algebra $SO(d, 1)$, namely $J^{\mu\nu}$. Now we want to identify the additional generators D, P^μ, and K^μ. By eyeball, we see that D is a scalar under $SO(d, 1)$, and so it can only be $J^{d,d+1}$. We identify $J^{d,d+1} = D$. Similarly, by eyeball, we see that P^μ and K^μ carry an index μ, and hence are vectors under $SO(d, 1)$. They could only be linear combinations of $J^{\mu,d}$ and $J^{\mu,d+1}$. So, let us make the educated guess $J^{\mu,d} = \frac{1}{2}(K^\mu + P^\mu)$ and $J^{\mu,d+1} = \frac{1}{2}(K^\mu - P^\mu)$.

We check only a few commutators to show that this assignment is correct. For example, (16) gives $[J^{\mu,d}, J^{\nu,d}] = -\eta^{dd} J^{\mu\nu} = J^{\mu\nu} = \frac{1}{4}[K^\mu + P^\mu, K^\nu + P^\nu] = \frac{1}{4}([K^\mu, P^\nu] - [K^\nu, P^\mu]) = \frac{1}{4}(4J^{\mu\nu})$, where in the last step we used (15). Similarly, $[J^{\mu,d+1}, J^{\nu,d+1}] = -\eta^{d+1,d+1} J^{\mu\nu} = -J^{\mu\nu} = \frac{1}{4}[K^\mu - P^\mu, K^\nu - P^\nu] = -\frac{1}{4}(4J^{\mu\nu})$. As another example, $[D, \frac{1}{2}(K^\mu + P^\mu)] = [J^{d,d+1}, J^{\mu,d}] = \eta^{dd} J^{d+1,\mu} = \frac{1}{2}(K^\mu - P^\mu)$.

The poor man now speaks up. "It is easier to see through all this if we pick a definite value of $d + 2$, say 6, and forget about signature, let it take care of itself. Just think about $SO(6)$." Evidently, $J^{\mu\nu}$, for $\mu, \nu = 1, 2, 3, 4$, generates the rotation algebra for $SO(4)$. In addition, we have $J^{\mu,5}$ and $J^{\nu,6}$, clearly vectors labeled by 5 and 6. For $\mu \neq \nu$, they commute with each other, while for $\mu = \nu$, they commute to produce J^{56}. Recall, as we learned in chapter I.3, that (16) merely says that J^{MN} and J^{PQ} commute with each other, unless a pair of indices, one from each of the Js, are equal, in which case the commutator is a J

carrying the remaining two indices. Thus, J^{56} commuted with $J^{\mu,5}$ and $J^{\mu,6}$ just turns one into the other.

Very good! In fact, an even simpler case is the conformal algebra of $(1+1)$-dimensional Minkowski spacetime, namely, the $SO(2,2)$ algebra.

You might feel that dilation obviously does not hold in physics, since there are fundamental length scales set by the masses of elementary particles. But in our present theoretical understanding, the building blocks of the universe, the quarks, the leptons, and all the rest, are postulated to be massless before the Higgs mechanism kicks in.

Appendix: $(1+1)$-dimensional Minkowski spacetime in light cone coordinates

It is instructive to work out the conformal algebra for a familiar spacetime written in not-so-familiar coordinates, namely, the $(1+1)$-dimensional Minkowski spacetime written in the light cone coordinates introduced in chapter VII.2. Define $x^\pm = t \pm x$. Then $ds^2 = dt^2 - dx^2 = dx^+ dx^- = \eta_{\mu\nu} dx^\mu dx^\nu$, which tells us that $\eta_{+-} = \eta_{-+} = \frac{1}{2}$ and $\eta^{+-} = \eta^{-+} = 2$. The other components, which we do not display, such as η^{++}, all vanish. Also, define $\partial_\pm = \frac{1}{2}\left(\frac{\partial}{\partial t} \pm \frac{\partial}{\partial x}\right)$, so that $\partial_+ x^+ = 1$ and $\partial_- x^- = 1$.

Then $P^\pm = \partial^\pm = 2\partial_\mp$, $D = x^+ \partial_+ + x^- \partial_-$, $J \equiv \frac{1}{2} J^{+-} = \frac{1}{2}(x^+ \partial^- - x^- \partial^+) = x^+ \partial_+ - x^- \partial_-$. Note that $D \pm J = 2x^\pm \partial_\pm$ works out nicely. (It is understood that the $\pm$ signs are correlated unless otherwise noted.) Can you guess what the conformal generators are? Let's find out; simply evaluate (14): $K^+ = x^2 \eta^{+-} \partial_- - 2x^+ (x^+ \partial_+ + x^- \partial_-) = -2(x^+)^2 \partial_+$, and similarly, $K^- = -2(x^-)^2 \partial_-$.

Rather elegantly, the six generators of $SO(2,2)$ can be taken to be

$$\partial^\pm, \quad x^\pm \partial_\pm, \quad \text{and} \quad -(x^\pm)^2 \partial_\pm \tag{17}$$

You can now check the algebra in (15). For example, (15) gives $[K^+, P^-] = 2(J^{+-} + \eta^{+-} D) = 4(J+D) = 8x^+ \partial_+$, and indeed, we compute $[K^+, P^-] = [-2(x^+)^2 \partial_+, 2\partial_+] = 8x^+ \partial_+$.

Interestingly, $SO(4)$ and its two analytically continued descendants, $SO(3,1)$ and $SO(2,2)$, have all appeared in this book.

Notes

1. See, for example, chapter IX.9 in G Nut.
2. See, for example, chapter I.5 in G Nut.
3. Some people have argued that this has led to a distorted view of the world with unfortunate consequences.
4. Or dilatation, if dilation is not academic enough for you.
5. This observation turns out to be of central importance in string theory. See J. Polchinski, String Theory, chapter 2.
6. Here I omit the overall factors of i commonly included in quantum mechanics. You and I live in free countries and, according to what is convenient in a given context, can include or omit overall factors at will.

O God, I could be bounded in a nutshell, and count myself a
king of infinite space—were it not that I have bad dreams.
  —Hamlet, speaking to Rosencrantz and Guildenstern

The everyday sphere as a coset manifold

We learned, back in chapter IV.7, that, in contrast to S^1 and S^3 (which are both group
manifolds), the familiar everyday sphere S^2 is a coset manifold: $S^2 = SO(3)/SO(2)$. You
might want to review the discussion in chapter IV.7 to recall how the coset construction
works, but a simple intuitive understanding is easy to acquire.

Pick any fiducial point, say, the south pole. To rotate it to an arbitrary point on the
sphere—say, your favorite city—you first rotate around the x-axis and bring the south
pole to the appropriate latitude, and then rotate around the z-axis to further bring it to the
appropriate longitude. In other words, an arbitrary point on the sphere is characterized
by the two angles characterizing these two rotations; the dimension of the coset manifold
G/H is given by the number of generators of G (three in the case of $SO(3)$) minus the
number of generators of H (one in the case of $SO(2)$).

More explicitly, we can write a general element of $SO(3)$ as $e^{i\varphi J_z}e^{i\theta J_x}e^{i\psi J_z}$. When this
rotation acts on the south pole, the angle ψ drops out (it "quotients out" or "mods out"),
and the general point on the sphere is characterized by θ and φ. The first rotation $R_z(\psi)$,
corresponding to the denominator $SO(2)$ in $SO(3)/SO(2)$, can be omitted.

One point to keep in mind for later use is that we enjoy considerable freedom in choosing
the "two-parameter net rotation" $e^{i\varphi J_z}e^{i\theta J_x}$, as long as the rotation on the right is not around
the z-axis and that the two rotations are not around the same axis. We can choose $e^{i\varphi J_y}e^{i\theta J_x}$,
or $e^{i\varphi(J_x+J_y)}e^{i\theta J_x}$, or $e^{i\varphi(J_x+J_y)}e^{i\theta(J_x-J_y)}$, or any number of other possibilities. The canonical
choice is convenient merely because θ and φ then correspond to spherical coordinates.

All this amounts to a quick summary of the discussion in chapter IV.7 (and goes back
conceptually to the first introduction of cosets in chapter I.2). Another way of looking at
this is to imagine ourselves living at the south pole. Then the invariance group of our local

flat space right at the south pole is in fact $SO(2)$, the denominator in the construction of the coset manifold.

Since the invariance group of our local flat spacetime is the Lorentz group $SO(3, 1)$, we are motivated to look at the coset manifold $SO(4, 1)/SO(3, 1)$. Amazingly, this leads us to an expanding universe that closely approximates the universe we live in.

The purpose of this chapter is to study the manifold[1] $SO(4, 1)/SO(3, 1)$, which, in analogy with the Euclidean sphere $SO(3)/SO(2)$, we may refer to as a Minkowskian sphere.

The group theory behind the exponentially expanding universe

Let us be a bit more general and discuss $dS^d \equiv SO(d, 1)/SO(d - 1, 1)$, known as the d-dimensional de Sitter spacetime. (The dimension of the coset manifold is equal to the difference in the number of generators of $SO(d, 1)$ and of $SO(d - 1, 1)$, namely, $\frac{1}{2}(d + 1)d - \frac{1}{2}d(d - 1) = d$.) We proceed by analogy with $S^d = SO(d + 1)/SO(d)$. In essence, it boils down to the plea "What are a few signs among friends?"

Picture $(d + 1)$-dimensional Minkowski spacetime defined by[2]

$$ds^2 = \eta_{MN}dX^M dX^N = -(dX^0)^2 + \sum_{i=1}^{d-1}(dX^i)^2 + (dX^d)^2 \tag{1}$$

Here the indices M and N range over $0, 1, 2, \cdots d$. We split the index range into time and space as follows: $M = (0, i, d)$, where the index $i = 1, 2, \cdots, d - 1$ takes on $d - 1$ possible values. (For the real world with $d = 4$, we are thus working in $(4 + 1) = 5$-dimensional Minkowski spacetime.)

Consider the d-dimensional manifold in this $(d + 1)$-dimensional spacetime defined by

$$\eta_{MN}X^M X^N = -(X^0)^2 + \sum_{i=1}^{d-1}(X^i)^2 + (X^d)^2 = 1 \tag{2}$$

As remarked earlier, this may be thought of as a Minkowski unit sphere. It is left invariant under $SO(d, 1)$ (just as the Euclidean unit sphere S^2 defined by $X^2 + Y^2 + Z^2 = 1$ is left invariant under $SO(3)$.) In fact, you have known the Lie algebra of $SO(d, 1)$ for a long time, ever since chapter I.3. In the notation used here, it reads

$$[J_{MN}, J_{PQ}] = i(\eta_{MP}J_{NQ} + \eta_{NQ}J_{MP} - \eta_{NP}J_{MQ} - \eta_{MQ}J_{NP}) \tag{3}$$

As in the $S^2 = SO(3)/SO(2)$ example, pick a reference point on the Minkowski unit sphere defined in (2), say, $X_* = (0, \vec{0}, 1)$. It is left invariant by $SO(d - 1, 1)$, namely, the set of transformations that do not touch the dth axis. Acting with $SO(d, 1)$ on X_*, we then map out the coset manifold.

Writing out $SO(d, 1)/SO(d - 1, 1)$ explicitly

In a sense, that is the end of the story. But as in the case of the S^2, the goal is to choose a convenient set of coordinates analogous to the familiar spherical coordinates. For ease of exposition, it is convenient to give names to the generators J_{MN}. (For S^2, this step is taken implicitly, since we already have the names J_x, J_y, and J_z branded into our brains virtually since childhood.) But wait, we already have a set of names ready made from chapter VIII.2, when we discussed the conformal algebra, namely, P, K, and D (the generators of translation, conformal transformation, and dilation, respectively) and the generators of Lorentz transformation. Even though we are not discussing conformal transformations here, we can use these names for free.

There is a slight catch: in chapter VIII.2 we talked about $SO(d, 2)$; here we are dealing with $SO(d, 1)$. But we are surely adult enough to deal with this minor adjustment in the range of indices. As already noted, we divide the index M into $M = (0, i, d)$, with $i = 1, 2, \cdots, d - 1$.

We now identify the $\frac{1}{2}(d + 1)d$ generators of $SO(d, 1)$. First, we have the generators of rotation J_{ij}. Next, recall from chapter VIII.2 that P and K are orthogonal combinations of the J_{i0} and $J_{d,i}$ (adapting the notation to suit the discussion here). So, let us denote

$$P_i \equiv J_{i0} + J_{d,i}, \qquad K_i \equiv J_{i0} - J_{d,i}, \qquad D \equiv J_{d,0} \tag{4}$$

Then, for example, $[P_i, P_j] = [J_{i0} + J_{d,i}, J_{j0} + J_{d,j}] = i(-J_{ij} + J_{ij} - \delta_{ij}D + \delta_{ij}D) = 0$. Note that to obtain this familiar result, we have to define translation P_i as a linear combination of a boost in the ith direction and a rotation in the (d, i)-plane. As another example, $[D, P_i] = [J_{d,0}, J_{i0} + J_{d,i}] = -i(J_{d,i} - J_{0i}) = -iP_i$.

Since the context is different from chapter VIII.2 (and also given our inclusion of factors of i here), we display here the $SO(d, 1)$ algebra, which you can also deduce from (3):

$$[P_i, P_j] = 0, \qquad [K_i, K_j] = 0,$$

$$[D, P_i] = -iP_i, \qquad [D, J_{ij}] = 0, \qquad [D, K_i] = iK_i,$$

$$[J_{ij}, P_k] = i(\delta_{ik}P_j - \delta_{jk}P_i), \qquad [J_{ij}, K_k] = i(\delta_{ik}K_j - \delta_{jk}K_i),$$

$$[P_i, K_j] = 2i\delta_{ij}D - 2iJ_{ij} \tag{5}$$

Perhaps it is worth emphasizing that the algebra $SO(d, 2)$ in chapter VIII.2 has two time coordinates, while the algebra $SO(d, 1)$ here has only one. We are merely borrowing the names D, P, K, and J.

These generators act linearly on the embedding coordinates X^M. As in chapter III.3, their action can be represented by $J_{MN} = i(X_M \partial_N - X_N \partial_M)$. Thus, each of these generators is represented by a $(d + 1)$-by-$(d + 1)$ matrix. We arrange the indices in the order $(0, \{i\}, d) = (0, 1, 2, \cdots, d - 1, d)$, as already indicated above. For example, the boost $D \equiv J_{d,0}$ in the dth direction is represented by

$$D = i \begin{pmatrix} 0 & 0 & -1 \\ \hline 0 & 0 & 0 \\ \hline -1 & 0 & 0 \end{pmatrix} \tag{6}$$

The notation is such that along the diagonal, in the upper left, the 0 represents a 1-by-1 matrix with entry equal to 0; in the center, the 0 represents a $(d - 1)$-by-$(d - 1)$ matrix with all its entries equal to 0; and finally, in the lower right, the 0 once again represents a 1-by-1 matrix with entry equal to 0. Exponentiating the generator D to obtain the group element, we obtain

$$
e^{iDt} = \left(\begin{array}{c|c|c}
\cosh t & 0 & \sinh t \\
\hline
0 & I & 0 \\
\hline
\sinh t & 0 & \cosh t
\end{array}\right)
\tag{7}
$$

In other words, this is a $(d+1)$-by-$(d+1)$ matrix with a $(d-1)$-by-$(d-1)$ identity matrix in its center. It is simply a boost in the dth direction, leaving the other $d-1$ spatial coordinates untouched.

Similarly, we have

$$
\vec{P} \cdot \vec{x} = i \left(\begin{array}{c|c|c}
0 & \vec{x}^T & 0 \\
\hline
\vec{x} & 0 & \vec{x} \\
\hline
0 & -\vec{x}^T & 0
\end{array}\right)
\tag{8}
$$

In the matrix, $\vec{x}$ is to be interpreted as a $(d-1)$-dimensional column vector (so that $\vec{x}^T$ is a $(d-1)$-dimensional row vector). (Notice that as a linear combination of a boost and a rotation, $\vec{P} \cdot \vec{x}$ is symmetric in its upper left corner and antisymmetric in its lower right corner, so to speak.) Exponentiating, you will find

$$
e^{i\vec{P}\cdot\vec{x}} = \left(\begin{array}{c|c|c}
1 + \frac{1}{2}\vec{x}^2 & \vec{x}^T & \frac{1}{2}\vec{x}^2 \\
\hline
\vec{x} & I & \vec{x} \\
\hline
-\frac{1}{2}\vec{x}^2 & -\vec{x}^T & 1 - \frac{1}{2}\vec{x}^2
\end{array}\right)
\tag{9}
$$

Notice that $(\vec{P} \cdot \vec{x})^3 = 0$, so that the exponential series terminates. You are invited to verify that $e^{i\vec{P}\cdot\vec{x}}e^{i\vec{P}\cdot\vec{y}} = e^{i\vec{P}\cdot(\vec{x}+\vec{y})}$.

Choice of coordinates on the coset manifold and observational astronomers

Our chosen reference point $X_* = (0, \vec{0}, 1)$ is not touched by J_{ij} and $J_{i0} = \frac{1}{2}(P_i + K_i)$. Thus, in analogy with the general rotation $e^{i\varphi J_z}e^{i\theta J_x}e^{i\psi J_z}$ in the case of the ordinary sphere, we can place the transformations generated by J_{ij} and $(P_i + K_i)$ on the right. They don't do anything. For the analog of the rotations $e^{i\varphi J_z}e^{i\theta J_x}$ on the left, which do the actual work, we have any number of choices, as explained above. Different choices correspond to different coordinates on the de Sitter spacetime dS^d.

With the benefit of hindsight, we choose $e^{i\vec{P}\cdot\vec{x}}e^{iDt}$. As we will see momentarily, the parameters $\vec{x}$ and t turn out to be what observational astronomers call space and time in an expanding universe. All that remains for us to do is plug in (7) and (9) and determine where X_* gets mapped to:

$$X = (gX_*) = e^{i\vec{P}\cdot\vec{x}}e^{iDt}(0, \vec{0}, 1) = \left(\sinh t + \frac{1}{2}e^t\vec{x}^2, e^t\vec{x}, \cosh t - \frac{1}{2}e^t\,\vec{x}^2 \right) \tag{10}$$

From (10), a straightforward computation gives

$$ds^2 = \eta_{MN}dX^MdX^N = -(dX^0)^2 + \sum_i (dX^i)^2 + (dX^d)^2$$

$$= -dt^2 + e^{2t}d\vec{x}^2 \tag{11}$$

This group theoretic derivation should make us a bit less amazed at the simplicity of the final result.

We recognize this as the metric of the exponentially expanding, spatially flat universe. Space is Euclidean, described by the Pythagorean $d\vec{x}^2$ multiplied by the scale factor e^{2t}, so that the distance between any two points in space is increasing like e^t.

I mentioned earlier that our observed universe is well approximated by this metric. Group theoretically, we coordinatize an event at $(t, \vec{x})$ by the group element $g(t, \vec{x}) = \exp(i\vec{P}\cdot\vec{x})\exp(iDt)$ needed to bring the reference point X_* on the Minkowskian sphere to our event.

Appendix: Helmholtz's remarkable insight

In 1876, Helmholtz understood[3] that space could be Euclidean, spherical, and hyperbolic using the language of coset manifolds. Imagine ourselves living in the late nineteenth century (and as smart as Helmholtz). Then we could have started with two empirical observations and arrived at these three possibilities. The isotropy of space implies that space is of the form $G/SO(3)$. The 3-dimensionality implies that G must have $3 + 3$ generators. There are three Lie groups with six generators, namely, $G = E(3)$ (that is, the Euclidean group consisting of rotations and translations), $G = SO(4)$, and $G = SO(3, 1)$. The resulting manifold $G/SO(3)$ corresponds to Euclidean, spherical, and hyperbolic, respectively. Note the appearance, in this context, of the Lorentz group long before special relativity!

Now let us generalize this discussion to spacetime. We know that spacetime is Lorentz invariant and 4-dimensional. Thus, if spacetime is homogeneous, it should be of the form $G/SO(3, 1)$, with G having $6 + 4 = 10$ generators. Again, there are three possibilities, namely, $G = E(3, 1)$ (that is, the Poincaré group consisting of Lorentz transformations and translations), $G = SO(4, 1)$, and $G = SO(3, 2)$. The resulting manifold $G/SO(3, 1)$ corresponds to Minkowski de Sitter, and anti de Sitter spacetime, respectively. We have not discussed anti de Sitter spacetime here; the reader is invited to work it out or to look it up.[4]

Notes

1. The material here is based on the work by S. Deser and A. Waldron.
2. This material is adapted from appendix 1 to chapter IX.10 in G Nut. Since in gravity, the metric choice $\eta_{00} = -1$, $\eta_{ij} = +1$ is almost standard (see G Nut, p. 866), we have flipped the overall sign of η. Again, I invoke Emerson here.
3. H. Helmholtz, "The Origin and Meaning of Geometrical Axioms," Mind, Vol. 1, No. 3 (Jul. 1876), pp. 301–321, http://www.jstor.org/stable/2246591.
4. One possible reference is chapter IX.11 in G Nut.

Part IX | The Gauged Universe

The modern understanding of the strong, weak, and electromagnetic interactions involves quarks and leptons interacting with one another via a web of gauge bosons. Group theory determines the interactions among the gauge bosons as well as the interactions of the gauge bosons with the quarks and leptons.

The discussion here emphasizes the group theoretic aspects rather than the field theoretic aspects of this understanding. My hope is that readers without a firm grounding in quantum field theory will still be able to appreciate the crucial role played by group theory. My goal is to show that the group theory you studied in this book suffices in this daring enterprise of grand unifying the three nongravitational interactions.

In the final chapter I describe the puzzle of why quarks and leptons come in three generations. We do not yet know how this family problem is to be resolved and whether group theory will play an essential role.

IX.1 | The Gauged Universe

The Creator likes group theory

When a clergyman asked J.B.S. Haldane[1] what he had learned about the Creator after a lifetime of studying Nature, he answered, "an inordinate fondness for beetles." If the clergyman were to ask a theoretical physicist like me a similar question, I would have answered, an inordinate fondness for group theory.[2]

Group theory governs the universe. Literally.

Group theory is useful not only in classifying the states of atoms and molecules, or in determining what types of crystals are possible. Sure, that's all important stuff, but group theory can do a lot more. Leaving aside gravity, which is not yet completely understood, we now know that the universe is governed by a web of gauge bosons, each corresponding to a generator of a Lie algebra.

Let that statement sink in for a second. It is almost as if the Creator of the universe understood Lie groups and decided to use them (a few of the simplest exemplars anyway) to construct the interactions among the fundamental particles that fill the universe. You might call Him or Her an applied group theorist.

To read this chapter, you need a rudimentary sense of what quantum field theory is. Since I will be emphasizing the group theoretic rather than the field theoretic aspects of the story, what knowledge I require is actually quite minimal. I also have to assume that you have heard of quarks and leptons, and about the four fundamental interactions among them—the strong, the electromagnetic, the weak, and gravity (which we are leaving aside, as just mentioned).

Quarks and leptons

So first we have to talk about the quarks and the leptons, known collectively as fermions in honor of Enrico Fermi. In school we learned that matter is composed of protons, neutrons, and electrons. The proton is now known to be made of two up quarks and a down quark, thus $P = (uud)$, and the neutron to be made of two down quarks and an up quark, thus $N = (ddu)$. In β-decay, a down quark inside the neutron is transformed by the weak interaction into an up quark and an electron and an antielectron-neutrino, in a process written as $d \rightarrow u + e^- + \bar{\nu}_e$. At this level, matter is composed of u, d, $\bar{\nu}_e$, and e^-. The electron carries electric charge -1, the up quark charge $+\frac{2}{3}$, the down quark charge $-\frac{1}{3}$, and the antineutrino charge 0. The photon couples to the various fermions with a strength proportional to the fermion's charge, thus generating the electromagnetic interaction we know, love, and depend on for livelihood and entertainment. It leaves the antineutrino alone.*

For a long time, the strong interaction gluing the quarks inside protons and neutrons together (not to mention the protons and neutrons inside nuclei) was not understood. Eventually, it was realized that the quarks carry a quantum number named color by Gell-Mann. Each quark comes in three color varieties, call them red, green, and yellow. For example, a red up quark u^r and a yellow down quark d^y could scatter off each other to become a yellow up quark u^y and a red down quark d^r. Processes such as $u^r + d^y \rightarrow u^y + d^r$ are then responsible for the strong interaction.

It would certainly seem to you and me that this motley crew of fermions ($u^r, u^g, u^y, d^r, d^g, d^y, e^-$, and ν_e), dancing to the tune of the four fundamental interactions, would suffice to make up an attractive and functioning universe, which in fact could closely approximate the actual universe we live in.

But how naive we are compared to the all-knowing Creator. Physicists later discovered that this crew actually appears in triplicate!

Corresponding to the up quark is the charm quark c; to the down quark, the strange quark s; to the electron, the muon;[†] and to the electron-neutrino, the muon-neutrino. We have what is sometimes called the second generation of fermions, consisting of $c^r, c^g, c^y, s^r, s^g, s^y, \mu^-$, and ν_μ. Utter amazement and disbelief among physicists.

To even more amazement and disbelief, experimentalists subsequently discovered a third generation, consisting of $t^r, t^g, t^y, b^r, b^g, b^y, \tau^-$, and ν_τ. The street names these guys are known by also got uglier: the top quark, the bottom quark, the tau, and the tau-neutrino.

Why matter comes triplicated is a totally unsolved puzzle in particle physics known as the family problem.[3]

So much for a lightning-quick inventory of the matter content of the universe. For the purpose of this chapter, we will simply ignore the existence of the second and third generations.[4] I need hardly say that in this chapter I will be necessarily cutting corners, omitting details, and ignoring subtleties. For example, what appears with the up quark in the first generation is not purely the down quark, but the down quark mixed in with a bit of the strange quark, and an even smaller bit of the bottom quark. As another example, before the year 2000 or so, neutrinos were believed to be strictly massless, but then they

---

* It goes without saying that each one of these statements represents decades of work, sweat mixed in, not with blood, but with serious headaches and puzzlements, by armies of nuclear and particle physicists, theorists, and experimentalists collaborating closely.

† To wit, the famous query by I. I. Rabi, "Who ordered the muon?"

were discovered to have tiny masses much less than that of the charged leptons (the electron, the muon, and the tau). Again, what appears in the first generation is not the electron neutrino ν_e but a linear combination of ν_e, ν_μ, and ν_τ, in a phenomenon known as neutrino mixing. None of this is understood, and group theory may or may not play a role in the pattern of masses and mixing.

A web of interacting gauge bosons

Now that we have gone through what matter is made of, we have to discuss the forces among these fundamental particles.

We now know that the strong, the electromagnetic, and the weak interactions are described by nonabelian gauge theories written down by Yang and Mills in 1954. We will refer to these theories[5] as gauge theories for short.

For any compact Lie group G, we can write down a gauge theory. For each generator T^a of the Lie algebra of G, there exists a gauge boson, namely, a spin 1 particle that obeys Bose-Einstein statistics. (The familiar photon is an example of a gauge boson.) The gauge bosons interact among themselves in a manner discovered by Yang and Mills and determined by the structure constants f^{ab}_{c} of the Lie algebra G. The precise nature of this interaction does not concern us here but is in fact essential for the properties of the strong, the electromagnetic, and the weak interactions.

The notion of gauge theories evolved out of our understanding of electromagnetism, which turns out to be a particularly elementary (once again, Nature is kind to physicists) gauge theory based on that simplest possible Lie group, $U(1)$. Since $U(1)$ has only one generator, there is only one gauge boson, namely, our much-loved photon, which bestows light on the world. And since the structure constant vanishes for $U(1)$, the photon does not interact with itself. It only interacts with particles that carry electric charge, and the photon itself carries no charge.[6]

Constructing gauge theories

Clearly, there is no way for me to explain gauge theory here. I refer you to various modern textbooks on quantum field theory.[7] I can only give you the recipe for constructing a gauge theory. Group theory is essential.

Here is the setup for the fun and games.

1. Pick a compact Lie group G. For our purposes, simply fixate on $SU(N)$.

2. The $\frac{1}{2}N(N-1)$ gauge bosons A^a_μ transform like the adjoint representation (since that is how the generators transform, as explained in chapter IV.1, for example).

3. The fermions (namely, the quarks and leptons) are put into various representations $\mathcal{R}$ of G, with a $d_\mathcal{R}$-dimensional representation accommodating $d_\mathcal{R}$ fermion fields. The interactions between the gauge bosons and the fermions are then fixed by the representation each fermion belongs to (similar to the way that the interactions between the photon and the

fermions are fixed by the charge assigned to each fermion; each of the representations of $U(1)$ is characterized by a number identified physically as the electric charge).

You see that you can't even begin to talk about fundamental physics without learning group theory. Group theory provides the language of fundamental physics.

Strong interaction and $SU(3)$

I will presently illustrate this recipe with an $SU(3)$ example, but before I do that I have to mention the Lorentz group. In chapter VII.3, you learned that each spin $\frac{1}{2}$ fermion is described by two Weyl fields, one left handed, the other right handed. Parity conservation requires that the left and right handed fields be put into the same representation $\mathcal{R}$ of the gauge group G. For example, the discovery that quarks come in three different color varieties means that they can be put in the fundamental 3-dimensional representation 3 of $SU(3)$. Parity then requires that the two sets of fields for the up quark—the left handed u^r_L, u^g_L, and u^y_L, and the right handed u^r_R, u^g_R, and u^y_R—are both put into the 3.

You also learned in chapter VII.3 that the conjugate of a left handed field transforms like a right handed field.[8] It turns out to be convenient for grand unified theorists to rewrite all the right handed fields in the gauge theory as conjugate left handed fields. In other words, instead of ψ_R, we could just as well write ψ^c_L. Thus, the fields u^r_L, u^g_L, and u^y_L are put into the 3, while the conjugate fields u^{cr}_L, u^{cg}_L, and u^{cy}_L are put into the 3* of $SU(3)$. Since everybody is left handed, henceforth we can omit the subscript L. The u^cs are the fields for the different color varieties of the anti-up quark.

Our task is to assign the fermion fields of the first generation, namely, the 15 (left handed) fields

$$u^r, u^g, u^y, u^{cr}, u^{cg}, u^{cy}, d^r, d^g, d^y, d^{cr}, d^{cg}, d^{cy}, e^-, e^+, \nu_e \tag{1}$$

into an assortment of representations of $SU(3)$.

We are now ready to describe the interaction of the gauge bosons with the fermions in this particular example. The eight gauge bosons A^a_μ ($a = 1, \cdots, 8$) couple to the generators T^a as represented in the representation $\mathcal{R}$ the fermions belong to. In other words, A^a_μ couples to the u_Ls via λ_a and to the u^c_Ls via λ^*_a. (Here λ_a denotes the Gell-Mann matrices in chapters V.2 and V.3.)

More precisely, for those readers who know a bit of quantum field theory, the relevant terms in the Lagrangian involving the u quark read

$$\mathcal{L} = gA^a_\mu(\bar{u}\lambda_a\gamma^\mu u + \bar{u}^c\lambda^*_a\gamma^\mu u^c) \tag{2}$$

(Those readers who do not know can safely ignore this more technical remark.) To read off the coupling of a given gluon to the quarks, we simply look up the λ matrices in chapter V.3 and plug in. With the names chosen arbitrarily here of red, green, and yellow, $u = \begin{pmatrix} u^r \\ u^g \\ u^y \end{pmatrix}$.

For example, the fourth gluon A^4_μ couples to $(\bar{u}^r\gamma^\mu u^y + \bar{u}^y\gamma^\mu u^r + \bar{u}^{cr}\gamma^\mu u^{cy} + \bar{u}^{cy}\gamma^\mu u^{cr})$: it turns a yellow up quark into a red up quark, and so on.

This being a book on group theory, we focus on the group theoretic, rather than the field theoretic, aspects of gauge theories. Suffice it to know the eight $SU(3)$ gauge bosons, called gluons, transform the three up quarks into one another and the three anti-up quarks into one another, according to the group properties of the 3 and 3* representations, respectively. Similarly for the down quarks and the antidown quarks. The process $u^r + d^y \rightarrow u^y + d^r$, mentioned earlier as responsible for the strong interaction, is then understood "more microscopically" as due to the effect of a gluon being exchanged between an up quark and a down quark: u^r becomes u^y by emitting an appropriate gluon, which when absorbed by d^y turns it into d^r. The resulting gauge theory, known as quantum chromodynamics, furnishes the modern theory of the strong interaction.

The strong interaction is explained in this way in terms of the gauge group, namely,* $SU(3)$, and the representations $\mathcal{R}$ the fermions are put into. We list the representations as follows, using a self-evident notation: $u \sim 3$, $u^c \sim 3^*$, $d \sim 3$, $d^c \sim 3^*$, $e \sim 1$, $e^c \sim 1$, and $v_e \sim 1$. The leptons, the electron, the antielectron, and the neutrino, are assigned to the singlet representation of $SU(3)$. Recall that by definition the singlet or trivial representation does not transform at all under the group. Thus, the assignment of the leptons to the 1 of $SU(3)$ just states mathematically the physical fact that they do not participate in the strong interaction. They are ignored by the gluons, which do not transform them at all.

This cold shoulder can be expressed by saying that we can obtain the term describing the coupling of the electron to the gluon analogous to those terms in the Lagrangian in (2) by replacing λ_a with a big fat 0; thus, $A_\mu^a(\bar{e}0e) = 0$, which is as 0 as any 0. In other words, the term coupling the electron to the gluon does not exist. Mathematically, the generator T_a of $SU(3)$ is represented in the trivial representation 1 by 0.

In summary, the strong interaction can be specified by listing the nontrivial irreducible representations of $SU(3)$. We replace the list in (1) by

$$(u^r, u^g, u^y), \ (u^{cr}, u^{cg}, u^{cy}), \ (d^r, d^g, d^y), \ (d^{cr}, d^{cg}, d^{cy}) \tag{3}$$

In other words, the gluons transfer the three fields within each pair of parentheses into one another. Note that e^-, e^+, and v_e are not listed. The group theoretic statement is that the fermion fields of the first generation are assigned to the reducible representation $3 \oplus 3^* \oplus 3 \oplus 3^* \oplus 1 \oplus 1 \oplus 1$ of $SU(3)$.

The neutrino has no conjugate partner

Another detail here—namely, that we write only v_e but not v_e^c—is of great importance in particle theory, but readers not into particle physics may safely ignore it, at least for now. In chapter VII.3, I mentioned that the profound discovery of parity is intimately connected with the neutrino having only a left handed component, thus satisfying the Weyl equation

---

* At the risk of confusing some readers, I should mention that this $SU(3)$, which is supposed to be exact and fundamental, is to be clearly distinguished from the $SU(3)$ discussed in chapter V.2, which is approximate and not associated with any gauge bosons. Under that $SU(3)$, the up, down, and strange quarks transform into one another.

rather than the Dirac equation. There is ν_L, but not ν_R, and hence there is no ν_L^c. (By the way, since we are considering only one generation of fermions here, we can also safely drop the subscript e on the neutrino field ν_e.)

Let us pause and count how many Weyl fields we have in one generation: $3 + 3 + 3 + 3 + 1 + 1 + 1 = 15$. That is a number to keep in mind.

Weak interaction and $SU(2)$

Now that we have gone through the strong interaction in record time, let us now tackle the weak interaction.[9] The β-decay $d \rightarrow u + e + \bar{\nu}$ which started the arduous century-long study of the weak interaction (which is still going on) will act as our guide.

This process would be explained if we propose an $SU(2)$ gauge theory in which the up and down quarks transform as a doublet, schematically $\sim \begin{pmatrix} u \\ d \end{pmatrix}$. The gauge theory would have $2^2 - 1 = 3$ gauge bosons called W_μ^a (with $a = 1, 2, 3$ here) coupled to the generators T^a of $SU(2)$, which by now you know all too well. In the doublet representation (namely, the 2 of $SU(2)$), T^a is represented by the Pauli matrices τ_a. Hence the linear combinations[10] $W_\mu^{1\pm i2}$ couple to $\tau^\mp = \frac{1}{2}(\tau_1 \mp i\tau_2)$. These two gauge bosons thus transform the up quark into the down quark and vice versa. Similarly, we will put the neutrino and the electron into a doublet $\sim \begin{pmatrix} \nu \\ e \end{pmatrix}$. The same two gauge bosons thus transform the neutrino into the electron and vice versa.

Direct product structure

We are obliged to mention a couple of details, even in such a broadbrush description of the gauge theories of the fundamental interactions. Since the gauge group $SU(2)$ and the Lorentz group $SO(3, 1)$ operate in different arenas—the former in an internal space (as alluded to in chapter V.1) and the latter in spacetime—they commute.

It follows that since the neutrino is left handed, the electron field in the doublet has to be left handed. The right handed electron field e_R (or equivalently, e_L^c) has no neutrino field to partner with. Thus, it has to be a singlet, that is, transforming like the 1 of $SU(2)$.

Decades of experiments on the weak interaction, particularly those focusing on parity violation, have shown that the u and d in the weak interaction doublet $\sim \begin{pmatrix} u \\ d \end{pmatrix}$ are left handed. The gauge bosons $W_\mu^{1\pm i2}$ leave the right handed u and d fields alone; in other words, the fields u^c and d^c, just like e^c, transform like the 1 of $SU(2)$, that is, they don't transform at all.

Another important feature of the weak interaction bosons $W_\mu^{1\pm i2}$ is that, while they transform u_L and d_L into each other, they do not touch the color the quarks carry. In other words, the strong and weak interactions do not talk to each other. Mathematically, the statement is that the gauge group has a direct product structure (as was first mentioned way back in chapter I.1), namely, $SU(3) \otimes SU(2)$.

Nature strikes me as being extraordinarily kind to theoretical physicists, as I have remarked elsewhere. I am not going to make you guys master some horribly large group, Nature says. Since you live in 3-dimensional space, surely you can figure out $SO(3)$, and eventually its double cover $SU(2)$. And once you learn that, you are smart enough to figure out $SU(3)$, no? Yes, I know that the direct product is a cheap way of making a bigger group out of smaller groups, but I like it, and also I want to make life simple for you, so then how about $SU(3) \otimes SU(2)$? Is that easy enough?

The bottom line is that our 15 Weyl fields are assigned to irreducible representations of $SU(3) \otimes SU(2)$ as follows:

$$\left(\begin{pmatrix} u^r \\ d^r \end{pmatrix}, \begin{pmatrix} u^g \\ d^g \end{pmatrix}, \begin{pmatrix} u^y \\ d^y \end{pmatrix} \right), \begin{pmatrix} v \\ e \end{pmatrix}, (u^{cr}, u^{cg}, u^{cy}), (d^{cr}, d^{cg}, d^{cy}), e^c \tag{4}$$

Stated more compactly, the fields of the first generation form the reducible representation $(3, 2) \oplus (1, 2) \oplus (3^*, 1) \oplus (3^*, 1) \oplus (1, 1)$, giving in total $3 \cdot 2 + 1 \cdot 2 + 3 \cdot 1 + 3 \cdot 1 + 1 \cdot 1 = 6 + 2 + 3 + 3 + 1 = 15$. These numbers are intended to test whether you are following the discussion. If they don't make sense, please go back some.

Thus far, our gauge theory of the universe minus gravity is based on $SU(3) \otimes SU(2) \otimes SO(3, 1)$. Henceforth, we will omit mentioning the Lorentz group; it is certainly understood that we ain't talking about nonrelativistic physics when we discuss the fundamental interactions.

Where is electromagnetism?

Confusio suddenly speaks up. "What about the gauge boson W^3_μ? Is it the photon?"

Confusio is not confused at all this time. Good question! The gauge boson W^3_μ is associated with the generator T^3. Thus, it couples to u and v with strength $+\frac{1}{2}$, to d and e with strength $-\frac{1}{2}$, and does not couple to u^c, d^c, or e^c at all. Hence this boson cannot be the photon, which could not couple to the neutrino at all, since it carries no electric charge. Also, since electromagnetism respects parity, the photon must couple to e_L and e_R with the same strength, or equivalently, to e and e^c with the opposite strength.

We thus conclude that at least one more gauge boson has to be introduced. Here the Gell-Mann Nishijima formula, $Q = I_3 + \frac{1}{2}Y$ (with Q the electric charge, I_3 the third component of isospin, and Y the hypercharge), introduced historically in a different context and discussed in chapter V.1, gives us a hint. Let us write, in the present context*

$$Q = T^3 + \frac{1}{2}Y \tag{5}$$

Introduce another gauge group $U(1)$ with $\frac{1}{2}Y$ as its generator. We could determine the values of $\frac{1}{2}Y$ as follows. Given an $SU(2)$ representation, the sum of T^3 for the fields in the

---

* It must be emphasized that T^3 is not the third component of isospin, but one of the generators of the $SU(2)$ gauge group introduced here.

representation must add up to 0 by definition (recall the S in $SU(2)$!). Hence (5) tells us that $\frac{1}{2}Y$ of the representation is equal to the average electric charge Q of the fields in the representation.

Recall that the electric charges of the up quark, the down quark, the neutrino, and the electron are equal to $+\frac{2}{3}$, $-\frac{1}{3}$, 0, and -1, respectively. We can then readily determine the hypercharges of the 15 fermion fields. Thus, in the list (4), the fields transforming like the singlet representation 1 of $SU(2)$ have hypercharge equal[*] to its electric charge; $\frac{1}{2}Y$ for the fields u^c, d^c, and e^c is equal to $-\frac{2}{3}$, $+\frac{1}{3}$, and $+1$, respectively. (The hypercharge of the quarks, just like electric charge, is independent of the quark's color.)

For the fields in (4) transforming like the 2 of $SU(2)$, $\binom{u}{d}$ and $\binom{v}{e}$ have $\frac{1}{2}Y$ equal to the average charges, $\frac{1}{2}\left(+\frac{2}{3}+\left(-\frac{1}{3}\right)\right) = \frac{1}{6}$ and $\frac{1}{2}(0+(-1)) = -\frac{1}{2}$, respectively.

The gauge group underlying the theory of the strong, weak, and electromagnetic interactions is thus extended to $SU(3) \otimes SU(2) \otimes U(1)$. The irreducible representations of this direct-product group are then identified by three numbers, specifying how the irreducible representation transforms under $SU(3)$, $SU(2)$, and $U(1)$, respectively. The third number is just the value of $\frac{1}{2}Y$ for the representation. For example, the neutrino electron doublet $\binom{v}{e}$ transforms like the irreducible representation $(1, 2, -\frac{1}{2})$. With this notation, the 15 Weyl fields of the first generation are then assigned to the reducible representation

$$\left(3, 2, \frac{1}{6}\right) \oplus \left(3^*, 1, -\frac{2}{3}\right) \oplus \left(3^*, 1, \frac{1}{3}\right) \oplus \left(1, 2, -\frac{1}{2}\right) \oplus (1, 1, 1) \tag{6}$$

The values of $\frac{1}{2}Y$ do not particularly concern you; the important point is that the interactions of the gauge boson associated with $U(1)$, call it B_μ, with the quarks and the leptons are completely determined.

The photon comes out to be a particular linear combination of the W_μ^3 and B_μ, coupling to electric charge as given by (5). The linear combination orthogonal to the photon is known as the Z. The discovery of a gauge boson with the predicted properties of the Z gave particle physicists tremendous confidence in the correctness of the $SU(3) \otimes SU(2) \otimes U(1)$ gauge theory outlined here.

The electron mass

Recall that the electron mass is described by the bilinear $e^c Ce$ given in chapter VII.5. In electrodynamics, the two fields e and e^c, having opposite electric charges, transform oppositely ($e \to e^{i\theta}e$, $e^c \to e^{-i\theta}e^c$), and thus the Dirac mass bilinear does not change under the $U(1)$ of electromagnetism.[†] But in the $SU(2) \otimes U(1)$ theory, e and e^c transform quite differently: e lives inside a doublet $l \equiv \binom{v}{e}$, but e^c is a singlet living by herself. Thus, under

---

[*] Since there is only one field in each singlet representation 1, we do not even have to average.

[†] If you confused Thomson's e with Euler's e at this stage, go back to square one.

$SU(2)$, $l \to e^{i\vec{\theta}\vec{\sigma}}l$, while $e^c \to e^c$. Similarly, under the $U(1)$, $l \to e^{-i\xi/2}l$ and $e^c \to e^{i\xi}e^c$. Thus, the mass bilinear e^cCe is not invariant under $SU(2) \otimes U(1)$.

This poses a serious problem, which Weinberg solved by introducing a scalar (that is, scalar under the Lorentz group) doublet* (doublet under $SU(2)$) $\varphi \equiv \left(\begin{smallmatrix} \varphi^0 \\ \varphi^- \end{smallmatrix} \right) = \left(\begin{smallmatrix} \varphi^1 \\ \varphi^2 \end{smallmatrix} \right)$, known as a Higgs field, which transforms just like l under $SU(2) \otimes U(1)$, namely, $\varphi \to e^{i\vec{\theta}\vec{\sigma}}\varphi$. Thus, as we learned in chapters IV.4 and IV.5, the combination $l^i\varphi^j\varepsilon_{ij}$ (with $i, j = 1, 2$) is a singlet under $SU(2)$. The idea is then to replace the Dirac mass bilinear for the electron by

$$e^cCe \to (e^cCl^i)\varphi^j\varepsilon_{ij} \tag{7}$$

Note that for the hypercharge $Y/2$ to add up correctly ($+1 - \frac{1}{2} - \frac{1}{2} = 0$), φ must have $Y/2 = -\frac{1}{2}$. In other words, φ has the same quantum numbers as l and so has the same electric charge content, as already indicated: the top component $\varphi^1 = \varphi^0$ carries no electric charge.

Since $\varphi^1(x)$ is a scalar field, it is a function of the spacetime coordinates x and thus the right hand side of (7) looks nothing like the left hand side. But now notice that if $\varphi^1(x) = v$ happens to equal a constant v while $\varphi^2(x) = 0$, then $(e^cCl^i)\varphi^j\varepsilon_{ij}$ becomes $(e^cCl^2)v\varepsilon_{21} \propto e^cCe$. We get what we had asked for: the Dirac bilinear for the electron mass.

Roughly speaking, we can understand what is going on by appealing to an analogy. That one component of φ can be a constant while the other component vanishes is not so different from the electric field $\vec{E}$ between two capacitor plates oriented perpendicular to the x-axis, say. Then E^1 is constant independent of $\vec{x}$, while E^2 and E^3 vanish.

In this breathless rush[11] through what is called the standard model of particle physics, we have, as I forewarned, by necessity omitted a number of important topics. In particular, we did not address how some gauge bosons become massive, while others (the gluons and the photon) remain massless. (I will leave it to you to figure out how the quarks get mass. Remarkably, you will discover that one single Higgs field transforming, like $(1, \frac{1}{2}, -\frac{1}{2})$ and its conjugate, can do the job for both the electron and the up and down quarks.) But as I said, I am focusing on the group theoretic, rather than the field theoretic, aspects of the story. Keep (7) in mind. When we do grand unified theory in chapter IX.2, we will need its analog.

Pieces of the jigsaw puzzle

In summary, our universe with gravity set aside is well described by a gauge theory based on the Lie group (or better,† algebra) $SU(3) \otimes SU(2) \otimes U(1)$. The quarks and leptons of

---

* Here (φ^0, φ^-) and (φ^1, φ^2) are just different names for the same two fields.

† As remarked long ago in chapter I.3, for most purposes, and certainly for the purposes of this chapter, it suffices to talk about the Lie algebra and avoid global issues.

each generation are placed in the following representations of this group:

$$\left(3, 2, \frac{1}{6}\right), \ \left(3^*, 1, -\frac{2}{3}\right), \ \left(3^*, 1, \frac{1}{3}\right), \ \left(1, 2, -\frac{1}{2}\right), \quad \text{and } (1, 1, 1) \tag{8}$$

Call this the master list of the jigsaw puzzle.

The notation has been explained. For example, $(3, 2, \frac{1}{6})$ is a 6-dimensional representation of the direct product group $SU(3) \otimes SU(2) \otimes U(1)$, transforming like a 3 of $SU(3)$, a 2 of $SU(2)$, with hypercharge $\frac{1}{6}$ (and hence electric charges $\frac{1}{2} + \frac{1}{6} = +\frac{2}{3}$ and $-\frac{1}{2} + \frac{1}{6} = -\frac{1}{3}$). These "quantum numbers" thus identify the six fields as u^r, u^g, u^y, d^r, d^g, and d^y. Another example: $(1, 1, 1)$ is a 1 under both $SU(3)$ and $SU(2)$, with hypercharge 1 and hence electric charge 1. It can only be e^+. Can you identify $(1, 2, -\frac{1}{2})$?

So, this is the most compact description of how the nongravitational physical world works. Matter is specified by a triplet of numbers. The gluons couple to the first number; the weak bosons $W^\pm$ and a linear combination of the photon γ and the Z boson couple to the second number; and the orthogonal linear combination of γ and Z couples to the third number. In a sense, Nature appears to have tricked the theoretical physicists: they might have thought that the photon should naturally couple to the $U(1)$ subgroup of $SU(2)$ and have the world be simply $SU(3) \otimes SU(2)$. We will see in chapter IX.2 that Nature has Her reason.

Could this list (8) be the end of the story?

Exercises

1 Work out the group theory behind the quark masses.

Notes

1. From Boilerplate Rhino, by D. Quammen, p. 242.
2. Fearful, p. 132.
3. It has been hoped that group theory might be of some help here. What we need is a group with 3-dimensional representations. For instance, the Frobenius groups and the tetrahedral group have been considered in the literature. See interludes II.i3 and IV.i1.
4. The terminology has not been codified. Sometimes what I call "generation" is referred to as a family. One lame joke is that the family problem occurs when three generations have to live together.
5. The word "theory" in this context represents a slight abuse of language; in fact, the standard model of the three nongravitational interactions is fairly well established, at least in broad outline.
6. See QFT Nut, p. 230.
7. For example, chapter IV.5 and part VII in QFT Nut.
8. In 1974, very few particle physicists knew this fact due to the predominance of Dirac's 4-component spinor formulation in all the leading textbooks.
9. For example, E. Commins and P. Buksbaum, Weak Interactions of Leptons and Quarks.
10. We ignore various factors, such as $1/\sqrt{2}$, which are not relevant for our purposes.
11. For a more detailed exposition of the standard model, see, for example, chapters VII.2 and VII.3 in QFT Nut.

Crying out for unification

In the standard model of particle physics, quarks and leptons are put into this collection of irreducible representations of $SU(3) \otimes SU(2) \otimes U(1)$:

$$\left(3, 2, \frac{1}{6}\right), \ \left(3^*, 1, -\frac{2}{3}\right), \ \left(3^*, 1, \frac{1}{3}\right), \ \left(1, 2, -\frac{1}{2}\right), \quad \text{and } (1, 1, 1) \tag{1}$$

This motley collection of representations practically cries out for further unification. Who would have constructed the universe by throwing this bizarre looking list down?

What we would like to have is a larger gauge group G containing $SU(3) \otimes SU(2) \otimes U(1)$, such that this laundry list of representations would be unified into (ideally) one great big representation. The gauge bosons in G (but not in $SU(3) \otimes SU(2) \otimes U(1)$, of course) would couple the representations in (1) to each other, for example, $(3, 2, \frac{1}{6})$ to $(1, 2, -\frac{1}{2})$.

Clues galore

There are many clues, at least in the glare of hindsight.

Imagine yourself a spy in some espionage movie finding a slip of paper with the list (1) written on it. What does it say to you? Here is one clue. Add up the hypercharges $\frac{1}{2}Y$ of these 15 first generation Weyl fields: $3 \cdot 2 \cdot \frac{1}{6} + 3 \cdot 1 \cdot (-\frac{2}{3}) + 3 \cdot 1 \cdot \frac{1}{3} + 1 \cdot 2 \cdot (-\frac{1}{2}) + 1 \cdot 1 \cdot 1 = 1 - 2 + 1 - 1 + 1 = 0$. Is that a coincidence or what?

Well, if the $U(1)$ is part of a simple Lie algebra, such as $SU(N)$ or $SO(N)$, then Y would be a generator and its trace would vanish; in other words, the sum of Y would equal 0. In fact, if you are good at this sort of thing (that is, grand unifying the universe), you might see that the sum $1 - 2 + 1 - 1 + 1$ can be arranged as the sum of two zeroes: $1 + 1 - 2 = 0$ and $1 - 1 = 0$. Is this significant? By the end of this chapter, you will see that it is.

Do-it-yourself grand unification

By now, you know enough group theory to construct the so-called grand unified theory, unifying the strong, weak, and electromagnetic interactions. Indeed, I set up this book in such a way so that you can do it now. You should try it before reading on.

To start, answer this question: What is the smallest group that contains $SU(3) \otimes SU(2) \otimes U(1)$?

The smallest group that contains $SU(3) \otimes SU(2) \otimes U(1)$ is $SU(5)$. By now you know all this stuff in your sleep after having gone through the discussion of $SU(N)$ in chapter IV.4. In particular, you know that $SU(5)$ is generated by the $5^2 - 1 = 24$ 5-by-5 hermitean traceless matrices acting on five objects we denote by ψ^μ with $\mu = 1, 2, \cdots, 5$, and which form the fundamental or defining representation of $SU(5)$.

Simply separate ψ^μ into two sets: ψ^α with $\alpha = 1, 2, 3$, and ψ^i with $i = 4, 5$. Those $SU(5)$ matrices that act on ψ^α define an $SU(3)$, and those $SU(5)$ matrices that act on ψ^i define an $SU(2)$. Indeed, of the 24 matrices that generate $SU(5)$, 8 have the form $\begin{pmatrix} A & 0 \\ 0 & 0 \end{pmatrix}$, and 3 the form $\begin{pmatrix} 0 & 0 \\ 0 & B \end{pmatrix}$, where A represents 3-by-3 hermitean traceless matrices (of which there are $3^2 - 1 = 8$, the Gell-Mann matrices of chapter V.3), and B represents 2-by-2 hermitean traceless matrices (of which there are $2^2 - 1 = 3$, namely, the Pauli matrices). Clearly, the former generate an $SU(3)$ and the latter an $SU(2)$. This specifies how $SU(3)$ and $SU(2)$ fit inside $SU(5)$. The matrices of $SU(3)$ and the matrices $SU(2)$ commute, and so the group so defined is actually $SU(3) \otimes SU(2)$.

Furthermore, the 5-by-5 hermitean traceless matrix

$$\frac{1}{2}Y = \begin{pmatrix} -\frac{1}{3} & 0 & 0 & 0 & 0 \\ 0 & -\frac{1}{3} & 0 & 0 & 0 \\ 0 & 0 & -\frac{1}{3} & 0 & 0 \\ 0 & 0 & 0 & \frac{1}{2} & 0 \\ 0 & 0 & 0 & 0 & \frac{1}{2} \end{pmatrix} \tag{2}$$

generates a $U(1)$. Without being coy about it, we have already called this matrix the hypercharge $\frac{1}{2}Y$.

And indeed, it is traceless.

At the end of the preceding chapter, some theoretical physicists were left wondering why Nature had to throw in that extra $U(1)$ in $SU(3) \otimes SU(2) \otimes U(1)$. We now see that the reason may be grand unification: $SU(5)$ naturally breaks up into $SU(3) \otimes SU(2) \otimes U(1)$, not $SU(3) \otimes SU(2)$.

A perfect fit

The three objects ψ^α transform like a 3-dimensional representation under $SU(3)$ and hence could be a 3 or a 3*. We have a choice here, which corresponds to choosing $Y/2$

as the matrix in (2) rather than minus that matrix. Let us choose ψ^α as transforming like the 3, which corresponds to ψ^μ transforming like the 5 of $SU(5)$. In other words, we take ψ^μ to furnish the defining irreducible representation of $SU(5)$.

The three objects ψ^α do not transform under $SU(2)$ and hence belong to the singlet 1 representation. Furthermore, they carry hypercharge $-\frac{1}{3}$, as we can read off from (2). To sum up, ψ^α transform like $(3, 1, -\frac{1}{3})$ under $SU(3) \otimes SU(2) \otimes U(1)$.

In contrast, the two objects ψ^i transform like 1 under $SU(3)$ and 2 under $SU(2)$, and carry, according to (2), hypercharge $\frac{1}{2}$, and thus transform like $(1, 2, \frac{1}{2})$.

This means that we embed $SU(3) \otimes SU(2) \otimes U(1)$ into $SU(5)$ by specifying how the defining representation 5 of $SU(5)$ decomposes into representations of $SU(3) \otimes SU(2) \otimes U(1)$:

$$5 \rightarrow \left(3, 1, -\frac{1}{3}\right) \oplus \left(1, 2, \frac{1}{2}\right). \tag{3}$$

Taking the conjugate,* we obtain

$$5^* \rightarrow \left(3^*, 1, \frac{1}{3}\right) \oplus \left(1, 2, -\frac{1}{2}\right) \tag{4}$$

Inspecting (1), we see that $(3^*, 1, \frac{1}{3})$ and $(1, 2, -\frac{1}{2})$ appear on the list. We are on the right track! The fields in these two representations fit snugly into 5^*.

This accounts for five of the fields contained in (1); we still have to find the remaining ten fields

$$\left(3, 2, \frac{1}{6}\right), \left(3^*, 1, -\frac{2}{3}\right), \quad \text{and } (1, 1, 1) \tag{5}$$

Consider the next representation of $SU(5)$ in order of size, namely, the antisymmetric tensor representation $\psi^{\mu\nu}$. Its dimension is $(5 \cdot 4)/2 = 10$. Precisely the number we want, if only the quantum numbers under $SU(3) \otimes SU(2) \otimes U(1)$ work out!

You should have the fun of working it out, and of course this stuff would work out if it is in textbooks by now.

Well, here we go. Since we know that $5 \rightarrow (3, 1, -\frac{1}{3}) \oplus (1, 2, \frac{1}{2})$, we simply work out (as in part IV, for example, chapter IV.4) the antisymmetric product of $(3, 1, -\frac{1}{3}) \oplus (1, 2, \frac{1}{2})$ with itself, namely, the direct sum of (where $\otimes_A$ denotes the antisymmetric product)

$$\left(3, 1, -\frac{1}{3}\right) \otimes_A \left(3, 1, -\frac{1}{3}\right) = \left(3^*, 1, -\frac{2}{3}\right) \tag{6}$$

$$\left(3, 1, -\frac{1}{3}\right) \otimes_A \left(1, 2, \frac{1}{2}\right) = \left(3, 2, -\frac{1}{3} + \frac{1}{2}\right) = \left(3, 2, \frac{1}{6}\right) \tag{7}$$

and

$$\left(1, 2, \frac{1}{2}\right) \otimes_A \left(1, 2, \frac{1}{2}\right) = (1, 1, 1) \tag{8}$$

---

* Note that there is no such thing as 2* of $SU(2)$; we learned back in chapter II.4 that 2 is pseudoreal.

(I will walk you through (6): in $SU(3)$ $3 \otimes_A 3 = 3^*$ (remember ε_{ijk}?), in $SU(2)$ $1 \otimes_A 1 = 1$, and in $U(1)$ the hypercharges simply add: $-\frac{1}{3} - \frac{1}{3} = -\frac{2}{3}$. Were you paying attention? How come for $SU(2)$ we don't add 1 and 1 to get 2?)

Thus

$$10 \to \left(3, 2, \frac{1}{6}\right) \oplus \left(3^*, 1, -\frac{2}{3}\right) \oplus (1, 1, 1) \tag{9}$$

Lo and behold, these $SU(3) \otimes SU(2) \otimes U(1)$ representations form exactly the collection of representations in (5).

The known quark and lepton fields in a given family fit perfectly into the 5^* and 10 representations of $SU(5)$!

I have just described the $SU(5)$ grand unified theory of Georgi and Glashow.[1] In spite of the fact that the theory has not been directly verified by experiments, it is extremely difficult for me and for many physicists not to believe that $SU(5)$ is at least structurally correct, in view of the perfect group theoretic fit.

It is often convenient to display the content of the representation 5^* and 10, using the names given to the various fields historically rather than, say,* $(3^*, 1, \frac{1}{3})$. We write 5^* as a column vector

$$\psi_\mu = \begin{pmatrix} \psi_\alpha \\ \psi_i \end{pmatrix} = \begin{pmatrix} d^c \\ d^c \\ d^c \\ v \\ e \end{pmatrix} \tag{10}$$

and the 10 as an antisymmetric matrix

$$\psi^{\mu\nu} = \{\psi^{\alpha\beta}, \psi^{\alpha i}, \psi^{ij}\}$$

$$= \begin{pmatrix} 0 & u^c & -u^c & d & u \\ -u^c & 0 & u^c & d & u \\ u^c & -u^c & 0 & d & u \\ -d & -d & -d & 0 & e^c \\ -u & -u & -u & -e^c & 0 \end{pmatrix} \tag{11}$$

(I suppressed the color indices on the right hand sides here.)

We now understand why the hypercharges of the 15 Weyl fields add up to zero in two separate sets, as described at the start of this chapter. Hypercharge is a generator of $SU(5)$, and so is traceless in any given representation. Note the electric charge assignment in the 5^*, as shown in (10) (namely, $Q = +\frac{1}{3}$ on ψ_α, 0 on ψ_4, and -1 on ψ_5), implies that $Q = -\frac{2}{3}$, $-\frac{1}{3}, +\frac{2}{3}, 1$ on $\psi^{\alpha\beta}, \psi^{\alpha 4}, \psi^{\alpha 5}$, and ψ^{ij}, respectively, as shown in (11). (For example, acting on $\psi^{\alpha 5}$, $Q = -\frac{1}{3} + 1 = +\frac{2}{3}$, and that identifies the up quark u in the first three rows and the fifth column.)

---

* You recognize this? Yes, the antidown quark.

Aside from its aesthetic appeal, grand unification based on $SU(5)$ deepens our understanding of physics enormously. Among other things, it explains (i) that electric charge is quantized and (ii) that the proton charge is exactly equal and opposite to the electron charge.[2]

Quark and lepton masses in grand unified theory

We discussed how the electron gets its mass in the $SU(3) \otimes SU(2) \otimes U(1)$ theory in chapter IX.1. Let us now see how quarks and leptons get their masses in the $SU(5)$ grand unified theory.

Consider the Dirac mass bilinear $u^c C u$ for the up quark. Both the Weyl fields u^c and u appear in the $10 \sim \psi^{\mu\nu}$, and so we have to search for the up quark mass in the bilinear $\psi^{\mu\nu} C \psi^{\rho\sigma}$. But we have known since chapter IV.4 on $SU(N)$ that this 4-indexed tensor is not invariant under $SU(5)$. This is precisely the same problem for the electron mass in the $SU(2) \otimes U(1)$ theory, as discussed in chapter IX.1.

We can solve the problem with the same "trick" as was used there: introduce a Lorentz scalar field φ^τ transforming like the 5 under $SU(5)$ and bring out the antisymmetric symbol. As promised, the group theory we learned in chapter IV.4 is just what is needed here. Add the $SU(5)$-invariant term $\varepsilon_{\mu\nu\rho\sigma\tau} \psi^{\mu\nu} \psi^{\rho\sigma} \varphi^\tau$ to the Lagrangian. The electric charges of the five fields contained in φ are the opposite of those in ψ_μ in (10); in particular, φ^4 is electrically neutral. Setting φ^4 to a constant (that is, independent of spacetime), we have

$$\varepsilon_{\mu\nu\rho\sigma4} \psi^{\mu\nu} C \psi^{\rho\sigma} \varphi^4 \rightarrow \psi^{12} C \psi^{35} \pm \text{permutations} \sim u^c C u \tag{12}$$

We thus obtain the up quark mass, as desired.

What about the masses of the down quark and of the electron? Since d^c and e belong to the 5*, while their partners d and e^c belong to the 10, these masses have to come from multiplying 5* and 10. Indeed, 5* ⊗ 10 does contain the 5, as we have known since chapter IV.4. Explicitly, in parallel to (12), we have

$$\psi_\mu C \psi^{\mu\nu} \varphi_\nu \rightarrow \psi_\mu C \psi^{\mu\nu} \varphi_4 \rightarrow \psi_\mu C \psi^{\mu 4} \sim d^c C d + e^c C e \tag{13}$$

Here φ_ν is the conjugate of φ^ν, as explained in chapter IV.4.

Again, as in $SU(3) \otimes SU(2) \otimes U(1)$, one single Higgs field transforming like the 5 does the job in $SU(5)$. This offers a clue that $SU(5)$ may not be the end of the story either.

Protons are not forever

I have to mention here one stunning prediction of the $SU(5)$ grand unified theory, namely, that protons, on which the world is founded, are not forever. I sketched in chapter IX.1 how the weak interaction, as described by $SU(2)$, causes the neutron to decay; a similar mechanism occurring in $SU(5)$ causes the proton also to decay.

Again, group theory rules. There are $5^2 - 1 = 24$ generators, and hence 24 gauge bosons, in the Lie algebra $SU(5)$, $3^2 - 1 = 8$ gauge bosons in $SU(3)$, $2^2 - 1 = 3$ gauge bosons in $SU(2)$, and 1 gauge boson in $U(1)$. Thus, the grand unified theory contains an additional[3] $24 - (8 + 3 + 1) = 12$ gauge bosons not in the standard $SU(3) \otimes SU(2) \otimes U(1)$ theory. You could well ask what these bosons do.

Back in part IV, you learned that the adjoint representation of $SU(N)$ is furnished by the traceless tensor with one upper index and one lower index. The 24 gauge bosons of $SU(5)$ can be identified with the 24 independent components of the traceless tensor A^μ_ν with $\mu, \nu = 1, 2, \cdots, 5$.

Recall that we decompose the $SU(5)$ index μ, into $\mu = \{\alpha, i\}$, with $\alpha = 1, 2, 3$ and $i = 1, 2$. The eight gauge bosons in $SU(3)$ transform an index of the type α into an index of the type α, while the three gauge bosons in $SU(2)$ transform an index of the type i into an index of the type i.

The fun comes with the $(2 \cdot 3) + (2 \cdot 3) = 6 + 6 = 12$ additional gauge bosons A^α_i and A^i_α, which transform the index α into the index i and vice versa. In other words, these gauge bosons have one foot in $SU(3)$ and the other foot in $SU(2)$. Since α is a color index pertaining to quarks, these previously unknown gauge bosons are capable of turning a quark into a lepton and vice versa. Referring to (11), we see that A^5_α can change

$$\psi^{\alpha 4} = d \to \psi^{54} = e^+ \tag{14}$$

for example. Group theoretically, the product $A^5_\alpha \psi^{\alpha 4}$ transforms just like ψ^{54}, as we can see simply by contracting indices. This is just how we multiply tensors in $SU(N)$, as we learned in chapter IV.4. That a generator of the Lie algebra transforms $\psi^{\alpha 4}$ into ψ^{54} inside an irreducible representation is realized physically in a gauge theory by a fermion absorbing a gauge boson to turn into another fermion. Mathematics is realized as physics.

Similarly, the gauge boson A^α_5 can change an up quark into an anti-up quark:

$$\psi^{5\beta} = u \to \psi^{\alpha\beta} = \bar{u} \tag{15}$$

(Again, group theoretically, this corresponds to $A^\alpha_5 \psi^{5\beta} \sim \psi^{\alpha\beta}$.) In quantum field theory, as mentioned in chapter IV.9, this can be described equivalently as an up quark changing into an anti-up quark upon emitting the gauge boson A^5_α, namely, $\psi^{5\beta} \sim A^5_\alpha \psi^{\alpha\beta}$.

Picture a proton $= uud$ (that is, a bag of two up quarks and one down quark) sitting there minding its own business. By emitting a gauge boson of the type A^5_α, one of the up quarks u becomes an anti-up quark $\bar{u}$, according to (15). This gauge boson, when absorbed by the down quark d, changes it into an antielectron e^+, also known as the positron, according to (14). Thus, this gauge boson, by being emitted and then absorbed, can generate the process $u + d \to \bar{u} + e^+$, causing $uud \to u\bar{u} + e^+$. Note that one of the two up quarks in uud sits around as a spectator when the other up quark combines with the down quark to produce an anti-up quark and a positron. The bag $u\bar{u}$, consisting of an up quark with an anti-up quark, describes a neutral pion π^0. At the level of hadrons, this process is thus observed as proton decay: $p \to \pi^0 + e^+$.

Proton decay has not yet been observed, but this could be due simply to the tank of water used to detect the process not being large enough. Note the important conceptual point that while group theory allows us to predict which processes are possible, it cannot possibly tell us what the actual decay rate is. For that we need quantum field theory.

The power of group theory as a constraint on the low energy effective theory

As I said, $SU(5)$ grand unification has not yet been confirmed, although the fit has been seamless. Quite impressively, group theoretic arguments can take us a long way in discussing proton decay. Here is another chance for me to impress on you the power of group theory.

Within $SU(3) \otimes SU(2) \otimes U(1)$, the proton has no inclination—indeed, has no ability[4]— to decay. It is only when we grand unify that the proton decays (as shown in the preceding section). Now suppose that there is indeed physics beyond the $SU(3) \otimes SU(2) \otimes U(1)$ gauge theory, be it grand unified theory or something else, and that it causes the proton to decay. But regardless of the details of this unknown physics, whatever it might be—let that "regardless" and the "whatever" sink in for a minute—we can use group theory to say something about proton decay. How is that possible? Read on.

Experimentally, there are in principle several observable decay modes, such as $p \to \pi^0 + e^+$, $p \to \pi^+ + \bar{\nu}$, and $p \to \pi^+ + \nu$. If we don't know in detail the physics causing proton decay, we are surely not able to calculate the rates for each of these processes. But remarkably, we are able to relate the various rates.[*]

The reason is that in the face of almost total ignorance, we are nevertheless armed with one crucial piece of knowledge, namely, the list (1). We know how the various quark and lepton fields involved transform under $SU(3) \otimes SU(2) \otimes U(1)$. Group theory thus determines a good deal of what the Lagrangian (known as the effective Lagrangian) responsible for proton decay must look like.[5]

The proton contains three quark fields and has spin $\frac{1}{2}$. Lorentz invariance insists that the initial state and the final state must match, or in other words, that the effective Lagrangian responsible for proton decay must transform as a singlet under $SO(3, 1)$. Since group theory tells us that $\frac{1}{2} \otimes \frac{1}{2} \otimes \frac{1}{2} \otimes \frac{1}{2}$ contains a singlet, but not $\frac{1}{2} \otimes \frac{1}{2} \otimes \frac{1}{2}$, to form a singlet, we need another spin $\frac{1}{2}$ field. Thus, the relevant terms in the effective Lagrangian must consist of three quark fields and one lepton field, with the schematic form $\sim qqql$. To form a Lorentz scalar, we have to bring the knowledge we gained in part VII to bear and tie up the spinor indices on these four Weyl fields by inserting various matrices (such as the C matrix of chapter VII.5). You are fully capable of doing this now, but this is not the point I want to emphasize here.

Instead, note that since $SU(3) \otimes SU(2) \otimes U(1)$ is a good symmetry at the energy scale[6] relevant for the physics of proton decay, the terms $\sim qqql$ must also not transform under $SU(3) \otimes SU(2) \otimes U(1)$, as well as under the Lorentz group. Another way of saying this

---

[*] This is somewhat reminiscent of the discussion in part III: group theory can tell us about the degeneracy and the pattern of the energy levels, but not the energies themselves.

is that the quark field q and the lepton field l carry $SU(3) \otimes SU(2) \otimes U(1)$ indices (as specified by (1)), and you must tie up these indices as well as the spinor indices. This ends up relating the various observable proton decay modes.

The stability of the world and conservation laws

That the proton cannot decay in $SU(3) \otimes SU(2) \otimes U(1)$, a profound fact that accounts for the stability of the world, can be put on a loftier footing by saying that baryon number B is conserved. As the name suggests, and as you might recall from chapter V.2, each baryon, such as the proton and the neutron, is simply assigned the number $B = +1$. The up quark and down quark each carry $B = \frac{1}{3}$. Since nothing in the $SU(3) \otimes SU(2) \otimes U(1)$ theory changes a quark into something else, the conservation of B follows essentially by definition. The conservation law is associated* with a global[7] $U(1)$ symmetry of the Lagrangian: transform every quark field by the same phase factor $q \to e^{i\alpha}q$, and the $SU(3) \otimes SU(2) \otimes U(1)$ Lagrangian is left unchanged. While this conservation law is of fundamental importance, the group theory behind the associated symmetry is elementary.

Similarly, leptons (such as the neutrino and the electron) are assigned a lepton number $L = +1$. Lepton number conservation is then associated with another $U(1)$ symmetry: transform every lepton field by the same phase factor $l \to e^{i\beta}l$, and the $SU(3) \otimes SU(2) \otimes U(1)$ Lagrangian is left unchanged.

Well, in $SU(5)$, neither B nor L is conserved, and the proton can decay, as already noted. But observe that the terms $\sim qqql$ describing proton decay still conserve $B - L$. The combination of fields $qqql$ has $B = \frac{1}{3} + \frac{1}{3} + \frac{1}{3} = 1$ and $L = 1$. Thus, $B - L = 1 - 1 = 0$.

Conservation or violation of $B - L$ in proton decay

Whether $B - L$ is conserved or violated is not merely for theoretical debate, but a matter of some importance for experimenters searching for proton decay. Possible decay modes include $p \to \pi^0 + e^+$ in the first case, and $p \to \pi^+ + \pi^+ + e^-$ in the second. Do you expect a positron or an electron? Looking back, we see that (14) is what decides the issue.

To achieve a deeper understanding[8] of why $B - L$ is still conserved while neither B nor L is conserved, and also for guidance in how to construct alternative theories that would violate $B - L$, let us go through an instructive little exercise in group theory. Go back to the two terms in (12) and (13) $\varepsilon_{\mu\nu\rho\sigma\tau}\psi^{\mu\nu}C\psi^{\rho\sigma}\varphi^\tau$ and $\psi_\mu C\psi^{\mu\nu}\varphi_\nu$ that we added to the Lagrangian. Write them group theoretically as $(10\ 10\ 5_\varphi)$ and $(5^*\ 10\ 5^*_\varphi)$. These two terms conserve a quantum number, call it X, because $3 - 2 = 1$; what I mean by that is that there are three numbers we can choose freely (namely, $X(10)$, $X(5^*)$, and $X(5_\varphi)$, the quantum number of the fermions in the 10, of the fermions in the 5*, and of the Higgs field φ, respectively), while there are only two constraints imposed by the

---

* The deep connection between symmetry and conservation laws was mentioned in chapter III.3.

two terms added to the Lagrangian. The two constraints are $X(10) + X(10) + X(5_\varphi) = 0$ and $X(5^*) + X(10) - X(5_\varphi) = 0$, with the solution[9] $X(10) = 1$ (this merely sets the overall normalization of X and is arbitrary), $X(5_\varphi) = -2$ (from the first equation), and $X(5^*) = -3$. In other words, under the $U(1)$ transformation $\psi^{\mu\nu} \to e^{i\theta} \psi^{\mu\nu}, \varphi^\tau \to e^{-2i\theta} \varphi^\tau$, $\psi_\mu \to e^{-3i\theta} \psi_\mu$, the Lagrangian (in particular the two terms in (12) and (13)) remains unchanged.

But you object. Since to generate quark and lepton masses, we set the component φ_4 of 5^*_φ to a constant, doesn't this break this $U(1)$ symmetry?

Excellent, it does. Looking at the charge (and hence hypercharge) content (10) of the fermions assigned to 5^*, we see that φ_4 has the same quantum numbers as the neutrino field ν, in particular, $Y/2 = -1/2$. Thus, the linear combination $X + 4(Y/2)$ is still conserved (since its value on φ_4 is equal to $2 + 4(-1/2) = 0$).

We conclude that $SU(5)$ grand unification conserves a mystery quantum number $X + 4(Y/2)$. What is it? It turns out that $\frac{1}{5}(X + 4(Y/2))$ is none other than $B - L$, baryon minus lepton number. You can verify this as an exercise.

Exercise

1   Show that $\frac{1}{5}(X + 4(Y/2)) = B - L$.

Notes

1. H. Georgi and S. L. Glashow, Phys. Rev. Lett. 32 (1974), p. 438. For this and other important papers on grand unified theory, see the collection of reprints in A. Zee, Unity of Forces in the Universe.
2. Since this is a textbook on group theory, I am content to merely mention these points, referring you to various standard sources for details. There exist any number of excellent texts on particle physics. See also chapter VII.5 in QFT Nut.
3. In other words, $5^2 - 3^2 - 2^2 + (-1 + 1 + 1 - 1) = 5^2 - 3^2 - 2^2 = 4^2 - 2^2 = (4+2)(4-2) = 12$.
4. An aside for the experts: I am leaving aside tiny nonperturbative effects, which cause protons to decay in sets of three.
5. See S. Weinberg, Phys. Rev. Lett. 43 (1979), p. 311; F. Wilczek and A. Zee, Phys. Rev. Lett. 43 (1979), p. 1571; H. A. Weldon and A. Zee, Nucl. Phys. B 173 (1980), p. 269. For more details, see chapter VIII.3 in QFT Nut.
6. I am glossing over a detail here, seeing that it pertains to quantum field theory rather than to group theory.
7. Unfortunately, it is beyond the scope of this book to discuss the distinction between global and local symmetries. See any modern quantum field theory textbook.
8. This is taken from the article "Conservation or Violation of $B - L$ in Proton Decay," F. Wilczek and A. Zee, Phys. Lett. 88 B (1979), p. 311.
9. We have in some sense come full circle from the first page of the linear algebra review.

IX.3 | From *SU*(5) to *SO*(10)

An entire generation of quarks and leptons into a single representation

Aesthetically, it seems rather unsatisfactory that the 15 quark and lepton Weyl fields in each generation are put into two irreducible representations, 5* and 10, of $SU(5)$.

An important hint comes from quantum field theory, which requires theories to pass a "health check," known as freedom from anomaly (which you may think of as a kind of disease). Each Weyl field in a gauge theory contributes to the anomaly a certain amount, which, roughly speaking, involves the cube[1] of various charges. All these contributions must add up to zero for the theory to pass muster. In $SU(5)$ theory, the contribution of the 5* and the contribution of the 10 happen to cancel each other exactly (see exercise 1).

This cancellation strongly suggests that $SU(5)$ unification is not the end of the story.

What larger[2] group G, with $SU(5)$ as a subgroup, would have a single irreducible representation that would break into the 5* and the 10 when we restrict G to $SU(5)$?

The spinor representation of $SO(10)$

Now that you have mastered the spinor representations of $SO(N)$ in chapter VII.1, you should be able to answer this question and to take the next crucial step. Recall that we can embed $SU(5)$ naturally into[3] $SO(10)$. Indeed, in chapter VII.1 you learned that the spinor of $SO(10)$ decomposes,[4] upon the restriction of $SO(10)$ to $SU(5)$, into the representations of $SU(5)$ as follows:

$$16^+ \to [0] \oplus [2] \oplus [4] = 1 \oplus 10 \oplus 5^* \tag{1}$$

The 5* and the 10 of $SU(5)$ fit inside the 16^+ of $SO(10)$! Again, the tight fit of the 5* and the 10 of $SU(5)$ inside the 16^+ of $SO(10)$ has convinced many physicists that it is surely right. Another benefit of $SO(10)$ grand unification* is that the anomaly vanishes automatically (see exercise 2).

One feature of $SO(10)$ grand unification appeals to me greatly: the quarks and leptons transform as spinors in both spacetime and in internal space. At present, we do not know quite what to make of this. So back to the fit.

Almost a perfect fit, but not perfect. The 16^+ breaks into $5^* \oplus 10 \oplus 1$. We have to throw in an additional Weyl field transforming as an $SU(5)$ singlet.

Who is this mysterious intruder?

The long lost antineutrino field

We don't have to guess his identity: group theory pins it down.

He is a singlet under $SU(5)$, and hence is a fortiori a singlet under $SU(3) \otimes SU(2) \otimes U(1)$. This Weyl field does not participate in the strong, weak, and electromagnetic interactions. In plain English, he is a lepton, with no electric charge, and is not involved in the known weak interaction. Thus, we identify the mysterious 1 in (1) as the "long lost" antineutrino field ν_L^c. This guy does not listen to any gauge bosons, known or unknown[†] to experiments.

We are using a convention in which all fermion fields are left handed, and hence we have written ν_L^c. By a conjugate transformation, as explained in chapter VII.5, this is equivalent to the right handed neutrino field ν_R, which was missing from the $SU(3) \otimes SU(2) \otimes U(1)$ theory. Why have experimentalists not seen it? The natural explanation is that this field is endowed with a large Majorana mass.

Majorana versus Dirac mass

Let us recall, summarize, and generalize the discussion of Majorana versus Dirac mass given in chapter VII.5. Majorana explained to us that a field without any electric charge can have a Majorana mass. In contrast, a charged field, such as the electron field, can only have a Dirac mass. Let us review.

Consider a Weyl field ψ_L, that is, something transforming like $(\frac{1}{2}, 0)$ under the Lorentz group. Group theory allows us to write a Majorana mass term of the form $\psi_L C \psi_L$, since $(\frac{1}{2}, 0) \otimes (\frac{1}{2}, 0)$ contains[‡] $(0, 0)$.

Suppose the field ψ_L also transforms under a nontrivial irreducible representation R under some internal (that is, not having to do with spacetime) symmetric group G, which is of course the group theoretic way of saying that ψ_L carries a charge. (The familiar electric charge corresponds to $G = U(1)$ and R equal to some nontrivial 1-dimensional

---

* Ironically, in a 1961 paper,[5] Glashow and Gell-Mann listed Lie algebras up to $SO(9)$, saying that it was "hard to imagine that any higher Lie algebras will be of physical interest."

† Even the gauge bosons of $SU(5)$ don't know about him.

‡ Although we are talking about the Lorentz group here, the fact that $\frac{1}{2} \otimes \frac{1}{2}$ contains 0 goes back to $SU(2)$.

representation characterized by $q \neq 0$.) The Majorana mass term $\psi_L C \psi_L$ transforms like $R \otimes R$. Thus, unless $R \otimes R$ contains the trivial identity representation in its decomposition, $\psi_L C \psi_L$ will not be invariant. The symmetry group G forbids the Majorana mass term. (In the simple case of the electric charge, this just says that unless $q + q = 0$, that is, unless $q = 0$, the Majorana term is not allowed.) The executive summary is that the Majorana term is possible only if ψ_L does not carry a charge. More precisely, a Majorana mass is allowed if

$$R \otimes R = 1 \oplus \cdots \tag{2}$$

in words, if $R \otimes R$ contains the singlet, that is, the identity representation.

Next, consider the Dirac mass. Suppose that in addition to ψ_L, we are given a Weyl field ψ_R, that is, a field transforming like $(0, \frac{1}{2})$ under the Lorentz group. Due to the magic of $J + iK \leftrightarrow J - iK$, the hermitean conjugate of ψ_R, namely $\bar{\psi}_R$ (the bar contains the conjugation and was introduced in chapter VII.4), transforms like $(\frac{1}{2}, 0)$. Something new, namely, a Dirac mass term (schematically of the form $\bar{\psi}_R \psi_L$) then becomes possible; it transforms like $(\frac{1}{2}, 0) \otimes (\frac{1}{2}, 0)$, which contains the singlet $(0, 0)$ under the Lorentz group.

The next twist in the story is that given a ψ_R, we can also form the charge conjugate field ψ_L^c, as explained in chapter VII.5. A key fact is that ψ_L^c transforms like $(\frac{1}{2}, 0)$, as the subscript L indicates. Thus, we can write the Dirac mass bilinear $\bar{\psi}_R \psi_L$ as $\psi_L^c C \psi_L$.

Here is a quick summary. In the all-left-handed fields formalism used here, the Dirac mass links a Weyl 2-component field with its conjugate. For example, the electron mass would have schematically the form $e^c C e$ in the Lagrangian. In contrast, a Majorana mass for the electron would have the form $e C e$, but this violates charge conservation, since this combination carries electric charge $-1 - 1 = -2$. In contrast, the Dirac mass has the form $e^c C e$ and hence carries electric charge $+1 - 1 = 0$. The electron can have a Dirac mass but not a Majorana mass.

The mass of the right handed neutrino

Under $SU(5)$, the left handed antineutrino (or equivalently, the right handed neutrino) transforms like a singlet. Thus, as far as $SU(5)$ is concerned, the right handed neutrino can have Majorana mass: $1 \otimes 1 = 1$ is a particularly forceful realization of (2). In other words, $SU(5)$ does not forbid the right handed neutrino from being massive.

Thus, the right handed neutrino can have a mass as high as the energy or mass scale M at which $SU(5)$ becomes a good symmetry. This scale is thought to be extremely high, many orders of magnitude beyond what present accelerators can reach.[6] Here we focus on the group theory.

In contrast, under $SO(10)$, the right handed neutrino transforms like the 16^+. We learned in (VII.1.50) that

$$16^+ \otimes 16^+ = 10 \oplus 120 \oplus 126 \tag{3}$$

Recall that the irreducible representations of 10, 120, and 126 are furnished by anti-symmetric tensors with 1, 3, and 5 indices, respectively.

The right hand side of (3) most definitely does not contain the 1 of $SO(10)$. Thus, $SO(10)$ forbids the right handed neutrino from having a Majorana mass. Since $SO(10)$ breaks into its subgroup $SU(5)$, the mass scale $\tilde{M}$ at which $SO(10)$ becomes a good symmetry is supposed to be higher than M. In other words, the right handed neutrino cannot have a Majorana mass larger than $\tilde{M}$.

Quark and lepton masses in $SO(10)$

In the previous two chapters we learned how to render quarks and leptons massive by setting a Higgs field to a constant. Recall that in $SU(5)$, the Higgs field transforms like a 5. In $SO(10)$, it follows from the group theory decomposition in (3) that the Higgs field φ needed transforms like a 10, which on restriction to $SU(5)$ transforms like $5 \oplus 5^*$. In other words, denoting the 16^+ by ψ, we simply add the term $\psi C \psi \varphi$ to the Lagrangian. The two terms (IX.2.12) and (IX.2.13) in the $SU(5)$ theory naturally combine into a single term in the $SO(10)$ theory. Everything fits together and make sense.

But what about the Majorana mass of the right handed neutrino, which (as just noted) transforms like a singlet under $SU(5)$? Where oh where can we find an $SU(5)$ singlet in the right hand side of (3)? Do you see it?

Yes, it is in the $126 = [5]$! Recall from chapter VII.1 that the $[5]$ is the self-dual 5-indexed tensor of $SO(10)$ (as the notation indicates). When restricted to $SU(5)$, it contains, among a whole load of other stuff, the 5-indexed antisymmetric tensor of $SU(5)$, but as we learned way back in chapter IV.4, this is nothing but the singlet 1 of $SU(5)$. Again, the group theory "conspires" to endow the right handed neutrino with a Majorana mass.

A binary code for the world

In the early 1970s, after $SU(2)$, $SU(3)$, and $SU(5)$, many theoretical physicists felt they had reached a new level of sophistication, moving from the orthogonal world, with its rotation and Lorentz transformation, to the more sophisticated unitary world, with two kinds of indices and what not. The sudden appearance of $SO(10)$ almost felt like the revenge of the orthogonals.

Group theory is trying to tell us something.

Go back to chapter VII.1. Because of the direct product form of the γ matrices, and hence of σ_{ij}, we can write the states of the spinor representations of $SO(2n)$ as $|\varepsilon_1 \varepsilon_2 \cdots \varepsilon_n\rangle$ where each of the εs takes on the values ± 1. The right handed spinor consists of those states $|\varepsilon_1 \varepsilon_2 \cdots \varepsilon_n\rangle$ with $(\Pi_{j=1}^n \varepsilon_j) = +1$, and the left handed spinor those states with $(\Pi_{j=1}^n \varepsilon_j) = -1$.

Thus, in $SO(10)$ unification the fundamental quarks and leptons are described by a 5-bit binary code, with states like $|+ + - - +\rangle$ and $|- + - - -\rangle$. Personally, I find this a rather pleasing picture of the world.

Let us work out the quark and lepton states explicitly, to clarify the group theory involved.

Start with the much simpler case of $SO(4)$. The spinor S^+ consists of $|++\rangle$ and $|--\rangle$, while the spinor S^- consists of $|+-\rangle$ and $|-+\rangle$. As discussed in chapter II.3, $SO(4)$ contains two distinct $SU(2)$ subgroups. Recall from chapter VII.1 that, under one $SU(2)$, $|++\rangle$ and $|--\rangle$ transform as a doublet, while $|+-\rangle$ and $|-+\rangle$ transform as two singlets. Or, we could choose the other $SU(2)$, under which $|++\rangle$ and $|--\rangle$ transform as two singlets, while $|+-\rangle$ and $|-+\rangle$ transform as a doublet. This is consistent with what we learned in chapter VII.1, that on the restriction of $SO(4)$ to $U(2)$, the spinors decompose as $S^+ \to [0] \oplus [2] = 1 \oplus 1$ and $S^- \to [1] = 2$.

Similarly, on the restriction of $SO(6)$ to $U(3)$, $4^+ \to [0] \oplus [2] = 1 \oplus 3^*$, and $4^- \to [1] \oplus [3] = 3 \oplus 1$. (Our choice of which triplet representation of $U(3)$ to call 3 or 3^* is made to conform to common usage, as we shall see presently.)

We are now ready to figure out the identity of each of the 16 states, such as $|++--+\rangle$ in $SO(10)$ unification. First of all, (VII.1.56) tells us that under the subgroup $SO(4) \otimes SO(6)$ of $SO(10)$ the spinor 16^+ decomposes as (since $\Pi_{j=1}^5 \varepsilon_j = +1$ implies $\varepsilon_1 \varepsilon_2 = \varepsilon_3 \varepsilon_4 \varepsilon_5$)

$$16^+ \to (2^+, 4^+) \oplus (2^-, 4^-) \tag{4}$$

We identify the natural $SU(2)$ subgroup of $SO(4)$ as the $SU(2)$ of the electroweak interaction and the natural $SU(3)$ subgroup of $SO(6)$ as the color $SU(3)$ of the strong interaction. Thus, according to the preceding discussion, $(2^+, 4^+)$ are the $SU(2)$ singlets of the standard $U(1) \otimes SU(2) \otimes SU(3)$ model, while $(2^-, 4^-)$ are the $SU(2)$ doublets. Here is the line-up (all fields left handed as usual):

$SU(2)$ doublets

$$\nu = |-+---\rangle$$
$$e^- = |+----\rangle$$
$$u = |-+++-\rangle, |-++-+\rangle, \text{ and } |-+-++\rangle$$
$$d = |+-++-\rangle, |+-+-+\rangle, \text{ and } |+--++\rangle$$

$SU(2)$ singlets

$$\nu^c = |+++++\rangle$$
$$e^+ = |--+++\rangle$$
$$u^c = |+++--\rangle, |++-+-\rangle, \text{ and } |++--+\rangle$$
$$d^c = |--+--\rangle, |---+-\rangle, \text{ and } |----+\rangle$$

I assure you that this is a lot of fun to work out, and I urge you to reconstruct this table without looking at it. Here are a few hints if you need help. From our discussion of $SU(2)$, I know that $\nu = |-+\varepsilon_3\varepsilon_4\varepsilon_5\rangle$ and $e^- = |+-\varepsilon_3\varepsilon_4\varepsilon_5\rangle$, but how do I know that $\varepsilon_3 = \varepsilon_4 = \varepsilon_5 = -1$? First, I know that $\varepsilon_3\varepsilon_4\varepsilon_5 = -1$. I also know that $4^- \to 3 \oplus 1$ on restricting $SO(6)$ to color $SU(3)$. Well, of the four states $|---\rangle, |++-\rangle, |+-+\rangle,$ and $|-++\rangle$, the odd man out is clearly $|---\rangle$. By the same heuristic argument, among the 16 possible states, $|+++++\rangle$ is the odd man out and so must be ν^c.

There are lots of consistency checks. For example, once I identified $\nu = |-+---\rangle$, $e^- = |+----\rangle$, and $\nu^c = |+++++\rangle$, I can figure out the electric charge Q, which— since it transforms as a singlet under color $SU(3)$—must have the value $Q = a\varepsilon_1 + b\varepsilon_2 + c(\varepsilon_3 + \varepsilon_4 + \varepsilon_5)$ when acting on the state $|\varepsilon_1\varepsilon_2\varepsilon_3\varepsilon_4\varepsilon_5\rangle$. The constants a, b, and c can be determined from the three equations $Q(\nu) = -a + b - 3c = 0$, $Q(e^-) = a - b - 3c = -1$, and $Q(\nu^c) = a + b + 3c = 0$. Thus, $Q = -\frac{1}{2}\varepsilon_1 + \frac{1}{6}(\varepsilon_3 + \varepsilon_4 + \varepsilon_5)$.

The $SU(5)$ singlet generator of $SO(10)$

In discussing $B - L$ conservation or violation in chapter IX.2, we encountered a mystery quantum number we called X, which is an $SU(5)$ singlet. In other words, when evaluating X on a given $SU(5)$ irreducible representation, we obtain a number. Using a rather circuitous physicist's route giving masses to the quarks and leptons, we found that

$$X(10) = 1, \qquad X(5^*) = -3 \tag{5}$$

(The overall normalization is arbitrary.) The experimentally important quantum number $B - L$ turns out to be a linear combination of X, which is outside $SU(5)$, and of hypercharge $Y/2$, which is of course inside. So is $B - L$ entirely inside $SO(10)$, or does it have a piece outside?

In chapter VII.1 we showed that the 45 generators of $SO(10)$ break up, on restriction to the natural subgroup $SU(5)$, into $24 \oplus 1 \oplus 10 \oplus 10^*$. I then challenged you to figure out what the singlet 1 of $SU(5)$ does to the irreducible representations of $SU(5)$ contained in the spinor 16 of $SO(10)$. Surely, you figured it out?

With this setup, it is almost signed sealed and delivered that the 1 contained in the 45 must have something to do with the mysterious X. This gives me another chance to show the power of the binary code formalism. Since 1 is an $SU(5)$ singlet, it can't tell the five εs apart; so up to an overall normalization, $X = \varepsilon_1 + \varepsilon_2 + \varepsilon_3 + \varepsilon_4 + \varepsilon_5$.

Let's check by evaluating X on the states listed in the previous section. Since $\varepsilon_1\varepsilon_2\varepsilon_3\varepsilon_4\varepsilon_5 = +1$, we can only have them equal to (i) all $+$s, in which case $X = 5$; (ii) three $+$s and two $-$s, in which case $X = 1$; and (iii) one $+$ and four $-$s, in which case $X = -3$. Referring to the list, we see that the states with one $+$ and four $-$s describe ν, e^-, and d^c, which form the 5^* of $SU(5)$, while the states with three $+$s and two $-$s describe u, d, u^c, d^c, and e^+, which form the 10 of $SU(5)$.

Lo and behold, this agrees with (5).

As a bonus we learned that $X = 5$ on the 1 contained in the spinor 16. Trivial check: $\text{tr } X = 10 \cdot 1 + 5 \cdot (-3) + 1 \cdot 5 = 0$, as expected.

So the X of chapter IX.2 is a generator of $SO(10)$, and since hypercharge $\frac{1}{2}Y$ is for sure a generator of $SO(10)$, this means that the physicist's $B - L$, being a linear combination of X and Y, is a generator of $SO(10)$. This provides another way to see the necessity of the right handed neutrino field: $\mathrm{tr}(B - L) = 0$ only with it included.

Lepton as the fourth color in a left right symmetric world

I interrupt this narrative stream to discuss an interesting idea, due to Jogesh Pati and Abdus Salam, which unifies quarks and leptons. Each quark comes in three colors. Could it be that the fourth color describes a lepton?

Starting from the $SU(3) \otimes SU(2) \otimes U(1)$ theory, Pati and Salam built up to $SU(4) \otimes SU(2) \otimes SU(2)$. To see how this works, let us go back to (IX.1.4), which I reproduce here for your convenience:

$$\left(\begin{pmatrix} u^r \\ d^r \end{pmatrix}, \begin{pmatrix} u^g \\ d^g \end{pmatrix}, \begin{pmatrix} u^y \\ d^y \end{pmatrix} \right), \begin{pmatrix} \nu \\ e \end{pmatrix}, \; (u^{cr}, u^{cg}, u^{cy}), \; (d^{cr}, d^{cg}, d^{cy}), \; e^c \tag{6}$$

The 15 Weyl fields of the first generation are listed as variously belonging to doublets and singlets of $SU(2)$.

Recall that in chapter IX.1 we conjugated the right handed fields to write all 15 Weyl fields as left handed fields. For the Pati-Salam theory, it is convenient to undo this and write, instead of (6),

$$\left(\begin{pmatrix} u^r \\ d^r \end{pmatrix}_L, \begin{pmatrix} u^g \\ d^g \end{pmatrix}_L, \begin{pmatrix} u^y \\ d^y \end{pmatrix}_L \right), \begin{pmatrix} \nu \\ e \end{pmatrix}_L; \; (u_R^r, u_R^g, u_R^y), \; (d_R^r, d_R^g, d_R^y), \; e_R \tag{7}$$

To lessen clutter, let us suppress the color indices r, g, and y and write (7) as

$$\underset{\frac{1}{6}}{\begin{pmatrix} u \\ d \end{pmatrix}_L}, \underset{-\frac{1}{2}}{\begin{pmatrix} \nu \\ e \end{pmatrix}_L}; \; \underset{\frac{2}{3}}{u_R}, \underset{-\frac{1}{3}}{d_R}, \underset{-1}{e_R} \tag{8}$$

where u_R, for instance, is shorthand for the three Weyl fields (u_R^r, u_R^g, u_R^y). Here we have listed the value of $\frac{Y}{2}$ under each set of fields.

Note that in (7) and (8) the left and right handed worlds are separated by a semicolon. Historically, before $SU(2) \otimes U(1)$ became firmly established, various authors proposed making the two worlds look similar by enlarging the electroweak gauge group from $SU(2) \otimes U(1)$ to $SU(2)_L \otimes SU(2)_R \otimes U(1)$.

The standard $SU(2)$ responsible for the weak interaction is renamed $SU(2)_L$. Correspondingly, the standard W bosons introduced in chapter IX.1, which transform u_L and d_L into each other, are renamed W_L.

An additional gauge group $SU(2)_R$ is introduced, under which u_R and d_R transform as a doublet $\begin{pmatrix} u \\ d \end{pmatrix}_R$. Now there are gauge bosons, call them W_R, which transform u_R and d_R into each other.

You might immediately object that the corresponding processes are unknown experimentally, but this can be explained by postulating that the W_R bosons are much heavier* than the W_L bosons. The suggestion is that the left and right handed worlds have the same structure and that the underlying left-right symmetry is hidden from us because the $SU(2)_R$ gauge bosons are much heavier than the $SU(2)_L$ gauge bosons. Thus, at this point, we have an $SU(3) \otimes SU(2)_L \otimes SU(2)_R \otimes U(1)$ gauge theory with quarks and leptons assigned to the following irreducible representation:

$$\begin{pmatrix} u \\ d \end{pmatrix}_L, \begin{pmatrix} v \\ e \end{pmatrix}_L; \begin{pmatrix} u \\ d \end{pmatrix}_R, \begin{pmatrix} v \\ e \end{pmatrix}_R \tag{9}$$

At the cost of introducing unobserved gauge bosons, we have made the world look more symmetrical: compare (8) with (9).

But now you have another serious objection. To have the desired left-right symmetry, we have to put e_R into an $SU(2)_R$ doublet, but there is no known field to partner it with. We are forced to introduce† a field v_R with the quantum numbers of the right handed neutrino, and to explain away the fact that no such particle is observed experimentally by giving it a very large Majorana mass, beyond the energies accessible to experimentalists.

The notion that the world is secretly left-right symmetric appeals to many theoretical physicists. But the pronouncement that the invented particles are all too massive to be seen also turns off many. Each to his or her own taste, and you should make up your own mind.

Looking at (9), Pati and Salam proposed unifying quarks and leptons by extending $SU(3)$ to $SU(4)$. This amounts to putting parentheses around the quarks and lepton doublets in (9), so that

$$\left(\begin{pmatrix} u \\ d \end{pmatrix}, \begin{pmatrix} v \\ e \end{pmatrix} \right)_L; \left(\begin{pmatrix} u \\ d \end{pmatrix}, \begin{pmatrix} v \\ e \end{pmatrix} \right)_R \tag{10}$$

Recall that the quark doublet is written in shorthand and actually represents three doublets (as written in (7)). So there are in fact four doublets in each of the pairs of big parentheses in (10), and we might as well further compactify the notation and write

$$\begin{pmatrix} U \\ D \end{pmatrix}_L; \begin{pmatrix} U \\ D \end{pmatrix}_R \tag{11}$$

by defining (U^A, D^A), $A = 1, 2, 3, 4$, with $(U^\alpha, D^\alpha) = (u^\alpha, d^\alpha)$, $\alpha = 1, 2, 3$, and $U^4 = v$, $D^4 = e$. In other words, the leptons are theorized to be quarks carrying a fourth color. In

---

* In quantum field theory, the quantum effect of a gauge boson becomes weaker as the mass of the gauge boson increases.

† This evasive ploy, of inventing new particles and then giving them huge masses to explain away the fact that they are unknown experimentally, unfortunately has been much abused in the subsequent decades.

$SU(4)$ there are then gauge bosons capable of changing a quark into a lepton and vice versa.

This scheme is sometimes referred to as partial unification, since the gauge theory is still based on a direct product group. Quarks and leptons are unified but not the gauge forces.

Different routes to our low energy world

Remarkably, group theory indicates that the Pati-Salam theory and the Georgi-Glashow theory, which at first sight look quite different, may end up being realizations of the same underlying theory. Starting with the grand unified world of $SO(10)$, there is another route to our low energy world other than $SO(10) \to SU(5) \to SU(3) \otimes SU(2) \otimes U(1)$.

We have discussed $SU(5)$ as a natural subgroup of $SO(10)$, but $SO(10)$ is large enough to have other interesting subgroups. It clearly contains $SO(6) \otimes SO(4)$, under which the defining vector representation $10 \to (6, 1) + (1, 4)$. In other words, we split the ten components of the vector ϕ^M, $M = 1, \cdots, 10$ trivially into the two sets ϕ^A, $A = 1, \cdots, 6$ and ϕ^P, $P = 7, \cdots, 10$.

Furthermore, from chapter VII.1 we know that $SO(6)$ is locally isomorphic to $SU(4)$, which contains $SU(3)$. We also know that $SO(4)$ is locally isomorphic to $SU(2) \otimes SU(2)$, which contains $SU(2) \otimes U(1)$. Thus, we can travel an alternate route to the low energy world, going through the following chain of subgroups:

$$SO(10) \to SO(6) \otimes SO(4) = SU(4) \otimes SU(2) \otimes SU(2)$$
$$\to SU(3) \otimes U(1) \otimes SU(2) \otimes U(1) \to SU(3) \otimes SU(2) \otimes U(1) \tag{12}$$

The same spinor 16^+ now decomposes quite differently than in (1):

$$16^+ \to (4^+, 2^+) \oplus (4^-, 2^-) = (4, 2, 1) \oplus (4^*, 1, 2)$$
$$\to (3, 2, 1) \oplus (1, 2, 1) \oplus (3^*, 1, 2) \oplus (1, 1, 2) \tag{13}$$

We explained, earlier in this chapter, that the right handed neutrino could acquire a large Majorana mass from the 126 of $SO(10)$. As a good exercise in group theory, you should be able to work out how the 126 decomposes under $SO(10) \to SO(6) \otimes SO(4) = SU(4) \otimes SU(2) \otimes SU(2) \to SU(3) \otimes SU(2) \otimes SU(2)$. Do it before reading on!

I need perhaps only remind you that the 126, being a 5-indexed antisymmetric tensor T^{ABCDE}, is self-dual (or antiself-dual). Here we denote the vector index of $SO(10)$ by $A = 1, 2, \cdots, 10$. Dividing $A = \{a, i\}$ into two sets with $a = 1, 2, 3, 4, 5, 6$ and $i = 7, 8, 9, 10$ pertaining to $SO(6)$ and $SO(4)$, respectively; keeping in mind the duality properties of the tensor T^{abcij}; and perhaps consulting the table in (VII.1.66), we obtain the decomposition:

$$126 \to (6, 1) \oplus (15, 4) \oplus (10, 3) \oplus (10, 3)$$
$$\to (6, 1, 1) \oplus (15, 2, 2) \oplus (10, 3, 1) \oplus (10^*, 1, 3)$$
$$\to (3^*, 1, 1) \oplus (3, 1, 1) \oplus (8, 2, 2) \oplus (3^*, 2, 2) \oplus (3, 2, 2) \oplus (1, 2, 2)$$
$$\oplus (6, 3, 1) \oplus (3, 3, 1) \oplus (1, 3, 1) \oplus (6^*, 1, 3) \oplus (3^*, 1, 3) \oplus (1, 1, 3) \tag{14}$$

In the first line, the 6 is the [1] of $SO(6)$ and [2] of $SU(4)$, respectively; the 15 is the [2] and the adjoint $(1, 1)$; and the 10 is the [3] and the $\{2\}$. In the third line, under $SU(4) \to SU(3)$, $6 \to 3^* \oplus 3$, $15 \to 8 \oplus 3^* \oplus 3 \oplus 1$, and $10 \to 6 \oplus 3 \oplus 1$. The usual check: $126 = 1 \cdot 6 + 4 \cdot 15 + 3 \cdot 10 + 3 \cdot 10 = 3 + 3 + 32 + 12 + 12 + 4 + 2(18 + 9 + 3)$.

Exercises

1 The contribution to the anomaly by a given irreducible representation is determined by the trace of the product of a generator and the anticommutator of two generators, namely, tr $T^A\{T^B, T^C\}$. Here T^A denotes a generator of the gauge group G. Show that for $G = SU(5)$, the anomaly cancels between the 5^* and the 10. Note that for $SU(N)$, we can, with no loss of generality, take A, B, and C to be the same, so that the anomaly is determined by the trace of a generator cubed (namely, tr T^3). It is rare that we get something cubed in physics (see endnote 1), and so any cancellation between irreducible representations can hardly be accidental.

2 Show that the anomaly vanishes for any representation of $SO(N)$ except for $SO(6)$. Why is $SO(6)$ exceptional?

3 Work out how the 3-indexed antisymmetric 120 decomposes on restriction to $SO(4) \otimes SO(6)$.

Notes

1. Yes, quantum field theory is very strange. See chapter IV.7 in G Nut.
2. Incidentally, your first guess might be $SU(6)$ with its 15-dimensional irreducible representation furnished by the 2-indexed antisymmetric tensor ψ^{AB} with the indices $A, B = 1, 2, \cdots, 6$. But on restriction to its natural $SU(5)$ subgroup, the 15 breaks into $\psi^{\mu\nu}$ and $\psi^{\mu 6}$, with $\mu, \nu = 1, 2, \cdots, 5$, corresponding to 10 and 5, not 10 and 5^*.
3. Howard Georgi told me that he actually found $SO(10)$ before $SU(5)$.
4. Recall also that the conjugate spinor 16^- breaks up into the conjugate of the representations in (1): $16^- \to [1] \oplus [3] \oplus [5] = 5 \oplus 10^* \oplus 1$.
5. S. L. Glashow and M. Gell-Mann, Ann. Phys. 15 (1961), p. 437.
6. The mass of the gauge bosons responsible for proton decay is intimately tied to M. In quantum field theory, the more massive a gauge boson, the weaker is the effect of that gauge boson. (As mentioned in chapter IX.1, this explains why the weak interaction is weak.) Since the lower bound on the proton lifetime is enormous (something like 10^{20} times the present age of the universe), M has to be huge. At low energies, the strong, the electromagnetic, and the weak couplings are very different. Grand unification can only happen at an energy scale M when the three couplings become comparable. In quantum field theory, how the couplings vary with the energy scale is known and calculable. For both of these points, see any modern textbooks on quantum field theory, for example, part VII in QFT Nut.

IX.4 | The Family Mystery

A speculation on the origin of three generations in a family

I now end this book on group theory with a profound mystery.

A great unsolved puzzle in particle physics is the family problem. Why does Nature repeat herself? Why do quarks and leptons come in three generations* $\{v_e, e, u, d\}$, $\{v_\mu, \mu, c, s\}$, and $\{v_\tau, \tau, t, b\}$?

The way we incorporate this experimental fact into our present day theory can only be described as pathetic: we repeat the fermionic sector of the Lagrangian three times without any understanding whatsoever. Fundamental forces are unified, but not fundamental fermions. Three generations living together gives rise to a nagging family problem. Group theory may or may not offer a solution.

Even before grand unification, at the level of $SU(3) \otimes SU(2) \otimes U(1)$, the family problem already poses a puzzle. The family problem may be a separate issue from grand unification, or the two might be inextricably linked. We don't know.

One natural approach that almost suggests itself is to introduce a family group[1] $\mathcal{F}$, under which the three generations transform as a 3-dimensional irreducible representation, and to extend the theory[2] to have the symmetry $SU(3) \otimes SU(2) \otimes U(1) \otimes \mathcal{F}$. With the advent of grand unification, theories based on $SU(5) \otimes \mathcal{F}$ and $SO(10) \otimes \mathcal{F}$ have all been explored. That the number of generations is equal to three is not explained at all in this approach; we merely look for an $\mathcal{F}$ with a 3-dimensional representation. Both continuous and finite groups have been considered, and you sure know enough group theory by now to be able to join in the fun. Continuous groups, such as $SO(3)$ or $SU(3)$, tend to be far too constraining: one difficulty is that the masses of the three generations are vastly different,[3] unlike the situations confronting our great predecessors Heisenberg and Gell-Mann. Perhaps finite groups offer a better bet, and a number of them have been explored, for example the tetrahedral group.[4]

---

* As mentioned in chapter IX.1.

Quarks and leptons as possibly composites

Another possibility, one that has been entertained for a long time but has not progressed very far, is that perhaps quarks and leptons are not elementary. If so, then group theory might be expected to play a role[5] as well, perhaps similar to the ways in which atomic states are organized, as was discussed in parts III and IV.

Living in the computer age, I find it intriguing that in $SO(10)$ the fundamental constituents of matter are coded by five bits. Our beloved electron is composed of the binary strings $+----$ and $--+++$. An attractive possibility[6] suggests itself, that quarks and leptons may be composed of five different species of fundamental fermionic objects.[7] We construct composites, writing down a $+$ if that species is present, and a $-$ if absent. For example, from the expression for Q given in chapter IX.3, we see that species 1 carries electric charge -1, species 2 is neutral, and species 3, 4, and 5 carry charge $\frac{1}{3}$. In other words, $Q = -n_1 + \frac{1}{3}(n_3 + n_4 + n_5)$ with $n_j = \frac{1}{2}(\varepsilon_j + 1)$. A more or less concrete model can even be imagined by binding these fundamental fermionic objects to a magnetic monopole.

The dream of one group, one irreducible representation

Behind the thrust that has successfully moved us from $SU(3) \otimes SU(2) \otimes U(1)$ to $SO(10)$ is the desire to have the fermions of one generation unified into one single irreducible representation R of a group G. We can now be more ambitious and continue that thrust. Imagine putting all the fundamental fermions, of all three known generations, into one single irreducible representation $\mathcal{R}$ of a group $\mathcal{G}$, such that when we restrict $\mathcal{G}$ to its subgroup G, the irreducible representation breaks up into several copies of R—hopefully, three copies in the form $R \oplus R \oplus R$, thus reproducing the repetitive structure seen in Nature.

A survey of all Lie algebras reveals that only the spinor representations of orthogonal groups come close* to having the desired decomposition property.[8] I remind you that the reason that a spinor decomposes into a direct sum of spinors essentially rests on two facts: (i) spinor representations exist because Clifford algebras exist, and (ii) Clifford algebras may be constructed iteratively.

Indeed, our binary code view of the world encodes this approach to the family problem: we add more bits, generalizing $|++--+\rangle$, for example, to $|++--+\varepsilon_6\varepsilon_7\cdots\rangle$. One

---

* To see what goes wrong with other groups, consider, as an illustration, the traceless antisymmetric tensor $T^{\mu\nu}_\rho$ of $SU(8)$. Break $SU(8)$ down to $SU(5)$ in the standard way, such that $8 \rightarrow 1 + 1 + 1 + 5$. To be specific, split the index set $\mu = \{i, A\}$ with $i = 1, 2, \cdots, 5$ and $A = 6, 7, 8$. Then the decomposition of $T^{\mu\nu}_\rho$ contains T^{AB}_i equal to three copies of the 5^* of $SU(5)$ and T^{ij}_A equal to three copies of the 10 of $SU(5)$. This would appear to account for the observed three generations, but unfortunately, we also obtain a bunch of unwanted fermions transforming like the 24 and 5. Of course, at the cost of being contrived, one could then try to find ways of hiding these unwanted fermions by giving them large masses, for example. But then, with $SO(10)$, we no longer have so much faith in $SU(5)$.

possibility is to "hyperunify" into an $SO(18)$ theory, putting all fermions into a single spinorial representation $S^+ = 256^+$, which on the breaking of $SO(18)$ to $SO(10) \otimes SO(8)$ decomposes as

$$256^+ \rightarrow (16^+, 8^+) \oplus (16^-, 8^-) \tag{1}$$

The good news and the bad news. First the good news: we get the 16^+s we want, repeated. The bad news: too many* of a good thing, eight of them instead of three. More bad news: group theory dictates that we also get a bunch of unwanted[†] 16^-s.

We are left with several possibilities, which are not necessarily mutually exclusive. (i) The idea of obtaining repetitive family structure from spinor representations may be altogether wrong. (ii) The idea of using spinor representations is correct. Two further possibilities. (iia) Perhaps fermions transforming like the 16^- will eventually be seen. (iib) The 16^- fermions, as well as the additional 16^+ fermions, are to be concealed somehow.[9]

Concealing unwanted particles

Two ways of concealing particles are known. One way, already mentioned in passing in chapter IX.3, is to give the unwanted fermions large masses by using the Higgs mechanism. This would appear somewhat ad hoc and contrived. A better way may be to take a hint from the way Nature conceals quarks.

After quarks were first proposed, physicists generally found the idea implausible. Why are these fractionally charged particles not seen? Eventually, it was theorized that quarks were permanently confined.[‡] The "dogma" that emerged, now widely accepted, holds that the color $SU(3)$ force is such that any state that does not transform as a color singlet cannot be liberated and exist in isolation. For instance, an up quark, which transforms as a 3, cannot exist in isolation and be observed. Similarly an antidown quark, which transforms as a 3*, cannot exist in isolation. But since $3 \otimes 3^* = 1 \oplus 8$ (as you learned in parts IV and V), a bound state of an up quark and an antidown quark that transforms as the 1 could exist; this is in fact the positively charged pion π^+. So you see that once again, group theory plays a leading role. As another example, the proton, consisting of two up quarks and a down quark, exists because $3 \otimes 3 \otimes 3$ contains the 1. In short, only objects that do not transform under color $SU(3)$ (sometimes referred to as colorless) can exist in isolation.[§]

In this picture, the observed strong interaction between, say, the pion and the nucleon (as discussed in chapter V.1) is but a pale shadow of the strong color force. A rough (but rather misleading[10]) analogy would be that if the electric force were much stronger than it is, then only electrically neutral atoms and molecules could exist. The ionization threshold

---

* In contrast, $SO(14)$ would give only two 16^+s, too few.

[†] The 16^-s lead to fields with opposite handedness from those that are observed. But for all we know, they could be just around the corner.

[‡] I have already alluded to quark confinement in chapter V.2.

[§] Are there some other ways of concealing particles?

may be unattainable. The observed interaction between atoms and molecules would then be a mere shadow of the underlying electric force.

Hypercolor

The hope is that Nature would once again show theoretical physicists her legendary kindness and use the same trick twice. Let us start with $SO(10 + 4k)$ and break it to $SO(10)$ $\otimes SO(4k)$. Suppose some subgroup of $SO(4k)$ remains unbroken down to low energy (just like the $SU(3)$ of color remains unbroken down to low energy[11]), and thus generates an analog of the strong color force, call it hypercolor. We then expect that only those fermions that transform like singlets under hypercolor can exist in isolation.[12]

Now the key point is that the two spinors $S^\pm$ of $SO(4k)$ can decompose quite differently under the hypercolor subgroup and thus will contain a different number of hypercolor singlets. An example would make this clear. Group theory tells us that the two spinors $S^\pm$ of $SO(8m)$ are real, as we learned in chapter VII.1, and so on restriction of $SO(8m)$ to $SU(4m)$, will in general decompose differently. In particular, under $SO(8) \to SU(4)$, we have $8^+ \to 1 \oplus 6 \oplus 1$, while $8^- \to 4 \oplus 4^*$. Thus, we will end up with two 16^+s and no 16^-s in the low energy world.

In the pantheon of attractive groups, $SO(8)$ is regarded by many as the most beautiful of all. We hope that Nature likes it too. In any case, $SO(8)$ has a strikingly symmetrical Dynkin diagram* with a 3-fold symmetry that arises because the two spinors $8^\pm$ have the same dimension as the vector 8^v. (The equation $2^{n-1} = 2n$ has the unique solution $n = 4$.) The algebra admits a transformation† that cyclically rotates these three representations 8^+, 8^-, and 8^v into one another.

Indeed, under the "natural" embedding of $SO(6) = SU(4)$ into $SO(8)$ (leaving two of the eight Cartesian axes untouched), the vector decomposes as $8_v \to 6 \oplus 1 \oplus 1$, while $8^\pm \to 4 \oplus 4^*$. Thus, we can understand the embedding used above (under which $8^+ \to 6 \oplus 1 \oplus 1$, while $8^- \to 4 \oplus 4^*$ and $8^v \to 4 \oplus 4^*$) as the natural embedding followed by an outer automorphism. Loosely speaking, the spinor 8^+ in the embedding used above is secretly the vector in the natural embedding. In other words, the embedding we want corresponds to the natural embedding followed by a "twist."

But now we see how to get three generations: we break the $SO(6)$ further down to an $SO(5) = Sp(4)$ of hypercolor, so that $8^+ \to 5 \oplus 1 \oplus 1 \oplus 1$. The 5 is confined by $SO(5)$, leaving us three $SO(5)$ singlets transforming like the 16^+ of $SO(10)$. Meanwhile, $8^- \to 4 \oplus 4^*$, and the 16^-s remain confined.

The trouble is that nothing prevents us from saying that $SO(5)$ breaks further to $SO(4) = SU(2) \otimes SU(2)$ of hypercolor, so that $8^+ \to 4 \oplus 1 \oplus 1 \oplus 1 \oplus 1$. We get four generations. One step further, with $SO(4) \to SO(3) = SU(2)$, and we get five generations.

---

* As shown in figure VI.5.8.
† Our good friend the jargon guy informs us that this is called an outer automorphism.

Thus, our nice little group theoretic game ends up predicting the number of 16^+ generations to be two, three, four, or five, respectively, according to whether $SO(8)$ is broken down to[13] $SO(6)$, $SO(5)$, $SO(4)$, or $SO(3)$. Unfortunately, our knowledge of the dynamics of symmetry breaking is far too paltry for us to make any further statements. It is as if our mastery of classical mechanics is so incomplete that we cannot predict where a rolling ball will end up, but can merely list the various possibilities.

It seems somehow appropriate, as we come to the end of this book, to hope that Nature will show her kindness once again. I have spoken about Nature's kindness at various points in this book and in my popular books. Indeed, the history of physics almost amounts to a series of beginner's luck. We got to practice classical mechanics before tackling quantum mechanics, Maxwell's theory before Yang-Mills theory, and so on. In group theory also, Nature, like a kindly pedagogue, first showed us $SO(3)$, then $SU(2)$. From $SU(2)$ and $SO(3)$, we graduated to $SO(4) = SU(2) \otimes SU(2)$, and then were able to move to $SO(3, 1)$ and $SU(3)$. Imagine being in another universe where we are slammed with $SO(9)$ from day one. Imagine encountering as our first internal symmetry group, not Heisenberg's $SU(2)$, but $G(2)$ or $E(8)$, say. Nature takes us by the hand, and lets us cut our teeth on representations with dimensions of 2 or 3, before moving onto larger irreducible representations.

Nature is kind.

Notes

1. There is by now a vast literature on the family group; an early paper is F. Wilczek and A. Zee, Phys. Rev. Lett. 42 (1979), p. 421.
2. The group $\mathcal{F}$ may or may not be gauged; people have considered both possibilities.
3. Feynman allegedly said, "Do you want to be famous? Do you want to be a king? Do you want more than the Nobel Prize? Then solve the mass problem."
4. This has been long advocated by E. Ma and collaborators (E. Ma and G. Rajasekaran, Phys. Rev. D 64 (2001) 113012; E. Ma, Mod. Phys. Lett. A 17 (2002), p. 289; hep-ph/0508099, and references therein). As you may recall from parts II and IV, this group has many attractive features. For instance, the existence of three inequivalent 1-dimensional representations might be relevant.
5. There is a vast literature on this. For one speculative attempt to make the neutrinos composite using $SU(2)$, see P. Kovtun and A. Zee, Phys. Lett. B 640 (2006), pp. 37–39.
6. For further details, see F. Wilczek and A. Zee, Phys. Rev. D 25 (1982), p. 553, section IV.
7. Recall the description of $SO(2n)$ spinors using fermions given in appendix 3 to chapter VII.1.
8. The fact that the group decomposition property of spinors is highly suggestive of the observed repetitive family structure was noted independently by Gell-Mann, Ramond, and Slansky, and by Wilczek and Zee. Some other authors have also studied the use of spinor representations to unify fermions. See the references given in F. Wilczek and A. Zee, ibid.
9. For a recent development on this front, see Y. BenTov and A. Zee, arXiv:1505.04312.
10. For one thing, the electrostatic potential goes like $V(r) \propto -1/r$, while the confining potential between quarks is theorized to grow like r.
11. In fact, down to zero energy.
12. This general picture was proposed by Gell-Mann, Ramond, and Slansky.
13. Note that these orthogonal subgroups are not embedded in $SO(8)$ in the standard way, however.

Epilogue

We have traveled a long road together. You ought to be impressed with yourself. Starting from a vague notion that an equilateral triangle is more symmetrical than an isosceles triangle, we saw the role of transformations, which led us to ponder what one transformation followed by another would produce. Concepts pop up one after another, often naturally suggesting themselves. We stand in awe of the beautiful results mathematicians have uncovered, such as the complete classification of Lie algebras.

Perhaps because of this sentiment, somebody famous once referred to group theory as an alluring temptress. But group theory is much much more than a temptress; it is our indispensable guide to a fundamental understanding of the universe. The Lorentz group determines for us how the fundamental fields should comport themselves in spacetime. Eventually, group theory, together with quantum field theory, shows us how three of the four fundamental interactions can be unified. And thus we arrive at the threshold of our understanding of the universe, confronted by dark but yet concrete mysteries, such as why the matter content of the universe is repeated three times. Cartoonists often imagine theorists pondering the basic laws standing before some massive computer, but in truth computations such as $16 \rightarrow 10 \oplus 5^* \oplus 1$ are more like it. Who would have guessed that figuring out what it is exactly that makes the circle so pleasing turns out to be the key first step in our quest to make sense of the universe? But such is the human mind and the creative spirit.

Timeline of Some of the People Mentioned

Pierre Fermat (1601 [1607/1608]–1665)

Gabriel Cramer (1704–1752)

Leonhard Euler (1707–1783)

Joseph-Louis Lagrange (born Giuseppe Lodovico Lagrangia) (1736–1813)

Pierre-Simon, marquis de Laplace (1749–1827)

Adrien-Marie Legendre (1752–1833)

Carl Gustav Jacob Jacobi (1804–1851)

William Rowan Hamilton (1805–1865)

Évariste Galois (1811–1832)

Auguste Bravais (1811–1863)

Hermann Ludwig Ferdinand von Helmholtz (1821–1894)

Arthur Cayley (1821–1895)

Charles Hermite (1822–1901)

Alfred Clebsch (1833–1872)

Paul Gordan (1837–1912)

Marius Sophus Lie (1842–1899)

William Kingdon Clifford (1845–1879)

Wilhelm Killing (1847–1923)

Ferdinand Georg Frobenius (1849–1917)

Hendrik Antoon Lorentz (1853–1928)

Henri Poincaré (1854–1912)

Élie Cartan (1869–1951)

Issai Schur (1875–1941)

Emmy Noether (1882–1935)

Hermann Weyl (1885–1955)

Paul Adrien Maurice Dirac (1902–1984)

Eugene Wigner (1902–1995)

Bartel Leendert van der Waerden (1903–1996)

Ettore Majorana (1906–disappeared at sea in 1938; probably dead after 1959)

Eugene Borisovich Dynkin (1924–2014)

Solutions to Selected Exercises

In the book of life, the answers aren't in the back.
—Charles M. Schulz, speaking through
Charlie Brown

A Brief Review of Linear Algebra

Exercises 1, 2, and 3 simply test if you understand how to multiply matrices.

1 $\begin{pmatrix} 0 & 1 \\ 1 & 0 \end{pmatrix} \begin{pmatrix} a & b \\ c & d \end{pmatrix} = \begin{pmatrix} c & d \\ a & b \end{pmatrix}.$

2 $\begin{pmatrix} s_1 & 0 \\ 0 & s_2 \end{pmatrix} \begin{pmatrix} a & b \\ c & d \end{pmatrix} = \begin{pmatrix} s_1 a & s_1 b \\ s_2 c & s_2 d \end{pmatrix}.$

3 $\begin{pmatrix} 1 & 0 \\ s & 1 \end{pmatrix} \begin{pmatrix} a & b \\ c & d \end{pmatrix} = \begin{pmatrix} a & b \\ sa+c & sb+d \end{pmatrix}.$

5 $\begin{pmatrix} a & b & c \\ d & e & f \\ g & h & i \end{pmatrix} \begin{pmatrix} 1 & 0 & 0 \\ 0 & 0 & 1 \\ 0 & 1 & 0 \end{pmatrix} = \begin{pmatrix} a & c & b \\ d & f & e \\ g & i & h \end{pmatrix}$ and $\begin{pmatrix} 1 & 0 & 0 \\ 0 & 0 & 1 \\ 0 & 1 & 0 \end{pmatrix} \begin{pmatrix} a & c & b \\ d & f & e \\ g & i & h \end{pmatrix} = \begin{pmatrix} a & c & b \\ g & i & h \\ d & f & e \end{pmatrix}.$

6   For example, to add the ith row to the jth row (for $i < j$), multiply M from the left by the matrix E defined as the identity matrix modified by adding a 1 to the entry in the jth row and the ith column.

8   Note that $\det M = \det M^T = \det(-M) = (-1)^n \det M$. For n odd, this vanishes.

9   The two eigenvalues are a and b, with the corresponding eigenvector $\begin{pmatrix} 1 \\ 0 \end{pmatrix}$ and $\begin{pmatrix} 1 \\ b-a \end{pmatrix}$. Thus, $S = \begin{pmatrix} 1 & 1 \\ 0 & b-a \end{pmatrix}$ with the inverse $S^{-1} = \frac{1}{b-a} \begin{pmatrix} b-a & -1 \\ 0 & 1 \end{pmatrix}$. You should check that $S^{-1} M S$ is diagonal. But if $a = b$, then S^{-1} fails to exist. For example, the matrix $M = \begin{pmatrix} 0 & 1 \\ 0 & 0 \end{pmatrix}$ cannot be diagonalized. (A quick way of seeing this is to note that $M^2 = 0$. Thus, if M could be diagonalized, its diagonal elements have to be 0, but then M would be the zero matrix, which it is not.)

10   Diagonalize M.

11   Note that λ solves the equation $\det(\lambda I - M) = \lambda^2 - (a^2 - b^2 + c^2) = 0$. Thus, $w = (a^2 - b^2 + c^2)^{\frac{1}{2}}$, which is either real or imaginary, according to whether b^2 is larger or smaller than $a^2 + c^2$. For $b = 0$ the eigenvalues are real, thus verifying that a real symmetric matrix must have real eigenvalues.

13 $A^{-1}(I + AB)A = (I + BA)$.

14   Take the transpose of $M^{-1}M = I$ to obtain $M^T(M^{-1})^T = I$, and thus the first identity follows. Similarly, complex conjugate $M^{-1}M = I$, and the second identity follows.

15   The expression $\Pi_{i<j}(x_i - x_j)$ follows instantly by noting that when $x_i = x_j$ for any pair of i and j, row i and row j in M are equal, and hence the determinant vanishes. (That the determinant cannot be equal to $\Pi_{i<j}(x_i - x_j)^s$ for some integer $s \neq 1$ follows from counting powers of the xs.)

16   You could simply evaluate the determinant by brute force and verify that it is equal to the area, as claimed. More elegantly, note that the area is invariant under translation and rotation of the triangle. Call the matrix in the statement of the exercise M. Adding a suitable multiple of the third row of M to its first and second row, we could set $x_1 = y_1 = 0$. This is of course just placing the origin at vertex 1 of the triangle. Then by multiplying the first row by a suitable factor and adding to it the second row multiplied by a suitable factor, we can change M to the form $\begin{pmatrix} 0 & b & x_3' \\ 0 & 0 & h \\ 1 & 1 & 1 \end{pmatrix}$ without changing the determinant. (The more advanced reader will realize that we are simply multiplying M by a 2-by-2 rotation matrix; the less advanced reader will have to wait until after he or she reads chapter I.3.) We then recover the familiar high school formula that the area of a triangle is equal to half of its base b multiplied by its height h.

I.1   Symmetry and Transformation

2   This is a consequence of the "once and only once rule."

I.2   Finite Groups

2   In the text it was shown that any permutation can be written as a product of 2-cycles. The permutations in A_n are even, and hence are equal to the product of an even number of 2-cycles. Using the result of the preceding exercise, we could write this product in the form $(1a)(1b)(1c)\cdots$ with an even number of 2-cycles. Each neighboring pair, for example, $(1a)(1b) = (a1)(1b) = (a1b)$, can be written as a 3-cycle. Thus, any element A_n can be written as a product of 3-cycles.
   A more laborious proof is to simply multiply two 3-cycles together. The possible cases are three letters in common, two in common, one in common, and none in common (which is a trivial case). Thus, $(123)(123) = (132)$, $(123)(234) = (12)(23)(23)(34) = (12)(34)$, $(123)(345) = (12345)$.

4 $1+1+1+1+1, 2+1+1+1, 2+2+1, 3+1+1, 3+2, 4+1, 5$.

5   The required number is just the number of ways of putting n objects into n_j boxes of length j each. To start with, we visualize that all the boxes are lined up in order from large to small and that the j objects in each box of length j are lined up in some order, as in $(xxxxx)(xxxxx)(xxxx)(xx)(xx)(xx)(x)(x)(x)(x)$. There are $n!$ ways of putting the n objects into the n available slots, but we overcount, since for each j the n_j boxes could be permuted around in $n_j!$ ways. Furthermore, in each box, the objects can be cyclically permuted without changing anything. For each box, there are j ways of ordering the j objects (by starting with any one of them at the "head" of the box). We should thus divide by j for each box, and thus, by j^{n_j} all together. The number of elements with a given cycle structure is then $n!/\Pi_j(j^{n_j}n_j!)$, thus verifying (4).

8   We have $(132)(13)(24)(123) = (14)(23)$. Thus, the elements of A_4 with the cycle structure $(xx)(xx)$ form an invariant subgroup.

10   If $f(h_1) = I$ and $f(h_2) = I$, then $f(h_1 h_2) = f(h_1)f(h_2) = I$. So the kernel of f forms a subgroup. Next, $f(g^{-1}hg) = f(g^{-1})f(h)f(g) = f(g^{-1})If(g) = f(g^{-1})f(g) = I$, and so the kernel is an invariant subgroup.

11   The elements consist of $(R^k, R^k r, k = 0, 1, \cdots, n-1)$. Calculate away! Thus, $\langle R^k, R^l \rangle = I$, $\langle R^k, R^l r \rangle = R^{-k}r^{-1}R^{-l}R^k R^l r = R^{-k}r R^k r = R^{-k}R^{-k} = R^{-2k} = R^{n-2k}$, and $\langle R^k r, R^l r \rangle = R^{2(k-l)}$.

     The derived subgroup consists of the elements $\{I, R^{n-2k}, R^{2(k-l)}\}$ with various possibilities for k and l. For $n = 4$ (for example), the derived subgroup $\{I, R^2\} = Z_2$. For $n = 5$, the derived subgroup $\{I, R, R^3, R^2, R^4\} = Z_5$. Convince yourself that the derived subgroup is $Z_{n/2}$ for n even and Z_n for n odd.

13   Divide the set of group elements into the two disjoint sets $\mathcal{I} = \{$all g such that $g^2 = I\}$ and $\mathcal{N} = \{$all g such that $g^2 \neq I\}$. Pick an element g_1 of $\mathcal{N}$. Since $g_1^2 \neq I$, $g_1^{-1} \neq g_1$, and $\mathcal{N}$ contains at least two distinct elements, namely g_1 and g_1^{-1}. If these exhaust $\mathcal{N}$, then, since by assumption G has an even number of elements, $\mathcal{I}$ has an even number of elements. But $\mathcal{I}$ cannot be null, since it contains the identity I. The theorem is proved. If g_1 and g_1^{-1} do not exhaust $\mathcal{N}$, then we pick another element g_2 and ask whether $\{g_1, g_1^{-1}, g_2, g_2^{-1}\}$ exhaust $\mathcal{N}$. We repeat this process until $\mathcal{N}$ is exhausted. Thus, $\mathcal{N}$ has an even number of elements, and so $\mathcal{I}$ has an even number of elements. The theorem is proved.

16   Given $(ab)^2 = abab = I$, we have $ba = ba(abab) = (baab)ab = ab$.

17   Since H is invariant, for any of its elements h_i, $gh_ig^{-1} = h_j$ for some h_j. Thus, the set $\{gh_1, \cdots, gh_i, \cdots\}$ is the same as the set $\{h_1 g, \cdots, h_i g, \cdots\}$ with the elements listed in a different order.

18   Label the six edges of the tetrahedron by $1, 2, \cdots, 6$. When each element of A_4 permutes the four vertices, it also permutes the six edges.

19   Explicitly, take two 2-cycles in S_n and calculate $\langle (xy), (yz) \rangle = (yx)(zy)(xy)(yz) = (xyz)(xyz)$. Now use the result of exercise 2 that the A_n is generated by 3-cycles. More quickly, given two permutations P and Q, $\det(P^{-1}Q^{-1}PQ) = \det(P^{-1}P)\det(Q^{-1}Q) = 1$, and so $P^{-1}Q^{-1}PQ$ is an element of A_n. Note that $S_n/A_n = Z_2$ is abelian.

I.3 Rotations and the Notion of Lie Algebra

4   Hermitean conjugate (18) to obtain $[J_i, J_j]^\dagger = -ic_{ijk}^* J_k$. Remembering that hermitean conjugation involves the transpose, which reverses the order in a product, the left hand side is equal to $-[J_i, J_j] = -ic_{ijk}J_k$. Therefore, $c_{ijk}^* = c_{ijk}$.

6   Two. We could choose, for example, J_{12} and J_{34}, since they commute with each other. This will be of use in part VI.

7   Define $f_i(\varphi) = e^{-i\varphi J_3} K_i e^{i\varphi J_3}$ for $i = 1, 2$. Differentiate to obtain $\frac{df_1}{d\varphi} = f_2(\varphi)$ and $\frac{df_2}{d\varphi} = -f_1(\varphi)$. Solve for $f_1(\varphi)$ with appropriate initial conditions at $\varphi = 0$.

II.2 Schur's Lemma and the Great Orthogonality Theorem

1 $u + v + w = 0$, $u + \omega v + \omega^* w = 0$, $u + \omega^* v + \omega w = 0$. Adding, we get $u = 0$. Subtracting the second equation from the third, we get $v - w = 0$, and hence $v = w = 0$.

2   As discussed in chapter I.2, the group D_5, the invariance group of the pentagon, consists of ten elements $\{I, R, R^2, R^3, R^4, r, Rr, R^2r, R^3r, R^4r\}$ falling into four classes. Call these classes I, R, R^2, and r. The class averages are $K(I) = I$, $K(R) = \frac{1}{2}(R + R^4)$, $K(R^2) = \frac{1}{2}(R^2 + R^3)$, and $K(r) = \frac{1}{5}\sum_{k=0}^{4} R^k r$. Then

$$K(R)K(R) = \frac{1}{4}K(R) + \frac{1}{2}K(R^2) \tag{31}$$

$$K(R)K(R^2) = \frac{1}{2}(K(R) + K(R^2)) \tag{32}$$

$$K(r)K(r) = \frac{1}{5}K(I) + \frac{2}{5}(K(R) + K(R^2)) \tag{33}$$

You could work out the rest.

II.3   Character Is a Function of Class

1   Again we look for fixed points, if any. The (123) leaves 4 untouched; so there is one fixed point, and hence $\chi = 1$. Explicitly, $(123) \sim \begin{pmatrix} 0 & 0 & 1 & 0 \\ 1 & 0 & 0 & 0 \\ 0 & 1 & 0 & 0 \\ 0 & 0 & 0 & 1 \end{pmatrix}$ with trace $= 1$. Similarly for (132). In contrast, (12)(34) has no fixed points and is represented by a matrix with trace $= 0$. Thus, the regular representation has characters given by $\begin{array}{|c|} \hline 4 \\ \hline 0 \\ \hline 1 \\ \hline 1 \\ \hline \end{array}$. Now we merely have to apply the orthogonality theorems. For example, $1 \cdot 4^2 + 3 \cdot 0^2 + 4 \cdot 1^2 + 4 \cdot 1^2 = 16 + 0 + 4 + 4 = 24 = 2(12)$, and hence the regular representation contains two irreducible representations. It is fairly clear which two these are. For example, calculate the orthogonality between the 4 and the 1: $1 \cdot 4 + 3 \cdot 0 + 4 \cdot 1 + 4 \cdot 1 = 4 + 0 + 4 + 4 = 12$, and thus the 4 contains 1 once. Indeed, by inspection $\begin{array}{|c|} \hline 4 \\ \hline 0 \\ \hline 1 \\ \hline 1 \\ \hline \end{array} = \begin{array}{|c|} \hline 1 \\ \hline 1 \\ \hline 1 \\ \hline 1 \\ \hline \end{array} + \begin{array}{|c|} \hline 3 \\ \hline -1 \\ \hline 0 \\ \hline 0 \\ \hline \end{array}$. Thus, $4 \to 1 + 3$.

3   Imposing various orthogonality theorems, obtain the character table as shown. Here are some selective checks:

Column orthonormality.
The irreducible representation 2: $1 \cdot 2^2 + 3 \cdot 2^2 + 8 \cdot (-1)^2 + 0 + 0 = 24$, ✓.
The irreducible representation 3: $1 \cdot 3^2 + 3 \cdot (-1)^2 + 0 + 6 \cdot 1^2 + 6 \cdot (-1)^2 = 24$, ✓.
Column orthogonality.
The irreducible representations 1 and $\bar{1}$: $1 \cdot 1 \cdot 1 + 3 \cdot 1 \cdot 1 + 8 \cdot 1 \cdot 1 + 6 \cdot 1 \cdot (-1) + 6 \cdot 1 \cdot (-1) = 1 + 3 + 8 - 6 - 6 = 0$, ✓.
The irreducible representations 1 and 2: $1 \cdot 1 \cdot 2 + 3 \cdot 2 + 8 \cdot 1 \cdot (-1) + 0 + 0 = 0$, ✓.
The irreducible representations 1 and 3: $1 \cdot 1 \cdot 3 + 3 \cdot 1 \cdot (-1) + 0 + 6 \cdot 1 \cdot 1 + 6 \cdot 1 \cdot (-1) = 0$, ✓.

Note that by subtracting the orthogonality between 1 and 3 and between $\bar{1}$ and 3, we can immediately conclude that the character of (12) and the character of (1234) in 3 (and in $\bar{3}$ also of course) have to be equal and opposite.

The irreducible representations 2 and 3: $1 \cdot 2 \cdot 3 + 3 \cdot 2 \cdot (-1) + 0 + 0 + 0 = 0$. ✓
Row orthogonality between the first and the second rows: $1 \cdot 1 + 1 \cdot 1 + 2 \cdot 2 + 3 \cdot (-1) + 3 \cdot (-1) = 0$. ✓

At this point, I will just let you go on.

6   Within S_n, any 3-cycle x is equivalent to (123): that is, there exists a permutation $s \in S_n$ such that $x = s^{-1}(123)s$. If s is also in A_n, then the claim stated in the exercise is true. If not, define $s = (45)t$; then $t \in A_n$. Now $s^{-1}(123)s = t^{-1}(54)(123)(45)t = t^{-1}(123)t = x$. Thus, t does the job. In A_4, (132) is manifestly not equivalent to (123).

7   As explained in chapter I.2, D_5 is presented by $D_5 = \langle R, r | R^5 = I, r^2 = I, Rr = rR^{-1} \rangle$. The 10 elements are divided into four equivalence classes, namely, $\{I\}$, $\{R, R^4\}$, $\{R^2, R^3\}$, and $\{r, Rr, R^2r, R^3r, R^4r\}$. Thus, there are four irreducible representations with dimensions fixed by $1^2 + 1^2 + 2^2 + 2^2$. The character table works out to be

D_5	n_c		1	1'	2	2'
	1	I	1	1	2	2
Z_5	2	R	1	1	$2\cos\theta$	$2\cos 2\theta$
Z_5	2	R^2	1	1	$2\cos 2\theta$	$2\cos\theta$
Z_2	5	r	1	-1	0	0

(26)

with $e^{5i\theta} = 1$. As usual, the first row and the first and second columns can be written down immediately. Column and row orthogonality fix the remaining entries.

II.4   Real, Pseudoreal, Complex Representations, and the Number of Square Roots

5   On the 3 of S_4, $\sum_g \chi(g^2) = 1 \cdot 3 + 3 \cdot 3 + 8 \cdot 0 + 6 \cdot 3 + 6 \cdot (-1) = 3 + 9 + 18 - 6 = 24$, and so the 3 is real.

6   Simply look up the character table of quarternionic group Q given in chapter II.3. Of the eight elements $\{1, -1, i, -i, j, -j, k, -k\}$, two square to 1 (namely, 1 and -1), while six square to -1 (namely, i, $-i$, j, $-j$, k, and $-k$). Thus, for the 2-dimensional irreducible representation, our trusty reality checker gives $\sum_g \chi^{(2)}(g^2) = 2 \cdot 2 + 6 \cdot (-2) = -8$, and so the 2 is pseudoreal.

   The neat thing is that we never have to explicitly show what the 2-dimensional irreducible representation actually is, as you might have noticed.

7   We know that $(123)^2 = (132)$. Evaluate $\sigma_{(123)} = 1 - 0 = 1$, ✓. Similarly, $\sigma_{(132)} = 1$, ✓.

8   For S_4,

$$\sigma_{(12)(34)} = 1 + 1 + 2 \cdot 2 - 3 - 3 = 0 \quad ✓ \tag{18}$$

$$\sigma_{(123)} = 1 + 1 - 1 = 1 \quad ✓ \tag{19}$$

$$\sigma_{(12)} = 1 - 1 + 0 + 1 - 1 = 0 \quad ✓ \tag{20}$$

$$\sigma_{(1234)} = 1 - 1 + 0 - 1 + 1 = 0 \quad ✓ \tag{21}$$

9   For A_5,

$$\sigma_I = 1 + 3 + 3 + 4 + 5 = 16 = 1 + 15 \quad ✓ \tag{22}$$

$$\sigma_{(12)(34)} = 1 - 1 - 1 + 1 = 0 \quad ✓ \tag{23}$$

$$\sigma_{(123)} = 1 + 0 + 0 + 1 - 1 = 1 \quad ✓ \tag{24}$$

$$\sigma_{(12345)} = \sigma_{(12354)} = 0 \quad ✓ \tag{25}$$

10   Calculate $\sum_g D^{(r)}(g^2)$ for A_4. The character table, from which the representation matrices for the 1-dimensional representations can be read off (of course), and the representation matrices for the 3-dimensional representation were all given in chapter II.3. The element (12)(34) squares to I, while the elements (123) and (132) square to each other. Thus, for 1', the sum equals $4 \cdot 1 + 4 \cdot (\omega + \omega^*) = 0$, which equals to $N(G)(\eta^{(r)}/d_r)I$ trivially, since $\eta^{1'} = 0$. Similarly for 1''. For the 3, the sum equals $4I$, because $c + r_1 cr_1 + r_2 cr_2 + r_3 cr_3 = 0$, and similarly for $c \to a$. In contrast, $N(G)(\eta^{(r)}/d_r)I = 12(1^2/3)I = 4I$.

11   Since A_4 has two real representations, we get, applying (17), $\tau_I = 12 \cdot 2 = 24$. To verify this, solve $f^2 g^2 = I$. Denote $\{I, \ (12)(34), \ (13)(24), \ (14)(23)\}$ by a_i, $i = 1, \cdots, 4$. Then $a_i^2 a_j^2 = I$. For examples, $I^2((12)(34))^2 = I$, $((12)(34))^4 = I$. This gives all together $4 \cdot 4 = 16$ solutions. Next, $(123)^2 (132)^2 = I$; there are $2 \cdot 4 = 8$ of these. All together $16 + 8 = 24$, ✓.

II.i2   Euler's φ-Function, Fermat's Little Theorem, and Wilson's Theorem

1   The group G_{10} has four elements: 1, 3, 7, 9. But in chapter I.1, we found all possible groups with four elements. Working out a few entries of the multiplication table (for example, $3^2 = 9$ (mod 10), $7^2 = 9$ (mod 10), $9^2 = 1$ (mod 10), $7 \cdot 3 = 1$ (mod 10), and $7 \cdot 9 = 3$ (mod 10)), we see that this is just Z_4 with the identification $1 \to 1$, $3 \to i$, $7 \to -i$, and $9 \to -1$.

3   Let's keep the elements G_{16} in front of us: 1, 3, 5, 7, 9, 11, 13, and 15. Again, by brute force, $3 = 3$, $3^2 = 9$, $3^3 = 11$, and $3^4 = 1$. The arithmetic here is mod 16. This generates a Z_4 subgroup. The elements yet unaccounted for are 5, 7, 13, and 15. Next, $5 = 5$, $5^2 = 9$, $3^3 = 11$, and $3^4 = 1$; again, this generates a Z_4 that intersects with the previous Z_4. Next, try $7 = 7$ and $7^2 = 1$, generating a Z_2, which does not intersect with the Z_4 generated by 3, for example. Hence we conclude $G_{16} = Z_4 \otimes Z_2$.

     You might wonder where the three elements 5, 13, and 15 are. In the notation of the direct product group $Z_4 \otimes Z_2$ they are (3, 7), (11, 7), and (13, 7), respectively. (In other words, $3 \cdot 7 = 21 = 5$, $11 \cdot 7 = 77 = 13$, and $9 \cdot 7 = 63 = 15$.)

III.2   Group Theory and Harmonic Motion: Zero Modes

1   Sure there is. The problem is still translation invariant: the two masses could be gliding along without stretching the spring. So we even know the eigenvector: it has to be (1, 1). With $m_1 \ddot{x}_1 = -(x_1 - x_2)$, and so forth, and defining $a_i = 1/m_i$, $i = 1, 2$, we end up with a nonsymmetric $H = \begin{pmatrix} a_1 & -a_1 \\ -a_2 & a_2 \end{pmatrix}$. Verify that a zero mode exists. The other eigenvector is $(a_1, -a_2)$, which is no longer orthogonal to (1, 1), with the corresponding eigenfrequency $\omega^2 = m_1^{-1} + m_2^{-1}$.

2   Setting $i = j$ in the Great Orthogonality theorem and summing, we obtain $\sum_g \chi^{(r)*}(g) D^{(s)}(g)^k_l = \frac{N(G)}{d_r} \delta^{rs} \delta^k_l$. The harmonic oscillator problem gives us a reducible representation $D(g)$ (for example, the 6-dimensional representation for the triangular molecule in the text). The operator $P_r \equiv \frac{d_r}{N(G)} \sum_g \chi^{(r)*}(g) D(g)$ thus projects out the irreducible representation r contained in $D(g)$.

IV.1   Tensors and Representations of the Rotation Groups $SO(N)$

5   It has only one component, which we can write as $T^{123} = \frac{1}{3!} \varepsilon^{ijk} T^{ijk}$. Under a rotation, $\varepsilon^{ijk} T^{ijk} \to \varepsilon^{ijk} R^{ii'} R^{jj'} R^{kk'} T^{i'j'k'} = (\det R) \varepsilon^{i'j'k'} T^{i'j'k'} = \varepsilon^{i'j'k'} T^{i'j'k'}$, where we used the definition of the determinant.

8   For example, for $D = 2$, evaluate $\varepsilon^{ij} R^{ip} R^{jq}$ for $p = 1$, $q = 2$. We have $\varepsilon^{ij} R^{i1} R^{j2} = \varepsilon^{12} R^{11} R^{22} + \varepsilon^{21} R^{21} R^{12} = \varepsilon^{12}(R^{11} R^{22} - R^{21} R^{12}) = \varepsilon^{12} \det R$.

10   To show the C indeed transforms like a vector, let $\varepsilon^{ijk} A^i B^j \to \varepsilon^{ijk} R^{im} R^{jn} A^m B^n = \varepsilon^{mnl} R^{kl} A^m B^n = R^{kl} C^l$. Thus, as expected, $C^k \to R^{kl} C^l$. An extreme nerd joke from the American Mathematical Society: What do you get when you cross a mosquito with a mountain climber? Nothing. You can't cross a vector with a scaler.

11   Simply note that both sides transform as invariant symbols with four indices, and the symmetry properties (such as under $i \leftrightarrow j$) of the two sides match. You can also prove this by trying it out for various values of i, j, l, n.

12   1, 5, 10, 14, and 30.

13   Simply repeat the reasoning in the text. The symmetric tensor has the form $S^{44\cdots 4xx\cdots x}$ (that is, among the indices are j xs, where x stands for 1, 2, or 3, with j ranging from 0 to h, and $h - j$ 4s). Thus, the total number of components is determined by $\sum_{j=0}^{h} \frac{1}{2}(j+1)(j+2) = \frac{1}{6}(h+1)(h+2)(h+3)$. Here we used $\sum_{j=0}^{h} j^2 = \frac{1}{6}h(h+1)(2h+1)$. (As an interim check, we have $\frac{1}{6}2 \cdot 3 \cdot 4 = 4$ for $h = 1$; and $\frac{1}{6}3 \cdot 4 \cdot 5 = 10$ for $h = 2$.) Next, impose the traceless condition: $\delta_{i_1 i_2} S^{i_1 i_2 \cdots i_j} = 0$. The left hand side here is a totally symmetric tensor carrying $(h-2)$ indices, which has $\frac{1}{6}(h-1)h(h+1)$ components. Therefore, the dimension is given by $d = \frac{1}{6}(h+1)(h+2)(h+3) - \frac{1}{6}(h-1)h(h+1) = (h+1)^2$.

IV.2   Lie Algebra of $SO(3)$ and Ladder Operators: Creation and Annihilation

1   For $j = \frac{1}{2}$, $J_+ = \begin{pmatrix} 0 & 1 \\ 0 & 0 \end{pmatrix}$, $J_- = J_+^T$, and $J_z = \frac{1}{2}\begin{pmatrix} 1 & 0 \\ 0 & -1 \end{pmatrix}$.

For $j = 1$, $J_+ = \begin{pmatrix} 0 & 1 & 0 \\ 0 & 0 & 1 \\ 0 & 0 & 0 \end{pmatrix}$, $J_- = J_+^T$, and $J_z = \begin{pmatrix} 1 & 0 & 0 \\ 0 & 0 & 0 \\ 0 & 0 & -1 \end{pmatrix}$.

For $j = 2$, $J_+ = \begin{pmatrix} 0 & 2 & 0 & 0 & 0 \\ 0 & 0 & \sqrt{6} & 0 & 0 \\ 0 & 0 & 0 & \sqrt{6} & 0 \\ 0 & 0 & 0 & 0 & 2 \\ 0 & 0 & 0 & 0 & 0 \end{pmatrix}$, $J_- = J_+^T$, and $J_z = \begin{pmatrix} 2 & 0 & 0 & 0 & 0 \\ 0 & 1 & 0 & 0 & 0 \\ 0 & 0 & 0 & 0 & 0 \\ 0 & 0 & 0 & -1 & 0 \\ 0 & 0 & 0 & 0 & -2 \end{pmatrix}$.

Verifying the commutation relations is a straightforward exercise involving multiplication and subtraction of matrices.

2   Write $T^{ijkl} = V^i V^j V^k V^l - A(\delta^{ij} V^k V^l + \delta^{jk} V^i V^l + \delta^{ki} V^j V^l + \delta^{il} V^k V^j + \delta^{jl} V^i V^k + \delta^{kl} V^j V^i) + B(\delta^{ij}\delta^{kl} + \delta^{jk}\delta^{il} + \delta^{ki}\delta^{jl})$. Impose the traceless condition to obtain $A = 1/7$, $B = 1/35$. We obtain $P_4(\cos\theta) = 35\cos^4\theta - 30\cos^2\theta + 3$.

IV.3   Angular Momentum and Clebsch-Gordan Decomposition

1, 2   Tables of Clebsch-Gordan coefficients are readily found on the web. For example, http://pdg.lbl.gov/2002/clebrpp.pdf.

5   Start with $\left| j = l + \frac{1}{2}, m = l + \frac{1}{2} \right\rangle = \left| l, s = \frac{1}{2}, l, \frac{1}{2} \right\rangle = \left| l, \frac{1}{2} \right\rangle$ and climb down by applying J_- repeatedly.

6   We have $3 \otimes 3 = 1 \oplus 3 \oplus 5$, and so $(3 \otimes 3) \otimes (3 \otimes 3)$ contains 1 three times. The three scalars are $(\vec{u} \cdot \vec{v})(\vec{w} \cdot \vec{z})$ and its two "cousins" by permutation. Thus, $(\vec{u} \times \vec{v}) \cdot (\vec{w} \times \vec{z})$ should be expressible in terms of these three guys, and indeed, it is equal to $(\vec{u} \cdot \vec{w})(\vec{v} \cdot \vec{z}) - (\vec{u} \cdot \vec{z})(\vec{v} \cdot \vec{w})$.

IV.4   Tensors and Representations of the Special Unitary Groups $SU(N)$

2   For $SU(2)$, they are simply given by the antisymmetric symbol ε^{abc}. For $SU(3)$, the answer is given in chapter V.2.

3   Write $U = e^{iH}$. Then $U^T = e^{iH^T} = U$ implies that $H^T = H$. Let $W = e^{\frac{1}{2}iH}$.

IV.5 $SU(2)$: Double Covering and the Spinor

2 The symmetric 2-indexed tensor T^{ij} has $2 \cdot 3/2 = 3$ components. By the "what-else-could-it-be" argument, it is clearly the vector of $SO(3)$. We can make this explicit as follows. Written as a 2-by-2 matrix, this tensor transforms as $T^{ij} \rightarrow U^{ik}U^{jl}T^{kl} = U^{ik}T^{kl}(U^T)^{lj} = (UTU^T)^{ij}$. The 3 components of the vector are then given by $\phi_a = \text{tr } T\sigma_2\sigma_a$. Since $U^T\sigma_2 = \sigma_2 U^\dagger$, we have, under an $SU(2)$ transformation, $\phi_a \rightarrow$ $\text{tr } UTU^T\sigma_2\sigma_a = \text{tr } T\sigma_2 U^\dagger\sigma_a U = R_{ab} \text{ tr } T\sigma_2\sigma_a = R_{ab}\phi_b$, as expected.

IV.7 Integration over Continuous Groups, Topology, Coset Manifold, and $SO(4)$

1 Trace to obtain $\text{tr } R(\vec{n}, \psi) = \text{tr } R(\vec{e}_z, \psi) = 1 + 2\cos\psi$. For the last equality, simply look at the form of $R(\vec{e}_z, \psi)$ given in chapter IV.3.

2 The axis $\vec{n}$ is defined by $R\vec{n} = \vec{n}$. First, note that, as is sensible, this equation determines $\vec{n}$ only up to an overall constant, which is fixed on imposing the normalization condition $\vec{n}^2 = 1$. Next, note that $R^T\vec{n} = R^T R\vec{n} = \vec{n}$, where we use the orthogonality of R. Subtracting, we obtain $(R - R^T)\vec{n} = 0$, that is, $\sum_j (R_{ij} - R_{ji})n_j = 0$. Setting $i = 1, 2$, we obtain two equations that determine n_1 and n_2 in terms of n_3, namely $(R_{12} - R_{21})n_2 = -(R_{13} - R_{31})n_3$ and $(R_{21} - R_{12})n_1 = -(R_{23} - R_{32})n_3$. This fixes the rotation axis.

IV.12 Crystal Field Splitting

2 We obtain $13 \rightarrow 3 \oplus 3 \oplus 3 \oplus 1 \oplus 1 \oplus 1' \oplus 1''$. The 1 appears twice, because $1 \cdot 1 \cdot 13 + 3 \cdot 1 \cdot 1 + 4(1 \cdot 1 + 1 \cdot 1) = 24 \Longrightarrow n_1 = 2$. A check: $13^2 + 3 \cdot 1^2 + 4(1^2 + 1^2) = 180 = 15(12)$, and $3^2 + 2^2 + 1^2 + 1^2 = 15$.

IV.13 Group Theory and Special Functions

2 We have (suppressing some not-so-relevant factors) $P_+ |pm\rangle = -ip |p, m+1\rangle = c_m \int(\cdots)e^{im\varphi}P_+ |p, \varphi\rangle$ $= c_m p \int(\cdots)e^{im\varphi}e^{i\varphi} |p, \varphi\rangle = (c_m p/c_{m+1}) |p, m+1\rangle$, where in the first equality we used (8). Thus, $c_{m+1} = ic_m$; and so $c_m = i^m$ if we fix $c_0 = 1$.

IV.14 Covering the Tetrahedron

1 From chapter II.3, we learned that c corresponds to a rotation through angle $2\pi/3$. Referring to the $SU(2)$ matrix U given in the text, we see that the character $= 2\cos\frac{\pi}{3} = 1$.

V.1 Isospin and the Discovery of a Vast Internal Space

1 The ratio is 2. The key observation is that the initial state is an $I = \frac{1}{2}$, while the final state has isospin $\frac{1}{2} \otimes 1$.

2 The relevant Clebsch-Gordan decompositions are

$$|\pi^+ p\rangle = \left|\frac{3}{2}, \frac{3}{2}\right\rangle$$

$$|\pi^- p\rangle = \sqrt{\frac{1}{3}}\left|\frac{3}{2}, -\frac{1}{2}\right\rangle - \sqrt{\frac{2}{3}}\left|\frac{1}{2}, -\frac{1}{2}\right\rangle$$

$$|\pi^0 n\rangle = \sqrt{\frac{2}{3}}\left|\frac{3}{2}, -\frac{1}{2}\right\rangle + \sqrt{\frac{1}{3}}\left|\frac{1}{2}, -\frac{1}{2}\right\rangle \tag{15}$$

Thus,

$$\langle\pi^+ p|\,\mathcal{T}\,|\pi^+ p\rangle = A_{\frac{3}{2}}$$

$$\langle\pi^- p|\,\mathcal{T}\,|\pi^- p\rangle = \frac{1}{3}A_{\frac{3}{2}} + \frac{2}{3}A_{\frac{1}{2}}$$

$$\langle\pi^0 n|\,\mathcal{T}\,|\pi^- p\rangle = \frac{\sqrt{2}}{3}\left(A_{\frac{3}{2}} - A_{\frac{1}{2}}\right) \tag{16}$$

Here $\mathcal{T}$ denotes some complicated strong interaction transition operator that is beyond our powers to compute analytically even today, and A the amplitude in the two isospin channels. According to the rules of quantum mechanics, $\sigma(\pi^+ p) = C|A_{\frac{3}{2}}|^2$ and

$$\sigma(\pi^- p) = C\frac{1}{9}\left\{(1+2)|A_{\frac{3}{2}}|^2 + (4+2)|A_{\frac{1}{2}}|^2\right\} \tag{17}$$

with C some kinematic factor we don't care about. For $A_{\frac{1}{2}} \simeq 0$, we obtain the stated result.

V.2 The Eightfold Way of $SU(3)$

2 The relevant identity is (note that none of the indices is summed over) $\varepsilon_{ikm}\varepsilon^{jln} \propto \delta^j_i\delta^l_k\delta^n_m \pm$ permutations. (We could of course work out the proportionality constant, but we won't need it.) This follows from the trivial observation that the left hand side vanishes unless (i, k, m) is a permutation of $(1, 2, 3)$, and similarly for (j, l, n). Thus, $\varepsilon_{ikm}\varepsilon^{jln}A_{jl}{}^{ik}$ is equal to various traces of the traceless tensor $A_{jl}{}^{ik}$ and hence vanishes.

3 Now that we listed all ten states for $n = 3$ in the text, we understand how it goes. Here we list the representative states and write the corresponding number of states and then total it up. For $n = 4$: (4,0,0), (3,1,0), (2,1,1), and (2,2,0), with, respectively, $3 + 3 \cdot 2 + 3 + 3 = 15 = \frac{1}{2}(5 \cdot 6)$. For $n = 5$: (5,0,0), (4,1,0), (3,2,0), (3,1,1), and (2,2,1), with, respectively, $3 + 3 \cdot 2 + 3 \cdot 2 + 3 + 3 = 21 = \frac{1}{2}(6 \cdot 7)$. See the pattern?

4 The $U(1)$ corresponds to multiplying each $a_i^\dagger$ by a common phase. It is none other than translation in time, according to the rules of quantum mechanics.

VI.2 Roots and Weights for Orthogonal, Unitary, and Symplectic Algebras

1 The first two are easy: $e^1 - e^3 = (e^1 - e^2) + (e^2 - e^3)$, and $e^1 + e^3 = (e^1 - e^2) + (e^2 + e^3)$. The other one is a bit harder and is best done in two steps: $(e^2 - e^3) + (e^2 + e^3) = 2e^2$, and $(e^1 - e^2) + e^2 = e^1$. Hence $e^1 + e^2$ is not simple.

VI.4   The Killing-Cartan Classification of Lie Algebras

1   Suppose that it contains three different lengths. Then, according to the table in this chapter, there are only a finite number of possibilities. For example, the three lengths (in some suitable units) could be 1, $\sqrt{2}$, and $\sqrt{3}$. But then this would contradict the table, since the ratio of the lengths of some roots would be $\sqrt{3/2}$.

VI.5   Dynkin Diagrams

1   $$A(F_4) = \begin{pmatrix} 2 & -1 & 0 & 0 \\ -1 & 2 & -2 & 0 \\ 0 & -1 & 2 & -1 \\ 0 & 0 & -1 & 2 \end{pmatrix}.$$

VII.1   Spinor Representations of Orthogonal Algebras

1   Because from (6) we would get $C\gamma_{2k}^* C^{-1} = (-1)^{n-k}\gamma_{2k}$, and the overall factor would depend on k.

VII.2   The Lorentz Group and Relativistic Physics

1   Suppress the coordinate z and write the Lorentz transformations as 3-by-3 matrices. Use an abbreviated notation $c = \cosh\varphi$, $c' = \cosh\varphi'$, etc. Then

$$L_x^{-1}L_yL_x = \begin{pmatrix} c & -s & 0 \\ -s & c & 0 \\ 0 & 0 & 1 \end{pmatrix}\begin{pmatrix} c' & 0 & s' \\ 0 & 1 & 0 \\ s' & 0 & c' \end{pmatrix}\begin{pmatrix} c & s & 0 \\ s & c & 0 \\ 0 & 0 & 1 \end{pmatrix} \tag{69}$$

As per the discussion of Lie algebra in chapter I.3, to extract the commutator we need only calculate the terms proportional to $\varphi\varphi'$, which fact allows us to simplify the calculation enormously. We could, for example, set c and c' effectively to 1. We obtain, almost immediately, $L_x^{-1}L_yL_x \sim \begin{pmatrix} 1 & 0 & 0 \\ 0 & 1 & -ss' \\ 0 & ss' & 1 \end{pmatrix}$, where "$\sim$" means effectively equal. We see explicitly that two boosts give a rotation: $[K_x, K_y] = -iJ_z$.

4   Setting $\sigma = \rho = 0$ in (38) gives $(L^0{}_0)^2 = 1 + \sum_i(L^i{}_0)^2 \geq 1$. Evidently, $\cosh\varphi \geq 1$ and $-\cosh\varphi \leq -1$.

$$\eta_{\mu\nu}L^\mu{}_\sigma L^\nu{}_\rho = \eta_{\sigma\rho} \tag{38}$$

VII.3 $SL(2, C)$ Double Covers $SO(3, 1)$: Group Theory Leads Us to the Weyl Equation

1   Boost in the z direction. Start with $iK_z = t\frac{\partial}{\partial z} + z\frac{\partial}{\partial t}$. In the notation used here, $\delta z = t$, $\delta t = z$, that is, $z' = z + \varphi t$, $t' = t + \varphi z$, or $z = z' - \varphi t = z' - \varphi t' + O(\varphi^2)$, $t = t' - \varphi z' + O(\varphi^2)$. Hence $\frac{\partial}{\partial t'} = \frac{\partial t}{\partial t'}\frac{\partial}{\partial t} + \frac{\partial z}{\partial t'}\frac{\partial}{\partial z} = \frac{\partial}{\partial t} - \varphi\frac{\partial}{\partial z}$, that is, $\delta\frac{\partial}{\partial t} = -\frac{\partial}{\partial z}$. Similarly, $\delta\frac{\partial}{\partial z} = -\frac{\partial}{\partial t}$.

2   Using the multiplication rule for $SU(2)$, we have $(\frac{1}{2}, \frac{1}{2}) \otimes (\frac{1}{2}, \frac{1}{2}) = (1 \oplus 0, 1 \oplus 0) = (1, 1) \oplus (1, 0) \oplus (0, 1) \oplus (0, 0)$. Repeating, we see that the direct product of h 4-vectors gives

$$\left(\frac{1}{2},\frac{1}{2}\right) \otimes \cdots \otimes \left(\frac{1}{2},\frac{1}{2}\right) = \left(\frac{1}{2}h \oplus \frac{1}{2}h - 1 \cdots, \frac{1}{2}h \oplus \frac{1}{2}h - 1 \cdots\right) = \left(\frac{1}{2}h, \frac{1}{2}h\right) \oplus \cdots$$

3   According to the preceding exercise, $(\frac{1}{2}h, \frac{1}{2}h)$ is the $SO(3, 1)$ tensor with h indices. The number of components is $(2 \cdot \frac{1}{2}h + 1)^2 = (h + 1)^2$, but according to exercise 13 you did in chapter IV.1, this is precisely the number of components contained in a symmetric traceless tensor of $SO(3, 1)$. Note that, although you did the counting for $SO(4)$, the number of components does not care about the analytic continuation from $SO(4)$ to $SO(3, 1)$.

5   Given $(E - \vec{\sigma} \cdot \vec{p})u = 0$, let us boost infinitesimally along the z direction and show that the variation of the left hand side of the Weyl equation vanishes. Use the notation in the text; in particular, $\vec{\sigma} \cdot \vec{p} = \sigma_1 p^x + \sigma_2 p^y + \sigma_3 p^z$, where I intentionally abuse notation. The variation is $(\delta E - \vec{\sigma} \cdot \delta \vec{p})u + (E - \vec{\sigma} \cdot \vec{p})\delta u = (p^z - \sigma_3 E)u + (E - \vec{\sigma} \cdot \vec{p})\frac{1}{2}\sigma_3 u$. Multiplying this by $2\sigma_3$ from the left, we obtain $-(E - \sigma_3 p^z)u + (\sigma_1 p^x + \sigma_2 p^y)u = -(E - \vec{\sigma} \cdot \vec{p})u = 0$.

VII.4   From the Weyl Equation to the Dirac Equation

1   Under parity, $(1, 0) \leftrightarrow (0, 1)$. The two irreducible representations correspond to $\vec{E} \pm i\vec{B}$.

4   We have

$$\bar{\psi}'(x') = \psi(x)^\dagger S(L)^\dagger \gamma^0 = \bar{\psi}(x)e^{+\frac{i}{4}\omega_{\mu\nu}\sigma^{\mu\nu}} \tag{40}$$

and thus, for example,

$$\bar{\psi}'(x')\psi'(x') = \bar{\psi}(x)e^{+\frac{i}{4}\omega\sigma}e^{-\frac{i}{4}\omega\sigma}\psi(x) = \bar{\psi}(x)\psi(x) \tag{41}$$

as claimed. The necessity for introducing $\bar{\psi}$ in addition to $\psi^\dagger$ in relativistic physics is traced back to the $(+, -, -, -)$ signature of the Minkowski metric. For other bilinears of the form $\bar{\psi}\Gamma\psi$, with Γ denoting a product of γ matrices, we encounter the expression $e^{+\frac{i}{4}\omega\sigma}\Gamma e^{-\frac{i}{4}\omega\sigma}$. Each of the γs contained in Γ transforms separately along the lines discussed in chapter VII.1.

5   For example, $(\bar{\psi}\psi)^\dagger = (\psi^\dagger\gamma^0\psi)^\dagger = \psi^\dagger\gamma^0\psi = \bar{\psi}\psi$.

8   $v^\dagger u = -iw^T\sigma_2 u$.

9   The $SO(2)$ in question rotates (W_1, W_2) as a vector. We find

$$\mathcal{L} = i\psi_+^\dagger\sigma^\mu\partial_\mu\psi_+ + i\psi_-^\dagger\sigma^\mu\partial_\mu\psi_- - m(\psi_+^T i\sigma_2\psi_- + h.c.)$$

which exhibits the $U(1)$ symmetry $\psi_\pm \to e^{\pm i\theta}\psi_\pm$.

VII.5   Dirac and Majorana Spinors: Antimatter and Pseudoreality

1   Let $\xi = \begin{pmatrix} \chi \\ \zeta \end{pmatrix}$. Then $\psi = \frac{1}{2}\begin{pmatrix} \chi + \sigma_2\zeta^* \\ \zeta - \sigma_2\chi^* \end{pmatrix}$. We check that indeed, $-\sigma_2(\chi + \sigma_2\zeta^*)^* = -\sigma_2\chi^* + \zeta$, in agreement with (27).

IX.1 The Gauged Universe

1 The mass terms for the up and down quarks transform like $(3, 2, \frac{1}{6}) \otimes (3^*, 1, -\frac{2}{3}) \sim (1, \frac{1}{2}, -\frac{1}{2})$ and $(3, 2, \frac{1}{6}) \otimes (3^*, 1, \frac{1}{3}) \sim (1, \frac{1}{2}, \frac{1}{2})$, respectively. Again, due to the defining representation of $SU(2)$ being pseudoreal, the representations $(1, \frac{1}{2}, -\frac{1}{2})$ and $(1, \frac{1}{2}, \frac{1}{2})$ are conjugate to each other, and one Higgs field does the job. One reason that some people do not like low-energy supersymmetry is that in that theory, separate Higgs fields have to be introduced for the up and down quarks.

IX.2 Grand Unification and $SU(5)$

1 First, $X(5^*) = -3$, while $(Y/2) = +\frac{1}{3}, -\frac{1}{2}$, and $-\frac{1}{2}$ on d^c, ν, and e, respectively. Thus, $\frac{1}{5}(X + 4(Y/2)) = \frac{1}{5}(-3 + 4/3) = -\frac{1}{3}$ on d^c and $= \frac{1}{5}(-3 - 2) = -1$ on ν and e. Next, $X(10) = 1$, while $(Y/2) = -2/3, 1/6, 1/6$, and 1 on u^c, u, d, and e^c, respectively. Thus, $\frac{1}{5}(X + 4(Y/2)) = \frac{1}{5}(1 - 8/3) = -\frac{1}{3}$ on u^c and $= \frac{1}{5}(1 + 2/3) = \frac{1}{3}$ on u, d, and $= \frac{1}{5}(1 + 4) = 1$ on e^c, respectively. We conclude that $\frac{1}{5}(X + 4(Y/2)) = B - L$, baryon minus lepton number.

IX.3 From $SU(5)$ to $SO(10)$

1 Clearly, to calculate tr T^3, our work is minimized by taking a diagonal T (for example, the hypercharge Y). To simplify the arithmetic, multiply Y by -3 and write $T = \text{diag}(2, 2, 2, -3, -3)$. Then tr T^3 evaluated on the 5^* gives $3(-2)^3 + 2(+3)^3 = -24 + 54 = 30$, and evaluated on the 10 it gives $3(+4)^3 + 6(-1)^3 + 1(-6)^3 = 192 - 6 - 216 = -30$.

2 The generators of $SO(N)$ are given by $T^{ij} = -T^{ji}$, with $i, j = 1, \cdots N$, as we learned way back when. The anomaly is determined by tr $T^{ij}\{T^{kl}, T^{mn}\}$. But this transforms like a 6-indexed tensor of $SO(N)$, with various symmetries, such as antisymmetry under the exchange of i and j. No such tensor exists except for $SO(6)$, in which case the antisymmetric symbol ε^{ijklmn} works. (This argument is due to H. Georgi and S. L. Glashow, Phys. Rev. D 6 (1972), p. 429.) This is all consistent, because, as you might recall from chapter VII.1, $SO(6)$ is locally isomorphic to $SU(4)$. Remarkable how group theory all hangs together.

3 $120 \rightarrow (20, 1) \oplus (15, 4) \oplus (6, 3) \oplus (6, 3) \oplus (1, 4)$; check: $120 = 20 + 60 + 18 + 18 + 4$.

Bibliography

Books on group theory

I do not necessarily recommend all these books as paragons of clarity.

M. A. Armstrong. Groups and Symmetry. Springer-Verlag, 1988.

R. D. Carmichael. Introduction to the Theory of Groups of Finite Order. Dover, 2000.

J.-Q. Chen. Group Representation Theory for Physicists. Second edition. World Scientific, 1989.

M. Gell-Mann and Y. Ne'eman. The Eightfold Way. Westview Press, 2000.

H. Georgi. Lie Algebras in Particle Physics: From Isospin to Unified Theories. Second edition. Westview Press, 1999.

R. Gilmore. Lie Groups, Lie Algebras, and Some of Their Applications. Dover, 2006.

G. G. Hall. Applied Group Theory. Elsevier, 1967.

M. Hamermesh. Group Theory and Its Application to Physical Problems. Dover, 1989.

B. Hayes. Group Theory in the Bedroom, and Other Mathematical Diversions. Hill and Wang, 2009.

H. F. Jones. Groups, Representations and Physics. Second edition. CRC Press, 1998.

H. J. Lipkin. Lie Groups for Pedestrians. Dover, 2002.

Z.-Q. Ma. Group Theory for Physicists. World Scientific, 2007.

R. McWeeny. Symmetry: An Introduction to Group Theory and Its Applications. Dover, 2002.

P. Ramond. Group Theory: A Physicist's Survey. Cambridge University Press, 2010.

M. Ronan. Symmetry and the Monster: One of the Greatest Quests of Mathematics. Oxford University Press, 2006.

J. S. Rose. A Course on Group Theory. Dover, 2012.

K. Tapp. Matrix Groups for Undergraduates. American Mathematical Society, 2005.

M. Tinkham. Group Theory and Quantum Mechanics. Dover, 2003.

W.-K. Tung. Group Theory in Physics. World Scientific, 1985.

B. G. Wybourne. Classical Groups for Physicists. John Wiley and Sons, 1974.

Books on related subjects

S. Gasiorowicz. Elementary Particle Physics. John Wiley and Sons, 1966.

L. Reimer and W. Reimer. Mathematicians Are People, Too: Stories from the Lives of Great Mathematicians. Dayle Seymour Publications, 1994.

J. J. Sakurai and J. J. Napolitano. Modern Quantum Mechanics. Second edition. Addison-Wesley, 2010.

S. Weinberg. Lectures on Quantum Mechanics. Cambridge University Press, 2012.

A. Zee. Unity of Forces in the Universe. Vols 1 and 2, World Scientific, 1982.

A. Zee. Fearful Symmetry: The Search for Beauty in Modern Physics. Princeton University Press, 1999.

A. Zee. Quantum Field Theory in a Nutshell. Princeton University Press, 2003.

A. Zee. Einstein Gravity in a Nutshell. Princeton University Press, 2013.

Index

Page numbers for entries occurring in notes are followed by an *n*.

The following is a loosely organized list of formulas used in this text, compiled here for the convenience of some readers.

Part I

Presentations

$$D_n = \langle R, r \,|\, R^n = I, r^2 = I, Rr = rR^{-1} \rangle$$

$$\langle a_1, a_2, \cdots, a_k \,|\, (a_i)^2 = I, (a_i a_j)^{n_{ij}} = I, n_{ij} \geq 2, \quad \text{with } i, j = 1, 2, \cdots, k \rangle$$

$$\langle a, b \rangle \equiv a^{-1} b^{-1} ab = (ba)^{-1}(ab)$$

$SO(3)$

$$[J_i, J_j] = i\epsilon_{ijk} J_k$$

$SO(N)$

$$[J_{(mn)}, J_{(pq)}] = i(\delta_{mp} J_{(nq)} + \delta_{nq} J_{(mp)} - \delta_{np} J_{(mq)} - \delta_{mq} J_{(np)})$$

$SO(4)$

$$[J_i, J_j] = i\varepsilon_{ijk} J_k$$

$$[J_i, K_j] = i\varepsilon_{ijk} K_k$$

$$[K_i, K_j] = i\varepsilon_{ijk} J_k$$

Part II

Great Orthogonality theorem

$$\sum_g D^{(r)\dagger}(g)^i{}_j D^{(s)}(g)^k{}_l = \frac{N(G)}{d_r} \delta^{rs} \delta^i{}_l \delta^k{}_j$$

Character orthogonality

$$\sum_c n_c (\chi^{(r)}(c))^* \chi^{(s)}(c) = N(G)\delta^{rs}$$

Decomposing a reducible representation

$$\sum_c n_c \chi^*(c)\chi(c) = N(G) \sum_r (n_r)^2$$

$$\sum_c n_c \chi^{*(r)}(c)\chi(c) = N(G)n_r$$

Dimensions of the irreducible representations

$$\sum_r d_r^2 = N(G)$$

Column orthogonality

$$\sum_c n_c (\chi^{(r)}(c))^* \chi^{(s)}(c) = N(G)\delta^{rs}$$

Row orthogonality

$$\sum_r \chi^{(r)}(c)^* \chi^{(r)}(c') = \frac{N(G)}{n_c} \delta^{cc'}$$

The character table is square

$$N(C) = N(R)$$

Reality check

$$\sum_{g \in G} \chi^{(r)}(g^2) = \eta^{(r)} N(G), \quad \text{with } \eta^{(r)} = \begin{cases} 1 & \text{if real,} \\ -1 & \text{if pseudoreal,} \\ 0 & \text{if complex} \end{cases}$$

Square roots

$$g^2 = f: \quad \sigma_f = \sum_r \eta^{(r)} \chi^{(r)}(f)$$

$$g^2 = I: \quad \sigma_I = \sum_r \eta^{(r)} d_r = \sum_{r=\text{real}} d_r - \sum_{r=\text{pseudoreal}} d_r$$

$$\sum_g D^{(r)}(g^2) = N(G)(\eta^{(r)}/d_r) I$$

$$f^2 g^2 = h^2 : \tau_h = N(G) \sum_r (\eta^{(r)})^2 \chi^{(r)}(h)/d_r$$

$$f^2 g^2 = I : \tau_I = N(G) \sum_r (\eta^{(r)})^2 = N(G) \sum_{\substack{r=\text{real or} \\ \text{pseudoreal}}} 1$$

Part III

$G \Longrightarrow$ degeneracy and $G \Longleftarrow$ degeneracy
$d =$ degrees of degeneracy $=$ dimension of irreducible representation

Euler-Lagrange equation

$$\frac{d}{dt}\left(\frac{\delta L}{\delta \frac{dq}{dt}}\right) - \frac{\delta L}{\delta q} = 0$$

Hamiltonian

$$H(p,q) = p\dot{q} - L(\dot{q}, q)$$

$$\dot{q} = \frac{\partial H}{\partial p}, \qquad \dot{p} = -\frac{\partial H}{\partial q}$$

Part IV

S and O

$$O^T O = 1$$

$$\det O = 1$$

Raising and lowering

$$J_\pm \equiv J_x \pm i J_y$$

$$[J_z, J_\pm] = \pm J_\pm, \qquad [J_+, J_-] = 2J_z$$

$$J_+ |m\rangle = c_{m+1} |m+1\rangle = \sqrt{(j+1+m)(j-m)} \, |m+1\rangle$$

$$J_- |m\rangle = c_m^* |m-1\rangle = \sqrt{(j+1-m)(j+m)} \, |m-1\rangle$$

$$J^2 |j, m\rangle = \left(\tfrac{1}{2}(J_+ J_- + J_- J_+) + J_z^2\right) |j, m\rangle = j(j+1) |j, m\rangle$$

$$j = \tfrac{1}{2} : J_+ \left|-\tfrac{1}{2}\right\rangle = \left|\tfrac{1}{2}\right\rangle, \quad J_- \left|\tfrac{1}{2}\right\rangle = \left|-\tfrac{1}{2}\right\rangle$$

$j = 1$:

$$J_+ \ket{-1} = \sqrt{2}\ket{0}, \qquad J_+ \ket{0} = \sqrt{2}\ket{0}$$

$$J_- \ket{1} = \sqrt{2}\ket{0}, \qquad J_- \ket{0} = \sqrt{2}\ket{-1}$$

$j = 2$:

$$J_+ \ket{-2} = 2\ket{-1}, \qquad J_+ \ket{-1} = \sqrt{6}\ket{0}, \qquad J_+ \ket{0} = \sqrt{6}\ket{1}, \qquad J_+ \ket{1} = 2\ket{2}$$

$$J_- \ket{2} = 2\ket{1}, \qquad J_- \ket{1} = \sqrt{6}\ket{0}, \qquad J_- \ket{0} = \sqrt{6}\ket{-1}, \qquad J_- \ket{-1} = 2\ket{-2}$$

Clebach-Gordon decomposition

$$j \otimes j' = (j + j') \oplus (j + j' - 1) \oplus (j + j' - 2) \oplus \cdots \oplus (|j - j'| + 1) \oplus |j - j'|$$

Angular momentum

$$L_x = i\left(z\frac{\partial}{\partial y} - y\frac{\partial}{\partial z}\right), \qquad L_y = i\left(x\frac{\partial}{\partial z} - z\frac{\partial}{\partial x}\right), \qquad L_z = i\left(y\frac{\partial}{\partial x} - x\frac{\partial}{\partial y}\right)$$

$$\vec{L}^2 = \frac{1}{2}(L_+L_- + L_-L_+) + L_z^2 = -\left(\frac{1}{\sin\theta}\frac{\partial}{\partial\theta}\left(\sin\theta\frac{\partial}{\partial\theta}\right) + \frac{1}{\sin^2\theta}\frac{\partial^2}{\partial\varphi^2}\right)$$

$$\vec{L}^2 Y_l^m(\theta, \varphi) = l(l + 1)Y_l^m(\theta, \varphi)$$

$$L_z Y_l^m(\theta, \varphi) = m Y_l^m(\theta, \varphi)$$

$$Y_l^m(\theta, \varphi) = N_l^m e^{im\varphi} P_l^m(\cos\theta), \qquad P_l(\cos\theta) \equiv P_l^{m=0}(\cos\theta)$$

Heisenberg and Dirac algebras

$$[q, p] = i: \quad a = \frac{1}{\sqrt{2}}(q + ip) \quad \text{and} \quad a^\dagger = \frac{1}{\sqrt{2}}(q - ip)$$

$$[a, a^\dagger] = 1$$

$$a\ket{n} = \sqrt{n}\ket{n-1} \quad \text{and} \quad a^\dagger\ket{n} = \sqrt{n+1}\ket{n+1}$$

$$H = \frac{1}{2}a^\dagger a + \frac{1}{2} = \frac{1}{2}(p^2 + q^2) = -\frac{1}{2}\frac{d^2}{dx^2} + \frac{1}{2}x^2$$

$$J_z \leftrightarrow \frac{1}{2}(a^\dagger a - b^\dagger b), \qquad J_+ \leftrightarrow a^\dagger b, \qquad J_- \leftrightarrow b^\dagger a$$

$$\ket{j, m} = \frac{1}{\sqrt{(j+m)!(j-m)!}}(a^\dagger)^{j+m}(b^\dagger)^{j-m}\ket{0}$$

$$J_z = \frac{1}{2}(\ket{+}\bra{+} - \ket{-}\bra{-}), \qquad J_+ = \ket{+}\bra{-}, \qquad J_- = \ket{-}\bra{+}$$

$\ket{J, M}$ in terms of $\ket{m, m'}$:

$$\ket{J, M} = \sum_{m=-j}^{j} \sum_{m'=-j'}^{j'} \ket{j, j', m, m'} \braket{j, j', m, m'|J, M}$$

$$\frac{1}{2} \otimes \frac{1}{2} = 1 \oplus 0$$

$$|1, 1\rangle = \left|\tfrac{1}{2}, \tfrac{1}{2}\right\rangle$$

$$|1, 0\rangle = \frac{1}{\sqrt{2}}\left(\left|-\tfrac{1}{2}, \tfrac{1}{2}\right\rangle + \left|\tfrac{1}{2}, -\tfrac{1}{2}\right\rangle\right); \qquad |1, 0\rangle = \frac{1}{\sqrt{2}}\left(\left|-\tfrac{1}{2}, \tfrac{1}{2}\right\rangle - \left|\tfrac{1}{2}, -\tfrac{1}{2}\right\rangle\right)$$

$$|1, -1\rangle = \left|-\tfrac{1}{2}, -\tfrac{1}{2}\right\rangle$$

$$1 \otimes 1 = 2 \oplus 1 \oplus 0$$

$J = 2$:

$$|2, 2\rangle = |1, 1\rangle$$

$$|2, 1\rangle = \frac{1}{\sqrt{2}}(|0, 1\rangle + |1, 0\rangle)$$

$$|2, 0\rangle = \frac{1}{\sqrt{6}}(|-1, 1\rangle + 2|0, 0\rangle + |1, -1\rangle)$$

$$|2, -1\rangle = \frac{1}{\sqrt{2}}(|0, -1\rangle + |-1, 0\rangle)$$

$$|2, -2\rangle = |-1, -1\rangle$$

$J = 1$:

$$|1, 1\rangle = \frac{1}{\sqrt{2}}(|0, 1\rangle - |1, 0\rangle)$$

$$|1, 0\rangle = \frac{1}{\sqrt{2}}(|-1, 1\rangle - |1, -1\rangle)$$

$$|1, -1\rangle = \frac{1}{\sqrt{2}}(|-1, 0\rangle - |0, -1\rangle)$$

$J = 0$:

$$|0, 0\rangle = \frac{1}{\sqrt{3}}(|-1, 1\rangle - |0, 0\rangle + |1, -1\rangle)$$

S and U

$$U^\dagger U = I$$

$$\det U = 1$$

$$U = e^{iH}$$

$$\psi^i \to \psi'^i = U^i_{\ j}\psi^j$$

$$\psi_i \to \psi'_i = \psi_j (U^\dagger)^j_{\ i}$$

$$\psi_i \equiv \psi^{i*}$$

$$\varepsilon_{i_1 i_2 \cdots i_N} U^{i_1}{}_{j_1} U^{i_2}{}_{j_2} \cdots U^{i_N}{}_{j_N} = \varepsilon_{j_1 j_2 \cdots j_N}$$

$$\varphi \to \varphi' = U \varphi U^\dagger$$

$$(SU(2) \otimes SU(2))/Z_2 = SO(4)$$

Pauli matrices

$$\sigma_1 = \begin{pmatrix} 0 & 1 \\ 1 & 0 \end{pmatrix}, \qquad \sigma_2 = \begin{pmatrix} 0 & -i \\ i & 0 \end{pmatrix}, \qquad \sigma_3 = \begin{pmatrix} 1 & 0 \\ 0 & -1 \end{pmatrix}$$

$$\sigma_a \sigma_b = \delta_{ab} I + i \varepsilon_{abc} \sigma_c$$

$$\{\sigma_a, \sigma_b\} = 2\delta_{ab}$$

$$\left[\frac{\sigma_a}{2}, \frac{\sigma_b}{2} \right] = i \varepsilon_{abc} \frac{\sigma_c}{2}$$

$$U = e^{i\vec{\varphi} \cdot \vec{\sigma}/2} = \cos \frac{\varphi}{2} I + i \hat{\varphi} \cdot \vec{\sigma} \sin \frac{\varphi}{2}$$

$$\sigma_2 \sigma_a^* \sigma_2 = \sigma_2 \sigma_a^T \sigma_2 = -\sigma_a$$

$$\sigma_2 (e^{i\vec{\varphi}\vec{\sigma}})^* \sigma_2 = e^{i\vec{\varphi}\vec{\sigma}}$$

$$U(N) = \left(SU(N)/Z_N \right) \times U(1)$$

$$\psi \to e^{i(2\pi)\sigma_3/2} \psi = \begin{pmatrix} e^{i\pi} & 0 \\ 0 & e^{-i\pi} \end{pmatrix} \psi = -\psi$$

Kramer's degeneracy

$$K U^\dagger \vec{\sigma} U K = \eta^* \eta K \sigma_2 \vec{\sigma} \sigma_2 K = -\vec{\sigma}$$

$$T^2 = \eta \sigma_2 K \eta \sigma_2 K = \eta \sigma_2 \eta^* \sigma_2^* K K = -1$$

Integration over group manifolds

$$\chi(j, \psi) = \frac{\sin(j + \frac{1}{2})\psi}{\sin \frac{\psi}{2}}$$

$$\int_{SO(3)} d\mu(g) F(g) = \int_0^\pi d\psi (\sin^2 \frac{\psi}{2}) F(\psi)$$

$$\chi(j) = (\zeta^{j+1} - \zeta^{-j})/(\zeta - 1)$$

$$\chi(k)\chi(j) = \chi(j+k) + \chi(j+k-1) + \cdots + \chi(j-k)$$

$$S^2 = SO(3)/SO(2)$$

$$SO(4): \quad W = t + i\vec{x} \cdot \vec{\sigma}; \ t^2 + \vec{x}^2 = 1, \qquad W \to U^\dagger W V$$

Symplectic algebra

$$Sp(2n, R): \quad R^T J R = J, \qquad J = \left(\begin{array}{c|c} 0 & I \\ \hline -I & 0 \end{array} \right)$$

$$USp(2n) = U(2n) \cap Sp(2n, C): \quad U^T J U = J$$

$$J = I \otimes i\sigma_2: \quad iA \otimes I, \qquad S_1 \otimes \sigma_1, \qquad S_2 \otimes \sigma_2, \qquad S_3 \otimes \sigma_3$$

Euclidean algebra

$$E(2): \quad [J, P_i] = i\varepsilon_{ij} P_j, \quad i = 1, 2, \qquad [P_1, P_2] = 0$$

$$P^2 |pm\rangle = p^2 |pm\rangle, \qquad J |pm\rangle = m |pm\rangle$$

$$\langle pm' | e^{-iaP_1} |pm\rangle = J_{m-m'}(pa)$$

Part V

Gell-Mann matrices

$$\lambda_1 = \begin{pmatrix} 0 & 1 & 0 \\ 1 & 0 & 0 \\ 0 & 0 & 0 \end{pmatrix}, \qquad \lambda_2 = \begin{pmatrix} 0 & -i & 0 \\ i & 0 & 0 \\ 0 & 0 & 0 \end{pmatrix}, \qquad \lambda_3 = \begin{pmatrix} 1 & 0 & 0 \\ 0 & -1 & 0 \\ 0 & 0 & 0 \end{pmatrix},$$

$$\lambda_4 = \begin{pmatrix} 0 & 0 & 1 \\ 0 & 0 & 0 \\ 1 & 0 & 0 \end{pmatrix}, \qquad \lambda_5 = \begin{pmatrix} 0 & 0 & -i \\ 0 & 0 & 0 \\ i & 0 & 0 \end{pmatrix}, \qquad \lambda_6 = \begin{pmatrix} 0 & 0 & 0 \\ 0 & 0 & 1 \\ 0 & 1 & 0 \end{pmatrix},$$

$$\lambda_7 = \begin{pmatrix} 0 & 0 & 0 \\ 0 & 0 & -i \\ 0 & i & 0 \end{pmatrix}, \qquad \lambda_8 = \begin{pmatrix} 1 & 0 & 0 \\ 0 & 1 & 0 \\ 0 & 0 & -2 \end{pmatrix}$$

$SU(3)$

$$3 \otimes 3^* = 8 \oplus 1$$

$$3 \otimes 3 = 6 \oplus 3^*$$

$$3 \otimes 6 = 10 \oplus 8$$

$$3 \otimes 3 \otimes 3 = (6 \oplus 3^*) \otimes 3 = (6 \otimes 3) \oplus (3^* \otimes 3) = 10 \oplus 8 \oplus 8 \oplus 1$$

$$8 \otimes 8 = 27 \oplus 10 \oplus 10^* \oplus 8 \oplus 8 \oplus 1$$

$$\dim(m, n) = \frac{1}{2}(m + 1)(n + 1)(m + n + 2)$$

$$f^{123} = 1$$

$$f^{147} = -f^{156} = f^{246} = f^{257} = f^{345} = -f^{367} = \frac{1}{2}$$

$$f^{458} = f^{678} = \frac{\sqrt{3}}{2}$$

Part VI

Math	Physics	Number of generators	Roots	Simple roots
A_{l-1}	$SU(l)$	$l^2 - 1$	$e^i - e^j$	$e^i - e^{i+1}$
B_l	$SO(2l+1)$	$l(2l+1)$	$\pm e^i \pm e^j,\ \pm e^i$	$e^{i-1} - e^i,\ e^l$
C_l	$Sp(2l)$	$l(2l+1)$	$\pm e^i \pm e^j,\ \pm 2e^i$	$e^{i-1} - e^i,\ 2e^l$
D_l	$SO(2l)$	$l(2l-1)$	$\pm e^i \pm e^j$	$e^{i-1} - e^i,\ e^{l-1} + e^l$

$$[H^i, H^j] = 0$$

$$[H^i, E_\alpha] = \alpha^i E_\alpha$$

$$[E_\alpha, E_\beta] = N_{\alpha, \beta} E_{\alpha + \beta}$$

$$[E_\alpha, E_{-\alpha}] = \alpha_i H^i$$

$$2\frac{(\vec{\alpha}, \vec{\beta})}{(\vec{\alpha}, \vec{\alpha})} = q - p \equiv n$$

$$2\frac{(\vec{\alpha}, \vec{\beta})}{(\vec{\beta}, \vec{\beta})} = q' - p' \equiv m$$

$$\cos^2 \theta_{\alpha\beta} = \frac{(\vec{\alpha}, \vec{\beta})^2}{(\vec{\alpha}, \vec{\alpha})(\vec{\beta}, \vec{\beta})} = \frac{mn}{4}$$

Weyl reflection

$$\vec{\beta}' = \vec{\beta} - 2\cos\theta_{\alpha\beta} |\beta| \hat{\alpha}$$

m	n	$\frac{(\vec{\alpha}, \vec{\alpha})}{(\vec{\beta}, \vec{\beta})}$	$\cos^2 \theta_{\alpha\beta}$	$\theta_{\alpha\beta}$
1	1	1	$\frac{1}{4}$	$60°$
2	1	2	$\frac{1}{2}$	$45°$
3	1	3	$\frac{3}{4}$	$30°$

$\theta_{\alpha\beta} = 90°$ implies that $(\vec{\alpha}, \vec{\beta}) = 0$ and $\rho_{\alpha\beta}$ is indeterminate.

Cartan matrix

$$A_{ij} \equiv 2\frac{(\alpha_i, \alpha_j)}{(\alpha_i, \alpha_i)}$$

Part VII

Clifford algebra

$$\{\gamma_i, \gamma_j\} = 2\delta_{ij}I$$

$$\gamma_j^{(n+1)} = \gamma_j^{(n)} \otimes \tau_3, \quad j = 1, 2, \cdots, 2n$$

$$\gamma_{2n+1}^{(n+1)} = I \otimes \tau_1$$

$$\gamma_{2n+2}^{(n+1)} = I \otimes \tau_2$$

Generating rotations

$$\sigma_{ij}^{(n+1)} = -i\gamma_i^{(n+1)}\gamma_j^{(n+1)} = \sigma_{ij}^{(n)} \otimes 1$$

$$\sigma_{i,2n+1}^{(n+1)} = \gamma_i^{(n)} \otimes \tau_2$$

$$\sigma_{i,2n+2}^{(n+1)} = -\gamma_i^{(n)} \otimes \tau_1$$

$$\sigma_{2n+1,2n+2}^{(n+1)} = 1 \otimes \tau_3$$

$$\gamma_F = (-i)^n \gamma_1 \gamma_2 \cdots \gamma_{2n}$$

$$\gamma_F \gamma_i = -\gamma_i \gamma_F \quad \text{for all } i$$

$$\psi_L \to (e^{\frac{i}{4}\omega_{ij}\sigma_{ij}} P_-)\psi_L = e^{\frac{i}{4}\omega_{ij}\sigma_{ij}}\psi_L \quad \text{and} \quad \psi_R \to (e^{\frac{i}{4}\omega_{ij}\sigma_{ij}} P_+)\psi_R = e^{\frac{i}{4}\omega_{ij}\sigma_{ij}}\psi_R$$

Conjugation

$$C_n^{-1}\gamma_i^{(n)\,T} C_n = (-1)^n \gamma_i^{(n)}$$

$$C^{-1}\sigma_{ij}^T C = -\sigma_{ij} = C^{-1}\sigma_{ij}^* C$$

$$C_n^{-1}\gamma_F C_n = (-1)^n \gamma_F$$

$$C_n^T = (-1)^{\frac{1}{2}n(n+1)} C_n$$

$$\gamma_i^{(n)} C_n = (-1)^{n+i+1} C_n \gamma_i^{(n)}$$

$SO(4k+2)$	Complex
$SO(8m)$	Real
$SO(8m+4)$	Pseudoreal

Spinor decomposition

$$2_+^{n+m-1} \to (2_+^{n-1}, 2_+^{m-1}) \oplus (2_-^{n-1}, 2_-^{m-1})$$

$$2_-^{n+m-1} \to (2_+^{n-1}, 2_-^{m-1}) \oplus (2_-^{n-1}, 2_+^{m-1})$$

Lorentz transformations

$$J_z = i \left(y \frac{\partial}{\partial x} - x \frac{\partial}{\partial y} \right), \qquad i K_x = t \frac{\partial}{\partial x} + x \frac{\partial}{\partial t}$$

$$[J_i, J_j] = i \varepsilon_{ijk} J_k$$

$$[J_i, K_j] = i \varepsilon_{ijk} K_k$$

$$[K_i, K_j] = -i \varepsilon_{ijk} J_k$$

$$J_{\pm, i} = \frac{1}{2} (J_i \pm i K_i)$$

$$x^\mu \to x'^\mu = L^\mu_{\ \nu} x^\nu, \qquad \eta_{\mu\nu} L^\mu_{\ \sigma} L^\nu_{\ \rho} = \eta_{\sigma\rho}$$

$$SL(2, C)/Z_2 = SO(3, 1)$$

$$x^\mu = (t, \vec{x}), \qquad x_\mu = (t, -\vec{x})$$

$$\partial_\mu x^\nu = \frac{\partial x^\nu}{\partial x^\mu} = \delta^\nu_\mu$$

$$\partial_\mu x_\nu = \eta_{\mu\nu}$$

$$J_{\mu\nu} = i(x_\mu \partial_\nu - x_\nu \partial_\mu)$$

Poincaré algebra

$$[J_{\mu\nu}, J_{\rho\sigma}] = -i(\eta_{\mu\rho} J_{\nu\sigma} + \eta_{\nu\sigma} J_{\mu\rho} - \eta_{\nu\rho} J_{\mu\sigma} - \eta_{\mu\sigma} J_{\nu\rho})$$

$$[J_{\mu\nu}, P_\rho] = -i(\eta_{\mu\rho} P_\nu - \eta_{\nu\rho} P_\mu)$$

$$[P_\mu, P_\nu] = 0$$

$$W_\sigma \equiv -\frac{1}{2} \varepsilon_{\mu\nu\rho\sigma} J^{\mu\nu} P^\rho$$

$$[W_\mu, W_\nu] = i \varepsilon_{\mu\nu\rho\sigma} W^\rho P^\sigma$$

$$[J_i, J_j] = i \epsilon_{ijk} J_k, \qquad [J_i, P_j] = i \epsilon_{ijk} P_k, \qquad [J_i, K_j] = i \epsilon_{ijk} K_k$$

$$[P_i, H] = 0, \qquad [J_i, H] = 0, \qquad [P_i, P_j] = 0$$

$$[K_i, H] = i P_i, \qquad [K_i, P_j] = 0, \qquad [K_i, K_j] = 0$$

$$(j^+, j^-) \to (j^+ + j^-) \oplus (j^+ + j^- - 1) \oplus (j^+ + j^- - 2) \oplus \cdots \oplus |j^+ - j^-|$$

Weyl spinors

$$u \to e^{i(\vec{\theta} - i\vec{\varphi}) \cdot \frac{\vec{\sigma}}{2}} u = e^{i\vec{\xi} \cdot \frac{\vec{\sigma}}{2}} u$$

$$\mathcal{L} = i u^\dagger \left(\frac{\partial}{\partial t} + \vec{\sigma} \cdot \vec{\nabla} \right) u$$

$$\sigma^\mu = (I, \vec{\sigma}) = (I, \sigma_1, \sigma_2, \sigma_3)$$

$$\sigma^\mu \partial_\mu u = 0 \quad \text{and} \quad \sigma^\mu p_\mu u = 0$$

$$\tilde{u} \equiv i\sigma_2 u^*: \quad \tilde{u} \to e^{i(\vec{\theta}+i\vec{\varphi})\cdot\frac{\vec{\sigma}}{2}}\tilde{u}$$

$$v \to e^{i(\vec{\theta}+i\vec{\varphi})\cdot\frac{\vec{\sigma}}{2}}v = e^{i\vec{\xi}^*\cdot\frac{\vec{\sigma}}{2}}v$$

$$\mathcal{L} = iv^\dagger \left(\frac{\partial}{\partial t} - \vec{\sigma}\cdot\vec{\nabla}\right)v$$

$$\bar{\sigma}^\mu \equiv (I, -\vec{\sigma})$$

$$\bar{\sigma}^\mu \partial_\mu v = 0 \quad \text{or} \quad \bar{\sigma}^\mu p_\mu v = 0$$

$$\mathcal{L} = iu^\dagger \left(\frac{\partial}{\partial t} + \vec{\sigma}\cdot\vec{\nabla}\right)u - \frac{1}{2}(mu^T\sigma_2 u + m^*u^\dagger\sigma_2 u^*)$$

$$i\left(\frac{\partial}{\partial t} + \vec{\sigma}\cdot\vec{\nabla}\right)u = m^*\sigma_2 u^*$$

Dirac spinors

$$\gamma^\mu \equiv \begin{pmatrix} 0 & \bar{\sigma}^\mu \\ \sigma^\mu & 0 \end{pmatrix}$$

$$(\gamma^\mu p_\mu - m)\psi(p) = 0$$

$$(i\gamma^\mu \partial_\mu - m)\psi(x) = 0$$

$$\sigma^{\mu\nu} \equiv \frac{i}{2}[\gamma^\mu, \gamma^\nu]$$

$$\bar{\psi} \equiv \psi^\dagger \gamma^0$$

$$\mathcal{L} = \bar{\psi}(i\not{\partial} - m)\psi$$

$$\gamma^0 = I \otimes \tau_1 = \begin{pmatrix} 0 & I \\ I & 0 \end{pmatrix} \quad \text{Weyl}, \quad \text{or} \quad \gamma^0 = I \otimes \tau_3 = \begin{pmatrix} I & 0 \\ 0 & -I \end{pmatrix} \quad \text{Dirac}$$

$$\gamma^i = \sigma_i \otimes i\tau_2 = \begin{pmatrix} 0 & -\sigma_i \\ \sigma_i & 0 \end{pmatrix} \quad \text{Weyl and Dirac}$$

$$\gamma_5 = I \otimes \tau_3 = \begin{pmatrix} I & 0 \\ 0 & -I \end{pmatrix} \quad \text{Weyl}, \quad \text{or} \quad \gamma_5 = -I \otimes \tau_1 = -\begin{pmatrix} 0 & I \\ I & 0 \end{pmatrix} \quad \text{Dirac}$$

Charge conjugation

$$\psi_c \equiv C\gamma^0\psi^*$$

$$C = \gamma^2\gamma^0$$

$$C = -\sigma_2 \otimes \tau_3 = \begin{pmatrix} -\sigma_2 & 0 \\ 0 & \sigma_2 \end{pmatrix} \quad \text{Weyl,} \quad \text{or} \quad C = \sigma_2 \otimes \tau_1 = \begin{pmatrix} 0 & \sigma_2 \\ \sigma_2 & 0 \end{pmatrix} \quad \text{Dirac}$$

$$C = C^{-1} = -C^* = -C^T = C^\dagger$$

$$\psi_c \equiv C\gamma^0 \psi^* = (\psi^*)^T (C\gamma^0)^T = \psi^\dagger \gamma^0 C^T = -\bar{\psi} C$$

$$\psi = \begin{pmatrix} u \\ v \end{pmatrix} \quad \text{and} \quad \psi_c = \begin{pmatrix} -\sigma_2 v^* \\ \sigma_2 u^* \end{pmatrix}$$

Majorana

$$i \not{\partial} \psi = m \psi_c$$

$$\mathcal{L} = i u^\dagger \sigma^\mu \partial_\mu u - \frac{1}{2} m (u^T \sigma_2 u + u^\dagger \sigma_2 u^*)$$

$$\mathcal{L} = \bar{\psi} i \gamma^\mu \partial_\mu \psi - \frac{1}{2} m (\psi^T C \psi + h.c.)$$

$$\psi = \begin{pmatrix} u \\ -\sigma_2 u^* \end{pmatrix}$$

Part VIII

Central extension

$$[K_i, P_j] = i M \delta_{ij}$$

$$K_x = -i t \frac{\partial}{\partial x} - M x$$

Conformal algebra

$$g'_{\rho\sigma}(x') = \Omega^2(x') g_{\rho\sigma}(x')$$

$$\partial_\rho \xi_\sigma + \partial_\sigma \xi_\rho = \frac{2}{d} \eta_{\rho\sigma} \partial \cdot \xi$$

$$[P^\mu, P^\nu] = 0, \qquad [K^\mu, K^\nu] = 0$$

$$[D, P^\mu] = -P^\mu, \qquad [D, J_{\mu\nu}] = 0, \qquad [D, K^\mu] = +K^\mu$$

$$[J^{\mu\nu}, P^\lambda] = -\eta^{\mu\lambda} P^\nu + \eta^{\nu\lambda} P^\mu, \qquad [J^{\mu\nu}, K^\lambda] = -\eta^{\mu\lambda} K^\nu + \eta^{\nu\lambda} K^\mu$$

$$[J^{\mu\nu}, J^{\lambda\rho}] = -\eta^{\mu\lambda} J^{\nu\rho} - \eta^{\nu\rho} J^{\mu\lambda} + \eta^{\nu\lambda} J^{\mu\rho} + \eta^{\mu\rho} J^{\nu\lambda}$$

$$[K^\mu, P^\nu] = 2(J^{\mu\nu} + \eta^{\mu\nu} D)$$

$$SO(2, 2): \quad \partial_\pm, \qquad x^\pm \partial_\pm, \quad \text{and} \quad -(x^\pm)^2 \partial_\pm$$

Expanding universe

$$X = (gX_*) = e^{i\vec{P}\cdot\vec{x}} e^{iDt}(0, \vec{0}, 1) = \left(\sinh t + \frac{1}{2}e^t\vec{x}^2, e^t\vec{x}, \cosh t - \frac{1}{2}e^t\vec{x}^2 \right)$$

$$ds^2 = \eta_{MN}dX^M dX^N = -dt^2 + e^{2t}d\vec{x}^2$$

Part IX

$$u^r, u^g, u^y, u^{cr}, u^{cg}, u^{cy}, d^r, d^g, d^y, d^{cr}, d^{cg}, d^{cy}, e^-, e^+, \nu_e$$

$$(3, 2, \tfrac{1}{6}) \oplus (3^*, 1, -\tfrac{2}{3}) \oplus (3^*, 1, \tfrac{1}{3}) \oplus (1, 2, -\tfrac{1}{2}) \oplus (1, 1, 1)$$

$$16^+ \rightarrow [0] \oplus [2] \oplus [4] = 1 \oplus 10 \oplus 5^*$$

$$16^+ \otimes 16^+ = 10 \oplus 120 \oplus 126$$

$SU(2)$ doublets

$$\nu = |-+---\rangle$$

$$e^- = |+----\rangle$$

$$u = |-+++-\rangle, \qquad |-++-+\rangle, \quad \text{and} \quad |-+-++\rangle$$

$$d = |+-++-\rangle, \qquad |+-+-+\rangle, \quad \text{and} \quad |+--++\rangle$$

$SU(2)$ singlets

$$\nu^c = |+++++\rangle$$

$$e^+ = |--+++\rangle$$

$$u^c = |+++--\rangle, \qquad |++-+-\rangle, \quad \text{and} \quad |++--+\rangle$$

$$d^c = |--+--\rangle, \qquad |---+-\rangle, \quad \text{and} \quad |----+\rangle$$